Handbook of Experimental Pharmacology

Volume 116

Springer
Berlin
Heidelberg
New York
Barcelona
Budapest
Hong Kong
London
Milan
Paris
Santa Clara
Singapore
Tokyo

The Pharmacology of Sleep

Contributors

J. Adrien, F. Albani, A. Baruzzi, M. Berger, E.O. Bixler
A.A. Borbély, D.G. Dikeos, R. Drucker-Colín, R. Fritsch Montero
Y. Hishikawa, S. Inoué, A. Kales, E. Lugaresi, H. Merchant-Nancy
J.M. Monti, W.E. Müller, O. Pompeiano, M. Radulovacki
A.E. Reinberg, D. Riemann, R. Riva, E. Sforza, C.R. Soldatos
G. Tononi, E. Van Cauter, A. Vela-Bueno, A.N. Vgontzas
A. Wauquier

Editor
Anthony Kales

Springer

ANTHONY KALES, M.D.
Professor and Chairman
Department of Psychiatry and
Director, Sleep Research and Treatment Center
Pennsylvania State University College of Medicine
M.S. Hershey Medical Center
500 University Drive
Hershey, PA 17033
USA

With 56 Figures and 26 Tables

ISBN 3-540-58961-9 Springer-Verlag Berlin Heidelberg New York

Library of Congress Cataloging-in-Publication Data. The pharmacology of sleep/contributors, J. Adrien . . . [et al.]: editor, Anthony Kales. p. cm. – (Handbook of experimental pharmacology: v. 116) Includes bibliographical references and index. ISBN 3-540-58961-9 (alk. paper) 1. Sleep – Physiological aspects. 2. Sleep disorders. 3. Hypnotics. 4. Neuropsychopharmacology. I. Kales, Anthony. II. Series. QP905.H3 vol. 116 [QP425] 615'.1 s – dc20 [612.8'21] 95-24466

Cover design: Springer-Verlag, Design & Production

Typesetting: Best-set Typesetter Ltd., Hong Kong

SPIN: 10126531 27/3136/SPS – 5 4 3 2 1 0 – Printed on acid-free paper

Preface

The last four decades have witnessed considerable advances in our knowledge of the pharmacology of sleep. Both basic and clinical pharmacology have made major contributions toward our current understanding of the complex mechanisms of sleep and wakefulness. In addition, these advances in our understanding of the pharmacology of sleep have benefited the treatment of sleep disorders and various neurologic and psychiatric conditions.

This volume is organized into three different parts. The first is a review of the basic mechanisms of sleep and wakefulness and the chronobiology of sleep. The second part reviews the basic pharmacology of the various neurotransmitter systems involved in sleep and wakefulness, while the third is clinically oriented and focuses on the effects of a variety of drugs on sleep and wakefulness.

The initial part begins with a historical review of the hypotheses of the mechanisms of sleep, evolving from passive to active regulation, and concepts involving sleep-related neurotransmitters and other sleep factors. Then regulation of sleep and wakefulness is discussed in terms of homeostatic, circadian, and ultradian processes. Also discussed is the fact that sleep homeostasis is not disrupted by the administration of hypnotic drugs. This part also reviews time-dependent properties of pharmacologic agents in relation to endogenous biologic rhythms and more specifically to chronopharmacologic changes.

The second part reviews the role of various neurotransmitter systems in the regulation of sleep and wakefulness. Intact monoaminergic (catecholaminergic, serotonergic, and histaminergic) systems are necessary for the normal expression of both sleep and wakefulness, as well as the different sleep stages. Intact catecholamine transmission is necessary for realization of REM sleep, while noradrenergic neurons play a modulatory role in wakefulness and in the occurrence of REM sleep. Similarly, serotonin acts as a neuromodulator with a very complex role. Specifically, during wakefulness, the activated serotoninergic system participates in creating the conditions for sleep occurrence, while during sleep the deactivated serotoninergic system facilitates in the occurrence of REM sleep. The histaminergic system has an important role in the regulation of the waking state as its pathways resemble those of the ascending noradrenergic and serotonergic components of the reticular activating system. In addition, histamine may act to modulate

REM sleep. Finally, it appears that the cholinergic system has a central role in at least two major functions: generation of REM sleep; and activation of the thalamus, cerebral cortex, and hippocampus during both wakefulness and REM sleep.

The second part continues with a description of the role GABAergic interneurons play in inhibiting most neuronal systems that underlie the sleep–wakefulness cycle. It appears that enhancement of central GABA-mediated inhibition by benzodiazepines is the primary mechanism for their hypnotic properties. Several CNS peptides have been shown to possess sleep-modulating activities, although their precise role in sleep regulation requires further investigation of the dynamic interactions among these and other substances. Current evidence indicates that both circadian rhythmicity and sleep inputs can be recognized in the 24-h profiles of all pituitary and pituitary-dependent hormones, while hormones which are not directly controlled by the hypothalamopituitary axis also show consistent changes during sleep and wakefulness. Hormones also may affect sleep and be of functional significance to the maintenance and quality of sleep. Finally, in several animal species, adenosine and adenosine analogs have been shown to produce a hypnotic effect and to affect sleep by increasing both slow-wave and REM sleep.

The third and final part of this volume reviews the clinical pharmacology of sleep. It begins with a review of the methodologic issues relevant for evaluation of hypnotic drugs and stresses the importance of sleep laboratory studies that provide objective and precise measurements throughout the night. Hypnotic efficacy, development of tolerance for hypnotic efficacy, adverse events and withdrawal effects associated with the use of benzodiazepines and related drugs are highly related to their pharmacokinetic and pharmacodynamic properties as well as other special characteristics related to their chemical structures. Various anxiolytic agents, including benzodiazepines, have also been assessed in the sleep laboratory for their efficacy and side effects.

Central stimulants have long been used to maintain wakefulness and alertness. Currently, these drugs are used for the treatment of disorders of excessive sleepiness such as narcolepsy, idiopathic hypersomnia, and other hypersomniac conditions. Most neuroleptics possess sedative effects; however, tolerance to this sedation develops quickly. While, traditionally, antihistamines have been associated with sedative effects, more recently, antihistamine drugs have been developed without this side effect. Most antidepressant drugs suppress REM sleep. If a cholinergic-aminergic neurotransmitter imbalance is assumed to cause depressive and manic states, then the REM sleep abnormalities often observed in depressed patients may be considered as the result of cholinergic hyperfunction or muscarinic supersensitivity. There is considerable information on the clinical pharmacology of antiepileptic drugs, which are a heterogeneous group. However, data on the effects of antiepileptic drugs on human sleep are relatively scarce, partly because of a complex interaction between sleep and epilepsy itself.

In the development of benzodiazepine dependence, dose and duration of administration are well-recognized factors while half-life, binding affinity, and specific chemical structure are less recognized but important factors. Finally, sleep disturbances can be caused by a large variety of therapeutic drugs. Dose and route of administration, as well as the pharmacokinetic and pharmacodynamic properties and chemical structures of various drugs, are major contributing factors to explain differences in the potential of drugs to produce sleep disturbances.

The editor would like to express gratitude to Springer-Verlag, who supported this volume and compiled the indices. In particular, I am indebted to Doris Walker for her tireless and skillful administrative assistance and guidance and to W. Shufflebotham for his excellent editorial assistance. Special thanks are expressed to my colleagues, Drs. Edward O. Bixler, Alexandros N. Vgontzas, and Antonio Vela-Bueno, who expertly assisted me in all phases of the editorial work. Finally, I am indebted to the authors for their extraordinary efforts in writing outstanding and careful reviews of the literature, thereby making this volume an important contribution to the field of the pharmacology of sleep and wakefulness.

Hershey
August 1995

Anthony Kales

List of Contributors

ADRIEN, J., INSERM, U288, CHU Pitié-Salpêtrière, 91, Boulevard de l'Hôpital, 75013 Paris, France

ALBANI, F., Istituto di Clinica Neurologica dell'Università di Bologna, Via Ugo Foscolo 7, 40123 Bologna, Italia

BARUZZI, A., Istituto di Clinica Neurologica dell'Università di Bologna, Via Ugo Foscolo 7, 40123 Bologna, Italia

BERGER, M., Psychiatrische Universitätsklinik, Albert-Ludwigs-Universität Freiburg, Hauptstraße 5, 79104 Freiburg, Germany

BIXLER, E.O., Department of Psychiatry and Sleep Laboratory, Sleep Research and Treatment Center, Pennsylvania State University College of Medicine, 500 University Drive, Hershey, PA 17033, USA

BORBÉLY, A.A., Institute of Pharmacology, University of Zürich, Winterthurevstrasse 190, 8057 Zürich, Switzerland

DIKEOS, D.G., Sleep Research Unit, Department of Psychiatry, Athens University Medical School, Eginition Hospital, 74, Vasilias Sofias, 11528 Athens, Greece

DRUCKER-COLÍN, R., Instituto de Fisiología Celular and Departamento de Fisiología, Facultad de Medicina, Universidad Nacional Autónoma de México, Apdo. Postal 70-250, 04510 México, D.F., México

FRITSCH MONTERO, R., Instituto Psichiatrico Jose Horwitz Barak, Servicio B, Av. La Paz 841, Santiago de Chile, Chile

HISHIKAWA, Y., Department of Neuropsychiatry, Akita University Medical School, Hondo 1-1-1, Akita, Japan 010

INOUÉ, S., Institute for Medical and Dental Engineering, Tokyo Medical and Dental University, Kanda-Surugadai 2-3-10, Chiyoda-ku, Tokyo 101, Japan

KALES, A., Department of Psychiatry and Sleep Research and Treatment Center, Pennsylvania State University College of Medicine, 500 University Drive, Hershey, PA 17033, USA

LUGARESI, E., Istituto di Clinica Neurologica dell'Università di Bologna, Via Ugo Foscolo 7, 40123 Bologna, Italia

MERCHANT-NANCY, H., Instituto de Fisiología Celular and Departamento de Fisiología, Faculdad de Medicina, Universidad Nacional Autónoma de México, Apdo, Postal 70-253, 04510 México, D.F., México

MONTI, J.M., Department of Pharmacology and Therapeutics, Clinics Hospital, J. Zudañez 2833/602, Montevideo 11300, Uruguay

MÜLLER, W.E., Zentralinstitut für Seelische Gesundheit, Abteilung Psychopharmakologie, Postfach 122120, 68159 Mannheim, Germany

POMPEIANO, O., Dipartimento di Fisiologia e Biochimica, Università di Pisa, Via S. Zeno 31, 56127, Italia

RADULOVACKI, M., Department of Pharmacology, College of Medicine, University of Illinois at Chicago, 901 S. Wolcott (M/C 868), Chicago, IL 60612, USA

REINBERG, A.E., UA CNRS 581, Chronobiology Unit, Foundation Adolphe de Rothschild, 29, Rue Manin, 75940 Paris Cedex, France

RIEMANN, D., Psychiatrische Universitätsklinik, Albert-Ludwigs-Universität Freiburg, Haupstraße 5, 79104 Freiburg, Germany

RIVA, R., Istituto di Clinica Neurologica dell'Università di Bologna, Via Ugo Foscolo 7, 40123 Bologna, Italia

SFORZA, E., Istituto di Clinica Neurologica dell'Università di Bologna, Via Ugo Foscolo 7, 40123 Bologna, Italia

SOLDATOS, C.R., Sleep Research Unit, Department of Psychiatry, Athens University Medical School, Eginition Hospital, 74, Vasilias Sofias, 11528 Athens, Greece

TONONI, G., Dipartimento di Fisiologia e Biochimica, Università di Pisa, Via S. Zeno 31, 56127, Italia

VAN CAUTER, E., Department of Medicine, MC 1027, University of Chicago, 5841 South Maryland Avenue, Chicago, IL 60637, USA

VELA-BUENO, A., Institute of Sleep Medicine and Biological Rhythms and Department of Psychiatry, Autonomous University, Ferraz 27 1° D[a], 28008 Madrid, Spain

VGONTZAS, A.N., Department of Psychiatry and Sleep Disorders Clinic, Sleep Research Treatment Center, Pennsylvania State University College of Medicine, 500 University Drive, Hershey, PA 17033, USA

WAUQUIER, A., Sleep Center, Department of Neurology, Texas Tech University Health Sciences Center, 3601 4th Street, Lubbock, TX 79430, USA

Contents

CHAPTER 4

Pharmacology of the Catecholaminergic System

CHAPTER 5

The Serotoninergic System and Sleep-Wakefulness Regulation

CHAPTER 9

Pharmacology of the CNS Peptides

S. Inoué. With 2 Figures . . . 243

CHAPTER 11

Pharmacology of the Adenosine System

CHAPTER 12

Methodological Issues in Pharmacological Studies of Sleep

CHAPTER 14

Anxiolytic Drugs

CHAPTER 15

Stimulant Drugs

CHAPTER 16

Neuroleptics, Antihistamines and Antiparkinsonian Drugs: Effects on Sleep

CHAPTER 18

Anticonvulsant Drugs

CHAPTER 19

Mechanisms of Benzodiazepine Drug Dependence

CHAPTER 20

Sleep Disturbances as Side Effects of Therapeutic Drugs

CHAPTER 1

Evolution of Concepts of Mechanisms of Sleep

R. DRUCKER-COLÍN and H. MERCHANT-NANCY

Theories may become obsolete, even while the validity of the experiments on which they are based, remains unchallenged.

(Moruzzi 1972)

A. Introduction

Even after more than 40 years of systematic research on sleep, there is little consensus among current workers concerning a specific mechanism of sleep regulation. Sleep is a behavioral state which interacts at various levels of organization with other higher brain functions; therefore "dissection" of the mechanisms of sleep has been difficult.

In this review, we attempt to describe the main concepts of mechanisms of sleep which have been put forward over the years. Because the theories of sleep regulation have been postulated according to the experimental data, which in turn have followed the technical advances in neuroscience research, the different sections of this article are divided according to the main techniques that have been utilized in sleep research.

B. Brief Historical Overview

Early attempts to explain sleep were a mixture of philosophical and scientific concepts. In ancient Greece, Empedocles believed that sleep followed a slight cooling of the blood contained in the heart. The father of medicine, Hippocrates, concluded from the cooling of the limbs of a sleeping person that sleep was caused by the retreat of blood and warmth into the inner regions of the body. Aristotle suggested that the immediate cause of sleep was to be found in the food we ate, which was supposed to give off fumes into the veins. These fumes then collected in the head causing sleepiness, cooled off the brain and sank back into the lower parts of the body, drawing heat away from the heart.

In the middle ages, Hildegard of Bingen, a German Benedictine nun, emphasized the parallelisms between sleep and feeding. She thought that human beings consisted of two parts, waking and sleeping, and that the human body was nourished by food and rest.

In the sixteenth century Paracelsus, a classical physician, thought that natural sleep lasted 6 h and its function was to eliminate the tiredness caused by working. Moreover, he recommended sleeping following sunset and neither too much nor too little.

The combination of physiological concepts and metaphysics was often utilized to explain sleep mechanisms during the seventeenth and eighteenth centuries. Alexander Stuart, a british physiologist and physician, thought that sleep resulted from a deficit of the spirits which worked and moved within the body during waking and eventually became exhausted. According to the dutch physician Boerhaave, sleep occurred because the fluids in the brain could not move freely and thus could no longer fill the small vessels and nerves that run from the brain to the sensory organs and voluntary muscles. In addition, the german physiologist Ackerman suggested that the inhalation of oxygen in the air gives off an "ether of life" which reaches the brain by way of the blood and is stored there. During tiredness there is a deficiency in the life of this ether, and only through sleep could the supply be replenished.

The nineteenth century was characterized by the growth of the natural sciences. The concepts of the mechanism of sleep began to have a more physiological and chemical basis, most theories postulating that sleep was exclusively the result of an absence or reduction of oxygen and blood supply to the brain (see BORBÉLY 1986; HORNE 1988).

C. Early Neurophysiological Concepts of Sleep (From Passive to Active)

In 1929 Hans Berger (see JOUVET and MORUZZI 1972) developed the first electroencephalograph. This opened a new era in the field of hypnic studies, with the demonstration that striking changes occurred in the electroencephalogram (EEG) during sleep. Before Berger's work, sleep studies were merely based on the observation of postures and movements, and on the study of a few changes occurring mainly in the autonomic sphere such as pupil diameter, blood pressure, heart rate and body temperature. Then, in the 1930s, BREMER studied the effects of two different transections in the cat brain stem on the sleep-wake cycle (BREMER 1935, 1938). The first, the *cerveau isolé* preparation, consisted of a transection at the intercollicular level, just behind the third cranial nerves. In this preparation, the sensory afferents to the isolated cerebrum were interrupted, with the exception of the olfactory and optic nerves, while the outflow was mediated exclusively by the motor neurons and the preganglionic parasympathetic nerve cells of the oculomotor nucleus. After this transection extreme myosis and a persistent slow-wave (6–10 Hz) rhythm of regular amplitude was observed, resembling that of a cat with slow-wave sleep. Moreover, throughout the short survival time of the *cerveau isolé* preparation, the alternation of sleep and wakefulness was abolished altogether. On the other hand, in the *encephale isolé* preparation, where a knife incision was made at the medulla oblongata-spinal cord junction (C1), animals showed a normal sleep-wake cycle. These experiments led to BREMER's conclusion that a tonic influence arising below the midbrain transection, but above the spinal cord segments, is necessary in order to maintain the sleep-wake cycle. From this, BREMER (1935) suggested, as KLEITMAN (1939) had done before him, that sleep is a passive state, occurring simply as a result of a slight reversible sensory deafferentation.

In the light of the landmark stimulation experiments of the reticular formation by MORUZZI and MAGOUN (1949), it was necessary to restructure the ideas of how sleep came about and BREMER's passive theory of sleep was no longer tenable. These investigators thought that the structures of critical importance for the maintenance of wakefulness were localized within both planes of the *encephale isolé* and *cerveau isolé* preparations. The now classical experiments performed in the *encephale isolé* preparation showed that stimulation of the medial reticular formation, the pontine and midbrain tegmentum and the dorsal hypothalamus and subthalamus elicited EEG arousal in synchronized basal conditions. MORUZZI and MAGOUN (1949) concluded that the sleep syndrome produced by the intercollicular transection was not due to sensory deafferentation in the strict sense of the word, but rather to the elimination of the waking influence of an ascending activating reticular system (AARS). They suggested then that the AARS through ascending projections exerted a tonic activating influence on subcortical and cortical forebrain structures in order to elicit and maintain a state of wakefulness manifested by tonic electrocortical activation. The existence of this system was confirmed when LINDSLEY et al. (1949) showed that medial mesencephalic lesions, interrupting most of the midbrain reticular formation, but leading to the classical ascending pathways, resulted in a synchronization of the EEG, while lateral lesions, sparing the reticular formation but interrupting the classical specific pathway, did not abolish the sleep-wake cycle.

Paralleling in time the above-mentioned observations, the anatomoclinical descriptions of VON ECONOMO (1929) provided the first evidence for an active mechanism of sleep regulation. He described, in patients affected by the epidemic of encephalitis lethargica which occurred in Europe around 1918, a striking correlation between inflammatory lesions of the anterior hypothalamic area and death in a state of insomnia, and between inflammatory lesions of the posterolateral hypothalamus and death in a state of somnolence.

In the 1930s RANSON (1939) reported the effects of hypothalamic lesions made in rhesus monkeys. He showed that bilateral lesions in posterolateral hypothalamic areas, extending to the caudal border of the mammillary bodies, were most effective in producing a sleep syndrome. However, when the lesions were placed in the basal forebrain region, the animal's sleep was not affected. Ranson concluded that excitation of the cerebral cortex by way of the hypothalamus is important, but not essential for maintaining the waking state, since posterolateral lesioned animals became progressively less sleepy.

A few years later Nauta (1946) described in albino rats that transverse sections situated in the vicinity of the mammillary bodies induced a sleep syndrome throughout a survival period of 8 days. This alteration in the sleep-wake cycle was attributed to the disconnection of a waking center from the underlying brain stem. In addition, Nauta performed transverse sections in the rostral half of the hypothalamus, observing then an opposite effect, i.e., insomnia. He thus concluded that the anterior hypothalamus, roughly corresponding to the suprachiasmatic nucleus and preoptic area, is the site of a regulatory structure of sleep. Interestingly, combined transections of both areas induced no changes in the sleep-wake cycle. These experiments generated the notion of two separate systems in the brain which controlled sleeping and waking. Moreover, Nauta further suggested an inhibitory action of the sleep center on the waking center, which would thus activate the cortex through the lateral hypothalamus.

At about the same time, the stimulation experiments of different central nervous system (CNS) areas that were reported by Hess (1944) confirmed the existence of brain structures that actively participated in the regulation of sleep.

It was in 1944 that Hess showed that low-rate, low-voltage stimulation of a region lying just lateral to the massa intermedia of the thalamus elicited a progressive decrease in activity, with clear behavioral signs of drowsiness and the induction of well-integrated physiological patterns of sleep. Subsequently, Monnier et al. (1960) showed that stimulation of the intralaminar thalamic nuclei induced EEG synchronization and physiological sleep, which he called orthodox sleep.

Sterman and Clemente (1962) described that unilateral low-rate stimulation of the lateral preoptic region and adjacent diagonal band of Broca, in the immobilized and freely moving cat, produced a rapid bilateral EEG synchronization. Moreover, diathermic warming of these structures produced sleep-like postural changes in the freely moving cat (Roberts and Robinson 1969).

Additionally, Moruzzi's group showed that low-rate electrical stimulation of the solitary tract nucleus of the medulla induced EEG synchronization in waking cats (Magnes et al. 1961). All these studies, together with the results obtained by Batini et al. (1959) from a pretrigeminal mediopontine preparation, in which cats presented a normal sleep-wake cycle, definitely ended support for the passive theory of sleep. It was then suggested that the reticular formation of the brain stem was a complex structure with both hypnogenic and waking functions. Sleep was no longer conclusively considered a passive phenomenon; however, as of today the concept of a tonic reticular activation which participates in the control of wakefulness and sleep can still be considered valid.

Interestingly, both the early passive and active theories of sleep regulation considered wakefulness as an opposing state to the condition of sleep, and sleep itself was viewed as a homogeneous state. The ASERINSKY and KLEITMAN (1953) discovery of rapid eye movement (REM) sleep, and its association with human dreaming marked the starting point of a series of studies where the monolithic idea of sleep was abandoned, and substituted by the dual concepts of active mechanisms regulating both slow-wave sleep (SWS) and REM sleep.

Since the time of this study, several lesion and transection studies have demonstrated that REM sleep is actively regulated by the brain stem. For example, transections at the pontomedullary junction produced a preparation without REM sleep signs frontal or caudal to the transection (WEBSTER et al. 1986). On the other hand, structures rostral to the rostro-pontine level produced a preparation with all the local signs of REM sleep caudal to the transection (JOUVET 1962). However, a mid-pontine transection induced the abolition of REM sleep. Since it was shown that only extensive electrolytic lesions of the pons in the cat permanently suppressed REM sleep (JOUVET 1962; CARLI and ZANCHETTI 1965), all these studies pointed towards the pontomedullary region as the critical place for REM sleep generation.

Recently, STERIADE et al. (1986; STERIADE and LLINÁS 1988) have suggested a dual "passive-active" mechanism for SWS. This idea was based on several studies of which only a few can be described. During wakefulness and REM sleep there is an activating influence from brain stem cholinergic (pedunculopontine tegmental and laterodorsal tegmental nuclei) and monoaminergic systems (norepinephrine from the locus coeruleus and serotonin from the raphe dorsalis) to the thalamus and cerebral cortex (see JONES 1990). These neurotransmitters lead to an increased excitability and a reduction or suppression of the long-lasting inhibitory processes in thalamocortical cells (STERIADE et al. 1990; MCCORMICK 1992). During SWS, or at least during the associated spindle rhythms, there is an "active" reduction of the influence of the ascending cholinergic and monoaminergic projections, which thus allows thalamocortical and reticular thalamic cells to oscillate and synchronize synaptic networks "passively" (STERIADE et al. 1990). The "passive" oscillatory nature of thalamic cells depends on their intrinsic properties and different ionic conductances, such as the very slowly inactivating Na^+ conductance, the low-threshold Ca^{2+} spike and the Ca^{2+}-dependent K^+ conductance (STERIADE and LLINÁS 1988). Moreover, these properties are subject to controlling influences from the gamma-aminobutyric acid (GABA) ergic reticular thalamic nuclei that function as a pacemaker, which, by virtue of their connections to all thalamic nuclei, synchronize the activity of thalamic neurons, inducing a coordinated long-lasting hyperpolarization.

On the other hand, the active participation of the anterior hypothalamic area on sleep induction, and the opposite role of the posterior hypothalamus implicated in wakefulness, was validated by recent studies in which injections of neurotoxic drugs in the medial preoptic area and diagonal band nuclei induced long-lasting insomnia (SALLANON et al. 1989). However, injections of the GABA agonist muscimol in the posterolateral hypothalamus of previously lesioned cats in the anterior hypothalamic area induced sleep recovery. These results were interpreted as suggesting that the preoptic area of the anterior hypothalamus is not required for sleep onset, but its sleep-promoting role results from the inhibition of posterior hypothalamic wakefulness-related histaminergic neurons (SAKAI et al. 1990).

In summary, the regulatory mechanisms of SWS are believed to reside within diencephalic structures, while those regulating REM sleep appear to reside mainly in the pontine brain stem.

D. Neurohumoral Concepts of Sleep Regulation

I. Progressive Development of an Old Idea

At the turn of the century, PIERON (1913) and LEGENDRE and PIERON (1913) in France furnished the first experimental evidence for the humoral or so-called wet nature of sleep. The hypothesis of these authors was based on experiments which showed that a period of wakefulness presumably caused the accumulation in the cerebrospinal fluid (CSF) of sleep-inducing products which they baptized "hypnotoxins." PIERON's experiments demonstrated that CSF obtained from sleep-deprived dogs was capable of inducing sleep in normal dogs. Similar experiments were performed by ISHIMORI (1909). Since these original observations, a steadily increasing number of experiments have yielded a series of substances with sleep-inducing properties which have been obtained from blood, CSF, urine or brain. To a greater or lesser degree all such experiments were influenced by PIERON's approach.

The new era for the search of sleep substances was probably begun by KORNMÜLLER et al. (1961), who reported the presence of sleep-promoting factors in the blood. These authors reported that, in two pairs of cats with crossed arterial circulation, sleep was induced in one of the animals using the Hess paradigm of low-frequency electrical stimulation in the medial thalamus. EEG recording demonstrated that 20–30 s after stimulation sleep also occurred in the unstimulated cat. These findings were extensively confirmed by MONNIER's group using the same low-frequency electrical stimulation of the mediocentral thalamus in rabbits (MONNIER et al. 1963). The cerebral venous blood of the stimulated animals was dialyzed during 80 min of continuous sleep, induced by repeated stimulation. Then the dialysate was injected intraventricularly (i.v.) to the recipient rabbits, which exhibited slow-wave EEG activity 10–15 min after the injection. The active dialysate was also effective after infusion into the third ventricle. Some years later, purification of the active dialysate fractions led to the identification of a nonapeptide, named the delta sleep-inducing peptide (DSIP) (SCHOENENBERGER et al. 1972, 1978).

Delta sleep-inducing peptide is one of the compounds which has been most extensively tested as a sleep-promoting substance. However, the somnogenic action of this peptide could not be confirmed in all studies. It is important to note that DSIP induces hyperthermia, which may cause sleep without having specific sleep-inducing properties (see BORBÉLY and TOBLER 1989).

In the 1960s, PAPPENHEIMER and coworkers initiated a series of studies that demonstrated the existence of sleep-promoting substances in the CSF. They used 72 h sleep-deprived goats as donor animals, to allow the repeated extraction of CSF from the cisterna magna. The infusion of CSF from sleep-deprived donors into the third ventricle of rats and cats produced behavioral sleep and a decrease in motor activity (PAPPENHEIMER et al. 1967). The effects were enhanced as a function of the duration of sleep deprivation in the donor animal. The sleep-inducing factor was characterized and designated as factor S (PAPPENHEIMER et al. 1975). After these early experiments factor S was not only purified and concentrated from CSF, but also from brain tissue of sleep-deprived goats, sheep, rabbits and cattle (KRUEGER et al. 1978). Factor S from human urine was also purified and showed similar sleep-inducing properties in rabbits as the factor obtained from brain. Moreover, a highly purified fraction of urinary factor S was administered by microinjection into several brain regions in rabbits. The most active sites were located in the basal forebrain (GARCIA-ARRARAS and PAPPENHEIMER 1983). Posteriorly, the analysis of urinary and brain factor S revealed that the major somnogenic constituent was a muramyl peptide with a molecular weight of 921 (MARTIN et al. 1984). A hyperthermic response that paralleled the somnogenic action was observed after the administration of different muramyl peptides purified from factor S (KRUEGER et al. 1987).

The hypnogenic action of muramyl peptides was not only due to substances isolated from urine or intestinal bacteria, but was also found to be the result of the normal immunoresponses of muramyl dipeptide (MDP), which include immunostimulatory and pyrogenic effects (RIVEAU et al. 1980). The MDP immunostimulatory effects are thought to be mediated through interleukin-1 (IL-1), which may contain an MDP-like structure with the capacity of inducing prolonged increases of SWS after intraperitoneal (i.p.). and i.v. administration (KRUEGER et al. 1984; KRUEGER and KARNOVSKY 1987). Moreover, administration of antipyretics simultaneously with MDP or IL-1 blocked febrile responses, but not the somnogenic one (KRUEGER et al. 1986). Since these immunostimulatory substances and others such as interferon-α_2 and the alpha-tumor necrosis factor (α-TNF) induce and increase SWS, KRUEGER and coworkers proposed that sleep may be involved in immunomodulatory processes, while the immunoresponses of the body which induce sleep could also have an adaptive function or serve as a defense function for the body.

Japanese workers led by Uchizono et al. (1975) started the study of endogenous sleep-promoting substances in the rat brain following sleep deprivation. The animals were sleep deprived for 24 h by exposure to electric foot shocks every 3 min for 1 min. Although the deprivation procedure induced stress (Inoue et al. 1985), these investigators suggested the presence of a somnogenic substance in the brain stem, mesencephalon and hypothalamus extracted from the sleep-deprived rats; injection of the homogenized, dialyzed and lyophilized brain material induced an increase in SWS in naive recipient rats. The somnogenic component contained in the brain extract was named sleep-promoting substance (SPS). Further purification of SPS allowed them to identify three different fractions with somnogenic activity, one of which was uridine, which presents strong somnogenic properties, increasing both SWS and REM sleep (Honda et al. 1984).

The study of sleep-promoting substances obtained from brain tissues was studied by our group, who demonstrated with the use of the push-pull cannula technique an increase in brain protein concentration in perfusates obtained from the mesencephalic reticular formation (MRF) during REM sleep (Drucker-Colín et al. 1975b). Moreover, perfusates obtained from the MRF of sleep-deprived cats, through a push-pull cannula, were capable of inducing sleep in normal recipient cats (Drucker-Colín 1973). In addition, antibodies against proteins obtained from MRF perfusates were quite effective in inhibiting REM sleep, when they were administered into the MRF of normal cats (Drucker-Colin et al. 1980).

These latter experiments suggested an important role for brain protein synthesis in REM sleep regulation. This idea was first formulated by Oswald in 1969, who proposed that during REM sleep there is an intense neuronal restoration mediated by an increase in brain protein synthesis. Several other studies support this notion:

1. A high correlation has been demonstrated between SWS and the release of growth hormone (GH) (Takahashi et al. 1968). Moreover, it has been suggested that since GH is an anabolic hormone the release of this substance during SWS induces an increase in the subsequent protein synthesis during REM sleep. In fact, GH induces REM sleep in rats (Drucker-Colín et al. 1975a), cats (Stern et al. 1975) and humans (Mendelson et al. 1980).
2. Systemic injection of different protein synthesis inhibitors, such as chloramphenicol, anisomycin and vincristine induce a selective inhibition of REM sleep without effects on SWS (Rojas-Ramírez et al. 1977; Drucker-Colín et al. 1979a).

Although the above-mentioned experiments are highly suggestive that proteins are involved in REM sleep modulation, the evidence is merely correlative and it is not clear which kinds of proteins are involved with REM regulation. More recently, however, a number of different neuropeptides have been implicated in the regulation of the sleep-wake cycle.

Studies on the effects of peptides derived from proopiomelanocortin (POMC) indicate that three of these compounds play a determining role on sleep-wake cycle regulation. The intracerebroventricular (i.c.v.) injection of adrenocorticotropic hormone (ACTH) induces a significant increase (64%) in the waking state, whereas the i.c.v. administration of two ACTH-derived peptides, α-melanocyte-stimulating hormone (α-MSH) and corticotropin-like intermediate-lobe peptide (CLIP), at 10 ng induces a 31% increase in SWS in the first case and a 59% increase in REM sleep, respectively (CHASTRETTE and CESPUGLIO 1985). However the α- and β-endorphins produce no change in sleep.

It has been observed that i.c.v. or i.p. injections of arginine vasotocin induce an increase in SWS and a suppression of REM sleep. These effects were prevented by lesions of the raphe dorsalis (GOLDSTEIN and PRATTA 1984).

The gastropancreatic peptides such as the vasoactive intestinal polypeptide (VIP) and the cholecystokinin octapeptide (CCK-8) produce a selective increase in REM sleep when they are i.c.v. administered in rats and cats (RIOU et al. 1982; DRUCKER-COLÍN et al. 1984; OBAL 1986). Moreover VIP induction of REM sleep is independent of its effects on temperature (OBAL et al. 1986). Since the incubation of CSF obtained from sleep-deprived cats with anti-VIP antibodies prevents the REM sleep-inducing effect of the CSF in paraclorophenylalanine (PCPA) insomniac cats, it has been suggested that VIP is a REM sleep factor that may accumulate in the CSF during wakefulness (DRUCKER-COLÍN et al. 1988, 1990). Moreover, radioimmunoassay (RIA) of VIP in CSF from sleep-deprived (SD) cats showed an increase in VIP-like substance almost proportional to the SD time (JIMÉNEZ-ANGUIANO et al. 1994).

It is important to indicate that several hormones and neuromodulators such as insulin, somatostatin and prostaglandins have been suggested to play an important role in sleep-wake cycle regulation. Central infusion of insulin enhanced SWS without affecting REM sleep in rats, while anti-insulin antibodies caused a decrease in SWS (DANGUIR and NICOLAÏDIS 1984). The same procedure with central infusions of somatostatin in the third ventricle induces a selective increase in REM sleep (DANGUIR 1986), whereas neutralization of central somatostatin by long-term i.c.v. administration of somatostatin antiserum resulted in a suppression of REM sleep, with no effect on SWS (DANGUIR and DE SAINT-HILAIRE-KAFI 1988). In addition, microinjections of prostaglandin D_2 (PGD_2) into the preoptic area of rats induced a dose-response increase in SWS, but also caused bradycardia and hypo- or hyperthermia (UENO et al. 1982). Moreover, continuous infusion of PGD_2 into the third ventricle produces a marked increase in SWS and REM sleep in rats and rhesus monkeys (*Macaca mulata*), with little alteration in body temperature and heart rate (ONEO et al. 1988). On the other hand, prostanglandins E_1 and E_2 induce a dose-dependent reduction of sleep (MASEK et al. 1976; MATSUMURA et al. 1988).

The list of sleep factor candidates and the methods used to detect their hypnogenic role have increased over the years. Moreover, the criteria for consideration of a substance as a sleep factor are very heterogeneous. However, the notion of "sleep factor" has a common implication: these substances are ascribed the intrinsic property of being the basis for the sleep mechanism. This implication has serious problems from a neurophysiological point of view. A different view of this problem (DRUCKER-COLÍN et al. 1985) suggests that all substances that either facilitate or disrupt sleep may do so because they impinge upon a yet unknown mechanism, which is responsible for producing sleep. Thus the reason for the existence of so many substances which affect sleep is that they all impinge upon this same mechanism, and that sleep factors in the strict Pieron sense do not really exist. Table 1 lists

Table 1. Peptides linked to sleep regulation

Substances		SWS	REM sleep
Neuropeptides			
VIP	+	No change	↑
	−	No change	↓
CCK-8	+	↑	↑
Endorphins, Enkephalins	±	No change	No change
Arginine-vasotocin	+	↑	No change
Angiotensin	+	↑	No change
Renin	+	No change	↓
DSIP	+	↑	No change
Interleukin-1	+	↑	No change
Interferon-α_2	+	↑	No change
Muramyl dipeptide	+	↑	No change
Substance P	+	No change	↓
Hormones and derivatives of GH	+	No change	↑
	−	No change	↓
GH-releasing factor	+	↑	↑
CLIP	+	No change	↑
Somatostatin	+	No change	↑
α-MSH	+	↑	No change
Insulin	+	↑	↑
	−	↓	No change
ACTH	+	↓	↓
Other substances			
CSF of sleep-deprived animals	+	No change	↑
	−	No change	↓
Prostaglandin D_2	+	↑	No change
	−	↓	↓
Uridine	+	↑	No change

all the peptides which have been studied in relation to sleep, regardless of whether they have been discussed in the text or not.

II. Sleep Factor Regulation Viewed Physiologically

The mechanism which triggers and/or maintains REM sleep could be closely related to the striking increase in brain stem excitability which occurs on the transition of an animal from SWS to REM sleep. The studies on the discharge rate of neurons from different brain regions have shown that there is widespread augmentation of neuronal activity during REM sleep. Such increases begin precisely at the transitional period from SWS to REM sleep (HUTTENLOCHER 1961; HOBSON et al. 1975; STERIADE et al. 1982; SAKAI 1988; DRUCKER-COLÍN et al. 1982; VERTES 1984).

On the other hand, it has been reported that an interesting characteristic of those REM sleep periods which are short or "abortive" is the absence of the gradual increase of unit activity which occurs at the transition time from SWS to REM sleep (DRUCKER-COLÍN et al. 1979b). Moreover, such abortive REM sleep periods can be induced by either systemic or local brain stem administration of chloramphenicol (DRUCKER-COLÍN et al. 1982) and this in turn is accompanied by the same absence of unit activity increase. By contrast, auditory stimulation during REM sleep which induces a major increase in REM sleep duration (DRUCKER-COLÍN and BERNAL-PEDRAZA 1983) is accompanied by an increase in the discharge rate of the pontine reticular formation (PRF) neurons and an augmentation of c-*fos* expression in several brain stem nuclei (DRUCKER-COLÍN et al. 1990; MERCHANT-NANCY et al. 1992). Such studies have recently been confirmed by YUMUY et al. (1993). Since the latter has been associated with excitability changes, it could be argued that the triggering and maintenance mechanism for REM sleep could simply occur as a result of changes in excitability levels in a widespread group of brain stem nuclei. This concept would suggest that REM sleep would function as a result of an excitostat system, whereby REM sleep is maintained if excitability is increased, while animals would awaken should such excitability diminish.

Another more general model to explain sleep mechanisms is the two-process model of sleep regulation of BORBÉLY (1982; BORBÉLY and TOBLER 1989), which suggests a direct role of sleep factors in the homeostatic component of sleep. According to this model, two separate processes underlie sleep regulation: a sleep-dependent or homeostatic process (S), and a sleep-independent or circadian process (C). The level of S is assumed to increase during waking and to decrease during sleep. The time course of process S was based on EEG slow-wave activity (power spectra of the delta band), which in animals (BORBÉLY et al. 1984) and man (BORBÉLY et al. 1981) increases as a function of prior waking and declines during sleep. The homeostatic component of sleep is independent of the circadian component, since rats with the circadian sleep-wake rhythm permanently abolished by suprachiasmatic bilateral lesions still show sleep rebound after 24 h of SD and show standard daily amounts of both SWS and REM sleep (Borbély and Tobler 1989). Process C reflects the circadian variation of sleep propensity, which is largely independent of the occurrence of sleep, since it persists even during prolonged sleep deprivation (Akerstedt and Gillberg 1981). The circadian aspect of sleep regulation is reflected by the oscillations of the sleep threshold and wake-up threshold, which interact with the S threshold to trigger sleep. Then according to Borbély and Tobler (1989) the sleep factors may define the process S threshold.

E. Neurochemical–Neurophysiological Concepts of Sleep Regulation

I. Classical Neurotransmitters

In the 1960s and 1970s Jouvet proposed the monoaminergic theory of the sleep-wake cycle, based on pharmacological, transection, lesion and neuroanatomical studies (see Jouvet and Moruzzi 1972). According to the monoaminergic theory the serotoninergic (5-HT) neurons are involved in the regulation of SWS, whereas noradrenergic neurons are responsible for both the executive mechanisms of REM sleep and the maintenance of behavioral EEG waking. This theory indicates that SWS is initiated by the release of 5-HT at some serotoninergic synapses. The 5-HT perikarya located in the anterior part of the raphe system, such as raphe dorsalis (RD) and centralis superior, would be responsible for the behavioral aspects of SWS including miosis and decrease in muscle tone and EEG synchronization. The caudal raphe system, including the raphe pontis, magnus, pallidus and obscurus, were proposed to be responsible for the priming of REM sleep, interacting directly with the executive mechanism of REM sleep in the locus coeruleus (LC) complex, probably through a cholinergic neuron.

Jouvet's theory suggests that the caudal two-thirds of the LC complex (nucleus locus coeruleus, subcoeruleus and possibly nucleus parabrachialis medialis) act as the trigger for the executive mechanisms of REM sleep. Moreover, the caudal third of this complex is responsible for the mechanisms involved in the control of the total inhibition of the muscle tone, whereas the medial third of the "coeruleus complex" would correspond to the pontine pacemaker of the pontogeniculo-occipital (PGO) activity and is responsible for both the phasic and tonic ascending components of REM sleep.

The evidence that supports the participation of the 5-HT-containing neurons of the raphe system in SWS generation mechanisms and in the "priming" mechanism of REM sleep can be summarized as follows: Inhibition of the synthesis of 5-HT by inhibition of tryptophan hydroxylation with *p*-cholorophenylalanine induces insomnia by decreasing both SWS and

REM sleep. This can be reversed back to normal sleep by the injection of small doses of 5-HTP, the immediate precursor of 5-HT (Mouret and Jouvet 1968). Destruction of the 5-HT-containing perikarya located in the raphe system induces insomnia, which is correlated with a decrease in cerebral serotonin (Pujol et al. 1969); the destruction of the anterior part of the raphe (raphe dorsalis and centralis) induces a state of permanent arousal, with REM sleep still appearing, whereas lesions of the caudal raphe (raphe pontis and magnus) are followed by a marked decrease in REM sleep frequency and a 40% decrease in SWS (Renault 1967). The increase in 5-HT turnover during instrumental REM sleep deprivation increases REM sleep frequency (Pujol 1970). Monoamine oxidase (MAO) inhibitors have the strongest suppressor effect upon REM sleep (Mouret et al. 1968; Jouvet and Delorme 1965).

The monoaminergic theory further suggests that the catecholaminergic system of the brain stem plays an important role in the REM sleep executive mechanism because: (a) the administration of dihydroxyphenylalanine (DOPA), in reserpine-pretreated cats, induces a decrease in REM sleep latency (Matsumoto and Jouvet 1964); (b) the α-methyl-DOPA that displaces norepinephrine (NE) from the synaptic stores suppresses REM sleep in the cat (Dusan-Peyrethon et al. 1968); (c) disulfiram, a blocker of NE synthesis, induces a decrease in REM sleep (Dusan-Peyrethon and Forment 1968); (d) α-receptor antagonist drugs suppress REM sleep (Matsumoto and Watanabe 1967); (e) bilateral lesions of the caudal part of the LC suppress the motor inhibition during REM, whereas more extensive bilateral lesions involving the caudal two-thirds of the LC and subcoeruleus nuclei suppress the occurrence of REM sleep, without changes in the PGO activity (Jouvet and Delorme 1965); (f) total lesions of the LC complex are followed by a permanent suppression of REM sleep, correlated with a significant decrease in NE in the mesencephalon and telediencephalon (Buguet 1969).

In addition, transection experiments indicate that the rostropontine preparation is characterized by normal REM sleep, caudal to the transection (Jouvet and Moruzzi 1972), and that the caudopontine preparation is followed by a loss of periodic muscular atonia, caudal to the section. However, the PGO activity associated with EEG desynchronization and REM sleep still persists (Jouvet 1962). Since the mediopontine preparation induces the suppression of REM sleep on both sides of the section, it was suggested that the pontine structures destroyed in this transection are related to the executive mechanism of REM sleep (Renault 1967).

Contemporary to the monoaminergic theory was Hernández-Peón's (1963, 1965) cholinergic theory of sleep-wake cycle regulation, based on discrete injections of acetylcholine (ACh) in a variety of brain sites. According to this theory the sleep-wake cycle is regulated by two antagonistic cholinergic systems: the sleep system and the waking system.

The sleep system is composed of two components. The descending component follows the trajectory of the medial forebrain bundle from the preoptic region through the lateral hypothalamus and into the limbic midbrain area. Local injections of ACh microcrystals into this circuit are followed by SWS, whereas lesions or atropine injections caudal to the injection of ACh suppress the sleep-inducing effect (VELLUTI and HERNÁNDEZ-PEÓN 1963). The descending system meets at the pontine level with an ascending component originating from the gray matter of the spinal cord (HERNÁNDEZ-PEÓN et al. 1963, 1965). ACh injections in the paramedian caudal medulla support this idea, since they are regularly followed by EEG synchronization, whereas ACh injections within the dorsal spinal cord are followed a short time later by all the signs of sleep (ROJAS-RAMÍREZ and DRUCKER-COLÍN 1973). The HERNANDEZ-PEÓN theory suggests that sleep is a unitary process, whereby SWS and REM sleep are not separate entities but are rather simply different manifestations of the same basic process. Thus, according to HERNÁNDEZ-PEÓN, the cholinergic sleep system operates through two hypnogenic stimuli: (a) the ascending component, which was postulated as the primary hypnogenic stimulus arising from the peripheral neurons of the ascending and descending segments and (b) the descending component, postulated as the conditioned stimulus arising from the neocortex. These influences converge, forming a final pathway of progressive inhibition. As the inhibition begins to ascend, mesencephalic neurons become inhibited, and then the thalamic recruiting neurons, now disinhibited, organize the thalamocortical activity which gives off spindles and slow-wave activities. Inhibition of these structures would release the activity of the neocortex which thus becomes rapid during REM sleep.

On the other hand, the cholinergic waking system corresponds to the ascending cholinergic reticular system described by SHUTE and LEWIS (1966). Local cholinergic stimulation of this system produces arousal (MORGANE 1969). Moreover, there is an increase in cortical ACh release during wakefulness or cortical arousal provoked by mediopontine transection and peripheral or central sensory stimulation (BARTHOLINI and PEPEU 1967; COLLIER and MITCHELL 1967). However, both sleep and waking systems are topographically intermixed in the medial forebrain bundle and the lateral hypothalamic area, since ACh injections in very restricted and closed areas produced lethargy or sleep behavior (BANDLER 1969).

The HERNÁNDEZ-PEÓN theory of sleep has not been pursued further mostly because there is an almost universal consensus for the independent nature of SWS and REM sleep at the neurophysiological, neurochemical and functional levels. However, the role of cholinergic neurons in the brain stem and the basal forebrain in the regulation of REM sleep has been completely confirmed (see below). In addition, the breakdown in the brain stem cholinergic system activity during SWS, the disinhibition of the thalamic pacemaker that induces subsequent oscillatory thalamocortical firing, and the mechanism of spindles and slow-wave activity have recently been studied by STERIADE et al. (1982) as described previously.

The monoaminergic theory was refuted by several electrophysiological, lesion and neuroanatomical studies. The first evidence against this theory came from the unitary recordings in the LC and raphe system, which demonstrated that the activity of the monoaminergic nuclei (LC and RD) show their highest frequency discharge during wakefulness. This activity progressively decreases upon the presentation of SWS, with almost complete inactivation during REM sleep (REM-OFF cells) (McGinty and Harper 1976; Hobson et al. 1975). Moreover, selective neurotoxic lesions with 6-hydroxydopamine in the LC (Laguzzi et al. 1972), or with 5,6-dihydroxytryptamine in the RD (Froment et al. 1974), do not abolish SWS or REM sleep, but rather produce the dissociated condition of permanent PGO activity throughout the sleep-wake cycle. In addition, pharmacological manipulations of brain NE or 5-HT do not provide consistent changes in SWS or REM sleep. It has been observed that long-term i.p. administration of parachlorophenyalanine (PCPA) in the cat induces only temporal insomnia; the sleep-wake cycle returned to basal levels by the 7th day of administration, although 5-HT was almost totally diminished by the 5th day (Dement et al. 1972). Moreover, biochemical studies demonstrate that brain serotonin release is lower during SWS than during wakefulness (Puizillout et al. 1979). On the other hand, several norepinephrine (NE) synthesis blockers, such as α-methyltyrosine, do not decrease REM sleep (Stern and Morgane 1973). Moreover, the administration of propranolol (a β-adrenergic antagonist), clonidine (an agonist of presynaptic α-adrenergic receptors) or amphetamine (a drug which enhances the release of synaptic concentrations of catecholamines) induces a significant decrease in SWS and REM sleep, and an increase in waking (Jacobs and Jones 1978; Hilakivi 1983), whereas the injection of phentolamine or prazocin (α-receptor antagonists) induces an increase in REM sleep (Hilakivi and Leppavuori 1984).

Recently Jouvet (1984; Sallanon et al. 1985) suggested that serotonin is a neuromodulator that may be responsible for the synthesis and accumulation of sleep factors during waking in neurons of the hypothalamus, to which raphe dorsalis neurons project. This hypothesis is supported by studies of PCPA-pretreated cats in which local injection of serotonin into the hypothalamus can restore sleep. Moreover, voltametric studies performed by Cespuglio et al. (1988) indicated that there are two different mechanisms of 5-HT release in the raphe dorsalis during the sleep-wake cycle. The axonal nerve release of 5-HT, measured in the hypothalamus, which is directly correlated with the neuronal activity of raphe system neurons, i.e., increases during the waking state, decreases during SWS and is even more reduced during REM sleep. This modality of release was interpreted as the signal for hypothalamic sleep factor accumulation during waking. The second modality of 5-HT release may be through local dentritic release in the raphe dorsalis, since a decrease in extracellular concentration of serotonin occurs during the waking state in this nucleus, while an increase is measured during SWS and REM sleep. The local dentritic release of 5-HT was hypothesized to be responsible for an auto-inhibitory process that induces the inhibition of 5-HT neurons during sleep (Cespuglio et al. 1990).

In an attempt to integrate the pharmacological, biochemical and electrophysiological studies of LC and NE in relation to sleep regulation, Siegel and Rogawski (1988) suggested that a function of REM sleep is the regulation of noradrenergic receptor sensitivity. According to this theory REM sleep serves to upregulate and/or prevent downregulation of brain NE receptors, since the LC neurons are inactive during REM and the NE availability is reduced. Moreover, REM sleep deprivation induces a decrease in β-adrenergic receptor binding and may be downregulating the NE receptors, confirming this hypothesis (Mogilnicka et al. 1986).

II. Neurotransmitters and Single-Unit Studies

In 1975, Hobson et al. proposed a reciprocal interaction model of sleep-wake cycle control to attempt to explain at the cellular level the basis of sleep control. In its original form the reciprocal interaction model postulated that the behavioral states constituting the sleep-wake cycle were a function of the out-of-phase discharge profiles displayed by two brain stem neuronal populations. This hypothesis was based on the findings that neurons from the PRF discharge selectively during REM sleep (REM-ON cells) (Hobson et al. 1974), whereas neurons from the RD and LC decrease or cease their discharge during REM sleep (REM-OFF cells). The model was supported by experiments of cholinergic injections which showed a striking increase in REM sleep after carbachol microinjections into the PRF (Baghdoyan et al. 1987).

According to the reciprocal interaction model, the generation of REM sleep depends on the cessation of REM-OFF cells, which were postulated to play a monoaminergic permissive role by disinhibiting the REM-ON cells. The cells that were disinhibited were supposed to play a cholinergically mediated role in the executive mechanism of REM sleep. This model suggested that the REM-ON cells present feedforward (to REM-OFF cells) and feedback (auto) excitatory loops, while the REM-OFF cells have feedforward (to REM-ON cells) and feedback (auto) inhibitory interconnections. A REM sleep period finishes when the excitatory influence of PRF to the REM-OFF cells reaches a level of monoaminergic activity in which the animal awakes or changes to a SWS state.

The original version of the reciprocal interaction model postulated that the PRF area was the REM sleep center controlling this state. However, the electrophysiological studies in the PRF were performed in restrained cats. Work carried out in freely moving cats showed that the PRF neurons that discharged during REM sleep also discharged during waking movements and could more likely be related to the startle reflex and motor activation during both REM sleep and wakefulness (Siegel and McGinty 1977). These latter studies suggested that the PRF activity was not exclusively related to REM sleep. Moreover, neurotoxic lesions of these structures did not modify the amount of REM sleep (DRUCKER-COLÍN and BERNAL-PEDRAZA 1983; STASTRE et al. 1981); however, under such conditions, carbachol was then unable to increase REM sleep (DRUCKER-COLÍN AND BERNAL-PEDRAZA 1983). It was thus suggested that PRF neurons are probably involved but are not essential for the REM sleep generation (DRUCKER-COLÍN and BERNAL-PEDRAZA 1983). The PRF REM sleep center notion was thus no longer tenable.

Based on recent electrophysiological, neuronanatomical, carbachol microinjection and lesion studies, the initial formulation of the reciprocal theory was altered but retained many of its original features (HOBSON et al. 1986; MCCARLEY and MASSAQUOI 1992). The concept that the sleep-wake cycle is generated by the interaction of multiple and widely distributed sets of anatomically distinct groups of neurons has gradually replaced the hypothesis that sleep is generated by a single, highly localized neuronal oscillator.

The present model of the reciprocal interaction hypothesis maintains the old postulate, which suggests the causal interdependence between the reciprocal discharge of cholinergic and cholinoceptive REM-ON cells and the putatively aminergic REM-OFF cell populations. However, the present model considers the existence of several cell populations distributed widely along the brain stem, rather than a unique REM-ON and REM-OFF structure. This consideration probably emerges from the observation that several nuclei such as the parabrachialis lateralis, laterodorsal (LDT) and pedunculopontine tegmental (PPT) and magnocellularis nuclei also demonstrate REM-ON cells, whereas the parabrachial nucleus and the raphe pontis, pallidus, medianus superioris and magnus contain REM-OFF cells (see HOBSON et al. 1986). The great number of multiple neuron populations that interact to regulate REM sleep probably explains why only large lesions of the brain stem suppress this behavioral state. A simplified account of the model's dynamics is that the REM-ON or REM executive neurons in the cholinergic mesopontine nuclei LDT and PPT activate reticular formation effector neurons in a positive feedback interaction to produce the onset of REM sleep. This also excites REM-OFF permissive or REM suppressive neurons in the raphe system, LC and peribrachial areas. As the REM-OFF neurons become active at the end of a REM sleep period due to their recruitment by REM-ON activity, they then terminate REM sleep because of their inhibition of REM-ON neurons. REM-OFF neuronal activity is maximal just after REM sleep and then decreases in the subsequent non-REM sleep period and becomes minimal at the onset of REM sleep due to a self-inhibitory feedback. This decreased REM-OFF activity disinhibits REM-ON cells and allows the onset of a REM sleep episode. Then the cycle repeats itself (MCCARLEY and MASSAQUOI 1992). There are recent neurobiological data relevant to the cell population interactions mentioned above.

There are strong in vitro and in vivo data indicating the excitatory effects of ACh on PRF neurons (see Hobson et al. 1986; Baghdoyan et al. 1987; Sakai 1988; Steriade and McCarley 1990), which is clear anatomical evidence of cholinergic projections from the LDT/PPT nuclei to both PRF and bulbar reticular formations (Shiromani et al. 1988). These projections depolarize and excite neurons in the effector systems important in REM sleep including those for REMs, PGO waves, muscle atonia and EEG desynchronization. The joint activation of these neuronal groups produces the state of REM sleep. In terms of modeling the positive feedback, a key consideration is the fact that ACh hyperpolarizes the cholinergic neurons (Luebke et al. 1992a), and thus the growth of REM sleep-ON neuronal activity cannot be due to a positive feedback within the LDT/PPT neuronal population alone. Instead the positive feedback for exponential growth in REM-ON activity likely stems from the inclusion of reticular formation neurons in the loop LDT/PPT – PRF – LDT/PPT. Additional positive feedback occurs via reticulo-reticular connections. In vitro data indicate excitatory effects of the excitatory amino acids (EAs) on LDT/PPT neurons (Leonard and Sanchez 1991; Sanchez et al. 1991) and anatomical data indicate reticulo-LDT/PPT projections (Higo et al. 1990). Moreover there is strong evidence that EAs are the principal excitatory transmitters of PRF (Stevens et al. 1990). Further support for the concept of PRF-LDT/PPT interaction comes from the data of Lydic et al. (1991), showing that excitation of PRF neurons increases the release of brain stem ACh, presumably as a result of PRF excitation of LDT/PPT neurons.

Because in some brain stem areas these cell groups are not anatomically segregated and may instead be neurochemically mixed or interpenetrated, as in the LC and peribrachial areas, the new model suggests a gradient of sleep-dependent membrane excitability changes that may be a function of the connectivity strength within an anatomically distributed network. The connectivity strength may be influenced by the degree of neurochemical interpenetration between the REM-ON and the REM-OFF cells. Although the latter hypothesis has yet to be proven, it nevertheless has an important heuristic value for further experiments.

There is anatomical evidence for cholinergic projections to both LC and RD (Jones 1990). The REM-ON neuronal excitation of RD neurons may be mediated through the reticular formation; there is evidence for an EA excitatory effect on both LC and RD neurons.

In vitro evidence indicates that most cholinergic neurons in the LDT are inhibited by serotonin (Luebke et al. 1992b). However, there are no data supporting the NE inhibitory effect on EA or cholinergic neurons in the PRF or the LDT/PPT nuclei, respectively.

Several studies have demonstrated the NE inhibition of LC neurons and of serotonergic inhibition of RD neurons, and anatomical studies indicate the presence of recurrent inhibitory collaterals (see Steriade and McCarley 1990). Moreover, the above-mentioned studies of Cespuglio et al. (1988,

1990) indicate that the RD may be autoinhibited during REM sleep by the increased dendritic release during this state.

Since there are connections from the suprachiasmatic nucleus (SNC) to both LC and RD (MOORE 1990), it has been suggested that the input from circadian systems (SNC) controls the circadian modulation of the decline in the activity of REM-OFF population at the onset of sleep (MCCARLEY and MASSAQUOI 1986).

Because the initial model of reciprocal interaction was hotly debated and many points were disclaimed on the basis of experimental data, the recent version has not produced a strong impact on the research community. However, this hypothesis has a strong heuristic value, and depending on its future development may be able to unify the concept of sleep mechanisms in the future.

Table 2 lists the neurotransmitters studied in relation to sleep regulation, regardless of mechanism or degree of assigned importance (see also Chaps. 4–8, this volume).

F. Final Considerations

One of the common features of the different concepts of mechanisms of sleep is the idea of a single form of SWS and a single form of REM sleep. Thus, most but not all theories postulate the unitary nature of the mechanisms which regulate SWS and REM sleep.

There are, however, alternative suggestions, as for example the sequential hypothesis of sleep functions (AMBROSINI et al. 1988), which suggests a dependence of the functions of the sleeping brain on the nature of the previous waking experience. This hypothesis, which is supported by studies

Table 2. List of the various neurotransmitters studied in relation to sleep

Substances		SWS	REM sleep
Neurotransmitters			
Serotonin	+	↑	No change
	–	↓	↓
Norepinephrine	+	No change	↑
	–	↓	↑ or ↓
Acetylcholine	+	↓	↑
	–	No change	↓
Dopamine	+	↓	↓
	–	↑	↑
Histamine	+	↓	No change
	–	↑	No change
GABA	+	↑	↑ or ↓

in which an increment of SWS or REM sleep was observed after the acquisition of a learning paradigm (see Smith 1985), in the final analysis, however, considers SWS and REM sleep as a unitary process, remarkably similar to the original Hernández-Peón postulate of sleep transitions.

Finally, a different alternative hypothesis which could explain the mechanisms of sleep has to do with the consideration that the many behavioral patterns of activities during waking could be responsible for triggering different kinds of SWS and REM sleep during the night. In other words, it is conceivable that sleep seen after feeding, stress, coitus, fatigue or infections, ect. is not necessarily triggered by the same mechanism and therefore we cannot talk about the same types of sleep behavior. A follow-up consideration of this hypothesis could then be that SWS and REM sleep are indeed not unitary phenomena, but they each could express different polygraphic, biochemical, physiological and functional manifestations depending on the previous waking activity. Thus, the consideration of multifactorial, but specific interactions between brain structure that are the substrate for the different types of sleep could explain the reason why so many sleep substances, neurotransmitters and brain structures are said to be involved in sleep regulation. This hypothesis suggests that the different physiological roles of sleep depend also on the nature of the previous waking experience. However, the definition of sleep as a heterogeneous behavioral state does not have as of today sufficient experimental support.

The various sleep theories discussed in the text are listed in Table 3 and are grouped as tightly as possible. It is likely that not all theories or hypotheses which exist today have been included; we apologize for such omissions, due mostly to space requirements. However, it should be noted that most "wet" or "dry" theories do not view the sleep-wake cycle as a unitary process. As a final reflection, it is also clear that, after more than 40 years of modern research on sleep, knowledge about the intrinsic mechanism

Table 3. Synopsis of the various theories or hypotheses of the mechanisms of sleep from early to more modern versions

1. Passive theory of sleep
2. Passive AARS theory
3. Active theory of sleep
4. Early humoral theory of sleep
5. Theory of sleep-promoting substances
6. Monoaminergic theory
7. Cholinergic theory
8. Reciprocal interaction model (early and more recent modified version)
9. The two-process model
10. The excitostat hypothesis
11. Hypothesis of intrinsic properties of thalamic neurons and the regulation of the EEG
12. Hypothesis of sleep as a result of previous waking experience (not one kind of sleep but several types)

of sleep still remains quite elusive, despite the great strides which have been made in that direction.

G. Summary

This chapter attempts to review in a succinct manner the most important hypotheses about the mechanisms of sleep from a historical perspective and the development of the various ontogenic concepts. It describes how the early concepts of sleep viewed this biological activity as merely the result of passive forces mostly linked to tiredness and loss of sensory influences, until several landmark observations demonstrated very clearly that sleep is the result of very elaborate active brain mechanisms which involve several brain structures. A review is also made of the development of the concepts involving neurotransmitters and the so-called sleep factors, most of which are of a peptidic nature. In this chapter, an attempt is made to reconcile the neuroanatomical localization of the cell groups involved in sleep regulation with particular neurotransmitters and/or peptides. Finally, in view of the ample number of nuclei and substances purported to be involved in sleep regulation, a hypothesis is presented suggesting that sleep depends on previous waking activity and that the events during waking induce on a day-to-day basis different chemical signals which activate different neuronal groups which in turn determine how we sleep.

References

Akerstedt T, Gillberg M (1981) The circadian variation of experimentally displaced sleep. Sleep 4:159–169

Ambrosini MV, Sadile AG, Gironi Carnevale U, Mattiaccio M, Giuditta A (1988) The sequential hypothesis of sleep function. I. Evidence that the structure of sleep depends on the nature of the previous waking experience. Physiol Behav 43:325–337

Aserinsky E, Kleitman N (1953) Regular occurring periods of eye motility and concomitant phenomena during sleep. Science 118:273–274

Baghdoyan H, Rodrigo-Angulo M, McCarly R, Hobson A (1987) A neuroanatomical gradient in the pontine tegmentum for the cholinoceptive induction of desynchronized sleep signs. Brain Res 414:245–261

Bandler R (1969) Aggression induced in rats by cholinergic stimulation of hypothalamus. Nature 224:1035–1036

Bartholini A, Pepeu G (1967) Investigations into the acetylcholine output from the cerebral cortex of the cat in the presence of hyoscine. Br J Pharmachol Chemother 31:66–74

Batini C, Moruzzi G, Palestini M, Rossi GF, Zanchetti A (1959) Effects of complete pontine transections on the sleep-wakefulness rhythm: the midpontine pretrigeminal preparation. Arch Ital Biol 97:1–12

Borbély A (1982) A two process model of sleep regulation. Hum Neurobiol 1: 195–204

Borbély A (1986) Secrets of sleep. Basic, New York

Borbély A, Tobler I (1989) Endogenous sleep-promoting substances and sleep regulation. Physiol Rev 69:605–670
Borbély A, Baumann AF, Brandeis D, Strauch I, Lehmann D (1981) Sleep-deprivation: effect on sleep stages and EEG power density in man. Electroencephalogr Clin Neurophysiol 51:483–493
Borbély A, Tobler I, Hanagasioglu M (1984) Effect of sleep deprivation on sleep and EEG power spectra in the rat. Behav Brain Res 14:171–182
Bremer F (1935) Cerveau "isolé" et physiologie du sommeil. C R Soc Biol (Paris) 118:1235–1241
Bremer F (1938) L'activité électrique de l'ecorce cérébrale et le problème physiologique du sommeil. Boll Soc Ital Biol Sper 13:271–290
Buguet A (1969) Monoamines et sommeils. V. Etude des relations entre les structures monoaminergiques du pont et les pointes pontogeniculo occipitales du sommeil. Thesis, University of Lyon, p 214
Carli G, Zanchetti A (1965) A study of pontine lesions suppressing deep sleep in the cat. Arch Ital Biol 103:751–789
Cespuglio R, Chastrette N, Jouvet M (1988) Opposite variations of 5-hydroxyindoleacetic acid (5-HIAA) extracellular concentrations, measured with voltammetry either in the axonal nerve endings or in the cell bodies of the nucleus raphe dorsalis, throughout the sleep-waking cycle. C R Acad Sci (Paris) 307:817–823
Cespuglio R, Chastrette N. Prevautel H, Jouvet M (1990) Serotonine and hypnogenic factors: functional relationship for sleep induction. In: Inoué S, Krueger JM (eds) Endogenous sleep factors. Academic, The Hague
Chastrette N, Cespuglio R (1985) Influence of propiomelanocortin derived peptides on the sleep-waking cycle of the rat. Neurosci Lett 62:365–370
Collier B, Mitchell JF (1967) The central release of acetylcholine during consciousness and after brain lesions. J Physiol (Lond) 210:424
Danguir J (1986) Intracerebroventricular infusion of somatostatin selectively increases paradoxical sleep in rats. Brain Res 367:26–30
Danguir J, De Saint-Hilaire-Kafi S (1988) Somatostatin antiserum blocks carbachol-induced increase of paradoxical sleep in the rat. Brain Res Bull 20:9–12
Danguir J, Nicolaïdis S (1984) Chronic intracerebroventricular infusion of insulin causes selective increase of slow wave sleep in rats. Brain Res 306:97–103
Dement WC, Mitler MM, Henriksen SL (1972) Sleep changes during chronic administration of parachlorophenylalanine. Rev Can Biol 31:239–246
Drucker-Colín R (1973) Crossed perfusion of a sleep inducing brain tissue substance in conscious cats. Brain Res 56:123–134
Drucker-Colín R, Bernal-Pedraza JG (1983) Kainic acid lesions of gigantocellular tegmental field (FTG) neurons does abolish REM sleep. Brain Res 272:387–391
Drucker-Colín R, Prospéro-García O (1990) Neurophysiology of sleep. In: Thorpy MJ (ed) Handbook of sleep disorders. Dekker, New York, pp 33–53
Drucker-Colín R, Spanis CW, Hunyadi J, Sassin JF, McGaugh JL (1975a) Growth hormone effects on sleep and wakefulness in the rat. Neuroendocrinology 18: 1–8
Drucker-Colín R, Spanis CW, Cotman CW, McGaugh JL (1975b) Changes in protein levels in perfusates of freely moving cats: relation to behavioral state. Science 187:963–965
Drucker-Colín R, Zamora J, Bernal-Pedraza J, Sosa B (1979a) Modification of REM sleep and associated phasic activities by protein synthesis inhibitors. Exp Neurol 63:458–467
Drucker-Colín R, Dreyfus-Cortés G, Bernal-Pedraza J (1979b) Differences in multiple unit activity discharge rate during short and long REM sleep periods: effects of protein synthesis inhibition. Behav Neural Biol 26:123–127
Drucker-Colín R, Tuena de Gómez-Puyou M, Gutiérrez MC, Dreyfus-Cortés G (1980) Immunological approach to the study of neurohumoral sleep factors:

effects on REM sleep of antibodies to brain stem proteins. Exp Neurol 69: 563–575
Drucker-Colín R, Bowersox SS, McGinty DJ (1982) Sleep and medial reticular unit responses to protein synthesis inhibitors: effects of chloramphenicol and thiamphenicol. Brain Res 252:117–127
Drucker-Colín R, Bernal-Pedraza J, Fernández-Cancino F, Oksenberg A (1984) Is vasoactive intestinal polypeptide (VIP) a sleep factor? Peptides 5:837–840
Drucker-Colín R, Aguilar-Roblero R, Arankowsky-Sandoval G (1985) Re-evaluation of the hypnogenic factor notion. In: Wauquier A, Gaillard JM, Monti JM, Radulovacki R (eds) Sleep: neurotransmitters and neuromodulators. Raven, New York, pp 291–304
Drucker-Colín R, Prospéro-García O, Arankowsky-Sandoval G, Pérez-Montfort R (1988) Gastropancreatic peptides and sensory stimuli as REM sleep factors. In: Inoué S, Schneider-Helmert D (eds) Sleep peptides: basic and clinical approaches. Japan Scientific Society, Tokyo; Springer, Berlin Heidelberg New York, pp 73–94
Drucker-Colín R, Arankowsky-Sandoval G, Prospéro-García O, Jiménez-Anguiano A, Merchant H (1990) The regulation of REM sleep: some considerations on the role of vasoactive intestinal peptide, acetylcholine and sensory modalities. In: Mancia M (ed) The diencephalon and sleep. Raven, New York, pp 313–330
Dusan-Peyrethon D, Froment JL (1968) Effets du disulfiram sur les états de sommeil chez le chat. C R Soc Biol (Paris) 162:2144–2145
Dusan-Peyrethon D, Peyrethon J, Jouvet M (1968) Suppression sélective du sommeil paradoxal chez le chat par alpha méthil-dopa. C R Soc Biol (Paris) 162:116–118
Froment JL, Petitjean F, Bertrand N, Jouvet M (1974) Effects delínjection intracérébrale de 5,6-hydroxytryptamine sur les monoamines cérebrals et les états de sommeils du chat. Brain Res 67:405–409
Garcia-Arraras JE, Pappenheimer JR (1983) Site of action of sleep-inducing muramyl peptide isolated from human urine: microinjections studies in rabbit brains. J Neurophysiol 49:528–533
Goldstein R, Pratta D (1984) Sleep in cat: raphe dorsalis and vasotocin. Sleep 7:373–379
Hernández-Peón R (1963) Sleep induced by localized electrical or chemical stimulation of the forebrain. Electroencephalogr Clin Neurophysiol Suppl 24:188–198
Hernández-Peón R (1965) A cholinergic hypnogenic limbic forebrainhindbrain circuit. In: Jouvet M (ed) Aspects anatomo-fonctionnels du sommeil. Centre National de la Recherche Scientifique, Paris
Hess WR (1944) Das Schlafsyndrom als Folge dienzephaler Reizung. Helv Physiol Pharmacol Acta 2:305–344
Higo S, Ito K, Fuchs D, McCarley RW (1990) Anatomical interconnections of the peductulopontine tegmental nucleus and the nucleus prepositus hypoglossi in the cat. Brain Res 536:79–85
Hilakivi I (1983) The role of beta- and alpha-adrenoreceptors in the regulation of the states of the sleep-waking cycle in the cat. Brain Res 277:109–118
Hilakivi I, Leppavuori A (1984) Effect of methoxamine, an alpha-1 adrenoreceptor agonist, and prazosin, an alpha-1 antagonist, on the stages of the sleep-waking cycle in the rat. Acta Physiol Scand 120:363–372
Hobson A, McCarley RW, Freedman R, Pivik RT (1974) Time-course of discharge rate changes in cat pontine brainstem neurons during the sleep cycle. J Neurophysiol 37:1297–1309
Hobson A, McCarley RW, Wyzinski PW (1975) Sleep cycle oscillation: reciprocal discharge by two brainstem neuronal groups. Science 189:55–58
Hobson A, Lydic R, Baghdoyan H (1986) Evolving concepts of sleep cycle generation: from brain centers to neuronal populations. Behav Brain Sci 9:371–448

Honda K, Komoda Y, Nishida S, Nagasaki H, Higashi A, Uchizono K, Inoué S (1984) Uridine as an active component of the sleep promoting substance: its effect on nocturnal sleep in rats. Neurosci Res 1:243–252
Horne J (1988) Why we sleep? Oxford University Press, Oxford
Huttenlocher PR (1961) Evoked and spontaneous activity in single units of medial brainstem during natural sleep and waking. J Neurophysiol 24:451–468
Inoué S, Honda K, Komoda Y (1985) Sleep-promoting substances. In: Wauquier A, Gaillard JM, Monti JM, Radulovacki M (eds) Sleep neurotransmitters and neuromodulators. Raven, New York, pp 305–318
Ishimori K (1909) True cause of sleep – a hypnogenic substance as evidenced in the brain of sleep-deprived animals. Tokyo Igakkai Zasshi 23:429–457
Jacobs BL, Jones BE (1978) The role of central monoamine and acetylcholine systems in sleep-wakefulness states: mediation or modulation? In: Bucher LL (ed) Cholinergic, monoaminergic interactions of the brain. Academic, New York, pp 271–290
Jiménez-Anguiano A, Báez-Saldaña A, Drucker-Colín R (1994) Cerebrospinal fluid (CSF) extracted immediately after REM sleep deprivation prevents REM rebound and contains vasoactive intestinal peptide (VIP). Brain Res 631:345–348
Jones BE (1990) Influence of the brain stem reticular formation, including intrinsic monoaminergic and cholinergic neurons, on forebrain mechanism of sleep and waking. In: Mancia M, Marini G (eds) The Diencephalon and sleep. Raven, New York
Jouvet M (1962) Recherches sur les structures nerveuses et les mécanismes responsables des différentes phases du sommeil physiologique. Arch Ital Biol 100:125–206
Jouvet M (1984) Neuromediateurs et facteurs hypnogenes. Rev Neurol (Paris) 140: 389–407
Jouvet M, Delorme JF (1965) Locus coeruleus et sommeil paradoxal. C R Soc Biol (Paris) 159:895–899
Jouvet M, Moruzzi G (1972) Neurophysiology and neurochemistry of sleep and wakefulness. Springer, Berlin, Heidelberg, New York
Kleitman N (1939) Sleep and wakefulness. University of Chicago Press
Kornmüller AE, Lux HD, Winkel K, Klee M (1961) Neurohumoral ausgelöste Schlafzustände an Tieren mit gekreuztem Kreislauf unter der Kontrolle von EEG-Ableitungen. Naturwissenschaften 48:503–505
Krueger J, Karnovsky ML (1987) Sleep and the immune response. Ann NY Acad Sci 496:510–516
Krueger J, Pappenheimer JR, Karnovsky ML (1978) Sleep-promoting factor S: purification and properties. Proc Natl Acad Sci USA 75:5235–5238
Krueger J, Walter J, Dinatello CA, Wolf SM, Chedid L (1984) Sleep-promoting effects of endogenous pyrogen (interleukin-1). Am J Physiol 246:994–999
Krueger JM, Kubillus S, Shoham S, Davenne D (1986) Enhancement of slow wave sleep by endotoxin and lipid. Am J Physiol 251:R591–R597
Krueger JM, Rosenthal RS, Martin SA, Walter J, Davenne D, Shoham S, Kubillus L Biemann K (1987) Bacterial peptidoglycans as modulators of sleep. I. Anhydro forms of muramyl peptides enhance somnogenic potency. Brain Res 403:249–257
Laguzzi R, Petitjean F, Pujol JF, Jouvet M (1972) Effets de l'injection intraventriculaire de 6-hydroxydopamine. II. Sur le cycle veille-sommeils du chat. Brain Res 48:295–310
Legendre R, Pieron H (1913) Rechercher sur le besoin de sommeil consécutif à une veille prolongée. Z Allg Physiol 14:235–262
Leonard CS, Sanchez R (1991) Synaptic potentials in mesopontine cholinergic neurons evoked by local stimulation in vitro. Soc Neurosci Abstr 17:1042

Lindsley DB, Bowden JW, Magoun HW (1949) Effect upon the EEG of acute injury to the brain stem activating system. Electroencephalogr Clin Neurophysiol 5: 295–296
Luebke JI, McCarley RW, Greene RW (1992a) Inhibitory action of the acetylcholine agonist on neurons of the rat laterodorsal tegmental nucleus in the in vitro brain stem slice. Sleep Res 21:32–34
Luebke JI, Greene RW, Semba K, Kamondi A, McCarley RW, Reiner PB (1992b) Serotonin hyperpolarizes cholinergic low threshold burst neurons in the rat laterodorsal tegmental nucleus in vitro. Proc Natl Acad Sci USA 89:743–747
Lydic R, Baghdoyan H, Lorine Z (1991) Microdialysis of cat pons reveals enhanced acetylcholine release during state dependent respiratory depression. Am J Physiol 261:766–770
Magnes J, Moruzzi G, Pompeiano O (1961) Synchronization of the EEG produced by low-frequency electrical stimulation of the regions of the solitary tract. Arch Ital Biol 99:33–67
Martin SA, Karnovsky ML, Krueger JM, Pappenheimer JR, Biemann K (1984) Peptidoglycans as promoters of slow-wave sleep. I. Structure of the sleep-promoting factor isolated from human urine. J Biol Chem 259:12652–12658
Masek K, Kadlecova O, Pöschlova N (1976) Effect of intracisternal administration of prostanglandin El, on waking and sleep in the rat. Neuropharmacology 16:491–494
Matsumoto J, Jouvet M (1964) Effecs de réserpine, DOPA et 5HTP sue les 2 états de sommeil. C R Soc Biol (Paris) 158:2137–2140
Matsumoto J, Watanabe S (1967) Paradoxical sleep: effects of adrenergic blocking agents. Proc Jpn Acad 43:680–683
Matsumura H, Goh Y, Ueno R, Sakai T, Hayaishi O (1988) Awaking effect of PGE_2 microinjected into the preoptic area of rats. Brain Res 444:265–272
McCarley RW, Massaquoi SG (1986) A limit cycle mathematical model of the REM sleep oscillator system. Am J Physiol 251:1011–1029
McCarley RW, Massaquoi SG (1992) Neurobiological structure of the revised limit cycle reciprocal interaction model of RE; cycle control. J Sleep Res 1:132–138
McCormick DA (1992) Cellular mechanisms underlying cholinergic and noradrenergic modulation of the neuronal firing mode in the cat and guinea pig dorsal lateral geniculate nucleus. J Neurosci 12:278–289
McGinty DJ, Harper RM (1976) Dorsal raphe neurons: depression of firing during sleep in cats. Brain Res 101:569–575
Mendelson WB, Seater S, Grold P, Gillin JC (1980) The effect of growth hormone administration on human sleep: a dose-response study. Biol Psychiatr 15:613–618
Merchant-Nancy H, Vázquez J, Aguilar-Roblero R, Drucker-Colín R (1992) c-fos proto-oncogene changes in relation to REM sleep duration. Brain Res 579: 342–346
Mogilnicka E, Przewlocka B, van Luijtelaar E, Klimek V, Coenen AM (1986) Effects of REM sleep deprivation on central alpha 1- and beta-adrenoceptors in rat brain. Pharmacol Biochem Behav 25:329–332
Monnier M, Kalbere M, Krupp P (1960) Functional antagonism between diffuse reticular and intralaminary recruiting projections in the medial thalamus. Exp Neurol 2:271–289
Monnier M, Koller T, Graber S (1963) Humoral influences of the induced sleep and arousal upon electrical brain activity of animals with crossed circulation. Exp Neurol 8:264–277
Moore RY (1990) The circadian system and sleep-wake behavior. In: Montplaisir J, Godbout R (eds) Sleep and biological rhythms: basic mechanisms and application to psychiatry. Oxford University Press, New York, pp 3–10
Morgane PJ (1969) Chemical mapping of hypnogenic and arousal systems in the brain. Psychophysiology 6:219

Moruzzi G, Magoun HW (1949) Brain stem reticular formation and activation of the EEG. Electroencephalogr Clin Neurophysiol 1:455–473
Moruzzi G (1972) The sleep waking cycle. Rev Physiol 64:1–165
Mouret JR, Jouvet M (1968) Insomnia following parachlorophenylalanine in the rat. Eur J Pharmacol 5:17–22
Mouret JR, Vilppula A, Frachon N, Jouvet M (1968) Effects d'un inhibiteur de la monoamine oxidase sur le sommeil du rat. C R Soc Biol (Paris) 162:914–917
Nauta WJH (1946) Hypothalamic regulation of sleep in rats. Experimental study. J Neurophysiol 9:285–316
Obal F (1986) Effects of peptides (DSIP, DSIP analogues, VIP, GRF, CCK) on sleep in the rat. Clin Neuropharmacol 9:459–461
Obal F, Sary G, Alfoidi P, Rubicsek G (1986) Vasoactive intestinal polypeptide promotes sleep without effects on brain temperature in rats at night. Neurosci Lett 64:236–240
Oneo H, Ueno R, Fujita I, Nishino H, Oomura Y, Hayaishi O (1988) Prostanglandin D_2 a cerebral sleep-inducing substance in monkeys. Proc Natl Acad Sci USA 85:4082–4086
Oswald I (1969) Human brain protein, durgs and dreams. Nature 223:893–897
Pappenheimer JR, Miller TB, Goodrich CA (1967) Sleep-promoting effects of cerebrospinal fluid from sleep-deprived goats. Proc Natl Acad Sci USA 58: 513–517
Pappenheimer JR, Koski G, Fencl V, Karnovsky ML, Krueger J (1975) Extraction of sleep-promoting factor S from cerebrospinal fluid and from brains of sleep-deprived animals. J Neurophysiol 38:1299–1311
Pieron H (1913) Le Problème physiologique du sommeil. Masson, Paris Prospéro-García O, Ott T, Durcker-Colín R (1987) Cerebroventricular infusion of cholecystokinin (CCK-8) restores REM sleep in parachlorophenylalanine (PCPA)-pretreated cats. Neurosci Lett 78:205–210
Puizillout JJ, Gaudin-Chazel G, Daszuta A (1979) Release of endogenous serotonin from "encephale isolé" cats. II. Correlations with raphe neuronal activity and sleep and wakefulness. J Physiol (Paris) 75:531–535
Pujol JF (1970) Contribution à l'étude des modifications de la régulation de métabolisme des monoamines centrales pendant le sommeil et la veille. Thesis, University of Paris, p 192
Pujol JF, Bobilllier P, Buguet A, Jones B, Jouvet M (1969) Biosynthèse de la sérotonine cérébrale: étude neurophysiologique et biochimique après p-cholophénylalanine et déstruction du système du raphé. C R Acad Sci (Paris) 268:100–102
Ranson SW (1939) Somnolence caused by hypothalamic lesions in the monkey. Arch Neurol Psychiatry 41:1–23
Renault J (1967) Monoamines et sommeil. Role du système du raphé et la sérotonine cérébrale dans l'endormissement. Thesis, University of Lyon, p 140
Riou F, Cespuglio R, Jouvet M (1982) Endogenous peptides and sleep in the rat. III. The hypnogenic properties of vasoactive intestinal polypeptide. Neuropeptides 2:265–277
Riveau G, Masek K, Parant M, Chedid L (1980) Central pyrogenic activity of muramyl dipeptide. J Exp Med 152:869–877
Roberts WW, Robinson TCL (1969) Relaxation and sleep induced by warming of preoptic region and anterior hypothalamus in cats. Exp Neurol 25:282–294
Rojas-Ramírez JA, Drucker-Colín RR (1973) Sleep induced by spinal cord cholinergic stimulation. Int J Neurosci 5:215–221
Rojas-Ramírez JA, Aguilar-Jiménez E, Posadas-Andrews A, Bernal-Pedraza JG Durcker-Colín R (1977) The effects of various protein synthesis inhibitors on the sleep-wake cycle of rats. Psychopharmacology (Berl) 53:147–150
Sakai K (1988) Executive mechanisms of paradoxical sleep. Arch Ital Biol 126: 239–257

Sakai K, El Mansari M, Lin JS, Zhang JG, Vanni-Mercier G (1990) The posterior hypothalamus in the regulation of wakefulness and paradoxical sleep. In: Mancia M, Marini G (eds) The diencephalon and sleep. Raven, New York, pp 171–198
Sallanon M, Buda C, Janin M, Jouvet M (1985) Implications of serotonin in sleep mechanisms: induction, facilitation. In: Wauquier A, Monti JM, Gaillard JM, Radulovacki M (eds) Sleep: neurotransmitters and neuromodulators. Raven, New York, pp 136–159
Sallanon M, Denoyer M, Kitahama K, Jouvet M (1989) Long-lasting insomnia induced by preoptic lesions and its transient reversal by muscimol injection into the posterior hypothalamus in the cat. Neuroscience 32:669–683
Sánchez R, Khateb A, Mühlethaler M, Leonard CS (1991) Glutamate and NMDA actions on mesopontine cholinergic neurons in vitro. Soc Neurosci Abstr 17:256
Sastre JP, Sakai K, Jouvet M (1981) Are the gigantocellular tegmental field neurons responsible for paradoxical sleep? Brain Res 229:147–161
Schoenenberger GA, Cueni LB, Hatt AM, Monnier M (1972) Isolation and physical-chemical characterization of a humoral, sleep inducing substance in rabbits (factor "delta"). Experientia 28:919–921
Schoenenberger GA, Maier PF, Tobler HJ, Wilson K, Monnier M (1978) The delta EEG (sleep)-inducing peptide (DSIP). XI. Amino-acid analysis, sequence, synthesis and activity of the monapeptide. Pflugers Arch 376:119–129
Shiromani PJ, Armstrong DM, Gillin JC (1988) Cholinergic neurons from the dorsolateral pons project to the medial pons: a WGA-HRP and choline acetyltransferase immunohistochemical study. Neurosci Lett 95:19–23
Shute CC, Lewis PR (1966) Cholinergic and monoaminergic systems of the brain. Nature 212:710–711
Siegel JM, McGinty DJ (1977) Pontine reticular formation neurons: relationship of discharge to motor activity. Science 196:678–680
Siegel JM, Rogawski MA (1988) A function for REM sleep: regulation of noradrenergic receptor sensitivity. Brain Res Rev 13:213–233
Smith C (1985) Sleep states and learning: a review of the animal literature. Neurosci Biobehav Rev 9:157–168
Steriade M, Llinás R (1988) The functional states of the thalamus and the associated neuronal interplay. Physiol Rev 68:649–742
Steriade M, McCarley RW (1990) Brainstem control of wakefulness and sleep. Plenum, New York
Steriade M, Oakson G, Robert N (1982) Firing rates and patterns of midbrain reticular neurons during steady and transitional states of the sleep-waking cycle. Exp Brain Res 46:37–51
Steriade M, Domich L, Oakson G (1986) Reticularis thalami neurons revisited: activity changes during shifts in states of vigilance. J Neurosci 6:68–81
Steriade M, Datta S, Paré D, Oakson G, Curró Dossi R (1990) Neuronal activities in brainstem cholinergic nuclei related to tonic activation processes in thalamocortical systems. J Neurosci 10:2541–2559
Sterman MB, Clemente CD (1962) Forebrain inhibitory mechanisms: cortical synchronization induced by basal forebrain stimulation. Exp Neurol 6:91–102
Stern WC, Morgane PJ (1973) Effects of alpha-methyltyrosine on REM sleep and brain amine levels in the cat. Biol Psychiatry 6:301–306
Stern WC, Jalowiec E, Shabshalowitz H, Morgane PJ (1975) Effects of growth hormone on sleep-waking patterns in cats. Horm behav 6:189–196
Stevens DR, Greene RW, McCarley RW (1990) Pontine reticular formation neurons: excitatory amino acid receptor-mediated responses. In: Horne J (ed) Sleep '90. Pontenagel, Bochum, pp 100–101
Takahashi Y, Kipnis DM, Danghaday WH (1968) Growth hormone secretion during sleep. J Clin Invest 47:2079–2090
Uchizono K, Inoué S, Iriki M, Ishikawa M, Komoda Y, Nagasaki H (1975) Purification of the sleep-promoting substances from sleep-deprived rat brain. In: Walter

R, Meienhofer J (eds) Peptides: chemistry, structure and biology. Ann Arbor Science, Ann Arbor, pp 667–671
Ueno R, Ishikawa V, Nakayama T, Thayaishi O (1982) Prostaglandin D_2 induces sleep when microinjected into the preoptic area of conscious rats. Biochem Biophys Res Commun 109:576–582
Velluti R, Hernández-Peón R (1963) Atropine blockade within a cholinergic hypnogenic circuit. Exp Neurol 8:20–29
Vertes RR (1984) Brain stem control of the events of REM sleep. Prog Neurobiol 22:241–287
Von Economo C (1929) Schlaftheorie. Ergeb Physiol 28:312–339
Webster HH, Friedman L, Jones Be (1986) Modification of paradoxical sleep following transections of the reticular formation at the pontomedullary junction. Sleep 9:1–23
Yumuy J, Mancillas JR, Morales FR, Chase MH (1993) C-fos expression in the pons and medulla of the cat during carbachol-induced active sleep. J Neurosci 13:2703–2718

CHAPTER 2

Principles of Sleep Regulation: Implications for the Effect of Hypnotics on Sleep

A.A. BORBÉLY

A. Basic Processes

Three basic processes underlie sleep regulation: (a) a *homeostatic* process determined by sleep and waking; (b) a *circadian* process, a clock-like mechanism defining the alternation of periods with high and low sleep propensity and which is basically independent of sleep and waking; and (c) an *ultradian* process occurring within sleep and represented by the alternation of the two sleep states, non-REM sleep and REM sleep (Fig. 1). This chapter focuses on sleep homeostasis and on the ultradian dynamics, whereas the circadian aspects are covered in Chap. 3 of this volume.

B. Sleep Homeostasis

Sleep reduction gives rise to increased sleep propensity, whereas excess sleep reduces the tendency for sleep. This general observation indicates that regulatory processes exist that counteract deviations from an average "reference level" of sleep. The term "sleep homeostasis" was coined (BORBÉLY 1980) to characterize the sleep-wake dependent aspect of sleep regulation. Homeostatic mechanisms counteract deviations from a physiological "reference level."

I. EEG Slow Waves as Indicators of "Sleep Pressure"

Non-REM sleep is not a homogeneous substate of sleep, but can be further subdivided according to the predominance of EEG slow-wave activity. For human non-REM sleep, the percentage of slow waves (frequency, 0–2 Hz; minimum amplitude, 75 μV) is the major criterion for scoring stages 2, 3 or 4 (RECHTSCHAFFEN and KALES 1968). Stages 3 and 4 are commonly referred to as slow-wave sleep. In animals, the term "slow-wave sleep" is often used to denote non-REM sleep. This confusing terminology should be abandoned, since a more detailed analysis demonstrated clearly that there are large and systematic variations in the predominance of slow waves also in the non-REM sleep of animals (see below).

Presently, EEG parameters can be assessed by computer-aided methods of signal analysis. One of the most important functional EEG parameters

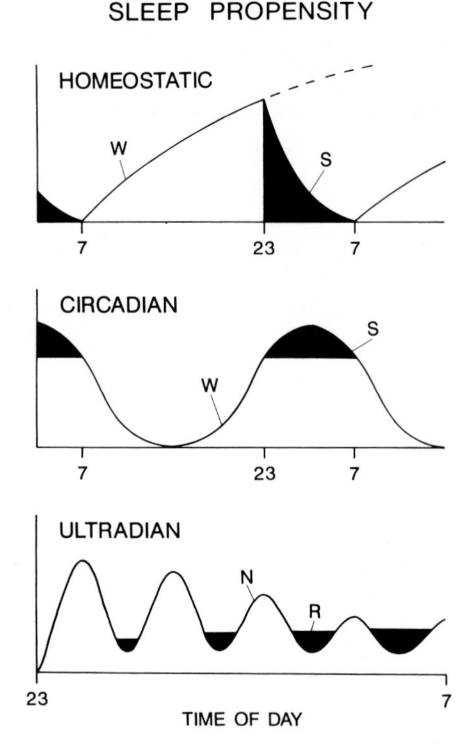

Fig. 1. Schematic representation of the three major processes underlying sleep regulation. *W*, waking; *S*, sleep; *N*, non-REM sleep; *R*, REM sleep. The progressive decline of non-REM sleep intensity is represented in both the *top and bottom diagrams* (decline of ultradian amplitude). The increase in the duration of successive REM sleep episodes is indicated. (From Borbély and Achermann 1992)

will be referred to as "slow-wave activity." It is equivalent to "delta activity" and encompasses components of the EEG signal in the frequency range of approximately 0.5–4.5 Hz. A quantitative measure of slow-wave activity can be obtained by spectral analysis (e.g., Borbély et al. 1981) or period amplitude analysis (e.g., Church et al. 1975). Although each method has its inherent limitations (Geering et al. 1993), they provide a quantitative measure of EEG components in the low-frequency range.

It was recognized as early as 1937 that the predominance of slow waves in the sleep EEG is an indicator of sleep intensity (Blake and Gerard 1937). Subsequent studies confirmed that the responsiveness to stimuli decreases as EEG slow waves become more predominant (e.g., Williams et al. 1964). Under physiological conditions, this EEG parameter can therefore be regarded as an indicator of the depth or the intensity of non-REM sleep. This statement applies not only to humans but also to animals. Thus in the rat, EEG slow-wave activity was recently shown to be inversely related to the rate of spontaneous brief awakenings from sleep (FRANKEN et al. 1991).

1. Global Changes During Baseline Sleep Episodes

Continuous spectral analysis of the sleep EEG made it possible to quantify the slow-wave activity and to delineate its time course during sleep in humans (BORBÉLY et al. 1981) and animals (BORBÉLY et al. 1984; TRACHSEL et al. 1988). Slow-wave activity typically shows a monotonic decline over the major sleep episode.

Figure 2 (upper part) shows the changes in mean EEG power density of human sleep over four cycles for the frequency range between 0.25 and 20 Hz. The values of each bin are expressed relative to the reference level of cycle 1 (100%). Note that although the largest changes occur in the low delta range, they also encompass the frequencies of the theta band.

The typical changes in the EEG spectrum are not restricted to human sleep but are also evident in other mammals. Thus in the rat, the application of quantitative EEG analysis revealed that EEG slow-wave activity as well as EEG amplitude progressively decline in the course of the light period, the animal's circadian sleep period, and increase during the dark period (ROSENBERG et al. 1976; BORBÉLY and NEUHAUS 1979; TRACHSEL et al. 1988). Similar results have been obtained in the hamster (TOBLER and JAGGI 1987).

2. Effect of Sleep Deprivation

It has been shown by various investigators that partial or total sleep deprivation enhances slow-wave sleep in the recovery night (see BORBÉLY 1982 for references). WEBB and AGNEW (1971) presented compelling evidence that slow-wave sleep increases as a function of prior waking. Also the quantitative assessment of slow-wave activity by all-night spectral analysis demonstrated that a night without sleep (i.e., 40.5 h of wakefulness) resulted in the enhancement of this EEG parameter during recovery sleep (BORBÉLY et al. 1981). Figure 2 (lower part) illustrates the changes in power density in the two recovery nights relative to the baseline level (100%). In the first recovery night, the largest increase was present in the low delta range, the part of the spectrum undergoing the most prominent changes in the course of baseline sleep (Fig. 2, upper part). The enhancement of slow-wave activity by sleep deprivation has been confirmed in several studies (for references see BORBÉLY and ACHERMANN 1992), and the extent of the increase was shown to be a function of the duration of prior waking (DIJK et al. 1990a).

There is strong evidence that it is the duration of prior waking per se which determines the level of EEG slow-wave activity, and not some secondary effect of sleep deprivation. The clearest demonstration derives from a study in which naps were recorded at various times during the day (BEERSMA et al. 1987; DIJK et al. 1987). Slow-wave propensity was shown to

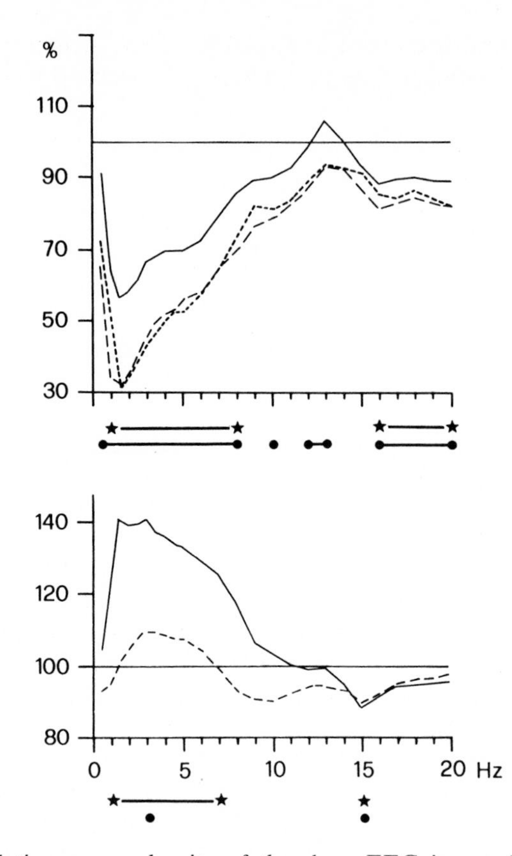

Fig. 2. *Upper* Relative power density of the sleep EEG in non-REM-REM sleep cycles 2(—), 3(--) and 4(. . .) of baseline nights. The values of 0.5-Hz bins (0.25–5.0 Hz) and 1.0-Hz bins (5.25–20.0 Hz) are expressed relative to the value in the first cycle (= 100%). Bins with significant differences between cycles 1 and 2 (*asterisks and connecting lines*) and between cycles 2 and 3 (*filled circles and connecting lines*) are indicated below the abscissa ($P < 0.05$; paired Wilcoxon test, two-sided). *Lower* Relative power density of the sleep EEG in recovery night 1 (*solid line*) and 2 (*interrupted line*) following sleep deprivation (40.5 h of waking). The values are plotted relative to the baseline values (= 100%). Bins with significant differences from baseline are indicated by *asterisks and connecting lines* (recovery 1) and *filled circles and connecting lines* (recovery 2) *below the abscissa* ($P < 0.05$; paired Wilcoxon test, two-sided; $N = 8$). (Modified from Borbély et al. 1981)

build up gradually during the day, its level being a function of the preceding waking period. If sleep is prevented at the time of its habitual occurrence, this buildup proceeded further. These results show that the deprivation procedure does not cause a qualitatively different kind of response, but merely augments a physiological trend.

Effects of sleep deprivation have also been studied in animals. An increase in EEG slow-wave activity after sleep deprivation was first demonstrated in the rabbit (Pappenheimer et al. 1975; Tobler et al. 1990) and was then extensively documented in the rat (Borbély and Neuhaus 1979; Friedman et al. 1979). By varying the duration of sleep deprivation it was shown that the increase in slow-wave activity is a function of the duration of prior waking (Tobler and Borbély 1986, 1990). Also sleep deprivation in other mammalian species (i.e., primates, dog, cat, chipmunk, guinea pig and dolphin) has been shown to increase slow-wave sleep or slow-wave activity (see references in Tobler 1985; also Dijk and Daan 1989; Tobler et al. 1992).

II. Ultradian Dynamics of Slow-Wave Activity and Spindle Activity

1. Baseline Sleep and Recovery Sleep After Sleep Deprivation

Not only the mean and peak level of slow-wave activity is determined by the duration of prior waking and sleep, but also the rise rate within single non-REM sleep episodes (Sinha et al. 1972; Achermann and Borbély 1987; Dijk et al. 1990b). It is evident from Fig. 3 that the rise rate in slow-wave activity decreases over the first three non-REM sleep episodes both under baseline conditions and during recovery from sleep deprivation. In addition, the effect of prolonged waking manifests itself in a more rapid buildup of slow-wave activity within the episodes.

Figure 3 illustrates not only the ultradian changes in slow-wave activity but also those of spindle frequency activity. The term "spindle frequency activity" is used to denote the power density in the frequency range of sleep spindles (13.25–15.0 Hz). There is a close correspondence between this measure and measures based on the occurrence of sleep spindles (Dijk et al. 1993). The time courses of slow-wave activity and spindle frequency activity differ in several respects. The global declining trend of slow-wave activity is not present in the spindle frequency range. Within non-REM sleep episodes, spindle frequency activity shows a bimodal pattern with an initial and a terminal peak. This gives rise to a U-shaped curve within the episode. The negative correlation between slow-wave activity and spindle frequency activity reported by Uchida et al. (1991) is present for the major part of the non-REM sleep episode (Aeschbach and Borbély 1993). However, in the early and late part of the episode, changes in the same direction are seen. The negative correlation which is most prominent in the middle part of the non-REM sleep episode decreases progressively in negativity over consecutive episodes (Aeschbach and Borbély 1993).

2. Neurophysiological Basis of Slow Waves and Sleep Spindles

Recent experiments revealed that hyperpolarized thalamic neurons exhibit fluctuations of the membrane potential in the slow-wave frequency range (Steriade et al. 1991; Leresche et al. 1991; Dossi et al. 1992; see also

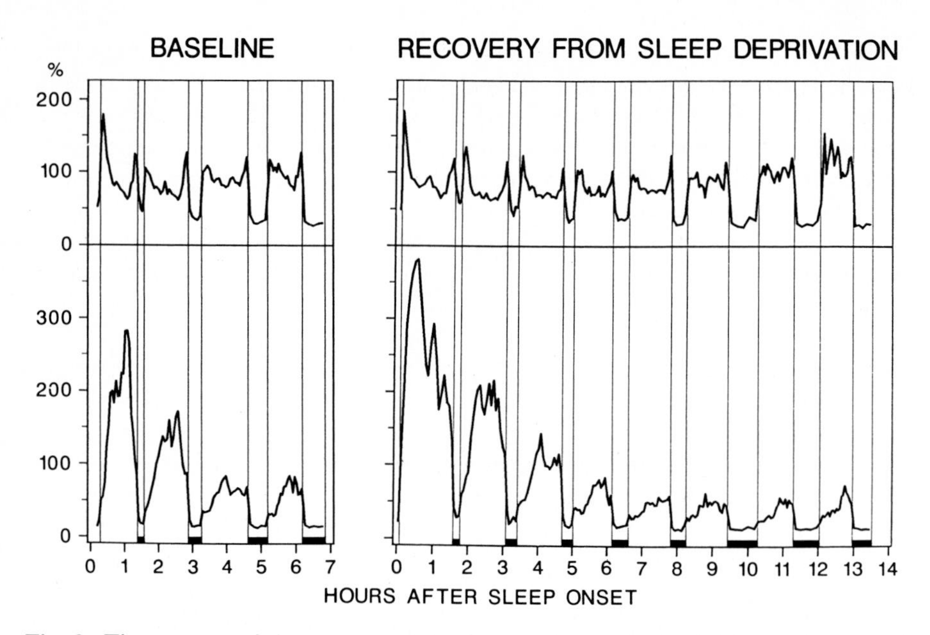

Fig. 3. Time course of slow-wave activity (power density in the 0.75- to 4.5-Hz band; *lower curves*) and activity in the spindle frequency range (13.25–15.0 Hz; *upper curves*) recorded under baseline conditions and after sleep deprivation (36 h of wakefulness). The non-REM sleep episodes were divided into 20 equal parts, the REM sleep episode into 5 equal parts. The curves represent mean percentile values ($N = 8$ except for cycle 8 of recovery sleep where $N = 6$) and are expressed relative to the mean slow-wave activity level in baseline non-REM sleep (= 100%). The mean timing of REM sleep episodes is delimited by *vertical lines and horizontal bars above the abscissa*. (Reanalysis of the data of Dijk et al. 1990b)

McCarley and Massaquoi 1992). At intermediate levels of hyperpolarization, the membrane fluctuates in the frequency range of sleep spindles. The progression of sleep stages from "spindle sleep" to slow-wave sleep (Fig. 3) therefore has a counterpart at the level of thalamocortical neurons. If, as these results seem to suggest, the fluctuations of neuronal membrane potentials are at the origin of the typical patterns in the sleep EEG, sleep homeostasis would be open for investigation at a cellular level.

III. Homeostasis of REM Sleep and Interactions with Non-REM Sleep

During recovery from total sleep deprivation, slow-wave sleep and EEG slow-wave activity exhibit an immediate rebound whereas the increase in REM sleep is delayed to subsequent nights or is not present at all (references in Borbély 1982). Selective REM sleep deprivation augments "REM sleep pressure," which is manifested by the increasing number of interventions required to prevent REM sleep episodes (see Borbély 1982 for references).

However, the occurrence of a REM sleep rebound during recovery sleep was found to be inconsistent (CARTWRIGHT et al. 1967). Thus, although human REM sleep responds to a sleep deficit, the results seem to indicate that it is not as finely regulated as slow-wave sleep. Recent findings contradict this notion. A REM sleep deprivation in the first 5 h of sleep entailed a REM sleep rebound in the subsequent 2.25 h (BEERSMA et al. 1990). A curtailment of sleep duration for 2 nights which induced a substantial REM sleep deficit was followed by a REM sleep rebound in the two recovery nights (BRUNNER et al. 1990). In both experiments, the REM sleep rebound occurred at a time when slow-wave pressure was either low at the end of sleep (BEERSMA et al. 1990) or was much less increased than "REM sleep pressure" (BRUNNER et al. 1990). These results suggest that REM sleep is also finely regulated but that the full manifestation of REM sleep homeostasis is prevented by an elevated slow-wave pressure.

An electrophysiological indicator of an intensity dimension of human REM sleep, comparable to slow-wave activity in non-REM sleep, has not been identified. The density of rapid eye movements is not associated with REM sleep pressure (ANTONIOLI et al. 1981), but is inversely related to slow-wave propensity (BORBÉLY and WIRZ-JUSTICE 1982; FEINBERG et al. 1987). However, a recent partial sleep deprivation study showed an intriguing association between REM sleep pressure and power density of sleep EEG in the upper delta and alpha range (BRUNNER et al. 1993). In the rat (BORBÉLY et al. 1984; FRANKEN et al. 1991) and rabbit (TOBLER et al. 1990), theta activity during REM sleep was enhanced following 24-h sleep deprivation. This change in the EEG could represent an electrophysiological correlate of an increased REM sleep intensity.

C. Effect of Hypnotics on Sleep and the Sleep EEG

I. Benzodiazepine Hypnotics Alter the Sleep EEG

All-night spectral analysis of the sleep EEG revealed that sleep after a hypnotic differs from drug-free, physiological sleep. The typical EEG changes consist in a depression of activity in the low-frequency range (delta and theta bands) and an increase in activity in the spindle frequency range (BORBÉLY et al. 1983, 1985; DIJK et al. 1989; AESCHBACH et al. 1994a). Figure 4 contrasts the changes in the sleep spectra induced by an increase in physiological "sleep pressure" (top) with those induced by a "pharmacological sleep intensification" (bottom). It is obvious that the two curves are almost mirror images of each other. Thus on the basis of the sleep EEG, benzodiazepine (BDZ) hypnotics do not mimick physiological sleep.

The BDZ-induced changes of the sleep EEG do not always parallel the time course of the hypnotic action. Thus after administration of short half-life hypnotics (i.e., triazolam or midazolam) the depression of slow-wave

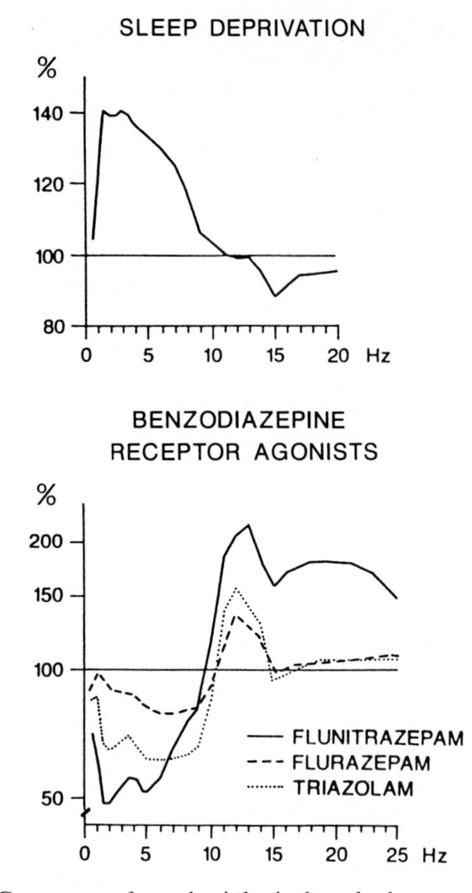

Fig. 4. Sleep EEG spectra after physiological and pharmacological "sleep intensification." The curves represent mean power density relative to baseline (*top*) and placebo (*bottom*). During recovery sleep from sleep deprivation, values in the low-frequency range are enhanced. Benzodiazepine hypnotics taken before bedtime reduce low-frequency activity and enhance activity in higher frequencies (doses: flunitrazepam, 2 mg; flurazepam, 30 mg; triazolam, 0.5 mg). (Modified from Borbély et al. 1981, 1983)

activity persisted unmitigated throughout the sleep period, while the hypnotic action declined (Borbély et al. 1985; Trachsel et al. 1990; Aeschbach et al. 1993). Moreover, the reduction in slow-wave activity was present even in the post-drug night, at a time when the hypnotic effect had vanished or was strongly attenuated (Borbély et al. 1985; Gaillard and Blois 1989). Other EEG effects of hypnotics such as the increase in EEG activity in the spindle frequency range were more closely related to their sleep-promoting action (Borbély et al. 1983; Johnson et al. 1983; Trachsel et al. 1990).

II. Benzodiazepine-Receptor Agonists Have Similar Effects on the Sleep EEG

The two new non-BDZ compounds zolpidem, an imidazopyridine, and zopiclone, a cyclopyrrolone, have recently been introduced in various countries as hypnotics. They both bind to the GABA-BDZ receptor complex and exert an agonistic action that can be antagonized by flumazenil. It has been emphasized that the compounds exhibit distinctive binding profiles to the receptor subunits. However, it is unclear whether their binding characteristics give rise to a specific pharmacological profile that distinguishes these substances in the pharmacotherapy of insomnia.

Zopiclone and zolpidem induced spectral changes that are very similar to those induced by BDZ hypnotics (Fig. 5) (Trachsel et al. 1990; Brunner et al. 1991), yet different from those induced by ethanol (Dijk et al. 1992). On the basis of these observations it has been proposed that a "spectral EEG signature" may reflect the agonistic effect of hypnotics on the GABA-BDZ-receptor complex (see also Steiger et al. 1993).

It was reported in early studies that the reduction in SWS persists after discontinuation of BDZ hypnotics (Gaillard et al. 1973), an effect that was thought to indicate a prolonged alteration of physiological sleep regulation (Borbély et al. 1983). The BDZ antagonist flumazenil was capable of reversing several effects of flunitrazepam on sleep (e.g., prolonged total sleep time, reduced waking after sleep onset, increased sleep efficiency, shortened sleep latency, prolonged REM sleep latency) and on the sleep EEG (enhanced alpha and spindle activity), but did not antagonize the BDZ-induced depression of slow-wave activity in the delta and theta range (Gaillard and Blois 1983, 1989). In fact, this antagonist actually reduced slow-wave activity when administered alone. These results indicate that the reduction in slow-wave activity by BDZ hypnotics is not simply a consequence of an action on the GABA-BDZ receptor complex but must be mediated by other, as yet undefined, processes.

III. Do Hypnotics Disrupt Sleep Homeostasis?

Hypnotics have been reported to exert prominent actions on the sleep stages. Typical changes induced by BDZ hypnotics include the reduction of slow-wave sleep and REM sleep, and the prolongation of REM sleep latency (Kales et al. 1970; Gaillard et al. 1973) (see also Chap. 13, this volume). While some investigators stated that the significance of changes in the amounts of various sleep stages was not known (Kales and Kales 1975), others concluded from such effects that BDZ hypnotics impair the recuperative aspects of sleep (Schneider-Helmert 1988). However, caution is indicated in drawing such conclusions, because the single administration of BDZ hypnotics (Borbély et al. 1985; Achermann and Borbély 1987) and non-BDZ hypnotics (Trachsel et al. 1990; Brunner et al. 1991) does not alter

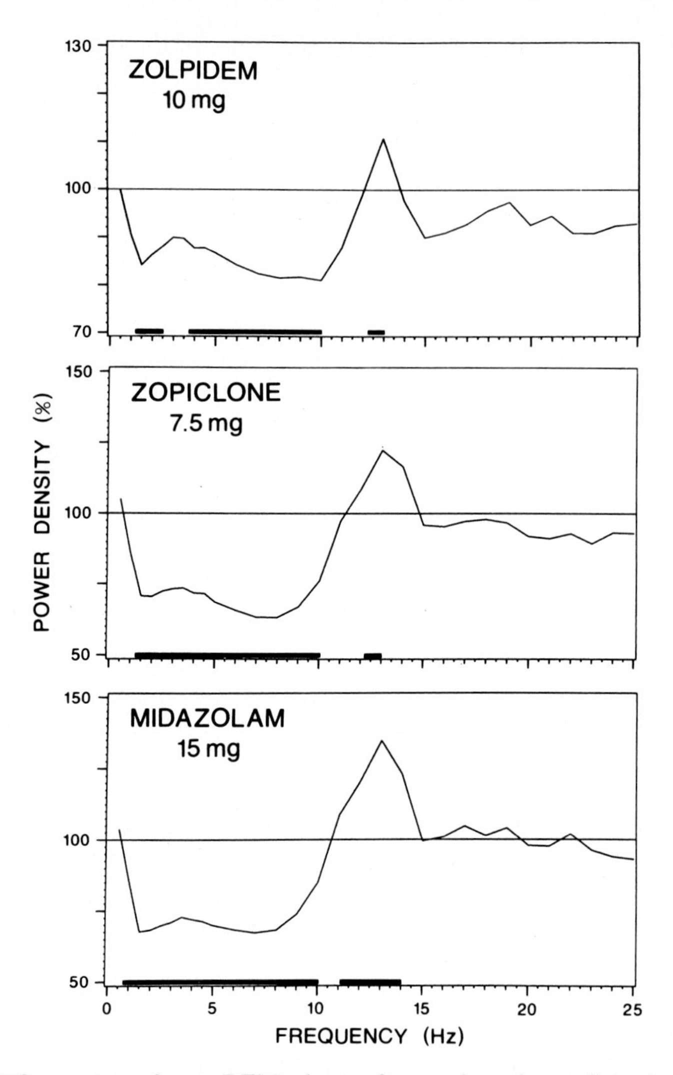

Fig. 5. EEG spectra of non-REM sleep after various benzodiazepine receptor agonists. The curves connect mean values ($N = 8$ for zolpidem; $N = 9$ for zopiclone and midazolam) for successive 0.5- or 1-Hz bins which are expressed as a percentage of the corresponding placebo reference value. *Lines above the abscissae* indicate the frequency ranges in which the values differed significantly from placebo ($P < 0.05$; paired t-test, two-tailed, on log-transformed values). (Modified from TRACHSEL et al. 1990; BRUNNER et al. 1991)

the salient features of sleep architecture. Thus, despite the clear depression of slow-wave activity by the BDZ hypnotic flunitrazepam (Fig. 6, interrupted curve), the declining trend of slow-wave activity over the cycles and the non-REM/REM sleep cyclicity persisted, and the intraepisodic rise and fall rates of slow-wave activity were little affected (BORBÉLY and

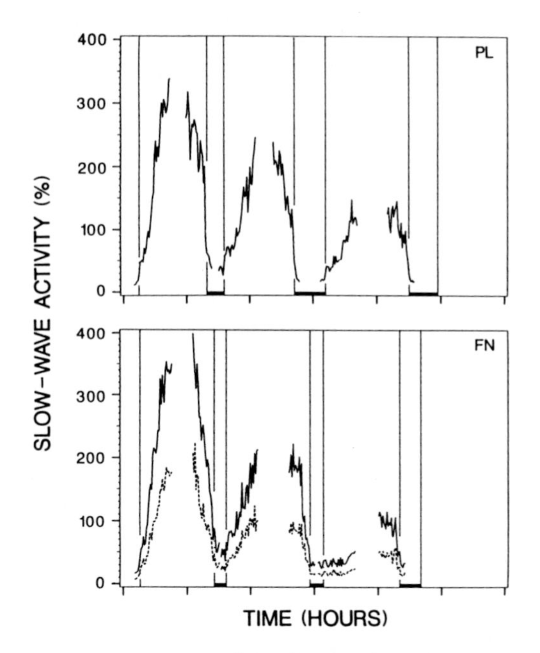

Fig. 6. Time course of slow-wave activity in the first three non-REM-REM sleep cycles during the placebo night (*PL*) and after intake of flunitrazepam (*FN*, 2 mg). The mean 1-min values are aligned relative to the first occurrence of stage 2 (first cycle) and the beginning and end of REM sleep. The curves represent the initial 30-min and the final 20-min intervals of the first three non-REM sleep episodes. In addition, the values for the 5 min preceding the onset and following the termination of episodes have been plotted. The mean times of onset and termination of REM sleep episodes are delimited by *vertical lines and black horizontal bars*. Two curves are shown for FN: The values indicated by *interrupted lines* are expressed in the same units as the PL night, whereas the values represented by the *continuous lines* were standardized to render the mean slow-wave activity equal to the mean PL level (= 100%). (Modified from BORBÉLY and ACHERMANN 1991)

ACHERMANN 1991). This is particularly evident from the continuous curves in the FN panel, which represents the values standardized to equal the placebo level. Moreover, a recent analysis has shown that the inverse relationship between slow-wave activity and activity in the spindle frequency range is largely preserved after the administration of a BDZ hypnotic (AESCHBACH et al. 1994b). From these results it appears that the drug-induced changes on the sleep EEG mainly reflect a pharmacological modulation of EEG-generating processes rather than an effect on sleep regulation. This interpretation is supported by the observation that the typical BDZ-induced effects on EEG spectra (i.e., reduction of slow-wave activity, enhancement of activity in the spindle frequency band) are seen in all sleep stages (BORBÉLY et al. 1985), although some stage-specific effects on alpha activity have been reported (GAILLARD and BLOIS 1989). As has been mentioned previously, the use of the conventional sleep stage criteria may result in fortuitous drug effects on the sleep stage distribution which are of little physiological relevance.

The prolongation of REM sleep latency by BDZ hypnotics (Gaillard and Blois 1983, 1989; Belyavin and Nicholson 1987) may be due, in part, to a drug-induced inhibition of EEG desynchronization rather than to the disruption of the REM sleep-generating processes (Borbély and Achermann 1991). Taken together, all these observations concur that the processes underlying sleep regulation are little affected by the usual doses of BDZ hypnotics.

IV. Effect of Benzodiazepine Hypnotics on Circadian Rhythms

When the main sleep episode is shifted to the circadian phase of habitual waking, sleep is likely to be disturbed and, consequently, sleepiness tends to be increased during the subsequent waking period. Drugs that would shift the phase of circadian rhythms would be expected to promote the adaptation to altered sleep-waking rhythms during shift work or after transmeridian travel. Studies in hamsters have suggested that the rapidly eliminated BDZ hypnotics, triazolam and midazolam, exert a chronopharmacological action by altering the phase of the circadian rest-activity rhythm (Turek and Losee-Olson 1986; Wee and Turek 1989). A phase-response curve has been established for triazolam (Turek and Losee-Olson 1986). However, since restraining the animals prevented the effect of triazolam (van Reeth and Turek 1989), the phase shifts seem to be due to a drug-induced motor activation of the hamsters (Mrosovsky and Salmon 1987) rather than to an effect of the BDZ on the circadian pacemaker. Experiments in humans failed to demonstrate an effect of triazolam on the circadian rhythm of sleep and other variables (Copinschi et al. 1990).

D. Concluding Remarks: Pitfalls in Pharmacological Sleep Studies

When sleep is disturbed, hypnotics may counteract the intrinsic or extrinsic sleep-disrupting influences by virtue of their permissive action on sleep, and thereby normalize the sleep pattern. Consequently, the percentages of slow-wave sleep and REM sleep may be increased. For example, when experimental subjects slept under non-sleep-conducive conditions, the administration of a BDZ hypnotic increased slow-wave sleep, although ordinarily the compound depresses slow-wave sleep (Balkin et al. 1989). A similar effect was observed in a protocol in which sleep was experimentally disturbed by having subjects sleep in a sitting position (Aeschbach et al. 1994a). These observations show that the action of a drug on sleep may be easily misinterpreted when its permissive action on sleep is not taken into consideration.

It has become increasingly clear that computer-aided methods for analyzing effects of hypnotics on electrophysiological sleep parameters and for relating them to functional aspects of sleep will be mandatory in future studies. Pharmacological effects cannot be adequately analyzed by relying exclusively on the conventional sleep scoring. One of the main problems derives from the arbitrary specification of scoring criteria. This is exemplified by the substates of the non-REM sleep states 2, 3 and 4, for which the major discriminating criterion is the abundance of EEG delta waves within a scoring epoch. The definition of the frequency range (0–2 Hz), the minimum amplitude of delta waves (75 μV) and the required prevalence in a scoring epoch (20%–50% for stage 3; >50% for stage 4) are all arbitrary. Consequently, variations that are unlikely to be of major physiological or pharmacological relevance (e.g., interindividual or age-related variations in EEG amplitude) may have large effects on the sleep stage distribution. Even more important, drugs affecting those EEG parameters that are critical for scoring (e.g., the amplitude of delta waves) give rise to prominent changes in sleep architecture, whereas equally potent drug effects on other EEG parameters (e.g., an augmentation of beta activity) do not affect the sleep scores and are therefore not reflected in the sleep stage distribution. For example, after administration of a BDZ hypnotic, the sleep EEG was still massively altered in the drug-free post-drug night, whereas sleep architecture as defined by the standard sleep scores was no longer significantly changed (Borbély et al. 1985). Therefore, studies based on the conventional scoring procedure alone may give rise to misleading results due to the overemphasis of drug-induced changes in sleep architecture as well as to the inadequate recognition of alterations of the sleep EEG.

E. Summary

A sleep-wake dependent homeostatic process, a circadian process and an ultradian process have been identified as underlying sleep regulation. EEG slow-wave activity is a marker of sleep homeostasis and an indicator of sleep intensity. Its level in the initial part of sleep and its rate of buildup within the non-REM sleep episode are a function of the duration of prior waking. Spindle frequency activity exhibits in part a negative correlation to slow-wave activity, which can be related to electrophysiological changes at the membrane level of thalamocortical neurons. Although REM sleep is also regulated by a homeostatic process, the manifestation of REM sleep homeostasis may be prevented when "slow-wave pressure" is high.

Benzodiazepine hypnotics have profound effects on the sleep EEG: They suppress activity in the low-frequency range (delta and theta band) and increase activity in the spindle frequency range. The time course of the effect on spindle-frequency activity is more closely related to the sleep-promoting action than the effect on low-frequency activity. The typical BDZ effect on the sleep EEG is induced also by the non-BDZ hypnotics zolpidem and zopiclone. Since these compounds also bind to the GABA-BDZ receptor complex, the distinct "spectral EEG signature" seems to reflect the agonistic action at this site. Despite the marked changes of the sleep EEG, the dynamics of slow-wave activity and spindle-frequency activity across the night are preserved after administration of hypnotics belonging to the class of BDZ-receptor agonists. This indicates that sleep homeostasis is not disrupted by these hypnotics.

The application of computer-aided methods of EEG analysis allows the analysis of effects of hypnotics in detail and enables them to be related to functional aspects of sleep. Studies based on the conventional sleep stages alone may give rise to misleading results.

Acknowledgements. I thank Dr. D.J. Dijk and Dr. I. Tobler for comments on the manuscript. The author's studies discussed in this chapter were supported by the Swiss National Science Foundation.

References

Achermann P, Borbély AA (1987) Dynamics of EEG slow wave activity during physiological sleep and after administration of benzodiazepine hypnotics. Hum Neurobiol 6:203–210

Aeschbach D, Borbély AA (1993) All-right dynamics of the sleep EEG. J Sleep Res 2:70–81

Aeschbach D, Cajochen C, Tobler I, Dijk DJ, Borbély AA (1994a) Sleep in a sitting position: effect of triazolam on sleep and EEG power spectra. Psychopharmacology (Berl) 114:209–214

Antonioli M, Solano L, Torre A, Violani C, Costa M, Bertini M (1981) Independence of REM density from other REM sleep parameters before and after REM deprivation. Sleep 4:221–225

Balkin TJ, O'Donnell VM, Kamimori GH, Redmond DP, Belenky G (1989) Administration of triazolam prior to recovery sleep: effects on sleep architecture, subsequent alertness and performance. Psychopharmacology 99:526–531

Beersma DGM, Daan S, Dijk DJ (1987) Sleep intensity and timing – a model for their circadian control. Lect Math Life Sci 19:39–62

Beersma DGM, Dijk DJ, Blok CGH, Everhardus I (1990) REM sleep deprivation during 5 hours leads to an immediate REM sleep rebound and to suppression of non-REM sleep intensity. Electroencephalogr Clin Neurophysiol 76:114–122

Belyavin A, Nicholson AN (1987) Rapid eye movement sleep in man: modulation by benzodiazepines. Neuropharmacology 26:485–491

Blake H, Gerard RW (1937) Brain potentials during sleep. Am J Physiol 119:692–703

Borbély AA (1980) Sleep: circadian rhythm versus recovery process. In: Koukkou M, Lehmann D, Angst J (eds) Functional states of the brain: their determinants. Elsevier, Amsterdam, pp 151–161

Borbély AA (1982) A two-process model of sleep. Hum Neurobiol 1:195–204

Borbély AA, Achermann P (1991) Ultradian dynamics of sleep after a single dose of benzodiazepine-hypnotics. Eur J Pharmacol 195:11–18

Borbély AA, Achermann P (1992) Concepts and models of sleep regulation: an overview. J Sleep Res 1:63–79

Borbély AA, Neuhaus HU (1979) Sleep-deprivation: effect on sleep and EEG in the rat. J Comp Physiol 133:71–87

Borbély AA, Wirz-Justice A (1982) Sleep, sleep deprivation and depression. A hypothesis derived from a model of sleep regulation. Hum Neurobiol 1:205–210
Borbély AA, Baumann F, Brandeis D, Strauch I, Lehmann D (1981) Sleep-deprivation: effect on sleep stages and EEG power density in man. Electroencephalogr Clin Neurophysiol 51:483–493
Borbély AA, Mattmann P, Loepfe M, Fellmann I, Gerne M, Strauch I, Lehmann D (1983) A single dose of benzodiazepine hypnotics alters the sleep EEG in the subsequent drug-free night. Eur J Pharmacol 89:157–161
Borbély AA, Tobler I, Hanagasioglu M (1984) Effect of sleep deprivation on sleep and EEG power spectra in the rat. Behav Brain Res 14:171–182
Borbély AA, Mattmann P, Loepfe M, Strauch I, Lehmann D (1985) Effect of benzodiazepine hypnotics on all-night sleep EEG spectra. Hum Neurobiol 4:189–194
Brunner DP, Dijk DJ, Tobler I, Borbély AA (1990) Effect of partial sleep deprivation on sleep stages and EEG power spectra: evidence for non-REM and REM sleep homeostasis. Electroencephalogr Clin Neurophysiol 75:492–499
Brunner DP, Dijk DJ, Münch M, Borbély AA (1991) Effect of zolpidem on sleep and sleep EEG spectra in healthy young man. Psychopharmacology (Berl) 104:1–5
Brunner DP, Dijk DJ, Borbély AA (1993) A repeated partial sleep deprivation progressively changes the EEG during sleep and waking. Sleep 16:100–113
Cartwright RD, Monroe LJ, Palmer C (1967) Individual differences in response to REM deprivation. Arch Gen Psychiatry 16:297–303
Church MW, March JD, Hibi S, Benson K, Cavness C, Feinberg I (1975) Changes in the frequency and amplitude of delta activity during sleep. Electroencephalogr Clin Neurophysiol 39:1–7
Copinschi G, Van Onderbergen A, L'Hermite-Balériaux M, Šzyper M, Caufriez A, Bosson D, L'Hermite M, Robyn C, Turek FW, Van Cauter E (1990) Effects of the short-acting benzodiazepine triazolam taken at bedtime on circadian and sleep-related hormonal profiles in normal men. Sleep 13:232–244
Dijk DJ, Daan S (1989) Sleep EEG spectral analysis in a diurnal rodent: Eutamias sibiricus. J Comp Physiol [A] 165:205–215
Dijk DJ, Beersma DGM, Daan S (1987) EEG power density during nap sleep: reflection of an hourglass measuring the duration of prior wakefulness. J Biol Rhythms 2:207–219
Dijk DJ, Beersma DGM, Daan S, van den Hoofdakker RH (1989) Effects of seganserin, a 5-HT_2 antagonist, and temazepam on human sleep stages and EEG power spectra. Eur J Pharmacol 171:207–218
Dijk DJ, Brunner DP, Beersma DGM, Borbély AA (1990a) Electroencephalogram power density and slow wave sleep as a function of prior waking and circadian phase. Sleep 13:430–440
Dijk DJ, Brunner DP, Borbély AA (1990b) Time course of EEG power density during long sleep in humans. Am J Physiol 258:R650–R661
Dijk DJ, Brunner DP, Aeschbach D, Tobler I, Borbély AA (1992) The effects of ethanol on human sleep EEG power spectra differ from those of benzodiazepine receptor agonists. Neuropsychopharmacology 7:225–232
Dijk DJ, Hayes B, Czeisler CA (1993) Analysis of spindle activity by transient pattern recognition software and power spectral analysis. Sleep Res 22:426
Dossi RC, Nunez A, Steriade M (1992) Electrophysiology of a slow (0.5–4 Hz) intrinsic oscillation of cat thalamocortical neurones in vivo. J Physiol (Lond) 447:215–234
Feinberg I, Floyd TC, March JD (1987) Effects of sleep loss on delta (0.3–3 Hz) EEG and eye movement density: new observations and hypotheses. Electroencephalogr Clin Neurophysiol 67:217–221
Franken P, Dijk DJ, Tobler I, Borbély AA (1991) Sleep deprivation in the rat: effects of electroencephalogram power spectra, vigilance states, and cortical temperature. Am J Physiol 261:R198–R208

Friedman L, Bergmann BM, Rechtschaffen A (1979) Effects of sleep deprivation on sleepiness, sleep intensity, and subsequent sleep in the rat. Sleep 1:369–391

Gaillard JM, Blois R (1983) Effect of the benzodiazepine antagonist Ro 15-1788 on flunitrazepam-induced sleep changes. Br J Clin Pharmacol 15:529–536

Gaillard JM, Blois R (1989) Differential effects of flunitrazepam on human sleep in combination with flumazenil. Sleep 12:120–132

Gaillard JM, Schulz P, Tissot R (1973) Effects of three benzodiazepines (nitrazepam flunitrazepam and bromazepam) on sleep of normal subjects studied with an automatic sleep scoring system. Pharmacopsychiatry 6:207–217

Geering BA, Achermann P, Eggimann F, Borbély AA (1993) Period-amplitude analysis and power spectral analysis: a comparison based on all-night sleep EEG recordings. J Sleep Res 2:121–129

Johnson LC, Spinweber CL, Seidel WF, Dement WC (1983) Sleep spindle and delta changes during chronic use of a short-acting and a long-acting benzodiazepine hypnotic. Electroencephalogr Clin Neurophysiol 55:662–667

Kales A, Kales JD, Scharf MB, Tan TL (1970) All-night EEG studies of chloral hydrate, flurazepam, and methaqualone. Arch Gen Psychiatr 23:219–225

Kales A, Kales JD (1975) Shortcomings in the evaluation and promotion of hypnotic drugs. NEJM 293:827–827

Leresche N, Lightowler S, Soltesz I, Jassik-Gerschenfeld D, Crunelli V (1991) Low-frequency oscillatory activities intrinsic to rat and cat thalamocortical cells. J Physiol (Lond) 441:155–174

McCarley RW, Massaquoi S (1992) Neurobiological structure of the revised limit cycle reciprocal interaction model of REM cycle control. J Sleep Res 1:132–137

Mrosovsky N, Salmon PA (1987) A behavioural method for accelerating re-entrainment of rhythms to new light-dark cycles. Nature 330:372–373

Pappenheimer JR, Koski G, Fencl V, Karnovsky ML, Krueger JM (1975) Extraction of sleep-promoting factor S from cerebrospinal fluid and from brains of sleep-deprived animals. J Neurophysiol 38:1299–1311

Rechtschaffen A, Kales AA (1968) Manual of standardized terminology techniques and scoring system for sleep stages of human subjects. Department of Health, Education and Welfare, Public Health Service, Bethesda

Reeth van O, Turek FW (1989) Stimulated activity mediates phase shifts in the hamster circadian clock induced by dark pulses or benzodiazepines. Nature 339:49–51

Rosenberg RS, Bergmann BM, Rechtschaffen A (1976) Variations in slow wave activity during sleep in the rat. Physiol Behav 17:931–938

Schneider-Helmert D (1988) Why low-dose benzodiazepine-dependent insomniacs can't escape their sleeping pills. Acta Psychiatr Scand 78:706–711

Sinha AK, Smythe H, Zarcone VP, Barchas JC, Dement WC (1972) Human sleep-electroencephalogram: a damped oscillatory phenomenon. J Theor Biol 35: 387–393

Steiger A, Trachsel L, Guldner J, Hemmeter U, Rothe B, Rupprecht R, Vedder H, Holsboer F (1993) Neurosteroid pregnenolone induces sleep-EEG changes in man compatible with inverse agonistic $GABA_A$-receptor modulation. Brain Res 615:267–274

Steriade M, Curró Dossi R, Nunez A (1991) Network modulation of a slow intrinsic oscillation of cat thalamocortical neurons implicated in sleep delta waves: cortically induced synchronization and brainstem cholinergic suppression. J Neurosci 11:3200–3217

Tobler I (1985) Deprivation of sleep and rest in vertebrates and invertebrates. In: Inoué S, Borbély AA (eds) Endogenous sleep substances and sleep regulation, vol 8. Scientific Societies Press, Tokyo, pp 57–66

Tobler I, Borbély AA (1986) Sleep EEG in the rat as a function of prior waking. Electroencephalogr Clin Neurophysiol 64:74–76

Tobler I, Jaggi K (1987) Sleep and EEG spectra in the Syrian hamster (Mesocricetus auratus) under baseline conditions and following sleep deprivation. J Comp Physiol [A] 161:449–459

Tobler I, Borbély AA (1990) The effect of 3-h and 6-h sleep deprivation on sleep and EEG spectra of the rat. Behav Brain Res 36:73–78

Tobler I, Franken P, Scherschlicht R (1990) Sleep and EEG spectra in the rabbit under baseline conditions and following sleep deprivation. Physiol Behav 38:121–129

Tobler I, Franken P, Trachsel L, Borbély AA (1992) Models of sleep regulation in mammals. J Sleep Res 1:125–127

Trachsel L, Tobler I, Borbély AA (1988) Electroencephalogram analysis of non-rapid eye movement sleep in rats. Am J Physiol 255:R27–R37

Trachsel L, Dijk DJ, Brunner D, Klene C, Borbély AA (1990) Effect of zopiclone and midazolam on sleep and EEG spectra in a phase-advanced sleep schedule. Neuropsychopharmacology 3:11–18

Turek FW, Losee-Olson S (1986) A benzodiazepine used in the treatment of insomnia phase-shifts the mammalian circadian clock. Nature 321:167–168

Uchida S, Maloney T, March JD, Azari R, Feinberg I (1991) Sigma (12–15 Hz) and delta (0.3–3 Hz) EEG oscillate reciprocally within NREM sleep. Brain Res Bull 27:93–96

Webb WB, Agnew HW (1971) Stage 4 sleep. Influence of time course variables. Science 174:1354–1356

Wee BE, Turek FW (1989) Midazolam, a short-acting benzodiazepine, resets the circadian clock of the hamster. Pharmacol Biochem Behav 32:901–906

Williams HL, Hammack JT, Daly RL, Dement WC, Lubin A (1964) Responses to auditory stimulation, sleep loss and the EEG stages of sleep. Electroencephalogr Clin Neurophysiol 16:269–279

CHAPTER 3

Principles of Chronopharmacology and the Sleep-Wake Rhythm

A.E. REINBERG

A. Introduction

The first studies showing that a given effect of a drug was dependent on time of dosing were carried out on narcotics. VIREY (1814) reported that following the recommendations of Thomas Sydenham he prescribed the use of opium (in the form of laudanum) as a narcotic during the late evening hours rather than during the morning or another time of day. Thus, even in 1814 VIREY considered the potential existence of biological clocks and emphasized that a change in the effectiveness of drugs as a function of the dosing time was related to changes in the subject's susceptibility. One hundred and fifty years later it was confirmed that some pharmacological agents able to induce sleep and related processes operate more efficiently when they are administered during evening or night hours in diurnally active subjects. This was the case for: halothane (FUKAMI et al. 1970), best efficiency between midnight and 6 a.m. (0600 h); ethanol (REINBERG et al. 1975), maximum self-rated inebriety and poorer psychological performance with ethanol (0.67 g/kg body weight) at 11 p.m. (2300 h) as compared with other dosing times (7 a.m., 11 a.m., 7 p.m.); and 3-alkyl pyrazolyl piperazine (SIMPSON et al. 1973), better subjective "quality" of sleep and less broken sleep than control (placebo) when administered in the morning.

With regard to animal experiments, the pioneer work of SCHEVING et al. (1968) clearly demonstrated (Fig. 1) that the duration of anesthesia induced by pentobarbital sodium (35 mg/kg) injected into adult Sprague-Dawley rats was time dependent for dosing. In animals synchronized with 12 h of light from 6 a.m. to 6 p.m. and with 12 h of darkness from 6 p.m. to 6 a.m. [light (L), 0600–1800 h; dark (D), 1800–0600 h], the largest duration of induced sleep corresponded to the treatment at 1900 h. In addition, SCHEVING et al. (1968) showed that this chronopharmacological phenomenon observed in LD persisted when the animals were kept in continuous light (LL) as well as in continuous darkness (DD).

The results of some other investigations in rodents are shown in Table 1. Statistically significant circadian (about 24 h) rhythms were detected in all of the chronopharmacological phenomena investigated. Differences in the circadian peak time of the same variable (e.g., duration of anesthesia) can be seen in the reported data. Such differences could be due to the time of

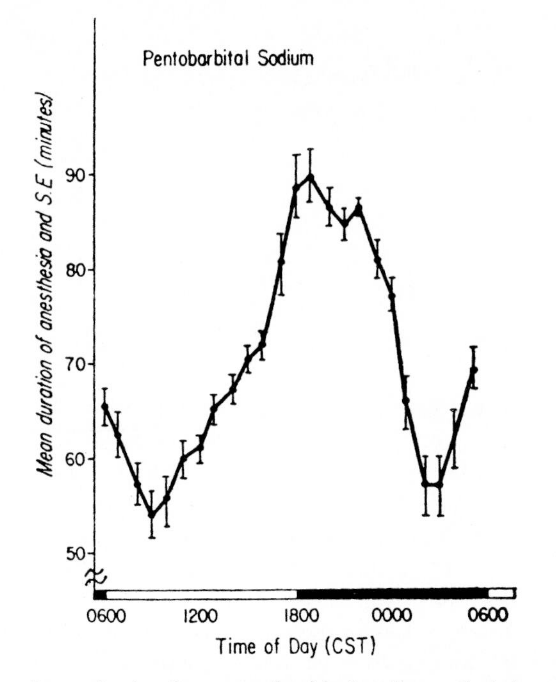

Fig. 1. Duration of anesthesia after pentobarbital sodium administration at different times of the day. Scheving et al. demonstrated that the duration of sleep induced by pentobarbital sodium injection (35 mg/kg) in the Sprague-Dawley adult rat was a function of the hour of administration (circadian-rhythm-phase-dependent). Animals were synchronized with light from 0600 to 1800 h and with darkness from 1800 to 0600 h. (After SCHEVING et al. 1968)

year (BOUYARD et al. 1974; BRUGUEROLLE et al. 1979), time of day (circadian rhythms) and conventional factors such as animal strain and age (REINBERG 1992a; REINBERG and ASHKENAZI 1993).

Experimental evidence of circadian changes in the biological responses of laboratory animals and human beings to various agents and/or drugs has been presented in several reviews (REINBERG and HALBERG 1971; REINBERG 1981, 1992a; LEMMER 1989). These reviews include those of: chronopharmacology in anesthesiology (BOURDALLE-BADIE et al. 1990); factors affecting circadian variation in responses to psychotropic drugs (REDFERN and MOSER 1988); and factors concerning drugs which are supposed to alleviate or prevent jet lag and intolerance of shift work (REINBERG et al. 1988). Therefore, within the space of a few pages, it is a challenge to give a complete overview of clinical chronopharmacology even if it is restricted to so-called sleeping pills and sleep-inducing drugs.

The aims of this chapter are primarily to: (a) outline briefly some basic definitions and concepts pertinent to clinical chronopharmacology, e.g., chronopharmacokinetics, chronesthesy, chronergy, chronoptimization; and

Table 1. Effects of drugs on induced sleep, general anesthesia and related phenomena according to circadian rhythm stage in the rat

Strain, age, sex	Chemical agent (fixed dose and multiple test times)	Variable investigated	Synchronization[a]	Chronopharmacologically induced changes	References
Sprague-Dawley, adult	Pentobarbital sodium	Duration of anesthesia	L: 0600–1800 D: 1800–0600 And continuous L or D	Longest duration (peak time at ~1900 h) Persistent rhythm in continuous L and/or D	Scheving et al. 1968
Sprague-Dawley, adult	Pentobarbital sodium	Duration of anesthesia	L: 0800–2000 D: 2000–0800	Longest duration of anesthesia at ~1700 h coinciding with trough time of brain amines	Walker 1974
Wistar, adult	Pentobarbital sodium	Induction time and duration of anesthesia	L: 0800–2000 D: 2000–0800	Induction time and duration of anesthesia peaks at ~2000 h	Simmons et al. 1974
Wistar, old	Hexobarbital sodium	Duration of anesthesia; hexobarbital oxidation in liver	L: 0700–1900 D: 1900–0700	Longest duration of anesthesia and lowest oxidation at ~1800 h	Müller 1974
Sprague-Dawley, adult	Hexobarbital sodium	Duration of anesthesia: hexobarbital oxidation in liver	L: 0600–1800 D: 1800–0600	Longest duration of anesthesia and lowest enzyme activity at ~1400 h	Nair 1974
Wistar, adult	Pentobarbital sodium	Induction time of anesthesia	L: 0600–1800 D: 1800–0600	Peak at ~1100 h (January–May)	Bruguerolle et al. 1979
Wistar, adult	Althesin[b]	Induction time of anesthesia	L: 0600–1800 D: 1800–0600	Peak at ~1600 h in October Peak at ~1000 h in March	Bouyard et al. 1974
Wistar, adult	Curarizing agents, e.g., pancuronium bromide	Neuromuscular activation	L: 0600–1800 D: 1800–0600	Peak of the curarizing effect at 0800 h in March and October	

[a] L, light; D, dark.
[b] An iso-osmotic solution containing alphaxone and alphadolone acetate in castor oil.

(b) give special attention to chronopharmacological factors concerned with drugs that presumably influence sleep and related processes.

B. Properties of Biological Rhythms

The quantitative study of biological rhythms shows that biophysical and biochemical processes vary with respect to time in a periodic, regular and predictable manner (Aschoff 1963; Bünning 1963; Halberg 1969). We know today that rhythmic activity is a fundamental property of living matter. Biological rhythms can in fact be demonstrated in all living beings, from nucleated unicellular organisms to man, and at all levels: entire organisms, organ systems, organs, tissues, cells and subcellular material. Biological rhythms exhibit similar basic properties in plants and animals: they are genetic in origin; they persist without time clue or cue; they can be characterized for a given species (e.g., rat, mice, man) although interindividual differences can be demonstrated (e.g., differences between strains of mice and monozygotic versus dizygotic twin studies in man); and they can be influenced by cyclic variations of certain environmental factors called synchronizers or zeitgebers.

For most plant and animal species alternation of the light:dark cycle with a period equal to 24 h (e.g., L:D/12:12) is the most powerful zeitgeber. In man the time constraints of our social life (e.g., hours of work and family life) seem to play the role of a major zeitgeber (Apfelbaum et al. 1969; Aschoff et al. 1971). However, several cyclical phenomena of environmental factors having synchronizing effects on rhythms have to be taken into account. Alternation of relative heat and cold, noise and silence and levels of odors may have a synchronizing effect. This is also the case for light and darkness in man when light intensity is considered. It seems that 2500 lux light is needed to obtain a synchronizing effect in man. This is also the amount of light which is needed to decrease the nocturnal secretion of melatonin in our species.

When animals are synchronized with LD it is important to know the hours of lights-on and lights-off respectively. In human beings waking time and lights-off are taken as indices of the subject's synchronization. This does not mean that the sleep-wake rhythm controls the temporal structure but that it is used as an index to assess how the subject is synchronized. In fact, the sleep-wake rhythm seems to possess properties similar to those of many other biological rhythms.

In mammals and certain species of birds the suprachiasmatic nuclei appear to act as a biological clock or circadian oscillator (Moore 1983; Stephan and Zucker 1972). Despite the fact that it is the only circadian oscillator which has been anatomically identified to date, other functional circadian oscillators are likely to exist in man including in the brain cortex (Ashkenazi et al. 1993; Reinberg et al. 1987). Environmental zeitgebers are used by the organism to synchronize all its biological clocks, which means calibrating the period to 24 h and resetting peaks and troughs of circadian rhythms in their respective physiological situations on the 24-h scale.

C. Biological Rhythms as Adaptive Phenomena to Predictable Changes in Environmental Factors

These bioperiodic changes result from adaptive phenomena to predictable variations of a set of factors directly related to the period (τ) of the earth's rotation around its axis ($\cong$24 h) and around the sun ($\cong$365.25 days). In fact, most examples of the temporal organization of living beings have been reported in circadian and circannual domains of biological rhythms (see also Chap. 2, this volume). Bioperiodic phenomena obviously not related to cosmic changes, with a periodicity of τ = 7 days, τ = 1 month, for example, have also been documented (ASCHOFF 1963; HALBERG 1969), but to a lesser extent than circadian and circannual rhythms.

Experimental data indicate that, for example, cells (from unicellular eukaryotes to mammals) are not able to perform all their potential functions at any one time during the 24-h scale. Thus the liver cell of mice and rats shows a temporal organization in its metabolic processes (RNA, DNA, enzymes, phospholipids, glycogen, etc.) (HAUS et al. 1974), as well as in its cytological reorganization associated with functional changes (VON MAYERSBACH 1978). Peak times (acrophases) of each type of activity are not randomly distributed on the 24-h scale; on the contrary, the distribution of crest times is genetically programmed.

The temporal organization at the cellular level can be viewed as an adaptive phenomenon. As an example, liver enzymes which might have competitive effects if their peak activity occurs at the same time are in fact programmed to vary their peak activity several hours (up to 12) apart. As a result a cooperative effect occurs. In addition, this programming in time may represent an optimal use of the available energy. In rodents synchronized with light from 0600 to 1800 h (rest) and with darkness from 1800 to 0600 h (activity), the glycogen synthesis peak in the liver cell occurs around the onset of light, whereas the albumin synthesis peak occurs 12 h later or earlier (HAUS et al. 1974; VON MAYERSBACH 1978).

It is not surprising, therefore, to observe: that the metabolic fate of a chemical agent (drug, nutrient, hormone, etc.) is not constant but varies as a function of time; and that the effects of the chemical agent at the target level vary according to which part of the metabolic program is in play when the agent reaches the cells concerned.

D. Temporal Distribution and Dosing Time of Therapies

According to time-honored traditions and conventions, patients are given therapeutic agents (chemical as well as physical) at specified hours on the 24-h scale. The temporal distribution of a therapy is frequently related to psychosocial considerations and less frequently to empirical observations, although sometimes reference is made to the homeostatic hypothesis. Before the 1960s no convincing experiment demonstrated objectively, for a given agent, the advantage of a well-defined chronotherapy (REINBERG 1978; HALBERG 1969).

Psychosocial Considerations. Instruction is often given to a patient to take a drug at mealtimes, with the hope that they will not forget to take it. However, that does not necessarily mean it is the time the medication will be the most effective.

Empirical Observations. VIREY's thesis (1814) is a good illustration of an empirical approach.

Homeostatic Hypothesis. The homeostatic hypothesis presumes that toxic, pharmacological and therapeutic effects of a given agent are constant as a function of time (see also Chap. 2, this volume). More precisely, desired and undesired effects are assumed to be identical at any time of the drug's administration: the clock hour on the 24-h scale, the day on the scale of the month and the month on the scale of the year. However, it has not been possible to validate this hypothesis even though it has been tested in many experiments (HAUS et al. 1974; REINBERG et al. 1965; REINBERG and HALBERG 1971; REINBERG 1992a; LEMMER 1989). It is, of course, interesting to note that the homeostatic approach for therapeutic methods is not only obsolete but dangerous. It is more important to know that a chronobiological approach may be useful for solving a number of therapeutic problems, e.g., reduction of the undesired effects of certain drugs.

Specific chronobiological methods are now available to "chronoptimize" the use of therapeutic means (REINBERG et al. 1991).

E. Concepts in Chronopharmacology

The rapidly growing number of biological rhythm-related changes in pharmacology has led to concepts such as chronopharmacokinetics, chronesthesy and chronergy (REINBERG 1978, 1992; REINBERG and SMOLENSKY 1981). To illustrate these concepts examples will be taken from commonly used sleep-inducing drugs.

I. Chronopharmacokinetics (or Chronokinetics) of a Drug

Chronokinetics is defined as rhythmic (e.g., circadian) changes in the bioavailability of a drug, its pharmacokinetics and/or its excretion in the urine or other routes (feces, sweat, saliva, etc.). When the timing of administration is changed (e.g., single daily dose, four different dosing times in the 24-h period, each being administered 1 week apart), statistically significant circadian rhythms can be demonstrated in the parameters of a given model used to characterize the pharmacokinetics of substances such as t_{max} (time to peak), C_{max} (maximum concentration at t_{max}), $t_{1/2}$ (half-life) and AUC (area under the time/concentration curve). Chronokinetic changes have been documented in adult subjects with regard to clorazepate dipotassium (AYMARD and SOULAIRAC 1979), hexobarbital (ALTMAYER et al. 1979), lithium carbonate (LAMBINET et al. 1982), diazepam (NAKANO et al. 1982) and ethanol (STURTEVANT 1976).

The metabolic fate of a pharmacological agent, as well as that of a nutrient, should not be expected to be independent of time. Indeed, experiments have clearly proved that rhythms strongly influence the metabolism of medication and nutrients. The metabolic pathways are open neither permanently nor with a constant patency on the 24-h scale.

II. Biomechanisms Presumably Involved in Chronokinetic Changes

Circadian changes in one or more conventional mechanisms help us to understand better the observed chronokinetics of a drug. In fact, circadian rhythms have been demonstrated in the following functions involved in drug chronokinetics (REINBERG and SMOLENSKY 1981; LEMMER 1989; REINBERG et al. 1986, 1991):

- Drug absorption (gastric pH, gastric emptying, enteric absorption)
- Drug transport (drug binding to proteins and erythrocytes)
- Drug distribution (cardiac output, capillary resistance, blood volume, peripheral blood flow, e.g., in liver, kidney and skin)
- Liver metabolisms (drug-metabolizing enzymes, microsomal oxydation glucurono- and sulfo-conjugation, enzymatic induction)
- Kidney metabolism (glomerular filtration rate, diuresis, urinary pH, tubular reabsorption)

Liver metabolism seems to play an important role in the chronokinetics of certain drugs which may act as sleep inducers. In adult Sprague-Dawley rats synchronized with L: 0600–1800 h and D: 1800–0600 h, NAIR (1974) has shown that the hexobarbital oxidation in the liver has a circadian rhythm with a peak around 2200 h and a trough around 1400 h. Hexobarbital sodium (150 mg/kg) was given intraperitoneally at six different time points on the 24-h scale. The hexobarbital oxidase activity rhythm varied in phase opposition with the duration of sleep (Fig. 2). BELANGER (1988) in a number of animal experiments demonstrated that barbiturate-related enzymatic induction in the liver has a circadian rhythmicity involving the cytochrome P-450 oxygenase system. Still in the rat, STURTEVANT and GARBER (1988) have shown that the circadian rhythm in the plasma ethanol disappearance rate (β-slope) was related to circadian changes in liver alcohol dehydrogenase. This was the case in experiments where effects of acute administration of ethanol were explored. However, in chronically (6 months)

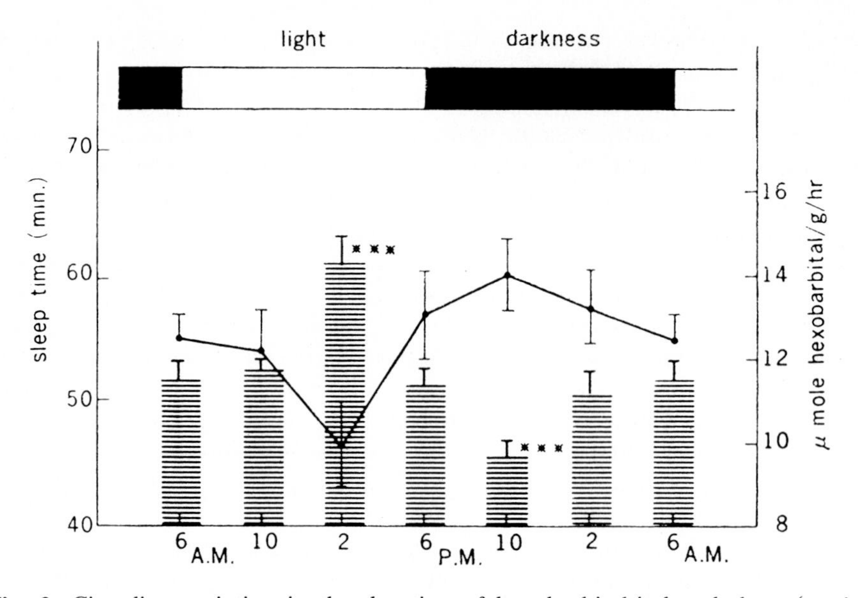

Fig. 2. Circadian variation in the duration of hexobarbital-induced sleep (*vertical bars*) compared with the circadian changes in hepatic hexobarbital oxidase activity (*solid line*). The duration of induced sleep at each clock hour of study represents the average duration for six to ten animals. Hexobarbital sodium (150 mg/kg) was given intraperitoneally. The enzyme activity is expressed as micromoles hexobarbital metabolized per gram tissue per hour; each figure represents the mean of six to eight animals. The *shorter vertical lines* represent the standard errors; $***P < 0.001$. The timing of the highest and lowest hexobarital oxidase activity at 2200 h and 1400 h coincides with the shortest and longest hexobartibal-induced sleep. (After NAIR 1974)

ethanol-treated rats, another enzymatic system (microsomal ethanol oxidative system) was involved.

It is of major interest to know that the circadian rhythmicity of drug-metabolizing liver enzymes is controlled by that of corticoadrenal secretion. In adrenalectomized animals the enzymatic rhythm is obliterated; it can be restored by corticosterone administration with an appropriate dosing time (RADZIALOWKSI and BOUSQUET 1968; BELANGER 1988).

III. Chronesthesy of a Biosystem

Let us assume that through an appropriate method (actual sustained release preparation or pump-delivering drug at a constant rate) the plasma concentration of a drug remains constant. Are circadian changes in drug effects still present under such circumstances? The answer is yes. In fact, the target system on which the drug is acting also exhibits circadian rhythmicity.

Chronesthesy is defined as the rhythmic changes in the susceptibility of a biosystem; it includes molecular and membrane phenomena, and related metabolic processes. The chronesthesy involves cells, tissues, organs and organ systems of the host as well as the susceptibility of parasites, bacteria, tumors, etc.

A biosystem can be completely unresponsive to a given drug at certain times on the 24-h scale, whereas the same dose is highly effective at other hours of the day. This chonesthesy can be expressed and quantified in terms of bioperiodic changes of receptors of a given system to a given drug.

Circadian changes in brain receptors to various substances including α- and β-adrenergic receptors as well as receptors to acetylcholine, opiates, benzodiazepines, nicotine, etc. have been documented experimentally (WIRZ-JUSTICE 1987). Several generalizations need to be made. Circadian rhythmicity is associated with the number of receptors rather than with their binding affinity. These circadian rhythms are likely to be obliterated by suppressing the suprachiasmatic nuclei. However, they are not influenced by sleep deprivation. In human subjects, circadian rhythms in binding sites to benzodiazepine have been demonstrated on platelets (LÉVI et al. 1986).

In some experimental situations (e.g., time of year) circadian changes in the number of receptors may be either nil or with an amplitude smaller than the observed effect. LEMMER et al. (1987) have provided experimentally an explanation for this intriguing phenomenon. Even when receptor rhythms exhibit only a small circadian amplitude, the intracellular cAMP rhythm maintains a large circadian amplitude. In fact, it is likely that the adenylate cyclase-phosphodiesterase system is acting as a rhythm "amplifier" located after the receptor.

The chonesthesy also involves circadian changes in membrane permeability and other processes which are of major interest in central nervous system activity (BOURDALLE-BADIE et al. 1990).

IV. Chronergy of a Chemical Agent

Chronergy is defined as rhythmic changes produced by a chemical or physical agent, either desired (chronoeffectiveness) or undesired (chronotoxicity). The chronergy of chemical agent involves its chronopharmacokinetics as well as the chronesthesy of a set of biosystems. The acrophase (peak time) of chronergy does not necessarily coincide with the acrophase of the drug level in the blood. Since both chronokinetics and chronesthesy involve several mechanisms, chronergy corresponds to a multifactorial phenomenon. The term chronergy is neutral since the desired effect of a drug relies upon the therapeutic aim. As an example, induced sleepiness may be the desired effect for a sleeping pill but the nondesired effect for a tranquillizer or an antihistamine agent. Likewise, the adrenal suppression resulting from a corticoid administration may be the desired effect (e.g., dexamethasone adrenal supression test) or the effect to be avoided (e.g., treatment of nocturnal asthma) (REINBERG et al. 1991).

V. Chronotherapy

Chronopharmacological knowledge is used to optimize the prescription of a given drug, which means both increasing its effectiveness and favouring its tolerance. Conventional attempts to reduce the most conspicuous side effects have been made by changing the molecular structure of the drug, its vehicle or its route of adminstration with reference to a homeostasis point of view of pharmacological processes. The ensuing results have not been encouraging. The chronopharmacological approach on the other hand has made significant progress. This is the case for corticosteroids, anticancer agents, nonsteroidal anti-inflammatory agents, theophylline, H_1 and H_2 antihistamines, cardiovascular drugs and a number of hormones (Lemmer 1989; Reinberg et al. 1986, 1991).

F. Drug Effects on Sleep, Anesthesia and Related Processes

This section aims to show the clinical importance of chronopharmacological studies in the use of drugs that influence sleep states. It is restricted to some illustrative examples.

I. Anesthetics

A circadian rhythm can be demonstrated in the effects of drugs used to anesthetize rodents as well as humans (Bourdalle-Badie et al. 1990), and a circannual rhythm of these effects should be considered (Bruguerolle et al. 1979). Circadian rhythmicity of substances such as anesthetics and curarizing drugs has also been documented (Bruguerolle et al. 1979). Such chronopharmacologically induced changes exhibit a time relationship and presumably a causal relationship with physiological circadian rhythms of brain amines (Walker 1974; Redfern and Moser 1988), liver enzymes (Nair 1974; Belanger 1988), neuromuscular excitability (Bouyard et al. 1974) and toxic effects (Scheving et al. 1968), among other cyclical phenomena. In fact, biochemical mechanisms playing a role in any chronopharmacological change in sleep or anesthesia involve many circadian rhythms (so-called component rhythms) at several levels of organization. However, the basic knowledge resulting from many experiments in animals has not been transferred as yet to clinical practice. This is mainly due to the fact that animal experiments are conducted to explore the chronopharmacology of one or two drugs while in clinical practice many agents are used simultaneously. It is therefore not possible to decide which agent is responsible for a given effect at a given time. In addition, it is difficult to state to what extent the temporal structure of a given patient to be anesthesized is or not altered. Methodological aspects as well as interindividual differences are some of the new problems that chronopharmacologists have to solve (Reinberg and Ashkenazi 1993).

II. Ethanol

Circadian rhythms of performance as well as subjective phenomena such as sleepiness, drowsiness and fatigue can be demonstrated and quantified. Conventional tests (eye-hand skill, random number addition test, reaction time) as well as visual analogue scales to self-rate subjective variables can be used iteratively. Using these methods it has been possible to demonstrate a circadian rhythm in performance (RUTENFRANZ and SINGER 1967; REINBERG 1992b) as well as self-rated variables. In subjects synchronized with diurnal activity and nocturnal rest, self-rated inebriety was greatest after ingestion (ethanol 0.67 g/kg of body weight) at 2300 h and lowest after ingestion at 1100 h. Speed decrement in performing both random number addition and eye-hand skill tests was greatest after ingestion at 2300 h with regard to performance. A statistically significant fall of about 20% was observed with the evening administration as compared with the morning one (driving simulation with or without signal cancellation). Lowered body temperature is considered a reliable clinical index of acute ethanol toxicity. The 24-h temperature mean was found to be lowest when ethanol was ingested at 0700 h. Such findings are consistent with those of experiments in rodents (mortality rate) (HAUS and HALBERG 1959) and with body temperature (LI et al. 1986), indicating that the peak time of acute ethanol toxicity occurs around the beginning of the activity phase in both mice and men. Validated time-dependent effects of ethanol ingestion were not related to drug pharmacokinetics. Chronopharmacological changes in effects of ethanol with regard to a set of psychological and physiological variables can therefore be suggested. However, this does not exclude that changes in the effects of ethanol of these variables may be related to the sleep-wake rhythm.

III. Drug-Induced Drowsiness

Chronobiological self-measurements and self-ratings were also used to study differences in sleepiness induced by antihistaminic agents in relation to the timing of drug administration (0700 h versus 1900 h in diurnally active healthy subjects). A double-blind, placebo, crossover study was performed (REINBERG et al. 1978). Clemastine (3 mg) enhanced sleepiness at all times of measurement (every 4 h). In contrast, terfenadine (60 mg) did not induce any sleepiness with regard to control (placebo). When ingested at 1900 h, terfenadine induced a decrease rather than an increase in sleepiness. These results are in good agreement with those of NICHOLSON and STONE (1979), who used a different methodology to study the sleepiness resulting from administration of several agents. Most antihistamines (e.g., clemastine) induce sleepiness, whereas both terfenadine and mequitazine (REINBERG et al. 1984) may enhance vigilance.

Experimental results obtained thus far in humans are in agreement with the first empirical observations of Syndenham and Virey. If narcosis, general anesthesia or drug-induced sleep is the desired effect, the agent should be administered not far from the anticipated beginning of spontaneous sleep (e.g., between 2200h and midnight). In fact, the existence of a "best time to sleep" has been confirmed by many experiments (AKERSTEDT and GILLBERT 1981; WEITZMAN et al. 1970). However, the coincidence between the Sydenham-Virey report and the best time to sleep awaits further investigation.

G. Shift Workers, Jet Lag and Sleeping Pills

Pertinent to the scope of this chapter is the problem of shift workers (those on night shifts associated with daytime sleep), jet lag resulting from crossing five or more time zones (transmeridian flights) and the use of sleeping pills. Chronological and other studies have been devoted to the adjustment to the "new" timing of the activity-rest rhythm after each shift or transmeridian flight (so-called resynchronization), to tolerance of shift work and to inter-individual differences in resetting biological rhythms to their physiological situation (FOLKARD and MONK 1985; REINBERG and SMOLENSKY 1992). Despite the fact that subjects exposed to both shift work and jet lag need to recover from a zeitgeber manipulation of at least 5h, the practical aims in both these situations differ. In the case of passengers after a given transmeridian flight, a rapid adjustment of circadian rhythms to local time is desirable, while in the case of shift workers, as well as pilots and flight attendants, it is in their best interest in most instances not to adjust their biological rhythms. Therefore, two strategies have to be considered.

I. Shift Workers and Sleeping Pills

Some subjects are able to tolerate shift work situations without medical complaint for their active lifetime. As a rule, tolerant subjects do not suffer from sleep disturbance; they have a large-amplitude body core temperature circadian rhythm; and they adjust slowly when the socioecological synchronizer is changed as a result of shift work schedules (ANDLAUER et al. 1979; REINBERG and SMOLENSKY 1992). On the other hand, some subjects do not tolerate shift work either immediately (i.e., after 3 months or less) or when reaching their forties or fifties, i.e., even after 15 or 20 years of shift work. Symptoms of intolerance are persistent fatigue (physiological fatigue disappears after regular rests), digestive troubles, irritability and alteration of mood, and sleep disturbances (poor subjective quality of sleep, difficulty falling asleep and disrupted sleep with frequent awakenings).

Clinical symptoms observed in nontolerant shift workers are in almost all cases associated with the use of sleeping pills (barbiturates, benzodiazepines, tranquilizers, antihistaminecs, various plant alkaloids, etc.). The regular use of sleeping pills by a shift worker is considered in occupational medicine (ANDLAUER et al. 1979; REINBERG and SMOLENSKY 1992) to be a clinical sign of intolerance. Furthermore, shift workers who become nontolerant have circadian rhythms of relatively small amplitude and they adjust quickly after a shift. In contrast, tolerant subjects have a large amplitude of circadian rhythms and adjust slowly after manipulation of their rest activity cycle.

Such facts provide an anwser, at least in part, to the question of solving persistent sleep disturbances by regular intake of drugs: their regular use is of no help in solving problems of intolerance of shift work.

II. Jet Lag, Sleeping Pills, Melatonin and Bright Light

A transmeridian flight across five or more time zones is usually associated with sleep deprivation. Experimental studies have produced interesting findings (FRÖBERG et al. 1972). Prolonged (75 h vigil) sleep deprivation has shown the persistence of all documented circadian rhythms. Epinephrine, norepinephrine and temperature rhythms maintained similar 24-h mean levels respectively while performance tests deteriorated and self-rated fatigue increased over time. Since sleep deprivation reduces performance and increases fatigue the use of sleeping pills under certain circumstances and for short time spans may help some passengers. But this has nothing to do with the adjustment of a passenger's biological rhythms to local time.

In fact, many investigators have been dreaming of a wonder drug for resetting all biological clocks (REINBERG et al. 1988). Such a drug does not exist as yet and it is difficult to believe that it ever will. It has been given the name of chronobiotic (SIMPSON et al. 1973). This name is presently used to designate a substance (or a therapeutic means) able to reset (or shift) a limited group of biological rhythms (REINBERG and ASHKENAZI 1993).

The use of melatonin has been proposed by ARENDT et al. (1986a,b) to alleviate jet lag symptoms. Melatonin is normally secreted during the night hours, an endogenous circadian rhythm controlled by the suprachiasmatic nucleus. However, bright light (>2500 lux) in man is able to turn down this nocturnal secretion. In addition, melatonin has a mild hypnotic effect and can be used as such. It may well be that the observed alleviation of jet lag by melatonin adminstration (at a specified dosing time) is due at least in part to its hypnotic effect. It also seems that physical activity is able to reduce melatonin secretion.

The use of bright light (again at specified times) may help subjects to adjust a set of biological rhythms as suggested by results obtained by LEWY et al. (1987).

These types of research aiming to control jet lag symptoms by appropriate timing of, e.g., administration of melatonin, bright light and physical exercise must be encouraged.

H. Summary

The understanding of dosing time dependencies of an agent's effects (including anesthetic, hypnotic, narcotic and sedative effects) is based on properties of endogenous biological rhythms and more specifically on chronopharmacologic changes.

Circadian rhythms (having a period $\tau \cong 24$ h) with similar basic properties can be demonstrated from nucleated unicellular organisms to man: they persist without time clue and cue; they are genetic in origin; circadian clocks or oscillators have been identified (e.g., the suprachiasmatic nuclei in mammals); and signals (e.g., dawn and dusk) from the cyclic variation of a set of environmental factors (e.g., day/night alternation; social constraints of our life) called synchronizers or zeitgebers are used by the organism to calibrate the period and reset its circadian clocks.

As a consequence of this temporal organization the metabolic fate of a drug as well as its effects (both desired and nondesired) are time dependent for dosing. All parameters used to characterize pharmacokinetics (e.g., t_{max}, C_{max}, half-life, area under the curve, clearance rate) may vary as a function of the time, on the 24-h scale, at which the agent was administered (chronopharmacokinetics). In addition, the target system on which the agent is acting at the cellular and subcellular level (e.g., receptors, membranes) also varies rhythmically and, therefore, with predictable changes (chronesthesy or chronopharmacodynamics of the target system). The chronopharmacologic knowledge is used to select the best dosing time of a drug in order to increase both its effectiveness and its tolerance.

Circadian changes in drug-induced sleep, anesthesia and related processes have been investigated for a set of agents in laboratory rodents as well as in clinical studies. Anesthetics, curarizing substances, and ethanol and drugs that may induce drowsiness (e.g., H_1-antihistamines) have been considered. Some chronopharmacologic mechanisms have already been investigated such as circadian changes in drug-metabolizing liver enzymes and number of receptors, among others. Problems related to the use of sleeping pills by subjects exposed to the effects of zeitgeber manipulations (e.g., shift workers on night work, pilots and passengers after transmeridian flights) are briefly reviewed.

References

Akerstedt T, Gillbert M (1981) Sleep disturbances and shift work. In: Reinberg A, Vieux N, Andlauer P (eds) Shift-work studies: a multidisciplinary approach. Pergamon, New York, pp 127–135

Altmayer P, Mayer D, von Mayersbach H, Lücker P, Rindt W, Weizelsberger F (1979) Circadian variations in pharmacokinetic parameters after oral application of hexobarbital. Chronobiologia 6:73

Andlauer P, Reinberg A, Fourre L, Battle W, Duverneuil G (1979) Amplitude of the oral temperature circadian rhythm and the tolerance to shift work. J Physiol (Paris) 75:507–512

Apfelbaum M, Reinberg A, Nillus P, Halberg F (1969) Rythmes circadiens de l'alternance veille-sommeil pendant l'insolement souterrain de sept jeunes femmes . Presse Med 77:879–882
Arendt J, Aldhous M, Marks V (1986a) Alleviation of jet lag by melatonin: preliminary results and controlled double blind trial. Br Med J 292:1170
Arendt J, Aldhous M, Marks V (1986b) Alleviation of jet lag by melatonin. Annu Rev Chronopharmacol 3:49–51
Aschoff J (1963) Comparative physiology: diurnal rhythms. Annu Rev Physiol 25:581–600
Aschoff J, Fatranska M, Giedke H (1971) Human circadian rhythms in continuous darkness: entrainment by social cues. Science 171:213–215
Ashkenazi IE, Reinberg A, Bicakova-Rocher A, Ticher A (1993) The genetic background of individual variations of circadian rhythm periods in healthy human adults. Am J Hum Genet 52:1250–1259
Aymard N, Soulairac A (1979) Chronobiological changes in pharmacokinetics of dipotassic clorazepate, a benzodiazepine. In: Reinberg A, Halberg F (eds) Chronopharmacology. Pergamon, New York, pp 111–116
Belanger P (1988) Chronobiological variations in the hepatic elimination of drugs and toxic chemical agents. Annu Rev Chronopharmacol 4:1–46
Bourdallé-Badie C, Bruguerolle B, Labreque G, Robert S, Erny P (1990) Biological rhythms in pain and anesthesia. Annu Rev Chronopharmacol 6:155–181
Bouyard P, Bruguerolle B, Jadot G, Mesdjian E, Valli M (1974) Mise en évidence d'un rythme nycthéméral dans le phénomène de la curarisation chez le rat. C R Soc Biol (Paris) 168:1005–1007
Bruguerolle B, Valli M, Rakoto JC, Jabot G, Bouyard P, Reinberg A (1979) Chronopharmacology in pancuronium in the rat: anaesthesia and seasonal influence? In: Reinberg A, Halberg F (eds) Chronopharmacology. Pergamon, New York, pp 117–123
Bünning E (1963) Die Physiologische Uhr. Springer, Berlin Gōttingen Heidelberg
Folkard S, Monk TH (1985) Hours of work. Wiley, Chichester
Fröberg J, Karlsson CG, Levi L, Lidberg L (1972) Circadian variations in performance, psychological ratings, catecholamine excretion and diuresis during prolonged sleep deprivation. Int J Psychobiol 2:23–36
Fukami N, Kotani T, Shimoji K, Moriaka T, Isa T (1970) Circadian rhythm and anaesthesia. Jpn J Anaesthesiol 19:1235–1240
Halberg F (1969) Chronobiology. Annu Rev Phsyiol 31:675–725
Haus E, Halberg F (1959) 24 hour rhythm in susceptibility of C mice to toxic dose of ethanol. J Appl Physiol 14:878–880
Haus E, Halberg F, Kuhl JFW, Lakatua D (1974) Chronopharmacology in animals. Chronobiologia 1 [Suppl 1]:122–156
Lambinet D, Aymard N, Soulairac A, Reinberg A (1982) Chronoptimization of lithium administration in five manic depressive patients. Int J Chronobiol 7:274
Lemmer B (1989) Chronopharmacology. Cellular and biochemical interactions. Dekker, New York
Lemmer B, Bärmeier H, Schmidt S, Lang PH (1987) On the daily variation in the beta-receptor-adenylate-cyclase-cAMP-phosphodiesterase system in rat forebrain. Chronobiol Int 4:469–475
Lévi F, Benavidès J, Touitou Y, Quateronnnet D, Canton T, Uzan a, Guérémy C, Le Fur G, Reinberg A (1986) Circadian rhythm in peripheral type benzodiazepine binding sites in human platelets. Biochem Pharmacol 35:2623–2625
Lewy AJ, Sack RL, Miller LS, Hoban TM, (1987) Antidepressant and circadian phase-shifting effects of light. Science 235:352–354
Li DH, Novak R, Brockway B (1986) Effects of time single low dose ethanol ingestions on the circadian rhythm of body temperature of rats. Annu Rev Chronopharmacol 3:257–260
Moore RY (1983) Organization and function of a central nervous system circadian oscillator the suprachiasmatic hypothalamic nucleus. Fed Proc 42:2783–2789

Müller O (1974) Circadian rhythmicity in response to barbiturate. In: Scheving LE, Halberg F, Pauly J (eds) Chronobiology. Igaku Shoin, Tokyo, pp 187–190
Nair V (1974) Circadian rhythm in drug action: a pharmacologic biochemical and electron microscopic study. In: Scheving LE, Halberg F, Pauly J (eds) Chronobiology. Iagaku Shoin, Tokyo, p 182
Nakano S, Watanabe H, Nagai K, Ogawa N (1984) Circadian stage dependent changes in diazepam kinetics. Clin Pharmacol Ther 36:271–277
Nicholson AB, Stone BM (1979) Hypnotic activity during the day of diazepam and its hydroxylated metabolites, 3-hydroxydiazepam (temazepam) and 3-hydroxy-*N*-desmethmyl-diazepam (oxazepam). In: Reinberg A, Halberg F (eds) Chronopharmacology. Pergamon, New York, pp 159–169
Radzialowski FM, Bousquet WF (1968) Daily rhythmic variations in hepatic drug metabolism in the rat and mouse. J Pharmacol Exp Ther 163:229–238
Redfern PH, Moser PC (1988) Factors affecting circadian variation in responses to psychotropic drugs. Annu Rev Chronopharmacol 4:107–137
Reinberg A (1978) Clinical chronopharmacology, an experimental basis for chronotherapy. Arzneimittelforschung 28:1861–1867
Reinberg A (1981) Biological rhythms, sleep and drugs. In: Wheatly D (ed) Psychopharmacology of sleep. Raven, New York, pp 73–93
Reinberg A (1992a) Concepts in chronopharmacology. Annu Rev Pharmacol Toxicol 32:51–66
Reinberg A (1992b) Circadian changes in psychologic effects of ethanol. Neuropsychopharmacology 7:149–156
Reinberg A, Ashkenazi I (1993) Interindividual differences in chronopharmacologic effects of drugs: a background for individualization of chronotherapy. Chronobiol Int 10:449–460
Reinberg A, Halberg F (1971) Circadian chronopharmacology. Annu Rev Pharmacol 11:455–492
Reinberg A, Smolensky M (1981) Biological rhythms in medicine. Springer, Berlin Heidelberg New York
Reinberg A, Smolensky M (1992) Night and shift work and transmeridian and space flights. In: Touitou Y, Haus E (eds) Biological rhythms in clinical and laboratory medicine. Springer, Berlin Heidelberg New York, pp 243–255
Reinberg A, Sidi E, Ghata J (1965) Circadian rhythms of human skin to histamine or allergen and the adrenal cycle. J Allergy 36:273–283
Reinberg A, Clench J, Aymard N, Gaillot M, Bourdon R, Gervais P, Abulker C, Dupont J (1975) Variations circadiennes des effects de l'ethanol et de l'éthanolémie chez l'homme adult sain (étude chronopharmacologique). J Physiol (Paris) 7:435–456
Reinberg A, Lévi F, Guillet P, Burke JT, Nicolai A (1978) Chronopharmacological study of antihistamines in man with special references to terfenadine. Eur J Clin Pharmacol 14:245–252
Reinberg A, Lévi F, Fourtillon JP, Peiffer C, Bicakova-Rocher A, Nicolai A (1984) Antihistamine and other effects of 5 mg mequitazine vary between morning and evening acute administration. Annu Rev Chronopharmacol 1:57–60
Reinberg A, Motohashi Y, Bourdeleau P, Andlauer P, Levi F, Bicakova-Rocher A (1987) Alteration of period and amplitude of circadian rhythms in shift works (with special reference to temperature, ritht and left grip strength). Eur J Appl Physiol 57:15–25
Reinberg A, Smolensky M, Labrecque G (1986) New aspects in chronopharmacology. Annu Rev Chronopharmacol 2:3–26
Reinberg A, Smolensky M, Labrecque G (1988) The hunting of a wonder pill for resetting all biological clocks. Annu Rev Chronopharmacol 4:171–200
Reinberg A, Labrecque G, Smolensky M (1991) Chronobiologie et chronotherapeutique. Flammarion, Paris
Rutenfranz J, Singer R (1967) Untersuchungen zur Frage einer Abhängigkeit der Alkoholwirkung von der Tagezeit. Int Z Angew Physiol 24:1–17

Scheving LE, Donald F, Vedral DF, Pauly JE (1968) A circadian susceptibility rhythm in rats to pentobarbitol sodium. Anat Rec 160:741–750
Simmons DJ, Lester PA, Sherman NE (1974) Induction of sodium pentobarbital anaesthesia: a circadian rhythm. J Interdiscip Cycle Res 5:71–77
Simpson HW, Bellamy N, Bohlen J, Halberg F (1973) Double blind trial of a possible chronobiotic (Quiadon R). Int J Chronobiol 1:287–297
Stephan FK, Zucker I (1972) Circadian rhythms in drinking behavior and locomotor activity of rats are eliminated by hypothalamic lesions. Proc Natl Acad Sci USA 69:1583–1586
Sturtevant FM (1976) Chronopharmacokinetics of ethanol. Chronobiologia 3:237–262
Sturtevant RP, Garber SL (1988) Chronopharmacology of ethanol. Acute and chronic adminstration in the rat. Annu Rev Chronopharmacol 4:47–76
Virey JJ (1814) Ephémérides de la vie humaine ou recherches sur la révolution journalière et la périodicité de ses phénomènes dans la santé et les maladies. Thesis, University of Paris
von Mayersbach H (1978) Die Zeitstruktur des Organismus. Auswirkungen auf zelluläre Leistungsfähigkeit und Medikamentenempfindlichkeit. Arzneimittelforschung 28:1824–1835
Walker C (1974) Implications of biological rhythms in brain amine concentrations and drug toxicity. In: Scheving LE, Halberg F, Paul J (eds) Chronobiology. Igaku Shoin, Tokyo, p 205
Weitzman ED, Kripke, DF, Goldmacher D, McGregor P, Nogeire C (1970) Acute reversal of the sleep-waking cycle in man. Arch Neurol 22:483–489
Wirz-Justice A (1987) Circadian rhythms in mammalian neurotransmitter receptors. Prog Neurobiol 29:219–259

CHAPTER 4

Pharmacology of the Catecholaminergic System

A. WAUQUIER

A. Introduction

About 20 years ago Swedish investigators developed the techniques and described the existence of catecholaminergic (CA) neurons in rat brain (e.g., ANDEN et al. 1966; DAHLSTRÖM and FUXE 1964a,b). Since then new technologies of mapping have emerged which have led to the description of catecholaminergic neurons in the primate and human brains (PEARSON et al. 1983; see FUXE et al. 1985 for a more complete description of dopaminergic systems). It was found that the majority of these neurons were situated in the lower brain stem with widespread connections to a variety of regions within the nervous system. Subsequently, a large number of studies emphasized the role of these neurons in the control of motor and sensory systems, higher functions and neuronal plasticity (see, e.g., review on the "Neurobiology of the Locus Coeruleus," BARNES and POMPEIANO 1991).

Originally JOUVET (1972) collected evidence, based on neurophysiological, neurochemical and neuropharmacological studies, which supported the hypothesis that noradrenergic neurons originating from the locus coeruleus were critically involved in wakefulness and paradoxical sleep (PS).

The ideas on the specific role of this system and of the neurons in the vicinity of the locus coeruleus have changed during the past 20 years (JACOBS 1985; JONES 1991; GAILLARD 1985). Presented in this chapter is a hypothesis that describes the function of the noradrenergic system in sleep and wakefulness as it stands today. Part of the evidence supporting such a role has been derived from pharmacological studies, described herein.

Although the role of the dopaminergic system in sleep was originally thought to be involved only in the maintenance of wakefulness, its role has now been expanded to incorporate the idea that dopamine (DA) plays an important role in the maintenance of paradoxical sleep (PS). The identification of different subtypes of DA receptors and the discovery of specific agonists at these receptors elicited new hypotheses on the function of DA. It is suggested that the different subtypes of DA receptors are essential in the maintenance of wakefulness and PS.

Both norepinephrine (NE) and DA have an active and a permissive role, albeit different, in wakefulness and PS, two states of vigilance. It goes without saying that the ideas presented here are views of today in a continually developing field of research. This will also be evident from the occasional opposing views on the role of catecholaminergic systems in sleep-wakefulness regulation as demonstrated by different authors.

We will describe in sequence how sleep and waking are affected by: various drugs which interfere with the synthesis of CA and CA-depleting substances; drugs acting on NE receptors; and drugs acting on DA receptors. At the end of each subsection a summarizing table is presented. For the sake of convenience, we will refer to paradoxical sleep as PS, a term consistently reserved for identifying REM sleep in nonhuman species. For sleep, we will use the abbreviation SWS (slow-wave sleep), bearing in mind, however, that the sleep in animals is not identical to the SWS in humans.

B. Amine-Depleting and Synthesis-Inhibiting Substances (Table 1)

Some compounds can reduce CA neurotransmission by a method other than direct receptor interaction. Reserpine is one of these compounds: it depletes presynaptic stores by releasing NE. Alpha-methylparatyrosine (α-MPT) interferes with the synthesis of CA by inhibiting tyrosine hydroxylase, the rate-limiting enzyme involved in CA synthesis. Other substances, such as disulfiram or Fla-63, inhibit DA-β-hydroxylase, preventing the conversion of DA to NE.

I. Catecholaminergic-Depleting Substances: Reserpine

Reserpine induces a pattern on the EEG of slow waves in different cortical regions, often with a slow onset of action and a suppression of the cortical activation evoked by sensory or electrical stimulation. The effects of reserpine on sleep have been studied extensively. REITE et al. (1969) found that reserpine increased the length and number of PS epochs, shortened the PS latency, but did not change the total time of wakefulness and sleep in monkey. However, a decrease in PS and SWS has been reported without causing a rebound effect in subsequent nights in all other species studied: rats (GOTTESMANN 1966), cats (HOFFMAN and DOMINO 1969) and rabbits (TABUSHI and HIMWICH 1969).

MATSUMOTO and JOUVET (1964) found a longer-lasting suppression of PS than SWS in cats treated i.p. with 0.25 and 0.5 mg/kg reserpine; a subsequent injection of 5-hydroxytryptophan (5-HTP) prevented the diminution of SWS, but suppressed PS for 3 days. KARCZMAR et al. (1970) observed that adding acetylcholinesterase inhibitor eserine (which elevates acetylcholine concentrations) to reserpine produced all the characteristics of PS in rabbit, cat and rat. This drug combination, referring to an interaction between catecholaminergic and cholinergic systems (also see Chap. 7, this volume), was referred to as a pharmacological model of PS.

Table 1. Substances interfering with catecholamine availability

Compound			Species	Sleep			Behavior	Reference
Generic name	Dose (mg/kg)	Route		W	SWS lSWS|dSWS	PS		
1. Catecholamine depletion								
Reserpine	0.25–0.5	i.p.	Cat		↘	↘		MATSUMOTO and JOUVET (1964)
	0.5–5.0	i.p.	Rat			(↘)	S	GOTTESMANN (1966)
	0.01–0.16	i.m.	Cat		↘	↘		HOFFMAN and DOMINO (1969)
	0.25	p.o.	Monkey			↗		REITE et al. (1969)
	0.05–1.0	i.v.	Rabbit		↘ → ↗	↘		TABUSHI and HIMWICH (1969)
	0.15	i.p.	Cat			↘		STERN and MORGANE (1973)
	0.125	i.c.v.				↗		STERN and MORGANE (1973)
	0.5	i.p.	Cat		↘			BUCKINGHAM and RADULOVACKI (1976)
2. DA-β-hydroxylase inhibitors								
Disulfiram	400	p.o.	Cat		(↗)	↘		PEYRETHON-DUSAN and FROMENT (1968)
Fla-63	25	i.p.	Cat	(↘)	(↗)	↘	S	KEANE et al. (1976)
3. Tyrosine hydroxylase inhibitor								
α-Methylparatyrosine	50–200	i.p.	Rat	0	0	0	S	BRANCHEY and KISSIN (1970)
	3.125–100	i.p.	Cat		↗	↗	S	KING and JEWETT (1971)
	3.0–3.65	i.p. p.o.	Monkey		↗	↘	S	WEITZMAN et al. (1969)
	75–100	i.p.	Cat	↘		↗↘		KAFI et al. (1977)
	50–75	i.p.	Rat			(↘)		HARTMANN et al. (1971)
	50–100	p.o.	Rat			↗		HARTMANN et al. (1971)
4. NE uptake blocker								
Prindamine	0.5–5.0	i.p.	Cat	↗		↘		HILAVIKI et al. (1987)

Route: i.p., intraperitoneal; p.o., oral; s.c., subcutaneous; i.c.v., intracerebroventricular; i.v., intravenous; W, wake; SWS, slow-wave sleep; lSWS, light SWS; dSWS, deep SWS, to indicate where a differentiation within sleep is made; PS, paradoxical sleep.
Arrows indicate the direction of the change in wake or sleep; when in brackets they indicate a slight or tentative change. Behavior is indicated whenever specifically described: S, sedative or sedation; A, activation; D, drowsiness; H, hyperactivity.

Buckingham and Radulovacki (1976) found that injections of reserpine (0.5 mg/kg, i.p.) in cats caused periods of slow-wave activity after 8 h and a return to SWS after 16 h. This coincided with a normalization of DA metabolites in cerebrospinal fluid, although levels of the 5-HT metabolite 5-hydroxyindoleacetic acid (5-HIAA) remained low. Many studies supported the dissociation between the levels of NE and 5-HT and the modulation of PS and SWS, which led to the conclusion that these levels are a poor measure of the functions of these systems (Stern and Morgane 1973).

A well-studied phasic event occurring during sleep and mainly observed in cats is pontogeniculo-occipital (PGO) waves. Reserpine appears to induce PGOs throughout wake and PS in cats (Delorme et al. 1965), although the PGO induction was produced in the absence of other PS characteristics (Stern and Morgane 1973). Dement (1973) suggested that, in spite of a PS loss, there was no urge for PS rebound.

Depoortere and Lloyd (1979), studying the interaction between reserpine-induced PGO waves and α-adrenergic compounds in the cat, concluded that the functional state of the presynaptic NE receptors is the critical factor for the generation of PGOs. However, it was reported earlier that PGOs can be suppressed by 5-HT in cats (Delorme et al. 1965).

In the cat, ergolene derivatives, including bromocriptine, methysergide, *d*-LSD and dihydroxyergotoxine, all inhibited reserpine-induced PGO waves (Burki et al. 1978). Although these drugs increased NE turnover, they also affected 5-HT and DA differently and, therefore, these experiments failed to establish which transmitter system was crucial for the generation of PGO waves. Brooks and Gershon (1977) concluded that PGOs occurring during sleep are regulated by both NE and 5-HT systems and by CA during wakefulness. The induction of PGOs during PS does not appear to be regulated by CA, since in reserpine-treated cats L-dopa failed to return to prior PS levels.

It should be clear from the above that many controversial findings are determined mainly by the lack of specificity of reserpine. Furthermore, not much is known about the precise relationship between the degree of depletion of transmitters and the consequent functional disturbances.

II. Dopamine-β-hydroxylase Inhibitors

A selective depletion of NE can be achieved by substances such as Fla-63 or disulfiram which inhibit DA decarboxylase and thus prevent the conversion of DA to NE. As a consequence, DA levels are elevated. An earlier study described a slight increase in SWS, a significant decrease in PS and a reduction of PGO waves in cats treated orally with 400 mg/kg disulfiram (Peyrethon-Dusan and Froment 1968). Other studies were performed on specific models or using various drug combinations. Keane et al. (1976) used the *encéphale isolé* preparation in cats and found that Fla-63 tended to increase SWS and decrease wakefulness. The lack of pronounced and significant effects was ascribed to the limited depletion of CA. KAFI and GAILLARD (1978) stated that Fla-63 prevented the PS-increasing effects of chlorpromazine in rats, which suggested that NE was also partially involved in controlling PS.

III. Tyrosine Hydroxylase Inhibitor

α-Methylparatyrosine (α-MPT) reduces brain CA by competitive inhibition with tyrosine hydroxylase, the rate-limiting step in the synthesis of CA from tyrosine. The effects of α-MPT on sleep are far from consistent. In monkeys, WEITZMAN et al. (1969) found that α-MPT whether given i.p. or by nasogastric tube decreased PS and caused a proportional increase in SWS without consistently changing wakefulness. In cats, KING and JEWETT (1971) found that α-MPT (3.125–100 mg/kg, i.p.) increased PS and at higher doses was correlated with a reduction in CA. Higher doses of α-MPT given in fractions in order to avoid toxic effects also increased SWS.

In the *encéphale isolé* cat (KEANE et al. 1976) and in rats (BRANCHEY and KISSIN 1970; MARANTZ and RECHTSCHAFFEN 1967) sleep was not changed in spite of a decrease in NE levels. TORDA (1968) reported a decrease in PS and HARTMANN et al. (1971) found a significant increase in PS when α-MPT was given orally (50–100 mg/kg). The failure to influence PS following an i.p. injection was said to be due to peritoneal irritation.

To resolve some of the problems involved in interpreting the contradictory findings and to avoid renal toxicity, KAFI et al. (1977) gave different doses in multiple fractions of 75 mg/kg α-MPT. They found that doses of 75–100 mg/kg α-MPT increased PS, which was associated with a slight reduction in histochemically determined CA fluorescence. The length of the PS periods did not change, but the PS periods appeared more frequently.

According to these authors, low doses of α-MPT released a PS trigger mechanism. Higher doses decreased PS and this correlated with a decrease in brain CA fluorescence. They suggested that an intact CA transmission is necessary for the realization of PS.

IV. Norepinephrine Uptake Blockers

Few specific NE uptake blockers exist. Prindamine is a compound which besides its NE uptake blocking effect has weak anticholinergic and serotonergic uptake inhibiting effects. Prindamine given i.p. (0.5–5 mg/kg) immediately before recording sleep wake patterns in cats consistently decreased PS (HILAKIVI et al. 1987). The α-blocker, phentolamine, antagonized prindamine's effects, but yohimbine and prazosine were ineffective. HILAKIVI et al. (1987) suggested that the latter results were due to an increased wakefulness by the antagonists.

V. In Conclusion

Paradoxical sleep decreased following the administration of reserpine and Fla-63, but inconsistent results were obtained with α-MPT. The contradictory and inconsistent findings are partly due to the incompleteness of transmitter inhibition, lack of functional predictability from transmitter levels and lack of specificity of the available components. The varying findings of these drugs make it difficult to study the relationship between monoamines and the stages of sleep and wakefulness. However, most of the studies lead to the conclusion that an intact CA transmission is necessary for the realization of PS.

C. Adrenergic Agonists and Antagonists (Table 2)

Norepinephrine (NE) is formed from either a direct conversion of tyrosine or via the following sequence: tyrosine $\rightarrow$ dihydroxyphenylalanine $\rightarrow$ dopamine, which is converted to NE by the enzyme dopamine-β-hydroxylase. NE neurons typically possess this enzyme.

In the periphery a distinction has been made between two types of receptor sites, designated as α and β, and a further subdivision into type 1 and 2. α_1-receptors predominate at postsynaptic effector sites of smooth muscles and glandular cells and the α_2-receptors mediate presynaptic feedback regulating the release of NE. The β_1-receptors predominate in cardiac tissues and the β_2-receptors predominate in smooth muscle cells and glandular cells.

In brain also such distinctions can be made: α_1-receptors are sites where agonists (methoxamine) or antagonists (phenoxybenzamine, prazosin) act at the postsynaptic level; and α_2-receptors are presynaptic or "autoreceptors" (LANGER 1974) localized on soma or dendrites of NE neurons at their origin such as the locus coeruleus (CEDARBAUM and AGHAJANIAN 1977) or at the terminal sites in cortex (STARKE 1977) which fulfill an inhibitory role in NE release (SVENSSON et al. 1975). These are sites where agonists (clonidine) and antagonists (yohimbine, phentolamine) act. β_1-Receptors are located on neurons, whereas β_2-receptors are probably located on non-neuronal elements.

I. α_1-Antagonists (Phenoxybenzamine, Prazosin)

Phenoxybenzamine (40 mg/kg i.p.) in rats moderately increased PS (HARTMANN and ZWILLING 1976). In cats, however, phenoxybenzamine clearly increased PS (LEPPÄVUORI and PUTKONEN 1980). In 1962 it was suggested that the preoptic area is a site where activation can elicit sleep (HERNANDEZ-PEON 1962). Few neuropharmacological studies explored these suggestions further. MOHAN KUMAR et al. (1984) injected phenoxybenzamine (1 mg) into the preoptic area of rats and found that the drug induced sleep

Table 2. Adrenergic agonists and antagonists

Compound			Species	Sleep				Behavior	Reference
Generic name	Dose (mg/kg)	Route		W	SWS lSWS	SWS dSWS	PS		
α_1-Antagonists									
Phenoxybenzamine	40	i.p.	Rat				(↗)		HARTMAN and ZWILLING (1976)
	10	i.p.	Cat				↗		LEPPÄVUORI and PUTKONEN (1980)
Prazosin	0.5–1	i.p.	Cat				↗		HILAKIVI et al. (1980)
	0.5	i.p.	Kitten (2 weeks)				↗		MIETTINEN (1981)
	2.0	i.p.	Kitten			↗	↘		MIETTINEN (1981)
	0.5–2	i.p.	Kitten (1 month)	↘		↗			MIETTINEN (1981)
	0.125–1	i.p.	Rat				↘		PELLEJERO et al. (1984)
	0.5–10	i.p.	Cat	↘		(↘)			HILAKIVI and LEPPÄVUORI (1984)
α_1-Agonists									
Methoxamine	0.66	i.v. i.c.v.	Dog	↗		↘	↘	A	PICKWORTH et al. (1977)
	400 μg	i.c.v.	Dog	↗		↘	↘	A	PICKWORTH et al. (1977)
	4–8	i.p.	Rat	↗	↘		↘		PELLEJERO et al. (1984)
	0.5–3	i.p.	Cat	↗			↘		HILAKIVI and LEPPÄVUORI (1984)
α_2-Antagonists									
Phentolamine	20	i.p.	Cat				↗	S	PUTKONEN and LEPPÄVUORI (1977)
	5–20	i.p.	Kitten (2 wks–1 mnth)				↗	S	PUTKONEN and LEPPÄVUORI (1977)
	6	s.c.	Dog			↘	(↘)		GORDON and LAVIE (1984)

Table 2. *Continued*

Compound			Species	Sleep				Behavior	Reference
Generic name	Dose (mg/kg)	Route		W	SWS lSWS	dSWS	PS		
Yohimbine			Cat-dog	↗				A	Guerrero-Figuero et al. (1972)
	2	i.p.	Cat	↗		↘	(↘)		Putkonen et al. (1977)
	2	i.p.	Cat		↘	↘			Leppävuori and Putkonen (1980)
	0.03	i.p.	Rat				↗		Kafi and Gaillard (1981)
	3	i.p.	Rat				↘		Kafi and Gaillard (1981)
	3	i.p.	Rat	↗	↘		↘		Pellejero et al. (1984)
α_2-Agonists									
Clonidine	0.2–1	p.o.	Rat Rabbit	↘	↗	↘	↘		Kleinlogel et al. (1975)
	0.005–0.02	i.p.	Cat			(↘)	↘	S	Putkonen et al. (1977)
	0.005–0.02	i.p.	Cat	↘	↗	↘	↘	S	Leppävuori and Putkonen (1980)
	0.1	i.v.	Rat						Depoortere (1988)
	0.5	p.o.							
β_1- and β_2-antagonists									
Propranolol	1–40	p.o.	Rat	0		0	0		Hartmann and Zwilling (1976)
	5	i.p.	Cat			↘	↘	D	Hilakivi et al. (1978)
	20	i.p.	Rat			↗	↘		Mendelson et al. (1980)
	5	i.p.	Cat		0	↘	↘		Mendelson et al. (1980)
	2–20	i.p.	Rat			0	↘		Lanfumey and Adrien (1982)
	2 × 3 days	p.o.	Dog	0		0	0	D	Gordon and Lavie (1984)

β_1- and β_2-antagonists								
Isoproteranol	2–20	i.p.	Rat		0	↘		LANFUMEY et al. (1985)
L-Isoproteranol	10 μg	i.c.v.	Rat	0	0	0		
β_1-Antagonists								
Metoprolol	10–50	i.p.	Cat	(↗)		(↘)	D	HILAKIVI (1983)
β_1-Angonists								
Acebutolol	10–50	i.p.	Rat	0	0	0		LANFUMEY et al. (1985)
	10 μg	i.c.v.	Rat			↘		LANFUMEY et al. (1985)
Prenalterol	20–40	i.p.	Cat			↗		HILAKIVI (1983)
	20 μg	i.c.v.	Rat	0	0	0		LANFUMEY et al. (1985)
β_2-Agonists								
Clenbuterol	0.001–0.010	i.p.	Rat	0	0	0		LANFUMEY et al. (1985)
	0.001–0.100	i.p.	Rat	0	0	0		DUGOVIC and ADRIEN (1985)
Salbutamol	40	i.p.	Cat			↘		HILAKIVI (1983)

Route: i.p., intraperitoneal; p.o., oral; s.c., subcutaneous; i.c.v., intracerebroventricular; i.v., intravenous; W, wake; SWS, slow-wave sleep; lSWS, light SWS; dSWS, deep SWS, to indicate where a differentiation within sleep is made; PS, paradoxical sleep.
Arrows indicate the direction of the change in wake or sleep; when in brackets they indicate a slight or tentative change. Behavior is indicated whenever specifically described: S, sedative or sedation; A, activation; D, drowsiness; H, hyperactivity.

after 10 min, an effect lasting for 55 min. Periods of PS were also observed. Because phenoxybenzamine evoked sleep in the awake animal, MOHAN KUMAR et al. (1984) suggested that cessation of NE activity is "responsible for the natural induction of sleep."

In kittens 2 weeks of age, a low dose (0.5 mg/kg) of prazosin, a specific α_1-antagonist (U'PRICHARD et al. 1978), increased PS and a higher dose (2 mg/kg) decreased PS while increasing SWS; in kittens 1 month of age, wakefulness decreased and SWS increased. According to Miettinen (1981), these results reflect dynamic changes in the catecholaminergic balance during development.

In cats 0.5–1.0 mg/kg i.p. prazosin largely increased PS, but a higher dose (10 mg/kg) produced short-lasting inhibition (HILAKIVI et al. 1980) and shortened deep SWS (HILAKIVI and LEPPÄVUORI 1984). In cats treated with the NE uptake blocker prindamine, prazosin (1 mg/kg i.p.) prolonged PS latency and reduced SWS (HILAKIVI et al. 1987). There was no direct antagonism of the suppressed PS sleep by the NE uptake blocker prindamine, which was suggested to be due to an enhancement of wakefulness after the combination of prazosin and prindamine. Thus, NE uptake inhibition decreases PS and α_1-antagonists tend to increase PS, although not at all doses.

II. α_1-Agonist

Following i.v. and i.c.v. injections of methoxamine in dogs, deep SWS and PS decreased and a proportional increase in light SWS was observed (PICKWORTH et al. 1977). At the highest dose (0.66 mg/kg, i.v.) PS was completely abolished for 2 h. The α_1-antagonist, phenoxybenzamine, antagonized both behavioral arousal and the reduced SWS and PS. The effects seen with methoxamine were interpreted to be due to a stimulation of α-receptors at the hypothalamic level, but there is no direct experimental evidence to sustain this hypothesis.

III. α_2-Antagonists (Phentolamine, Yohimbine)

Though phentolamine lacks specificity, it can be considered as an α_2-antagonist (NICHERSON and COLLIER 1975). In rats it prolonged amobarbital-induced hypnosis (COHN et al. 1975) and in cats it produced behavioral sedation (PUTKONEN and LEPPÄVUORI 1977). Phentolamine, in kittens 2 weeks and 1 month of age (MIETTINEN 1981) and in cats (PUTKONEN and LEPPÄVUORI 1977), increased PS and antagonized the reduction of PS produced by α-methyldopa (LEPPÄVUORI and PUTKONEN 1978) or by clonidine (PUTKONEN 1979). Phentolamine (10 mg/kg i.p.) partially antagonized the inhibition of PS induced by the NE uptake blocker, prindamine (HILAKIVI et al. 1987).

Yohimbine at a low dose (30 μg/kg) increased PS in rats, which was prevented by the catecholamine depletor, α-methylparatyrosine; at a high dose (3 mg/kg) it decreased PS, an effect which was potentiated by α-methylparatyrosine (KAFI and GAILLARD 1981). In cats no such increase was found (LEPPÄVUORI and PUTKONEN 1980), but at the dose of 2 mg/kg i.p. it decreased light and deep SWS by more than 50%. These effects were interpreted by the authors as being due to interference from an increased NE transmission by blocking autoinhibitory presynaptic receptors. However, the contradictory results preclude a firm relationship between a particular effect and a presynaptic receptor blockade.

In the presence of the NE uptake blocker prindamine, yohimbine (1 mg/kg i.p.) decreased PS (HILAKIVI et al. 1987). This was suggested to be due to an enhancement of active wakefulness when the drugs were combined.

IV. α_2-Agonist

The α_2-agonist, clonidine, possesses a wide variety of peripheral and central actions. The sedative effect is associated with cortical slow waves in rabbits and sleep spindles in rat (DEPOORTERE 1981; KLEINLOGEL et al. 1975) and cat (HUKUHARA et al. 1968). Clonidine appears to reduce wakefulness and SWS and increase drowsiness in rats (KLEINLOGEL et al. 1975) and cats (PUTKONEN et al. 1977). In adult rats (KLEINLOGEL et al. 1975), in kittens 2 weeks and 1 month of age (MIETTINEN 1981) and in cats (LEPPÄVUORI and PUTKONEN 1980) PS was suppressed by dose-related amounts of clonidine.

In cats, PGO waves appear during PS. These PGO waves can be induced by reserpine and parachlorophenylalanine and are suppressed by clonidine (DEPOORTERE 1981; RUTH-MONACHON et al. 1976). DEPOORTERE et al. (1977) suggested that the ability to reduce these drug-induced PGO waves is positively correlated with the PS suppression by clonidine. This further reinforces the suggestion that clonidine suppresses neurons of the pontine reticular formation, which in the cat are involved in the generation of PGO waves (RUTH-MONACHON et al. 1976).

The slow waves produced by clonidine in rats, cats and rabbits are prevented by the α_2-antagonists, phentolamine and yohimbine, and not by the α_1-antagonist, phenoxybenzamine (FLORIO et al. 1975). The reduced PS in rats and cats is effectively antagonized by yohimbine and phentolamine (LEPPÄVUORI and PUTKONEN 1980; PUTKONEN and LEPPÄVUORI 1977; HOLMAN et al. 1971); PGO waves in the cat are antagonized by yohimbine, but not by prazosin and phenoxybenzamine (DEPOORTERE and LLOYD 1979). At variance with the above, phentolamine does not appear to antagonize the cortical spindles induced in rats (DEPOORTERE 1981), cats and rabbits (TRAN QUANG LOC et al. 1974).

V. β-Adrenergic Antagonists

The clinical use of propanol in humans is often associated with the side effects of insomnia (KALES et al. 1979). Propranolol given over a large dose range (1–40 mg/kg, orally) in rats caused no significant change in wakefulness or sleep (HARTMANN and ZWILLING 1976). In dogs, it did not change sleep-wake patterns (GORDON and LAVIE 1984). L-Propranolol (20 mg/kg i.p.) in rats (MENDELSON et al. 1980) suppressed the total percentage of PS when given during the day. When given during the night it prolonged PS latency and increased SWS. According to these authors this effect is caused by a blockage of melatonin synthesis, which increases during the night and in darkness and is associated with wakefulness. Isoproterenol (5–20 μl, i.c.v. infusion) in rats (LANFUMEY and ADRIEN 1982; LANFUMEY et al. 1985) reversed the decrease in PS caused by *dl*-propranolol. The same reversal was found with another β_1-agonist, prenalterol. The β_2-agonists, salbutamol and clenbuterol, could not reverse the propranolol-induced insomnia. However, HILAKIVI et al. (1978) found that propranolol (5 mg/kg, i.p.) in cats increased drowsiness, specifically decreasing deep SWS and PS without changing the light SWS.

HILAKIVI et al. (1987) found that propranolol (5 mg/kg i.p.) decreased PS in cats and partially reduced the increased PS latency caused by citalopram, a 5-HT uptake blocker. Such results might be related to the β-adrenergic effect of propranolol, whereas the interaction might be related to a direct action of propranolol on 5-HT receptors.

The effects on PS by propranolol might partially be due to indirect action on other neurotransmitter systems. LANFUMEY and ADRIEN (1982) strongly argue against this interpretation, since for example quipazine, a 5-HT agonist, was unable to reverse the effects of propranolol. It is suggested, therefore, that β_1-receptors might be involved in PS regulation.

VI. β-Adrenergic Agonists

Following an i.c.v. injection of *l*-isoproterenol in rats, PS increased during the 4 h of recording. The insomnia produced by an i.p. injection of 10 mg/kg propranolol was reversed by *l*-isoproterenol. In the rats which were subjected to a neonatal lesion of the locus coeruleus, restoration of PS was more efficient. This study demonstrated an interaction between a β-agonist and -antagonist, which enhanced and diminished the amount of PS respectively. It was also found that the more pronounced effects in previously lesioned animals were possibly due to supersensitivity of the receptors.

VII. In Conclusion

The above studies support the hypothesis that NE neurons aid in modulating wakefulness (EEG and behavioral aspects) and play a role in the occurrence of PS (GAILLARD 1985). The α_1-agonist, methoxamine, decreased PS, and the reverse was observed with the antagonist, phenoxybenzamine. These results, however, might be indirectly caused by a predominantly stimulatory effect of methoxamine on wakefulness. α_1-Antagonists, such as prazosin, increased PS and this was overcome by an α_2-antagonist but not by an α_1-antagonist or β-blocker. Because of its presynaptic action, clonidine inhibits the release of NE perhaps via an action on the locus coeruleus. Furthermore, the correlation between the inhibition of the PGO waves and the inhibition of PS suggests that they involve NE. These experiments reveal that interference with NE systems can change the amount of PS, but they did not implicate the NE locus coeruleus system as the essential substrate of PS.

The sleep "induction" with clonidine appears to be associated with light sleep, the occurrence of spindles and inhibiting deep SWS. The induction of this spindle could not be antagonized by the α_2-antagonist, phentolamine, in cats. The α_1-agonist, methoxamine, also reduced SWS, and this can be reversed by the α-blocker, phenoxybenzamine. However, there is not much evidence supporting the suggestion that α_1-receptors are essential or exclusive in the regulation of SWS.

The involvement of β-adrenergic receptors in the regulation of sleep is controversial. However, based predominantly on the effects of β-receptor agonists and the reversal of propranolol-induced insomnia, ADRIEN et al. (1985) and STENBERG and HILAKIVI (1985) accumulated evidence for the idea that these receptors may be involved in the regulation of PS. This would suggest that the effects are related to activation of β_1-receptors (ADRIEN et al. 1985).

D. Dopamine Agonists and Antagonists (Table 3)

The predominant action of DA at the receptor level as defined in electrophysiological studies using the iontophoretic technique is inhibition, though some excitatory effects have also been found in the striatum (YORK 1970). On the basis of pharmacological studies in the cat and electrophysiological studies on ganglia of *Helix aspersa* (e.g., COOLS and VAN ROSSUM 1976), two types of DA receptors were proposed, one excitatory and one inhibitory. Both receptors are thought to have a distinct localization in the caudate nucleus, be sensitive to different types of agonists and antagonists and be functionally different though interconnected. It was also suggested that stimulation of the D_1 receptor caused activation of adenylyl cyclase. The D_2 receptor was described as not being coupled or inhibitory to adenylyl cyclase. Since then, many different D-receptor subtypes have been described going beyond the D_1, D_2 classification. The heterogenicity was revealed by the cloning of genes from several members of the large family of G-protein-coupled receptors (e.g., ANDERSEN et al. 1990; STRANGE 1991; SIBLEY 1991). At present, but this could rapidly change, five subtypes have been described: D_1; D_2 (short) and D_2 (long), the latter identified by gene cloning and derived from alternative splicing of a common gene (STRANGE 1991); D_3 and D_4, subtypes related to the D_2 (SIBLEY 1991); and a D_5-receptor subtype, a D_1-like receptor. For the sake of simplicity and because of the limited

Table 3. Dopamine agonists and antagonists

Compound			Species	Sleep				Behavior	Reference
Generic name	Dose (mg/kg)	Route		W	SWS lSWS	dSWS	PS		
1. DA precursor									
L-Dopa	25[a]	i.p.	Cat	↗				H	Jones (1972)
	80	s.c.	Dog	↗	↘	↘	↘	A	Wauquier (1985)
2. D_2-agonists									
Apomorphine	0.43–4.9	i.p.	Rat	↗		↘	↘	H	Kafi and Gaillard (1976)
	≤0.1	s.c.	Rat			↗	↗	D	Mereu et al. (1979)
	≥0.1	s.c.	Rat				↘	A	Mereu et al. (1979)
	0.0125–0.05	s.c.	Rat	↘		↗	↗	D	Wauquier (1985)
	0.50	s.c.	Rat	↗		↘	↘	A	Wauquier (1985)
	0.04–1.25	s.c.	Dog	↗	↘	↘	↘	H	Wauquier et al. (1980)
	0.025–0.05	i.p.	Rat	↘		↗	↗		Monti et al. (1988)
	1.0–2.0	i.p.	Rat	↗		↘			Monti et al. (1988)
	0.5	s.c.	Rat			↘	↘		Chastrette et al. (1990)
Bromocriptine	0.25–6.0	i.p.	Rat	↘			(↗↘)		Morti et al. (1988)
Pergolide	0.05	i.p.	Rat			↗	↘		Monti et al. (1988)
	0.01–0.5	i.p.	Rat	↗		↘	↘		Monti et al. (1988)
Quinpirole	0.15–0.030	i.p.	Rat	↘			↘		Monti et al. (1989)
	0.25–1.0	i.p.	Rat	↗	↘	↘	↘		Monti et al. (1989)

3. D_2-antagonists									
Haloperidol	1–4	i.p.	Cat	↘	↗		↘		Monti (1968)
	2	p.o.	Rat	↘	↗	↘	↘		Stille and Lauener (1974)
	0.2–0.16	i.p.	Rat	↘	(↗)	↗	↘		Monti (1979)
	0.3–1.0	i.p.	Cat	(↗)		↘	↘		Polc et al. (1979)
	10	i.p.	Cat		(↗)	↗	↘	D	Tsuchiya et al. (1979)
	0.01–0.16	s.c.	Dog	↘	(↗)	↗	(↗)		Wauquier et al. (1980)
	0.02–0.16	i.p.	Rat	↘					Monti et al. (1988)
	0.1–3.0	p.o.	Rat	↘		↗			Trampus and Orgini (1990)
Pimozide	0.4–1	i.p.	Rat	↘	(↘)				Monti (1979)
	⩽0.016	p.o.	Dog	↗	↘	↘	(↘)		Wauquier et al. (1980)
	⩾0.063	p.o.	Dog	↘	↗	↗	↗		Wauquier et al. (1980)
Risperidone[b]	0.01–0.16	i.p.	Rat		↘	↗	↘		Dugovic et al. (1989)
	0.63–2.5	i.p.	Rat		↗	↘	↘		Dugovic et al. (1989)
4. D_1-agonist									
SKF 38393	0.1–4.0	i.p.	Rat				(↘)		Monti et al. (1990)
5. D_1-angagonist									
SCH 23390	0.1–2.0	i.p.	Rat	↘		↗	↘		Monti et al. (1990)
	0.003–0.03	s.c.	Rat	↘		↗	↗		Trampus and Orgini (1990)

Route: i.p., intraperitoneal; p.o., oral; s.c., subcutaneous; i.c.v., intracerebroventricular; i.v., intravenous; W, wake; SWS, slow-wave sleep; lSWS, light SWS; dSWS, deep SWS, to indicate where a differentiation within sleep is made; PS, paradoxical sleep.
Arrows indicate the direction of the change in wake or sleep; when in brackets they indicate a slight or tentative change. Behavior is indicated whenever specifically described: S, sedative or sedation; A, activation; D, drowsiness; H, hyperactivity.
[a] Effects obtained with combination of L-dopa and 50 mg/kg tropolone (an inhibitor of COMT).
[b] Risperidone also has potent 5-HT_2-receptor-blocking activity.

number of specific drugs available, the following description will pertain to D_1- and D_2-agonists and -antagonists only.

I. Dopamine$_2$-Agonists

In rats, apomorphine, a D_2 agonist, reduced sleep at doses causing stereotypic behavior (KAFI and GAILLARD 1976). In dogs, apomorphine produced a dose-related increase in wakefulness and a decrease in SWS and PS, specifically during the first hours following the injection (WAUQUIER et al. 1981). In rats, however, low doses of apomorphine produced hypomotility (STRÖMBOM 1976). The possible hypnogenic effects of apomorphine were studied by MEREU et al. (1979) in rats. Below a dose of 100 μg/kg the total amount of sleep increased from 18.2% to 54.4% of the recording time, which was mainly due to an increase in SWS, though PS also significantly increased. WAUQUIER (1985) found that apomorphine's effects were independent of the time of the recording and of the amount of sleep normally observed during control recordings.

Thus, apomorphine displays a biphasic effect on sleep in rats; stereotypogenic doses cause a marked reduction of sleep, whereas low doses are hypnotic. MONTI et al. (1988) studied different D_2 agonists on sleep-wake behavior in rats: apomorphine, bromocriptine and pergolide. All of these drugs produced a biphasic effect. However, pergolide suppressed PS sleep irrespective of the dosage. Such findings may further reinforce the idea of an independent mechanism for PS and SWS, but it does not explain why pergolide exerts this particular effect on PS. The D_2 antagonist, haloperidol, at doses preferentially acting upon presynaptic sites, reversed the effects of low doses of the D_2 agonists on wake and sleep (MONTI et al. 1988).

WAUQUIER et al. (1980) studied the interaction between different doses of apomorphine and a dose of a D_2-antagonist, pimozide (0.063 mg/kg p.o.), which did not influence sleep in dogs, and of domperidone (a peripherally acting DA antagonist) at a dose (0.16 mg/kg, p.o.) antagonizing the emetic effects of apomorphine.

The changes in the sleep-wakefulness patterns induced by apomorphine were not affected by the peripheral D_2-antagonist, domperidone, though domperidone completely antagonized the emetic effects of apomorphine. Thus, the effects are centrally mediated and obviously not related to the emetic properties of apomorphine.

Pimozide partially antagonized PS reduction caused by 0.02 mg/kg apomorphine. It also completely antagonized the increased wakefulness, decreased SWS and partially antagonized the PS caused by 0.16 mg/kg apomorphine. A partial antagonism was found against the highest dose of apomorphine. It is suggested that DA plays a prominent role in the organization of both PS and wakefulness.

In a further study MONTI et al. (1989) investigated the selective D_2-receptor agonist, quinpirole, in rats. Again, a biphasic affect was found.

Using another D_2 antagonist, YM-09151-2, at doses acting at presynaptic sites, reversed the suppressant effect of a low dose of quinpirole on wake and sleep. Using a dose blocking postsynaptic D_2 receptors, the depressed PS sleep was not affected. The latter is indicative of an active role of DA in the control of wakefulness. A presynaptic D_2-receptor activation appears more related to SWS and PS induction. BAGETTA et al. (1988) suggested that the site of action for the "hypnotic" effects of D_2 agonists is the ventral tegmental area. They derived this conclusion from their studies using bilateral injections of apomorphine (0.01–1.0 nmol) into the ventral tegmental area. Treatment with the D_1-receptor antagonist, SCH 23390, and an α_2-adrenoceptor antagonist, yohimbine, were ineffective. There are now experimental drugs with a purported specific action as agonists or antagonists at the D_1 receptor. A specific D_1-agonist, SKF 38393, was given i.p. at doses ranging from 0.1 to 4 mg/kg to rats and only significantly decreased PS at the highest dose and markedly increased PS latency at the doses of 0.25 and 4 mg/kg (MONTI et al. 1990). Such a weak effect could be related to poor brain penetration of the drug. Also it could not be concluded from these experiments that there is a relationship between D_1-receptor activation and the inhibition of PS, for a combination of the D_1 antagonist, SCH 23390, with the D_1 agonist, SCH 38393, led to an enhanced suppression of PS.

II. Dopamine$_2$-Antagonists

Although neuroleptics have D_2-antagonistic properties, most are not devoid of unrelated pharmacological activity. Interestingly, some newer compounds have primarily an antiserotonergic activity combined with a D_2-antagonistic property, such as risperidone. Such combined effects have numerous clinical advantages, particularly with respect to Parkinson's disease-like side effects.

The effects of haloperidol on sleep are species dependent: in rats (MONTI 1979) and cats (MONTI 1968; TAKEUCHI 1973) SWS is increased and PS is decreased: in dogs wakefulness is decreased, and SWS and PS are increased (WAUQUIER et al. 1980). In rats, STILLE and LAUENER (1974) found an inhibition of deep SWS, but an increase in spindle sleep and an increase in PS. This was confirmed by TSUCHIYA et al. (1979), who found that different neuroleptics increased spindle sleep and decreased PS.

The effects on sleep are also dependent upon the type of neuroleptic studied. In rats, for example, α-flupentixol decreased wakefulness, increased SWS and had no significant effect on PS. In addition, the effects were more pronounced during the dark phase than during the light period (FORNAL et al. 1982). Whereas a PS-suppressant effect was induced by loxapine in cats (STILLE and LAUENER 1974; SCHMIDEK et al. 1974), only the former authors found a strong suppressant effect on deep SWS.

As of yet, there has been no interpretation of these different effects; however, it is clear that they depend on the specificity of the DA antagonist studied and the doses utilized. It is, therefore, of importance to observe the effects with the specific D_2 antagonist, primozide, at different dose levels. WAUQUIER et al. (1980) found that in dogs pimozide induced a biphasic effect, which is the opposite of that found with apomorphine in the rat by MEREU et al. (1979). The lowest dose of pimozide increased wakefulness and decreased SWS and PS; the intermediate dose did not alter the sleep pattern; and the highest dose used decreased time awake and increased both SWS and PS.

It appears that a DA antagonist more effectively reduces wakefulness during the day-night cycle when DA turnover is lowest (FORNAL et al. 1982), suggesting that DA neurons are more specifically involved in wakefulness (JOUVET 1972). Our studies and those of MEREU et al. (1979) also suggest that DA is implicated in the occurrence of PS.

Risperidone, a new neuroleptic with potent 5-HT_2 and D_2-antagonistic properties, was given i.p. to rats (dose range, 0.01–2.50 mg/kg) at the onset of the light period (DUGOVIC et al. 1989). During the first 4 h low doses (0.01–0.16 mg/kg) increased deep sleep and decreased wakefulness, light sleep and PS. Higher doses (0.03–2.5 mg/kg) increased light sleep and decreased deep sleep, whereas the PS reduction persisted. The enhanced deep sleep at the low dose, which is in contrast to the effects of specific D_2 antagonist, is suggested to be due to a 5-HT_2 receptor antagonism. Similarly, no increase in PS was observed when high doses were administered.

III. Dopamine$_1$-Antagonists

Currently, there exist several studies on SCH 23390, a specific D_1-receptor antagonist. Studying behavioral and EEG changes in rats, GESSA et al. (1985) found that SCH 23390 prevented the stereotypy and EEG desynchronization. They suggested that D_2 receptors mediate sedation and sleep.

TRAMPUS and ORGINI (1990) found that SCH 23390 given s.c. at doses ranging from 0.0003 to 0.03 mg/kg increased SWS and produced an even more pronounced increase in PS, mainly due to an increase in the number of PS episodes. The fact that the proportion of SWS/PS increased after SCH 23390 and decreased after haloperidol suggested that the effects of PS are unrelated to the effects on SWS, which is a suggestion that was formulated previously by WAUQUIER et al. (1981). It was found that the latency to the first PS episode was not altered. It is, therefore, more likely that the D_1-receptor has a more specific effect on the process of PS.

MONTI et al. (1990) found a significant decrease in wakefulness and PS, no effect on PS latency and an increase in SWS. The effects were maximal at a dose of 0.25 mg/kg. The increase in SWS and decrease in wakefulness by SCH 23390 was prevented by the D_1 agonist SKF 38393, but not by the PS suppression. The biphasic effects seen by DA autoreceptors and drugs acting on postsynaptic D_2 receptors were not seen following administration of SCH 23390. Furthermore, a blockade of $5HT_2$ receptors could be partially responsible for the effects of SCH 23390 on sleep.

Some discrepancy exists between the interpretations of the effects of SCH 23390 on sleep mechanisms by different authors. TRAMPUS and ORGINI (1990) found an increase in PS, whereas MONTI et al. (1990) reported a decrease. This difference is probably related to the doses used: the former authors used doses ranging from 0.003 to 0.03 mg/kg given s.c., whereas the latter used doses ranging from 0.1 to 2.0 mg/kg given i.p. This would suggest that a D_1 antagonist, SCH 23390, in a similar way to D_2 agonists, induces a biphasic effect on PS.

IV. In Conclusion

Both D_2 agonists and antagonists have a biphasic effect on sleep-wakefulness, yet their effects are in opposition. Low doses of the D_2 agonists and high doses of the D_2 antagonists decrease wakefulness and increase both SWS and PS. Conversely, high doses of the D_2 agonists and low doses of the D_2 antagonists increase wakefulness and decrease both SWS and PS. In addition, there is a mutual interaction between D_2 agonists and antagonists.

There does not yet exist a potent, selective and centrally acting D_1 agonist. However, it is proposed that SKF 38393 has D_1-agonistic properties. It was found that SKF 38393 slightly decreases PS. The D_1-antagonist, SCH 23390, decreases wakefulness and increases SWS. Again a biphasic effect on PS could be suggested to exist: a low dose increased and a high dose decreased PS. However, it is evident that a D_1 antagonist does not affect PS latency, suggesting that the involvement of the D_1 receptor is in maintenance rather than initiation of PS.

E. Differential Regulation of Waking and PS by Catecholamines

In the past it was believed that neurons of the pontomesencephalic tegmentum, later identified as noradrenergic neurons, were essential to the initiation and maintenance of wakefulness and PS (JOUVET 1972). Later studies have suggested that the involvement of NE in waking and its involvement in PS are in opposition to each other: during waking there is activity of NE neurons, whereas PS activity is decreased, or even prevented (JONES 1991; GAILLARD 1985). With respect to wakefulness, taking the presented pharmacological studies into account, but also lesioning and electrophysiological studies (JONES 1991), noradrenergic neurons appear to play a modulatory role in wakefulness and are involved in cortical activation of waking, but they do not play an essential role in the maintenance of waking.

Paradoxical sleep is a much more vulnerable state, influenced by α- and β-adrenergic agonists and antagonists. Basically α_1- and α_2-agonists and β_1-antagonists decrease PS, whereas α_1- and α_2-antagonists and β_1-agonists increase PS.

It is thus suggested that α-adrenergic activation and β_1-antagonism is inhibitory to PS. Again lesion studies and electrophysiological recordings of locus coeruleus noradrenergic neurons showed that these neurons are turned off during PS (JACOBS 1985). This is in contrast to the previous hypothesis as formulated by GAILLARD (1985) that NE synaptic activity is "a requisite condition for the appearance of these states" (i.e., waking and PS). Pharmacological studies appear to provide a more subtle picture, but are more in agreement with the lesion studies. A good summary of the involvement of the β_1-receptor as part of the regulatory mechanisms of PS is found in ADRIEN et al. (1985). That neurons other than those of the locus coeruleus are responsible for the generation of PS has proven to be true. Yet, the neurons are cholinergic as suggested in the earlier pharmacological studies by JOUVET (1972) and are in contrast to the later studies by GAILLARD (1985). In fact, cholinergic neurons appear to be essential for the timing of PS. Finally, PGO waves, although influenced by adrenergic and by serotonergic drugs (see also Chaps. 5, 7, this volume), are rather dependent upon cholinergic activity (WEBSTER and JONES 1988).

It was suggested that DA has a permissive and active role on sleep-wakefulness (WAUQUIER 1985). The development of drugs acting at different subtypes of the DA receptor allows a refinement of the proposition without altering it. Waking is suggested to be a state sustained by D_2 activation, whereas decreased dopaminergic activity and inhibition of motor activity create the condition favoring sleep. Activation of dopaminergic autoreceptors (e.g., by low doses of D_2-agonists) is sleep inhibiting by inducing a postsynaptic disinhibition. The decreased waking observed with the D_1 antagonist, SCH 23390, is suggested to be due to an indirect activation of the D_2 receptor. In fact, waking and SWS are not significantly modified by D_1 activation (MONTI 1990). In addition, SCH 23390 prevented motor activity and stereotypy induced by the D_2 agonist, apomorphine, suggesting that D_1 inactivation leads to D_2 inactivation (GESSA et al. 1985).

Low doses of the D_1 antagonist SCH 23390 largely increase PS, which was suggested to be a specific effect (TRAMPUS and ORGINI 1990). In particular the D_1 receptor might be involved in the maintenance of PS, not the initiation, for there is no effect of these drugs on PS latency, but a specific effect on the number of PS periods. However, a more cautious conclusion is that D_1 receptors are involved in the regulation of PS, in particular its maintenance.

F. Summary

The present chapter provides a summary of the effects of catecholaminergic drugs on sleep and wakefulness: (a) drugs interfering with catecholaminergic transmission either by depleting or by inhibiting the synthesis of catecholamines; (b) α_1- and α_2-agonists and antagonists and β-adrenergic agonists and antagonists; and (c) dopamine 1 and 2 agonists and antagonists. From these pharmacological experiments in conjunction with other studies, hypotheses can be formulated as to the role of norepinephrine and dopamine on sleep-wakefulness regulation and organization.

Most studies on drugs interfering with catecholamine transmission lead to rather contradictory and inconsistent findings. However, these studies may also suggest that an intact catecholamine transmission is necessary for the realization of paradoxical sleep.

Studies of drugs acting on adrenergic receptors are more conclusive in that they lead to the hypothesis that noradrenergic neurons play a modulatory role in wakefulness (EEG and behavioral aspects) and in the occurrence of paradoxical sleep. In particular, β_1-receptors are involved; some studies suggest that paradoxical sleep is regulated by activation of β_1-receptors. Although there are sleep changes induced by drugs acting on α-receptors, it can be concluded that α_1-receptors are not essential, and neither are these receptors exclusive in the regulation of slow-wave sleep. In spite of the many studies demonstrating that drugs acting on the adrenergic system change sleep-wakefulness patterns, these studies do not imply that the noradrenergic–locus coeruleus system is an essential substrate of paradoxical sleep.

Dopamine has a permissive and active role on sleep-wakefulness regulation. Drugs acting as agonists or antagonists of the different receptor subtypes have biphasic and opposite effects on wakefulness and sleep. Wakefulness is sustained by dopamine 2 activation and a decrease of dopamine 2 activation may promote sleep. Dopamine 1 receptors are possibly involved in the maintenance of paradoxical sleep, not in its initiation and timing.

References

Adrien J, Lanfumey L, Dugovic C (1985) Effects of beta-adrenergic antagonists on sleep. In: Wauquier A, Gaillard JM, Monti JM, Radulovacki M (eds) Sleep: neurotransmitters and neuromodulators. Raven, New York, pp 93–106

Anden NE, Dahlstrom A, Fuxe K, Larsson K, Olson L, Ungerstedt U (1966) Ascending monoamine neurons to the telencephalon and diencephalon. Acta Physiol Scand 67:313–326

Andersen PH, Gingrich JA, Bates MD, Dearry A, Falardeau P, Senogles SE, Caron MG (1990) Dopamine receptor subtypes beyond the D1/D2 classification. Trends Pharmacol Sci 11:231–236

Bagetta G, DeSarro G, Priolo E, Nistico G (1988) Ventral tegmental area: site through which dopamine D2-receptor agonists evoke behavioral and electrocortical sleep in rats. Br J Pharmacol 95:860–866

Barnes CD, Pompeiano O (eds) (1991) Neurobiology of the locus coeruleus. Elsevier, Amsterdam

Branchey M, Kissin B (1970) The effects of alpha-methyl-paratyrosine on sleep and arousal in the rat. Psychosom Sci 19:281–282

Brooks DC, Gershon MD (1977) Amine repletion in the reserpinized cat: effect upon PGO waves and REM sleep. Electroencephalography 42:35–47

Buckingham RC, Radulovacki M (1976) The effects of reserpine, L-dopa and 5-hydroxytryptophan on 5-hydroxyindole acetic and homovanillic acids in cerebrospinal fluid, behavior and EEG in cats. Neuropharmacology 15:383–392
Burki HR, Asper H, Ruch W, Zuger PE (1978) Bromocriptine, dihydroergotoxine, methysergide, d-LSD, CF 25-397 and 29-712: effects on the metabolism of the biogenic amines in the brain of the rat. Psychopharmacology (Berl) 57:227–237
Cedarbaum JM, Aghajanian GK (1977) Catecholamine receptors on locus coeruleus neurons: pharmacological characterization. Eur J Pharmacol 44:375–385
Chastrette N, Cespuglio R, Lin YL, Jouvet M (1990) Propiomelanocortin (POMC) derived peptides and sleep in the rat. II. Aminergic regulatory processes. Neuropeptides 15:75–88
Cohn ML, Cohn M, Taylor FH (1975) Effects of phentolamine on dibutyryl cyclic AMP and norepinephrine in rats anesthestized with amobarbital. Arch Int Pharmacodyn 217:80–85
Cools AR, Van Rossum JM (1976) Excitation-mediating and inhibition-mediating dopamine receptors: a new concept towards a better understanding of electrophysiological, biochemical, pharmacological, function and chemical data. Psychopharmacologia 45:243–254
Dahlström A, Fuxe K (1964a) Evidence for the existence of monoamine-containing neurons in the central nervous system. Acta Physiol Scand 62 [Suppl]:232
Dahlström A, Fuxe K (1964b) A method for the demonstration of monoamine-containing fibers in the nervous system. Acta Physiol Scand 60:293–295
Delorme F, Jeannerod M, Jouvet M (1965) Effects remarquables de la réserpine sur l'activité EEG phasique ponto-géniculo-occipitale. C R Soc Biol (Paris) 159:900–903
Dement WC (1973) The biological role of REM sleep. In: Webb WB (ed) Sleep – an active process. Scott, Foresman, Chicago, pp 33–58
Depoortere H (1981) Sleep induction and central effects of some α-adrenoceptor agonists. In: Koella WP (ed) Sleep 1980. Karger, Basel, pp 283–286
Depoortere H, Lloyd KG (1979) Ponto-geniculo-occipital spikes: a model for the study of the central action of noradrenergic compounds. Adv Biosci 18:173–178
Depoortere H, Honore L, Jalfre M (1977) EEG effects of various imidazoline derivatives in experimental animals. In: Koella WP, Levin P (eds) Sleep 1976. Karger, Basel, pp 358–361
Dugovic C, Adrien J (1985) Effects of clenbuterol on sleep in the rat. In: Koella WP, Rüther, Schultz H (eds) Sleep. Fischer, Stuttgart, pp 332–334
Dugovic C, Wauquier A, Janssen PAJ (1989) Differential effects of the new antipsychotic risperidone on sleep and wakefulness in the rat. Neuropharmacology 28:1431–1433
Florio V, Bianchi L, Longo VG (1975) A study of the central effects of sympathomimetic drugs: EEG and behavioral investigations on clonidine and naphazoline. Neuropharmacology 14:707–714
Fornal C, Wojcik WS, Radulovacki M (1982) α-Flupenthixol increases slow-wave sleep in rats: effect of dopamine receptor blockade. Neuropharmacology 21: 323–325
Fuxe K, Agnati LF, Kalia M, Goldstein M, Anderson K, Harfstrand A (1985) Dopaminergic systems in the brain and pituitary. In: Halasz B, Fuxe K, Agnati LF, Kalia M, Goldstein M, Andersson K, Hàrfstrand A, Clark B (eds) The dopaminergic system. Springer, Berlin Heidelberg New York
Gaillard JM (1985) Involvement of noradrenaline in wakefulness and paradoxical sleep. In: Wauquier A, Gaillard JM, Monti JM, Radulovacki M (eds) Sleep: neurotransmitters and neuromodulators. Raven, New York, pp 57–67
Gessa GL, Porceddu ML, Collu M, Mereu G, Serra M, Orgini E, Baggio G (1985) Sedation and sleep induced by high doses of apomorphine after blockade of D-1 receptors by SCH 23390. Eur J Pharmacol 109:269–274
Gordon CR, Lavie P (1984) Effect of adrenergic blockers on the dog's sleep-wakefulness. Physiol Behav 32:345–350

Gottesmann C (1966) Réserpine et vigilance chez le rat. C R Soc Biol (Paris) 160: 2056–2061
Guerrero-Figuero R, Gallant DM, Galatas RF, Rye MM (1972) Effects of yohimbine on CNS structures: neurophysiological and behavioral correlations. Psychopharmacologia 25:133–145
Hartmann E, Zwilling G (1976) The effect of alpha and beta adrenergic receptor blockers on sleep in the rat. Pharmacol Biochem Behav 5:135–138
Hartmann E, Bridwell TJ, Schildkraut JJ (1971) Alpha-methyl paratyrosine and sleep in the rat. Psychopharmacologia 21:157–164
Hernandez-Peon R (1962) Sleep induced by localized electrical or chemical stimulation of the forebrain. Electroencephalagr Clin Neurophysiol 14:423–424
Hilakivi I (1983) The role of β- and α-adrenoceptors in the regulation of the stages of the sleep-waking cycle in the cat. Brain Res 277:109–118
Hilakivi I, Leppävuori A (1984) Effects of methoxamine, an alpha-1 adrenoceptor agonist and prazosin, an alpha-1 antagonist, on the stages of the sleep-waking cycle in the cat. Acta Physiol Scand 120:363–372
Hilakivi I, Mäkelä J, Leppävuori A, Putkonen PTS (1978) Effects of two adrenergic β-receptor blockers on the sleep cycle of the cat. Med Biol 56:138–143
Hilakivi I, Leppävuori A, Putkonen PTS (1980) Prazosin increases paradoxical sleep. Eur J Pharmacol 65:417–420
Hilakivi I, Kovala T, Leppavuori A, Shvaloff A (1987) Effects of serotonin and noradrenaline uptake blockers on wakefulness and sleep in cats. Pharmacol Toxicol 60:161–166
Hoffman JS, Domino EF (1969) Comparative effects of reserpine on the sleep cycle of man and cat. J Pharmacol Exp Ther 170:190–198
Holman RB, Shillito EE, Vogt M (1971) Sleep produced by clonidine (2-(2,6-chlorophenylamino)-2-imidazoline hydrochloride). Br J Pharmacol 43:685–695
Hukuhara T, Otsuka Y, Takeda R, Sakai F (1968) Synchronization of EEG and increased spindle activity following clonidine in cats. Arzneimittelforschung 18:1147–1153
Jacobs BL (1985) Overview of the activity of brain monoaminergic neurons across the sleep-wake cycle. In: Wauquier A, Gaillard JM, Monti JM, Radulovacki M (eds) Sleep: neurotransmitters and neuromodulators. Raven, New York, pp 1–14
Jones BE (1972) The selective involvement of noradrenaline and its deaminated metabolites in waking and paradoxical sleep: a neuropharmacological model. Brain Res 39:121–136
Jones BE (1991) The role of noradrenergic locus coerulus neurons and neighboring cholinergic neurons of the pontomesencephalic tegmentum in sleep-wake states. In: Barnes CD, Pompeiano O (eds) Neurobiology of the locus coeruleus. Elsevier, Amsterdam, pp 533–543
Jouvet M (1972) The role of monoamines and acetylcholine containing neurons in the regulation of the sleep-waking cycle. Ergeb Physiol 64:166–307
Kafi S, Gaillard JM (1976) Brain dopamine receptors and sleep in the rat: effects of stimulation and blockade. Eur J Pharmacol 38:357–364
Kafi S, Gaillard JM (1978) Biphasic effect of chlorpromazine on rat paradoxical sleep: a study of dose-related mechanisms. Eur J Pharmacol 49:251–257
Kafi S, Gaillard JM (1981) Pre- and postsynaptic effect of yohimbine on rat paradoxical sleep. In: Koella WP (ed) Sleep 1980. Karger, Basel, pp 292–293
Kafi S, Bouras C, Constantinidis J, Gaillard JM (1977) Paradoxical sleep and brain catecholamines in the rat after single and reported administration of alpha-methyl-paratyrosine. Brain Res 135:123–134
Kales A, Soldatos CR, Cadieux R, Bixler EO, Tan T-L, Scharf MB (1979) Propranolol in the treatment of narcolepsy. Ann Intern Med 91:742–743
Karczmar AG, Longo VG, Scotti de Carolis A (1970) A pharmacological model of paradoxical sleep. The role of cholinergic and monoamine systems. Physiol Behav 5:175–182

Keane PE, Candy JM, Bradley PB (1976) The role of endogenous catecholamines in the regulation of electrocortical activity in the encéphale isolé cat. Electroencephalogr Clin Neurophysiol 41:561–570

King CD, Jewett RD (1971) The effects of α-methyltyrosine on sleep and brain norepinephrine in cats. J Pharmacol Exp Ther 177:188–194

Kleinlogel H, Scholtysik G, Sayers AC (1975) Effects of clonidine and BS-100–141 on the EEG sleep pattern in rats. Eur J Pharmacol 33:159–163

Lanfumey L, Adrien J (1982) Regulation of sleep after neonatal coeruleus lesion: functional evidences of β-adrenergic suprasensitivity. Eur J Pharmacol 79:251–264

Lanfumey L, Dugovic C, Adrien J (1985) β_1 and β_2 adrenergic receptors: their role in the regulation of paradoxical sleep in the rat. Electroencephalogr Clin Neurophysiol 60:558–567

Langer SZ (1974) Presynaptic regulation of catecholamine release. Biochem Pharmacol 23:1793–1800

Leppävuori A, Putkonen PTS (1978) Evidence for central alpha adrenoceptor stimulation as the basis of paradoxical sleep suppression by alpha methyldopa. Neurosci Lett 9:37–43

Leppävuori A, Putkonen PTS (1980) Alpha-adrenoceptive influences on the control of the sleep-waking cycle in the cat. Brain Res 193:95–115

Marantz R, Rechtschaffen A (1967) Effect of alpha-methylparatyrosine on sleep in the rat. Percept Motor Skills 25:805–808

Matsumoto J, Jouvet M (1964) Effects de réserpine, DOPA et 5-HTP sur les deux états de sommeil. C R Soc Biol (Paris) 158:2137–2140

Mendelson WB, Gillin JC, Dawson SD, Lewy A, Wyatt RJ (1980) Effect of melatonin and propranolol on sleep of the rat. Brain Res 201:240–244

Mereu GP, Scarnati E, Paglietti E, Pellegrini Quarantotti B, Chessa P, Dihiara G, Gessa GL (1979) Sleep induced by low doses of apomorphine in rats. Electroencephalogr Clin Neurophysiol 46:214–219

Miettinen MVJ (1981) α-Adrenergic functions and sleep in kittens. In: Koella WP (ed) Sleep 1980. Karger Basel, pp 287–289

Mohan Kumar V, Data S, China GS, Ghandi N, Singh B (1984) Sleep-wake responses elicited from medial preoptic area on application of norepinephrine and phenoxybenzamine in free moving rat. Brain Res 322:322–325

Monti JM (1968) The effect of haloperidol on the sleep cycle of the cat. Experientia 24:1143

Monti JM (1979) The effects of neuroleptics with central dopamine and noradrenaline receptor blocking properties in the L-Dopa and (+) amphetamine-induced waking EEG in the rat. Br J Pharmacol 67:87–91

Monti JM, Hawkins M, Jantos H, D'Angelo L, Fernandez M (1988) Biphasic effects of dopamine D-2 receptor agonists on sleep and wakefulness in the rat. Psychopharmacology (Berl) 95:395–400

Monti J, Jantos H, Fernandez M (1989) Effects of the selective dopamine D-2 receptor against quinpirole on sleep and wakefulness in the rat. Eur J Pharmacol 169:61–66

Monti JM, Fernandez M, Jantos H (1990) Sleep during acute dopamine D1 agonist SKF 38393 or D1 antagonist SCH 23390 administration in rats. Neuropsychopharmacology 3:153–162

Nicherson M, Collier B (1975) Drugs inhibiting adrenergic nerves and structures innervated by them. In: Goodman LS, Gilman A (eds) The pharmacological basis of therapeutics, 5th edn. McMillan, New York, pp 533–564

Pearson J, Goldstein M, Markey K, Brandeis L (1983) Human brainstem catecholamine neuronal anatomy is indicated by immunocytochemistry with antibodies to tyrosine hydroxylase. Neuroscience 8:3–32

Pellejero T, Monti JM, Baglietto J, Jantos H, Pazos S, Cichevski K, Hawkins M (1984) Effects of methoxamine and α-adrenoceptor antagonists, prazosin and yohimbine, on the sleep-wake cycle of the rat. Sleep 7:365–372

Peyrethon-Dusan D, Froment JL (1968) Effects du disulfiram sur les états de sommeil chez le chat. C R Soc Biol (Paris) 162:2141–2145
Pickworth WB, Sharpe LG, Nozaki M, Martin WR (1977) Sleep suppression induced by intravenous and intraventricular infusions of methoxamine in the dog. Exp Neurol 57:999–1011
Polc P, Schneeberger J, Haefely W (1979) Effects of several centrally active drugs on the sleep-wakefulness cycle of cats. Neuropharmacology 18:259–267
Putkonen PTS (1979) α- and β-adrenergic mechanisms in the control of sleep stages. In: Priest RG, Plitscher A, Ward J (eds) Sleep research. MTP Press, Lancaster, pp 19–34
Putkonen PTS, Leppävuori A (1977) Increase in paradoxical sleep in the cat after phentoloamine and alpha-adrenoceptor antagonist. Acta Physiol Scand 100: 488–490
Putkonen PTS, Leppävuori A, Stenberg D (1977) Paradoxical sleep inhibition by central alpha-adrenoceptor stimulant clonidine antagonized by alpha-receptor blocker yohimbine. Life Sci 21:1059–1066
Reite M, Pegram GV, Stephens LM, Bixler EC, Lewis OL (1969) The effect of reserpine and monoamine oxidase inhibitors on paradoxical sleep in the monkey. Psychopharmacologia 14:12–17
Ruth-Monachon M, Jaffre M, Halfely W (1976) Drugs and PGO waves in the lateral geniculate body of a curarized cat. III. PGO wave activity and brain catecholamines. Arch Int Pharmacodyn Ther 219:287–307
Schmidek WR, Timo-Iara C, Schmidek M, Krakowiak M, Alves MR, Delmutti EE (1974) Influence of loxapine on the sleep-wakefulness cycle of the rat. Pharmacol Biochem Behav 2:747–751
Sibley DB (1991) Cloning of a "D3" receptor subtype expands dopamine receptor family. Trends Pharmacol Sci 12:7–9
Starke K (1977) Regulation of noradrenaline release by presynaptic receptor systems. Rev Physiol Biochem Pharmacol 77:1–124
Stenberg D, Hilakivi I (1985) Alpha-1 and alpha-2-adrenergic modulation of vigilance and sleep. In: Wauquier A, Gaillard JM, Monti JM, Radulovacki M (eds) Sleep: neurotransmitters and neuromodulators. Raven, New York, pp 69–77
Stern WC, Morgane PJ (1973) Effects of reserpine on sleep and brain biogenic amine levels in the cat. Psychopharmacologia 28:275–286
Stille G, Lauener H (1974) Die Wirkung von Neuroleptika im chronischen pharmacologischen Experiment. Arzneimittelforschung 24:1292–1294
Strange PG (1991) Interesting times for dopamine receptors. Trends Neurosci 14:43–45
Strömbom U (1976) Catecholamine receptor agonists. Effects on motor activity and rate of tyrosine hydroxylation in mouse brain. Naunyn Schmiedebergs Arch Exp Pathol Pharmakol 292:167–176
Svensson TH, Buney DS, Aghajanian GK (1975) Inhibition of both noradrenergic and serotonergic neurons in brain by the α-adrenergic agonist clonidine. Brain Res 92:291–306
Tabushi K, Himwich HE (1969) The acute effects of reserpine on the sleep-wakefulness cycle in rabbits. Psychopharmacologia 16:240–252
Takeuchi O (1973) Influences of psychotropic drugs (chlorpromazine, imipramine and haloperidol) on the sleep wakefulness mechanisms in cats. Psychiatr Neurol Jpn 75:424–459
Torda C (1968) Effects of changes of brain norepinephrine content on sleep cycle in rat. Brain Res 10:200–207
Trampus M, Orgini E (1990) The D1 dopamine receptor antagonist SCH 23390 enhances REM sleep in the rat. Neuropharmacology 29:889–893
Tran Quang Loc D, Tsoucaris-Kupfer Y, Bogaievsky D, Delbarre B, Schmitt H (1974) Antagonisme de l'action sédative de la clonidine par quelques α-adrénolytiques: étude électrocorticographique et comportementale chez le lapin et le chat. J Pharmacol (Paris) 5:51–55

Tsuchiya T, Tami S, Fokushima H (1979) Analysis of the dissociation between the neocortical and hypocortical EEG activity induced by neuroleptics. Psychopharmacology (Berl) 63:179–185
U'Prichard DC, Charness ME, Robertson D, Snyder SH (1978) Prazosin: differential affinities for two populations of α-noradrenergic receptor binding sites. Eur J Pharmacol 50:87–89
Wauquier A (1985) Active and permissive roles of dopamine in sleep-wakefulness regulation. In: Wauquier A, Gaillard JM, Monti JM, Radulovacki M (eds) Sleep: neurotransmitters and neuromodulators. Raven, New York, pp 107–120
Wauquier A, Van den Broeck WAE, Janssen PAJ (1980) Biphasic effects of pimozide on sleep-wakefulness in dogs. Life Sci 27:1469–1475
Wauquier A, Van den Broeck WAE, Niemegeers CJE, Janssen PAJ (1981) On the antagonistic effects of pimozide and domperidone on apomorphine-disturbed sleep-wakefulness in dogs. In: Koella WP (ed) Sleep 1980. Karger, Basel, pp 279–282
Webster HH, Jones BE (1988) Neurotoxic lesions of the dorsolateral pontomesencephalic tegmentum-cholinergic area in the cat. II. Effects upon sleep-waking states. Brain Res 458:285–302
Weitzman E, McGregor P, Moose C, Jacoby J (1969) The effect of alpha-methylparatyrosine on sleep patterns of the monkey. Life Sci 8:751–757
York DH (1970) Possible dopaminergic pathway from substantia nigra to putamen. Brain Res 20:233–249

CHAPTER 5

The Serotoninergic System and Sleep-Wakefulness Regulation

J. Adrien

A. The Serotoninergic System

The anatomical and biochemical aspects of the serotoninergic system have been widely studied for many years, while recently extensive research has been undertaken on the receptors of the system. This chapter aims to provide an understanding of the functional role of serotonin in sleep, by briefly reviewing the various properties of the serotoninergic system.

I. Serotoninergic Nuclei and Pathways

The serotoninergic system is composed of several neuronal groups (named B1–9) the somas of which are located in a medial sagittal plane extending from the brain stem to the midbrain, and which form the raphe nuclei (Fig. 1). The serotoninergic cells lying within the most caudally located raphe nuclei (magnus, pallidus and obscurus) give rise to a prominent bulbospinal serotoninergic system (Dahlström and Fuxe 1965; Bjorklund et al. 1971). The serotoninergic cells present in the rostral raphe nuclei (dorsalis B7, medianus and centralis superior B8) are the main source of the serotoninergic innervation of the forebrain (Anden et al. 1966; Azmitia and Segal 1978; Dahlström and Fuxe 1964; Fuxe and Jonsson 1974). The midbrain raphe nuclei also gives rise to an extensive supra- and subependymal plexus of serotoninergic axons in the walls of the ventricles. Thus, some central serotoninergic neurons may form a crucial link between the cerebrospinal fluid and the central nervous system.

One of the main features of the serotoninergic system is its widespread (and highly collateralized) distribution through the neuraxis. Indeed, a single serotoninergic raphe neuron may send divergent axon collaterals to a number of remote forebrain areas. The serotoninergic axonal varicosities generally also do not display the membrane differentiation of typical synaptic terminals, but form a sort of "floating terminal" within the neural tissue (Descarries et al. 1975; Chan-Palay 1975). Such morphological properties suggest two distincts types of action for serotonin (5-HT). In the synaptic configuration, 5-HT probably binds directly to postsynaptic receptors. In the "nonsynaptic" case, the 5-HT released may act on more distant receptors. These two characteristics, divergent axons and "nonsynaptic" terminals,

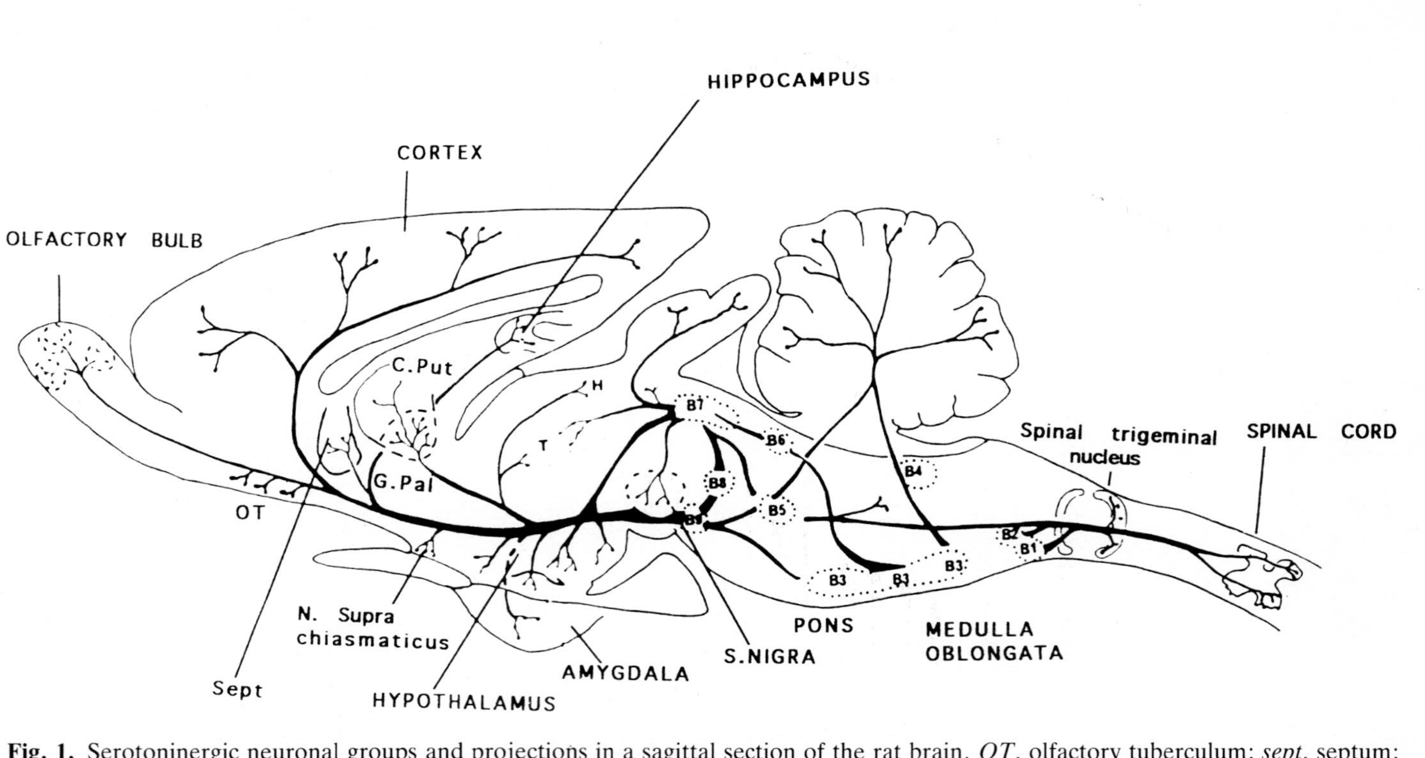

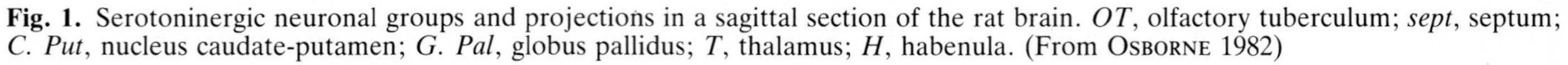

Fig. 1. Serotoninergic neuronal groups and projections in a sagittal section of the rat brain. *OT*, olfactory tuberculum; *sept*, septum; *C. Put*, nucleus caudate-putamen; *G. Pal*, globus pallidus; *T*, thalamus; *H*, habenula. (From OSBORNE 1982)

suggest that a single serotoninergic neuron could exert a fairly widespread and diffuse influence over vast neuronal populations.

II. Serotoninergic Metabolism and Neurotransmission

1. Synthesis of Serotonin

In the central nervous system, 5-HT is synthesized in two enzymatic steps: the natural precursor found in the blood, tryptophan, is hydroxylated in its free form by a rate-limiting enzyme, tryptophan hydroxylase. This takes place within serotoninergic neurons (GRAHAME-SMITH 1967). The product, 5-hydroxytryptophan (5-HTP), is then decarboxylated into serotonin by the enzyme 5-HTP decarboxylase, a reaction which may take place outside serotoninergic neurons, and possibly in catecholaminergic neurons, at least when 5-HTP is administered exogenously (JOHNSON et al. 1968). Thus, any pharmacological action aiming at enhancing 5-HT synthesis through an increase in tryptophan levels will be more specific than one involving an exogenous increase in 5-HTP levels.

2. Release and Inactivation of Serotonin

In nerve endings, 5-HT may be stored in the vesicles or catabolized into 5-hydroxyindoleacetic acid (5-HIAA) by monoamine oxidases. As a neurotransmitter, 5-HT is released at the terminal and possibly somatodendritic levels, and is secondarily inactivated through reuptake into the serotoninergic terminal itself, or into glial cells. In the latter case, it is directly catabolized into 5-HIAA, whereas in nerve endings it might be either stored again in the vesicles or catabolized as described above. The release and reuptake of 5-HT is controlled at the presynaptic level by autoreceptors localized on the teminal or on the soma of serotoninergic neurons.

3. Impairment of the Synthesis, Release and Catabolism of Serotonin

p-Chlorophenylalanine (*p*-CPA) has been used widely as a pharmacological means of preventing the synthesis of 5-HT. This compound inhibits the rate-limiting enzyme, tryptophan hydroxylase. Thus, under *p*-CPA treatment, only the administration of exogenous 5-HTP will restore potential for 5-HT synthesis.

Conversely, in order to enhance the levels of endogenous 5-HT, inhibitors of 5-HT reuptake have been commonly used. Under the action of such blockers, 5-HT accumulates in the extraneuronal space. In turn, this accumulation activates a negative feedback loop triggered by specific presynaptic autoreceptors, and the result of such activation is a reduction of 5-HT neurotransmission, concomitant with a cessation of activity of serotoninergic neurons. Such specific autoreceptors are located either at the terminal level on axonal varicosities, or at the somatodendritic level (Fig. 2).

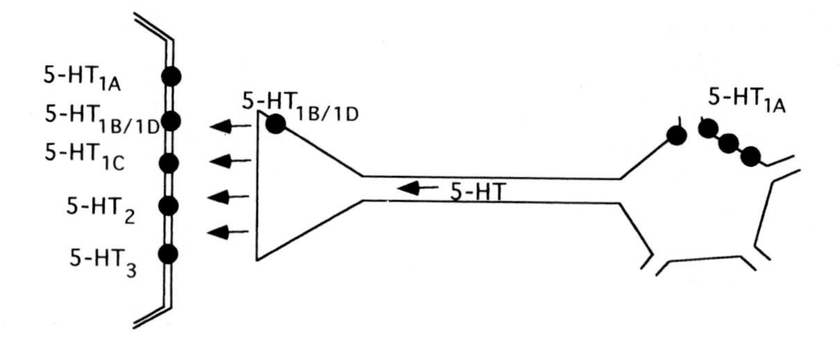

Fig. 2. Pre- and postsynaptic serotoninergic receptor types in relation to the serotoninergic neuron. 5-HT_{1A} receptors are found in the presynaptic position on the soma and dendrites of the serotoninergic neuron itself, and in the postsynaptic position at the axonal terminal level. 5-HT_{1B} (in the rat) and 5-HT_{1D} (in humans) are observed at the terminal level, in the pre- and postsynaptic position. 5-HT_{1C}, 5-HT_2, 5-HT_3 (and most probably 5-HT_4) receptors are found only in the postsynaptic position

Finally, another means of enhancing 5-HT levels is to prevent its catabolism, which is achieved by inhibitors of monoamine oxidase.

In all events, activation of 5-HT transmission usually induces an inhibition of the target cell in the telencephalon, whereas many of the pontomedullary and spinal neurons are excited (BLOOM et al. 1972; SEGAL 1980).

III. Serotoninergic Receptors

The binding sites for 5-HT in the central nervous system are heterogeneous and are classically differentiated into at least four classes, from 1 to 4 (but new classes have since been described), and into several types or subtypes: A, B, C, D, etc. (HOYER 1991; PEROUTKA 1988; SANDERS-BUSH 1988). This classification is based essentially on pharmacological data, but the various serotoninergic receptors also correspond to the respective affinities for 5-HT and for selective ligands, as well as to different transduction mechanisms in the membrane.

From a general point of view, receptors for neurotransmitters are located either on the neuron itself (presynaptic position) or on the target cells of this neuron (postsynaptic position) (Fig. 2). In terms of neurotransmission, the localization of a given receptor is particularly relevant because activation of the autoreceptor (presynaptic) induces an overall decrease in neurotransmitter release and thus reduced synaptic efficacy. Conversely, activation of postsynaptic receptors mimics an enhancement of neurotransmission.

1. $5\text{-}HT_{1A}$ Receptors

The $5\text{-}HT_{1A}$ type of receptor is located presynaptically on the serotoninergic neurons of the nucleus raphe dorsalis (NRD), where it acts as an autoreceptor (VERGE et al. 1986), as well as postsynaptically, in all other brain areas (in particular in the limbic system) where the serotoninergic system sends projections. The $5\text{-}HT_{1B}$ (in the rat) and $5\text{-}HT_{1D}$ (in other mammals, including humans) types of receptors are usually found on postsynaptic cells, but also in a presynaptic position on axonal terminals of serotoninergic neurons. They are distributed at a high density in the globus pallidus and the pars reticulata of the substantia nigra. The $5\text{-}HT_{1C}$ receptors are present essentially in the choroid plexus and, although at lower densities, in the limbic system and in structures associated with motor behavior.

2. $5\text{-}HT_2$ Receptors

Receptors of the $5\text{-}HT_2$ type are located in the postsynaptic position with regard to the serotoninergic neuron and are found essentially in the cortex, concentrated for the most part in laminae I and V.

3. $5\text{-}HT_3$ Receptors

$5\text{-}HT_3$ receptors are also postsynaptic and are present at a high desity in the brain stem (nucleus tractus solitarius, nucleus of the vagus nerve, area postrema).

4. $5\text{-}HT_4$ Receptors

Finally, $5\text{-}HT_4$ receptors are probably also located in the postsynaptic position and are observed in the colliculi and hippocampus, but their distribution in the brain remains largely unknown.

Selective ligands (agonists and antagonists) of various types and subtypes of serotoninergic receptors are being developed more and more (Table 1). They represent very useful tools for investigating the respective involvement of the various serotoninergic receptor systems in the regulation of sleep and wakefulness; their effects are described below. However, with regard to $5\text{-}HT_4$ receptors, to date no report of the effects of their ligands on behavior has been published.

IV. Conclusion

The pharmacological study of the relationship of the serotoninergic system to sleep and wakefulness is possible thanks to specific chemical tools that act upon the neurotransmitter itself and/or upon its receptors within the brain. However, the serotoninergic projections are ubiquitous, and the serotoninergic receptors in these targets are not evenly distributed. Thus,

Table 1. Selectivity of various serotoninergic ligands for the different serotoninergic receptor types whose effects on sleep-wakefulness have been studied

	5-HT_{1A}	$5\text{-HT}_{1B/1D}$	5-HT_{1C}	5-HT_2	5-HT_3
Agonists					
8-OH-DPAT	+				
Buspirone	+				
Ipsapirone	+				
CM 57493	+				
RU 24969	+	+			
m-CPP		+	+		
DOI, DOM			+	+	
2-Methyl-5-HT					+
Phenylbiguanide					+
Antagonists					
Cianserin				+	
Ritanserin			+	+	
Ondansetron					+
MDL 72222					+
ICS 205930					+

the interpretation of data must take into account not only the chemical specificity of the various compounds used, but also the brain area(s) involved in their mode of action.

B. Serotonin and Sleep

There is much evidence that disruption of serotoninergic function (by lesions or pharmacological means) is followed by insomnia of various degrees (references in JOUVET 1972). Concomitantly, there are some data which indicate that an increase in serotoninergic activity induces the appearance of some sleep signs, or at least sedation, i.e., a decrease in waking intensity (references in KOELLA 1982). In order to interpret the data obtained on sleep-wakefulness regulation after pharmacological manipulation of the serotoninergic system, the spontaneous fluctuations of the activity of the latter during sleep-wakefulness cycles should be considered.

I. Spontaneous Activity of the Serotoninergic System During Sleep and Wakefulness

Electrophysiological recording of serotoninergic neurons in the NRD of the cat has indicated intense activity during wakefulness (W), a progressive decrease in neuronal firing during slow-wave sleep (SWS) and complete cessation of activity during paradoxical sleep (PS) (MCGINTY and HARPER 1976; TRULSON and JACOBS 1979).

Studies of 5-HT or 5-HIAA release in brain areas such as the caudate, the cortex and the basal hypothalamus confirm earlier electrophysiological data indicating that the activity of the serotoninergic system is maximal during W, reduced during SWS and minimal during PS (PUIZILLOUT et al. 1979; CESPUGLIO et al. 1990; HOUDOUIN et al. 1991).

However, as measured in the NRD, the level of 5-HIAA, the catabolite of 5-HT which normally reflects the activity of the serotoninergic system, fluctuates during the sleep-wakefulness cycles in an inverse relationship to its level in forebrain areas (CESPUGLIO et al. 1992). These data indicate the pattern of 5-HT release during sleep-waking cycles to be opposite at terminal axonal and at somatodendritic levels. This finding is puzzling, and it awaits confirmation in other species and with more specific techniques such as direct measurement of 5-HT levels by microdialysis. Nevertheless, such a differential pattern of 5-HT release is consistent with the data on neurochemical regulation of serotoninergic neuronal activity: during W, the neurons are active and thus 5-HT is released at the terminal level. In contrast, during sleep, and in particular during PS, the activity of serotoninergic raphe neurons is reduced or even totally abolished as a consequence of increased amounts of 5-HT in the somatodendritic environment, which activates a negative feedback loop through 5-HT_{1A} somatodendritic autoreceptors (SPROUSE and AGHAJANIAN 1987).

II. Pharmacological Impairment of Serotoninergic Neurotransmission (Table 2)

1. Enhancement of Serotonin Levels in the Brain

The central endogenous amounts of 5-HT can be increased pharmacologically by the action at different levels of the previously described serotoninergic metabolism, i.e., synthesis, reuptake or catabolism.

a) Synthesis

Systemic administration of the natural precursors of 5-HT, tryptophan and 5-hydroxytryptophan, both of which cross the blood-brain barrier, whereas 5-HT itself does not, is followed by an increase in 5-HT levels in the forebrain (MOIR and ECCLESTON 1968). The effects of such treatment on sleep have generally been described as a deactivation of the waking system (see references in KOELLA 1982; SPINWEBER et al. 1983; URSIN 1976).

α) Tryptophan. The administration of tryptophan was reported to induce in the cat an increase in drowsiness, sometimes an enhancement of high-voltage synchronous cortical waves and a significant decrease in PS (URSIN 1976). However, this might indicate nonspecific effects on sleep because, at the same doses (200 and 300 mg/kg i.p.), tryptophan systematically induced vomiting. Indeed, in such cases, the drowsiness and decrease in PS might be only secondary to general behavioral impairment. In the rat, no major modifications of sleep-wakefulness patterns were observed after tryptophan

Table 2. Effects on wakefulness and sleep of various compounds which impair the serotoninergic neurotransmission

Compound	Species	Dose	W	SWS_1	SWS_2	PS
Enhancement of 5-HT levels						
Tryptophan	Cat	200–300 mg/kg i.p.	–	+	·	·
	Rat	Up to 600 mg/kg i.p.	·	·	·	·
	Human	1–15 g p.o.	·	·	(+)	(+)
5-HTP	Cat	40 mg/kg i.p.	·	+	·	·
	Rat	25–100 mg/kg i.p.	·	·	·	·
		0.5–1.0 μg i.c.v.	·	·	·	·
	Human	1.5–3 g p.o.	·	+	·	·
Blockers of 5-HT reuptake						
Citalopram, Zimelidine, Alaproclate, Indalpine, Fluoxetine, Paroxetine	Cat	5–10 mg/kg i.p.	·	·	(+)	–
	Rat	1–20 mg/kg i.p.	·	·	(+)	–
	Human	10–100 mg p.o.	·	·	·	–
Inhibitors of monoamine oxidase	Cat	200 mg/kg i.p.	·	·	·	–
	Rat	400 mg/kg i.p.	·	·	–	–
	Human	15–400 mg p.o.	·	·	·	–
Decrease in 5-HT levels						
Reserpine	Cat	0.2–0.4 mg/kg i.p.	+	–	–	–
p-CPA	Cat	100 mg/kg i.p.	+	–	–	–
	Rat	200–400 mg/kg i.p.	+	–	–	–
	Rabbit					
	Monkey					
Combined treatment						
p-CPA + 5-HTP	Cat	400 + 5 mg/kg i.p.	Restoration of all sleep states			
	Rat	500 + 50 mg/kg i.p.				

+, increased amounts; –, decreased amounts; (+) and (–), increase and decrease found by only some authors.

treatment, even at doses of up to 600 mg/kg i.p., except for a slight reduction in sleep latency (Hartmann and Chung 1972; Wojcik et al. 1980). In contrast, in human volunteers and in insomniac patients, L-tryptophan (1–15 mg per os) was found by some groups to induce sleep and to promote SWS at the expense of PS (Hartmann et al. 1974; Spinweber et al. 1983; Wyatt et al. 1971). Other groups reported no effects of L-tryptophan on either state of sleep (Adam and Oswald 1979; Brezinova et al. 1972).

These discrepancies in the reported effects of L-tryptophan on sleep are most probably due to the different doses used and to the time of the day-night cycle the experiments were performed. Finally, there is general agreement that at moderate doses L-tryptophan promotes cortical synchronization in humans and in animals, mainly by lowering arousal levels and thus setting the stage for more rapid sleep onset.

β) 5-Hydroxytryptophan. The natural precursor of 5-HT, 5-hydroxytryptophan, administered at 40 mg/kg i.p. in cats, induces an enhancement of the drowsy state and of light SWS (SWS_1) with no significant modification of deep SWS (SWS_2), and provokes total suppression of PS for 6 h after treatment (URSIN 1976). However, as in the case of tryptophan injection, 5-HTP induced vomiting in all animals. In other species such as rats, rabbits and dogs, a significant reduction in PS and the presence of synchronized cortical slow waves have been described after 5-HTP administration. It remains unknown whether this cortical activity reflects actual SWS or rather a state of drowsiness associated with synchronized cortical waves (ZOLOVIC et al. 1973; JOUVET 1967; URSIN 1976).

In patients suffering from insomnia due to neurological defects, two studies have reported a beneficial effect of 5-HTP administration (FISHER-PERROUDON et al. 1974; GUILLEMINAULT et al. 1973).

γ) Specificity. Finally, even though the data obtained with tryptophan may be more specific than those obtained with 5-HTP, because the latter may be decarboxylated in nonserotoninergic cells (JOHNSON et al. 1968; PUJOL et al. 1971), both compounds usually induce PS suppression and eventually an enhancement of cortical synchronization. With regard to PS inhibition, the data are consistent with the reciprocal interaction model of PS regulation, which implies than an activation of the serotoninergic system inhibits the executive mechanisms of PS (STERIADE and MCCARLEY 1990). With regard to induction of cortical synchronization, it is now broadly agreed that the main effect of 5-HT precursors on sleep-wakefulness cycles is a general deactivation of the waking state rather than a specific sleep-producing effect.

b) Reuptake

Many compounds are able selectively to block the reuptake of 5-HT and thus to induce an enhancement of 5-HT concentration in the synaptic cleft. Generally, the authors report that inhibitors of the 5-HT reuptake induce an inhibition of PS, whereas other sleep states are affected differently according to the various studies. In cats and rats, 5-HT uptake inhibitors, such as citalopram (HILAKIVI et al. 1987), zimelidine and alaproclate (SOMMERFELT and URSIN 1991; SOMMERFELT et al. 1987; URSIN et al. 1989), indalpine (KAFI DE SAINT HILAIRE et al. 1984), fluoxetine and paroxetine (KLEINLOGEL and BURKI 1987; PASTEL and FERNSTROM 1987), all induce a dose-dependent PS suppression. With regard to SWS_2, most reports indicate a weak suppressing effect. However, one group consistently found an increase (though delayed) in the amounts of SWS in cats and rats after zimelidine injections, associated with enhanced delta activity (SOMMERFELT and URSIN 1991; SOMMERFELT et al. 1987; URSIN et al. 1989).

In humans, 5-HT uptake inhibitors, such as fluoxetine and paroxetine (NICHOLSON and PASCOE 1988; OSWALD and ADAM 1986; SALETU et al. 1991; SLATER et al. 1978), fluvoxamine (HARTMANN and SPINWEBER 1979; KUPFER et al. 1991) and zimelidine (SHIPLEY et al. 1984), induce a reduction of PS which is accompanied by an increase in wakefulness. The latter phenomenon should be underlined since a similar PS inhibition is induced by norepinephrine uptake blockers, but this is accompanied by sedation (NICHOLSON and PASCOE 1986). However, when large doses of 5-HT uptake inhibitors were used (i.e., 40 mg/kg paroxetine), they induced drowsiness and sometimes even nausea (SALETU et al. 1991).

c) Inactivation

Inhibitors of monoamine oxydases (IMAOs) such as nialamide induce in cats and rats (MOURET et al. 1968b) and in humans (AKINDELE et al. 1970) an almost total suppression of PS. In rats, a concomitant reduction in the amounts of SWS is also reported. However, interpretation of these data in terms of the serotoninergic system would be biased since IMAOs are not selective compounds for the latter, but also act on the noradrenergic and dopaminergic systems. Because of this lack of specificity very few studies, except for the above-cited reports, have been devoted to this particular topic.

2. Decrease in Serotonin Levels in the Brain

a) Reserpine Treatment

Treatment with reserpine induces a drastic depletion of central 5-HT levels, resulting in total insomnia which can last for several hours. Cats treated with reserpine (0.2–0.4 mg/kg i.p.) exhibit a waking-like state with cortical activation and high muscle tone, associated with ongoing pontogeniculo-occipital (PGO) activity (BROOKS and GERSHON 1972; RUCH-MONACHON et al. 1976). Such PGO waves were first described in the cat in the pontine reticular formation, in the lateral geniculate nucleus, and in the occipital cortex (MOURET et al. 1963). They occur predominantly during PS and are initiated in the pons, from which they are widely distributed into several areas of the central nervous system (references in SAKAI 1980).

It seems that the pontine generator of PGO activity is under strong tonic inhibitory control of the serotoninergic system (SAKAI 1980; RUCH-MONACHON et al. 1976). Thus inactivation of the latter by reserpine has a releasing effect on PGO waves, and this effect can in turn be reversed by various agonists of the serotoninergic receptors such as 8-hydroxy-2-(di-*n*-propylamino)tertralin (8-OH-DPAT, DEPOORTERE and RIOU-MERLE 1988), CM 57 493 (ADRIEN et al. 1989) and ipsapirone (Fig. 3).

In sum, depletion of serotoninergic terminals with reserpine induces drastic insomnia and continuous PGO activity. However, knowing that this compound induces depletion of not only 5-HT but also norepinephrine and dopamine, another method for depleting 5-HT in the brain is described.

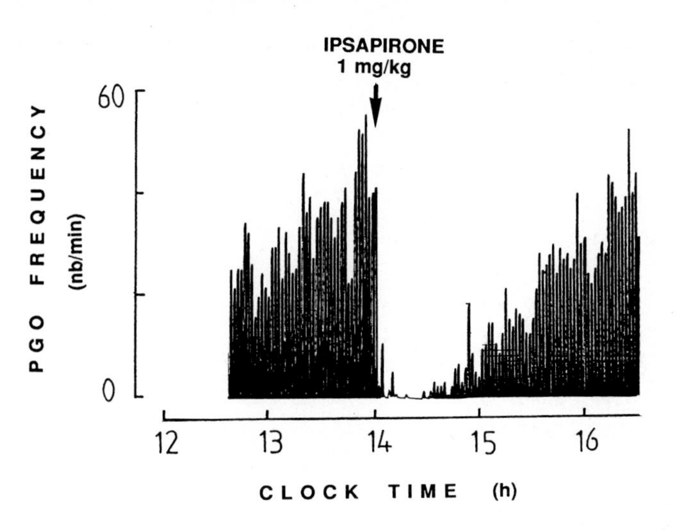

Fig. 3. Inhibitory effect of the 5-HT_{1A} agonist ipsapirone on PGO activity in the lateral geniculate nucleus of reserpinized cats. Ipsapirone (1 mg/kg i.p.) was injected 3 h after reserpine treatment (0.4 mg/kg i.p.). PGO density, expressed as number of spikes per minute, was drastically reduced for 30 min to 1 h under the action of the 5-HT_{1A} agonist. A progressive return to reserpine baseline occurred within the subsequent 2 h

b) para-Chlorophenylalanine Treatment

The endogenous levels of 5-HT can be reduced drastically by treatment with *para*-chlorophenylalanine (*p*-CPA), an inhibitor of tryptophan hydroxylase, the rate-limiting enzyme of 5-HTP formation. Concomitantly, *p*-CPA induces insomnia in the rat (MOURET et al. 1968a; ADRIEN et al. 1980), in the rabbit (FLORIO et al. 1968), in the monkey (WEITZMAN et al. 1968) and in the cat (JOUVET 1969; KOELLA et al. 1968; PETITJEAN et al. 1985). In rats, 300 mg/kg *p*-CPA given i.p. on two consecutive days induces a major reduction, but not total inhibition, of both SWS and PS for 2–3 days, with a gradual return to normal levels after 6–7 days (ADRIEN et al. 1980). In cats, a single dose of 100–400 mg/kg i.p. *p*-CPA produces a long-lasting reduction in sleep which is initiated about 24 h after treatment and which lasts for several days. In particular, SWS_2 and PS are totally inhibited for 2–4 days, and only some episodes of SWS_1 are present. In addition, as in the case of treatment with reserpine, continuous PGO activity is observed. These data suggest that normal 5-HT metabolism is a prerequisite for normal occurrence of sleep and PGO activity. However, when *p*-CPA was administered to cats for 7–10 consecutive days at a daily dose of 100 mg/kg i.p., recovery of behavioral and sleep states was observed during the treatment, even though the levels of 5-HT remained drastically low (DEMENT et al. 1972). Such adaptations, which were also observed after lesions of the raphe nuclei

(PETITJEAN et al. 1978), have not been studied in depth (ADRIEN et al. 1980), and the only interpretation given to data is based on possible functional receptor supersensitivity (BORBELY et al. 1981).

c) *Restoration of Serotonin Levels After p-CPA Treatment*

Restoration of normal 5-HT levels after *p*-CPA treatment can be achieved by administration of 5-HTP, the direct precursor of 5-HT. In cats, during the period of total *p*-CPA-induced insomnia, systemic administration of 2–5 mg/kg 5-HTP suppresses within 1–2 min the spontaneous occurrence of PGO waves, and restores within 30–60 min a normal alternation of SWS and PS for 6–8 h (JOUVET 1969; PETITJEAN et al. 1985) (Fig. 4). The same restoration of sleep is observed in rats (LAGUZZI and ADRIEN 1980). These data were used as strong evidence that there is need of 5-HT for sleep and that this transmitter plays a direct role in sleep production (JOUVET 1969). However, the delay observed in the restoration of sleep after 5-HTP injection is puzzling. It cannot be accounted for by peripheral mechanisms because direct infusion of 5-HTP in relevant brain sites (the basal hypothalamus) also requires a 1-h delay for sleep induction (DENOYER et al. 1989). Such a delay, in contrast to PGO waves, is not compatible with classical neurotransmission. It has been proposed that this time lapse corresponds to the synthesis of some hypnogenic substance(s) of a peptidic nature, which is in accordance with the finding that 5-HT release at the terminal level is maximal during W and minimal during sleep. During wakefulness, 5-HT released at various target sites (such as the basal hypothalamus) would induce the synthesis of hypnogenic substance(s) which would trigger sleep secondarily (references in CESPUGLIO et al. 1992).

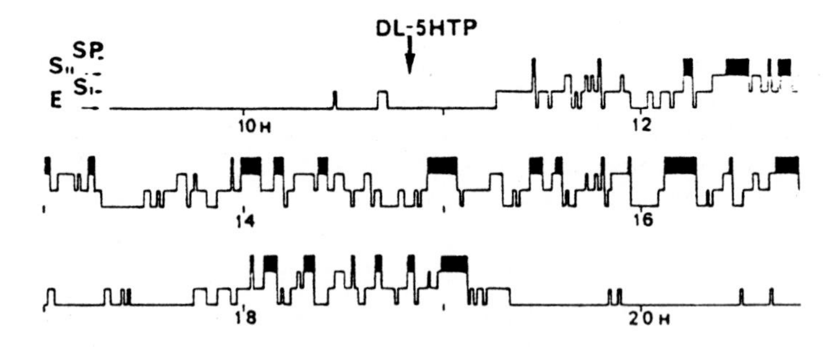

Fig. 4. Effect of 5-HTP on sleep-wakefulness cycles of a *p*-CPA-treated cat. The animal had received *p*-CPA (400 mg/kg i.p.) on the previous day, which induced almost total insomnia. After the injection of 5-HTP (5 mg/kg i.p.), all sleep states reappeared within 30 min to 1 h, and the sleep-wakefulness cycles were close to those observed under physiological conditions. (From PETITJEAN et al. 1985)

III. Pharmacological Impairment of Receptor Function

The ongoing development of numerous ligands of the various receptor types makes it possible to investigate their respective roles in the regulation of each state of vigilance (Table 3).

1. 5-HT_{1A} Receptors

a) 5-HT_{1A} Receptor Stimulation

Stimulation of 5-HT_{1A} receptors by 8-OH-DPAT, a selective ligand of these receptor sites (DEPOORTERE and RIOU-MERLE 1988; DZOLJIC et al. 1987; MONTI and JANTOS 1992), but also with buspirone (LERMAN et al. 1986), ipsapirone (ADRIEN et al. 1990; TISSIER et al. 1993) and eltoprazine (QUATTROCHI et al. 1993), induces quite characteristic modifications of the sleep-wakefulness cycles. In moderate and high doses, 5-HT_{1A} agonists induce an initial decrease in PS amounts and a concomitant increase in W, followed by a second period when PS amounts exhibit a rebound. The latter was described particularly in the rat with 8-OH-DPAT (DEPOORTERE and RIOU-MERLE 1988) and ipsapirone (TISSIER et al. 1993). In the cat, an identical deficit of PS concomitant with a suppression of PGO activity is observed for 1–3 mg/kg i.p. ipsapirone (ADRIEN et al. 1990) and for 2–4 mg/kg i.p. eltoprazine, where the reduction in PS is balanced by an increase in SWS amounts (QUATTROCHI et al. 1993).

In low doses, 8-OH-DPAT (0.01 mg/kg s.c.) was shown in the rat to induce effects opposite to those of large doses, at least with regard to SWS_2, i.e., an increase in the latter state 1–2 h after treatment and a concomitant decrease in PS (MONTI and JANTOS 1992).

It is important to note that, as was shown in the rat for sleep-wakefulness regulation but also for other physiological variables (references in TISSIER et al. 1993), the doses of 5-HT_{1A} agonists which induce sleep modifications depend on their route of administration. Indeed, subcutaneous and intravenous injections of 8-OH-DPAT or ipsapirone are five to eight times more potent than doses by the intraperitoneal route (DEPOORTERE and RIOU-MERLE 1988; MONTI and JANTOS 1992; TISSIER et al. 1993).

Because 5-HT_{1A} receptors are located not only on the somas and dendrites of serotoninergic neurons but also in the postsynaptic position at the terminal level (Fig. 2), the question arises of the respective involvement of either in the effects of 5-HT_{1A} agonists on sleep.

Indeed the postsynaptic 5-HT_{1A} receptors are most probably responsible for the effects of 5-HT_{1A} agonists on PS (initial inhibition and secondary rebound) because they persist after subtotal destruction of the somato-dendritic 5-HT_{1A} autoreceptors of the NRD (TISSIER et al. 1993). In addition, the latter study indicates that the effects of ipsapirone on PS were even enhanced after such a lesion of the NRD. This suggests that a phenomenon of denervation supersensitivity of 5-HT_{1A} receptors has occurred in lesioned

Table 3. Effects on wakefulness of the various serotoninergic ligands

Compound	Species	Dose	W	SWS_1	SWS_2	PS
$5\text{-}HT_1$ ligands						
Agonists 1A						
8-OH-DPAT	Rat	4 μg in NRD				
		0.01 mg/kg s.c.			+	−
		0.5–0.3 mg/kg i.p., s.c.	+		(−)	−
Buspirone	Rat	0.1–10 mg/kg i.p.	+		(−)	−
Ipsapirone		0.1–1 mg/kg i.p.				
Eltoprazine	Cat	0.06–4 mg/kg i.p.	+		(−)	−
Agonists 1B						
RU 24969	Rat	0.5–2 mg/kg s.c.	+		−	−
m-CPP	Human	0.5 mg/kg p.o.		+	−	−
Antagonists (nonselective)						
Pindolol	Rat	1–4 mg/kg s.c.	+		−	−
Propranolol	Rat	1–10 mg/kg i.p.	+		−	−
Combined treatment						
Pindolol +		1 mg/kg s.c.				
8-OH-DPAT	Rat	0.01 mg/kg s.c.	+		−	−
$5\text{-}HT_2$ ligands						
Agonists						
DOM	Rat	0.16–2.5 mg/kg i.p.			−	−
DOI		0.25–0.4 mg/kg i.p.			−	−
Antagonists						
Cinanserin	Rat	2.5–5 mg/kg i.p.			+	−
Ritanserin		0.1–0.5 mg/kg i.p.			+	−
	Cat	0.5–2.5 mg/kg i.p.			(−)	−
	Human	1–30 mg			+	−
ICI 169.369		40 mg/kg p.o.				−
ICI 170.809	Rat	40 mg/kg p.o.			+	−
RP 62 203		0.5–2 mg/kg p.o.			+	−
	Human	100 mg			+	−
Combined treatments						
Ritanserin + DOM	Rat	0.63 mg/kg i.p. + 0.63 mg/kg i.p.	Antagonists prevent the action of agonists on SWS_2 only			
$5\text{-}HT_3$ ligands						
Agonists						
Phenylbiguanide	Rat	12.5–50 μg i.c.v.	+		−	−
Antagonists						
MDL 72 222	Cat					
Ondansetron	Rat	0.01–10 mg/kg i.p.				(+)
ICS 205.930	Human	2–25 mg p.o.				
Combined treatments						
ICS 205.930 + Phenylbiguanide	Rat	0.5 mg/kg s.c. +50 μg i.c.v.	Antagonists prevent the action of agonists on SWS_2 and PS			

For explanation of symbols see Table 2.

animals, and it is a further argument to support the hypothesis of the involvement of primarily the postsynaptic 5-HT_{1A} receptors in the control of PS.

With respect to SWS, MONTI and JANTOS (1992) indicated that the effect observed with 8-OH-DPAT at a low dose could depend on the activation of 5-HT_{1A} receptors located within the NRD. However, the very large intracerebral doses of 8-OH-DPAT used in the latter study do not allow a definite conclusion to be drawn regarding this question.

b) 5-HT_{1A} Receptor Blockade

To date, no selective antagonist of 5-HT_{1A} receptors has been available. However, since most β-adrenergic antagonists are also blockers at 5-HT_{1A} sites, their action on sleep-wakefulness regulation has been tested against that of the agonist 8-OH-DPAT.

When injected alone, β_1-antagonists, such as propranalol (2–10 mg/kg i.p., LANFUMEY et al. 1985) and pindolol (1–4 mg/kg s.c., MONTI and JANTOS 1992), induce a decrease in PS and in SWS_2, associated with an increase in W for 1–4 h postinjection. Administered together with 8-OH-DPAT (0.01 mg/kg s.c.), pindolol (1 mg/kg s.c.) blocks the SWS_2-enhancing action of the 5-HT_{1A} agonist, but it induces a further increase in W and a further decrease in PS (MONTI and JANTOS 1992). Since pindolol is an antagonist at both the 5-HT_{1A} and the β-adrenergic receptors, it is difficult to draw definite conclusions from these studies.

Finally, the specific involvement of 5-HT_{1A} receptors in the regulation of sleep and wakefulness may well be clarified in the near future thanks to the development of selective ligands of these receptors, as well as to studies in which direct local infusions of such ligands are performed at various brain sites.

2. 5-HT_{1B} and 5-HT_{1C} Receptors

Agonists at 5-HT_{1B} sites, such as RU 24969 (which also binds to 5-HT_{1A} receptors) and *m*-chlorophenylpiperazine (*m*-CPP, a mixed 5-HT_{1B}/5-HT_{1C} agonist), induce an increase in the amounts of W and a decrease in those of PS (references in ADRIEN 1992). These effects were interpreted as being primarily enhancing on W, but until truly selective antagonists of the 5-HT_{1B} receptors are available no further conclusions can be drawn. With regard to 5-HT_{1C} ligands, because most of them also to 5-HT_2 receptors, their effects on sleep will be considered in the next section.

3. 5-HT_2 Receptors

Ligands of 5-HT_2 sites are only relatively selective with respect to the multiple serotoninergic receptors, and most of them bind with high affinity to 5-HT_{1C} sites. In fact, 5-HT_2 receptors are closely related to the latter type (HOYER 1991).

a) 5-HT$_2$ Receptor Stimulation

Stimulation of 5-HT$_2$ receptors was performed using the mixed 5-HT$_2$/5-HT$_{1C}$ agonists, methoxyphenethylamines DOM and DOI [1-(2,5-dimethoxy-4-methylphenyl)-2-aminopropane and 1-(2,5-dimethoxy-4-iodophenyl)-2-aminopropane], which are also (though with a 50-fold lower affinity) ligands at histaminergic H$_1$-, dopaminergic D$_2$- and adrenergic α_1-receptor sites.

These compounds, at the dose of 0.16 to 2.5 mg/kg i.p., induce in the rat a deficit of both SWS$_2$ and PS, concomitant with an increase in the amounts of W (Dugovic 1992; Dugovic et al. 1989b).

b) 5-HT$_2$ Receptor Antagonists

Antagonists at 5-HT$_2$ receptors are more selective than agonists as tools for investigating the functional role of these receptors in sleep and wakefulness. This is particularly the case with cinanserin, which exhibits a relatively high affinity for 5-HT$_2$ receptors, but also with ritanserin, even though it is a mixed 5-HT$_2$/5-HT$_{1C}$ antagonist (Hoyer 1991).

In the rat, cinanserin has no effect on sleep patterns at 2.5 mg/kg i.p., but it induces, at higher doses (5 mg/kg), a decrease in PS (Dugovic et al. 1989b). In contrast, ritanserin induces at doses of 0.5–2.5 mg/kg i.p. an increase in the amounts of SWS$_2$ and in EEG slow-wave activity, concomitant with a decrease in PS (Bjorvatn and Ursin 1990; Borbely et al. 1988; Dugovic et al. 1989b). No SWS-enhancing effect is obtained in cats (references in Ursin et al. 1992), but in humans 10 mg ritanserin given in the morning induces during the following night a twofold increase in deep SWS at the expense of light SWS, with no effect on rapid eye movement (REM) sleep (Idzikowski et al. 1986; references in Dugovic 1992). A similar action of several other mixed 5-HT$_2$/5-HT$_{1C}$ antagonists was observed, for example ICI 170.809 (20 mg/kg i.p.) and RP 62203 (0.5–2 kg/kg per os) in the rat, or ICI 169.369 (100 mg) in humans (Stutzmann et al. 1990; references in Dugovic 1992).

c) Agonist and Antagonist Combined Treatment

Combined treatments with agonists and antagonists at 5-HT$_2$ receptors were performed in order to test the specific involvement of the latter in the regulation of sleep and wakefulness. In fact, pretreatment with ritanserin (0.16–0.63 mg/kg i.p.) and cinanserin (2.5–5 mg/kg i.p.) prevents, in a dose-related manner, the deficit in SWS$_2$ and the enhancement of W induced by the 5-HT$_2$ agonist DOM at the dose of 0.63 mg/kg i.p., whereas it does not modify the effects on PS (Dugovic 1992; Dugovic et al. 1989b).

From these data, it might be said that receptors of the 5-HT$_2$/5-HT$_{1C}$ type are probably involved specifically in the regulation of SWS$_2$ rather than in that of PS or W.

d) Circadian Modulation of 5-HT_2 Influences

In nocturnal rodents such as the rat or the mouse, the amounts of sleep follow a circadian variation with more sleep during the light phase and more wakefulness during the dark. With regard to 5-HT_2 receptors, it was observed both in rats (DUGOVIC 1992; DUGOVIC et al. 1989a) and in humans (DIJK et al. 1989) that the action of 5-HT_2 ligands on sleep and wakefulness depended on the time of day the compound was administered. More precisely, in the rat, the SWS_2-enhancing and PS-inhibiting action of ritanserin (0.63 mg/kg i.p.) is observed when the compound is injected at the beginning of the light phase. In contrast, when ritanserin is administered at the onset of darkness, no enhancement of SWS_2 is observed, even at large doses (2.5 mg/kg i.p.). Similarly, the 5-HT_2 agonist DOM has a differential action depending on the time of treatment. When administered during the light phase, DOM induces a deficit in SWS and PS, whereas when injected during the dark period it provokes a sleep deficit only initially, whereas a compensatory rebound is observed secondarily (DUGOVIC et al. 1992).

It appears that these fluctuations could be related to a change in the functional sensitivity of 5-HT_2 receptors (MOSER and REDFERN 1985), with more receptor activity during the light phase than during the dark one. They could also be related to circadian variations in the endogenous production of melatonin because serotonin is the precursor of melatonin and because melatonin levels depend on the light-dark rhythm. In fact, 5-HT_2 ligands elicit modifications of sleep and wakefulness when they are administered during the light phase, when the melatonin levels are low. In addition, when melatonin is administered during the light phase together with 5-HT_2 ligands, it counteracts the normally observed effects of the latter on sleep (DUGOVIC 1992).

In humans, however, the relationship observed between efficiency of 5-HT_2 ligands and melatonin levels is the inverse of that in the rat, the maximal effect being observed during the night (references in DUGOVIC 1992). Thus, it could be the natural sleep period rather than the light/dark cycle per se which is relevant in the relationship between 5-HT_2 regulation of sleep and the circadian rhythm.

4. 5-HT_3 Receptors

a) 5-HT_3 Agonists

The 5-HT_3 agonists that are available to date do not cross the blood-brain barrier and thus have to be administered intracerebroventricularly.

In the rat, the 5-HT_3 agonist, *m*-chlorophenylbiguanide, infused into the lateral ventricle at 12.5–50 μg induces a dose-related increase in W and reduction in both SWS_2 and PS (PONZONI et al. 1993).

b) 5-HT$_3$ Antagonists

5-HT$_3$ antagonists have very weak effects on the regulation of sleep and wakefulness. In the rat, ondansetron, MDL 72222 and ICS 205930 (0.05–10 mg/kg i.p.), induce no impairment of the sleep-wakefulness cycles, except at certain doses. In particular 0.01 mg/kg ondansetron promotes slightly the state of PS (ADRIEN 1992), and 0.1 mg/kg s.c. MDL 72222 provokes a decrease in PS latency (PONZONI et al. 1993). In the cat, MDL 72222 does not modify sleep (ADRIEN 1992); neither does ICS 205930 in humans (references in ADRIEN 1992).

c) Combined Treatments with 5-HT$_3$ Ligands

Pretreatment with the 5-HT$_3$ antagonist MDL 72222 (0.1–0.5 mg/kg s.c.) was found recently in the rat to prevent the action of the agonist *m*-chlorophenylbiguanide (50 μg i.c.v.), i.e., the increase in W and the concomitant reduction in SWS$_2$ and PS (PONZONI et al. 1993).

5-HT$_3$ receptors seem to be only slightly involved in the regulation of sleep-wakefulness cycles. Their activation appears to enhance W and to decrease sleep. However, this action probably does not result from the activation of 5-HT$_3$ receptors themselves, but rather from the enhancement of the release of 5-HT and of dopamine which is observed after stimulation of these receptors (references in PONZONI et al. 1993).

5. Interaction Between the Various Serotoninergic Receptor Types

In the control of sleep and wakefulness the interaction between the various serotoninergic receptor types has been tested using 5-HT$_1$ agonists in conjunction with 5-HT$_2$ or 5-HT$_3$ antagonists. In the rat, pretreatment with the 5-HT$_2$ antagonists, ritanserin or cinanserin, does not modify the sleep impairments induced by the 5-HT$_{1A}$ agonist, 8-OH-DPAT, by the mixed 5-HT$_{1B}$/5-HT$_{1A}$ agonist, RU 24969, or by the mixed 5-HT$_{1C}$/5-HT$_{1B}$ agonist, *m*-chlorophenylpiperazine (references in DUGOVIC 1992). In contrast, in the cat, pretreatment with the 5-HT$_3$ antagonist MDL 72222 (0.1 mg/kg i.p.) prevents the action of the 5-HT$_{1A}$ agonist, ipsapirone (1 mg/kg), on W, PS and PGO activity (ADRIEN 1992).

To conclude, it is most probable that the various receptor types are differentially involved in the regulation of sleep and wakefulness. Today, it is thought that 5-HT$_2$ receptors are involved in the control of SWS$_2$, by exerting a tonic inhibitory influence on this state. With respect to 5-HT$_1$ receptors, the 5-HT$_{1A}$ and 5-HT$_{1B}$ type (essentially the postsynaptic ones) may play an active role in the inhibition of PS and in the promotion of W. Finally, 5-HT$_3$ receptors may favor W by an indirect route involving, in particular, the dopaminergic system.

IV. How Is Serotonin Involved in Sleep-Wakefulness Regulation?

The role of serotonin in the regulation of sleep and wakefulness appears more and more complex as investigations progress. The concept of serotonin as a sleep-promoting neurotransmitter (JOUVET 1972) or as an "antiwaking" agent (KOELLA 1982) has evolved into a more subtle role for both neurotransmitter and neurohormone. Hence, the serotoninergic system is active during W and may participate in building up the conditions of sleep occurrence, in particular through the synthesis of sleep-promoting substances (Fig. 5). This step may require the participation of 5-HT_1 receptors, and in particular the postsynaptic 5-HT_{1A} and 5-HT_{1B} types, eventually at the hypothalamic level. Concomitantly, when the serotoninergic neurons are activated, they inhibit both sleep states: SWS_2 through 5-HT_2 receptors, and PS through the 5-HT_{1A} receptors located within pontine structures, which are "executive" of this state of sleep (STERIADE and MCCARLEY 1990).

The conditions for sleep are met when the serotoninergic system becomes inactive, thus disinhibiting: (a) the mechanisms of SWS, in particular through 5-HT_2 receptors, and (b) cholinergic pontine neurons responsible for PS production (see also Chap. 7, this volume), through deactivation of the 5-HT_{1A} receptors in this area.

Further studies are needed to understand the mechanisms underlying these various steps, in which most probably the serotoninergic system should function within certain limits for the preparation, induction and maintenance of wakefulness and of the different states of sleep.

C. Summary

It is well established that serotonin plays a major role in the regulation of sleep-wakefulness cycles, and that any disruption of serotoninergic function induces modifications of sleep and waking.

The brain serotoninergic system arises from a limited number of neurons whose cell bodies are located in the midbrain and brain stem, and its projections are widespread through the neuraxis. Under physiological conditions, the activity of the serotoninergic neurons is at a maximum during wakefulness, reduced during slow-wave sleep, and almost totally abolished during paradoxical sleep.

Impairment of serotoninergic metabolism, or action of various ligands at serotoninergic receptor sites, has specific effects on the states of vigilance. Schematically, the pharmacologically induced increase of serotoninergic levels in the brain by using serotoninergic precursors results in a reduction of the waking state. Conversely, serotonin reuptake inhibitors mainly induce a suppression of paradoxical sleep. The prevention of serotonin synthesis in the central nervous system by para-chlorophenylalanine is followed by a drastic and long-lasting insomnia. The latter can be reversed by treatment

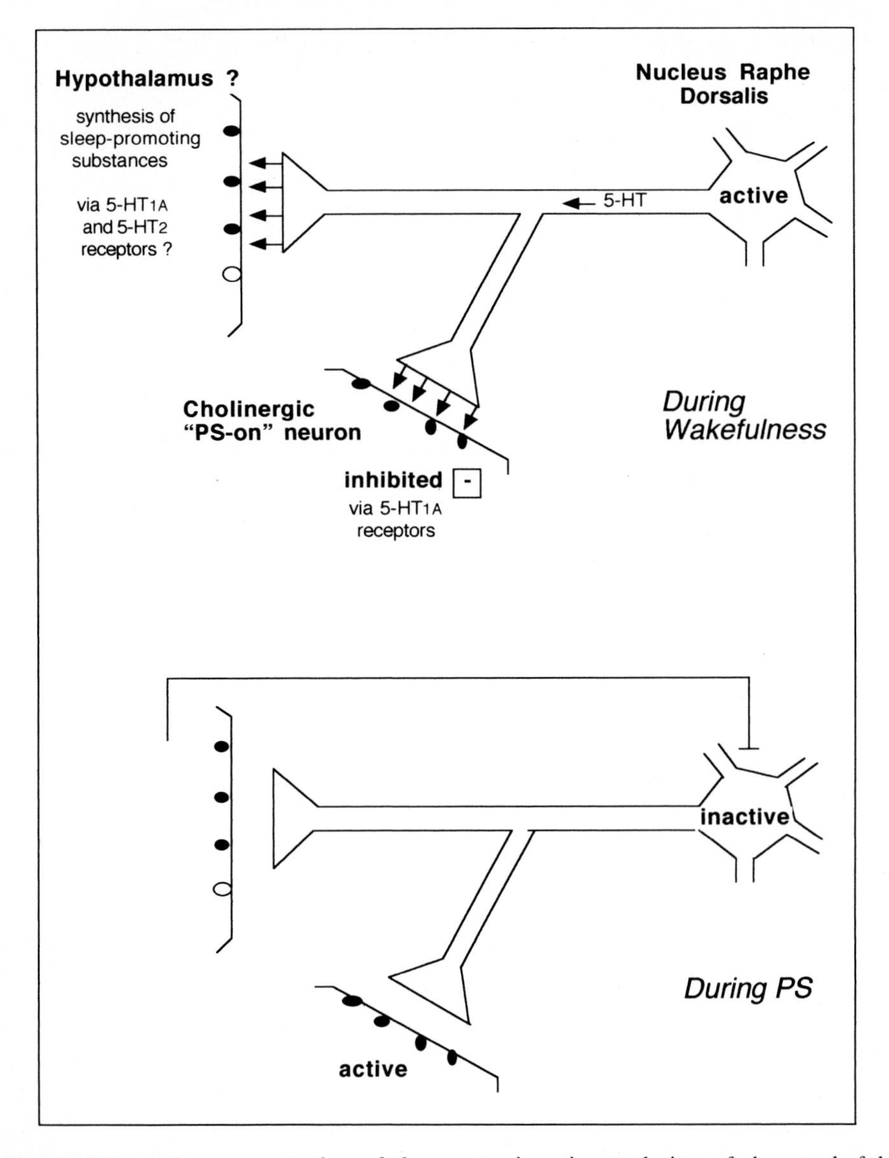

Fig. 5. Schematic representation of the serotoninergic regulation of sleep-wakefulness cycles. During waking, the serotoninergic neurons of the NRD are active. Thus 5-HT is liberated at the terminal level and activates postsynaptic serotoninergic receptors. This activation results in both the building up of conditions for the occurence of sleep at least at the hypothalamic level (CESPUGLIO et al. 1992) and the inhibition of cholinergic structures responsible for PS at the pontine level (STERIADE and MCCARLEY 1990). During sleep (and in particular during PS), the serotoninergic neurons of the NRD are inactive, possibly due to a negative feedback loop via sleep-promoting substances. In terms of serotoninergic receptors, this results in a reduction in both 5-HT_2 influence, therefore facilitating the occurrence of SWS_2, and 5-HT_{1A} inhibitory control of cholinergic "PS-on" neurons, therefore facilitating PS occurrence.

with the direct precursor of serotonin, 5-hydroxytryptophan, which restores the synthesis of serotonin.

The development of numerous selective ligands at the various serotoninergic receptor sites has allowed the investigation of their respective roles in the regulation of sleep-wakefulness cycles. To date, three main classes of receptors have been studied. It was found by several groups in various species that activation of the 5-HT_{1A} type induces a dose-dependent decrease in the amounts of paradoxical and slow-wave sleep, and a concomitant enhancement of wakefulness. This action probably involves the postsynaptic 5-HT_{1A} receptors rather than the somatodendritic ones. Activation of the 5-HT_{1C} type of receptors is followed by increased wakefulness and decreased paradoxical sleep. With regard to 5-HT_2 receptors, their blockade induces an enhancement of deep slow-wave sleep and an inhibition of paradoxical sleep. The effect on deep slow-wave sleep can be reversed by agonists at 5-HT_2 receptors, indicating that this receptor type would be specifically involved in the control of the latter sleep state. Finally, the activation of 5-HT_3 receptors results in an enhancement of wakefulness and a reduction of all sleep states, which are prevented by pre-treatment with a 5-HT_3 antagonist, but these effects would result from an action on the dopaminergic system.

Serotonin is no longer considered as a "sleep-promoting neurotransmitter" or an "antiwaking" agent, but as a neuromodulator with a very complex role. A simplified model of its involvement in sleep-wakefulness regulations has been proposed: during wakefulness, the activated serotoninergic system participates in building up the conditions for sleep occurrence (possibly in the hypothalamus); during sleep, the deactivated serotoninergic system controls the production of paradoxical sleep (probably at the brain stem level).

References

Adam K, Oswald I (1979) One gram of L-tryptophan fails to alter the time taken to fall asleep. Neuropsychopharmacology 18:1025–1027

Adrien J (1992) Are other 5-HT receptors besides 5-HT_{1A} and 5-HT_2 involved in the regulation of sleep and wakefulness? J Sleep Res 1:176–177

Adrien J, Laguzzi R, Bourgoin S, Hamon M (1980) Le sommeil du rat à lésion précoce du raphé: pharmacologie du système sérotoninergique. Waking Sleeping 4:119–129

Adrien J, Lanfumey L, Gozlan H, Fattaccini CM, Hamon M (1989) Biochemical and electrophysiological evidence for an agonist action of CM 57493 at pre- and postsynaptic 5-$hydroxytryptamine_{1A}$ receptors in brain. J Pharmacol Exp Ther 248:1222–1230

Adrien J, Davenne D, Jolas T, Hamon M (1990) Interaction of 5-HT_{1A} and 5-HT_3-receptor-active drugs in the control of PGO activity and sleep in the cat. In: Horne J (ed) Sleep '90. Pontenagel, Bochum, pp 123–125

Akindele MO, Evans JI, Oswald I (1970) Mono-amine oxydase inhibitors, sleep and mood. Electroencephalogr Clin Neurophysiol 29:47–56

Andén NE, Dalström A, Fuxe K, Olson L, Ungerstedt U (1966) Mapping out of catecholamine and 5-hydroxytryptamine neurons innervating the telencephalon and diencephalon. Acta Physiol Scand 67:313–326
Azmitia E, Segal M (1978) An autoradiographic analysis of the differential ascending projections of the dorsal and median raphe nuclei in the rat. J Comp Neurol 179:641–668
Bjorklund A, Falck B, Stenevi U (1971) Classification of monoamine neurons in the rat mesencephalon: distribution of a new monoamine neuronal system. Brain Res 32:269–285
Bjorvatn B, Ursin R (1990) Effects of zimelidine, a selective 5-HT reuptake inhibitor, combined with ritanserin, a selective 5-HT_2 antagonist, on waking and sleep stages in rats. Behav Brain Res 40:239–246
Bloom FE, Hoffer BJ, Siggins GR, Barker JL, Nicoll RA (1972) Effects of serotonin on central neurons: microiontophoretic administration. Fed Proc 31:97–106
Borbely AA, Neuhaus HY, Tobler I (1981) Effect of p-chloro-phenylalanine and tryptophan on sleep, EEG and motor activity in the rat. Behav Brain Res 2:1–22
Borbély A, Trachsel L, Tobler I (1988) Effects of ritanserin on sleep stages and sleep EEG in the rat. Eur J Pharmacol 156:275–278
Brezinova V, Loudon J, Oswald I (1972) Tryptophan and sleep. Lancet 2:1086–1087
Brooks DC, Gershon MD (1972) An analysis of the effect of reserpine upon pontogeniculo-occipital wave activity in the cat. Neuropharmacology 11: 499–510
Cespuglio R, Sarda N, Gharib A, Houdouin F, Rampin C, Jouvet M (1990) Voltametric detection of the release of 5-hydroxyindole compounds throughout the sleep-waking cycle of the rat. Exp Brain Res 80:121–128
Cespuglio R, Houdouin F, Oulerich M, El Mansari M, Jouvet M (1992) Axonal and somato-dendritic modalities of serotonin release: their involvement in sleep preparation, triggering and maintenance. J Sleep Res 1:150–156
Chan-Palay V (1975) Fine structure of labelled axons in the cerebellar cortex and nuclei of rodents and primates after intraventricular infusions with tritiated serotonin. Anat Embryol (Berl) 148:235–265
Dahlström A, Fuxe K (1964) Evidence for the existence of monoamines containing neurons in the central nervous system. I. Demonstration of monoamines in the cell bodies of brainstem neurons. Acta Physiol Scand 62:1–55
Dahlström A, Fuxe K (1965) Evidence for the existence of monoamines containing neurons in the central nervous system. II. Experimentally induced changes in the intraneuronal amine levels of bulbospinal neuron systems. Acta Physiol Scand 64 [Suppl 247]:1–36
Dement W, Mitler M, Henricksen S (1972) Sleep changes during chronic administration of parachlorophenylalanine. Rev Can Biol Suppl 31:239–246
Denoyer M, Sallanon M, Kitahama M, Aubert C, Jouvet M (1989) Reversibility of p-chlorophenylalanine induced insomnia by intrahypothalamic micro-injection of L-5-hydroxytryptophan. Neuroscience 28:83–94
Depoortere H, Riou-Merle F (1988) Electropharmacological characterization of serotonin receptors by PGO spikes in cats and by sleep-wakefulness cycle in rats. In: Koella W (ed) Sleep '86. Fischer, Stuttgart, p 349
Descarries L, Beaudet A, Watkins KC (1975) Serotonin nerve terminals in adult rat neocortex. Brain Res 100:563–588
Djik DJ, Beersma DG, Daan S, Van den Hoofdakker RH (1989) Effects of seganserin, a 5-HT_2 antagonist, and termazepam on human sleep stages and EEG power spectra. Eur J Pharmacol 171:207–218
Dugovic C (1992) Functional activity of 5-HT_2 receptors in the modulation of sleep/wakefulness states. J Sleep Res 1:163–168
Dugovic C, Leysen JE, Wauquier A (1989a) Melatonin modulates the sensitivity of 5-Hydroxytryptamine-2 receptor-mediated sleep-wakefulness regulation in the rat. Neurosci Lett 104:320–325

Dugovic C, Wauquier A, Leysen JE, Marannes R, Janssen PAJ (1989b) Functional role of 5-HT_2 receptors in the regulation of sleep and wakefulness in the rat. Psychopharmacology (Berl) 97:436–442
Dzoljic MR, Saxena PR, Ukponmwan OE (1987) 5-HT receptor agonists enhance alertness. Sleep Res 16:88
Fisher-Perroudon C, Mouret M, Jouvet M (1974) Sur un cas d'agrypnie (4 mois sans sommeil) au cours d'une maladie de Morvan. Effet favorable du 5-hydroxytryptophane. Electroencephalogr Clin Neurophysiol 36:1–18
Florio V, de Carolis AS, Longo VG (1968) Observations on the effect of DL-parachlorophenylalanine on the electroencephalogram. Physiol Behav 3:861–863
Fuxe K, Jonsson G (1974) Further mapping of central 5-hydroxytryptamine neurons: studies with the neurotoxic dihydroxytryptamine. Adv Biochem Psychopharmacol 10:1–12
Grahame-Smith DG (1967) The biosynthesis of 5-hydroxytryptamine in brain. Biochem J 105:351–360
Guilleminault C, Cathala JP, Castaigne P (1973) Effects of 5-hydroxytryptophan on sleep of patient with a brainstem lesion. Electroencephalogr Clin Neurophysiol 34:177–194
Hartmann E, Chung R (1972) Sleep inducing effects of L-tryptophan. J Pharm Pharmacol 24:252–253
Hartmann E, Spinweber C (1979) Fluvoxamine – a serotonin reuptake blocker: effects on sleep. Sleep Res 8:78
Hartmann E, Cravens J, List S (1974) Hypnotic effects of L-tryptophan. Arch Gen Psychiatry 31:394–397
Hilakivi I, Kovala T, Leppävuori A, Shvaloff A (1987) Effects of serotonin and noradrenaline uptake blockers on wakefulness and sleep in cats. Pharmacol Toxicol 60:161–166
Houdouin F, Cespuglio R, Gharib A, Sarda N, Jouvet M (1991) Detection of the release of 5-hydroxyindole compounds in the hypothalamus and the n. raphe dorsalis throughout the sleep-waking cycle and during stressful situations in the rat. A polygraphic and voltametric approach. Exp Brain Res 85:153–162
Hoyer D (1991) The 5-HT receptor family: ligands, distribution and receptor-effector coupling. In: Rodgers RJ, Cooper SJ (eds) 5-HT_{1A} agonists, 5-HT_3 antagonists and benzodiazepines: their comparative behavioural pharmacology. Wiley, New York, p 31
Idzikowski C, Mills FJ, Glennard R (1986) 5-Hydroxytryptamine-2 antagonist increases human slow wave sleep. Brain Res 378:164–168
Johnson GA, Kim EG, Boukma SJ (1968) Mechanism of norepinephrine depletion by 5-hydroxytryptophan. Proc Soc Exp Biol 128:509–512
Jouvet M (1967) Neurophysiology of the states of sleep. Physiol Rev 47:117–177
Jouvet M (1969) Biogenic amines and the states of sleep. Science 163:32–41
Jouvet M (1972) The role of monoamines and acetylcholine-containing neurons in the regulation of the sleep-waking cycle. Ergeb Physiol 64:166–307
Kafi de Saint-Hilaire S, Merica H, Gaillard JM (1984) The effect of indalpine – a selective inhibitor of 5-HT uptake – on rat paradoxical sleep. Eur J Pharmacol 98:413–418
Kleinlogel H, Burki HR (1987) Effects of the selective 5-hydroxytryptamine uptake inhibitor paroxetine and zimelidine on EEG sleep and waking stages in the rat. Neuropsychobiology 17:206–212
Koella WP (1982) A modern neurobiological concept of vigilance. Experientia 38:1426–1437
Koella WP, Fedlstein A, Czicman JS (1968) The effect of parachlorophenylalanine on the sleep of cats. Electroencephalogr Clin Neurophysiol 25:481–490
Kupfer DJ, Perel JM, Pollock BG, Nathan RS, Grochocinski VJ, Wilson MJ, McEachran AB (1991) Fluvoxamine versus desipramine: comparative polysomnographic effects. Biol Psychiatry 29:23–40

Laguzzi RF, Adrien J (1980) Inversion de l'insomnie produite par la para-chlorophenylalanine chez le rat. Arch Ital Biol 118:109–123
Lanfumey L, Dugovic C, Adrien J (1985) β_1 and β_2 adrenergic receptors: their role in the regulation of paradoxical sleep in the rat. Electroenceph Clin Neurophysiol 60:558–567
Lerman JA, Kaitin KI, Dement WC, Peroutka SJ (1986) The effects of buspirone on sleep in the rat. Neurosci Lett 72:64–68
McGinty D, Harper RM (1976) Dorsal raphe neurons: depression of firing during sleep. Brain Res 101:567–575
Moir ATB, Eccleston D (1968) The effects of precursor loading in the cerebral metabolism of 5-hydroxyindoles. J Neurochem 15:1093–1108
Monti J, Jantos H (1992) Dose-dependent effects of the 5-HT_{1A} receptor agonist 8-OH-DPAT on sleep and wakefulness in the rat. J Sleep Res 1:169–175
Monti J, Pineyro G, Orellana C, Boussard M, Jantos H, Labraga P, Olivare S, Alvarino F (1990) 5-HT receptor agonists DOI and 8-OH-DPAT increase wakefulness in the rat. Biogenic Amines 7:145–151
Moser PC, Redfern PH (1985) Circadian variation in behavioural response to central 5-HT receptor sitmulation in the mouse. Psychopharmacology (Berl) 86:223–227
Mouret J, Jeannerod M, Jouvet M (1963) L'activité électrique du système visuel au cours de la phase paradoxale du sommeil chez le chat. J Physiol (Paris) 55: 305–306
Mouret J, Bobillier P, Jouvet M (1968a) Insomnia following para-chlorophenylalanine in the rat. Eur J Pharmacol 5:17–22
Mouret J, Vilppula A, Frachon N, Jouvet M (1968b) Effets d'un inhibiteur de la monoamine oxydase sur le sommeil du rat. C R Soc Biol (Paris) 162:914–917
Nicholson AN, Pascoe PA (1986) 5-hydroxytryptamine and noradrenaline uptake inhibition: studies on sleep in man. Neuropharmacology 25:1079–1083
Nicholson AN, Pascoe PA (1988) Studies on the modulation of the sleep-wakefulness continuum in man by fluoxetine, a 5-HT uptake inhibitor. Neuropharmacology 27:597–602
Osborne NN (1982) Biology of serotoninergic transmission. Wiley, Chichester
Oswald I, Adam K (1986) Effects of paroxetine on human sleep. Br J Clin Pharmacol 22:97–99
Pastel RH, Fernstrom JD (1987) Short-term effects of fluoxetine and trifluomethylphenylpiperazine on electroencephalographic sleep in the rat. Brain Res 436: 92–102
Peroutka SJ (1988) 5-Hydroxytryptamine receptor subtypes. Annu Rev Neurosci 11:45–60
Petitjean F, Buda C, Janin M, Sakai K, Jouvet M (1978) Patterns of sleep alternation following selective raphe nuclei lesions. Sleep Res 8:40
Petitjean F, Buda C, Janin M, Sallanon M, Jouvet M (1985) Insomnie par administration de parachlorophenylalanine: réversibilité par injection périphérique ou centrale de 5-hydroxytryptophane et de sérotonine. Sleep 8:56–67
Ponzoni A, Monti J, Jantos H (1993) The effects of selective activation of the 5-HT_3 receptor with m-chlorophenylbiguanide on sleep and wakefulness in the rat. Eur J Pharmacol 249:259–264
Puizillout JJ, Gaudin-Chazal G, Daszuta A, Seyfritz N, Ternaux JP (1979) Release of endogenous serotonin from "encéphale isolé" cats. II. Correlations with raphe neuronal activity and sleep and wakefulness. J Physiol (Paris) 75:531–538
Pujol JF, Buguet A, Froment JL, Jones B, Jouvet M (1971) The central metabolism of serotonin in the cat during insomnia. A neurophysiological and biochemical study after administration of p-chlorophenylalanine or destruction of the raphe system. Brain Res 29:195–212
Quattrochi J, Mamelak A, Binder D, Williams J, Hobson JA (1993) Dose-related suppression of REM sleep and PGO waves by the serotonin-1 agonist Eltoprazine. Neuropsychopharmacology 8:7–13

Ruch-Monachon MA, Jalfre M, Haefely W (1976) Drugs and PGO waves in the lateral geniculate body of the curarized cat. V. Miscellaneous compounds. Arch Int Pharmacodyn Ther 219:326–346

Sakai K (1980) Some anatomical and physiological properties of ponto-mesencephalic tegmental neurons with special reference to the PGO waves and postural atonia during paradoxical sleep in the cat. In: Hobson JA, Brazier MAB (eds) The reticular formation revisited. Raven, New York, p 427

Saletu B, Frey R, Krupka M, Anderer P, Grünberger J, Wolf RS (1991) Sleep laboratory studies on single-dose effects of serotonin reuptake inhibitors paroxetine and fluoxetine on human sleep and awakening qualities. Sleep 14: 439–447

Sanders-Bush E (1988) the serotonin receptors. Humana, Clifton

Segal M (1980) The action of serotonin in the rat hippocampal slice preparation. J Physiol (Lond) 303:423–439

Shipley JE, Kupfer DJ, Griffin SJ, Dealy RS, Coble PA, McAechran AB, Grachocinski VJ (1984) Differential effects of amitriptyline and of zimelidine on the sleep electroencephalogram of depressed patients. Clin Pharmacol Ther 34:251–259

Slater IH, Jones GT, Moore RA (1978) Inhibition of REM sleep by fluoxetine, a specific inhibitor of serotonin uptake. Neuropharmacology 17:383–389

Sommerfelt L, Ursin R (1991) Behavioral, sleep-waking and EEG power spectral effects following the two specific 5-HT uptake inhibitors zimelidine and alaproclate in cats. Behav Brain Res 45:105–115

Sommerfelt L, Hauge ER, Ursin R (1987) Similar effect on REM sleep but differential effect on slow wave sleep of the two 5-HT uptake inhibitors zimelidine and alaproclate in cats and rats. J Neural Transm 68:127–144

Spinweber CL, Ursin R, Hilbert RP, Hilderbrand RL (1983) L-Tryptophan: effects on daytime sleep latency and the waking EEG. Electroencephalogyr Clin Neurophysiol 55:652–661

Sprouse JS, Aghajanian GK (1987) Electrophysiological responses of serotoninergic dorsal raphe neurons to 5-HT_{1A} and 5-HT_{1B} agonists. Synapse 1:3–9

Steriade M, McCarley R (1990) REM sleep as a biological rhythm. In: Steriade M, McCarley K (eds) Brainstem control of wakefulness and sleep. Plenum, New York, pp 363–393

Stutzmann JM, Eon B, Roux M, Lucas M, Blanchard JC, Laduron PM (1990) RP62203, a 5-HT_2 antagonist, enhances slow wave sleep in rats. Eur J Pharmacol 183:1394

Tissier MH, Lainey E, Fattaccini CM, Hamon M, Adrien J (1993) Effects of ipsapirone, a 5-HT_{1A} agonist, on sleep-wakefulness cycles: probable postsynaptic action. J Sleep Res 2:103–109

Trulson ME, Jacobs BJ (1979) Raphe unit activity in freely moving cats: correlation with the level of behavioural arousal. Brain Res 163:135–150

Ursin R (1976) The effects of 5-hydroxytryptophan and L-tryptophan on wakefulness and sleep patterns in the cat. Brain Res 106:105–115

Ursin R, Bjorvatn B, Sommerfelt L, Underland G (1989) Increased waking as well as increased synchronization following administration of selective 5-HT uptake inhibitors to rats. Behav Brain Res 34:117–130

Ursin R, Bjorvatn B, Sommerfelt L, Neckelmann D, Bjorkum AA (1992) Studies on sleep/wake effects of serotonin reuptake inhibitors and receptor subtype involvement. J Sleep Res 1:157–162

Vergé D, Daval G, Marcinkiewicz M, Patey A, El Mestikawy S, Gozlan H, Hamon M (1986) Quantitative autoradiography of multiple 5-HT_1 receptor sybtypes in the brain of control and 5,7-DHT treated rats. J Neurosci 6:3474–3482

Weitzman ED, Rapport MM, McGregor P, Jawby T (1968) Sleep patterns of the monkey and brain serotonin concentration: effect of p-chlorophenylalanine. Science 160:1361–1363

Wojcik WJ, Fornal C, Radulovacki M (1980) Effect of tryptophan on sleep in the rat. Neuropharmacology 19:163–167

Wyatt RS, Zarcone V, Engelman K, Dement WC, Snyder F, Sjoerdsma A (1971) Effects of 5-hydroxytryptophan on the sleep of normal human subjects. Electroencephalogr Clin Neurophysiol 30:505–509

Zolovic AJ, Stern WC, Panksepp J, Jalowiec JE, Morgane PJ (1973) Sleep-waking patterns in cats after administration of fenfluramine and other monoaminergic modulating drugs. Pharmacol Biochem Behav 1:41–16

CHAPTER 6

Pharmacology of the Histaminergic System

J.M. Monti

A. Introduction

Histamine (HA) was first synthesized by Winders and Vogt in 1907, but its pharmacological activity was recognized only a few years later when it was shown to stimulate smooth muscle (Barger and Dale 1910; Dale and Laidlaw 1910). Its presence in the body was established when HA was isolated from liver and lung (Best et al. 1927).

Although much of the attention since has been given to HA in peripheral tissues, allergic reactions and injury, knowledge of its presence in the central nervous system (CNS) goes back as far as that of norepinephrine.

Kwiatkowski (1943) was the first to detect the amine in the brain, mainly in gray matter, while Harris et al. (1952) found relatively higher amounts of HA in the hypothalamus than in other brain areas. The successful synthesis of HA in the brain from labeled histidine (White 1959) and the development of compounds with HA-blocking properties and marked sedative effects, later to be called H_1 antagonists (Bovet 1950), strongly suggested that the amine might be in histaminergic neurons.

Further advances in our understanding of HA as a neurotransmitter in the CNS were made possible by: (a) the development of more sensitive assays for both HA and L-histidine decarboxylase (Schwartz et al. 1970); (b) the finding that the amine occurs in neurons and axons (Panula et al. 1984) and is released by depolarization (Atack and Carlsson 1972); (c) the characterization in brain of three distinct subclasses of HA receptors, H_1, H_2 and H_3 (Ash and Schild 1966; Black et al. 1972; Arrang et al. 1983); and (d) the availability of new pharmacological tools capable of interfering with HA synthesis or of activating or selectively blocking HA receptors (Monti 1990).

B. Histamine Synthesis and Metabolism

Labeled HA does not cross the blood-brain barrier, indicating that the CNS depends upon local neuronal biosynthesis.

Histamine in brain is formed from L-histidine, this being actively transported into the brain. Histidine is decarboxylated by the enzyme L-histidine decarboxylase (HD). The distribution of HD immunoreactivity in the brain is not uniform and roughly parallels that of HA and of the metabolites tele-methylhistamine and tele-methylimidazolacetic acid (Table 1).

Histidine decarboxylase is not saturated, and consequently administration of L-histidine increases brain HA levels (Lee, et al. 1981). Compared with norepinephrine and dopamine, HA levels in brain are relatively low. HA content is highest in the hypothalamus, intermediate in basal ganglia and thalamus and lowest in the brain stem and cerebellum (Brownstein 1981; Oishi et al. 1983).

Because no high-affinity uptake system seems to exist for HA, its inactivation occurs solely by catabolic pathways. Two main catabolic pathways have been described at peripheral sites, which depend on the enzymes diamine oxidase and HA-*N*-methyltransferase. Almost no diamine oxidase is present in mammalian brain, which makes methylation by HA-*N*-methyltransferase at the tele position to tele-methylhistamine the only effective route of HA degradation. The methyl group is transferred from *S*-adenosyl-*L*-methionine. Tele-methylhistamine is further deaminated by monoamine oxidase (MAO)-B into tele-methylimidazolacetic acid. The metabolites of HA have a large concentration gradient between cisternal and lumbar cerebrospinal fluid in man and rhesus monkey, which indicates that they are derived from brain (Prell et al. 1989).

Studies using either the rate of accumulation of brain tele-methylhistamine levels after administration of the MAO inhibitor pargyline, or the rate of decline of HA levels after irreversible inhibition of HD by the compound α-fluoromethylhistidine, tend to indicate that brain HA turnover is rapid. In the rat, values for hypothalamus of 1.33 nmol/g per hour have been reported (Garbarg et al. 1980; Hough et al. 1984; Oishi et al. 1984).

In brain, HA is contained in both a neuronal and a nonneuronal pool. The latter corresponds to the mast cells that make only a minor contribution to HA at central sites. HA in the neuronal pool appears during the first few days of the postnatal period. In 1-day-old rats, ^{14}C-HA is slightly released

Table 1. Pathway of histamine synthesis and metabolism in the CNS

Compound	Enzyme
Histidine	
↓	Histidine decarboxylase
Histamine	
↓	HA-*N*-methyltransferase
N-Methyl histamine	
↓	MAO-B
N-Methylimidazolacetic acid	

by K^+ stimulation. The maximum release of HA occurs in 30-day-old rats, and is prevented when the incubation is performed in a Ca^{2+}-free superfusion medium (WATABE and ISHIKAWA 1984). HA in the neuronal pool is highly associated with HD activity. Its half-life calculated from the turnover rate is 7.7 min in the striatum, and about 50 min in the hypothalamus and thalamus (SCHWARTZ et al. 1979, 1981).

Histamine contained in mast cells comprises only 20% of the HA found in brain. It appears before birth, is associated with HD activity to a very limited extent, is released by mast cell degranulators such as compound 48/80 and its half-life is several days (GARBARG et al. 1980; PICAPOSTE et al. 1977).

There are diurnal and age-related changes in brain HA. In rats kept under controlled environmental conditions, a significantly higher HA content was found in hypothalamus at the end of the light period, then a decline during the dark hours followed by an increase during daytime. Histidine decarboxylase and HA-*N*-methyltransferase activities were higher during darkness. In guinea pig hypothalamus HA content was significantly higher in the morning at the beginning of the light period, while HA-*N*-methyltransferase showed no significant changes (ORR and QUAY 1975; TUOMISTO and TUOMISTO 1982).

The opposite changes in HA, its biosynthetic enzyme and its inactivation enzyme in rat hypothalamus during the light-dark cycle indicate the existence of 24-h rhythms. The lower level of HA during the dark period, when rats remain awake significantly longer, could be related to increased release leading to an increase in metabolite levels.

The effect of age on brain HA levels and its inactivation enzyme was investigated in hypothalamus, midbrain and cortex of 3-month-old and 12-month-old rats. HA concentration was higher and HA-*N*-methyl-transferase activity lower in the older animals (MAZURKIEWICZ-KWILECKI and PRELL 1984), which indicates that, similarly to other central neurotransmitters, the central histaminergic system is affected by the aging process.

C. Histaminergic Nuclei and Pathways

Histamine-containing neurons and their fibers cannot be visualized using the formaldehyde- or orthophthaldehyde-induced fluorescence because of lack of sensitivity. Also, autoradiography is of no use because no high-affinity uptake system exists for HA. The location of histaminergic neurons and their projection axons was determined by immunofluorescence techniques. This was achieved after the development of specific and sensitive antibodies to HD and HA (PANULA et al. 1984; FUKUI et al. 1980; WATANABE et al. 1984). With these methods histaminergic neurons were identified in the tuberal region of the posterior hypothalamus of the rat. The histaminergic neuronal system consists of a single group of diffusely distributed neurons in the tuberomammillary nucleus, where WATANABE et al. (1984), BLEIER et al. (1979) and STEINBUSCH and MULDER (1984) proposed the existence of three main subgroups of cells. The most prominent of the three subgroups was found in the caudal magnocellular nucleus, which lies close to the basal surface of the brain. A second subgroup was observed in the tuberal magnocellular nucleus, which is located in the dorsal tip of the third ventricle. The third cluster of HA-containing neurons was identified within the postmammillary caudal magnocellular nucleus. More recently, ERICSON et al. (1987) contended that the tuberomammillary nucleus consists only of two subgroups of HA neurons. The largest one, distinguished as the ventral tuberomammillary subgroup, comprises a rostral aspect which includes the caudal magnocellular nucleus, and a caudal aspect which corresponds to the posterior caudal magnocellular nucleus. The second cluster, identified as the medial tuberomammillary subgroup, was subdivided into a dorsal part which is equivalent to the tuberal magnocellular nucleus, and a ventral part which is in the proximity of the medial mammillary body.

Histamine immunoreactive neurons were found also in the tuberomammillary complex of the guinea pig, cat, tree shrew and man. In the guinea pig HA neurons were more numerous than in the rat, being also found between the medial and lateral mammillary nuclei. In the tree shrew the majority of cell bodies were located laterally in the ventral part of the tuberomammillary nucleus. In man distinct clusters of HA neurons were seen between the ventromedial nucleus and the third ventricle, and in the basal hypothalamus in areas corresponding to the tuberomammillary nucleus (AIRAKSINEN and PANULA 1988; AIRAKSINEN et al. 1989; PANULA et al. 1989, 1990).

Many of the HD-containing neurons also exhibit immunoreactivity for adenosine deaminase, L-glutamate decarboxylase, the neuropeptides met-enkephalin-arg-phe, galanin, and substance P (VINCENT et al. 1983; NAGY et al. 1984; KÖHLER et al. 1985; STAINES et al. 1986). In addition, the distribution of MAO-B activity in the posterior hypothalamus is very similar to the distribution of HA in the magnocellular neurons of this area (PANULA et al. 1984; MAEDA et al. 1984; LIN et al. 1992).

Initially, it was proposed by STEINBUSCH and MULDER (1985) that caudal, tuberal and postmammillary caudal magnocellular nuclei had distinct projection fields. However, recent retrograde labeling studies (ERICSON et al. 1987) showed projections from all parts of the tuberomammillary nucleus to all major areas of the telencephalon, rhinencephalon, diencephalon, mesencephalon and rhombencephalon of the rat (Table 2).

Using a new immunohistochemical method to study the distribution of histaminergic neuronal fibers and terminals in the rat brain, PANULA et al. (1988) recognized one descending and two ascending pathways. The ventral ascending pathway runs close to the major hypothalamic nuclei towards the nucleus of the diagonal band, the medial septal nucleus and the olfactory tubercle and bulb. The dorsal ascending pathway runs along the lateral side

Table 2. Projections of histaminergic axons in the rat. (From WATANABE et al. 1984; STEINBUSCH and MULDER 1985; PANULA et al. 1989; SCHWARTZ et al. 1991)

I. Ascending
1. Telencephalon and rhinencephalon
   - Cerebral cortex – Most neocortical and allocortical areas
   - Limbic system – Medial septal nucleus
     – Basal hippocampus, subiculum, dentate gyrus
     – Amygdaloid complex
   - Neostriatum – Caudate putamen, nucleus accumbens
2. Diencephalon
   - Thalamus – Periventricular areas, posterolateral nuclei
   - Hypothalamus – Medial preoptic area
     – Periventricular nucleus
     – Supraoptic and suprachiasmatic nuclei
     – Paraventricular nucleus
     – Dorsomedial and ventromedial nuclei
     – Arcuate nucleus
     – Basal hypothalamus (mammillary recess)

II. Descending
1. Mesencephalon – Pars compacta and pars reticulata of the substantia nigra
   – Interpeduncular nucleus
   – Superior and inferior colliculi
   – Mesencephalic central gray
2. Rhombencephalon – Raphe nuclei
   – Medial and lateral ventral gray at the pontine level
   – Locus coeruleus
   – Solitary tract nucleus
3. Cerebellum – Cerebellar cortex
4. Spinal cord – Dorsal horn

of the third ventricle and sends branches to the thalamic nuclei and rostral forebrain structures including the limbic system and cerebral cortex. These tracts remain largely ipsilateral, although there is a crossing over at the level of the retrochiasmatic area, the optic chiasma and in the supramammillary region. The extent of contralateral fibers reaching a given structure varies according to the species and the structure involved. The descending pathway reaches the central gray and dorsal longitudinal fasciculus. Some of these fibers project as far as the caudal medullary nuclei and the spinal cord (Table 2).

As in the rat, histaminergic fibers in the guinea pig and the tree shrew project to nearly all parts of the brain. However, the density of histaminergic fibers in the hindbrain of the tree shrew is higher than that in rodents (AIRAKSINEN and PANULA 1988; AIRAKSINEN et al. 1989).

A preliminary study in human brain revealed the presence of HA-immunoreactive fibers in the frontal and temporal cortex, with the densest network of fibers being located in lamina I (PANULA et al. 1990).

Several laboratories have shown afferents to HA-immunoreactive neurons from prefrontal cortex, septal nuclei, olfactory tubercle, hippocampus, medial septal area and dorsal tegmental region (Swanson and Cowan 1979; Groenewegen and Van Dijk 1984; Wouterlood et al. 1987; Wouterlood and Gaykema 1988).

D. Histamine Receptors

Histamine-related functions in the CNS are regulated at postsynaptic sites by the H_1 and the H_2 receptor. In contrast, the H_3 receptor shows the features of a presynaptic autoreceptor, mediating the synthesis and release of HA. The three types of receptors differ in their molecular properties, distribution in the CNS, electrophysiological responses and affinity for HA and synthetic agonists and antagonists.

The mapping of HA H_1 receptors has been established by autoradiography with [^{3}H]-mepyramine and [^{125}I]-iodobolpyramine (Hill et al. 1978; Bouthenet et al. 1988). The H_1 receptor is widely distributed in the CNS. However, its density and regional distribution vary between species. In the guinea pig the H_1 receptor is present in: (a) all areas and layers of the cerebral cortex with a higher density in the deep layers; (b) the limbic system, including the hippocampus (dentate gyrus, subiculum), amygdala (medial group of nuclei) and medial and lateral septal nuclei; (c) the caudate-putamen and the nucleus accumbens, which are moderately and highly labeled, respectively; (d) the thalamus, with a higher distribution of receptors in the anterior, median and lateral nuclei; (e) the hypothalamus, including the medial preoptic area, dorsolateral and ventromedial nuclei and the tuberomammillary complex; (f) the mesencephalon and lower brain stem, where H_1 receptors predominate in the nuclei of origin of cranial nerves, dopamine-, norepinephrine- and serotonin-containing neurons, cerebellum and the area postrema; and (g) the spinal cord, predominantly the dorsal horn (Bouthenet et al. 1988; Pollard and Bouthenet 1992; Schwartz et al. 1991).

The areas with the highest density of H_1 receptors in guinea pig brain, nucleus accumbens, thalamus and cerebellum, are scarcely labeled in rat and cat brain. The distribution of H_1 receptors in these two species is highest in the hypothalamus and tuberomammillary nucleus. Intermediate levels of binding are observed in the cerebral cortex, limbic system, corpus striatum, colliculi and medulla. The lowest binding is observed in the pons and spinal cord (Hill and Young 1980; Taylor et al. 1982; Schwartz et al. 1986).

Using [^{125}I]-iodobolpyramine to label the H_1 receptor, in the human and rhesus monkey brain, Martínez-Mir et al. (1990) found the receptor to be particularly abundant in the neocortex and hippocampus. Villemagne et al. (1991) and Yanai et al. (1990) used [^{11}C]-mepyramine as a radioligand to investigate H_1 receptors in the human brain with position emission tomography (PET). [^{11}C]-Radioactivity was observed at high concentrations in the frontal and temporal cortex, hippocampus and thalamus. The frontal, parietal and temporal cortices showed age-related decreases in binding of approximately 13% per decade (YANAI et al. 1990).

The H_1 receptor is related to the enzyme phospholipase C via a GTP-binding protein and acts primarily by increasing the Ca^{2+} concentration in the target cell. The process is linked to the hydrolysis of inositol phospholipids by the enzyme phospholipase C with the production of inositol-1,4,5-triphosphate, which mobilizes Ca^{2+}, and 1,2-diacylglycerol, which activates protein kinase C (TIMMERMAN 1989; HAAKSMA et al. 1990).

The H_1 receptor also stimulates the accumulation of cAMP mediated by the H_2 receptor. This is an indirect effect which is significantly reduced in the absence of Ca^{2+} (SCHWARTZ et al. 1991; HILL and DONALDSON 1992).

The distribution of HA H_2 receptors has been determined by autoradiographic studies using [^{125}I]-iodoaminopotentidine as a selective ligand (RUAT et al. 1990). The H_2 receptor is distributed extensively and in a heterogeneous fashion. In the rodent brain the H_2 receptor is present in: (a) the superficial layers of the cerebral cortex; (b) the limbic system including the hippocampus (subicular complex, particularly the CA1 region) and the amygdala (bed nucleus of the stria terminalis and basal group of nuclei); (c) the basal ganglia, where very high densities have been detected in the nucleus accumbens, the caudate-putamen and the olfactory tubercle; (d) the thalamus (medial group of nuclei); (e) the hypothalamus, where the density is very low; (f) the mesencephalon, where relatively high densities have been found at the level of the superior and inferior colliculi, central gray matter and substantia nigra; and (g) the lower brain stem, with the highest density being observed in the raphe nuclei (SCHWARTZ et al. 1991). In human and rhesus monkey brain the H_2 receptor sites are predominantly localized in the basal ganglia, although they are also present in the cerebral cortex (MARTINEZ-MIR et al. 1990).

Histamine, acting through the H_2 receptor, activates the stimulatory G protein (Gs), which in turn stimulates adenylate cyclase. Increased cAMP levels lead to phosphorylation of intracellular proteins and to a physiological response. The increase in cAMP levels is stimulated by selective H_2 agonists (HEGSTRAND et al. 1976; JOHNSON 1992).

The distribution of the H_3 receptor in the rodent brain is highly heterogeneous and not exactly the same as that of histaminergic terminals, which is in accordance with its presence on nonhistaminergic nerve endings. Thus, it has been localized on histaminergic axons, but also on serotonergic and noradrenergic nerve endings (SCHLICKER et al. 1988, 1989). HA H_3 receptors have been mapped with the potent and selective H_3 ligand R-α-[3H]-methylhistidine (ARRANG et al. 1987).

In rodents the H_3 receptor is found in: (a) all layers of the cerebral cortex, especially in the deep layers; (b) structures corresponding to the limbic system, including the hippocampus (molecular layer of the dentate gyrus, subicular complex), the amygdala (bed nucleus of the stria terminalis and basolateral, lateral and central nuclei); (c) the striatum, nucleus accumbens, anterior olfactory nuclei and olfactory tubercles; (d) the thalamus, mainly its medial part; (e) the hypothalamus at the level of the tuberomammillary nucleus; (f) the mesencephalon, where a relatively high density is found in the pars reticulata of the substantia nigra, the ventral tegmental area and the superior and inferior colliculi; (g) the lower brain stem, where low to moderate densities are found in the locus coeruleus, the raphe nuclei and the vestibular nuclei (SCHWARTZ et al. 1991; ARRANG et al. 1992).

In human and rhesus monkey brain H_3 receptors predominate in the basal ganglia, mainly the globus pallidus.

The H_3 receptor is coupled to its effector system via a G protein. The mechanisms involved in the signal transfer at the H_3 receptor are still unknown.

E. Effects of Histamine at the Cellular Level

The effect of HA and of H_1, H_2 or H_3 receptor agonists and antagonists on CNS neurons has been studied in tissue culture, tissue slices and anesthetized and unanesthetized decerebrate animals. Iontophoretic administration of HA reveals excitatory and inhibitory effects. The excitatory responses to HA are due in most cases to H_1 receptor activation. They manifest as depolarization, increase in firing rate and facilitation of signal transmission. In mammals these effects seem to be related to a reduced potassium conductance (HAAS 1985, 1992). Excitatory actions mediated via H_1 receptors have been found in the brain stem (medial pontine reticular formation, medial vestibular nucleus, locus coeruleus), hypothalamus (supraoptic, suprachiasmatic and ventromedial nuclei), hippocampus (CA1 and CA2 areas) and cerebral cortex (HAAS and BUCHER 1975; ARMSTRONG and SLADEK 1985; GERBER et al. 1990; GREENE and HAAS 1990; PHELAN et al. 1990). These effects are obtained with HA and are blocked by mepyramine.

Inhibitory effects of HA in the CNS are mediated by H_2 receptors. The effect is due to hyperpolarization and decrease in the firing rate of the corresponding neurons. They are reproduced by the H_2 receptor agonists 4-methylhistamine and impromidine, potentiated by phosphodiesterase inhibitors and blocked by the H_2 receptor antagonists metiamide and cimetidine (HAAS 1992).

Inhibitory actions mediated via H_2 receptors have been found in the spinal cord, brain stem (medullary reticular formation, vestibular neurons and dorsal raphe nucleus), hypothalamus (ventromedial nucleus, preoptic area), hippocampus, nucleus accumbens and cerebral cortex (HAAS 1979; CHRONISTER et al. 1982; LAKOSKI et al. 1984).

The H_3 receptor mediates the inhibitory response of HA on the tuberomammillary nucleus. The effect could be related to the inhibition of an inward Ca^{2+} current (ARRANG et al. 1990).

F. Drugs Acting at the Histamine Receptors

Agonists and antagonists are now available which act on H_1, H_2 or H_3 receptors (Table 3).

I. H_1 Receptor

Replacing the imidazole ring of HA by other heterocycles reduces HA-like activity but introduces selectivity at HA receptors. The heterocyclic analog 2-(2-thiazolyl)ethylamine, which retains the ring-nitrogen atom adjacent to the side chain, acts predominantly as H_1 receptor agonist. Its relative potency at H_1 sites (guinea pig ileum assay) is 26% that of HA, while it is only 2% (guinea pig atrium assay) at the H_2 sites (GANELLIN 1982, 1992).

Recently, thienyl analogs of 2-benzyl and 2-phenylhistamine with selective high affinity for the H_1 receptor were synthesized. The meta-bromo analog of 2-phenylhistamine was equipotent with HA, while meta-trifluoromethylphenylhistamine was significantly more potent (guinea pig ileum assay) (ZINGEL et al. 1990; LESCHKE et al. 1992). For many years, the more active H_1 receptor antagonists were structurally characterized by the presence of a tertiary amino group linked by a two- or three-atom chain to two aromatic substituents. They include mepyramine, diphenhydramine and chlorpheniramine, which after their administration are widely distributed throughout the body, including the CNS.

There are now new H_1 receptor antagonists which show high potency and do not fit this general structure. Many of them, including terfenadine,

Table 3. Pathway of histamine synthesis and metabolism and possible sites of drug action

Histidine

| Histidine decarboxylase (inhibitor: α-fluoromethylhistidine)

Histamine H_1 agonist: 2-thiazolylethylamine
antagonists: mepyramine, diphenhydramine, chlorpheniramine
H_2 agonist: dimaprit
antagonists: cimetidine, ranitidine, zolantidine
H_3 agonist: (R)-α-methylhistamine
antagonist: thioperamide

| *N*-Methyltransferase (inhibitors: homodimaprit, metoprine)

N-Methylhistamine

| MAO-B (inhibitor: pargyline)

N-Methyl imidazoleacetic acid

astemizole, loratadine and temelastine, are nonsedative H_1 receptor antagonists. This property is related to their poor penetration through the blood-brain barrier (SORKIN and HEEL 1985; JANSSENS et al. 1986; ROMAN et al. 1986; CALCUTT et al. 1987).

II. H_2 Receptor

A series of tautomeric amidine derivatives including dimaprit and impromidine act as highly selective H_2 receptor agonists. Dimaprit is an isothiourea derivative with a relative potency in vitro at H_2 sites (guinea pig right atrium assay) of 71% that of HA. Impromidine is an imidazolyalkylguanidine, which was found to be 48 times more potent than HA. It is devoid of activity at H_1 sites. However, it behaves as a potent H_3 receptor antagonist and also inhibits the enzyme HA-*N*-methyltransferase (PARSONS et al. 1977; GANELLIN 1982, 1992). Neither compound crosses the blood-brain barrier.

GALLENIN (1992) considers four groups of H_2 receptor antagonists: (a) imidazole derivatives (metiamide, cimetidine, oxmetidine); (b) aminomethylfuran compounds (ranitidine, nizatidine); (c) guanidinothiazole derivatives (tiotidine, famotidine); and (d) aminomethylphenoxy compounds (lamtidine, loxtidine). Although most of these derivatives are potent and selective H_2 receptor antagonists, they are polar molecules that cross the blood-brain barrier to a very limited extent. To overcome this drawback, YOUNG et al. (1988) designed zolantidine, which is a specific, reversible, brain-penetrating H_2 receptor antagonist.

III. H_3 Receptor

R-α-Methylhistamine (AMH) acts as a highly potent and selective ligand on presynaptic HA H_3 autoreceptors (ARRANG et al. 1987). However, it penetrates poorly through the blood-brain barrier. In order to increase AMH penetration through the blood-brain barrier, STARK et al. (1992) designed a series of azomethine derivatives which are highly lipophilic and behave as pro-drugs of the H_3 receptor agonist.

Several H_2-receptor antagonists show H_3-receptor-blocking properties. However, the first highly selective H_3-receptor antagonist was the imidazole-piperidine derivative thioperamide (LIPP et al. 1992).

G. Histamine Control of the Waking State

The proposal that HA plays a neuroregulatory role in the waking state is derived: (a) from electrophysiological studies where HA neurons were recorded during the sleep-wakefulness continuum; (b) by the characterization of circadian rhythms for HA, HD and HA-*N*-methyltransferase; (c) from the finding that lesions aimed at HA-containing areas profoundly affect wakefulness; and (d) from the results of pharmacological studies where drugs that interfere with HA synthesis and metabolism, or modify HA "tonus" at central sites, decrease or increase the occurrence of wakefulness.

I. Electrophysiological Studies

Single-unit extracellular recordings in the ventrolateral posterior hypothalamus of the freely moving cat or tuberomammillary nucleus of the anesthetized rat define a population of histaminergic cortically projecting neurons which show a relatively long-lasting action potential and a slow conduction velocity (VANNI-MERCIER et al. 1985; REINER and MCGEER 1987).

Histaminergic neurons display a slow and regular discharge during quiet waking ($\bar{x}$ = 1.4 spikes/s) in the cat. When the animal is moving (active waking) the mean discharge rate increases to 2.3 spikes/s. As the cat enters slow-wave sleep (SWS) the mean discharge rate shows a progressive decrease (0.43 spikes/s). During deep SWS and REM sleep all the neurons become silent (VANNI-MERCIER et al. 1985). According to SAKAI et al. (1990), these neurons are very similar to noradrenergic and serotonergic "PS off" cells involved in the mechanisms of REM sleep generation (see also Chaps. 4, 5, this volume).

II. Circadian Rhythms

Histamine levels reach a minimum during the dark phase followed by an increase during the light period, in rats and guinea pigs kept under controlled environmental conditions. The lower levels of HA during the dark phase, when rats are more active and sleep less, could be tentatively ascribed to an increased turnover (ORR and QUAY 1975; TUOMISTO and TUOMISTO 1982). In support of this contention is the finding by SCHWARTZ et al. (1982) that HA synthesis in hypothalamic slices is significantly higher when rats are put to death during the behaviorally active dark period. Results from microdialysis carried out in the posterior hypothalamus of rhesus monkeys confirm the early reports indicating increased turnover during the period of maximum wakefulness (ONOE et al. 1992). Thus, concentrations of HA in dialysates were much higher during daytime than nighttime.

III. Lesion Studies

Long before it was known that HA-immunoreactive neurons are located in the tuberomammillary nucleus, it was established that damage, lesioning or cooling of the posterior hypothalamus of several species, including man, gives rise to a state of somnolence or hypersomnia.

In the study of the Vienna epidemic of encephalitis lethargica, VON ECONOMO (1926) established that somnolence was present in only those patients with lesions of the caudal part of the hypothalamus. The experiments by RANSON (1939) on monkeys showed that bilateral lesions in the area of the mammillary bodies caused the same marked somnolence observed in patients with epidemic encephalitis. NAQUET et al. (1965) and SWETT and HOBSON (1968) found that electrolytic lesions of the posterior lateral hypothalamus in cats were followed by a behavioral state of somnolence; the animals had at least twice as much SWS as those with medial lesions. Temporary inhibition of the posterior hypothalamus by means of a thermode permitting a light progressive cooling gave way to slowing of cortical rhythms and to the appearance of bilateral spindles accompanied by behavioral sleep (NAQUET et al. 1966). Relevant to our topic was the report by SALLANON et al. (1987) that insomnia induced in cats after neurotoxic lesions of the paramedial preoptic area with ibotenic acid could be reversed by pharmacological inactivation of the posterior hypothalamus with the GABAergic agonist, muscimol.

In the rat complete bilateral transection of the posterior hypothalamus interfered with the maintenance of the waking state (NAUTA 1946). Accordingly, the animals slept uninterruptedly for several days in the absence of external stimuli.

Bilateral electrolytic lesions of the posterior hypothalamus and adjacent subthalamic area in rats produced a continuous sleep-like state which lasted from 1 to 4 days. EEG patterns were of large amplitude and low frequency. REM sleep was completely suppressed during the continuous phase of somnolence. Recovery from somnolence was gradual but complete, thus suggesting the rapid development of compensatory mechanisms (MCGINTY 1969).

In conclusion, electrolytic lesions, complete bilateral transection, light, progressive cooling or pharmacological inactivation of the posterior hypothalamus in cats and rats induced behavioral sleep, and in those studies where the EEG was recorded an increase in number of spindles and slow wave activity. Although the experimental procedures always included the area where HA-immunoreactive neurons are located, nonspecific histopathological changes in nonhistaminergic structures cannot be excluded following such extensive lesions.

IV. Pharmacological Studies

1. Intracerebroventricular Administration of Histamine

BOVET et al. (1958) and GOLDSTEIN et al. (1963) first reported that i.v. administration of HA caused desynchronization of cortical EEG of the anesthetized rat or rabbit. Since HA does not cross the blood-brain barrier following i.v. injection, the waking effect could be due to a reflex activation of the reticulocortical projections through nociceptive or chemoceptive afferent systems. In favor of this proposal was the finding by MONNIER et al. (1970) that pretreatment with acetylsalicylic acid markedly decreased the activating effect of i.v. infused HA.

Histamine infusion (150–300 μg) into the third ventricle of the conscious rabbit elicited a marked electrographic arousal reaction, and a decrease in spectral power in the delta band. Analgesia with acetylsalicylic acid did not reduce the response. On the other hand, pretreatment with mepyramine (1.5 mg/kg) abolished the cortical EEG activation (MONNIER et al. 1970; WOLF and MONNIER 1973). This observation led MONNIER et al. (1970) to postulate that HA may have a role in the modulation of EEG arousal. Histamine was also administered into the lateral cerebral ventricle of the conscious and pentobarbital-anesthetized rat. In the freely moving rat HA (0.25–25 μg) induced a significant increase in spontaneous motor activity; in the anesthetized rat, it induced a dose-related decrease in narcosis duration. Histamine-induced behavioral arousal was blocked by i.p. administration of the H_1 receptor antagonists, chlorpheniramine (5 mg/kg) or diphenhydramine (5 mg/kg).

Intracerebroventricular (i.c.v.) injection of the H_2 receptor antagonist, cimetidine did not alter HA-induced behavioral response or antagonism of pentobarbital (KALIVAS 1982).

TASAKA et al. (1989) found that i.c.v. administration of HA (0.5–5 μg) suppressed the increase in power in the low-frequency bands (0–6 Hz) induced by low-frequency electrical stimulation of the midbrain reticular formation of conscious rats. This activating effect was antagonized by i.c.v. mepyramine (5.1 μg) or diphenhydramine (4.6–15 μg), while cimetidine (11.4 μg) or ranitidine (14.2 μg) were ineffective in this respect. Injection of HA (5–60 μg) into the ventrolateral posterior hypothalamus of the cat, where HA immunoreactive neurons have been identified, also caused an increase in wakefulness (W) and a decrease in deep SWS and REM sleep. The effects of HA were blocked by a small dose of mepyramine (1 mg/kg) (LIN et al. 1988).

All these findings support the postulate that HA may have a physiological function in modulating arousal.

Histamine has also been microinjected into the nucleus accumbens of the rat, a structure that plays an important role in the control of motor activity (BRISTOW and BENNETT 1988a). H_2 and H_3 receptors are particularly abundant in the rat's nucleus accumbens, while H_1 receptors are scarcely labeled (YANAI et al. 1990; SCHWARTZ et al. 1991). Microinjection of HA (1–200 μg) into the nucleus accumbens induced dose-dependent, biphasic changes in rat motor activity, which comprised initial hypoactivity followed by marked hyperactivity that was attenuated by mepyramine (10 μg). The hyperactivity response was also found after administration of 2-thiazolylethylamine (100–200 μg), but not after administration of dimaprit (20–50 μg). The H_3 receptor agonist, AMH, induced only a decrease in spontaneous motor activity (BRISTOW and BENNETT 1988b). All these results tend to indicate that hyperactivity after HA injection into the nucleus accumbens is mediated by H_1 receptors, while hypoactivity could depend on the inhibitory H_3 receptor.

Microinjection of HA (1 μg) into the caudal hippocampus of the rat inhibited locomotor and rearing activity, two behavioral parameters associated with exploratory behavior in the rat (ALVAREZ and GUERRA 1982).

2. Histamine Synthesis Inhibition

α-Fluoromethylhistidine (α-FMH) is a highly specific, irreversible inhibitor of HD (KOLLONITSCH et al. 1978). α-FMH does not inhibit other decarboxylases, such as dopa and glutamate decarboxylase, or the HA-metabolizing enzyme HA-*N*-methyltransferase. Single injections of α-FMH (10–50 mg/kg) produce an 80%–95% inhibition of HD activity at central sites within 90–120 min. Restoration of HD activity to control levels occurs within 3 days in the hypothalamus, but is incomplete in cerebral cortex after 4 days (GARBARG et al. 1980).

In the studies by MONTI et al. (1985, 1988), where rats were given α-FMH 50 mg/kg, i.p., and recorded for 3 days, 10 h/day in the light period, wakefulness (W) and light sleep (high-voltage slow waves interrupted by low-voltage fast EEG activity) were significantly decreased, while SWS and REM sleep showed opposite changes.

In rats housed under 16 h light/8 h dark compared with 12 h light/12 h dark conditions, W increased significantly during the dark period (HALPERIN et al. 1981). Under these conditions α-FMH decreased W and increased SWS during the dark period (MONTI et al. 1988). In the study by KIYONO et al. (1985), where rats were given α-FMH 100 mg/kg early in the light period and recorded for 24 h, W and NREM sleep (light sleep + SWS) were increased during the dark period, while REM sleep was augmented only during the light phase. The results of this study are somewhat surprising in view of the rapid decrease in HA levels to less than 10% that of controls during the first 6 h after α-FMH injection.

Bilateral injection of α-FMH (50 μg) into the ventrolateral posterior hypothalamus of the cat resulted in a significant increase in deep SWS and decrease in W, without a change in light sleep or REM sleep; these effects appeared as early as the 2nd h after injections (LIN et al. 1988). Similar results were obtained when α-FMH (20 mg/kg) was given intraperitoneally, the effect being apparent early after drug administration.

3. H_1 Receptor Agonists and Antagonists

The relatively specific H_1-receptor agonist, 2-thiazolylethylamine (64.5–258 μg dose), dependently increased W and decreased NREM sleep and REM sleep when delivered by the i.c.v. route in rats prepared for chronic sleep recordings. The H_1 receptor antagonist mepyramine (1–2 mg/kg), prevented both the increase in W and decrease in SWS (MONTI et al. 1986).

The H_1 receptor antagonists, mepyramine (1–5 mg/kg), diphenhydramine (1.6–10 mg/kg) and chlorpheniramine (10 mg/kg), decreased W and REM sleep and increased NREM sleep in rats and dogs (MONTI et al. 1985,

1986; WAUQUIER 1983). A decrease in W and increase in SWS in cats followed bilateral injection of mepyramine (120 μg) into the ventrolateral posterior hypothalamus. Although mepyramine has anticholinergic and membrane-stabilizing effects, the results obtained after direct administration into the mammillary nucleus do not seem to be related to nonspecific mechanisms, since neither atropine nor tetracaine reproduced the actions of the H_1 receptor antagonist on sleep variables (LIN et al. 1988).

Astemizole, terfenadine or loratadine (3 mg/kg, p.o.) given to cats and dogs at doses similar to those effective as antihistamines in man were devoid of any significant effect on the sleep-wakefulness pattern (ONGINI et al. 1987). Even slightly larger amounts of astemizole (10 mg/kg, p.o.) did not alter sleep variables in dogs (WAUQUIER 1983; WAUQUIER et al. 1981, 1984). On the other hand, doses of astemizole or terfenadine in the order of 30 mg/kg p.o. reduced SWS and/or REM sleep in cats (ONGINI et al. 1987).

The lack of an effect of astemizole, terfenadine or loratadine on sleep-waking patterns of laboratory animals after doses similar to those effective in man could be related to their poor penetration through the blood-brain barrier and weaker affinity for central H_1 receptors (NIEMEGEERS et al. 1986; AHN and BARNETT 1986).

Conflicting results have been reported on the effects of H_1 antagonists on sleep in man.

Methapyrilene 25 mg with 0.25 mg scopolamine was administered to insomniac subjects (KALES et al. 1971). This dose was found not to produce any favorable effects in terms of inducing sleep. Similarly, mepyramine (50–100 mg) given to healthy subjects did not influence W during sleep or total sleep time (NICHOLSON 1985).

BORBELY and YOUMBI-BALDENER (1988) found no significant differences in a series of subjective sleep parameters including sleep latency, number of nocturnal awakenings and duration of wake time between a group of healthy adults who received diphenhydramine (50–75 mg) before bedtime and those who received placebo.

In another study, which was double-blind and placebo controlled, healthy volunteers received diphenhydramine 50 mg before bedtime. The compound did not modify sleep induction or maintenance compared with baseline. However, in regard to sleep architecture, diphenhydramine significantly increased stage 4 NREM sleep and REM sleep latency. Topographic pharmaco-EEG showed a decrease in total power and alpha activity as well as an increase in theta and slow beta activity, and an acceleration in the centroid of the combined delta-theta activities (SALETU et al. 1987, 1988).

In another study, RICKELS et al. (1983) found diphenhydramine 50 mg to improve various sleep parameters including sleep latency, number of awakenings and duration, depth and quality of sleep in a group of moderately insomniac patients.

Diphenhydramine was also evaluated in patients with a diagnosis of insomnia and generalized anxiety or major depression (KUDO and KURIHARA

1990). The H_1 receptor antagonist improved the severity of insomnia after doses of 12.5–50 mg administered for a 2-week period. Dose-dependent effects were found only in those patients who had not been previously treated for insomnia.

Moreover, ROEHRS et al. (1984), GENGO et al. (1989) and ROTH et al. (1992) observed that diphenhydramine in doses of 25–100 mg given during daytime to normal volunteers produced significant feelings of drowsiness, mental impairment, performance decrements in a divided attention and complex reaction time task, and a decrease in the mean latency to stage 1 evaluated by the multiple sleep latency test (MSLT).

The H_1 antagonists, tripolidine (10–20 mg), brompheniramine (4–8 mg) or promethazine (20 mg), given to healthy adults or poor sleepers did not significantly modify wake time after sleep onset or total sleep time but reduced the percentage of REM sleep (NICHOLSON et al. 1985; ADAM and OSWALD 1986).

It can be concluded that H_1 receptor antagonists which cross the blood-brain barrier have little effect on W in sleep laboratory controlled studies, when given before bedtime to normal subjects. However, in equal doses given during the day, they induce drowsiness, an increased tendency to sleep and impaired performance.

Sleep laboratory studies have shown that H_1 receptor antagonists whose structure is incompatible with the ability to easily cross the blood-brain barrier, including astemizole (10–20 mg), cetirizine (10–20 mg), temelastine (100 mg), terfenadine (60–120 mg) and loratadine (10–40 mg), are devoid of sleep-inducing or -maintaining properties (NICHOLSON et al. 1985; ADAM and OSWALD 1986; HINDMARCH and BHATTI 1987; SEIDEL et al. 1990). Moreover, terfenadine (120 mg) did not induce daytime sleepiness in healthy volunteers as measured by the MSLT (MURRI et al. 1991).

Analysis of the EEG spectrum in healthy male volunteers showed terfenadine to increase slow waves and to decrease activity in the alpha band. On the other hand, no variation in spectral parameters was observed after administration of cetirizine or loratadine (PECHADRE et al. 1988; RAMAEKERS et al. 1992).

4. H_2 Receptor Agonists and Antagonists

Following i.c.v. administration of either the H_2 receptor agonist, dimaprit, or the H_2 receptor antagonists, metiamide, cimetidine or ranitidine, hypersynchronous electrocortical activity was induced in the frontal and occipital areas of the rat cortex. The cortical discharges showed sustained high-voltage spikes which lasted 10 s to 5 min. These electrographic changes were not accompanied by motor manifestations (MONTI et al. 1986). When cimetidine was given by the i.p. route, sleep and W values showed slight but inconsistent changes which were not significant. It is possible that factors other than those related to H_2 receptors contribute to effects of these drugs on rat cortical EEG activity.

The novel benzthiazole derivative, zolantidine, which is a potent antagonist of H_2 receptors, penetrates the brain (CALCUTT et al. 1988). Zolantidine (0.25–8.0 mg/kg) had no significant effects on sleep parameters in rats implanted for chronic sleep recordings (MONTI et al. 1990). Although these results tend to suggest that H_2 receptors are not involved in sleep regulation, the absence of selective, brain-penetrating H_2 receptor agonists leaves the matter of their effects on sleep and W unresolved.

Cimetidine (200–400 mg) or ranitidine (150–300 mg) given to healthy volunteers did not affect wake time after sleep onset, total sleep time or REM sleep. However, cimetidine increased SWS and the number of movements during sleep (NICHOLSON 1985; NICHOLSON et al. 1985).

Although the CSF concentration of ranitidine is lower than the corresponding CSF concentration of cimetidine, the CSF/plasma ratios of these drugs are almost equal (JONSSON et al. 1984). Thus, the difference in effects on sleep architecture cannot be due to small differences in CSF concentration. Tentatively, it could be related to interaction with other receptor types including serotonin and GABA-benzodiazepine-binding sites (LAKOSKI and AGHAJANIAN 1983; LAKOSKI et al. 1983).

5. H_3 Receptor Agonists and Antagonists

The effect of the H_3 receptor agonist, AMH, was compared with that of the H_3 receptor antagonist, thioperamide, in freely moving rats and cats. R-α-methylhistamine (1–4 μg) injected bilaterally into the premammillary area of the rat, where HA immunoreactive neurons are located, increased SWS, whereas W and REM sleep were decreased. No significant effects were observed when AMH (1–8 mg/kg) was administered intraperitoneally, which could be related to its poor penetration through the blood-brain barrier (MONTI et al. 1990, 1991; STARK et al. 1992).

Thus, it is puzzling that oral administration of the H_3 receptor agonist (10–20 mg) caused a significant increase in deep SWS in cats (LIN et al. 1990).

Thioperamide (1–20 mg/kg) increased W and decreased SWS and REM sleep in rats and cats. Pretreatment with thioperamide (4 mg/kg) prevented the effect of AMH (2 μg) on SWS and W in rats (MONTI et al. 1991). Moreover, the arousal effect of thioperamide (2 mg/kg) was prevented by pretreatment with AMH (20 mg/kg) or mepyramine (1 mg/kg) in cats (LIN et al. 1990).

Results obtained after administration of the H_3-receptor agonist or antagonist have been ascribed to a decrease and increase in HA release, respectively. The attenuation of the thioperamide-induced increase in the waking state by mepyramine is in keeping with results obtained after i.c.v.

HA administration (MONNIER et al. 1970) and indicates its dependence on the activation of the H_1 receptor.

H. Summary

Neuroanatomical, neurochemical and neuropharmacological evidence presently indicates a role for HA in the control of the waking state. The known neuroanatomical connections of the histaminergic pathways resemble those of the ascending noradrenergic and serotonergic components of the reticular activating system. Also, the arousing effect of i.c.v. HA administration indicates an important role for HA in the regulation of this system as a major determinant of the waking state. This is further supported by recent findings in which 2-thiazolylethylamine, the predominantly H_1-receptor agonist, and thioperamide, the H_3-receptor antagonist, increase W while the HA synthesis inhibitor, α-FMH, the H_1-receptor antagonists, mepyramine, diphenhydramine and chlorpheniramine, and the H_3-receptor agonist, AMH, produce opposite effects (Table 4).

It has been proposed that HA may act to modulate REM sleep, such that inhibition of HA functional activity would be followed by increased amounts of REM sleep. Histamine synthesis inhibition is related to a REM sleep increase. However, stimulation or blockade of the H_1- or H_3-receptor suppresses REM sleep. These seemingly conflicting sets of data could be related to the lack of specificity of drugs which modify HA transmission. However, experimental manipulations involving direct interactions with receptors may not necessarily have the same consequences for REM sleep as would manipulations that result in reduced HA synthesis.

Table 4. Effects on sleep and waking of drugs influencing the Histaminergic system in the brain

Compound	Effect on sleep and waking
α-Fluoromethylhistidine	Decrease in W and LS Increase in SWS and REM sleep
2-Thiazolylethylamine	Increase in W Decrease in SWS and REM sleep
Mepyramine, diphenhydramine, chlorpheniramine	Decrease in W and REM sleep Increase in SWS
Zolantidine	No effects on sleep or W
(R)-α-Methylhistamine	Decrease in W and REM sleep Increase in SWS
Thioperamide	Increase in W Decrease in SWS and REM sleep

References

Adam K, Oswald I (1986) The hypnotic effects of an antihistamine: promethazine. Br J Clin Pharmacol 22:715–717

Ahn HS, Barnett A (1986) Selective displacement of ^{3}H-mepyramine from peripheral vs. central nervous system receptors by loratadine, a non-sedating antihistamine. Eur J Pharmacol 127:153–155

Airaksinen MS, Panula P (1988) The histaminergic system in the guinea-pig central nervous system: an immunocytochemical mapping using an antiserum against histamine. J Comp Neurol 273:163–186

Airaksinen MS, Flugge G, Fuchs E, Panula P (1989) The histaminergic system in the tree shrew brain. J Comp Neurol 286:289–310

Alvarez EO, Guerra FA (1982) Effects of histamine microinjections into the hippocampus on open-field behavior in rats. Physiol Behav 28:1035–1040

Armstrong WE, Sladek CD (1985) Evidence for excitatory actions of histamine on supraoptic neurons in vitro: mediation by an H_1-type receptor. Neuroscience 16:307–322

Arrang JM, Garbarg M, Schwartz JC (1983) Autoinhibition of brain histamine release mediated by a novel class (H_3) of histamine receptor. Nature 302:832–837

Arrang JM, Garbarg M, Lancelot JC, Lecomte JM, Pollard H, Robba M, Schunack W, Schwartz JC (1987) Highly potent and selective ligands for histamine H_3-receptors. Nature 327:117–123

Arrang JM, Roy J, Morgat JL, Schunack W, Schwartz JC (1990) Histamine H^3 receptor binding sites in rat brain membranes: modulations by guanine nucleotides and divalent cations. Eur J Pharmacol 188:219–227

Arrang JM, Garbarg M, Schwartz JC (1992) H_3-receptor and control of histamine release In: Schwartz JC, Haas HL (eds) The histamine receptor. Wiley-Liss, New York, p 145

Ash ASF, Schild HO (1966) Receptors mediating some actions of histamine. Br J Pharmacol Chemother 27:427–439

Atack C, Carlsson A (1972) In vitro release of endogenous histamine, together with noradrenaline and 5-hydroxytryptamine, from brain slices of mouse cerebral hemispheres. J Pharmacol 24:990–992

Barger G, Dale HH (1910) The presence in ergot and physiological activity of β-imidazolylethylamine. J Physiol 40:38–40

Best CG, Dale HH, Dudley HW, Thorpe WV (1927) The nature of the vasodilator constituents of certain tissue extracts. J Physiol 62:397–417

Black JW, Duncan WAM, Durant CJ, Ganellin CR, Parsons ME (1972) Definition and antagonism of histamine H_2-receptors. Nature 236:385–390

Bleier R, Cohn P, Siggelkow TR (1979) A cytoarchitectonic atlas of the hypothalamus and hypothalamic third ventricle of the rat. In: Morgane PJ, Panksepp J (eds) Handbook of the hypothalamus. Anatomy of the hypothalamus. Dekker, New York, p 137

Borbély AA, Youmbi-Baldener G (1988) Effect of diphenhydramine on subjective sleep parameters and on motor activity during bedtime. Int J Clin Pharmacol Ther Toxicol 26:392–396

Bouthenet ML, Rual M, Salès N, Schwartz JC (1988) A detailed mapping of histamine H_1-receptors in guinea-pig central nervous system established by autoradiography with [^{125}I]iodobolpyramine. Neuroscience 26:553–600

Bovet D (1950) Introduction to antihistamine agents and antergan derivatives. Ann NY Acad Sci 50:1089–1126

Bovet D, Kohn R, Mardita M, Silvestrini B (1958) Some effects of histamine on normal and Haemophilus pertussis vaccinated rat. Br J Pharmacol Chemother 13:74–83

Bristow LJ, Bennett GW (1988a) Biphasic effects of intra-accumbens histamine administration on spontaneous motor activity in the rat; a role for central histamine receptors. Br J Pharmacol 95:1292–1302

Bristow LJ, Bennett GW (1988b) A role for histamine H_3 receptors in histamine induced hypoactivity in the rat. Br J Pharmacol 94:319P
Brownstein MJ (1981) Serotonin, histamine and the purines. In: Siegel GJ, Albers RM, Agranoff BW, Katzman R (eds) Basic neurochemistry. Little Brown, Boston, p 224
Calcutt CR, Ganellin CR, Jackson B, Leigh BK, Owen DAA, Smith IR (1987) Evidence for low brain penetration by the H_1-receptor antagonist temelastine (SK&F 93944). Eur J Pharmacol 133:65–74
Calcutt CR, Ganellin CR, Griffiths R, Leigh BK, Maguire JP, Mitchel RC, Mylek ME, Parsons ME, Smith IR, Young RC (1988) Zolantidine (SK&F 95282) is a potent selective brain-penetrating histamine H_2-receptor antagonist. Br J Pharmacol 93:69–78
Chronister RB, Palmer GC, Defrance JK, Sikes RW, Hubbard JI (1982) Histamine: correlative studies in nucleus accumbens. J Neurobiol 13:23–37
Dale HH, Laidlaw PP (1910) The physiological action of β-imidazolylethylamine. J Physiol 41:318–344
Ericson H, Watanabe T, Kohler C (1987) Morphological analysis of the tuberomammillary nucleus in the rat brain: delineation of subgroups with antibody against L-histidine decarboxylase as a marker. J Comp Neurol 263:1–24
Fukui H, Watanabe T, Wada H (1980) Immunochemical cross reactivity of the antibody elicited against L-histidine decarboxylase purified from the whole bodies of fetal rats with the enzyme from rat brain. Biochem Biophys Res Commun 93:333–339
Ganellin CR (1982) Chemistry and structure-activity relationships of drugs acting at histamine receptors. In: Ganellin CR, Parsons ME (eds) Pharmacology of histamine receptors. Wright-PSC, Bristol, p 10
Ganellin CR (1992) Pharmacochemistry of H_1 and H_2 receptors. In: Schwartz JC, Haas HL (eds) The histamine receptor. Wiley-Liss, New York, p 1
Garbarg M, Barbin G, Rodergas E, Schwartz JC (1980) Inhibition of histamine synthesis in brain by α-fluoromethylhistidine, a new irreversible inhibitor: in vitro and in vivo studies. J Neurochem 35:1045–1052
Gengo F, Gabos C, Miller JK (1989) The pharmacodynamics of diphenhydramine-induced drowsiness and changes in mental performance. Clin Pharmacol Ther 45:15–21
Gerber U, Greene RW, Haas HL, McCarley RW (1990) Excitation of brain stem neurones by noradrenaline and histamine. J Basic Clin Pharmacol 1:71–76
Goldstein L, Pfeiffer CC, Muñoz C (1963) Quantitative EEG analysis of the stimulant properties of histamine and histamine derivatives. Fed Proc 22:424
Greene RW, Haas HL (1990) Effects of histamine on dentate granule cells in vitro. Neuroscience 34:299–303
Groenewegen HJ, Van Dijk C (1984) Efferent connections of the dorsal tegmental region in the rat, studied by means of anterograde transport of the lecitin Phaseolus vulgaris leucoagglutinin (PHA-L). Brain Res 455:170–176
Haaksma EEJ, Leurs R, Timmerman H (1990) Histamine receptors: subclasses and specific ligands. Pharmacol Ther 47:73–104
Haas HL (1979) Histamine and noradrenaline are blocked by amitryptiline on cortical neurones. Agents Actions 9:83–84
Haas HL (1985) Histamine. In: Rogawski MA, Barker JL (eds) Neurotransmitter actions in the vertebrate nervous system. Plenum, New York, p 321
Haas HL (1992) Electrophysiology o histamine-receptors. In: Schwartz JC, Haas HL (eds) The histamine receptor. Wiley-Liss, New York, p 161
Haas HL, Bucher UM (1975) Histamine H_2-receptors on single central neurons. Nature 255:634–635
Halperin JM, Miller D, Iorio LC (1981) Sleep-inducing effects of three hypnotics in a new model of insomnia in rats. Pharmacol Biochem Behav 14:811–814

Harris GW, Jacobson D, Kahlson G (1952) The occurrence of histamine in the cerebral regions related to the hypophysis. In: Wolstenholme GEW (ed) CIBA Foundation colloquia on endocrinology. Blakiston, New York, p 186
Hegstrand LR, Kanof PD, Greengard P (1976) Histamine sensitive adenylate cyclase in mammalian brain. Nature 260:163–165
Hill SJ, Donaldson J (1992) The H_1-receptor and inositol phospholipid hydrolysis. In: Schwartz JC, Haas HL (eds) The histamine receptor. Wiley-Liss, New York, p 109
Hill SJ, Young JM (1980) Histamine H_1-receptors in the brain of the guinea-pig and the rat: differences in ligand binding properties and regional distribution. Br J Pharmacol 68:687–696
Hill SJ, Emson PC, Young JM (1978) The binding of [^{3}H]mepyramine to histamine H_1 receptors in guinea-pig brain. J Neurochem 31:997–1004
Hindmarch I, Bhatti JZ (1987) Psychomotor effects of astemizole and chlorpheniramine, alone and in combination with alcohol. Int Clin Psychopharmacol 2:117–119
Hough LB, Khandelwal JK, Green JP (1984) Histamine turnover in regions of rat brain. Brain Res 291:103–109
Janssens F, Jansen MAC, Awouters F, Niemegeers CJE, Bussche GV (1986) Chemical development of astemizole-like compounds. Drug Dev Res 8:27–36
Johnson CL (1992) Histamine receptors and cyclic nucleotides. In: Schwartz JC, Haas HL (eds) The histamine receptor. Wiley-Liss, New York, p 129
Jonsson KA, Eriksson SE, Kagevi I, Norlander B, Bodemar G, Walan A (1984) No detectable concentrations of oxmetidine but measurable concentrations of cimetidine in cerebrospinal-fluid (CSF) during multiple dose treatment. Br J Clin Pharmacol 17:781–782
Kales JD, Tan T-L, Swearingen C, Kales A (1971) Are over-the-counter sleep medications effective? All night EEG studies. Curr Ther Res 13:143–151
Kalivas PW (1982) Histamine-induced arousal in the conscious and pentobarbital-pretreated rat. J Pharmacol Exp Ther 222:37–42
Kiyono S, Seo ML, Shibagaki M, Watanabe T, Maeyama K, Wada H (1985) Effects of α-fluoromethylhistidine on sleep-waking parameters in rats. Physiol Behav 34:615–617
Köhler C, Swanson L, Haglund L, Wu JY (1985) The cytoarchitecture, histochemistry and projections of the tuberomammillary nucleus in the rat brain. Neuroscience 16:85–110
Kollonitsch J, Patchett AA, Marburg S, Maycock AL, Perkins LM, Doldouras GA, Duggan DE, Aster SD (1978) Selective inhibitors of biosynthesis of aminergic neurotransmitters. Nature 274:906–908
Kudo Y, Kurihara M (1990) Clinical evaluation of diphenhydramine hydrochloride for the treatment of insomnia in psychiatric patients: a double-blind study. J Clin Pharmacol 30:1041–1048
Kwiatkowski H (1943) Histamine in nervous tissues. J Physiol 102:32–41
Lakoski JM, Aghajanian GK (1983) Effects of histamine H_1- and H_2-receptor antagonists on the activity of serotonin-containing neurons in the dorsal raphe. J Pharmacol Exp Ther 227:517–523
Lakoski JM, Aghajanian GK, Gallager DW (1983) Interaction of histamine H_2-receptor antagonists with GABA and benzodiazepine binding-sites in the CNS. Eur J Pharmacol 88:241–246
Lakoski JM, Gallagher DW, Aghajanian GK (1984) Histamine-induced depression of serotonergic dorsal neurons: antagonism by cimetidine, a reevaluation. Eur J Pharmacol 103:153–156
Lee NS, Fitzpatrick D, Meier E, Fisher H (1981) Influence of histidine on tissue histamine concentration, histidine decarboxylase and histamine methyltransferase activity in the rat. Agents Actions 11:307–311

Leschke C, Elz S, Schunack W (1992) Synthesis and H_1 agonistic properties of 2-(3-bromophenyl)histamine and 2-thienyl analogues. 12th International Symposium on Medicinal Chemistry, Basel
Lin JS, Sakai K, Jouvet M (1988) Evidence for histaminergic arousal mechanisms in the hypothalamus of cat. Neuropharmacology 27:111–122
Lin JS, Sakai K, Vanni-Mercier G, Arrang JM, Garbarg M, Schwartz JC, Jouvet M (1990) Involvement of histaminergic neurons in arousal mechanisms demonstrated with H_3-receptor ligands in the cat. Brain Res 523:325–330
Lin JS, Fort P, Kitahama K, Panula P, Denney RM, Jouvet M (1992) Immunohistochemical evidence for the presence of monoamine oxidase B in histamine-containing neurons in the posterior hypothalamus of cats. J Sleep Res 1 [Suppl] 1:581
Lipp R, Stark H, Schunack W (1992) Pharmacochemistry of H_3-receptors. In: Schwartz JC, Haas HL (eds) The histamine receptor. Wiley-Liss, New York, p 57
Maeda T, Kimura H, Nagai T, Imai H, Arai R, Sakumoto T, Sakai K, Kitahama K, Jouvet M (1984) Histochemistry of the magnocellular neurons in the posterior hypothalamus, with special reference to MAO activity and ability of 5-HT uptake and decarboxylation. Acta Histochem Cytochem (Jena) 17:179–183
Martinez-Mir Ml, Pollard H, Moreau J, Arrang JM, Ruat M, Traiffort E, Schwartz JM, Palacios JM (1990) Three histamine receptors (H_1, H_2 and H_3) visualized in the brain of human and non-human primates. Brain Res 526:322–327
Mazurkiewicz-Kwilecki IM, Prell GD (1984) Age-related changes in brain histamine. Agents Actions 14:554–557
McGinty DJ (1969) Somnolence, recovery and hyposomnia following ventromedial diencephalic lesions in the rat. Electroencephalogr Clin Neurophysiol 26:70–79
Monnier M, Sauer R, Hatt AM (1970) The activating effect of histamine on the central nervous system. Int Rev Neurobiol 12:265–305
Monti JM (1990) The histaminergic system: sleep and wakefulness. In: Meyer P, Elghozi J-L, Quera Salva A (eds) Etat de veille et de sommeil Masson, Paris, p 25
Monti JM, Pellejero T, Jantos H, Pazos S (1985) Role of histamine in the control of sleep and waking. In: Wauquier A, Gaillard JM, Monti JM, Radulovacki M (eds) Sleep – neurotransmitters and neuromodulators. Raven, New York, p 197
Monti JM, Pellejero T, Jantos H (1986) Effects of H_1- and H_2-histamine receptor agonists and antagonists on sleep and wakefulness in the rat. J Neural Transm 66:1–11
Monti JM, D'Angelo L, Jantos H, Pazos S (1988) Effects of α-fluoromethyl-histidine on sleep and wakefulness in the rat. J Neural Transm 72:141–145
Monti JM, Orellana C, Boussard M, Jantos H, Olivera S (1990) Sleep variables are unaltered by zolantidine in rats: are histamine H_2-receptors not involved in sleep regulation? Brain Res Bull 2:229–231
Monti JM, Jantos H, Boussard M, Altier H, Orellana C, Olivera S (1991) Effects of selective activation or blockade of the histamine H_3 receptor on sleep and wakefulness. Eur J Pharmacol 205:283–287
Murri L, Massetani R, Krause M, Iudice A (1991) Daytime sleepiness induced by antihistamines: a multiple sleep latency test (MSLT) study with terfenadine. Sleep Res 20A:165
Nagy J, LaBella L, Buss M, Daddona P (1984) Immunohistochemistry of adenosine deaminase: implications for adenosine neurotransmission. Science 224:166–168
Naquet R, Denavit M, Lanoir J, Albe-Fessard D (1965) Altérations transitoires ou définitives de zones diencéphaliques chez le chat. Leurs effects sur l'activité électrique corticale et le sommeil. In: Jouvet M (ed) Aspects anatomofonctionnels de la physiologie du sommeil. CNRS, Paris, p 107
Naquet R, Denavit M, Albe-Fessard D (1966) Comparaison entre le rôle du subthalamus et celui des différentes structures bulbomésencéphaliques dans le maintien de la vigilance. Electroencephalogr Clin Neurophysiol 20:149–164

Nauta WJH (1946) Hypothalamic regulation of sleep in rats. An experimental study. J Neurophysiol 9:285–316
Nicholson AN (1985) Histaminergic systems: daytime alertness and nocturnal sleep. In: Wauquier A, Gaillard JM, Monti JM, Radulovacki M (eds) Sleep – neurotransmitters and neuromodulators. Raven, New York, p 211
Nicholson AN, Pascoe PA, Stone BM (1985) Histaminergic systems and sleep. Studies in man with H_1 and H_2 antagonists. Neuropharmacology 24:245–250
Niemegeers CJE, Awouters F, Janssen PAJ (1986) The pharmacological profile of a specific, safe, effective and non-sedative anti-allergic, astemizole. Agents Actions 18:141–144
Oishi R, Nishibori M, Saeki K (1983) Regional distribution of histamine and tele-methylhistamine in the rat, mouse and guinea-pig brain. Brain Res 280:172–175
Oishi R, Nishibori M, Saeki K (1984) Regional differences in the turnover of neuronal histamine in the rat brain. Life Sci 34:691–699
Ongini E, Marzanatti M, Guzzon V (1987) Comparative effects of loratadine and selected antihistamines on sleep-waking patterns in the cat. Drug Dev Res 10:75–83
Onoe H, Yamatodani A, Watanabe Y, Ono K, Mochizuki T, Wada H, Hayaishi O (1992) Prostaglandin E_2 and histamine in the posterior hypothalamus. J Sleep Res 1 [Suppl 1]:166
Orr EL, Quay WB (1975) Hypothalamic 24-hour rhythm in histamine, histidine decarboxylase and histamine-N-methyltransferase. Endocrinology 96:941–945
Panula P, Yang HYT, Costa E (1984) Histamine-containing neurons in the rat hypothalamus. Proc Natl Acad Sci USA 81:2572–2576
Panula P, Häppölä O, Airaksinen MS, Auvinen S, Virkamäki A (1988) Carbodiimide as a tissue fixative in histamine immunocytochemistry and its application in developmental neurobiology. J Histochem Cytochem 36:259–269
Panula P, Pirvola U, Auvinen S, Airaksinen MS (1989) Histamine immunoreactive nerve fibers in the rat brain. Neuroscience 28:585–610
Panula P, Airaksinen MS, Pirvola U, Kotilainen E (1990) A histamine-containing neuronal system in human brain. Neuroscience 34:127–132
Parsons ME, Owen DAA, Ganellin CR, Durant GJ (1977) Dimaprit – [S-[3-(N,N-dimethylamino)propyl]isothiourea] – a highly specific histamine H_2-receptor agonist. I. Pharmacology. Agents Actions 7:31–37
Pechadre JC, Vernay D, Trolese JF, Bloom M, Dupont P, Rihoux JP (1988) Comparison of the central and peripheral effects of cetirizine and terfenadine. Eur J Clin Pharmacol 35:255–259
Phelan KD, Nakamura J, Gallagher JP (1990) Histamine depolarizes rat medial vestibular nucleus neurons recorded intracellularly in vitro. Neurosci Lett 109:287–292
Picaposte F, Blanco I, Palacios JM (1977) The presence of two cellular pools of rat brain histamine. J Neurochem 29:735–738
Pollard H, Bouthenet ML (1992) Autoradiographic visualization of the three histamine receptor subtypes in the brain. In: Schwartz JC, Haas HL (eds) The histamine receptor. Wiley-Liss, New York, p 179
Prell GD, Khandelwal JK, LeWitt PA, Green JP (1989) Rostral-caudal concentration gradients of histamine metabolites in human cerebrospinal fluid. Agents Actions 26:267–272
Ramaekers JG, Uiterwijk MMC, O'Hanlon JF (1992) Effects of loratadine and cetirizine on actual driving and psychometric test performance, and EEG during driving. Eur J Clin Pharmacol 42:363–369
Ranson SW (1939) Somnolence caused by hypothalamic lesion in the monkey. Arch Neurol Psychiatry 41:1–2
Reiner PB, McGeer EG (1987) Electrophysiological properties of cortically projecting histamine neurons of the rat hypothalamus. Neurosci Lett 73:43–47

Rickels K, Morris RJ, Newman H, Rosenfeld H, Schiller H, Weinstock R (1983) Diphenhydramine in insomniac family practice patients: a double-blind study. J Clin Pharmacol 23:235–242

Roehrs TA, Tiez EI, Zorick FJ, Roth T (1984) Daytime sleepiness and antihistamines. Sleep 7:137–141

Roman IJ, Kassem N, Gural RP, Herron J (1986) Suspension of histamine-induced wheel response by loratidine (SCH 29851) over 28 days in man. Ann Allergy 57:253–256

Roth T, Wittek TJ, Merlotti L, Roehrs TA, Fortier J, Riker DK (1992) Dose related effect of diphenhydramine on daytime sleepiness. Sleep Res 21:71

Ruat M, Traiffort E, Bouthenet ML, Schwartz JC, Hirschfeld J, Buschauer A, Schunack W (1990) Reversible and irreversible labeling and autoradiographic localisation of the cerebral histamine H_2-receptor using [^{125}I]iodinated probes. Proc Natl Acad Sci USA 87:1658–1662

Sakai K, El Mansari M, Lin JS, Zhang JG, Vanni-Mercier G (1990) The posterior hypothalamus in the regulation of wakefulness and paradoxical sleep. In: Mancia M, Marini G (eds) The diencephalon and sleep. Raven, New York, p 171

Saletu B, Grünberger J, Krupka M, Schuster P (1987) Comparative double-blind, placebo-controlled sleep laboratory studies of a combination of lorazepam and diphenhydramine (SM-1014) and its single components. Curr Ther Res 42:1037–1058

Saletu B, Grünberger J, Anderer P, Barbanoj MJ (1988) Pharmacodynamic studies of a combination of lorazepam and diphenhydramine and its single components: electroencephalographic brain mapping and safety evaluation. Curr Ther Res 44:909–937

Sallanon M, Aubert C, Denoyer M, Kitahama K, Jouvet M (1987) L'insomnie consécutive à la lésion de la région préoptique paramédiane est réversible par inactivation de l'hypothalamus postérieur chez le chat. C R Acad Sci [III] 305:561–567

Schlicker E, Betz R, Gothert M (1988) Histamine H_3 receptor-mediated inhibition of serotonin release in the rat brain cortex. Naunyn Schmiedebergs Arch Pharmacol 337:588–590

Schlicker E, Fink K, Hinterthaner M, Gothert M (1989) Inhibition of noradrenaline release in the rat brain cortex via presynaptic H_3-receptors. Naunyn Schmiedebergs Arch Pharmacol 340:633–638

Schwartz JC, Lampart C, Rose C (1970) Properties and regional distribution of histidine decarboxylase in rat brain. J Neurochem 17:1527–1534

Schwartz JC, Barbin G, Baudry M, Garbarg M, Martres MP, Pollard H, Verdiere M (1979) Metabolism and functions of histamine in the brain. In: Essman WB, Valzelli L (eds) Current developments in psychopharmacology. Spectrum, New York, p 173

Schwartz JC, Barbin G, Duchemin AM, Garbarg M, Pollard H, Quach TT (1981) Functional role of histamine in the brain. In: Palmer GC (ed) Neuropharmacology of central nervous system and behavioral disorders. Academic, New York, p 539

Schwartz JC, Garbarg M, Lebrecht U, Nowak J, Pollard E, Rodergas C, Rose C, Quach TT, Morgat JL, Roy J (1982) Histaminergic systems in brain: studies on localisation and actions. In: Uvnas B, Tasaka K (eds) Advances in histamine research. Pergamon, Oxford, p 71

Schwartz JC, Arrang JM, Garbarg M, Korner M (1986) Properties and roles of the three subclasses of histamine receptors in brain. J Exp Biol 124:203–224

Schwartz JC, Arrang JM, Garbarg M, Pollard H, Ruat M (1991) Histaminergic transmission in the mammalian brain. Physiol Rev 71:1–51

Seidel WF, Cohen S, Gourash Bliwise N, Dement WC (1990) Direct measurement of daytime sleepiness after administration of cetirizine and hydroxyzine with a

standardized electroencephalographic assessment. J Allergy Clin Immunol 86:1029–1033
Sorkin EM, Heel RC (1985) Terfenadine: a review of its pharmacodynamic properties and therapeutic efficacy. Drugs 29:34–56
Staines WA, Yamamoto T, Daddona PE, Nagy JI (1986) Neuronal colocalization of adenosine deaminase, monoamine oxidase, galanin and 5-hydroxytryptophan uptake in the tuberomammillary nucleus of the rat. Brain Res Bull 17:351–365
Stark H, Arrang JM, Garbarg M, Rouleau A, Lecomte JM, Lipp R, Schwartz JC, Schunack W (1992) Prodrugs of histamine H_3-agonists for improved drug penetration through blood-brain barrier. 12th International Symposium on Medicinal Chemistry, Basel
Steinbusch HWM, Mulder AH (1984) Immunohistochemical localization of histamine in neurons and mast cells in the rat brain. In: Björklund A, Hökfelt T, Kuhar MJ (eds) Handbook of chemical neuroanatomy. Classical transmitter receptors in the CNS, part II. Elsevier, Amsterdam, p 126
Steinbusch HWM, Mulder AH (1985) Localization and projections of histamine-immunoreactive neurons in the central nervous system of the rat. Adv Biosc 51:119
Swanson LW, Cowan WM (1979) The connections of the septal region in the rat. J Comp Neurol 186:621–656
Swett CP, Hobson JA (1968) The effects of posterior hypothalamic lesions on behavioral and electrographic manifestations of sleep and waking in cats. Arch Ital Biol 106:279–283
Tasaka K, Chung YH, Sawada K (1989) Excitatory effect of histamine on EEGs of the cortex and thalamus in rats. Agents Actions 27:127–130
Taylor JE, Yaksh TL, Richelsen E (1982) Histamine H_1-receptors in the brain and spinal cord of the cat. Brain Res 243:391–394
Timmerman H (1989) Histamine receptors in the central nervous system. Pharm Weekbl 11:146–150
Tuomisto L, Tuomisto J (1982) Diurnal variations in brain and pituitary histamine and histamine-N-transferase in the rat and guinea pig. Med Biol 60:204–209
Vanni-Mercier G, Sakai K, Salvert D, Jouvet M (1985) Waking-state specific neurons in the posterior hypothalamus of the cat. In: Koella WP, Rüther E, Schulz H (eds) Sleep '84. Fischer, Stuttgart, p 238
Villemagne VL, Dannals RF, Sánchez-Roa PM, Ravert HT, Vazquez S, Wilson AA, Natarajan TK, Wong DF, Yanai K, Wagner HN (1991) Imaging histamine H1 receptors in the living human brain with carbon-11-pyrilamine. J Nucl Med 32:308–311
Vincent S, Hökfelt T, Skirboll L, Wu JI (1983) Hypothalamic γ-aminobutyric acid neurons project to the neocortex. Science 220:1309–1310
Von Economo C (1926) Die Pathologie des Schlafes. In: Berthe A, von Bergmann G, Embden G, Ellinger A (eds) Handbuch der normalen und pathologischen Psychologie, vol 17. Springer, Berlin, p 591
Watabe S, Ishikawa K (1984) Comparative study on the ontogenesis of the uptake and release of histamine and γ-aminobutyric acid in the rat hypothalamus. Biogenic Amines 1:329–339
Watanabe T, Taguchi Y, Shiosaka S, Tanaka J, Kubota H, Terano Y, Tohyama M, Wada H (1984) Distribution of the histaminergic neuron system in the central nervous system of rats; a fluorescent immunohistochemical analysis with histidine decarboxylase as a marker. Brain Res 295:13–25
Wauquier A (1983) Drug effects on sleep-wakefulness patterns in dogs. Neuropsychobiology 10:60–64
Wauquier A, Van den Broeck WAE, Awouters F, Janssen PAJ (1981) A comparison between astemizole and other antihistamines on sleep-wakefulness cycles in dogs. Neuropharmacology 20:853–859

Wauquier A, Van den Broeck WAE, Awouters F, Janssen PAJ (1984) Further studies on the distinctive sleep-wakefulness profiles of antihistamines (astemizole, ketotifen, terfenadine) in dogs. Drug Dev Res 4:617–625
White T (1959) Formation and catabolism of histamine in brain tissue in vitro. J Physiol (Lond) 149:34–42
Winders A, Vogt W (1907) Synthese des imidazolylethylamins. Ber Dtsch Chem Ges 40:3691–3695
Wolf P, Monnier M (1973) Electroencephalographic, behavioral and visceral effects of intraventricular infusion of histamine in the rabbit. Agents Actions 3:196
Wouterlood FG, Gaykema RPA (1988) Innervation of histaminergic neurons in the posterior hypothalamic region by medial preoptic neurons. Anterograde tracing with Phaseolus vulgaris leucoagglutinin combined with immunocytochemistry of histidine decarboxylase in the rat. Brain Res 455:170–176
Wouterlood FG, Steinbusch HWM, Luiten PGM, Bol JGJM (1987) Projection from the prefrontal cortex to histaminergic cell groups in the posterior hypothalamic region of the rat. Anterograde tracing with Phaseolus vulgaris leucoagglutinin combined with immunocytochemistry of histidine decarboxylase. J Histochem Cytochem 406:330–336
Yanai K, Yagi N, Watanabe T, Itoh M, Ishiwata K, Ido T, Matsuzawa T (1990) Specific binding of [^{3}H]pyrilamine to histamine H_1 receptors in guinea pig brain in vivo: determination of binding parameters by a kinetic four-compartment model. J Neurochem 55:409–420
Young RC, Mitchell RC, Brown TH, Ganellin CR, Griffiths R, Jones M, Rana KK, Saunders D, Smith IR, Sore NE, Wilks TJ (1988) Development of a new physicochemical model for brain penetration and its application to the design of centrally acting H_2-receptor histamine antagonists. J Chem Med 31:656–671
Zingel V, Elz S, Schunack W (1990) Histamine analogues. 33rd communication: 2-phenylhistamines with high histamine H_1-agonistic activity. Eur J Med Chem 25:673–680

CHAPTER 7
Pharmacology of the Cholinergic System

G. TONONI and O. POMPEIANO

A. Introduction

As long ago as 1931 HESS suggested that the parasympathetic or cholinergic system was involved in sleep promotion. Since then, the evidence for a role of the cholinergic system in the regulation of wakefulness (W) and sleep, particularly rapid eye movement (REM) sleep, has increased enormously and at the same time has become more specific. Reviewed in this chapter are: the biochemistry of the cholinergic system; the anatomy of cholinergic cell bodies and terminations in the central nervous system; the nature and localization of the various kinds of cholinergic receptors; and the postsynaptic effects of cholinergic receptor activation. After discussing the physiology of the cholinergic system in relation to sleep-waking states, the chapter reviews the effects on sleep of various manipulations of the cholinergic system, through systemic or local injections of agonists and antagonists, or through lesions of cholinergic cell groups. Also mentioned is the evidence linking acetylcholine, sleep, and some diseases of the brain. Two major functions subserved by the cholinergic system emerge from this evidence: an involvement in thalamic, cortical, and hippocampal activation, and a central role in the generation of REM sleep. In addition, we will briefly discuss the involvement of the cholinergic system in postural and vegetative control.

B. Biochemistry, Anatomy, and Physiology of the Cholinergic System

I. Biochemistry

1. Synthesis

Acetylcholine (ACh) is synthesized by the enzyme, choline acetyltransferase (ChAT) from acetyl coenzyme A (acetyl-CoA) and choline in the cytoplasm of cholinergic nerve terminals. ChAT is synthesized in the cell body and reaches nerve terminals by slow axoplasmic flow (2–17 mm/day in mammals). The enzyme is principally present as a soluble pool, but there is also a membrane-bound pool of ChAT whose function is as yet unclear. Acetyl-CoA is synthesized in the mitochondria, while choline comes from the blood as a dietary constituent or from the hydrolysis of ACh by the degradative enzyme acetylcholinesterase (AChE). More than 50% of choline from released Ach is reutilized.

2. Storage and Release

ACh is stored in electronlucent synaptic vesicles with ATP and a soluble protein called vesiculin (for a review see Zimmermann 1988). Such vesicles are released in a quantal fashion by exocytosis in response to depolarization of nerve terminals in the presence of calcium. There is a direct relationship between firing rate and ACh release. For example, ACh release in the cerebral cortex is maximal in W and REM sleep, when cortical activity is high, and minimal in slow-wave sleep (SWS), when cortical activity is low (Celesia and Jasper 1966; Szerb 1967; Jasper and Tessier 1971). Cholinergic nerve terminals contain many presynaptic autoreceptors (M_1 muscarinic receptors) and heteroreceptors that generally decrease ACh release. ATP, when coreleased from cholinergic terminals, is hydrolyzed to adenosine and participates in a negative feedback loop to inhibit further ACh and ATP release (see Agoston 1988 for a review). The amount of ACh synthesized and stored, moreover, influences the amount of ACh that can be released.

ACh can coexist and be coreleased with other potentially neuroactive molecules, such as vasoactive intestinal polypeptide (VIP), enkephalin, calcitonin gene-related peptide (CGRP), or galanin. In unitransmitter neurons, ACh generally acts as a classical ionotropic transmitter (fast opening of ionic gates in postsynaptic membranes). Multitransmitter cholinergic neurons, which release ACh and neuropeptides, also have a metabotropic action on target cells (long-lasting effects, primarily on metabolism). The release of neuropeptides from cholinergic multitransmitter neurons has a different stimulus frequency dependence than ACh release. Cholinergic multitransmitter neurons may therefore release different combinations of ACh and neuropeptides depending on the stimulus frequency.

At the concentrations of choline normally present in cholinergic terminals, ChAT is not saturated, and the availability of ACh for release is determined largely by the levels of choline coming from a specific reuptake system. In addition to the widespread low-affinity carrier for choline, cholinergic neurons have a specific high-affinity active transport system that transfers back choline into the presynaptic terminals. This sodium-dependent transport system seems to be the rate-limiting step controlling the level of ACh in the cell, although acetyl-CoA may also play some role. For a systematic review on this topic see, e.g., Tucek (1990).

II. Anatomy

1. Cholinergic Cells

Despite the vast body of knowledge on cholinergic synapses in the peripheral nervous system, for many years technical difficulties have confined the identification of cholinergic pathways in the central nervous system to the collaterals of motoneurons to Renshaw cells (POMPEIANO 1984). During the 1980s, new authoradiographic (WAMSLEY et al. 1981) and immunohistochemical (KIMURA et al. 1981; GEFFARD et al. 1985) methods were introduced to study systematically cholinergic receptors and neurons. It is generally accepted that detection of ChAT in a cell is a criterion sufficient to classify that cell as cholinergic. Monoclonal antibodies raised against ChAT (CRAWFORD et al. 1982; ECKENSTEIN and THOENEN 1982; LEVEY et al. 1983) have therefore been essential in clarifying the anatomy of the central cholinergic system in rat (ARMSTRONG et al. 1983; MESULAM et al. 1983a), cat (KIMURA et al. 1981; VINCENT and REINER 1987), and macaque monkey (MESULAM et al. 1984). ChAT-positive cells generally also contain the degradative enzyme AChE, and appear to be cholinergic if tested for choline uptake and ACh release. AChE staining, although often used previously to identify cholinergic cells (SHUTE and LEWIS 1967), is still valuable but clearly less reliable than ChAT, since several noncholinergic cell groups, particularly catecholaminergic and indolaminergic cells of the brain stem, are rich in AChE activity. Recently, it has been observed that cholinergic cells in the brain stem can be selectively stained by the NADPH-diaphorase technique for neurons containing nitric oxide synthase (VINCENT et al. 1983).

Central cholinergic neurons are found from the spinal cord to the forebrain. They can be subdivided into: (a) projection neurons, which include cells in the medial septum, diagonal band nuclei, nucleus basalis, medial habenula, parabigeminal nucleus, pedunculopontine and laterodorsal tegmental nuclei, cranial nerve nuclei, somatic motoneurons and preganglionic projection neurons in the spinal cord; and (b) cholinergic interneurons, which have been described in the caudate putamen, ventral striatum, cerebral cortex, hippocampal formation, amygdaloid complex, olfactory bulb, and hypothalamus (SALVATERRA and VAUGHN 1989). Even if they belong to physiologically different subsystems, all cholinergic neurons share some morphological and hodological characteristics probably fundamental to their ability to mediate complex behavior: together they innervate virtually every region of the brain, they receive input from areas of multimodal convergence as well as from systems related to homeostatic functions, and they are strongly interconnected among themselves in reentrant circuits (see WOOLF 1991 for a review). These characteristics provide the cholinergic system with the ability to mediate global functions in the brain. Here we focus on two cholinergic projection groups, the basal forebrain group and the pontomesencephalic group, that have been clearly implicated in the regulation of wakefulness and sleep.

2. Basal Forebrain Cholinergic Groups

In the telencephalon there are two subsystems of cholinergic cells, the striatal subsystem (islands of Calleja, olfactory tubercle, nucleus accumbens, and caudate-putamen complex), which consists of interneurons, and the projection neurons of the basal forebrain group. The latter contains, according to the classification (Ch1–6) proposed by MESULAM et al. (1983a,b; MUFSON et al. 1986), four subgroups of cholinergic neurons (Ch1–4). Ch1, the most rostral subgroup, corresponds to the medial septal nucleus. Ch2 corresponds to the vertical limb of the diagonal band of Broca, and Ch3 to the horizontal limb. In all these structures, a variable proportion of the cells (20%–60%) are cholinergic, and they are intermingled with GABAergic and peptidergic neurons (BRASHEAR et al. 1986; ONTENIENTE et al. 1987). They project principally to allocortical areas such as hippocampus, entorhinal, cingulate, retrosplenial and subicular cortex, to the amygdaloid complex, and to the olfactory bulb. Ch1–3 receive reciprocal projections from many of these areas; other afferents originate in the lateral septum, medial habenula, periventricular hypothalamic nucleus, ventral tegmental area, pedunculopontine and laterodorsal tegmental nucleus, dorsal and median raphe, and locus coeruleus (WOOLF 1991).

The cholinergic subgroup of the basal forebrain most directly involved in sleep regulation is Ch4. The anterior component of Ch4 in the monkey corresponds to the magnocellular preoptic area; the intermediate and posterior components correspond to the nucleus basalis (NB) and substantia innominata in the cat, and to the NB, substantia innominata, and nucleus ansa lenticularis in the rat (WOOLF 1991). The majority (90%) of cells in Ch4 are large-sized cholinergic neurons, projecting with a topographically organized pattern to the entire neocortex. The NB also projects to the reticular nucleus of the thalamus, but only a few of these projecting neurons are cholinergic (ASANUMA 1989). Cortical afferents to basal cholinergic neurons are still a matter of discussion. Ch1–4 receive some projections from cortical areas related to the limbic system (prepyriform cortex, orbitofrontal cortex, anterior insula, temporal pole, entorhinal cortex, and medial temporal cortex). Projections from many isocortical areas, in particular from primary sensory cortices, have not been unequivocally demonstrated. Ch4 receives a substance P input possibly from midbrain or upper pons, and a GABAergic input from either the striatum or the basal forebrain itself. Other afferents to Ch4 are from intralaminar nuclei of thalamus, zona incerta, ventral tegmental area, compact part of the substantia nigra, dorsal and medial raphe, pedunculopontine nucleus, and locus coeruleus. In many cholinergic neurons of the basal forebrain ACh colocalizes with galanin-like immunoreactivity.

3. Brain Stem Cholinergic Groups

There are four major brain stem cholinergic subgroups. The medial habenular nucleus (Ch7), located in the dorsomedial part of the thalamus, projects to the interpeduncular nucleus. The parabigeminal nucleus (Ch8) is located in the lateral part of the midbrain tegmentum. It projects to the superior colliculus (principally the contralateral one), and to the dorsal lateral geniculate nucleus. The other two subgroups, the pedunculopontine nucleus (PPT, Ch5) and the laterodorsal tegmental nucleus (LDT, Ch6), are particularly important for REM sleep generation.

The PPT extends from the caudal pole of the red nucleus to the parabrachial nucleus and is strictly associated with the ascending limb of the superior cerebellar peduncle. It includes a lateral compact part (PPTc), with a majority of cholinergic cells, and a medial, more diffuse part (PPTd), where the cholinergic cells are less abundant. In the rat the PPT consists almost exclusively of cholinergic neurons (RYE et al. 1987), while in the cat the peribrachial area contains a mixture of cholinergic, catecholaminergic, and other neurons (JONES and BEAUDET 1987). Caudally the medial PPTd merges with the LDT in the central gray matter. It should be noted that in Ch5–6 of rat, 15%–30% of cholinergic neurons also contain substance P, corticotropin-releasing factor, bombesin-gastrin-releasing peptide, and atriopeptin-like immunoreactivity (VINCENT et al. 1983; SAPER et al. 1985; STANDAERT et al. 1986). Bombesin and CRF may have a role in thalamo-cortical activation, because they can induce EEG and behavioral signs of arousal (SUTTON et al. 1982; EHLERS et al. 1983; RASLER 1984).

Together, Ch5–6 project to virtually all subcortical brain areas. These diffuse projections are in general not overlapping, because Ch6 project principally to medial, limbic-associated nuclei, and Ch5 to more lateral nuclei. In contrast with monoaminergic brain stem nuclei, Ch5–6 do not have direct cortical projections, save for a few exceptions such as LDT projections to infralimbic and anterior cingulate cortices (see WOOLF 1991). On the other hand, the abundant projections of Ch5–6 to the thalamus may be of primary importance in the process of thalamocortical activation during W and REM sleep. Ch5 has been shown to project to all thalamic nuclei, and Ch6 mainly to nuclei associated with the limbic system, such as the anteroventral, anteromedial, laterodorsal, habenular, mediodorsal, central medial, and ventromedial nuclei (DE LIMA and SINGER 1987; HALLANGER et al. 1987; PARÉ et al. 1988; SMITH et al. 1988; STERIADE et al. 1988). Some cholinergic neurons of Ch5–6 simultaneously innervate the central-lateral nucleus and the dorsolateral geniculate nucleus, possibly providing the anatomical basis for the simultaneous occurrence of cortical desynchronization and PGO waves during REM sleep (SHIROMANI et al. 1990b). The reticular thalamic nucleus is unique in that it receives a projection from both Ch4 and Ch5–6. Other projections of Ch5–6 include the hypothalamus, septum, substantia nigra, basal forebrain, basal ganglia, interpeduncular nucleus, raphe nuclei, locus coeruleus, sensory cranial nerve nuclei, reticular formation, and deep cerebellar nuclei.

Of special significance are the projections of Ch5–6 to the medial pontine reticular formation (Mitani et al. 1988; Shiromani et al. 1988b) and to the paramedian reticular nucleus of the caudal medulla (Shiromani et al. 1990a). In particular, dense fibers originating in PPT and LDT innervate the dorsal pontine tegmentum, including the locus coeruleus (LC), LCα, subcoeruleus, and lateral tegmental field. Through the tegmentoreticular tract, cholinergic fibers from the PPT/LDT reach the cholinosensitive area of the medullary inhibitory reticulospinal system (Lai and Siegel 1988), where they are involved in mediating the postural atonia of REM sleep. Cholinergic projections from the caudal part of the medial and descending vestibular nuclei reach the flocculonodular lobe of the cerebellum (Barmack et al., 1992a,b), while PPT/LDT project to corticocerebellar areas such as the anterior vermis (Cirelli et al. 1994).

The afferent projections to the LDT and PPT appear to originate mainly in the brain stem reticular formation, the midbrain central gray, and the lateral hypothalamus-zona incerta region. Some of the reticular connections are serotoninergic or noradrenergic, others are longitudinal or commissural cholinergic interconnections. Less prominent input originates in various limbic, oculomotor, extrapyramidal, and sensory structures (Semba and Fibiger 1992).

III. Cholinergic Receptors

ACh acts on target cells via two classes of cholinergic receptors, muscarinic (mAChRs) and nicotinic (nAChRs). mAChRs are selectively activated by muscarine and blocked by atropine, and nAChRs are activated by nicotine and blocked by curare.

1. Nicotinic Receptors

Because of their low density, strong evidence for the presence and function of central nicotinic synapses has been obtained at only two sites in the mammal CNS, Renshaw cells in the spinal cord and ganglion cells in the retina. However, many other central neurons are sensitive to nicotinic agonists and antagonists, implying the presence of nicotinic receptors. In the thalamus (Andersen and Curtis 1964; Phillis 1971), cerebral cortex (Stone 1972), cerebellar cortex (Andre et al. 1992a, 1993a), and the hypothalamic supraoptic nucleus (Dreifuss and Kelly 1972), nicotinic receptors are mixed with muscarinic receptors. Nicotinic receptors are also present on several types of presynaptic terminals, including cholinergic ones (Chesselet 1984) where, in contrast to presynaptic muscarinic receptors, they enhance ACh release.

2. Muscarinic Receptors

The first classical subsclassification of muscarinic receptors proposed by HAMMER and GIANCHETTI (1982) differentiated muscarinic M_1 and M_2 receptors on the basis of their high and low affinity for pirenzepine, respectively. More recently, the differential sensitivity to several antagonists has been used to pharmacologically define four muscarinic subtypes: M_1, M_2, M_3, and M_4 (for a review see HULME et al. 1990). Five different genes encoding proteins with characteristics of muscarinic receptor sites have been cloned and sequenced (see BONNER 1989 for a review). Their protein products are called *m1*, *m2*, *m3*, *m4*, and *m5*. The pharmacological M_1, M_2, and M_3 subtypes correspond to cloned *m1*, *m2*, and *m3* receptors (BUCKLEY et al. 1989). M_4 subtype probably corresponds to *m4* cloned receptor (MICHEL et al. 1989; VILARÓ et al. 1991). Even if each receptor subtype can couple with different effector systems, generally *m1*, *m3*, and *m5* gene products stimulate phosphoinositide hydrolysis, while *m2* and *m4* cloned receptors inhibit adenylate cyclase (BONNER et al. 1988; FUKUDA et al. 1988; PERALTA et al. 1988).

The distribution of the different muscarinic receptor subtypes is very heterogeneous. *m1*, *m3*, and *m4* subtype mRNAs exhibit a similar, prevalently telencephalic distribution (BUCKLEY et al. 1988). *m1* receptors are abundant in cerebral cortex, striatum, and hippocampus; *m3* is also present in several thalamic and brain stem nuclei; *m4* is prevalently expressed in the striatum, but is also present in cortex and hippocampus. Very low levels of *m5* transcripts are present in hippocampus and in some brain stem nuclei. M_2 receptors are prevalent in the brain stem and are associated throughout the brain with cholinergic cell groups and cholinergic projection areas (MASH and POTTER 1986; SPENCER et al. 1986; REGENOLD et al. 1989; WANG et al. 1989; LI et al. 1991; GITLER et al., 1992; VILARÓ et al. 1992). M_2 receptors on cholinergic neurons may function as autoreceptors, participating in the control of ACh release in brain (LAPCHAK et al. 1989a). M_2 receptors are also expressed in many noncholinergic neurons that receive cholinergic projections, such as in the olfacotry bulb, cerebral cortex, thalamic nuclei, amygdala, superior colliculus, and cerebellum, where they are probably located postsynaptically. Autoradiographic and in situ hybridization studies in the rat have demonstrated a high density of *m2* muscarinic receptors in the pons (MASH and POTTER 1986; SPENCER et al. 1986; VILARÓ et al. 1992), which is significant in view of their role in REM sleep generation.

IV. Postsynaptic Effects

The global impact of ACh on the cerebral cortex, the hippocampus, and the thalamus can be described as facilitatory with respect to ongoing synaptic transmission and magnitude of neuronal activity. The action of ACh on cortical neurons is complex, via both mAChRs and nAChRS, with both excitatory and inhibitory effects. The excitatory effect is mediated by M_1 receptors in part through a decrease in a voltage-dependent potassium conductance (M current) (HALLIWELL and ADAMS 1982; COLE and NICOLL 1983). In general, the resulting slow excitatory postsynaptic potential does not directly drive cortical neurons, but lowers the resting potential so that other excitatory inputs are more likely to reach threshold. This slow excitation is preceded by a fast inhibition, presumably indirect, via activation of γ-aminobutyric acid (GABA) interneurons mediated by M_2 receptors. Both mAChRs and nAChRS may be involved in these processes, because in cholinoceptive cortical neurons the two classes of cholinergic receptors are highly colocalized (VAN DER ZEE et al. 1992). ACh, together with norepinephrine (NE) and histamine (MCCORMICK and PRINCE 1985; MADISON and NICOLL 1986), also reduces the calcium-activated potassium conductance that underlies the afterhyperpolarization following a burst of action potentials in pyramidal cells, which has been assumed to play an important role in the generation of cortical slow waves (BUSZAKI et al. 1988). As discussed below, these cellular effects of the cholinergic (and noradrenergic) projections to the cortex are consistent with their role in cortical activation. The action of ACh on hippocampal neurons is not dissimilar from that on cortical neurons, and it consists largely of a direct excitatory effect coupled with a reduction of their inhibitory input (see VANDERWOLF 1988).

Cholinergic neurons increase thalamic excitability by directly exciting thalamocortical cells and by inhibiting GABAergic neurons in the reticular nucleus of the thalamus (STERIADE and MCCARLEY 1990). The excitation of thalamocortical cells is due to an early nicotinic component followed by a long-lasting muscarinic depolarization (CURRÓ DOSSI et al. 1990; STERIADE et al. 1991; MCCORMICK 1992). The net result of such depolarization is the switching between the oscillatory, burst mode of thalamic firing associated with SWS and the tonic, single-spike mode associated with cortical activation.

ACh tends to excite noncholinergic neurons in the pontine reticular formation (PRF), while it inhibits, presumably through autoreceptors, cholinergic neurons in the LDT/PPT. Two-thirds of PRF neurons are strongly depolarized by carbachol; the effect is direct, nonsynaptically mediated, and muscarinic in nature, because it can be blocked by atropine (GREENE et al. 1989a; GREENE and MCCARLEY 1990). This depolarization is associated with an increase in neuronal excitability as measured by an increased response to the same depolarizing current or to the electrical stimulation of the contralateral PRF. If applied to cholinergic neurons, cholinergic agonists produce the opposite effect, presumably through M_2 autoreceptors. In vitro, both PPT neurons (LEONARD and LLINAS 1989) and LDT neurons (LUEBKE et al. 1993) are hyperpolarized by cholinergic agonists through the activation of a non-M_1 receptor. In urethane-anesthetized rats ACh inhibits about half of the LDT neurons, and has no effects on the other ones (KOYAMA and KAYAMA 1992). For the effects of ACh on Purkinje cells in the cerebellum, see Sect. D.IV.

C. Cholinergic System and Sleep

I. Spontaneous Activity of Cholinergic Neurons

1. Nucleus Basalis

SZYMUSIAK and MCGINTY (1986; MCGINTY and SZYMUSIAK 1989) studied the spontaneous activity of the NB in relation to the sleep-waking cycle. The majority of the recorded neurons were more active during W and REM sleep, like neurons in most other brain areas. Twenty-four percent of the recorded neurons, on the other hand, increased their discharge rate selectively during sleep onset and tonically during SWS, some of them 10–15 s before the initial appearance of EEG synchrony. DETARI and VANDERWOLF (1987; DETARI et al. 1984) recorded basal forebrain cells projecting to the neocortex, and thus possibly cholinergic, with a firing rate fivefold higher during low-voltage fast-activity periods than during periods of large slow waves, while cells that did not project to the cortex were most active during SWS (DETARI et al. 1984; SZYMUSIAK and MCGINTY 1986). BUSZAKI et al. (1988) observed a negative correlation between the activity of NB neurons and the power of slow activity in the cortex. The neurochemical identity of neurons associated respectively with cortical activation and deactivation, however, is not conclusively established. On the basis of conduction velocity it seems more likely that sleep-active neurons were GABAergic and not cholinergic. The fact that more ACh is released during cortical activation (see Sect. C.II) also suggests that neurons active during W and REM sleep may be cholinergic. A recent study of the electrophysiological properties of basal forebrain neurons in guinea pig brain slices, however, opens up a different possibility (MUEHLTHALER et al. 1992). Identified cholinergic neurons from the substantia innominata have the ability to fire in a slow rhythmic bursting mode (2–6 Hz) when they are hyperpolarized (e.g., by serotonin and ACh) and in tonic mode (maximum 15 Hz) when they are depolarized (e.g., by NE and histamine) (MUEHLTHALER 1994).

2. LDT/PPT

According to several extracellular recording studies, in the dorsal pontine tegmentum of unanesthetized animals there is a subpopulation of neurons that fire at higher rates during W and REM sleep, when the EEG is desynchronized, than during SWS (EL MANSARI et al. 1989; STERIADE et al. 1990a; KAYAMA et al. 1992), and another subpopulation that fires most specifically during REM sleep. Many of these neurons fire tonically, but may show burst discharges in relation to PGO waves (see Sect. C.V.5). EL MANSARI et al. (1989) described two types of tonic neurons in freely moving cats. Tonic type I neurons showed a tonic pattern and higher rates of discharge slightly prior to and during both W and REM sleep, while tonic type II neurons showed a tonic pattern of discharge highly specific to periods of REM sleep. Some type I neurons (I-S) had a long spike duration, a low conduction velocity, and were inhibited by the local injection of carbachol, and may thus be cholinergic cells endowed with muscarinic inhibitory autoreceptors. STERIADE et al. (1990a), recording extracellularly in unanesthetized, chronically implanted cats, also described a group of PPT/LDT neurons that discharged at a higher rate during W and REM sleep, and another group that increased its firing rate from W to SWS to REM sleep. Similar findings have been reported by KAYAMA et al. (1992) in the LDT of undrugged rats.

These results do not prove conclusively the cholinergic nature of the recorded neurons. Intracellular recordings of LDT neurons in slices of rat brain using the whole-cell patch-clamp technique show that a class of neurons, identified as cholinergic by combining intracellular deposition of biocytin with NADPH diaphorase histochemistry, exhibit both single-spike mode and burst mode of firing depending upon their membrane potential (KAMONDI et al. 1992; see, however, LEONARD and LLINÁS 1987; KANG and KITAI 1990, who identified nonbursting neurons as cholinergic, and bursting neurons as noncholinergic). Although conclusive studies will require the recording of identified cholinergic neurons across behavioral states, it was suggested that during W, when noradrenergic and serotonergic discharge is high (see also Chaps. 4, 5, this volume), cholinergic neurons would be hyperpolarized and would fire only in short bursts. During REM sleep, when monoaminergic systems fall silent, they would fire tonically. Burst discharge would still be possible, but only at the offset of a long period of hyperpolarization due, for example, to inhibitory input from the substantia nigra pars reticulata (DATTA et al. 1991).

A new approach based on the expression of immediate early genes should also prove valuable in clarifying, with the use of double-labeling techniques, the neurochemical identity of neurons activated by sleep deprivation or which are particularly active during different sleep-waking states in regions that are known to contain cholinergic neurons (M. POMPEIANO et al. 1992, 1994). In a recent study, after the induction of prolonged REM-like episodes through carbachol microinjections into the dorsal pons, numerous neurons in that area showed Fos staining, and some of them, corresponding to the LDT/PPT, were identified immunohistochemically as cholinergic (SHIROMANI et al. 1992).

II. Spontaneous Release of ACh

The spontaneous release of ACh in different brain regions has been measured by cups placed over the cortex and filled with perfusion liquid containing neostigmine or eserine (see, for example, CELESIA and JASPER 1966) or by push-pull cannulae (COLLIER and MITCHELL 1967), and more recently by microdialysis (KODAMA et al. 1990; LYDIC et al. 1991a,b, 1992).

Early studies provided evidence that cortical release of ACh is related to the activity and arousal state of the animal. ACh release increases when the cortex is activated by electrical stimulation of the reticular formation and it is high during W and REM sleep and low in SWS (KANAI and SZERB 1965; CELESIA and JASPER 1966; COLLIER and MITCHELL 1967; SZERB 1967; PHILLIS et al. 1968; JASPER and TESSIER 1971). A similar variation in ACh release has been demonstrated in subcortical structures such as striatum (GADEA-CIRIA et al. 1973), hippocampus (KAMETANI and KAWAMURA 1990), thalamus (PHILLIS et al. 1968; WILLIAMS et al. 1994), and cerebral ventricles (HARANATH and VENKATAKRISHNA-BHATT 1973). More recently, by employing in vivo microdialysis, it has been shown that the release of ACh in the PRF is about two times higher during REM sleep than during SWS and W, and it begins to increase during the transition period from SWS to REM sleep (KODAMA et al. 1990). ACh release in the PRF increases also during carbachol-induced REM sleep-like states (LYDIC et al. 1991a,b), or after electrical stimulation of the PPT (LYDIC et al. 1992). During REM sleep ACh release is also increased in the nucleus paramedianus of the medulla (KODAMA et al. 1992), which has been implicated in mediating muscle atonia (see Sect. C.V.4).

III. Availability of Cholinergic Receptors and AChE

In addition to sleep-related changes in ACh release, some authors have demonstrated changes in the level of cholinergic receptors. In rats killed during different sleep waking states the number of muscarinic receptors is higher in REM sleep and in W than in SWS in a brain stem region including the dorsal pontine tegmentum (POMPEIANO and TONONI 1990). The number of muscarinic receptors stays constant in other brain areas, such as cortex or cerebellum, and their affinity does not vary (M. POMPEIANO et al. 1990). Other, indirect evidence links REM sleep with high levels of muscarinic receptors. BOEHME et al. (1984) and KILDUFF et al. (1986) described an increased number of muscarinic receptors in the medial PRF of narcoleptic dogs. SUTIN et al. (1986) found an increase in muscarinic receptors in several brain areas of rats, associated with an increase in REM sleep percentage, during the withdrawal from a 7-day chronic scopolamine treatment. FSL rats, a line of rats selectively bred for central-cholinergic hypersensitivity, exhibit an increase in REM sleep compared with control rats both in physiological conditions and after brief periods of sleep deprivation (SHIROMANI et al. 1988a, 1991). The increase in REM sleep observed in FSL rats seems to be due to an increased concentration of muscarinic receptors in several brain regions, including the pons, rather than to changes in levels or turnover of ACh (SCHILLER et al. 1989; SHIROMANI et al. 1991). On the other hand, the number of muscarinic receptors in rat cerebral cortex is reduced by REM sleep deprivation (SALIN-PASCUAL et al. 1992). Recently, some changes associated with REM sleep deprivation have also been demonstrated for the membrane-bound form of AChE in the brain stem (MALLICK and THAKKAR 1992).

IV. Lesions of the Cholinergic System

1. Lesions and Stimulation of Basal Forebrain

Several studies have dealt with the effects on cortical activity of basal forebrain lesions, but until a selective lesion of cholinergic cells is achieved, their interpretation remains difficult. Electrolytic lesions induce a reduction in cortical activity (LO CONTE et al. 1982). Large kainic acid lesions of the NB, substantia innominata, and neighboring areas (STEWART et al. 1984) produce slowing of cortical activity and loss of vigilance; small ibotenic acid lesions are associated with localized increased delta power (BUSZAKI et al. 1988). Some electrical or chemical stimulation experiments are consistent with lesion results, indicating short-latency activation of cortical areas (FUNAHASHI 1983; STERN and PUGH 1984), long-lasting desynchronization (BELARDETTI et al. 1977), and increased release of ACh in the cortex (CASAMENTI et al. 1986). Conversely, the release of ACh from the neocortex is reduced by the injection of muscimol (a GABA agonist) in the basal forebrain (CASAMENTI et al. 1986), where it is known to inhibit cortically projecting cholinergic cells (LAMOUR et al. 1986). Reduction of the activity of the magnocellular preoptic area of Ch4 also facilitated the induction of sleep (TICHO et al. 1988).

On the other hand, some authors have observed a disruption of SWS after lesion of the NB (SZYMUSIAK and MCGINTY 1986; MCGINTY and SZYMUSIAK 1989). Consistent with the latter results, there are sites within this region where electrical stimulation may induce SWS (STERMAN and CLEMENTE 1962; MCGINTY and SZYMUSIAK 1989; SZYMUSIAK and MCGINTY 1986).

2. Lesions of LDT/PPT

The rostral pontine tegmentum has long been implicated in both sustaining and facilitating cortical activation (MORUZZI 1972), and in the generation of REM sleep (JOUVET 1972). Because this area is now known to contain both cholinergic and noradrenergic cell bodies, it becomes essential to ascertain the specific contribution of cholinergic cells to both cortical activation and REM sleep. Although a sufficiently specific neurotoxin for cholinergic cells in the LDT/PPT has not yet been employed, JONES and WEBSTER (1988) and WEBSTER and JONES (1988) used kainic acid, which appears to be more toxic to cholinergic than to noradrenergic cells. Kainic acid lesions of the dorsolateral pontomesencephalic tegmentum disrupted REM sleep and accompanying signs for several weeks; within this area, the amount of residual REM sleep and the rate of pontogeniculo-occipital (PGO) waves were positively correlated with the number of residual cholinergic but not with that of noradrenergic neurons (WEBSTER and JONES 1988). However, such lesions did not affect waking or cortical activation, indicating that pontine cholinergic cells must be considered, alongside with noradrenergic cells, as only one component of the ascending reticular activating system. Another study employing bilateral electrolytic or radio frequency lesions of PPT (SHOUSE and SIEGEL 1992) confirmed the importance of this region for REM sleep and PGO waves, but indicated that it is not necessary for muscle atonia. On the other hand, cytotoxic lesions by quisqualate of the medial medullary reticular formation suggest a facilitatory role of cholinergic cells in this area in the motor inhibition associated with REM sleep (HOLMES and JONES 1994).

V. Pharmacological Manipulations of the Cholinergic System

Systemic and local injections of cholinergic agonists and antagonists offered the first indications that the cholinergic system is involved in the regulation of sleep and wakefulness (JOUVET 1975). Such injections have revealed a significant association between ACh and REM sleep on the one hand, and between ACh and arousal accompanied by cortical activation and hippocampal theta rhythm on the other hand.

In order to facilitate the review of pharmacological studies, a brief summary follows of the substances most frequently used. Carbachol is a mixed muscarinic and nicotinic agonist. Nicotine and mecamylamine are used as nicotinic agonists and antagonists, respectively. Arecoline, pilocarpine, RS 86, oxotremorine, and betanechol are muscarinic agonists, while atropine and scopolamine are muscarinic antagonists. McN 343-A is an M_1 agonist; pirenzepine and biperidene are antagonists. Oxotremorine-M and cismethyldioxolane are M_2 agonists, 4-DAMP and gallamine are antagonists (gallamine is also a nicotinic antagonist). Physostigmine (eserine), neostigmine, and galantamine are used as reversible AChE inhibitors, while di-isopropyl-fluorophosphate is irreversible. Hemicolinium is used as a reuptake inhibitor.

1. Cortical Activation

Cortical activation, which implies a readiness to respond in an effective way to incoming stimuli, has been traditionally associated with low-voltage fast activity (LVFA) or "desynchronized" EEG signals. In general, a desynchronized EEG is correlated with behavioral arousal, but there are several exceptions, the most notable being REM sleep. LVFA is associated with a tonic irregular pattern of firing in neocortical cells (and thalamic cells) depending on the relative depolarization of their membrane potential (STERIADE et al. 1990a). Conversely, synchronized sleep is characterized by two EEG rhythms: sleep spindles, around 7–14 Hz, and delta or slow waves, from 0.5 to 4 Hz, the latter associated with deeper sleep. Thalamo-cortical and corticothalamic projection neurons fire short, high-frequency bursts separated by long silent periods.

It is clearly established that cholinergic mechanisms play an important role in the production of LVFA in the neocortex. ACh and cholinergic agonists are able to induce LVFA when applied via several different routes, and these effects are antagonized by atropine (see Domino et al. 1968; Jouvet 1975 for references). Physostigmine can produce cortical desynchronization also during behavioral sleep. Intraventricular administration of hemicolinium-3, an inhibitor of choline uptake, or of atropine, a muscarinic antagonist, produces a decrease in LVFA in EEG during waking and also during REM sleep (e.g., Domino et al. 1968; Hazra 1970; Longo 1966; Henriksen et al. 1972).

On the other hand, LVFA can persist, under certain conditions, despite muscarinic blockade (atropine-resistant LVFA). Vanderwolf (1988) has distinguished, in rodents, between two types of behavior: voluntary or type 1 behavior is characterized by turning the head, walking, running, jumping, or manipulating objects with the forelimbs. Type 2 behavior includes alert immobility in any posture and automatic behavior such as shivering, licking, and face washing. In the cortex of a waking animal, LVFA is always present during type 1 behavior, and often present during type 2 behavior as well. Atropine abolishes LVFA during type 2 behavior but not during type 1 behavior. LVFA during type 1 behavior is also unaffected by lesions of the substantia innominata, which reduce the high-affinity uptake of choline and the release of ACh in the neocortex. Vanderwolf (1992) suggests that the cholinergic NB and the brain stem serotoninergic system are both involved in the activation of the neocortex. The atropine-sensitive component of LVFA is responsible for all LVFA during type 2 behavior, while during type 1 behavior an atropine-resistant, probably serotonergic system is also implicated. When both the cholinergic and serotoninergic system are pharmacologically blocked, large-amplitude, irregular slow waves appear in the electrocorticogram. The animals are aroused, but their behavior is severely disorganized, indicating that normal cortical functioning is impaired.

2. Hippocampal Theta Rhythm

Cortical LVFA is often associated with hippocampal rhythmic slow activity (RSA) or theta rhythm. The theta rhythm is a pattern of high-amplitude regular waves recorded in the hippocampal formation with a frequency of 4–7 Hz in the cat and 6–10 Hz in the rat. In rodents, it has been clearly related to states of brain activation, as during voluntary or type 1 behavior and during REM sleep. During type 2 behavior a large-amplitude irregular activity is generally present in the hippocampus. In some cases, however, RSA can be elicited by a variety of sensory stimuli even if no visible movements occur.

Systemic administration of cholinergic agonists such as eserine or pilocarpine in waking or anesthetized animals produces hippocampal RSA, and ACh induces RSA when injected in the carotid artery of anesthetized animals (STUMPF 1965). Conversely, systemically given antimuscarinic drugs such as atropine or scopolamine are able to abolish all RSA in anesthetized animals (e.g., KRAMIS et al. 1975; STEWART et al. 1984; LEUNG 1985) and RSA during type 2 behavior in waking animals (KRAMIS et al. 1975; VANDERWOLF 1975). However, the same antimuscarinic drugs that abolish RSA during anesthesia and type 2 behavior are ineffective on RSA during type 1 behavior. According to VANDERWOLF (1988), the atropine-sensitive RSA is dependent on a muscarinic activation of the hippocampus; the atropine-resistant input may be dependent on serotoninergic mechanisms.

Cholinergic agonists infused directly in the hippocampus also produce RSA, in both freely moving and in anesthetized animals, and they produce RSA-like waves even in hippocampal slices. Both muscarinic and nicotinic receptors have been described in the hippocampus, and hippocampal neurons are sensitive to iontophoretically applied ACh (see VANDERWOLF 1988 for references). The contribution of nicotine to the cholinergic type of RSA is not clear because systemic injections of nicotine produce brief periods of RSA (STUMPF 1965), but local injections in the hippocampus are ineffective (OTT et al. 1983). The cholinergic input to the hippocampus originates largely in the medial septal nucleus and in the nucleus of the diagonal band. Selective lesions of the cholinergic neurons in the medial septal area caused by using saporin strongly reduce the power of hippocampal theta (LEE et al. 1994). Stimulation of the medial septal nucleus produces a release of ACh in the hippocampus (DUDAR 1975, 1977). Cholinergic input from the brain stem excites these septal neurons and thereby indirectly contributes to hippocampal RSA. Accordingly, stimulation of the brain stem reticular formation produces a release of ACh in the hippocampus (DUDAR 1975, 1977) and induces RSA activity (VERTES 1982).

3. REM Sleep

ACh has been the first neurotransmitter implicated in REM sleep generation. In 1962 JOUVET observed that, after section of the neuraxis at the midbrain level, signs of REM sleep were present just caudal to the cut, and could be suppressed by atropine and enhanced by eserine. Since then, significant insights into the role of the cholinergic system in the regulation of sleep-waking states have come from the use of systemic and particularly local injections of cholinergic agonists and antagonists.

a) Systemic Injections

Interference with brain cholinergic mechanisms has been achieved by injecting nicotinic and muscarinic agonists or antagonists, cholinesterase inhibitors, and choline uptake inhibitors. Although there are exceptions, a potentiation of cholinergic mechanisms tends to decrease REM latency and to increase the percentage of time spent in REM sleep.

An increase in REM sleep can be induced in cats by intravenous (DOMINO and YAMAMOTO 1965) or subcutaneous administration of nicotine (JEWETT and NORTON 1986), and blocked by the nicotine antagonist, mecamylamine. Arecoline, a muscarinic agonist, induced REM sleep in cats and chickens (DOMINO et al. 1968). In humans, arecoline (SITARAM et al. 1978) and RS 86 (SPIEGEL 1984) reduce REM latency. Pilocarpine, another muscarinic agonist, shortens REM latency in cats (MATSUZAKI et al. 1968).

In decerebrate cats lacking forebrain input to the brain stem, the reversible AChE inhibitor, physostigmine (eserine), precipitates REM sleep or some of its components, although in the intact animal it induces behavioral arousal (MATSUZAKI et al. 1968; MAGHERINI et al. 1971; POMPEIANO and HOSHINO 1976a,b; POMPEIANO 1980); both effects are blocked by atropine. The irreversible AChE inhibitor di-isopropyl-fluorophosphate increases REM sleep in rats (GNADT and PEGRAM 1986). In humans, systemic administration of cholinesterase inhibitors can produce an increase in REM sleep (DUFFY et al. 1979; METCALF and HOLMES 1969) and a shortening of REM latency (SITARAM et al. 1976). Infusions of eserine seem to produce variable effects depending on the state of vigilance of the subject: in SWS eserine decreases the latency to REM sleep, while in W the same drug induces arousal and increases sleep latency (SITARAM et al. 1976). On the other hand, galanthamine, another cholinesterase inhibitor, has been shown to produce synchronizing effects on healthy humans (HOLL et al. 1992).

Consistent with an increase in REM sleep after activation of the cholinergic system, its blockade is generally followed by a decrease in REM sleep. Low doses of hemicholinium (a choline uptake inhibitor) injected in the fourth ventricle (HAZRA 1970) and in the lateral ventricles of the cat (DOMINO and STAWINSKI 1970) suppress REM sleep. The muscarinic antagonists, atropine and scopolamine, reduce and delay REM sleep in humans. In the cat, atropine is able to inhibit REM sleep and its various components both in the intact animal and in the decerebrate, reserpinized, or *p*-chlorophenylalanine (PCPA)-treated animal (see DOMINO et al. 1968; JOUVET 1975; POMPEIANO 1980).

b) Local Microinjections

Local microinjection experiments have been particularly useful in order to precisely identify key areas involved in the regulation of sleep and to avoid confounding effects due to peripheral or diffuse receptor activation or blockade. In general, microinjection experiments have concentrated on the role of ACh in REM sleep generation. To date, the cardinal role of the cholinergic system in REM sleep generation has been definitely demonstrated by different experimental approaches. In a pioneering study in 1963, HERNÁNDEZ-PEÓN et al. induced a REM-like state after application of minute crystals of ACh into several regions of the limbic forebrain-limbic midbrain circuit. Also in 1963, CORDEAU et al. observed that a solution of ACh directly injected into various areas of the brain stem reticular formation of the cat was generally able to induce cortical synchronization and behavioral sleep. In some cases, following injection in the rostral pontine and bulbar tegmentum, they observed brief periods of cortical desynchronization and hippocampal synchronization, which were interpreted as REM sleep episodes. GEORGE et al. in 1964 discovered that the injection of carbachol in the PRF of the cat induced desynchronization of EEG, muscular atonia, and synchronization of hippocampal activity. In some cases the total picture was indistinguishable from a REM sleep episode, while in others the cats were alert despite the absence of muscular tone. Atonia after carbachol injection in the pontine tegmentum was also described by MITLER and DEMENT (1974) and VAN DONGEN (1980). KATAYAMA et al. (1984) characterized an inhibitory cholinoceptive area in the dorsal pontine tegmentum of the cat where carbachol induced a suppression not only of postural somatomotor, but also of sympathetic visceromotor and nociceptive somatosensory functions. A cholinoceptive, REM-triggering area in the dorsal pontine tegmentum also exists in the rat (GNADT and PEGRAM 1986). A dose-dependent and site-specific induction of REM sleep has been obtained in many other local injection studies using not only carbachol (among others: AMATRUDA et al. 1975; SILBERMAN et al. 1980; SHIROMANI and MCGINTY 1986; BAGHDOYAN et al. 1987; VANNI-MERCIER et al. 1989), but also selective muscarinic agonists such as oxotremorine (GEORGE et al. 1964) and bethanechol (HOBSON et al. 1983). Moreover, local infusion of muscarinic receptor antagonists such as atropine and scopolamine can prevent pharmacologically induced REM sleep and reduce physiological REM sleep (VELLUTI and HERNÁNDEZ-PEÓN 1963; SHIROMANI and FISHBEIN 1986; BAGHDOYAN et al. 1989). The anti-cholinesterase, neostigmine, is also able to trigger a REM sleep episode when microinjected in the PRF, suggesting that physiological REM sleep is equally mediated by an increase in the secretion of ACh in the pons (BAGHDOYAN et al. 1984a). Finally, there is evidence that two-thirds of PRF neurons are strongly depolarized by carbachol through muscarinic receptors (GREENE et al. 1989a; GREENE and MCCARLEY 1990). This depolarization is associated with an increase in neuronal excitablity that resembles the changes in membrane potential and excitability of medial PRF neurons occurring during natural REM sleep.

c) Critical Regions

Many studies have been devoted to the identification of the most sensitive zone for cholinoceptive triggering of REM sleep. Several investigators (BAGHDOYAN et al. 1984b; SHIROMANI and FISHBEIN 1986; VANNI-MERCIER et al. 1989) have observed that while cholinomimetic injections within the pontine and caudal mesencephalic tegmentum may trigger REM sleep, injections into the rostral midbrain or the medullary reticular formation induce arousal. There is now a consensus, despite some terminological discrepancies, that the optimal region for short-latency REM sleep induction is the rostrodorsal medial pons, at least in the cat. BAGHDOYAN et al. (1987) described a gradient of effectiveness in the PRF, with the anterodorsal sites being the most effective ones. Other studies by VANNI-MERCIER et al. (1989) and YAMAMOTO et al. (1990a,b) restricted the most sensitive cholinoceptive zone to a site of the anterodorsal pontine tegmentum (L 2.0 to 3.0, P 1.0 to 3.5, and H −4.0 to −5.5), which approximately corresponds to the LCα and the peri-LCα in the nomenclautre of SAKAI (1980). This area does not contain ChAT-positive cells (SAKAI et al. 1986; JONES and BEAUDET 1987; SHIROMANI et al. 1988c; JONES 1991a; SAKAI 1991), but it is critically located to receive cholinergic input from LDT and PPT nuclei (MITANI et al. 1988; SHIROMANI et al. 1988b), as well as noradrenergic and serotoninergic input from LC and raphe nuclei. Some neurons in this area fire specifically during REM sleep episodes (REM-on cells), and at least some of them are activated by local injections of carbachol in parallel with the appearance of REM sleep (EL MANSARI et al. 1990; YAMAMOTO et al. 1990a,b; see also SHIROMANI and McGINTY 1986). Since local ACh release is increased during REM sleep (LYDIC et al. 1991a) as well as after stimulation of the PPT (LYDIC et al. 1992), the cholinergic projections to this area could be crucial for the physiological induction of REM sleep.

Near the sensitive zone there are sites where the injection of carbachol can separately induce only one or few polygraphic signs of REM sleep, such as atonia with arousal, REM with arousal, or pontogeniculo-occipital (PGO) waves (see HOBSON et al. 1986 for references). According to VANNI-MERCIER et al. (1989), the different syndromes triggered by carbachol injections within the anterodorsal pons can be attributed to slightly different injection sites. LOPEZ-RODRIGUEZ et al. (1994), however, showed that a crucial factor is the state of the animal during the injection: if the animal (cat) is awake, carbachol induces a dissociated state in which waking with EEG desynchronization is associated with atonia and sometimes PGOs, while when the cat is in quiet sleep carbachol induces a full-fledged REM sleep episode. In addition, there is some evidence that, even if the dorsal pons is necessary for the induction of REM sleep, it may not be sufficient. The contribution of the medulla is indicated by the finding that in cats transected at prebulbar level neither spontaneous nor carbachol-induced REM signs were observed (VANNI-MERCIER et al. 1991).

While the short-latency site for the induction of REM sleep in the rostrodorsal medial pons can be considered a cholinoceptive executive area, recent microinjection experiments show that cholinergic/cholinoceptive cells located in the PPT/LDT have a long-lasting and persistent effect on REM sleep generation (CALVO et al. 1992). A single unilateral injection of carbachol into the pontine peribrachial nucleus, a region near the PPT that does contain ChAT-positive cells and that has been implicated in PGO wave generation, produces state-independent PGOs persisting for 3–4 days, and a long-lasting increase in REM sleep, beginning the 2nd day and persisting for more than 6 days. The authors suppose that immediate PGO enhancement is a direct consequence of the membrane activation of cholinoceptive PGO burst neurons by carbachol, while the subsequent REM sleep enhancement is a consequence of metabolic activation of endogenous cholinergic neurons that would be responsible for REM sleep induction by acting on the cholinoceptive region around the LCα.

d) Receptors

REM sleep induction by cholinergic stimulation in the PRF is probably mediated by the non-M_1 subtype of muscarinic receptors (HOBSON et al. 1983; VELÁZQUEZ-MOCTEZUMA et al. 1991). McN 343-A, a selective M_1 agonist, does not produce any modification of sleep percentages after local infusion into the medial pontine reticular formation of cats (VELÁZQUEZ-MOCTEZUMA et al. 1989). Pirenzepine, a selective M_1 antagonist, does not block the increase in REM sleep induced by the (quite) selective M_2 agonist, cismethyldioxolane, when locally infused into the PRF (VELÁZQUEZ-MOCTEZUMA et al. 1991), although it can interfere with the carbachol-induced REM increase (ZOLTOSKI et al. 1991). Moreover, the density of M_1 receptors in the pons is very low (SPENCER et al. 1986). Some involvement of M_1 forebrain receptors in REM sleep regulation cannot be ruled out, however, because REM sleep can be induced by cholinergic stimulation of anterior areas (HERNANDEZ-PEON et al. 1963), and pirenzepine or biperiden, another M_1 antagonist, decreases REM sleep percentage when intracerebroventricularly administered in rats and cats, respectively SALIN-PASCUAL et al. 1991; IMERI et al. 1992). However, these effects are probably not mediated by M_1 receptors in the brain stem since pontine microinjection of pirenzepine induces no changes in the sleep-waking cycle (IMERI et al. 1994). Little is known about M_3 receptors, except that the M_3-specific blocking agent *p-F*-HHSiD (*para*-fluoro-hexahydro-sila-difenidol) seems to increase SWS at the expense of W, with no modification of REM sleep (IMERI et al. 1992).

On the other hand, compelling evidence suggests that the M_2 subtype is the most closely involved in REM sleep regulation, although a highly selective M_2 antagonist is not yet available. The M_2-selective agonists, oxotremorine-M and cismethyldioxolane, are able, with a fourfold lower dose than carbachol, to induce the same increase in REM sleep (VELÁZQUEZ-MOCTEZUMA et al. 1989). Atropine, a mixed M_1–M_2 muscarinic antagonist, but not pirenzepine, a M_1-muscarinic antagonist, block the cismethyldioxolane-induced increase in REM sleep (VELÁZQUEZ-MOCTEZUMA et al. 1991). Gallamine, a mixed nicotinic and M_2 muscarinic antagonist, tends to block the effect of cismethyldioxolane (VELÁZQUEZ-MOCTEZUMA et al. 1991). The selective M_2 antagonist, 4-DAMP, however, does not inhibit REM sleep when infused in the lateral ventricle of rats (ZOLTOSKI et al. 1991, 1992).

Although George et al. (1964) concluded that pontine microinjections of nicotine were ineffective on REM sleep, there is some recent evidence for a nicotinic component in the pontine cholinoceptive mechanism of REM sleep generation. Velázquez-Moctezuma et al. (1990) obtained an increase in time spent in REM sleep and a decrease in REM latency at the expense of W and drowsiness after microinjection of nicotine in the medial PRF. This effect was, however, much less remarkable than that produced by carbachol or cismethyldioxolane. In vitro studies in rats demonstrated the presence on PRF neurons of nicotinic receptors mediating excitatory effects (Birnstiel et al. 1991). In addition, it is known that the nicotine antagonist, mecamylamine, can block the PGO waves, one of the main components of REM sleep, in the cat lateral geniculate nucleus (Hu et al. 1988).

e) Cholinergic-Noradrenergic Interactions

Despite the abundant evidence linking the cholinergic system and REM sleep generation, it is clear that other neurochemical systems also play a role. The interaction between cholinergic and aminergic systems is particularly relevant (see also Chap. 4, this volume). For example, systemic injection experiments had demonstrated that eserine produced arousal, instead of REM sleep, unless the animal had been previously depleted of monamines with reserpine (Karczmar et al. 1970). Based on the fundamental observation that noradrenergic cells in the LC cease firing as the animal approaches and enters REM sleep, one of the most influential models of REM sleep generation, the reciprocal interaction model (see Hobson et al. 1986 for a review), postulated that cholinergic/cholinoceptive neurons in the PRF would trigger REM sleep, but only if released from noradrenergic inhibition. While the cholinergic side of the model is well supported by the above evidence, the postulated noradrenergic inhibition of REM-executive cells is still controversial (see Gaillard 1985; Stenberg and Hilakivi 1985; Tononi et al. 1991b). The model predicts that noradrenergic potentiation in the PRF should suppress REM sleep, while interference with noradrenergic inhibition of REM-on cells should facilitate REM sleep. Microinjection experiments into the dorsal PRF aimed at examining noradrenergic influences on REM sleep generation have demonstrated, in agreement with the reciprocal interaction model, that REM sleep is suppressed by adrenergic α_1- (Cirelli et al. 1993) and β-agonists (Tononi et al. 1988b, 1989) and enhanced by β-antagonists (Tononi et al. 1988a, 1989). These effects were attributed either to α_1-adrenergic activation of local GABAergic interneurons, inhibiting PRF neurons, or to a direct β-adrenergic inhibition of these pontine neurons. α_2-Agonists, which are known to suppress LC discharge, also reduce REM sleep, an effect which was linked to a direct inhibitory action of the injected substances on REM-executive neurons (Tononi et al. 1991a). Experiments using whole-cell patch-clamp recordings in slices, associated with intracellular labeling of mesopontine cholinergic neurons, have indeed shown that activation of α_2-adrenoceptors hyperpolarized these cholinergic neurons (WILLIAMS and REINER 1993).

4. ACh and Atonia

Postural atonia represents one of the main tonic manifestations of REM sleep. Observations made in the early 1960s by testing excitability changes of α-extensor (and flexor) motoneurons to orthodromic, antidromic, and direct stimulations during the sleep-waking cycle recorded in unrestrained, unanesthetized cats had shown that the suppression of postural activity during REM sleep was not due to a reduced discharge of supraspinal descending excitatory systems leading to a motoneuronal disfacilitation. Rather it was due to an increased discharge of supraspinal descending inhibitory systems, leading to postsynaptic inhibition of α-motoneurons (POMPEIANO 1967). Similar conclusions were reached by recording intracellularly the membrane potential change occurring in limb extensor motoneurons during REM sleep (MORALES and CHASE 1978; MORALES et al. 1987).

Atonia can be dissociated from other REM sleep signs and selectively induced by cholinergic stimulation of two different regions of the brain stem, an area immediately ventral to the LC (MITLER and DEMENT 1974; VAN DONGEN et al. 1978; BAGHDOYAN et al. 1984b; KATAYAMA et al. 1984; SHIROMANI et al. 1986) and an area in the medial bulbar reticular formation, corresponding to the region where MAGOUN and RHINES (1946) had obtained atonia in the decerebrate cat after electrical stimulation. Both areas receive dense cholinergic innervation from the LDT/PPT, the latter through the tegmentoreticular tract. Stimulation of this medullary area produces inhibitory postsynaptic potentials in both extensor and flexor motoneurons (JANKOWSKA et al. 1968). Similarly, inhibitory postsynaptic potentials were recorded from spinal motoneurons following local injection of carbachol in the dorsal pontine tegmentum of decerebrate cats (MORALES et al. 1987).

The role of the cholinoceptive area ventral to the LC (termed LCα and peri-LCα by SAKAI) in mediating muscle tone suppression during REM sleep has been confirmed by recording (for references see HOBSON and STERIADE 1986) and lesion studies (JOUVET and DELORME 1965; HENDRICKS et al. 1982; CHASE 1983; AMINI-SERESHKI and MORRISON 1986). When it is lesioned, the syndrome of REM sleep without atonia ensues (HENLEY and MORRISON 1974; SAKAI 1980).

Only the caudal portion of the bulbar inhibitory area of Rhines and Magoun (nucleus paramedianus) produces atonia when injected with cholinergic agonists, while in the more rostral portion (nucleus magnocellularis) non-NMDA excitatory amino acids are effective, but not cholinergic agonists (LAI and SIEGEL 1988, 1991; LAI et al. 1993). It has recently been demonstrated that spontaneous ACh release increases in the nucleus paramedianus during REM sleep in unanesthetized, freely moving cats (KODAMA et al.

1992). Because ACh and glutamate are colocalized in some neurons of the PPT (CLEMENTS and GRANT 1990), it is possible that the corelease of ACh and glutamate is necessary for the full expression of atonia. It is also of interest that in other regions of the brain, such as the cerebellar cortex, a major role of the cholinergic muscarinic system is to enhance the signal-to-noise ratio of the response to Purkinje cells to pulses of glutamate (ANDRE et al. 1993b). It is thus possible that even at the level of the medullary inhibitory area ACh may act by increasing both the steady background discharge and the recruitment of inhibitory reticulospinal neurons.

5. ACh and PGO Waves

Several studies have provided evidence that brain stem cholinergic neurons in PPT and LDT are involved in transferring PGO signals to the thalamocortical system. Electrical stimulation of these nuclei induce a sharp PGO wave in the lateral geniculate thalamic nucleus (SAKAI et al. 1976). Microinjections of carbachol into the peribranchial nucleus of the cat induce state-independent PGO waves lasting for several days (DATTA et al. 1992). Excitotoxic lesions of the PPT/LDT area, leading to a loss of 60% of ChAT-positive cells, reduce by about 75% the amount of PGO waves during REM sleep (WEBSTER and JONES 1988). Several classes of PGO-related cells have been recorded in the PPT/LDT area (SAITO et al. 1977; MCCARLEY et al. 1978; SAKAI and JOUVET 1980; STERIADE et al. 1990b). Some of them fire high-frequency spike bursts prior to geniculate PGO waves on a background of tonically increased activity during REM sleep. Because PGO waves in the thalamus are strongly depressed by nicotinic antagonists (HU et al. 1988), at least some of the PGO-on cells are likely to be cholinergic. Consistent with this, membrane properties of identified cholinergic cells projecting to the thalamus would permit both tonic and/or burst firing (LEONARD and LLINAS 1990; KAMONDI et al. 1992).

6. Diseases Involving ACh and Sleep

Additional evidence for an involvement of the cholinergic system in sleep-wakefulness regulation comes from several pathological conditions that show a combined alteration of sleep patterns and of the cholinergic system. There are many disorders with such characteristics, including myasthenia, Alzheimer's disease, and several psychiatric conditions. Here we will mention the two disorders that are most strictly associated with both REM sleep and cholinergic anomalies, namely depression and narcolepsy.

a) Depression

In patients with depression several abnormalities of sleep are well documented, in particular short REM latency, increased REM percentage and density, multiple awakenings, and reduction of SWS (stages 3 and 4) and of total sleep time (for a review see GILLIN 1989; VOGEL 1989). Even if not specific, short REM latency is an important state marker in depression, varying with the severity of the episode and returning toward normal with clinical recovery. These abnormalities of sleep patterns seem to be associated, at least in part, with an altered functioning of the cholinergic system, including a possible upregulation of muscarinic receptors. Based on many clinical and pharmacological findings, such as the fact that drugs with cholinomimetic or antiaminergic properties produced depression-like symptoms that can be antagonized by atropine, JANOWSKY et al. (1972) hypothesized that an increased ratio of cholinergic to aminergic neurotransmission plays a role in the pathophysiology of depression. As we have seen in Sect. C.V.3.e, such an increased ratio of cholinergic to aminergic transmission was recognized early on (HOBSON and MCCARLEY 1974) as a central characteristic of REM sleep. Consistent with this, some evidence has been found for an association between cholinergic supersensitivity and sleep disorders in depression. Studies in volunteers have shown that scopolamine-induced muscarinic supersensitivity produces sleep disorders mimicking those associated with depression. In the cholinergic REM induction test (CRIT) the latency to the second REM period is measured after infusion of arecoline or physostigmine during the second non-REM period. Patients with major depressive disorder enter REM sleep significantly faster than normal controls. These findings have also found some therapeutic counterpart: REM sleep deprivation via arousals at the onset of REM episodes was found to improve endogenous depression and effective antidepressant drugs produce a similar REM sleep deprivation (see GILLIN 1989 for references).

b) Narcolepsy

Narcolepsy is a complex disorder characterized by excessive daytime sleepiness, hypnagogic hallucinations, sleep onset REM episodes, sleep paralysis, and cataplexy. Cataplexy, a sudden loss of muscle tone during active waking that can be considered a sort of "atonia without REM sleep," is most clearly related to cholinergic mechanisms. The cholinesterase inhibitor, physostigmine salicylate, which potentiates both muscarinic and nicotinic action of ACh and crosses the blood-brain barrier, induces cataplexy in narcoleptic dogs, while atropine sulfate and scopolamine hydrobromide block spontaneous attacks in these animals. Neostigmine methylsufate, which has primarily peripheral actions, does not induce cataplexy. The muscarinic cholinomimetic, arecholine hydrochloride, increases the duration of cataplectic attacks, while nicotinic agonists are not effective. These findings suggest that the effects of cholinergic drugs on cataplexy are mediated by central muscarinic receptors. In humans, tricyclic antidepressant agents are useful against cataplexy, and their effectiveness is partially due to their anticholinergic properties (for a review of pharmacological studies in cataplexy see GUILLEMINAULT 1989). An increased number of muscarinic receptors have been described in the medial PRF and medial medulla of narcoleptic dogs (Boehme et al. 1984; Kilduff et al. 1986). A noticeable upregulation of cholinergic receptors in the medial mesopontine region has been demonstrated also in narcoleptic humans (Aldrich et al. 1990).

Microdialysis studies show that ACh release in the medial pontine reticular formation is enhanced during cataplexy (Reid et al. 1994a). Furthermore, cataplexy can be precipitated in narcoleptic dogs by pontine injections of carbachol, of the M_2 agonist, oxotremorine, but not of the M_1 agonist, McN-A-343 (Reid et al. 1994b). Both REM sleep atonia and cataplexy are therefore associated with an enhanced release of ACh in medial PRF and a subsequent M_2 receptor stimulation. In narcoleptic dogs, carbachol can induce status cataplecticus and long-lasting muscle atonia with EEG desynchronization also when injected into the lateral ventricle and after local injection in the diagonal bundle of the basal forebrain, but not in the lateral preoptic area, lateral septum, and amygdala (Nishino et al. 1992). This effect, but not the carbachol-induced muscle atonia after injection into the medial PRF, is antagonized by yohimbine.

D. Cholinergic System and Postural Mechanisms

I. Effects on Decerebrate Rigidity

It was mentioned previously that dorsal pontine reticular structures such as the peri-LCα and adjacent areas exert an excitatory influence on the inhibitory areas of the medullary reticular formation (Magoun and Rhines 1947), thus leading to suppression of posture (Foote et al. 1983; Pompeiano 1980; Sakai 1980, 1988; Hobson and Steriade 1986). This effect is apparently mediated by descending tegmentoreticular projections ending in two distinct regions of the medial medulla: the nucleus magnocellularis of the rostral medulla (Sakai et al. 1979; Jones and Yang 1985; Luppi et al. 1988) and the nucleus paramedianus of the caudal medulla (Lai and Siegel 1988; Rye et al. 1988; Shiromani et al. 1990a). Both nuclei give rise to reticulospinal (RS) projections (Ohta et al. 1988), which inhibit spinal motoneurons (Jankowska et al. 1968), by activating, in part at least, Renshaw cells (Kanamori et al. 1980; Pompeiano 1984). However, while the nucleus paramedianus is made up of cholinosensitive neurons (Lai and Siegel 1988), thus receiving an excitatory input from cholinergic pontine neurons (Rye et al. 1988; Shiromani et al. 1990a), the nucleus magnocellularis contains glutamate-sensitive neurons, receiving an excitatory input from glutamatergic neurons (Lai and Siegel 1988).

In contrast to these dorsal pontine structures which exert an inhibitory influence on posture, the LC complex exerts a facilitatory influence on posture. There is in fact evidence that the noradrenergic and NE-sensitive neurons (Foote et al. 1983; Ennis and Aston-Jones 1986a) send projections to the peri-LCα and the adjacent dorsal pontine reticular formation (SAKAI et al. 1977; JONES and YANG 1985; see also JONES and FRIEDMAN 1983) on which they exert a tonic inhibitory influence (FOOTE et al. 1983; HOBSON and STERIADE 1986; POMPEIANO 1980; SAKAI 1980, 1988). Moreover, in addition to this indirect projection, the LC complex gives rise to a direct coeruleospinal (CS) projection which exerts either a direct excitatory influence on spinal motoneurons (FUNG et al. 1991) or a suppressive influence on the related inhibitory Renshaw cells (FUNG et al. 1988). Moreover, iontophoretic application of NE not only depolarizes the α-motoneurons (WHITE et al. 1991), but also inhibits the corresponding Renshaw cell activity (see POMPEIANO 1984 for references).

Several lines of evidence indicate that the decerebrate rigidity is modulated by the activity of the cholinoceptive dorsal pontine tegmental structures indicated above. In fact, lesion of the peri-LCα and the dorsal PRF, which suppressed the episodes of REM sleep in intact cats (JOUVET 1972; HENLEY and MORRISON 1974; SAKAI 1985; WEBSTER and JONES 1988), reduced or suppressed the episodes of postural atonia induced by systemic injection of an anticholinesterase in decerebrate cats (D'ASCANIO et al. 1985b; cf. POMPEIANO 1985) as described in previous studies (MATSUZAKI et al. 1968; MATSUZAKI 1969; MERGNER et al. 1976; cf. POMPEIANO 1976, 1980). On the other hand, microinjection of the cholingergic agonist, carbachol, in that area, which induced postural atonia and/or REM sleep in intact animals (KATAYAMA et al. 1984; SAKAI 1988; VANNI-MERCIER et al. 1989; cf. SHIROMANI and MCGINTY 1986), decreased (BARNES et al. 1987; MORI 1989) or suppressed (D'ASCANIO et al. 1988; LAI and SIEGEL 1988) the postural activity in decerebrate cats, due to a tonic inhibitory influence on extensor motoneurons (MORALES et al. 1987). In line with these findings are the results of experiments showing that anatomical or functional inactivation of the noradrenergic LC complex elicited either by electrolytic lesion of this structure (D'ASCANIO et al. 1985a, 1989c) or by local injection into the LC complex of the α_2-adrenergic agonist, clonidine, (POMPEIANO et al. 1987) or the β-adrenergic agonist, isoproterenol, (D'ASCANIO et al. 1989a), which inhibited the discharge of the NE-containing neurons (ABERCROMBIE and JACOBS 1987b; FOOTE et al. 1983). In return this released from inhibition the cholinoceptive pontine reticular neurons and greatly descreased or suppressed the postural activity particularly in the ipsilateral limbs of decerebrate cats. Similar results were also obtained after microinjection into the dorsal PRF either of the β-noradrenergic antagonist, propranolol, (D'ASCANIO et al. 1989b) or else of the α_1-noradrenergic antagonist, prazosin, (CIRELLI et al. 1993), which blocked the inhibitory influence exerted by the noradrenergic LC neurons on the cholinoceptive pontine reticular neurons. It is likely that these effects were mediated either directly, through β-adrenoceptors, or indirectly by activating through α_1-adrenoceptors GABAergic interneurons inhibiting the cholinoceptive PRF neurons. On the other hand, an increase in postural activity affected the ipsilateral limbs of decerebrate cats after microinjection into the LC complex of a cholinergic agonist, carbrachol, (Stampacchia et al. 1987). There is, in fact, evidence that cholinergic fibers and terminals ending on the LC neurons probably originate from the dorsolateral pontine tegmentum (Horn et al. 1987) and that ACh activates the LC neurons through muscarinic receptors (Engberg and Sevensson 1980; Egan and North 1985).

The notion that the cholinoceptive neurons located in the peri-LCα and the adjacent PRF, as well as the related noradrenergic and NE-sensitive LC neurons, are critically involved in the maintenance of decerebrate rigidity is supported by experiments of unit recording performed in decerebrate cats, which showed that the postural activity is present as long as the LC complex neurons fire regularly, thus keeping under inhibitory control the cholinoceptive PRF neurons (Hoshino and Pompeiano 1976; Pompeiano and Hoshino 1976a,b; cf. Pompeiano 1976; Pompeiano 1980). However, as soon as the noradrenergic LC neurons ceased firing, as during the episodes of postural atonia induced by systemic injection of an anticholinesterase, the firing rate of the dorsal PRF neurons (Hoshino and Pompeiano 1976) as well as of the presumably inhibitory medullary RS neurons (Srivastava et al. 1982) increased. Similar changes in firing rate were also obtained by recording the unit discharge from the two populations of units either in intact animals during waking or during REM sleep (Sakai 1980; Foote et al. 1983; Hobson and Steriade 1986). It is of interest that the mean resting discharge of the LC neurons, which is very low (1–2 impulses/s) in intact animals during quiet waking (Sakai 1980; Foote et al. 1983; Hobson and Steriade 1986), due to self-inhibitory synapses acting on α_2-adrenoceptors through mechanisms of recurrent and/or lateral inhibition (Ennis and Aston Jones 1986a; Foote et al. 1983), increased to about 10 impulses/s after decerebration (Pompeiano and Hoshino 1976a,b; Oz. Pompeiano et al. 1990). This finding has been attributed to interruption of a supramesencephalic descending pathway exerting a tonic inhibitory influence on the LC neurons.

The hypothesis that the resulting increase in resting discharge of the LC-complex neurons contributes to the γ-rigidity which occurs after decerebration (Granit 1970) is supported by the fact reported above, i.e., that the LC neurons exert not only a direct facilitatory influence on α-extensor motoneurons (Fung et al. 1991), but also an inhibitory influence on Renshaw cells (Fung et al. 1988). This would lead to disinhibition and thus to an increased discharge of both the γ-motoneurons and the small-size tonic α-motoneurons innervating the extensor muscles (Pompeiano 1984). In addition to this finding, the increased discharge of the LC neurons would reduce the activity of the cholinoceptive PRF neurons and the related medullary inhibitory RS neurons, which may act on extensor motoneurons by activating Renshaw cells (Pompeiano 1984). No experiment has been performed so far to identify the source of origin of the supramesencephalic structures exerting an inhibitory influence on the LC neurons. This pathway could be constituted either by inhibitory neurons projecting directly on the LC or by excitatory neurons acting through GABAergic inhibitory interneurons. There is in fact evidence that GAD-immunoreactive cells are located close to or within the LC complex (JONES 1991b). The possibility that the involved structures include the substantia nigra (SN) pars reticulata, whose neurons are GABAergic, is supported by the facts that: (a) this part of the SN projects to the pontine tegmentum (see SCHWARZ et al. 1984 for references); (b) selective kainic acid destruction of this structure of one side produces in the rat a contralateral turning behavior (DI CHIARA et al. 1977), due in part at least to increased postural activity in the ipsilateral limbs; and (c) electrical stimulation of this structure reduces the discharge of the γ-static motoneurons (SCHWARZ et al. 1984), which are responsible for the γ-rigidity (GRANIT 1970). The possible involvement of the SN, pars reticulata, and the related noradrenergic and cholinergic systems during Parkinson's disease may account for the appearance of the γ-rigidity which occurs in parkinsonian patients. The crucial point in this hypothesis would be to verify whether the SN and pars reticulata inhibits the LC neurons, including the coeruleospinal neurons, thus disinhibiting the PRF neurons and the related medullary inhibitory RS neurons.

II. Effects on Vestibulospinal Reflexes

In addition to static changes in posture, the dorsal pontine tegmental structures described above may also exert a prominent influence on the postural changes induced by labyrinth stimulation. It is known that in decerebrate cats slow sinusoidal rotation about the longitudinal axis of the whole animal, leading to stimulation of macular (gravity) receptors, produced vestibulospinal (VS) reflexes characterized by a contraction of limb extensors, during ipsilateral (side-down) tilt and a relaxation during side-up tilt (α-responses). This effect, which involved not only forelimb (MANZONI et al. 1983a; SCHOR and MILLER 1981; LINDSAY et al. 1976) but also hindlimb extensors (MANZONI et al. 1984; D'ASCANIO et al. 1985a,b; POMPEIANO et al. 1991), is apparently mediated by the lateral vestibular nucleus (LVN), which exerts a direct excitatory influence on ipsilateral limb extensor motoneurons (LUND and POMPEIANO 1968). In fact, most of the lateral VS neurons, including those projecting to the lumbosacral segments of the spinal cord, were excited during side-down and depressed during side-up tilt (BOYLE and POMPEIANO 1980; SCHOR and MILLER 1982; MARCHAND et al. 1987). Experiments were performed to investigate: (1) whether presumably inhibitory RS neurons and LC-complex neurons, including the CS neurons, having the characteristics attributed to noradrenergic neurons, responded to sinusoidal stimulation of labyrinth receptors; and (2) whether these responses contributed to the postural changes of limb extensors during the VS reflexes.

1. The main result of the first group of experiments was that most of the RS neurons located in the inhibitory area of the medullary reticular formation (Manzoni et al. 1983b), as well as the CS neurons projecting to the lower segments of the spinal cord (Pompeiano et al. 1990), responded to roll tilt of the animal with a response pattern which was just opposite to that of the VS neurons, being excited during side-up tilt and depressed during side-down tilt (β-responses). These findings were attributed to the fact that the macular input of one side, responsible for the α-responses to tilt of the VS neurons, is transmitted via a crossed pathway (Pompeiano 1979) to the ventral part of medullary reticular formation (Kubin et al. 1980; Manzoni et al. 1983b), the neurons of which may project either directly to the spinal cord as RS neurons, or to the LC complex (Aston-Jones et al. 1986), where they exert a prominent excitatory influence (Ennis and Aston-Jones 1986b). One of the main channels transmitting the macular input of one side to the contralateral medullary reticular structures is represented by the lateral VS tract acting on neurons of the crossed spinoreticular pathway (Pompeiano 1979).

2. After these findings had been obtained, experiments were performed to investigate whether changes in the resting discharge of the medullary inhibitory RS neurons or else of the excitatory CS neurons modified the gain of the VS reflexes. The influence of these two populations of neurons on the VS reflexes can be understood only if we consider that the corresponding populations of units fire out of phase with respect to the VS neurons, and that, due to reciprocal interaction between the noradrenergic and the cholinoceptive PRF neurons, the activity of the excitatory CS neurons may from time to time predominate over that of the inhibitory RS neurons and vice versa. It is known for instance that, in precollicular decerebrate cats, the high resting discharge of the LC neurons (Pompeiano and Hoshino 1976a,b; O. Pompeiano et al. 1990) tonically inhibits that of the dorsal PRF neurons and the related medullary inhibitory RS neurons (Hoshino and Pompeiano 1976; Srivastava et al. 1982). In these instances the amplitude of the EMG modulation and thus the response gain of the limb extensors to labyrinth stimulation was quite small in the forelimbs (Manzoni et al. 1983a) and almost absent in the hindlimbs (Manzoni et al. 1984; D'Ascanio et al. 1985a,b; Pompeiano et al. 1991). This finding can be understood if we consider that the contraction of limb extensors during side-down animal tilt depends on the increased discharge of the VS neurons (Boyle and Pompeiano 1980; Schor and Miller 1982; Marchand et al. 1987). However, the reduced discharge of the excitatory CS neurons for the same direction of animal orientation would lead to disfacilitation of the extensor motoneurons, thus giving rise to a small gain of response of limb extensors to labyrinth stimulation. A further decrease in gain of the VS reflexes occurred after activation of the LC neurons elicited by local injection into the LC of the cholinergic agonist carbachol (Stampacchia et al. 1987). In contrast to these findings there were other conditions in which the activity of the PRF

neurons and the related medullary inhibitory RS neurons predominated over that of the LC and the corresponding CS neurons. In these instances two levels of activity of the cholinergic system were recognized, according to whether the increased discharge of these pontine and medullary reticular neurons was either moderate to decrease postural activity or so prominent as to suppress posture completely. We shall refer to these stages as the primary or the secondary range of operation of the system, respectively.

If the increased discharge of the pontine and medullary reticular neurons was only moderate so as to decrease postural activity (*primary range*), the amplitude of modulation and thus the response gain of the limb extensors to labyrinth stimulation increased. These findings were obtained particularly in decerebrate cats after systemic injection of small doses of the anticholinesterase, eserine sulfate, and were suppressed after i.v. administration of the muscarinic blocker, atropine sulfate, (POMPEIANO et al. 1983). Moreover, electrolytic lesion of the dorsal PRF neurons of one side suppressed the postural and reflex changes induced by systemic injection of the anticholinesterase (D'ASCANIO et al. 1985; cf. POMPEIANO 1985). In line with these findings are the results of experiments (BARNES et al. 1987) showing that unilateral microinjection into the peri-LCα and the adjacent dorsal PRF of a cholinergic agonist like carbachol which acts on both muscarinic and nioctinic receptors, or bethanechol which acts on muscarinic receptors, while activating the corresponding cholinoceptive neurons, decreased the postural activity in the ipsilateral limbs but greatly increased the gain of the VS reflexes. These effects were site specific and dose dependent and were suppressed after local injection into the same region of small doses of the muscarinic blocker, atropine sulfate.

Because, as reported above, the inhibitory RS neurons fire out-of-phase with respect to the excitatory VS neurons acting on ipsilateral limb extensor motoneurons, it appears that the increased contraction of ipsilateral limb extensor motoneurons during side-down tilt depends not only upon an increased discharge of excitatory VS neurons but also on a reduced discharge of inhibitory RS neurons. It appeared, therefore, that for a given labyrinth signal, the higher the firing rate of these RS neurons in the animal at rest, the greater the disinhibition affecting the extensor motoneurons during side-down animal tilt. These motoneurons would then respond more efficiently to the same excitatory VS volleys elicited by given parameters of stimulation, thus giving rise to an increased gain of the EMG responses of forelimb extensor muscles to labyrinth stimulation. Because the cholinoceptive pontine neurons located in the peri-LCα and the dorsal PRF are tonically inhibited by the LC, it was thought that inactivation of the noradrenergic LC neurons produced changes in posture and gain of the VS reflexes similar to those obtained by direct activation of the dorsal reticular system. Indeed, experiments performed in decerebrate cats have shown that unilateral microinjection into the dorsal PRF of the nonselective β-adrenergic antagonist, propranolol, (D'ASCANIO et al. 1989b) or else of the α_1-adrenergic antagonist, prazosin, (CIRELLI et al. 1993), which blocked the postsynaptic inhibitory influence exerted by the noradrenergic LC neurons on the pontine reticular neurons (see Sect. C.I), not only reduced the postural activity in the ipsilateral limb extensors, but also increased the amplitude of modulation and thus the response gain of the ipsilateral triceps brachii to labyrinth stimulation. Similar results were also obtained after unilateral electrolytic lesion of the LC (d'ACSANIO et al. 1985) or after local injection into the LC of one side either of the α_2-adrenergic agonist, clonidine, (POMPEIANO et al. 1987) or of the β-adrenergic agonist, isoproterenol, (D'ASCANIO et al. 1989a), which inactivated the noradrenergic and NE-sensitive LC neurons.

The possibility that in decerebrate cats clonidine acted on α_2-adrenoceptors located either presynaptically on cholinergic terminals, thus inhibiting the release of ACh (BEANI et al. 1978; BIANCHI et al. 1979; VIZI 1980; MORONI et al. 1983), or postsynaptically on presumptive cholinergic and cholinoceptive PRF neurons, thus suppressing their activity (GREENE et al. 1989b; BIER et al. 1990), was excluded by the fact that the postural and reflex changes induced by local injection of clonidine into the LC complex and the adjacent dorsal PRF closely resembled those elicited in previous experiments by local administration of cholinergic agonists into the peri-LCα and the adjacent dorsal PRF (BARNES et al. 1987). Moreover, the induced changes were suppressed by local injection into the same pontine region of the muscarinic blocker, atropine sulfate, (POMPEIANO et al. 1987).

If the increased discharge of the dorsal PRF neurons and the medullary RS neurons was prominent (*secondary range*), one observed a complete suppression not only of posture, but also of the VS reflexes. These findings were originally obtained in decerebrate cats after systemic injection of appropriate doses of an anticholinesterase (MATSUZAKI 1969; MATSUZAKI et al. 1968; MERGNER et al. 1976; POMPEIANO 1976, 1980). Moreover, transient episodes of postural atonia, associated with a suppression of the EMG responses of forelimb extensors to animal tilt, occurred after electrolytic lesion of the LC (D'ASCANIO et al. 1989a) or local injection into this structure either of the α_2-adrenergic agonist, clonidine, (unpublished observations) or the β-adrenergic agonist, isoproterenol, (D'ASCANIO et al. 1989). Similar results were also obtained after injection into the peri-LCα and the adjacent dorsal PRF of the β-adrenergic antagonist, propranolol, (D'ASCANIO et al. 1989b) was well as of the cholinergic agonist, carbachol, (BARNES et al. 1987; D'ASCANIO et al. 1988). These effects were attributed either to prominent disinhibition or to direct excitation of the dorsal pontine reticular neurons and the related medullary RS neurons, leading to intense postsynaptic inhibition of the extensor α-motoneurons (MORALES et al. 1987).

In conclusion, it appears that while an increased discharge of the LC neurons, such as occurs after decerebration, reduces the gain of the VS reflexes; a reduced discharge of the same neurons would either increase or decrease the gain according to the "primary" or the "secondary" range of operation of the system. This operational model suggests that the neuronal circuit made by the noradrenergic LC neurons and the related cholinergic and/or cholinoceptive PRF neurons acts as a variable gain regulator. This system may actually adapt the gain of the VS reflexes to the animal state. Experiments of unit recording, referred to above, have in fact shown that the resting discharge of the LC neurons, which is low during quiet waking, increases during alertness (FOOTE et al. 1983) or stress (ABERCROMBIE and JACOBS 1987a), but decreases and then disappears during REM sleep (SAKAI, 1980, 1985, 1988, 1991; ASTON-JONES 1985; FOOTE et al. 1983; HOBSON and STERIADE 1986; JACOBS 1986). In this extreme condition, the arrest of the tonic discharge of the LC neurons would lead to a prominent increase in activity of the cholinoceptive pontine reticular neurons (SAKAI 1980, 1988, 1991; FOOTE et al. 1983; HOBSON and STERIADE 1986) and the related medullary inhibitory RS neurons (SIEGEL et al. 1979; KANAMORI et al. 1980; CHASE et al. 1981) to suppress both posture (POMPEIANO 1967; JOUVET 1972) and the VS reflexes.

III. Effects on Postural Adjustments During Cortically Induced Movements

Both the cholinergic and the noradrenergic systems located in the dorsolateral pontine tegmentum intervene not only in the control of posture and the vestibulospinal (VS) reflexes, but also in the regulation of the postural adjustments induced by stimulation of the motor cortex. Electrical stimulation of the motor cortex performed in unanesthetized cats, at an intensity sufficient to elicit flexion of one limb, produces a diagonal pattern of postural adjustments, characterized by a decreased force by the limb diagonally opposite to the moving one and an increased force by the other two (GAHERY and NIEOULLON 1978). Latency measurements exclude that these changes depend on the peripheral feedback associated with the performance of the movement, because they parallel and actually precede the onset of movement (GAHERY and NIEOULLON 1978). Thus the diagonal pattern of postural adjustments results from a central command.

It was postulated recently (GAHERY et al. 1984) that the lateral vestibular nucleus (LVN), which projects to all the segments of the spinal cord, and exerts a direct excitatory influence on ipsilateral limb extensor motoneurons (LUND and POMPEIANO 1968), could be influenced by descending projections from the motor cortex acting either directly or indirectly through connections with precerebellar structures, such as the lateral reticular nucleus (LRN) and the inferior olive (IO) (ITO 1984). These structures would than act on Purkinje cells of the paramedial zone B of the cerebellar vermis which in turn project to the LVN (CORVAJA and POMPEIANO 1979), where they exert a GABA-mediated inhibitory (ITO 1984), Indeed there is evidence that a large proportion of LVN neurons, some of which are antidromically identified as VS neurons, responded to electrical stimulation of the contralateral motor cortex (Licata et al. 1990; Sarkisian and Fanardjian 1992); this effect, however, could not be detected after partial cerebellectomy (Gildenberg and Hassler 1971). The hypothesis that the LVN is involved in the execution of the postural adjustments during cortically induced limb movements is supported by the results of experiments performed in unanesthetized cats, in which the vertical force exerted by the limbs on four plates connected to strain gauges was evaluated following stimulation of the motor cortex with 20-ms trains (at 300/s, 1 ms, 1–5 V) (Luccarini et al. 1992). Unilateral microinjection into the LVN of a GABA-A or GABA-B agonist did not modify the threshold, latency or amplitude of the flexor responses in the performing fore- or hindlimb when the corresponding region of the motor cortex was stimulated. However, the early component of the responses which affected the three postural limbs and was considered of central origin decreased in amplitude, while the late component of reflex origin increased, due to the instability resulting from the GABA-induced depression of the early postural responses driven by the central command. Just the opposite results were obtained following microinjection into the LVN of one side of a GABA-A or GABA-B antagonist. These results were dose dependent and site specific (Pompeiano et al. 1993). It is likely, therefore, that descending projections originating for instance from the forelimb region of the motor cortex diverge at medullary level to influence portions of the LRN and the IO as well as the related corticocerebellar areas, in such a way as to decrease the discharge of the VS neurons which control the extensor activity in the hind limb diagonally opposite to the moving limb, but to increase the discharge of the VS neurons which control the extensor activity in the remaining two limbs. Inactivation of the LVN of one side would then decrease the effectiveness of the cortical input on the VS neurons of that side as well as, through crossed inhibitory interconnections, the VS neurons of the opposite side. While the LVN is apparently involved in the execution of the postural adjustments accompanying cortically induced limb movement, recent evidence indicates that the cholinoceptive PRF area may exert a permissive role on these postural changes induced by cortical stimulation. Experiments performed in unanesthetized cats (Luccarini et al. 1990) have in fact shown that bilateral microinjections into the peri-LC region and the surrounding dorsal PRF of the muscarinic agonist, bethanechol, produced a short-lasting episode of postural atonia followed by a period during which the cats were still able to stand on the measurement platforms. These microinjections did not modify the threshold, latency and amplitude of the flexion response of the performing limb, but greatly decreased or suppressed the early component of the postural responses in the remaining limbs, which is considered to be centrally triggered. On the other hand, the late component of the responses which is of reflex origin increased in amplitude, due to the instability resulting from the depression of the early postural responses. An additional finding was that the slope of the response curve of the moving limb remained unmodified, while that of the anticipatory response of the limbs involved in the postural adjustments decreased following stimulation of the motor cortex at different stimulus intensities. The effects described above were site specific and dose dependent. It is known that the cholinoceptive pontine reticular neurons exert a tonic excitatory influence on the medullary reticular region (SAKAI 1980, 1981; POMPEIANO 1985; HOBSON and STERIADE 1986), from which the inhibitory RS system originates (JANKOWSKA et al. 1968; MAGOUN and RHINES 1974), and that an increased discharge of these PRF neurons elicited by local injection of cholinergic agonists into this pontine region (VIVALDI et al. 1980; BARNES et al. 1987; VANNI-MERCIER et al. 1989) suppresses the postural activity, due to postsynpatic inhibition of limb extensor motoneurons (MORALES et al. 1987). There is now evidence (LUCCARINI et al. 1990) that changes in the postural adjustments opposite in sign to those reported above were obtained after local injection into the LC or bethanechol, which activates through muscarinic receptors the noradrenergic neurons (ENGBERG and SVENSSON 1980; EGAN and NORTH 1985), thus suppressing the discharge of the PRF neurons. It has been postulated that the medullary inhibitory RS neurons, whose excitability is tonically increased by the cholinergic agonist acting on the PRF, could be appropriately used through direct or indirect corticoreticular projections in such a way as to decrease the early component of the postural responses not only in the limbs whose vertical force increased during the cortical movement, but also in the limb diagonally opposed to the performing limb, whose vertical force decreased during the same cortical movement. Because the cholinoceptive pontine tegmental neurons are responsible for the spontaneous fluctuations in posture related to the sleep-waking cycle, they may intervene in order to adapt to the animal state the amplitude of the postural adjustments during cortically induced limb movements.

IV. Cerebellar Regulation of Vestibular Reflexes

In addition to the classical mossy fibers and climbing fibers, which utilize excitatory amino acids as neurotransmitters, the cerebellar cortex receives other afferent systems of different neurochemical specificity. One of these systems consists of cholinergic fibers. In particular, immunohistochemical studies have shown a ChAT immunoreactivity in subpopulations of mossy fibers ending with glomerular rosettes in the granular layer (KAN et al. 1978; OJIMA et al. 1989; ILLING 1990; IKEDA et al. 1991; BARMACK et al. 1992a,b) and parallel fibers (IKEDA et al. 1991), as well as in thin varicose fibers closely associated with Purkinje (P) cell bodies and their initial dendritic shafts in the lower part of the molecular layer (OJIMA et al. 1989; ILLING 1990; IKEDA et al. 1991; BARMACK et al. 1992a). Moreover, different types of muscarinic (BUCKLEY et al. 1988; NEUSTADT et al. 1988; ANDRE et al. 1992a) and nicotinic receptors (LAPCHAK et al. 1989b; SWANSON et al. 1987; ANDRE et al. 1993a) have been identified in the cerebellar cortex. In particular in rodents, muscarinic receptors showed a higher density in the granular and P-cell layers than in the molecular layer (NEUSTADT et al. 1988), some of these receptors being of the m2 subtype (ANDRE et al. 1992a for references). The nicotinic effects, however, were located in the granular layer (SWANSON et al. 1987). While the cholinergic fibers to the flocculonodular lobe originate from the medial and descending vestibular nuclei as well as from the praepositus hypoglossi (BARMACK et al. 1992), possible sources of cholinergic fibers to the paleo- and neocerebellum could be the cerebellar nuclei (IKEDA et al. 1991) and the lateral reticular nucleus (BARMACK et al. 1992; TATEHATA et al. 1987). Most interestingly, cholinergic neurons projecting to the cerebellar vermis (lobules V–VII) were also found in the dorsal pontine tegmentum of kittens (CIRELLI et al. 1994), namely in the PPT and LDT, as well as within the LC complex, where ChAT-positive neurons (JONES 1991b) appeared to be intermingled with noradrenergic LC neurons also projecting to the same corticocerebellar area (FOOTE et al. 1983). These findings are of interest since both populations of cholinergic and noradrenergic systems projecting to the cerebellum were found in the same dorsal tegmental regions which are involved in the regulation of the sleep-waking state (see Sect. C.V.3.c).

Let us describe first the action of the cholinergic system on the cerebellar cortex. It appeared from previous experiments that the physiological action of ACh on individual neurons in the cerebellar cortex, as well as the nature of the receptors involved, is rather controversial, some authors describing excitation and others describing inhibition of P cells (see ANDRE 1992a, 1993 for references). Moreover, the cholinergic system may act on the cerebellar cortex not only postsynaptically, but also presynaptically by utilizing nicotinic receptors (LAPCHAK et al. 1989b). Quite recently the effects of ACh at the cellular level have been reinvestigated in anesthetized rats and attempts made to study the possible modulatory role of the cholinergic system in the cerebellum (ANDRE et al. 1993b, 1994). In particular, the effects of microiontophoretic application of a muscarinic agonist were tested not only on the spontaneous discharge of P cells but also on their responses to pulses of glutamate, the putative transmitter of parallel fibers, or GABA, the neurotransmitter of stellate and basket cells. Microiontophoretic application of the muscarinic agonist bethanechol for 5 min produced inconsistent effects on the spontaneous discharge of P cells, which remained either unchanged or slightly decreased or increased. Independently of these actions, bethanechol increased on average the simple spike responses of P cells to glutamate by up to 186% of the control values. Similarly the muscarinic agonist increased the GABA-induced inhibition of the simple spike discharge of the P cells by up to 170% of the control values. These effects, which appeared towards the end of the bethanechol application, reached a peak within 5–15 min and slowly declined 30–45 min after the end of bethanechol application. This long-lasting effect was not due to the persistent action of the drug on the receptors, because a similar time course was observed with

ACh, which is rapidly inactivated. Moreover, the potentiation exerted by bethanechol in the glutamate and the GABA-induced responses was prevented by combined microiontophoresis of the muscarinic antagonist, scopolamine.

In conclusion, it appears that the action of the muscarinic receptors is to increase the signal-to-noise ratio of the responses of P cells to excitatory or inhibitory amino acids, which are the neurotransmitters of the conventional afferent system, thus improving information transfer within local circuits. This modulatory influence of muscarinic receptors in the cerebellar cortex, which facilitates the responses of P cells to regular neurotransmitters without necessarily affecting the neuronal discharge at rest, is consistent with earlier findings showing a facilitatory action of ACh on excitatory responses of neocortical neurons (McCormick and Prince 1986; McKenna et al. 1988; Metherate et al. 1988; see Andre 1993 for references). In particular, ACh would regulate cortical information processing by improving the signal-to-noise ratio of neuronal responses to sensory inputs. These effects were in part at least of muscarinic origin, because they could be produced by selective agonists (McKenna et al. 1988) and blocked by the corresponding antagonist (Metherate et al. 1988). The action of ACh at the cortical level has been attributed to a reduction in voltage-dependent K^+ current (I_m) activated by membrane depolarization, as well as in slow Ca^{2+}-dependent K^+ current responsible for the afterhyperpolarization. Blockade of either one of these conductances at the corticocerebellar level would increase the response of the P cells to any excitatory input, including pulses of glutamate. An additional mechanism mediating the potentiation of glutamate response could be the increase in membrane resistance as observed in P cells after bath or iontophoretic in vitro application of ACh (Crepel and Dhanjal 1982). Second messengers can also be implicated in the long-term increase in responsiveness of P cells to excitatory or inhibitory neurotransmitters. Indeed, some of the long-term effects of ACh on membrane resistance can be mimicked by cyclic guanosine monophosphate, a putative cholinergic second messenger (Woody et al. 1978).

The possible involvement of the cholinergic system in the control of cerebellar functions is supported by the results of experiments showing that local application into the cerebellar vermis of cholinergic agents may modify posture as well as the gain of the VS reflex. It is known that in decerebrate cats the P cells located in the paramedial zone B of the cerebellar anterior vermis, which projects directly to the LVN (Corvaja and Pompeiano 1979), on which they exert an inhibitory influence (Ito 1984), respond to roll tilt of the animal, leading to sinusoidal stimulation of labyrinth receptors. (Denoth et al. 1979). However, the P cells fired out of phase with respect to the VS neurons, which exert an excitatory influence on the ipsilateral limb extensor motoneurons. It appeared, therefore, that the activation of the VS neurons (Boyle and Pompeiano 1980; Marchand et al. 1987; Schor and Miller 1982), and thus the related contraction of limb extensors during ipsilateral tilt (MANZONI et al. 1983a; SCHOR and MILLER 1981), was due not only to an increased discharge of ipsilateral labyrinthine receptors, but also to a reduced discharge of P cells of the cerebellar vermis, leading to disinhibition of the corresponding VS neurons. The conclusion of this study, i.e., that the cerebellar vermis contributes positively to the gain of the VS reflex, was proved by the fact that local microinjection into the cerebellar vermis of GABA-A or GABA-B agonists, which inactivate the P cells of the cerebellar vermis, decreased the amplitude of the EMG responses of limb extensors to animal tilt (ANDRE et al. 1992b). Experiments performed in decerebrate cats (ANDRE et al. 1992a, 1993a) have now shown that unilateral injection into zone B of the cerebellar anterior vermis of cholinergic muscarinic and nicotinic agonists (e.g., bethanechol and nicotine) decreased the postural activity in the ipsilateral limbs, probably due to a slightly increased activity of the corresponding P cells. However, the gain of the VS reflexes significantly increased on both sides, while the phase angle of the responses, which was related to the extreme animal position, remained unmodified. Just the opposite changes in posture and gain of the VS reflexes were obtained after unilateral local injection into the same corticocerebellar area of the muscarinic antagonist, scopolamine, or the nicotinic antagonists, hexametonium or D-tubocurarine.

As reported above, the increased discharge of the VS neurons and thus the contraction of the corresponding limb extensors during side-down tilt depends not only upon an increased excitatory input originating from ipsilateral labyrinth receptors, but also on disinhibition of the same neurons resulting from a reduced discharge of the overlying P cells. It is likely, therefore, that the cholinergic system enhances the depth of modulation of the P cells of the cerebellar vermis to given parameters of labyrinth stimulation, thus increasing the gain of the VS reflex. This hypothesis is supported by the results of the experiments reported above showing that microiontophoretic application of bethanechol increases the signal-to-noise ratio of the response of P cells to glutamate and GABA, i.e., to the excitatory and inhibitory neurotransmitters which are used by the cerebellar network to determine the output of P cells to the labyrinth signals (ANDRE et al. 1993b, 1994). The fact the cholinergic receptors are also located on granule cells, which transmit the afferent signals from one side of the cerebellar vermis to the P cells of the opposite side, may explain why local administration of cholinergic agents affects the response gain not only of the ipsilateral but also of the contralateral limb extensors to labyrinth stimulation.

It is of interest that, similar to the increase in gain of the VS reflex which occurred after local injection of cholinergic agonists into the zone B of the vermis (ANDRE et al. 1992a, 1993), bilateral microinjection of the same agents into the flocculus of the rabbit increased the gain of the vestibuloocular reflex (VOR), by utilizing not only muscarinic but probably also nicotinic receptors (TAN and COLLEWJIN 1991). Further experiments have shown that a bilateral floccular microinjection of the aselective cholinergic agonist, carbachol, increased in the rabbit the gain of the optokinetic response to a sinusoidal motion stimulus (TAN and COLLEWJIN 1991) and accelerated the buildup of slow-phase velocity of optokinetic nystagmus (OKN) in either direction (TAN et al. 1992). On the other hand, unilateral floccular injection specifically enhanced the buildup of OKN slow-phase velocity only in the direction towards the injected side (ipsiversive) (TAN et al. 1993a). In addition to these findings, a bilateral injection of carbachol in the flocculus shortened the duration of the optokinetic afternystagmus (TAN et al. 1992), as well as the duration of the postrotatory vestibular nystagmus (TAN et al. 1993b).

In addition to cholinergic afferents, the dorsal pontine tegmentum gives rise to noradrenergic afferents impinging on the cerebellar cortex. These fibers originate from the LC and make synaptic contacts primarily on P-cell dendrites in the molecular layer and to a lesser extent on the P-cell body and superficial granule cell layers (FOOTE et al. 1983). The neurochemical specificity of the NE afferents to the cerebellar cortex is supported by the results of in vitro autoradiographic studies showing in rats a high level of α_1-adrenoceptors in the molecular layer (JONES et al. 1985; PALACIOS et al. 1987), a low level of α_2-adrenoceptors in the granular layer (BRUNING et al. 1987; UNNERSTALL et al. 1984) and β-adrenoceptors in the molecular layer (LORTON and DAVIS 1987; RAINBOW et al. 1984) as well as in "patches" surrounding small groups of P-cell somata (SUTIN and MINNEMAN 1985). The possibility that β-adrenoceptors, which are mainly of the β_2-subtype (see RAINBOW et al. 1984 for references), are not located postsynaptically on P cells, but rather presynaptically on the terminals of parallel fibers, is documented by the fact that the neurons expressing β_2-adrenoceptors correspond to the granule cells, as shown by using in situ hybridization (PALACIOS 1988).

As reported for ACh, even the physiological action of NE on the spontaneous discharge of corticocerebellar units is rather controversial (see ANDRE et al. 1991 for references). More consistent, however, are the results of NE on the induced discharge of P cells. In particular, there is evidence that iontophoretic application of NE, while depressing the spontaneous activity of P cells, enhanced the responses of these cells to both excitatory (mossy fibers and climbing fibers) and inhibitory (basket or stellate cells) inputs (MOISES et al. 1990), as well as to the corresponding excitatory (glutamate, aspartate) and inhibitory (GABA) transmitters (MOISES et al. 1979; YEH et al. 1981; WATERHOUSE et al. 1982). It appears, therefore, that one of the main functions of the NE input in cerebellar operation is to augment target neuron responsiveness to conventional afferent systems, thus increasing the signal-to-noise ratio of the evoked versus spontaneous activity (WATERHOUSE et al. 1988). The same input could also act to gate the efficacy of subliminal synaptic inputs conveyed by the classical afferent systems (WATERHOUSE et al. 1988; MOISES et al. 1990).

The involvement of the noradrenergic system in the control of cerebellar functions is supported by the results of experiments showing that local application into the cerebellar vermis of noradrenergic agents modifies both posture and gain of the VS reflex. Experiments performed in decerebrate cats (see ANDRE et al. 1991 for references) have in fact shown that unilateral microinjection into the vermal cortex of the cerebellar anterior lobe of the α_1- and α_2-adrenergic agonists, metoxamine and clonidine, as well as of the β-adrenergic agonist, isoproterenol, reduced the postural activity in the ipsilateral limbs, while that of the contralateral limbs either remained ummodified or slightly increased. Just the opposite changes in posture were obtained in other experiments after injection of α_1-, α_2-, and β-adrenergic antagonists (i.e., prazosin, yohimbine, and propranolol, respectively). Moreover, local administration of the selective α_1- (metoxamine) or the α_2-adrenergic agonist (clonidine) increased the gain of the VS reflexes of both sides. Bilateral effects similar in sign but smaller in amplitude than those induced by clonidine were also elicited by the aselective β-adrenergic agonist, isoproterenol, which acts on both β_1- and β_2-adrenoceptors. In these experiments, there were only slight changes in the phase angle of the responses, which always remained related to the extreme animal position. Moreover, the specificity of the results was supported by the fact that the changes in the VS reflexes induced by a given adrenergic agonist were greatly reduced or suppressed by previous injection into the cerebellar vermis of the corresponding antagonist.

In order to account for these findings we postulated microinjection of the noradrenergic agonists may, similar to the injection of cholinergic agonists, increase the amplitude of modulation of the P cells of the cerebellar vermis to labyrinth stimulation, thus increasing the gain of the VS reflex. There are, however, some differences in the relative role that the cholinergic and the noradrenergic systems exert in the cerebellar control of the vestibular reflex. In fact, while the cholinergic system acts by increasing the gain of both the VS reflex (ANDRE et al. 1992a, 1993a) and the VOR (TAN and COLLEWJIN 1991), the noradrenergic system increases the VS reflex gain (see ANDRE et al. 1991 for references) but not of the VOR (VAN NEERVEN et al. 1990; TAN et al. 1991). However, the β-noradrenergic system increases the adaptation of both the VS reflex (POMPEIANO et al. 1991) and the VOR (VAN NEERVEN et al. 1990).

Because noradrenergic (FOOTE et al. 1983) and cholinergic afferents to the cerebellar cortex (CIRELLI et al. 1994) may originate from the same dorsal pontine tegmental structures as those involved in the regulation of the sleepwaking state, such as the LC complex and the dorsal pontine tegmental area, we conclude that one of the main functions of these systems is to act as a variable gain regulator, which adapts the amplitude of both the VS reflex and the VOR to the animal state. This would ensure the development of appropriate vestibular reflex adjustments according to the state of the animal.

E. Cholinergic System and Respiratory Mechanisms

In addition to the postural changes which occur during natural REM sleep, this phase of sleep is accompanied by changes in the activity of various respiratory muscles, which differ among the various muscles studied (see REMMERS 1981 for review). Rib cage muscles are not uniformly affected; the intercostal muscles become markedly hypotonic throughout REM sleep (PARMEGGIANI and SABATTINI 1972; DURON and MARLOT 1980; MEGIRIAN et al. 1987), whereas the interchondral, triangular sterni, and levator costae muscles maintain their inspiratory activity (DURON and MARLOT 1980; MEGIRIAN et al. 1987). Also the activity of upper airway muscles can be affected differently; pharyngeal and laryngeal abductors become atonic or hypotonic (SAUERLAND and HARPER 1976; MEGIRIAN et al. 1985; SHERREY et al. 1986), but the activity of certain constrictors is maintained (SHERREY et al. 1986). Although the diaphragm is spared by the intense descending inhibition observed in postural muscles during REM sleep (REMMERS et al. 1976; DURON and MARLOT 1980; REMMERS 1981; PHILLIPSON and BOWES 1986), recent findings indicate that there is some suppression of activity, especially during episodes of phasic REM (OREM 1986b; SIECK et al. 1984; KLINE et al. 1986; HENDRICKS et al. 1990). To explain these variable effects of REM sleep on respiratory muscle activity, it has been postulated that motoneurons receive converging inputs, excitatory from respiratory premotor neurons of the medulla and inhibitory from areas of the brain stem reticular formation regulating posture and muscle tone, the strength of each varying with the importance of the muscle's functions in respiratory and postural mechanisms (DURON and MARLOT 1980; OREM 1980a; REMMERS 1981).

Recent experiments have shown that a depression of respiratory muscle activity similar to that obtained during natural REM sleep could also be elicited in chronically implanted intact cats following carbachol microinjections into the dorsal pontine tegmentum (LYDIC and BAGHDOYAN 1989; LYDIC et al. 1989, 1991a). Moreover, this depression was also associated with a significant decrease in respiratory rate, so that the decrease in ventilation was more pronounced than in natural REM sleep (LYDIC and BAGHDOYAN 1989).

It has already been reported in Sect. C.I that carbachol microinjection into the pontine tegmentum of decerebrate cats evokes a postural atonia that resembles the atonia of natural REM sleep. KIMURA et al. (1990) have recently used this preparation to study the changes in respiratory neuronal activity that accompany the atonia. The activity of representative respiratory motor nerves – phrenic, intercostal, and hypoglossal – and that of a motor branch of C4 was recorded in decerebrate, vagotomized, paralyzed, and artificially ventilated cats. After the microinjection of carbachol, there was a profound suppression of activity in all the nerves and a decrease in respiratory rate. The magnitude of the suppression of respiratory-related activity was phrenic (65% that of the control) < inspiratory intercostal (50%) < hypoglossal (10%) < expiratory intercostal (5%), while the decrease in respiratory rate corresponded to 70% that of the control. A reversal of the effects described above was elicited by the microinjection of atropine into the same sites as the carbachol injection. The sites, which upon injection of carbachol produced the effects described above, were located in the pontine reticular formation, ventral or ventromedial to the LC, thus corresponding to the site giving origin to many other signs of REM sleep in intact animals (KATAYAMA et al. 1984; VANNI-MERCIER et al. 1989; YAMAMOTO et al. 1990a,b). Effects similar to those described by KIMURA et al. (1990) were also obtained in a different model of REM sleep by KAWAHARA et al. (1988a,b, 1989a,b), who used electrical stimulation to provoke the atonia in decerebrate cats. It appears, therefore, that the cholinoceptive pontine reticular region has the potential to depress powerfully and differentially various respiratory motoneuronal pools and to reduce the respiratory rate.

It is of interest that while the inactivation of lumbar motoneurons during both natural REM sleep and the state of carbachol-induced, REM sleep-like atonia is due to a postsynaptic inhibitory mechanism involving glycine as a neurotransmitter (MORALES et al. 1987; CHASE et al. 1989), the suppression of activity of the hypoglossal motoneurons in the same experimental conditions was apparently not mediated by inhibitory amino acids (SOJA et al. 1987; KUBIN et al. 1993). Moreover, the tonic depressant effects on respiration occurring during the carbachol-induced REM episodes differed from the changes that occur during natural REM sleep (REMMERS 1981; PHILLIPSON and BOWES 1986) or the REM sleep-like state induced in chronically implanted cats by pontine carbachol injections (LYDIC and BAGHDOYAN 1989; LYDIC et al. 1991a–c), in that they were strong and not accompanied by irregularities in the respiratory pattern.

Unfortunately, in the original study by KIMURA et al. (1990), the use of paralyzed animals precluded study of the eye movements and thus relation of the changes in the respiratory activity to the occurrence of REM. In a later study, however, TOJIMA et al. (1992) reported the effects of pontine injections of carbachol on ventilation and the respiratory motor output in acutely decerebrated, spontaneously breathing cats in which the electrooculogram was also recorded. It was found in this study that phrenic nerve activity was not depressed in the steady-state phase of the carbachol response, unlike in the artificially ventilated cats, whereas the decreases in respiratory rate were similarly profound and long-lasting in the two preparations and paralleled the decrease in postural muscle tone and the activity of upper airway and rib cage muscles. Similar results were also elicited by pontine carbachol in spontaneously breathing decerebrate rats (TAGUCHI et al. 1992). An additional finding in these studies was that the breathing after carbachol was quite regular even in decerebrate animals showing vigorous eye movements in response to carbachol injection. This is in striking contrast to the pattern of breathing typically observed during natural REM sleep (OREM 1980b; KLINE et al. 1986; PHILLIPSON and BOWES 1986). The regularity of breathing described above can be attributed to the fact that decerebration interrupts ascending pathways which may, through cortical or subcortical (hypothalamic) loops, modify the activity of the pontine and medullary respiratory centers.

In conclusion, the mechanisms underlying the respiratory and postural effects described in decerebrate animals are likely to be common to both systems, because the effects were produced with similar timing through cholinoceptive neurons located in a pontine area that has been implicated in the generation of REM sleep and atonia (SIEGEL 1989). Moreover, the pattern of this depression was similar to that seen during natural REM sleep, in that phrenic nerve activity was depressed the least, and the upper airway (hypoglossal) and internal intercostal activities the most. There is, therefore, a potential for tonic respiratory inhibition to occur in the decerebrate preparations, which is suppressed and overridden by excitatory inputs during natural REM sleep.

The final question raised now is whether the atonia-related depression of activity in respiratory muscles, which occurs during natural REM sleep in intact animals or following pontine injections of carbachol in decerebrate animals, can be due to a state-specific inhibition or disfacilitation that acts directly on the motoneurons of the respiratory pump and/or to a reduced phasic respiratory drive that reaches these motoneurons through specific respiratory bulbospinal pathways. Observations made by OREM (1980a) in chronically implanted cats have shown that the activity of the premotor neurons of the ventral respiratory group (VRG) of the medulla was in general decreased during tonic REM sleep, whereas facilitatory effects dominated during the phasic events of REM sleep (but see TROTTER and OREM 1991).

This approach was extended by KUBIN et al. (1992), who studied the effect of pontine carbachol on the activity of VRG neurons in decerebrate vagotomized, paralyzed, and artificially ventilated cats. Carbachol injected in the pontine tegmentum produced a depression of the peak firing rate in most of the inspiratory and expiratory VRG neurons, as well as in the simultaneously recorded phrenic and intercostal, inspiratory and expiratory activity. However, the demonstration that phrenic nerve activity could be depressed to some degree even without changes in VRG cell activity and that intercostal activity was consistently depressed more than that of the phrenic nerve (KIMURA et al. 1990) suggested that there were additional inhibitory (and/or disfacilitatory) inputs responsible for the depression of intercostal motoneuronal activity during the REM sleep-like atonia (DURON and MARLOT 1980; OREM 1980b; REMMERS 1981). The carbachol-injected decerebrate animal may thus provide a suitable model for studying the tonic changes in respiration and upper airway muscle tone that are directly related to the mechanisms underlying the postural atonia of REM sleep, and perhaps other states that involve a generalized postural atonia, such as cataplexy (MITLER and DEMENT 1974) and concussion-induced behavioral suppression (HAYES et al. 1984).

It is outside the aims of this review to investigate whether the cholinergic pontine system is also involved in the changes in cardiovascular activity that occur during natural REM sleep. It is known that the sympathetic activity is in the cat maintained by the tonic discharge of excitatory neurons located in the rostral portion of the ventrolateral medulla (RVLM) (CALARESU and YARDLEY 1988) and that the basal activity of these RVLM neurons is sustained by neurons located in the lateral tegmental field (BARMAN and GEBBER 1987; GEBBER and BARMAN 1988; HAYES and WEAVER 1992).

In a recent study, using electrical stimulation within the dorsal pontine tegmentum of decerebrate cats, sites were found from which a bilateral decrease in muscle tone and a parallel drop in blood pressure could be elicited (IWAHARA et al. 1991). Similarly, in decerebrate cats, pontine carbachol produced a decrease in blood pressure that developed in parallel to the postural atonia and respiratory depression (KIMURA et al. 1990; LAI and SIEGEL 1990). These responses may thus result from cholinergic excitation of cells or pathways that may also be activated during REM sleep. It should also be mentioned that the noradrenergic LC complex may influence the sympathetic activity not only by supressing the discharge of the cholinosensitive area located in the dorsal pontine tegmentum (see Sect. I.C), but also by utilizing a direct coeruleospinal projection acting on the intermediate-lateral column of the spinal cord (see GUYENET 1991 for references).

F. Conclusions

The evidence reviewed above indicates that ACh is involved in several functions related to the regulation of sleep and wakefulness: the generation of REM sleep, the modulation of thalamic, hippocampal, and cortical activation, and the control of postural adjustments induced by reflex and/or central mechanisms. To conclude, the main results from the previous sections are summarized in the light of these general functions.

I. Cholinergic System and Cortical Activation

In their seminal work, MORUZZI and MAGOUN (1949) established that cortical activation can be reliably produced by stimulating the brain stem reticular formation, and they introduced the concept of the "ascending reticular activating system" or ARAS. Lesion studies indicated that a critical node of this system is the region found between the midpontine pretrigeminal transections, resulting in persistent wakefulness, and low collicular transections, resulting in persistent sleep (MORUZZI 1972). This region contains the PPT/

LDT cholinergic cell groups, the LC noradrenergic neurons, and several other cell types. Many studies have tried to establish a neurochemical counterpart for the physiological concept of the ARAS. The original hypothesis that the ARAS would be cholinergic (SHUTE and LEWIS 1967), which was based on AChE histochemistry, is still consistent with many recent findings. As we have seen, ChAT immunochemistry combined with retrograde tracers has confirmed a strong innervation of the thalamus on the part of LDT/PPT. Other data indicate that application of ACh and stimulation of the LDT/PPT can directly depolarize thalamic neurons (and inhibit reticular thalamic neurons), decouple synchronized neurons, disrupt the slow oscillations that support slow waves and spindles, enhance 40-Hz rhythms, and facilitate the transmission of incoming signals (STERIADE et al. 1991), and that these effects are blocked by nicotinic and/or muscarinic antagonists. LDT/PPT neurons discharge at higher rates during activated states, namely W and REM sleep, and increase their discharge 30–60 s earlier than EEG activation (STERIADE et al. 1990c). Finally, there is evidence that the cholinergic system is also implicated in the generation of the hippocampal theta rhythm, a slow rhythm that is associated with cortical activation both in waking and in REM sleep.

Despite the positive evidence for the involvement of the LDT/PPT in cortical activation, the brain stem cholinergic system is only part of the picture. Identified cholinergic LDT/PPT neurons are inhibited by NE and 5-HT and, when hyperpolarized, they tend to fire in bursts. Thus, their contribution to desynchronization and waking, when NE and 5-HT levels are high, may be phasic rather than tonic (KAMONDI et al. 1992). More importantly, lesions of the cholinergic LDT/PPT are not followed by cortical deactivation during waking (JONES and WEBSTER 1988). Finally, the ascending cholinergic system located in the LDT/PPT does not project to the cortex to any significant degree, and can thus contribute to cortical activation only in an indirect way, by modulating the activity of noncholinergic thalamocortical pathways.

On the other hand, there is now much evidence for a second cholinergic system that is involved in cortical activation but does not reside in the brain stem. This system, originating in the NB and neighboring areas, projects to the entire cortex, seems to be more active during activated states than during SWS, is responsible for the increased release of ACh during cortical activation, and its lesion interferes with cortical activation. ACh depolarizes cortical neurons as it does with thalamic neurons and, together with NE and 5-HT (MCCORMICK and WILLIAMSON 1989; MCCORMICK and PAPE 1990), it could interfere with slow-wave generation at the cortical level. Interestingly, the brain stem cholinergic group participates with cholinergic activation of the neocortex by projecting to the cholinergic basal forebrain. Basal forebrain structures are excited by ACh and cholinergic agonists, and the stimulation of the reticular formation increases the release of ACh from the neocortex.

Finally, the simultaneous involvement in cortical activation of neuromodulatory systems using other neurotransmitters should be underlined. Other ascending systems, notably the LC and the raphe dorsalis, release substances (NE and 5-HT) that depolarize thalamic neurons and dampen slow oscillations by decreasing various K conductances, thus contributing to cortical activation (McCORMICK and PAPE 1990). The noradrenergic and serotoninergic systems also discharge at higher rates during waking than during SWS. Unlike the cholinergic system, however, they are totally silent during REM sleep, and therefore the cholinergic system is the only one associated with that special case of activation. Thus, EEG desynchronization would be prevalently aminergic during W and cholinergic during REM sleep. The close opportunities for interactions between the cholinergic and the aminergic systems in the dorsal pontine tegmentum are therefore of great theoretical and experimental significance (JONES 1991a,c).

II. Cholinergic System and REM Sleep

As outlined above, early pharmacological experiments based on systemic injections had suggested the notion that the cholinergic system is involved in promoting REM sleep. To date, cholinergic agonists and cholinesterase inhibitors are still the most effective agents in inducing REM sleep, particularly after local microinjections. Further evidence has pointed to cholinergic cells located in the dorsal pontine tegmentum as playing a crucial role in REM sleep generation. The two cholinergic cell groups in the region, LDT and PPT, lie just caudal to the transection level above which REM sleep is abolished, and below which it persists (JOUVET 1962). Large electrolytic, radiofrequency or neurotoxic lesions of this region, the latter affecting preferentially cholinergic cells, are most effective in disrupting REM sleep and some of its components. Recording studies indicate that REM-on and/or PGO-on cells are found in this region, and are presumed to be cholinergic. Anatomically, cholinergic projections from this area are in an ideal position to orchestrate both tonic and phasic aspects of REM sleep by innervating the relevant executive structures. Finally, the importance of cholinergic cells in the peribrachial area is underlined by the recent discovery that stimulation of presumably cholinergic cells produces a delayed but persistent increase in REM sleep as well as state-independent PGO waves.

The mechanism and the pathways by which cholinergic neurons act to trigger REM sleep, however, are complex. The most effective/shortest-latency site for carbachol induction of REM sleep is a cholinoceptive area in the dorsal PRF, corresponding to the LCα/periLCα and neighboring structures, which receives a major cholinergic projection from PPT and LDT. Some neurons in this area fire specifically during REM sleep episodes, and many of them are activated by local injections of carbachol in parallel with the appearance of REM sleep. Since ACh release is increased in the PRF during REM sleep as well as after stimulation of the PPT, a possible interpretation is that the cholinergic PPT/LDT controls physiological REM sleep generation while the cholinoceptive LCα represents an executive area.

III. Cholinergic System and Postural Mechanisms

Observations made in the early 1960s had shown that the postural atonia which occurs during REM sleep was not due to a reduced discharge of supraspinal descending excitatory systems, leading to motoneuronal disfacilitation, but rather to an increased discharge of supraspinal descending inhibitory systems, leading to postsynaptic inhibition of α-motoneurons (POMPEIANO 1967). This conclusion was supported by the results of experiments in which the membrane potential change occurring in limb extensor motoneurons during REM sleep was recorded intracellularly (MORALES and CHASE 1978; MORALES et al. 1987). After the demonstration that the RS inhibitory system was under the tonic excitatory control of presumably cholinergic and/or cholinoceptive PRF neurons, and that these neurons were under the tonic inhibitory influence of NE LC neurons, evidence has been presented indicating that these dorsal pontine tegmental structures exert a prominent influence on decerebrate rigidity (Sect. D.I). In particular, it appears that the increase in postural activity which occurs after decerebration is due to an increased discharge of NE LC neurons, which are disinhibited by a mesencephalic transection (POMPEIANO et al. 1990). These neurons would then exert a facilitatory influence on posture not only by utilizing the direct CS projection (FUNG et al. 1991) but also by suppressing the activity of the PRF neurons and the related medullary inhibitory RS system (POMPEIANO et al. 1991). This interpretation of the experimental findings modifies the classical notion, i.e., that the decerebrate rigidity results from interruption of a corticoreticular pathway exerting an excitatory influence on the medullary inhibitory RS system (MAGOUN and RHINES 1947). In addition to static changes in posture, the dorsal pontine tegmental structures described above greatly contribute to the regulation of postural adjustments induced by labyrinth stimulation (Sect. D.II). The postural changes which occur by changing the position of the head in space are primarily due to activation of the three-neuronal VS reflexes, made by primary vestibular afferents, second-order VS neurons, and α-extensor motoneurons (BOYLE and POMPEIANO 1980; SCHOR and MILLER, 1982). There is also evidence, however, that both the medullary inhibitory RS neurons (MANZONI et al. 1983b) and the excitatory CS neurons (O. POMPEIANO et al. 1990) respond to natural stimulation of macular (gravity) receptors. Moreover, it appears that the neuronal circuit made by the NE LC neurons and the related cholinergic and/or cholinoceptive PRF neurons acts as a variable gain regulator, which adapts the gain of the VS reflexes to the animal state (POMPEIANO et al. 1991). The cholinergic, as well as the noradrenergic, systems located in the dorsal pontine tegmentum intervene also in the control of the postural adjustments induced by stimulation of the motor cortex (Sect. D.III).

Electrical stimulation of the motor cortex performed in unanesthetized cats at an intensity sufficient to elicit flexion of one limb produces a diagonal pattern of postural adjustments, characterized by a decreased force by the limb diagonally opposite to the moving one and an increased force by the other two. Latency measurements indicated that this diagonal pattern of postural changes results from a central command (GAHERY and NIEOULLON 1978). We have recently shown that, while the lateral vestibular nucleus is involved in the execution of the postural adjustments induced by cortical stimulation (LUCCARINI et al. 1992; POMPEIANO et al. 1993), the cholinoceptive PRF plays a prominent role in the gain regulation of these postural changes. In particular, bilateral microinjections into the peri-LC region and the surrounding dorsal PRF of the muscarinic agonist, bethanecol, which did not modify the flexion response of the performing limb, selectively decreased or suppressed the early component of the postural response driven by the cortical command (LUCCARINI et al. 1990). Since the cholinoceptive pontine tegmental neurons are responsible for the spontaneous fluctuations in posture related to the sleep-waking cycle, they may intervene in order to adapt to the animal state the amplitude of the postural adjustments during cortically induced limb movements. The cholinergic system may control not only spinal, but also cerebellar functions (Sect. D.IV). Experiments performed in the cerebellar cortex of rats (ANDRE et al. 1993b, 1994) have shown that microiontophoretic application of the muscarinic agonist, bethanechol, which produced inconsistent changes in the spontaneous discharge of Purkinje (P) cells, greatly increased the simple spike responses of these cells to both excitatory (glutamate) and inhibitory (GABA) neurotransmitters. This potentiation, which could also be duplicated by ACh, was prevented by combined microiontophoresis of the muscarinic antagonist, scopolamine. This finding clearly indicates that the action of the muscarinic receptors is to increase the signal-to-noise ratio of the responses of the P cells to excitatory or inhibitory amino acids, which are the neurotransmitters of the conventional afferent systems, thus improving the information transfer within local circuits. This conclusion is supported by the results of experiments showing that local injection into the cerebellar vermis of cholinergic (muscarinic and nicotinic) agents modifies the gain of both the VS reflex (ANDRE et al. 1992a, 1993a) and the VOR (TAN and COLLEWJIN 1991; TAN et al. 1992, 1993a,b). It is of interest that, in addition to cholinergic afferents, the cerebellar cortex also receives NE afferents. Moreover, there is evidence that iontophoretic application of NE in the cerebellar cortex of rats, while depressing the spontaneous activity of P cells, enhances the responses of these cells to conventional excitatory and inhibitory inputs, not only by increasing the signal-to-noise ratio of the evoked versus spontaneous activity, but also by acting to gate the efficacy of subliminal synaptic inputs conveyed by the classical afferent systems (WATERHOUSE et al. 1988). Experiments performed in different preparations have also shown that appropriate injections into the cerebellar cortex of α_1-, α_2-, and β-noradrenergic agents modified the basal gain of the VS reflex (ANDRE et al. 1991) but not of the VOR (VAN NEERVEN et al. 1990; TAN et al. 1991). Most interestingly, the β-noradrenergic system increased the adaptive change in gain of both the VS reflex (O. POMPEIANO et al. 1992) and the VOR (VAN NEERVEN et al. 1990). Because noradrenergic (FOOTE et al. 1983) and at least some of the cholinergic afferents to the cerebellar cortex (CIRELLI et al. 1994) originate from the dorsal pontine tegmentum structures which are involved in the regulation of the sleep-waking cycle (such as the LC complex and the dorsal pontine tegmental area), we conclude that these systems may contribute to adapt the amplitude of the VS reflexes and the VOR to the different states of the animal.

A final comment concerns the results of experiments showing that the same cholinergic and/or cholinoceptive neurons located in the dorsal pontine tegmental area and implicated in the generation of REM sleep and atonia intervene in the depression of some respiratory muscle activity as shown in both intact and decerebrate animals (see Sect. E).

In conclusion, while many aspects of the role of the cholinergic system in sleep are still provisional, the picture we have today is much more complete and precise than just a few years ago. In the near future, our understanding will be enhanced by more selective cholinergic neurotoxins, unequivocal identification of the transmitters contained in recorded neurons, new techniques such as microdialysis, and functional-anatomical correlations made possible by double-labeling studies with markers for ACh and indicators of neuronal activity such as immediate early genes.

G. Summary

This chapter reviews the biochemistry, anatomy, and physiology of the central cholinergic system in relation to sleep-wakefulness states. Acetylcholine was the first neurotransmitter to be implicated in the control of wakefulness and sleep, and these early pharmacological studies have been fundamental for our understanding of such control. New findings have greatly extended our knowledge of the location of cholinergic centers and pathways, of the regional distribution and molecular identity of cholinergic receptors, and of the cellular effects of cholinergic agonists on target neurons. The activity of cholinergic cells during sleep-wakefulness states, the variations in the release of acetylcholine, and the changes in the availability of enzymes or receptors are also being investigated. These recent results are useful for the interpretation of the effects on sleep of various manipulations of the cholinergic system, such as systemic or local injections of agonists and antagonists, or lesions of cholinergic nuclei.

The picture that emerges indicates that the cholinergic system plays a central role in at least two major functions. The generation of REM sleep is largely orchestrated by a collection of cholinergic cells in the dorsal pontine tegmentum, corresponding to the laterodorsal tegmental and pedunculopontine nuclei. The activation of the thalamus, the cerebral cortex, and the hippocampus during both waking and REM sleep are mediated to a significant extent by ascending cholinergic influences arising respectively in the dorsal pontine tegmentum, in the cholinergic nucleus basalis, and in the medial septal nucleus. Evidence is also presented indicating that the cholinergic system intervenes in the control of posture and in the regulation of postural adjustments induced by reflex and central mechanisms. Although by no means the only one, acetylcholine is to date the neuromodulator that is most unequivocally involved in the regulation of sleep-wakefulness states.

References

Abercrombie ED, Jacobs BL (1987a) Single-unit response of noradrenergic neurons in the locus coeruleus of freely moving cats. I. Acutely presented stressful and non-stressful stimuli. J Neurosci 7:2837–2843

Abercrombie ED, Jacobs BL (1987b) Microinjected clonidine inhibits noradrenergic neurons of the locus coeruleus in freely moving cats. Neurosci Lett 76:203–298

Agoston DV (1988) Cholinergic co-transmitters. In: Whittaker VP (ed) Handbook of experimental pharmacology, vol 86. Springer, Berlin Heidelberg New York, pp 479–533

Aldrich MS, Frey K, Albin RL, Penney JB (1990) Muscarinic receptor autoradiographic studies in postomortem human narcoleptic brain. Sleep Res 19:179

Amatruda TT, Black DA, McKenna TM, McCarley RW, Hobson JA (1975) Sleep cycle control and cholinergic mechanisms: differential effects of carbachol injections at pontine brainstem sites. Brain Res 98:501–515

Amini-Sereshki L, Morrison AR (1986) Effects of pontine tegmental lesions that induce paradoxical sleep without atonia on thermoregulation in cats during wakefulness. Brain Res 384:23–29

Andersen P, Curtis DR (1964) The pharmacology of the synaptic and acetylcholine-induced excitation of ventrobasal thalamic neurons. Acta Physiol Scand 61: 100–120

Andre P, D'Ascanio P, Pompeiano O (1991) Noradrenergic agents into the cerebellar anterior vermis modify the gain of vestibulospinal reflexes in the cat. Prog Brain Res 88:463–484

Andre P, Pompeiano O, Manzoni D, D'Ascanio P (1992a) Muscarinic receptors in the cerebellar vermis modulate the gain of the vestibulospinal reflexes in decerebrate cats. Arch Ital Biol 130:213–245

Andre P, D'Ascanio P, Manzoni D, Pompeiano O (1992b) Depression of the vestibulospinal reflexes by intravermal microinjection of GABA-A and GABA-B agonists in decerebrate cats. Pflugers Arch 420:51

Andre P, D'Ascanio P, Manzoni D, Pompeiano O (1993a) Nicotinic receptors in the cerebellar vermis modulate the gain of the vestibulospinal reflexes in decerebrate cats. Arch Ital Biol 13:1–24

Andre P, Pompeiano O, White SR (1993b) Activation of muscarinic receptors induces a long-lasting enhancement of Purkinje cell responses to GABA. Brain Res 617:28–36

Andre P, Fascetti F, Pompeiano O, White SR (1994) The muscarinic agonist bethanechol enhances GABA-induced inhibition of Purkinje cells in the cerebellar cortex. Brain Res 637:1–9

Armstrong DM, Saper CB, Levery AI, Wainer BH, Terry RD (1983) Distribution of cholinergic neurons in rat brain: demonstrated by the immunocytochemical localization of choline acetyltransferase. J Comp Neurol 216:53–68

Asanuma C (1989) Basal forebrain projections to the reticular thalamic nucleus in rat. Proc Natl Acad Sci USA 86:4746–4750
Aston-Jones G (1985) Behavioral function of locus coeruleus derived from cellular attributes. Physiol Psychol 13:118–126
Aston-Jones G, Ennis M, Pieribone VA, Nickell WT, Shipley MT (1986) The brain nucleus locus coeruleus: restricted afferent control of a broad efferent network. Science 234:734–737
Baghdoyan HA, Monaco AP, Rodrigo-Angulo ML, Assens F, McCarley RW, Hobson JA (1984a) Microinjection of neostigmine into the pontine reticular formation of cats enhances desynchronized sleep signs. J Pharmacol Exp Ther 231:173–180
Baghdoyan HA, Rodrigo-Angulo ML, McCarley RW, Hobson JA (1984b) Site-specific enhancement and suppression of desynchronized sleep signs following cholinergic stimulation of three brainstem regions. Brain Res 306:39–52
Baghdoyan HA, Rodrigo-Angulo ML, McCarley RW, Hobson JA (1987) A neuroanatomical gradient in the pontine tegmentum for the cholinoceptive induction of desychronized sleep signs. Brain Res 414:245–261
Baghdoyan HA, Lydic R, Callaway CW, Hobson JA (1989) The carbachol induced enhancement of desynchronized sleep signs is dose dependent and antagonized by centrally administered atropine. Neuropsychopharmacology 2:67–79
Barmack NH, Baughman RW, Eckenstein FP (1992a) Cholinergic innervation of the cerebellum of rat, rabbit, cat and monkey as revealed by choline acetyltransferase activity and immunohistochemistry. J Comp Neurol 317:233–249
Barmack NH, Baughman RW, Eckenstein FP, Shojaku H (1992b) Secondary vestibular cholinergic projection to the cerebellum of rabbit and rat as revealed by choline acetyltransferase immunohistochemistry, retrograde and orthograde tracers. J Comp Neurol 317:250–270
Barman SM, Gebber GL (1987) Lateral tegmental field neurons of cat medulla: a source of basal activity of ventrolateral medullospinal sympathoexcitatory neurons. J Neurophysiol 57:1410–1424
Barnes CD, D'Ascanio P, Pompeiano O, Stampacchia G (1987) Effects of microinjections of cholinergic agonists into the pontine reticular formation on the gain of the vestibulospinal reflexes in decerebrate cats. Arch Ital Biol 125:71–105
Baxter BL (1969) Induction of both emotional behavior and a novel form of REM sleep by chemical stimulation applied to cat mesencephalon. Exp Neurol 23: 220–229
Beani L, Bianchi C, Giacomelli A, Tamberi F (1978) Noradrenaline inhibition of acetylcholine release from guinea-pig brain. Eur J Pharmacol 48:179–193
Belardetti F, Borgia R, Mancia M (1977) Prosencephalic mechanisms of ECoG desynchronization in cerveau isolé cats. Electroencephalogr Clin Neurophysiol 42:213–225
Bianchi C, Spidalieri G, Guandalini P, Tanganelli S, Beani L (1979) Inhibition of acetylcholine outflow from guinea-pig cerebral cortex following locus coeruleus stimulation. Neurosci Lett 14:97–100
Bier MJ, Greene RW, McCarley RW (1990) Norepinephrine mediated responses in the pontine reticular formation: in vitro and in vivo studies. 10th European Congress on Sleep Research May 20–25, Strasbourg, p 333
Birnstiel S, Gerber U, Stevens DR, Greene RW, McCarley RW (1991) Intracellular investigation of nicotinic actions in the rat medial pontine reticular formation in vitro. Sleep Res 20:15
Boehme RE, Baker TL, Mefford IN, Barchas JD, Dement WC, Ciaranello RD (1984) Narcolepsy: cholinergic receptor changes in an animal model. Life Sci 34:1825–1828
Bonner TI (1989) The molecular basis of muscarinic receptor diversity. Trends Neurosci 12:148–151
Bonner TI, Young AC, Brann MR, Buckley NJ (1988) Cloning and expression of the human and rat m5 muscarinic acetylcholine receptor genes. Neuron 1: 403–410

Boyle R, Pompeiano O (1980) Reciprocal responses to sinusoidal title of neurons in Deiters' nucleus and their dynamic characteristics. Arch Ital Biol 118:1–32

Brashear HR, Zaborski L, Heimer L (1986) Distribution of GABAergic and cholinergic neurons in the rat diagonal band. Neuroscience 17:439–451

Bruning G, Kaulen P, Baumgarten HG (1987) Quantitative autoradiographic localization of α_2-antagonist binding sites in rat brain using [^{3}H]-idazoxan. Neurosci Lett 83:333–337

Buckley NJ, Bonner TI, Brann MR (1988) Localization of a family of muscarinic receptor mRNAs in rat brain. J Neurosci 8:4646–4652

Buckley NJ, Bonner TI, Buckley CM, Brann MR (1989) Antagonist binding properties of five cloned muscarinic receptors expressed in CHO-K1 cells. Mol Pharmacol 35:469–476

Buzsaki G, Bickford RG, Ponomareff G, Thal LJ, Mandel R, Gage FH (1988) Mucleus basalis and thalamic control of neocortical activity in the freely moving rat. J Neurosci 8:4007–4026

Calaresu FR, Yardley CP (1988) Medullary basal sympathetic tone. Annu Rev Phys 50:511–524

Calvo JM, Datta S, Quattrochi J, Hobson JA (1992) Cholinergic microstimulation of the peribrachial nucleus in the cat. II. Delayed and prolonged increases in REM sleep. Arch Ital Biol 130:285–301

Casamenti F, Deffenu G, Abbamondi AL, Pepeu G (1986) Changes in cortical acetylcholine output induced by modulation of the nucleus basalis. Brain Res Bull 16:689–695

Celesia GG, Jasper HH (1966) Acetylcholine released from cerebral cortex in relation to state of activation. Neurology (Minneap) 16:1053–1064

Chase MH (1983) Synaptic mechanisms and circuitry involved in motoneuron control during sleep. Int Rev Neurobiol 24:213–258

Chase MN, Enomoto S, Murakami T, Nakamura Y, Taira M (1981) Intracellular potential of medullary reticular neurons during sleep and wakefulness. Exp Neurol 71:226–233

Chase MH, Oja PJ, Morales FR (1989) Evidence that glycine mediates the postsynaptic potentials that inhibit lumbar motoneurons during the atonia of active sleep. J Neurosci 9:743–751

Chesselet M-F (1984) Presynaptic regulation of neurotransmitter release in the brain: facts and hypothesis. Neuroscience 12:347–375

Cirelli C, Tononi G, Pompeiano M, Pompeiano O, Gennari A (1992) Modulation of desynchronized sleep through microinjection of α_1-adrenergic agonists and antagonists in the dorsal pontine tegmentum of the cat. Pflugers Arch 422: 273–279

Cirelli C, D'Ascanio P, Horn E, Pompeiano O, Stampacchia G (1993) Modulation of vestibulospinal reflexes through microinjection of an α_1-adrenergic antagonist in the dorsal pontine tegmentum of decerebrate cats. Arch Ital Biol 131:275–302

Cirelli C, Lin R, Fung S, Barnes CD, Pompeiano O (1994) Cholinergic projections of the dorsal pontine tegmentum to the vermal cortex of the kitten. Pflugers Arch 428:199

Clements JR, Grant SJ (1990) Glutamate and acetylcholine are colocalized in the laterodorsal tegmental and pedunculopontine nuclei. Soc Neurosci Abstr 16:1189

Cole AE, Nicoll RA (1983) Acetylcholine mediates a slow synaptic potential in hippocampal pyramidal cells. Science 221:1299–1301

Collier B, Mitchell JF (1967) The central release of acetylcholine during consciousness and after brain lesions. J Physiol (Lond) 188:83–98

Cordeau JP, Moreau A, Beaulnes A, Laurin C (1963) EEG and behavioral changes following microinjections of acetylcholine and adrenaline in the brain stem of cats. Arch Ital Biol 101:30–47

Corvaja N, Pompeiano O (1979) Identification of cerebellar corticovestibular neurons retrogradely labeled with horseradish peroxidase. Neuroscience 4: 507–515
Crawford GD, Correa L, Salvaterra PM (1982) Interaction of monoclonal antibodies with mammalian choline acetyltransferase. Proc Natl Acad Sci USA 79:7031–7035
Crepel F, Dhanjal SS (1982) Cholinergic mechanisms and neurotransmission in the cerebellum of the rat. An in vitro study. Brain Res 244:59–68
Curró Dossi R, Paré D, Steriade M (1990) Short-lasting nicotinic and long-lasting muscarinic depolarizing responses of thalamocortical neurons to stimulation of mesopontine cholinergic nuclei. J Neurophysiol 65:393–406
D'Ascanio P, Bettini E, Pompeiano O (1985a) Tonic inhibitory influences of locus coeruleus on the response gain of limb extensors to sinusoidal labyrinth and neck stimulations. Arch Ital Biol 123:69–100
D'Ascanio P, Bettini E, Pompeiano O (1985b) Tonic facilitatory influences of dorsal pontine reticular structures on the response gain of limb extensors to sinusoidal labyrinth and neck stimulations. Arch Ital Biol 123:101–132
D'Ascanio P, Pompeiano O, Stampacchia G, Tononi G (1988) Inhibition of vestibulospinal reflexes following cholinergic activation of the dorsal pontine reticular formation. Arch Ital Biol 126:291–316
D'Ascanio P, Horn E, Pompeiano O, Stampacchia G (1989a) Injections of β-sdrenergic substances in the locus coeruleus affect the gain of vestibulospinal reflexes in decerebrate cats. Arch Ital Biol 127:187–218
D'Ascanio P, Horn E, Pompeiano O, Stampacchia G (1989b) Injections of a β-adrenergic antagonist in pontine reticular structures modify the gain of vestibulospinal reflexes in decerebrate cats. Arch Ital Biol 127:275–303
D'Ascanio P, Pompeiano M, Tononi G (1989c) Inhibition of vestibulospinal reflexes during the episodes of postural atonia induced by unilateral lesion of the locus coeruleus in the decerebrate cat. Arch Ital Biol 127:63–79
Datta S, Dossi RC, Pare D, Oakson G, Steriade M (1991) Sunstantia-nigra reticulata neurons during sleep waking states: relation with pontogeniculooccipital waves. Brain Res 566:344–347
Datta S, Calvo JM, Quattroohi J, Hobson JA (1992) Cholinergic microstimulation of the peribrachial nucleus in the cat. I. Immediate and prolonged increases in ponto-geniculo-occipital waves. Arch Ital Biol 130:263–284
Do Lima AD, Singer W (1987) The brainstem projection to the lateral geniculate nucleus in the cat: identification of cholinergic and monoaminergic elements. J Comp Neurol 259:92–121
Denoth F, Magherini PC, Pompeiano O, Stanojevic M (1979) Responses of Purkinje cells of the cerebellar vermis to neck and macular vestibular inputs. Pflugers Arch 381:87–98
Detari L, Vanderwolf CH (1987) Activity of identified cortically projecting and other basal forebrain neurons during large slow waves and cortical activation in anaesthetized rats. Brain Res 437:1–10
Detari L, Juhasz G, Kukorelli T (1984) Firing properties of cat basal forebrain neurones during sleep-wakefulness cycle. Electroencephalogr Clin Neurophysiol 58:362–368
Di Chiara G, Olianas M, Del Fiacco M, Spano PF, Tagliamonte A (1977) Intranigral kainic acid is evidence that nigral non-dopaminergic neurones control posture. Nature 268:743–745
Domino EF, Yamamoto K (1965) Nicotine: effect on the sleep cycle of the cat. Science 150:637–638
Domino EF, Stawinski M (1970) Effect of cholinergic antisynthesis agent HC-3 on the awake-sleep cycle in the cat. Psychophysiology 7:315–316
Domino EF, Yamamoto K, Dren AT (1968) Role of cholinergic mechanisms in states of wakefulness and sleep. Prog Brain Res 128:113–133

Dreifuss JJ, Kelly JS (1972) The activity of identified supraoptic neurones and their response to acetylcholine applied by iontophoresis. J Physiol (Lond) 220: 105–118

Dudar JD (1975) The effect of septal nuclei stimulation on the release of acetylcholine from the rabbit hippocampus. Brain Res 83:123–133

Dudar JD (1977) The role of the septal nuclei in the release of acetylcholine from the rabbit cerebral cortex and dorsal hippocampus and the effect of atropine. Brain Res 129:237–246

Duffy FH, Burchfiel P, Bartels M, Guon M, Sim VM (1979) Long-term effects of organophosphates upon the human electroencephalogram. Toxicol Appl Pharmacol 47:161–176

Duron B, Marlot D (1980) Intercostal and diaphragmatic electrical activity during wakefulness and sleep in normal unrestrained adult cats. Sleep 3:269–280

Eckenstein F, Thoenen H (1982) Production of specific antisera and monoclonal antibodies to choline acetyltransferase: characterization and use for identification of cholinergic neurons. EMBO J 1:363–368

Egan TM, North RA (1985) Acetylcholine acts on m2-muscarinic receptors to excite rat locus coeruleus neurons. Br J Pharmacol 85:733–735

Ehlers C, Hendricksen SJ, Wang M, Rivier J, Vale W, Bloom FE (1983) Corticotropin releasing factor produces increases in brain excitability and convulsive seizures in rats. Brain Res 278:332–336

El Mansari M, Sakai K, Jouvet M (1989) Unitary characteristics of presumptive cholinergic tegmental neurons during the sleep-waking cycle in freely moving cats. Exp Brain Res 76:519–529

El Mansari M, Sakai K, Jouvet M (1990) Responses of presumed cholinergic mesopontine tegmental neurons to carbachol microinjections in freely moving cats. Exp Brain Res 83:115–123

Engberg G, Svensson TH (1980) Pharmacological analysis of a cholinergic receptor mediated regulation of brain norepinephrine neurons. J Neural Transm 49: 137–150

Ennis M, Aston Jones G (1986a) Evidence for self- and neighbor-mediated postactivation inhibition of locus coeruleus neurons. Brain Res 374:299–305

Ennis M, Aston Jones G (1986b) A potent excitatory input to nucleus locus coeruleus from the ventrolateral medulla. Neurosci Lett 71:299–305

Foote SL, Bloom FE, Aston-Jones G (1983) Nucleus locus coeruleus: new evidence of anatomical and physiological specificity. Physiol Rev 63:844–914

Fukuda K, Higashida H, Kubo T, Maeda A, Akiba I, Bujo H, Mishina M, Numa S (1988) Selective coupling with K^+ currents of muscarinic acetylcholine receptor subtypes in NG108-15 cells. Nature 335:355–358

Funahashi S (1983) Responses of monkey prefrontal neurons during a visual tracking task reinforced by substantia innominata self-stimulation. Brain Res 276: 267–276

Fung SJ, Pompeiano O, Barnes CD (1988) Coeruleospinal influence on recurrent inhibition of spinal motonuclei innervating antagonistic hindleg muscles of the cat. Pflugers Arch 412:346–353

Fung SJ, Manzoni D, Chan JYH, Pompeiano O, Barnes CD (1991) Locus coeruleus control of spinal motor output. Prog Brain Res 88:395

Gadea-Ciria M, Stadler H, Lloyd KG, Bartholini G (1973) Acetylcholine release within the cat striatum during the sleep-wakefulness cycle. Nature 243:518–519

Gahery Y, Nieoullon A (1978) Postural and kinetic coordination following cortical stimuli which induce flexion movements in the cat's limbs. Brain Res 149:23–37

Gahery Y, Pompeiano O, Coulmance M (1984) Changes in posturo-kinetic limb responses to cortical stimulation following unilateral neck deafferentation in the cat. Arch Ital Biol 122:129–154

Gaillard JM (1985) Involvement of noradrenaline in wakefulness and paradoxical sleep. In: Wauquier A, Gaillard JM, Monti JM, Radulovacki M (eds) Sleep, neurotransmitters and neuromodulators. Raven, New York, pp 57–67

Gebber GL, Barman SM (1988) Studies on the origin and generation of sympathetic nerve activity. Clin Exp Hypertens [A] 10 [Suppl 1]:33–44
Geffard M, McRae-Degueurce A, Souan ML (1985) Immunocytochemical detection of acetylcholine in the rat central nervous system. Science 229:77–79
George R, Haslett W, Jenden D (1964) A cholinergic mechanism in the brainstem reticular formation: induction of paradoxical sleep. Int J Neuropharmacol 3:541–552
Gildenberg PL, Hassler R (1971) Influence of stimulation of the cerebral cortex on vestibular nuclei units in the cat. Exp Brain Res 14:77–94
Gillin JC (1989) Sleep and affective disorders: theoretical perspectives. In: Kryger MH, Roth T, Dement WC (eds) Principles and practice of sleep medicine. Saunders, Philadelphia, pp 420–422
Gitlor MS, Reba RC, Cohen VI, Rzeszotarski WJ, Baumgold J (1992) A noval m2-selective muscarinic antagonist: binding characteristics and autoradiographic distribution in rat brain. Brain Res 582:253–260
Gnadt JW, Pegram CV (1986) Cholinergic brainstem mechanisms of REM sleep in the rat. Brain Res 384:29–41
Granit R (1970) The basis of motor control. Academic, New York
Greene RW, McCarley RW (1990) Cholinergic neurotransmission in the brainstem: implications for behavioral state control. In: Steriade M, Biesold D (eds) Brain cholinergic systems. Oxford University Press, New York, pp 224–235
Greene RW, Cerber U, McCarley RW (1989a) Cholinergic activation of medial pontine reticular formation neurons in vitro. Brain Res 476:154–159
Greene RW, Gerber U, Haas HL, McCarley RW (1989b) Noradrenergic actions on neurons on the medial pontine reticular formation in vitro. Sleep Res 18:11
Guilleminault C (1989) Narcolepsy syndrome. In: Kryger MH, Roth T, Dement WC (eds) Principles and practice of sleep medicine. Saunders, Philadelphia, pp 338–346
Guyenet PG (1991) Central noradrenergic neurons: the autonomic connection. Prog Brain, Res 88:365–380
Hallanger AE, Levey AI, Lee HJ, Rye DB, Wainer BH (1987) The origins of cholinergic and other subcortical afferents to the thalamus in the rat. J Comp Neurol 262:105–124
Halliwell JV, Adams PR (1982) Voltage-clamp analysis of muscarinic excitation in hippocampal neurons. Brain Res 250:71–92
Hammer R, Gianchetti A (1982) Muscarinic receptor subtypes: M1 and M2 biochemical and functional characterization. Life Sci 31:2991–2998
Haranath PS, Venkatakrishna-Bhatt H (1973) Release of acetylcholine from perfused cerebral ventricles in unanesthetized dogs during waking and sleep. Jpn J Physiol 23:241–250
Hayes K, Weaver LC (1992) Tonic sympathetic excitation and vasomotor control from pontine reticular neurons. Am J Physiol 263 32:H1567–H1575
Hayes RL, Pechura CM, Katayama Y, Povlishock JT, Giebel ML, Becker DP (1984) Activation of ponteine cholinergic sites implicated in unconsciousness following cerebral concussion in the cat. Science 223:301–303
Hazra J (1970) Effect of hemicholinium-3 on slow wave and paradoxical sleep of cat Eur J Pharmacol 11:395–397
Hendricks JC, Morrison AR, Mann GL (1982) Different behaviors during paradoxical sleep without atonia depend on pontine lesion site. Brain Res 239:81–105
Hendricks JC, Kline LR, Davies RO, Pack AT (1990) The effect of dorsolateral pontine lesions on diaphragmatic activity during REMS. J Appl Physiol 68: 1435–1442
Henley K, Morrison AR (1974) A re-evaluation of the effects of lesions of the pontine tegmentum and locus coeruleus on phenomena of paradoxical sleep in the cat. Acta Neurobiol Exp 34:215–232
Henriksen SJ, Jacobs BL, Dement WC (1972) Dependence of REM sleep PGO waves on cholinergic mechanisms. Brain Res 48:412–416

Hernández-Peón R, Chávez-Ibarra G, Morgane PJ, Timo-Iaria C (1963) Limbic cholinergic pathways involved in sleep and emotional behavior. Exp Neurol 8:93–111
Hess WR (1931) Le sommeil. C R Soc Biol (Paris) 107:1333–1366
Hobson JA, McCarley RW (1974) Discharge patterns of cat pontine brain stem neurons during desynchronized sleep. J Neurophysiol 38:751–766
Hobson JA, Steriade M (1986) Neuronal basis of behavioral state control. In: Bloom FE (ed) Intrinsic regulatory system of the brain. American Physiological Society, Bethesda, pp 701–823 (Handbook of physiology, sect 1: The nervous system, vol 4)
Hobson JA, Goldbcrg M, Vivaldi E, Riow D (1983) Enhancement of desynchronized sleep signs after pontine microinjection of the muscarinic agonist bethanechol. Brain Res 275:127–136
Hobson JA, Lydic R, Baghdoyan HA (1986) Evolving concepts of sleep cycle generation: from brain centers to neuronal populations. Behav Brain Sci 9: 371–448
Holl G, Straschill M, Thomsen T, Fischer JP, Kewitz H (1992) Effect of the cholinesterase inhibiting substance galanthamine on human EEG and visual evoked-potentials. Electroencephalogr Clin Neurophysiol 82:445–452
Holmes CJ, Jones BE (1994) Importance of cholinergic, gabaergic, serotonergic, and other neurons in the medial medullary reticular formation for sleep-wake states studied by cytotoxic lesions in the cat. Neuroscience 62:1179–1200
Horn E, D'Ascanio P, Pompeiano O, Stampacchia G (1987) Pontine reticular origin of cholinergic excitatory afferents to the locus coeruleus controlling the gain of vestibulospinal and cervicospinal reflexes in decerebrate cats. Arch Ital Biol 125:273–304
Hoshino K, Pompeiano O (1976) Selective discharge of pontine neurons during the postural atonia produced by an anticholinesterase in the decerebrate cat. Arch Ital Biol 114:244–277
Hu B, Bouhassira D, Steriade M, Deschenes M (1988) The blockage of ponto-geniculo-occipital waves in the cat lateral geniculate nucleus by nicotinic antagonists. Brain Res 473:394–397
Hulme EC, Birdsall NJM, Buckley NJ (1990) Muscarinic receptor subtypes. Annu Rev Pharmacol Toxicol 30:633–673
Ikeda M, Houtani T, Ueyama T, Sugimoto T (1991) Choline acetyltransferase immunoreactivity in the cat cerebellum. Neuroscience 45:671–690
Illing RB (1990) A subtype of cerebellar Golgi cells may be cholinergic. Brain Res 522:267–274
Imeri L, Bianchi S, Angeli P, Mancia M (1992) M1 and M3 muscarinic receptors: specific roles in sleep regulation. Neuroreport 3:276–278
Imeri L, Bianchi S, Angeli P, Mancia M (1994) Selective blockade of different brain stem muscarinic receptor subtypes: effects on the sleep-wake cycle. Brain Res 636:68–72
Ito M (1984) The cerebellum and neural control. Raven, New York, p 580
Iwahara T, Wall, PT, Carcia-Rill E, Skinner RD (1991) Stimulation-induced setting of postural muscle tone in the decerebrate rat. Brain Res 557:331–335
Jacobs BL (1986) Single unit neurons in behaving animals. Prog Neurobiol 27: 183–194
Jankowska E, Lund S, Lundberg A, Pompeiano O (1968) Inhibitory effects evoked through ventral reticulospinal pathways. Arch Ital Biol 106:124–140
Janowsky DL, El-Yousef MK, Davis JM (1972) A cholinergic-adrenergic hypothesis of mania and depression. Lancet 2:632–635
Jasper HH, Tessier J (1971) Acetylcholine liberation from cerebral cortex during paradoxical (REM) sleep. Science 172:601–602
Jewett RE, Norton S (1986) Effects of some stimulants and depressant drugs on the sleep cycle of cats. Exp Neurol 15:463–474

Jones BE (1985) Immunohistochemical study of choline acetyltransferase immunoreactive processes and cells innervating the pontomedullary reticular formation in the rat. J Comp Neurol 295:485–514
Jones BE (1991a) Paradoxical sleep and its chemical/structural substrates in the brain. Neuroscience 40:637–656
Jones BE (1991b) Noradrenergic locus coeruleus neurons: their distant connections and their relationship to neighbouring (including cholinergic and GABAergic) neurons of the central gray and reticular formation. Prog Brain Res 88:15–30
Jones BE (1991c) The role of noradrenergic locus coeruleus neurons and neighbouring cholinergic neurons of the pontomesencephalic tegmentum in sleep-wake states. Prog Brain Res 88:533–543
Jones BE, Beaudet A (1987) Distribution of acetylcholine and catecholamine neurons in the cat brain stem studied by choline acetyltransferase and tyrosin hydroxylase immunohistochemistry. J Comp Neurol 261:15–32
Jones BE, Friedman L (1983) Atlas of catecholamine pericarya varicosities and pathways in the brainstem of the cat. J Comp Neurol 215:382–396
Jones BE, Webster HH (1988) Neurotoxic lesions of the dorsolateral pontomesencephalic tegmentum cholinergic area in the cat. I. Effects upon the cholinergic innervation of the brain. Brain Res 451:13–32
Jones BE, Yang T-Z (1985) The efferent projections from the reticular formation and the locus coeruleus studied by anterograde and retrograde axonal transport in the rat. J Comp Neurol 242:56–92
Jones LS, Gauger LL, Davis JN (1985) Anatomy of brain alpha-adrenergic receptors: in vitro autoradiography with [^{125}I]-HEAT. J Comp Neurol 231:190–208
Jouvet M (1962) Recherches sur les structures nerveuses et les mécanismes responsables des différentes phases du sommeil physiologique. Arch Ital Biol 100:125–206
Jouvet M (1965) Cholinergic mechanisms and sleep. In: Waser PG (ed) Cholinergic mechanisms. Raven, New York
Jouvet M (1972) The role of monoamine and acetylcholine-containing neurons in the regulation of the sleep-waking cycle. Ergeb Physiol 64:166–307
Jouvet M, Delorme JF (1965) Locus coeruleus et sommeil paradoxal. C R Soc Biol (Paris) 159:895–899
Kametani H, Kawamura H (1990) Alterations in acetylcholine release in the rat hippocampus during sleep-wakefulness detected by intracerebral dialysis. Life Sci 47:421–426
Kamondi A, Williams JA, Hutcheon B, Reiner PB (1992) Membrane properties of mesopontine cholinergic neurons studied with the whole-cell patch-clamp technique: implications for behavioral state control. J Neurophysiol 68:1359–1372
Kan K-SK, Chao L-P, Eng LF (1978) Immunohistochemical localization of choline acetyl-transferase in rabbit spinal cord and cerebellum. Brain Res 146:221–229
Kanai T, Szerb JC (1965) Mesencephalic reticular activating system and cortical acetylcholine output. Nature 205:80–82
Kanamori N, Sakai K, Jouvet M (1980) Neuronal activity specific to paradoxical sleep in the ventromedial medullary reticular formation of unrestrained cats. Brain Res 189:251–255
Kang Y, Kitai ST (1990) Electrophysiological properties of pedunculopontine neurons and their postsynaptic responses following stimulation of substantia nigra reticulata. Brain Res 535:79–95
Karezmar AG, Longo VG, DeCarolis AS (1970) A pharmacological model of paradoxical sleep: the role of cholinergic and monoamine systems. Physiol Behav 5:175–182
Katayama Y, DeWitt DS, Becker DP, Hayes RL (1984) Behavioral evidence for a cholinoceptive pontine inhibitory area: descending control of spinal motor output and sensory input. Brain Res 296:241–262

Kawahara K, Nakazono Y, Kumagai S, Yamauchi Y, Miyamoto Y (1988a) Parallel suppression of extensor muscle tone and respiration by stimulation of pontine dorsal tegmentum in decerebrate cats. Brain Res 473:81–90
Kawahara K, Nakazono Y, Kumagai S, Yamauchi Y, Miyamoto Y (1988b) Neuronal origin of parallel suppression of postural tone and respiration elicited by stimulation of midpontine dorsal tegmentum in the decerebrate cat. Brain Res 474: 403–406
Kawahara K, Nakazono Y, Kumagai S, Yamauchi Y, Miyamoto Y (1989a) Inhibitory influences on hypoglossal neural activity by stimulation of midpontine dorsal tegmentum in decerebrate cats. Brain Res 479:185–189
Kawahara K, Nakazono Y, Miyamoto Y (1989b) Depression of diaphragmatic and external intercostal muscle activities elicited by stimulation of midpontine dorsal tegmentum in decerebrate cats. Brain Res 491:180–184
Kayama Y, Ohta M, Jodo E (1992) Firing of "possibly" cholinergic neurons in rat laterodorsal tegmental nucleus during sleep and wakefulness. Brain Res 569: 210–220
Kilduff TS, Bowersox S, Kaitin K, Baker TL, Giaraneloo RD, Dement WC (1986) Muscarinic cholinergic receptors and the canine model of narcolepsy. Sleep 9:102–106
Kimura H, McGeer PL, Peng JH, McGeer EG (1981) The central cholinergic system studied by choline acectyltransferase immuno-histochemistry in the cat. J Comp Neurol 200:151–201
Kimura H, Kubin L, Davies RO, Pack AI (1990) Cholinergic stimulation of the pons depresses respiration in decerebrate cats. J Appl Physiol 69:2280–2289
Kline RL, Hendricks JC, Davies RO, Pack AI (1986) Control of activity of the diaphragm in rapid-eye-movement sleep. J Appl Physiol 61:1293–1300
Kodama T, Takahashi Y, Honda Y (1990) Enhancement of acetylcholine release during paradoxical sleep in the dorsal tegmental field of the cat brain stem. Neurosci Lett 114:277–282
Kodama T, Lai YY, Siegel JM (1992) Enhancement of acetylcholine release during REM sleep in the caudomedial medulla as measured by in vivo microdialysis. Brain Res 580:348–350
Koyama Y, Kayama Y (1992) Discrete influences of acetylcholine on the cholinergic, noradrenergic and serotonergic neurons in the anesthetized rat brainstem. Soc Neurosci Abstr 18:880
Kramis R, Vanderwolf CH, Bland BH (1975) Two types of hippocampal rhythmical slow activity in both the rabbit and the rat: relations to behavior and effects of atrophine, diethyl ether, urethane, and pentobarbital. Exp Neurol 49:58–85
Kubin L, Magherini PC, Manzoni D, Pompeiano O (1980) Responses of lateral reticular neurons to sinusoidal stimulation of labyrinth receptors in decerebrate cats. J Neurophysiol 44:922–936
Kubin L, Kimuar H, Tojima H, Pack AI, Davies RO (1992) Behavior of VRG neurons during the atonia of REM sleep induced by pontine carbachol in decerebrate cats. Brain Res 592:91–100
Kubin L, Kimura H, Tojima H, Davies RO, Pack AI (1993) Suppression of hypoglossal motoneurons during the carbachol-induced atonia of REM sleep is not caused by fast synaptic inhibition. Brain Res 611:300–312
Lai YY, Siegel JM (1988) Medullary regions mediating atonia. J Neurosci 8:4790–4796
Lai YY, Siegel JM (1990) Cardiovascular and muscle tone changes produced by microinjection of cholinergic and glutaminergic agonists in dorsolateral pons and medial medulla. Brain Res 514:27–36
Lai YY, Siegel JM (1991) Pontomedullary glutamate receptors mediating locomotion and muscle tone suppression. J Neurosci 11:2931–2937
Lai YY, Clements JR, Siegel JM (1993) Glutamatergic and cholinergic projections to the pontine inhibitory area identified with horseradish peroxidase retrograde transport and immunohistochemistry. J Comp Neurol 336:321–330

Lamour Y, Dutar P, Rascol O, Jobert A (1986) Basal forebrain neurons projecting to the rat fronto-parietal cortex: electrophysiological and pharmacological properties. Brain Res 362:122–131
Lapchak PA, Araujo D, Quirion R, Collier B (1989a) Binding sites for [^{3}H]AF-DX 116 and effect of AF-DX 116 on endogenous acetylcholine release from rat brain slices. Brain Res 496:285–294
Lapchak PA, Aranjo DM, Quirion R, Collier B (1989b) Presynaptic cholinergic mechanisms in the rat cerebellum: evidence for nicotinic, but not muscarinic autoreceptors. J Neurochem 53:1843–1851
Lee MG, Chrobak JJ, Sik A, Wiley RG, Buzsaki G (1994) Hippocampal theta activity following selective lesion of the septal cholinergic system. Neuroscience 62:1033–1047
Leonard CS, Llinás R (1987) Low threshold calcium conductance in parabrachial reticular neurons studied in vitro and its blockade by octanol. Soc Neurosci Abstr 13:1012
Leonard CS, Llinás R (1990) Electrophysiology of mammalian pedunculopontine and laterodorsal tegmental neurons in vitro: implications for the control of REM sleep. In: Steriade M, Biesold D (eds) Brain cholinergic systems. Oxford University Press, New York, pp 205–223
Leung LWS (1985) Spectral analysis of hippocampal EEG in the freely moving rat: effects of centrally active drugs and relations to evoked potentials. Electroencephalogr Clin Neurophysiol 60:65–77
Levey AI, Armstrong DM, Atweh SF, Terry RD, Wainer BH (1983) Monoclonal antibodies to choline acetyltransferase: production, specificity, and immunohistochemistry. J Neurosci 3:1–9
Li M, Yasuda RP, Wall SJ, Wellstein A, Wolfe BB (1991) Distribution of m2 muscarinic receptors in rat brain using antisera selective for m2 receptors. Mol Pharmacol 40:28–35
Licata F, Li Volsi G, Maugeri G, Santangelo F (1990) Effects of motor cortex and single muscle stimulation on neurons of the lateral vestibular nucleus in the rat. Neuroscience 34:379–390
Lindsay KW, Roberts TDM, Rosenberg JR (1976) Asymmetric tonic labyrinth reflexes and their interaction with neck reflexes in the decerebrate cat. J Physiol (Lond) 261:583–601
Lo Conte G, Gasamenti F, Bigi V, Milaneschi E, Pepeu G (1982) Effect of magnocellular forebrain nuclei lesions on acetylcholine output from the cerebral cortex, electrocorticogram and behavior. Arch Ital Biol 120:176–188
Longo VG (1966) Contributo allo studio dell' azione centrale dell' atropina. Bull Soc Ital Biol Sper 42:97–99
Lopez-Rodriguez F, Kohlmeier K, Morales FR, Chase MH (1994) State dependency of the effects of microinjection of cholinergic drugs into the nucleus pontis oralis. Brain Res 649:271–281
Lorton D, Davis JN (1987) The distribution of beta-1 and beta-2 adrenergic receptors of normal and reeler mouse brain: an in vitro autoradiographic study. Neuroscience 23:199–210
Luccarini P, Gahery Y, Pompeiano O (1990) Cholinoceptive pontine reticular structures modify the postural adjustments during the limb movements induced by cortical stimulation. Arch Ital Biol 128:19–45
Luccarini P, Gahery Y, Blanchet G, Pompeiano O (1992) GABA receptors in Deiters nucleus modulate posturokinetic responses to cortical stimulation in the cat. Arch Ital Biol 130:127–154
Luebke JI, McCarley RW, Greene RW (1993) Inhibitory action of muscarinic agonists on neurons in the rat laterodorsal tegmental nucleus in vitro. J Neurophysiol 70:2128–2135
Lund S, Pompeiano O (1968) Monosynaptic excitation of α-motoneurons from supraspinal structures in the cat. Acta Physiol Scand 73:1–21

Luppi P-H, Sakai K, Fort P, Salvert D, Jouvet M (1988) The nuclei of origin of monoaminergic, peptidergic and cholinergic afferents to the cat nucleus reticularis magnocellularis: a double-labeling study with cholera toxin as a retrograde tracer. J Comp Neurol 277:1–20
Lydic R, Baghdoyan HA (1989) Cholinoceptive pontine reticular mechanisms cause state-dependent respiratory changes in the cat. Neurosci Lett 102:211–216
Lydic R, Baghdoyan HA, Zwillich CW (1989) State-depdent hypotonia in posterior cricoarytenoid muscles of the larynx caused by cholinoceptive reticular mechanisms. FASEB J 3:1625–1631
Lydic R, Baghdoyan HA, Lorinc Z (1991a) Microdialysates from the medial pontine reticular formation (mPRF) reveal increased acetylcholine (ACh) release during the carbachol-induced REM sleep-like state (DCARB). Sleep Res 20:25
Lydic R, Baghdoyan HA, Lorinc Z (1991b) Microdialysis of cat pons reveals enhanced acetylcholine release during state-dependent respiratory depression. Am J Physiol 261:R766–R770
Lydic R, Baghdoyan HA, Wertz R, White DP (1991c) Cholinergic reticular mechanisms influence state-dependent ventilatory response to hypercapnia. Am J Physiol 261:R738–R746
Lydic R, Baghdoyan HA, Lorinc Z (1992) Microdialysis of the medial pontine reticular formation (mPRF) reveals increased acetylcholine release caused by electrical stimulation of the pedunculopontine tegmental nucleus (PPT). Sleep Res 21:10
Madison DV, Nicoll RA (1986) Actions of noradrenaline recorded intracellularly in rat hippocampal CA1 neurones in vitro. J Physiol (Lond) 372:221–244
Magherini P, Pompeiano O, Thoden U (1971) The neurochemical basis of REM sleep: a cholinergic mechanism responsible for rhythmic activation of the vestibulo-oculomotor system. Brain Res 35:565–569
Magoun HW, Rhines R (1946) An inhibitory mechanism in the bulbar reticular formation. J Neurophysiol 9:165–171
Magoun HW, Rhines R (1947) Spasticity. The stretch-reflex and extrapyramidal systems. Thomas, Springfield, p 59
Mallick BN, Thakkar M (1992) Effect of REM sleep deprivation on molecular forms of acetylcholinesterase in rats. Neuroreport 3:676–678
Manzoni D, Pompeiano O, Srivastava UC, Stampacchia G (1983a) Responses of forelimb extensors to sinusoidal stimulation of macular labyrinth and neck receptors. Arch Ital Biol 121:205–214
Manzoni D, Pompeiano O, Stampacchia G, Srivastava UC (1983b) Responses of medullary reticulospinal neurons to sinusoidal stimulation of labyrinth receptors in decerebrate cat. J Neurophysiol 50:1059–1079
Manzoni D, Pompeiano O, Srivastrava UC, Stampacchia G (1984) Gain regulation of vestibular reflexes in fore- and hindlimb muscles evoked by roll tilt. Boll Soc Ital Biol Sper 60 [Suppl 3]:9–10
Marchand AR, Manzoni D, Pompeiano O, Stampacchia G (1987) Effects of stimulation of vestibular and neck receptors on Deiters' neurons projecting to the lumbosacral cord. Pflugers Arch 409:13–23
Mash DC, Potter LT (1986) Autoradiographic localization of M1 and M2 muscarine receptors in the rat brain. Neuroscience 19:551–564
Matsuzaki M (1969) Differential effects of sodium butyrate and physostigmine upon the activities of para-sleep in acute brain stem preparations. Brain Res 13:247–265
Matsuzaki M, Okada Y, Shuto S (1968) Cholinergic agents related to para-sleep state in acute brainstem preparation. Brain Res 9:253–267
McCarley RW, Hobson JA (1975) Neuronal excitability modulation over the sleep cycle: a structural and mathematical model. Science 189:55–58
McCarley RW, Nelson JP, Hobson JA (1978) Ponto-geniculo-occipital (PGO) burst neurons: correlative evidence for neuronal generators of PGO waves. Science 201:269–272

McCormick DA (1992) Cellular mechanisms underlying cholinergic and noradrenergic modulation of neuronal firing mode in the cat and guinea-pig dorsal lateral geniculate nucleus. J Neurosci 12:278–289

McCormick DA, Pape H-C (1990) Noradrenergic and serotonergic modulation of a hyperpolarization-activated cation current in thalqmic relay neurons. J Physiol (Lond) 431:319–342

McCormick DA, Prince DA (1985) Two types of muscarinic response to acetylcholine in mammalian cortical neurons. Proc Natl Acad Sci USA 82:6344–6348

McCormick DA, Prince DA (1986) Mechanism of action of acetylcholine in guinea pig cerebral cortex in vitro. J Physical (Lond) 36:169–194

McCormick DA, Williamson A (1989) Convergence and divergence of neurottansmitter action in human cerebral cortex. Proc Natl Acad Sci USA 86:8098–8102

McGinty D, Szymusiak R (1989) The basal forebrain and slow wave sleep: mechanistic and functional aspects. In: Wauquier A. Dugovic C, Radulovacki M (eds) Slow wave sleep: physiological, pathophysiological and functional aspects. Raven, New York, pp 61–74

McKonna TM, Ashe JH, Hui CK, Weinberger NM (1988) Muscarinic agonist modulates spontaneous and evoked unit discharge in auditory cortex of cat. Synapse 2:54–68

Megirian D, Hinrichsen CFL, Herrey JH (1985) Respiratory roles of genioglossus, sternothyroid and sternohyoid muscles during sleep. Exp Neurol 90:118–128

Megirian D, Pollard JM, Sherrey JH (1987) The labile respiratory activity of ribeage muscles of the rat during sleep. J Physiol (Lond) 389:99–110

Mergner T, Magherini PC, Pompeiano O (1976) Temporal distribution of rapid eye movements and related monophasic potentials in the brain stem following injection of an anticholinesterase. Arch Ital Biol 114:75–99

Mesulam M-M, Mufson EJ, Wainer BH, Levey AI (1983a) Central cholinergic pathways in the rat: an overview based on an alternative nomenclature (Ch1–Ch6). Neuroscience 10:1185–1201

Mesulam M-M, Mufson EJ, Levey AI, Wainer BH (1983b) Cholinergic innervation of cortex by the basal forebrain: cytochemistry and cortical connections of the septal area: diagonal band nuclei, nucleus basalis (substantia innominata), and hypothalamus in the rhesus monkey. J Comp Neurol 214:170–197

Mesulam M-M, Mufson EJ, Levey AI, Wainer BH (1984) Atlas of cholinergic neurons in the forebrain and upper brainstem of the macaque based on monoclonal choline acetyltransferase immunohistochemistry and acetylcholinesterase histochemistry. Neuroscience 12:669–686

Metcalf DR, Holmes JH (1969) EEG, psychological and neurological alterations in humans with organophosphate exposure. Ann NY Acad Sci 160:357–365

Metherate R, Trembley N, Dykes RW (1988) Transient and prolonged effects of acetylcholine on responsiveness of cat somatosensory cortical neurons. J Neurophysiol 59:1253–1275

Michel AD, Stefanich F, Whiting RL (1989) PC12 phaeochromocytoma cells contain an atypical muscarinic receptor binding site. Br J Pharmacol 97:914–920

Mitani A, Ito K, Hallanger AE, Wainer BH, Kataoka K, McCarley RW (1988) Cholinergic projections from the laterodorsal and pedunculopontine tegmental nuclei to the pontine gigantocellular tegmental field in the cat. Brain Res 451:397–402

Mitler MM, Dement WC (1974) Catapletic-like behavior in cats after microinjections of carbachol in pontine reticular formation. Brain Res 68:335–343

Moises HC, Woodward DJ, Hoffer BJ, Freedman R (1979) Interactions of norepinephrine with Purkinje cell responses to putative amino acid neurotransmitters applied by microiontophoresis. Exp Neurol 64:493–515

Moises HC, Burne RA, Woodward DJ (1990) Modification of the visual response properties of cerebellar neurons by norepinephrine. Brain Res 514: 259–275

Morales FR, Chase MH (1978) Intracellular recording of lumbar motoneuron membrane potential during sleep and wakfulness. Exp Neurol 62:821–827
Morales FR, Engelhardt JK, Soja PJ, Pereda AE, Chase MH (1987) Motoneurons properties during motor inhibition produced by microinjection of carbachol into the pontine reticular formation of the decerebrate cat. J Neurophysiol 57: 1118–1129
Mori S (1989) Contribution of postural muscle tone full expression of posture and locomotor movements: multifaceted analyses of its setting brainstem-spinal cord mechanisms in the cat. Jpn J Physiol 39:785–809
Moroni F, Tanganelli S, Antonelli T, Carla V, Bianchi C, Beani L (1983) Modulation of cortical acetylcholine and amino-butyric acid release in freely moving guinea pigs: effects of clonidine and other adrenergic drugs. J Pharmacol Exp Ther 227:435–440
Moruzzi G (1972) The sleep-waking cycle. Ergeb Physiol 64:1–165
Moruzzi G, Magoun HW (1949) Brain stem reticular formation and activation of the EEG. Electroencephalogr Clin Neurophysiol 1:455–473
Mufson EJ, Martin TL, Mash DC, Wainer BH, Mesulam MM (1986) Cholinergic projections from the parabigeminal nucleus (Ch8) to the superior colliculus in the mouse: a combined analysis of horseradish peroxidase transport and choline acetyltransferase immunohistochemistry. Brain Res 370:144–148
Muehlthaler M (1994) Electrophysiological properties and neuromodulation of cholinergic and non-cholinergic basal forebrain neurons. J Sleep Res [Suppl 1]:173
Muehlthaler M, Khateb A, Fort P, Jones BE, Alonso A (1992) Forty Hz membrane potential oscillations and theta-like activity in basal forebrain neurones. Soc Neurosci Abstr 18:197
Neustadt A, Frostholm A, Rotter A (1988) Topographical distribution of muscarinic cholinergic receptors in the cerebellar cortex of the mouse, rat, guinea pig and rabbit: a species comparison. J Comp Neurol 272:317–330
Nishino S, Shelton J, Reid MS, Siegel JM, Dement WC, Mignot E (1992) A cholinoceptive site in the basal forebrain is involved in canine narcolepsy. Soc Neurosci Abstr 18:880
Ohta Y, Mori S, Kimura H (1988) Neuronal structures of the brainstem participating in postural suppression in cats. Neurosci Res 5:181–202
Ojima H, Kawajiri S-I, Yamasaki T (1989) Cholinergic innervation of the rat cerebellum: qualitative and quantitative analyses of elements immunoreactive to a monoclonal antibody against choline acetyltransferase. J Comp Neurol 290: 41–52
Onteniente B, Geffard M, Campistron G, Calas A (1987) An ultrastructural study of GABA-immunoreactive neurons and terminals in the septum of the rat. J Neurosci 7:48–54
Orem J (1980a) Medullary respiratory neuron activity: relationship to tonic and phasic REM sleep. J Appl Physiol 48:54–65
Orem J (1980b) Neuronal mechanisms of respiration in REM sleep. Sleep 3:251–267
Ott T, Malisch R, Destrade C (1983) Nachweis der transmitterspezifischen Modulation der langsamen rhytmischen Aktivität des Hippokampus. Biomed Biochim Acta 42:96–99
Palacios JM (1988) Mapping brain receptors by autoradiography. ISI Atlas Sci Pharmacol 2:71–77
Palacios JM, Moyer D, Cortes R (1987) α_1-Adrenoceptors in the mammalian brain: similar pharmacology but different distribution in rodents and primates. Brain Res 419:65–75
Paré D, Smith Y, Parent A, Steriade M (1988) Projections of brainstem core cholinergic and non-cholinergic neurons of cat to intralaminar and reticular thalamic nuclei. Neuroscience 25:69–86

Parmeggiani PL, Sabattini L (1972) Electromyographic aspects of postural, respiratory and thermoregulatory mechanisms in sleeping cats. Electroencephalogr Clin Neurophysiol 33:1–13
Peralta FG, Ashkenazi A, Wislow JW, Ramachandran J, Capon DJ (1988) Differential regulation of PI hydrolysis and adenylyl cyclase by muscarinic receptor subtypes. Nature 334:434–437
Phillipson EA, Bowes G (1986) Control of breathing during sleep. In: The respiratory system control of breathing. Am Physiological Society, Bethesda, pp 649–689 (Handbook of physiology, sect 3, vol 2)
Phillis JW (1971) The pharmacology of thalamic and geniculate neurons. Int Rev Neurobiol 14:1–48
Phillis JW, Tebecis AK, York DH (1968) Acetylcholine release from the feline thalamus. J Pharm Pharmacol 20:476–478
Pompeiano M, Tononi G (1990) Changes in pontine muscarinic receptor binding during sleep-waking states in the rat. Neurosci Lett 109:347–352
Pompeiano M, Tononi G, Galbani P (1990) Muscarinic receptor binding in forebrain and cerebellum during sleep-waking states in the rat. Arch Ital Biol 128:77–79
Pompeiano M, Cirelli C, Tononi G (1992) Effects of sleep deprivation on fos-like immunoreactivity in the rat brain. Arch Ital Biol 130:325–335
Pompeiano M, Cirelli C, Tononi G (1994) Immediate-early genes in spontaneous wakefulness and sleep: Expression of c-fos and NGFI-A mRNA and protein. J Sleep Res 3:80–96
Pompeiano O (1967) The neurophysiological mechanisms of the postural and motor events during desynchronized sleep. Res Publ Assoc Nery Ment Dis 45:351–423
Pompeiano O (1976) Mechanisms responsible for spinal inhibition during desynchronized sleep: experimental study. Adv Sleep Res 3:411–449
Pompeiano O (1979) Neck and macular labyrinthine influences on the cervical spinoreticulocerebellar pathway. Prog Brain Res 50:501–514
Pompeiano O (1980) Cholinergic activation of reticular and vestibular mechanisms controlling posture and eye movements. In: Hobson JA, Brazier MAB (eds) The reticular formation revisited. Raven, New York, pp 473–512 (IBRO monograph series, vol 6)
Pompeiano O (1984) Recurrent inhibition. In: Davidoff RA (ed) Handbook of the spinal cord, vol 2, 3. Dekker, New York, pp 461–557
Pompeiano O (1985) Cholinergic mechanisms involved in the gain regulation of postural reflexes. In: Wauquier A, Gaillard JM, Monti JM, Radulovacki M (eds) Sleep: neurotransmitters and neuromodulators. Raven, New York, pp 165–184
Pompeiano O, Hoshino K (1976a) Tonic inhibition of dorsal posntine neurons during the postural atonia produced by an anticholinesterease in the decerebrate cat. Arch Ital Biol 114:310–340
Pompeiano O, Hoshino K (1976b) Central control of posture: reciprocal discharge by two pontine neuronal groups leading to suppression of decerebrate rigidity. Brain Res 116:131–138
Pompeiano O, Manzoni D, Srivastava UC, Stampacchia G (1983) Cholinergic mechanisms controlling the response gain of forelimb extensor muscles to sinusoidal stimulation of macular labyrinth and neck receptors. Arch Ital Biol 121:285–303
Pompeiano O, D'Ascanio P, Horn E, Stampacchia G (1987) Effects of local injection of the α_2-adrenergic agonist clonidine into the locus coeruleus complex on the gain of vestibulospinal and cervicospinal reflexes in decerebrate cats. Arch Ital Biol 125:225–269
Pompeiano O, Manzoni D, Barnes CD, Stampacchia G, D'Ascanio P (1990) Responses of locus coeruleus and subcoeruleus neurons to sinusoidal stimulation of labyrinth receptors. Neurocience 35:227–248

Pompeiano O, Horn E, D'Ascanio P (1991) Locus coeruleus and dorsal pontine reticular influences on the gain of vestibulospinal reflexes. Prog Brain Res 88:435–462

Pompeiano O, Andre P, D'Ascanio P, Manzoni D (1992) Local injections of β-noradrenergic substances in the cerebellar anterior vermis of cats affect adaptation of the vestibulospinal reflex (VSR) gain. Soc Neurosci Abstr 18: 215–216

Pompeiano O, Luccarini P, Gahery Y, Blanchet G (1993) Somatotopical effects of local microinjection of GABAergic agents in Deiters nucleus on posturokinetic responses to cortical stimulation. J Vest Res 3:391–407

Rainbow TC, Parsons B, Wolfe BB (1984) Quantitative autoradiography of α_1- and α_2-adrenergic receptors in rat brain. Proc Natl Acad Sci USA 81:1585–1589

Rasler FE (1984) Behavioral and electrophysiological manifestations of bombesin: excessive grooming and elimination of sleep. Brain Res 321:187–198

Regenold W, Araujo DM, Quirion R (1989) Quantitative autoradiographic distribution of [^{3}H]AF-DX 116 muscarinic-M2 receptor binding sites in rat brain. Synapse 4:115–125

Reid MS, Tafti M, Nishino S, Siegel JM, Dement WC, Mignot E (1994a) Cholinergic regulation of catplexy in canine narcolepsy in the pontine reticular formation is mediated by M2 muscarinic receptors. Sleep 17:424–435

Reid MS, Tafti M, Geary JN, Siegel JM, Dement WC, Mignot E (1994b) Cholinergic mechanisms in canine narcolepsy: I. Modulation of cataplexy by local drug administration into the pontine reticular formation. Neuroscience 59:511–522

Remmers JE (1981) Control of breathing during sleep. In: Hornbein TF (ed) Regulation of breathing. Dekker, New York, pp 1197–1249

Remmers JE, Bartlett D Jr, Putnam MD (1975) Changes in the respiratory cycle associated with sleep. Respir Physiol 28:227–238

Rye DB, Saper CB, Lee HJ, Wainer BH (1987) Pedunculopontine tegmental nucleus of the rat: cytoarchitecture, cytochemistry, and some extrapyramidal connections of the mesopontine tegmentum. J Comp Neurol 259:483–528

Rye DB, Lee HJ, Saper CB, Wainer BH (1988) Medullary and spinal efferents of the pedunculopontine tegmental nucleus and adjacent mesopontine tegmentum in the rat. J Comp Neurol 269:315–341

Saito H, Sakai K, Jouvet M (1977) Discharge patterns of the nucleus parabrachialis lateralis neurons of the cat during sleep and waking. Brain Res 134:59–72

Sakai K (1980) Some anatomical and physiological properties of pontomesencephalic tegmental neurons with special reference to the PGO waves and postural atonia during paradoxical sleep. In: Hobson JA, Brazier MAB (eds) The reticular formation revisited. Raven, New York, pp 427–447 (IBRO monograph series, vol 6)

Sakai K (1985) Anatomical and physiological basis of paradoxical sleep. In: McGinty DJ, Drucker-Colin R, Morrison A, Parmeggiani PL (eds) Brain mechanisms of sleep. Raven, New York, pp 111–137

Sakai K (1988) Executive mechanisms of paradoxical sleep. Arch Ital Biol 126: 239–257

Sakai K (1991) Physiological properties and afferent connections of the locus coeruleus and adjacent tegmental neurons involved in the generation of paradoxical sleep in the cat. Prog Brain Res 88:31–45

Sakai K, Jouvet M (1980) Brain stem PGO-on cells projecting directly to the cat dorsal lateral geniculate nucleus. Brain Res 194:500–505

Sakai K, Petitjean F, Jouvet M (1976) Effects of ponto-mesencephalic lesions and electrical stimulation upon PGO waves and EMPs in unanesthetized cats. Electroencephalogr Clin Neurophysiol 41:49–63

Sakai K, Touret M, Salvert D, Leger L, Jouvet M (1977) Afferent projections to the cat locus coeruleus as visualized by horseradish peroxidase technique. Brain Res 119:21–41

Sakai K, Sastre J-P, Salvert D, Touret M, Tohyama M, Jouvet M (1979) Tegmentoreticular projections with special reference to the muscular atonia during paradoxical sleep in the cat. An HRP study. Brain Res 176:233–254
Sakai K, Sastre J-P, Kanamori N, Jouvet M (1981) State-specific neurons in the ponto-medullary reticular formation with special reference to the postural atonia during paradoxical sleep in the cat. In: Pompeiano O, Ajmone Marsan C (eds) Brain mechanism of perceptual awareness and purposeful behavior. Raven, New York, pp 405–429 (IBRO monograph series, vol 8)
Sakai K, Luppi PH, Salvert D, Kimura H, Maeda T, Jouvet M (1986) Localization of cholinergic neurons in the cat lower brain stem. C R Acad Sci (Paris) 303:317–324
Salin-Pascual RJ, Jimenez-Anguiano A, Granados-Fuentes D, Drucker-Colin R (1991) Effects of biperiden on sleep at baseline and after 72 h of REM sleep deprivation in the cat. Sleep Res 20:83
Salin-Pascual RJ, Ortega-Soto H, Huerto-Delgadillo L, Chavez JL, Granados-Fuentes D (1992) Changes in the cholinergic system as a result of rapid eye movement sleep deprivation in the rat: behavioral and biochemical evidences. Sleep Res 21:324
Salvaterra PM, Vaughn JE (1989) Regulation of choline acetyltransferase. Int Rev Neurobiol 31:81–143
Saper CB, Standaert DG, Currie MG, Schartz D, Geller DM, Needleman P (1985) Atriopeptin-immunoreactive neurons in the brain: presence in cardiovascular regulatory areas. Science 227:1047–1049
Sarkisian VH, Fanardjian V (1992) Neuronal mechanisms of interaction of Deiters nucleus with the cerebral cortex. Arch Ital Biol 130:113–126
Sauerland EK, Harper RM (1976) The human tongue during sleep; electromyographic activity of the genioglossus muscle. Exp Neurol 51:160–170
Schiller GD, Byrne C, Orbach J, Overstreet DH (1989) Regulation of muscarinic receptors in two lines of rats selectively bred for differential cholinergic sensitivity. Neurosci Lett 34:S147
Schor RH, Miller AD (1981) Vestibular reflexes in neck and forelimb muscles evoked by roll tilt. J Neurophysiol 46:167–178
Schor RH, Miller AD (1982) Relationship of cat vestibular neurons to otolith-spinal reflexes. Exp Brain Res 47:137–144
Schwarz M, Sontag K-M, Wand P (1984) Non-dopaminergic neurones of the reticular part of the substantia nigra can gate static fusimotor action onto flexors in cat. J Physiol (Lond) 354:333–344
Semba K, Fibiger HC (1992) Afferent connections of the laterodorsal and the pedunculopontine tegmental nuclei in the rat. A retrograde and anterograde transport and immunohistochemical study. J Comp Neurol 323:387–410
Sherrey JH, Pollard MJ, Megirian D (1986) Respiratory functions of the inferior pharyngeal constrictor and ternohyoid muscles during sleep. Exp Neurol 92: 267–277
Shiromani PJ, Fishbein W (1986) Continuous pontine cholinergic micronifusion via mini-pump induces sustained alterations in rapid eye movement (REM) sleep. Pharmacol Biochem Behav 25:1253–1261
Shiromani PJ, McGinty DJ (1986) Pontine neuronal response to local cholinergic microinfusion: relation to REM sleep. Brain Res 386:20–31
Shiromani PJ, Siegel JM, Tomaszewski KS, McGinty DJ (1986) Alterations in blood pressure and REM sleep after pontine carbachol microinfusion. Exp Neurol 91:285–292
Shiromani PJ, Overstreet DH, Levy D, Goodrich G, Campbell SS, Gillin JC (1988a) Increased REM sleep in rats genetically bred for cholinergic hyperactivity. Neuropsychopharmacology 1:127–133
Shiromani PJ, Armstrong DM, Gillin JC (1988b) Cholinergic neurons from the dorsolateral pons project to the medial pons: a WGA-HRP and choline acetyltransferase immunohistochemical study. Neurosci Lett 95:19–23

Shiromani PJ, Armstrong DM, Berkowitz A, Jeste DV, Cillin JC (1988c) Distribution of choline acetyltransferase immunoreactive somata in the feline brainstem: implication for REM sleep generation. Sleep 11:1–16
Shiromani PJ, Lai YY, Siegel JM (1990a) Descending projections from the dorsolateral pontine tegmentum to the paramedian reticular nucleus of the caudal medulla in the cat. Brain Res 517:224–228
Shiromani PJ, Floyd C, Veláquez-Moctezuma J (1990b) Pontine cholinergic neurons simultaneously innervate two thalamic targets. Brain Res 532:317–322
Shiromani PJ, Velazquez-Moctezuma J, Overstreet DH, Shalauta M, Lucero S, Floyd C (1991) Effects of sleep deprivation on sleepiness and increased REM sleep in rats selectively bred for cholinergic hyperactivity. Sleep 14:116–120
Shiromani PJ, Kilduff TS, Bloom FE, McCarley RW (1992) Cholinergically induced REM-sleep triggers Fos-like immunoreactivity in dorsolateral pontine regions associated with REM-sleep. Brain Res 580:351–357
Shouse MN, Siegel JM (1992) Pontine regulation of REM sleep components in cats: integrity of the pedunculopontine tegmentum (PPT) is important for phasic events but unnecessary for atonia during REM sleep. Brain Res 571:50–63
Shute CCD, Lewis PR (1967) The ascending cholinergic reticular system: neocortical, olfactory and subcortical projections. Brain 90:497–520
Sieck GC, Trelease RB, Harper RM (1984) Sleep influences on diaphragmatic motor unit discharge. Exp Neurol 85:316–335
Siegel JM (1989) Brainstem mechanisms generating REM sleep. In: Kryger M, Roth T, Dement WC (eds) Principles and practice of sleep medicine. Saunders, Philadelphia, pp 104–120
Siegel JM, Wheeler RL, McGinty DJ (1979) Activity of medullary reticular formation neurons in the unrestrained cat during waking and sleep. Brain Res 1(9): 49–60
Silberman E, Vivaldi E, Garfield J, McCarley RW, Hobson JA (1980) Carbachol triggering of desynchronized sleep phenomena: enhancement via small volume intusions. Brain Res 191:215–224
Sitaram N, Wyatt RJ, Dawson S, Gillin JC (1976) REM sleep induction by physostigmine infusion during sleep. Science 191:1281–1283
Sitaram N, Moore AM, Gillin JC (1978) Induction and resetting of REM sleep rhythm in normal man by arecholine: blockade by scopolamine. Sleep 1:83–90
Smith Y, Paré D, Deschênes M, Parent A, Steriade M (1988) Cholinergic and noncholinergic projections from the upper brainstem core to the visual thalamus in the cat. Exp Brain Res 70:166–180
Soja PJ, Finch DM, Chase MH (1987) Effect of inhibitory amino acid antagonists on masseteric reflex suppression during active sleep. Exp Neurol 96:178–193
Spencer DG Jr, Horváth E, Traber J (1986) Direct autoradiographic determination of M1 and M2 muscarinic acetylcholine receptor distribution in the rat brain: relation to cholinergic nuclei and projections. Brain Res 380:59–68
Spiegel R (1984) Effects of RS 86, an orally active cholinergic agonist on sleep in man. Psychiatry Res 11:1–13
Srivastava UC, Manzoni D, Pompeiano O, Stampacchia G (1982) State-dependent properties of medullary reticular neurons involved during the labyrinth and neck reflexes. Neurosci Lett 10:S461
Stampacchia G, Barnes CD, D'Ascanio P, Pompeiano O (1987) Effects of microinjection of a cholinergic agonist into the locus coeruleus on the gain of vestibulospinal reflexes in decerebrate cats. Arch Ital Biol 125:107–138
Standaert DG, Saper CB, Rye DB, Wainer BH (1986) Colocalization of atriopeptin-like immunoreactivity with choline acetyltransferase and substance P-like immunoreactivity in the pedunculopontine and laterodorsal tegmental nuclei in the rat. Brain Res 382:163–168
Stenberg D, Hilakivi I (1985) α_1- and α_2-adrenergic modulation of vigilance and sleep. In: Wauquier A, Gaillard JM, Monti JM, Radulovacki M (eds) Sleep, neurotransmitters and neuromodulators. Raven, New York, pp 69–77

Steriade M, McCarley RW (1990) Brainstem control of wakefulness and sleep. Plenum, New York

Steriade M, Pare D, Parent A, Smith Y (1988) Projections of cholinergic and noncholinergic neurons of the brainstem core to relay and associational thalamic nuclei in the cat and macaque monkey. Neuroscience 25:47–67

Steriade M, Datta S, Pare D, Oakson G, Dossi RC (1990a) Neuronal activities in brain-stem cholinergic nuclei related to tonic activation processes in thalamocortical systems. J Neurosci 10:2541–2559

Steriade M, Pare D, Datta S, Oakson G, Curró Dossi R (1990b) Different cellular types in mesopontine cholinergic nuclei related to ponto-geniculo-occipital waves. J Neurosci 10:2560–2579

Steriade M, Jones EG, Llinas RR (1990c) Thalamic oscillations and signalling. Wiley, New York

Steriade M, Curró Dossi R, Nunez A (1991) Network modulation of a slow intrinsic oscillation of cat thalamocortical neurons implicated in sleep delta waves: cortically induced synchronization and brainstem cholinergic suppression. J Neurosci 11:3200–3217

Sterman MB, Clemente CD (1962) Forebrain inhibitory mechanisms: sleep patterns induced by basal forebrain stimulation. Exp Neurol 6:103–117

Stern WC, Pugh WW (1984) Responses of frontal cortex single units to magnocellular basal forebrain stimulation. Soc Neurosci Abstr 10:9

Stewart DJ, Macfabe DF, Vanderwolf CH (1984) Cholinergic activation of the electrocorticogram: role of the substantia innominata and effects of atropine and quinuclidinyl benzilate. Brain Res 322:219–232

Stone TW (1972) Cholinergic mechanisms in the rat somatosensory cerebral cortex. J Physiol (Lond) 225:485–499

Stumpf CH (1965) Drug action on the electrical activity on the hippocampus. Int Rev Neurobiol 8:77–138

Sutin J, Minneman KP (1985) Adrenergic beta receptors are not uniformly distributed in the cerebellar cortex. J Comp Neurol 236:547–554

Sutin EL, Shiromani PJ, Kelsoe JR, Storch FI, Gillin JC (1986) Rapid eye movement sleep and muscarinic receptor binding in rats are augmented during withdrawal from chronic scopolamine treatment. Life Sci 39:2419–1986

Sutton RE, Koob GF, Le Moal M, Rivier J, Yale W (1982) Corticotropin releasing factor produces behavioral activation in rats. Nature 297:31–33

Swanson LW, Simmons DM, Whiting PJ, Lindstrom J (1987) Immunohistochemical localization of neuronal nicotinic receptors in the rodent central nervous system. J Neurosci 7:3334–3342

Szerb JC (1967) Cortical acetylcholine release and electroencephalographic arousal. J Physiol (Lond) 192:329–343

Szymusiak R, McGinty D (1986) Sleep-related neuronal discharge in the basal forebrain of cats. Brain Res 370:82–92

Taguchi O, Kubin L, Pack AI (1992) Evocation of postural atonia and respiratory depression by pontine carbachol in the decerebrate rat. Brain Res 595:107–115

Tan HS, Collewjin H (1991) Cholinergic modulation of optokinetic and vestibulocular responses: a study with microinjections in the flocculus of the rabbit. Exp Brain Res 85:475–481

Tan HS, van Neerven J, Collewjin H, Pompeiano O (1991) Effects of α-noradrenergic substances on the optokinetic and vestibulo-ocular responses in the rabbit: a study with systemic and intrafloccular injections. Brain Res 562:207–215

Tan HS, Collewjin H, Van der Steen J (1992) Optokinetic nystagmus in the rabbit and its modulation by bilateral microinjection of carbachol in the cerebellar flocculus. Exp Brain Res 90:456–468

Tan HS, Collewjin H, Van der Steen J (1993a) Unilateral cholinergic stimulation of the rabbit's cerebellar flocculus: asymmetric effects on optokinetic responses. Exp Brain Res 92:375–384

Tan HS, Collewjin H, Van der Steen J (1993b) Shortening of vestibular nystagmus in response to velocity steps by microinjection of carbachol in the rabbit's cerebellar flocculus. Exp Brain Res 92:385–390

Tatehata T, Shiosaka S, Wanaka A, Rao ZR, Tohyama M (1987) Immunocytochemical localization of the choline acetyltransferase containing neuron system in the rat lower brain stem. J Hirnforsch 28:707–716

Ticho SR, Virus RM, Radulovacki M (1988) Intracerebral administration of adenosine to medial preoptic area enhances sleep in rats. Soc Neurosci Abstr 14:1307

Tojima H, Kubin L, Kimura H, Daview RO (1992) Spontaneous ventilation and respiratory motor output during carbachol-induced atonia of REM sleep in the decerebrate cat. Sleep 15:404–414

Tononi G, Pompeiano M, Gianni S, Pompeiano O (1988a) Enhancement of desynchronized sleep signs after microinjection of the β-adrenergic antagonist propranolol in the dorsal pontine tegmentum. Arch Ital Biol 125:119–122

Tononi G, Pompeiano M, Pompeiano O (1988b) Desynchronized sleep suppression after microinjection of the β-adrenergic agonist isoproterenol in the dorsal pontine tegmentum. Arch Ital Biol 125:123–125

Tononi G, Pompeiano M, Pompeiano O (1989) Modulation of desynchronized sleep through microinjection of β-adrenergic agonists and antagonists in the dorsal pontine tegmentum of the cat. Pflugers Arch 415:142–149

Tononi G, Pompeiano M, Cirelli C (1991a) Suppression of desynchronized sleep through microinjection of the α-adrenergic agonist clonidine in the dorsal pontine tegmentum of the cat. Pflugers Arch 418:512–518

Tononi G, Pompeiano M, Cirelli C (1991b) Effects of local pontine injection of noradrenergic agents on desynchronized sleep in the cat. Prog Brain Res 85: 545–553

Trotter R, Orem J (1991) Increased excitability of medullary respiratory neurons in REM sleep. FASEB J 5:A734

Tucek S (1990) The synthesis of acetylcholine: twenty years of progress. Prog Brain Res 84:467–477

Unnerstall JR, Kopajtic TA, Kuhar MJ (1984) Distribution of α_2-agonist binding sites in the rat and human central nervous system: analysis of some functional, anatomic correlates of the pharmacological effects of clonidine and related adrenergic agents. Brain Res Rev 7:69–101

Van der Zee EA, Streefland C, Strosberg AD, Schröder H, Luiten PGM (1992) Visualization of cholinoceptive neurons in the rat neocortex: colocalization of muscarinic and nicotinic acetylcholine receptors. Mol Brain Res 14:326–336

Van Dongen PAM (1980) Locus coeruleus region: effects on behavior of cholinergic, noradrenergic and opiate drugs injected intracerebrally into freely moving cats. Exp Neurol 6:52–78

Van Dongen PAM, Broekkamp CLE, Cools AR (1978) Atonia after carbachol microinjections near the locus coeruleus in cats. Pharmacol Biochem Behav 8:527–532

Van Neerven J, Pompeiano O, Collewjin H, Van der Steen J (1990) Injections of β-noradrenergic substances in the flocculus of rabbits affect adaptation of the VOR gain. Exp Brain Res 79:249–260

Vanderwolf CH (1975) Nocortical and hippocampal activation in relation to behavior: effects of atropine, eserine, phenotiazines, and amphetamine. J Comp Physiol Psychol 88:300–323

Vanderwolf CH (1988) Cerebral activity and behavior: control by cholinergic and serotonergic systems. Int Rev Neurobiol 30:225–340

Vanderwolf CH (1992) The electrocorticogram in relation to physiology and behavior: a new analysis. Electroencephalogr Clin Neurophysiol 82:165–175

Vanni-Mercier G, Sakai K, Lin JS, Jouvet M (1989) Mapping of cholinoceptive brainstem structures responsible for the generation of paradoxical sleep in the cat. Arch Ital Biol 127:133–164

Vanni-Mercier G, Sakai K, Lin JS, Jouvet M (1991) Carbachol microinjections in the mediodorsal pontine tegmentum are unable to induce paradoxical sleep after caudal pontine and prebulbar transection in the cat. Neurosci Lett 130:41–45

Velázquez-Moctezuma J, Gillin JC, Shiromani PJ (1989) Effect of specific M1, M2 muscarinic receptor agonists on REM sleep generation. Brain Res 503:128–131

Velazquez-Moctezuma J, Shalauta MD, Gillin JC, Shiromani PJ (1990) Microinjections of nicotine in the medial pontine reticular formation elicits REM sleep. Neurosci Lett 115:265–268

Velazquez-Moctezuma J, Shalauta MD, Gillin JC, Shiromani PJ (1991) Cholinergic antagonists and REM sleep generation. Brain Res 543:175–179

Velluti R, Hernández-Peón R (1963) Atropine blockade within a cholinergic hypnogenic circuit. Exp Neurol 8:20–29

Vertes RP (1982) Brain stem generation of the hippocampal EEG. Prog Neurobiol 19:159–186

Vilaro MT, Wiederhold KH, Palacios JM, Mengod G (1991) Muscarinic cholinergic receptors in the rat caudate-putamen and olfactory tubercle belong predominantly to the m4 class: in situ hybridization and receptor autoradiography evidence. Neuroscience 40:159–167

Vilaró MT, Wiederhold KH, Palacios JM, Mengod G (1992) Muscarinic M2 receptor mRNA expression and receptor binding in cholinergic and non-cholinergic cells in the rat brain: a correlative study using in situ hybridization histochemistry and receptor autoradiography. Neuroscience 47:367–393

Vincent SR, Reiner PB (1987) The immunohistochemical localzation of choline acetyltransferase in the cat brain. Brain Res Bull 18:371–415

Vincent SR, Satoh K, Armstrong DM, Fibiger HC (1983) NADPH-diaphorase: a selective histochemical marker for the cholinergic neurons of the pontine reticular formation. Neurosci Lett 43:31–36

Vivaldi E, McCarley RW, Hobson JA (1980) Evocation of desynchronized sleep signs by chemical microstimulation of the pontine brainstem. In: Hobson JA, Brazier MAB (eds) The reticular formation revisited. Raven, New York, pp 513–529 (IBRO monography series, vol 8)

Vizi ES (1980) Modulation of cortical release of acetylcholine by noradrenaline released from nerves arising from the rat locus coeruleus. Neuroscience 5: 2139–2144

Vogel G (1989) Sleep variables and the treatment of depression. In: kryger MH, Roth T, Dement WC (eds) Principles and practice of sleep medicine. Saunders, Philadelphia, pp 419–420

Wamsley JK, Lewis MS, Young WS, Kuhar MJ (1981) Autoradiographic localization of muscarinic cholinergic receptors in rat brainstems. J Neurosci 1:176–191

Wang JX, Roeske WR, Hawkins KN, Gehlert DR, Yamamura HI (1989) Quantitative autoradiography of M2 muscarinic receptors in the rat brain identified by using a selective radioligand [^{3}H]AF-DX 116. Brain Res 477:322–326

Waterhouse BD, Moises HC, Yeh HH, Woodward JD (1982) Norepinephrine enhancement of inhibitory synaptic mechanisms in cerebellum and cerebral cortex: mediation by beta adrenergic receptors. J Pharmacol Exp Ther 221: 495–506

Waterhouse DB, Sessler FM, Cheng JT, Woodward JD, Azizi SA, Moises HC (1988) New evidence for a gating action of norepinephrine in central neuronal circuits of mammalian brain. Brain Res Bull 21:425–432

Webster HH, Jones BE (1988) Neurotoxic lesions of the dorsolateral pontomesencephalic tegmentum cholinergic area in the cat. II. Effects upon sleep-waking states. Brain Res 458:285–302

White SR, Fung SJ, Barnes CD (1991) Norepinephrine effects on spinal motoneurons. Prog Brain Res 88:343–350

Williams JA, Reiner PB (1993) Noradrenaline hyperpolarizes identified rat mesopontine cholinergic neurons in vitro. J Neurosci 13:3878–3883

Williams JA, Comisarow J, Day J, Fibiger HC, Reiner PB (1994) State-dependent release of acetylcholine in rat thalamus measured by in vivo microdialysis. J Neurosi 14:5236–5242
Woody CD, Swartz BE, Gruen E (1978) Effects of acetylcholine and cyclic GMP on input resistance of cortical neurons in awake cats. Brain Res 158:373–395
Woolf NJ (1991) Cholinergic systems in mammalian brain and spinal cord. Prog Neurobiol 37:475–524
Yamamoto K, Mamelak AN, Quattrochi JJ, Hobson JA (1990a) A cholinoceptive desynchronized sleep induction zone in the anterodorsal pontine tegmentum: locus of the sensitive region. Neuroscience 39:279–293
Yamamoto K, Mamelak AN, Quattrochi JJ, Hobson JA (1990b) A cholinoceptive desynchronized sleep induction zone in the anterodorsal pontine tegmentum: spontaneous and drug-induced neuronal activity. Neuroscience 39:295–304
Yeh HH, Moises HC, Waterhouse BD, Woodward DJ (1981) Modulatory interactions between norepinephrine and taurine, beta-alanine, gamma-aminobutyric acid and Neuropharmacology 20:549–560
Zimmermann H (1988) Cholinergic synaptic vesicles. In: Whittaker VP (ed) Handbook of experimental pharmacology, vol 86. Springer, Berlin Heidelberg New York, pp 349–382
Zoltoski RK, Floyd C, Gillin JC (1991) The effects of centrally adminstered pirenzepine and 4-DAMP on sleep in rats. Sleep Res 20:93
Zoltoski RK, Kleiger ER, Gillin JC (1992) Brainstem M1 muscarinic receptor subtype in REM sleep. Sleep Res 21:21

CHAPTER 8

Pharmacology of the GABAergic/ Benzodiazepine System

W.E. Müller

A. Introduction

Although γ-aminobutyric acid (GABA) was discovered only as recently as 1950, today we have convincing evidence that GABA represents the most important inhibitory neurotransmitter of the mammalian central nervous system (CNS). Estimations of the percentage of all nerve endings that are GABAergic synapses range from 50% (forebrain) to about 30% for the whole mammalian CNS. These histological data are supported by iontophoretic studies indicating that nearly all central neurons can be affected by GABA (and therefore might possess GABA receptors). Thus, on a quantitative basis, GABA probably represents the most important neurotransmitter of our brain (for reviews, see Snodgrass 1983; Roberts 1984).

GABAergic inhibitory neurotransmission is mediated in the mammalian CNS by several kinds of GABAergic neurons and by two different types of neuronal responses. If GABAergic neurons synapse presynaptically to axons of other neurons (axoaxonic synapses), GABA acts as a presynaptic inhibitory transmitter, probably by depolarizing the respective axon (Fig. 1). This kind of presynaptic inhibition has been found on the endings of primary afferents of spinal or cranial nerves (Fig. 2). All other kinds of GABAergic synapses are axosomatic or axodentritic (Fig. 1), where the GABAergic neurons may be small interneurons (collateral inhibition or recurrent inhibition) or projecting principal neurons (Fig. 1). All of this second type of GABAergic synapses mediate so-called postsynaptic inhibition and account for the majority of GABAergic synapses in the CNS (Simmonds 1984). A schematic representation of the hyperpolarization induced by GABA and its postsynaptically inhibitory effect is given in Fig. 3. Very interestingly, both kinds of GABAergic inhibition (pre- and postsynaptic) are mediated by the same $GABA_A$ receptor subtype, which operates by gating a chloride channel of the neuronal membrane. The electrophysiological response for presynaptic inhibition (depolarization), however, will be different from that in the case of postsynaptic inhibition (hyperpolarization) due to the different intracellular chloride concentrations of the receptive neurons (Simmonds 1984; Snodgrass 1983).

Beside its well-documented $GABA_A$ receptor response, some synaptic effects of GABA are also mediated by a second GABA receptor type

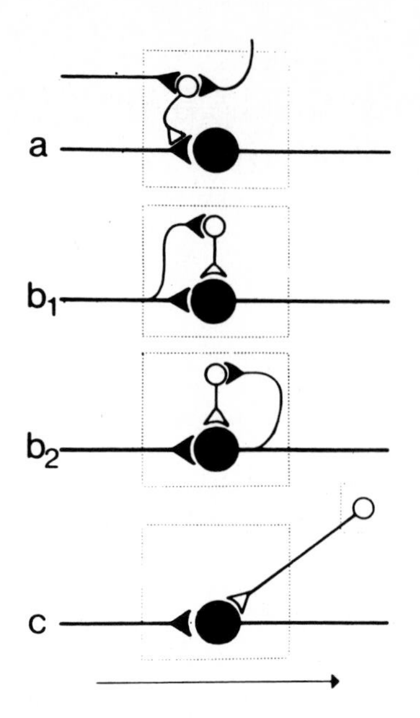

Fig. 1. Schematic diagrams of inhibitory GABAergic circuits: *a*, presynaptic inhibition by axoaxonic synapse; b_1, postsynaptic inhibition, collateral inhibition; b_2, postsynaptic inhibition, recurrent inhibition; *c*, postsynaptic inhibition by a projecting neuron. GABAergic neurons are shown *in white*, excitatory neurons *in black*. (From HAEFELY 1984)

($GABA_B$). $GABA_B$ receptors (via G-protein coupling) inhibit adenylatcyclase activity or enhance K^+ conductance (BOWERY et al. 1990). $GABA_B$ receptors are relevant for the drug baclofen, which acts as an agonist (BOWERY et al. 1990), but are not considered relevant for the pharmacology of sedative-hypnotic drugs.

Contrary to the contemporary, broad functional role of GABA in the mammalian CNS, for many years very little was known about the pharmacological relevance of GABA. Accordingly, the first reports about a possible involvement of GABA in the mechanism or action of benzodiazepines were accepted very enthusiastically, because for the first time they provided a basis for the combination of the broad spectrum of pharmacological and therapeutic properties of the benzodiazepines (Table 1) with a specific neurotransmitter system. It was only subsequently that the mechanisms of action of other hypnotics like the barbiturates were also linked to GABA. However, it took some time before the presumed GABAmimetic properties of the benzodiazepines and other hypnotics could be explained at the level of GABAergic synapses of the CNS.

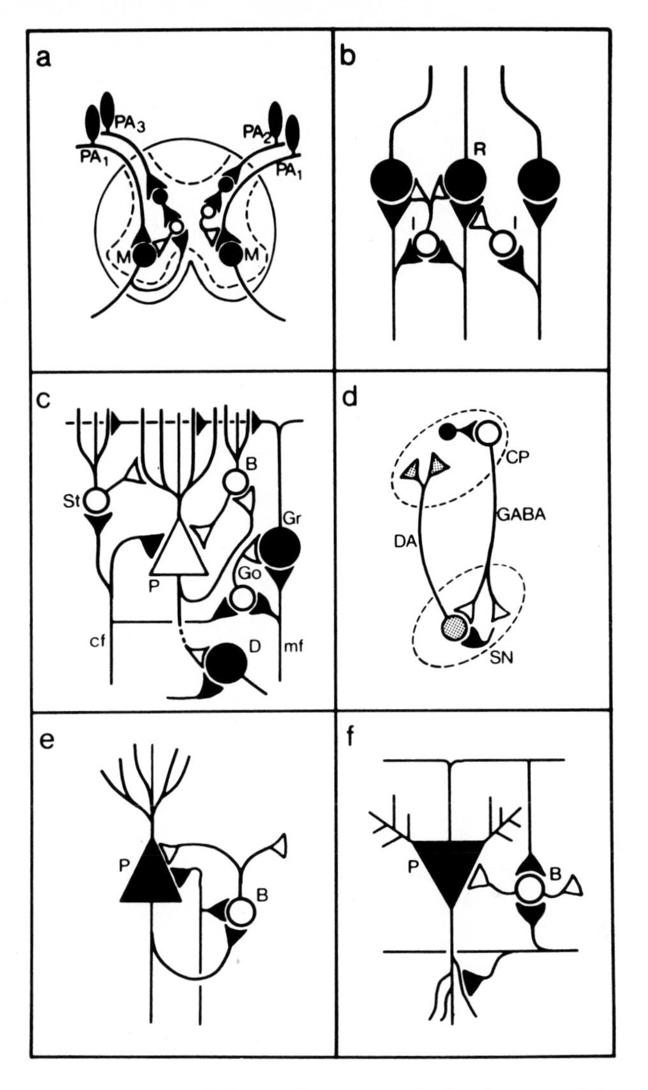

Fig. 2. Schematic diagrams of the main neuronal circuits containing GABAergic synapses, where benzodiazepines have been found to enhance transmission. GABAergic neurons are shown *in white*, excitatory neurons *in black*. *a*, spinal cord (*M*, motoneuron; *PA*, primary afferents); *b*, dorsal column nuclei (*R*, relay cell; *I*, interneuron); *c*, cerebellar cortex (*P*, Purkinje cell; *Gr*, granule cell; *Go*, Golgi cell; *B*, basket cell; *St*, stellate cell; *D*, output neuron of Deiter's nucleus; *cf*, climbing fiber; *mf*, mossy fiber): *d*, neostriatum and substantia nigra (*CP*, caudate putamen; *SN*, substantia nigra; *DA*, nigrostriatal dopamine pathway; *GABA*, GABAergic striatonigral pathway); *e*, cerebral cortex (*P*, pyramidal cell; *B*, basket cell; *f*, hippocampus (*P*, pyramidal cell; *B*, basket cell). (From HAEFELY 1984)

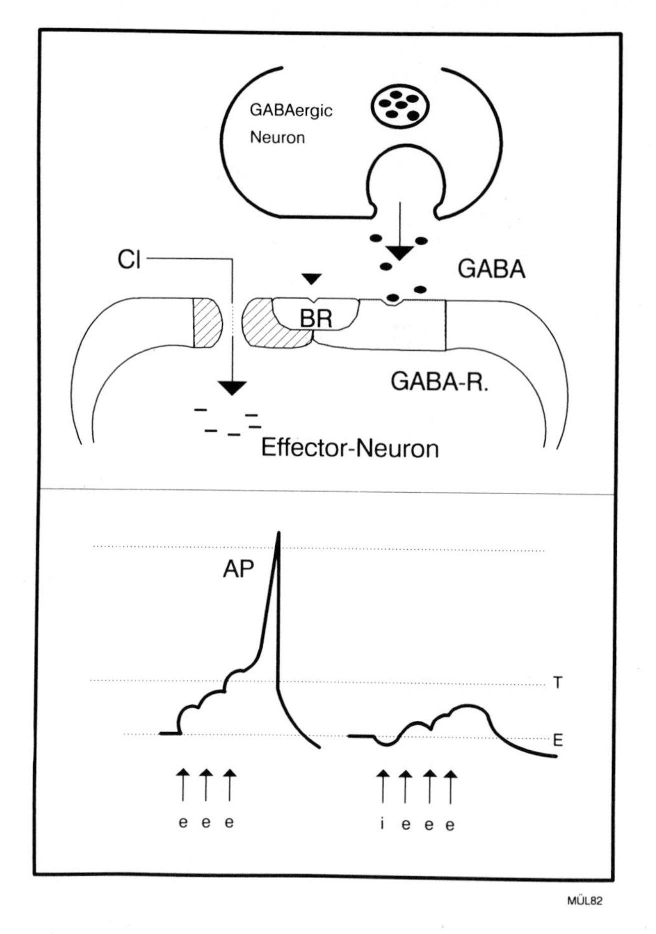

Fig. 3. Schematic presentation of the $GABA_A$ receptor complex (*upper*) and the postsynaptic inhibition mediated by GABA (*lower*). After GABA-induced hyperpolarization (*i*) the same excitatory input (*e*) is no longer sufficient to reach the threshold potential (*T*) and to trigger an action potential (*AP*)

B. Functional Evidence for the GABA-mimetic Properties of Benzodiazepines

Even the first experiments with chlordiazepoxide demonstrated quite readily that this type of drug exhibited sedative-hypnotic, anxiolytic, anticonvulsive, and centrally muscle relaxant properties (see SCHALLEK et al. 1979). For more than 10 years of extensive use of the benzodiazepines very little was discovered about the biological or biochemical basis or their activity. The fourth edition of Goodman and Gilman's standard textbook of pharmacology printed in 1970 stated that "the effects of this compound (chlordiazepoxide)

Table 1. Major pharmacological properties of the benzodiazepines and their corresponding therapeutic uses

Pharmacological actions	Therapeutic indications
Antipunishment and antifrustration activity, behavioral disinhibition	Aniety, anxious depression
Arousal reduction, antiaggressive activity	Hyperemotional states
Facilitation of sleep	Insomnia
Anticonvulsant action	Various forms of epileptiform activity
Attenuation of centrally mediated autonomic nervous and endocrine response to emotions and to excessive afferent stimuli	Psychosomatic disorders (cardiovascular, gastrointestinal, hormonal)
Central muscle relaxation	Somatic and psychogenic muscle spasms, tetanus
Potentiation of the activity of centrally depressant agents, anterograde amnesia	Surgical anesthesia

upon the brain are not well known." Ten years later, two major advances occurred in respect of our understanding of benzodiazepine action.

1. Evidence from biochemical, neuropharmacological, and electrophysiological experiments indicated that the action of the benzodiazepines at the neuronal level was in some way linked to that of the major inhibitory neurotransmitter GABA (Costsa et al. 1975; Haefely et al. 1975; Fuxe et al. 1975).
2. Specific benzodiazepine-binding sites were identified in mammalian brain, specifically located within GABAergic synapses and acting as the primary target of benzodiazepine drugs (Squires and Braestrup 1977; Möhler and Okada 1977).

Both findings finally led to our present concept of a specific benzodiazepine-binding site as a regulating subunit of the postsynaptic $GABA_A$-receptor complex.

As already mentioned above, the first evidence for the concept of benzodiazepines acting as GABA-mimetic drugs had already been reported several years before the discovery of the benzodiazepine receptor. However, the possible mechanism of this GABA-mimetic activity was completely unknown, because benzodiazepines did not act via indirect mimetic mechanisms already known from other neurotransmitters, such as enhancement of neuronal release, or a blockade of the synaptic degradation of the neurotransmitter. Starting with the discovery of the benzodiazepine receptor, enormous progress has been made, and today we have a fairly clear picture of how benzodiazepines enhance inhibitory GABAergic transmission.

However, a number of mainly electrophysiological observations are not consistent with the assumption that all properties of the benzodiazepines are mediated via GABA-mimetic effects (POLC 1988). One example is the effect of several benzodiazepines at nanomolar concentrations on K^+ conductance (CARLEN et al. 1983), which can be blocked by the benzodiazepine receptor antagonist flumazenil. Because some of these effects occur at very low concentrations of benzodiazepines, it remains to be demonstrated whether there are flumazenil-sensitive benzodiazepine receptors which are not part of the $GABA_A$-receptor complex (POLC 1988).

I. Electrophysiology

In most cases of GABA-mediated responses in the mammalian CNS, benzodiazepines have been shown to enhance the effects of GABA in spite of GABA's pre- or postsynaptic locus of action (see the schematic diagrams of neuronal circuits containing inhibitory GABAergic neurons in Fig. 2) (SIMMONDS 1984).

Because GABAergic inhibition is mediated by changes in the chloride conductance of the neuronal membrane, other studies have concentrated on the effects of benzodiazepines on the GABA-gated chloride channel of the neuronal membrane. Such studies (see BARKER et al. 1984 for a review) have indicated that, of the possible parameters of the chloride channel at the microscopic level, benzodiazepines only increase the frequency of chloride channel openings, while the lifetime of opening events and the channel conductance are not altered.

II. Pharmacology and Biochemistry

Pharmacological and biochemical evidence also points to an involvement of GABA in the mechanism of action of the benzodiazepines. Because these data contribute little further information when compared with the electrophysiological data, only a very short update will be given (for further details see the reviews of SCHALLEK et al. 1979; HAEFELY et al. 1981). Benzodiazepines are usually more potent antagonists against convulsions induced by agents impairing GABAergic neurotransmission than against convulsions induced by agents acting via other neuronal system (e.g., strychnine as a glycine receptor antagonist). Moreover, a large variety of effects of benzodiazepines at the electrophysiological, biochemical, and pharmacological level can be blocked by relatively low doses of GABA antagonists such as bicuculline and picrotoxin. Finally, in a variety of experimental settings, benzodiazepine activity is profoundly lowered when the brain levels of GABA are reduced by inhibitors of GABA synthesis. All these data suggest indirect GABA-mimetic properties of the benzodiazepines, but without indicating the possible mechanism of action.

III. Histochemistry

The hypothesis of a GABA-mimetic activity of the benzodiazepines was strongly supported by histological data indicating that the presence of the benzodiazepine receptor in all brain regions was strongly correlated with the presence of GABAergic nerve terminals. One of the first reports in this respect was the findings of PLACHETA and KAROBATH (1979) of a comparable regional distribution of $GABA_A$ receptors and benzodiazepine receptors, with the consistent finding of more GABA receptors than benzodiazepine receptors in all brain regions studied. Moreover, by using the combination of benzodiazepine receptor autoradiographic techniques with immunohistochemical methods for glutamic acid decarboxylase, the marker enzyme of GABAergic neurons (MÖHLER and RICHARDS 1983; KUHAR 1983), strong evidence for a coexistence of benzodiazepine receptors with GABAergic synapses could be demonstrated at the light microscopic level as well as the electron microscopic level. Final evidence for the colocalization of the benzodiazepine receptor and GABA receptor comes from data about the isolation and purification of a synaptic macromolecule containing the benzodiazepine and the GABA receptor (SCHOCH et al. 1985; HÄRING et al. 1985) and the identification of its subunit composition (LÜDDENS and WISDEN 1991; MÖHLER et al. 1993) (see Sect. C.II). In conclusion, strong evidence exists that most if not all benzodiazepine receptors are colocalized with $GABA_A$ receptors. Up to now, little evidence exists for the presence of benzodiazepine receptors that are not biochemically linked to GABA receptors. On the other hand, because our CNS contains many more $GABA_A$ receptors than benzodiazepine receptors, it is generally assumed that $GABA_A$ receptors are a long way short of all being linked to benzodiazepine receptors (BOWERY et al. 1984). A specific example might be a diazepam-insensitive $GABA_A$ receptor on cerebellar granule cells (LÜDDENS et al. 1990).

The GABA receptors associated with the benzodiazepine receptors are always the $GABA_A$ subclass, which can be specifically antagonized by bicuculline. No evidence exists that $GABA_B$ receptors (which can be specifically activated by the drug baclofen) are associated with benzodiazepine receptors (BOWERY er al. 1984, 1990).

Even in their first publication on the presence of specific benzodiazepine-binding sites in rat brain, BRAESTRUP and SQUIRES (1977) also demonstrated the presence of specific and high-affinity benzodiazepine-binding sites in several peripheral organs. However, these sites exhibited a different substrate specificity, since the highly potent benzodiazepine derivative, clonazepam, showed only a very weak affinity for these sites. Similar sites have subsequently been identified in most peripheral tissues investigated including platelets and white blood cells. All these early studies used tritiated diazepam or tritiated flunitrazepam as radioligands. However, after the introduction of tritiated 4′-chlordiazepam (^{3}H-Ro 5-4864), a specific ligand for these sites became available with negligible affinity for the brain-specific benzodiazepine receptor. By the use of this radioligand it was possible to identify so-called peripheral benzodiazepine-binding sites not only on peripheral organs but also in CNS tissues of many species. These sites differ from the benzodiazepine receptor in respect to regional distribution, cellular distribution with a possibly specific association of these sites with glial cells, and phylogenetic development, because these sites are probably specific for mammalian vertebrates. The most important difference between the central benzodiazepine receptor and the peripheral benzodiazepine-binding sites is the completely different substrate specificity, which, however, is typical for these sites and does not differ between the different organs and between the different species investigated. Very importantly, there is no correlation between affinity for these sites and benzodiazepine activity in animals and man, as has been demonstrated for the central benzodiazepine receptor (see Sect. C.VI). There is presently little doubt that peripheral benzodiazepine-binding sites are not involved in any of the relevant (Table 1) pharmacological or clinical properties of the benzodiazepines (for reviews see SANNO et al. 1989; KRUEGER 1991; GAVISH et al. 1992).

IV. Receptor Interactions: Effects of GABA Receptor-Agonists on Benzodiazepine-Receptor Binding and Vice Versa

Strong evidence for a close functional relationship between the GABA receptor and the benzodiazepine receptor also comes from binding experiments (see BRAESTRUP et al. 1983 for a review). Activation of the GABA receptor by agonists increases agonist binding to the benzodiazepine receptor, strongly suggesting allosteric interactions between both receptors. This effect, usually called GABA shift or GABA ratio, is very specific for benzodiazepine receptor agonists and is not observed for antagonists. Accordingly, the GABA ratio represents an excellent in vitro test system for differentiating benzodiazepine receptor agonists, antagonists, and inverse agonists (see Sect. C.IV). However, the GABA shift cannot explain the GABA-mimetic properties of the benzodiazepines.

Very interestingly, an increase in the affinity of agonist binding to the GABA receptor in the presence of benzodiazepines has also been described (JOHNSTON and SKERRITT 1984). Theoretically, this effect could much better explain the GABA-mimetic properties of benzodiazepines. However, it is also not the final explanation, because the potency of benzodiazepines to enhance GABA receptor binding does not correlate at all with their pharmacological potency or their affinity for the benzodiazepine receptor.

C. $GABA_A$ Receptor–Benzodiazepine Receptor–Chloride Channel Complex

I. Functional Aspects

Taking all the evidence (see above) together, the following functional model of the $GABA_A$ receptor–benzodiazepine receptor complex can be given

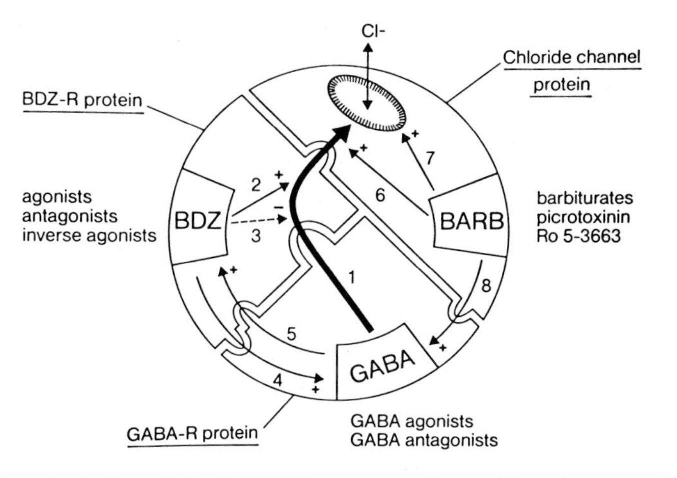

Fig. 4. Hypothetical model of GABA receptor–benzodiazepine receptor–chloride channel complex of the GABAergic synapse. For explanation of *numbers 1–8* see text. (From POLC et al. 1981)

(Fig. 4). Although most parts of this model have been identified in biochemical experiments (see Sect. C.II), this rather hypothetical model is still useful today in understanding the pharmacology of the $GABA_A$ receptor.

This model assumes the presence of a multifunctional macromolecular complex, consisting of the chloride channel, the $GABA_A$ receptor, the benzodiazepine receptor, and a third recognition site binding drugs such as barbiturates as agonists and some convulsants such as picrotoxin as antagonists. All four functional units can interact with each other, finally resulting in a very complicated pattern of responses.

1. Activation of the GABA receptor by an agonist leads to a conductance change of the chloride channel. Although this represents the physiologically most important part of the model, the biochemical mechanism of the GABA-induced gating of the chloride channel is still not yet finally understood.
2. Agonist binding to the benzodiazepine receptor enhances the effect of GABA agonists at the chloride channel by increasing the frequency of channel-opening events. Again, the biochemical mechanism is not yet known.
3. Binding of inverse agonists (see Sect. C.IV) to the benzodiazepine receptor has the opposite effect (decrease in the frequency of channel-opening events).
4. Agonist binding to the benzodiazepine receptor enhances agonist affinity at the GABA receptor.
5. Agonist binding to the GABA receptor enhances agonist binding to the benzodiazepine receptor (GABA ratio).

6. Agonists at the picrotoxin-binding site (e.g., barbiturates) at low concentrations also enhance the GABA receptor mediated activation of the chloride channel (see Sect. D.II).
7. Barbiturates at high (possibly anesthetic) concentrations seem to activate the chloride channel directly (hyperpolarization) (see Sect. D.II).
8. Barbiturates at low concentrations can also enhance agonist binding to the benzodiazepine receptor.

The effects of agonist at all three recognition sites can be prevented by specific antagonists, e.g., bicuculline at the $GABA_A$ receptor, flumazenil at the benzodiazepine receptor, and picrotoxin at the barbiturate recognition site.

This model not only summarizes the effects of benzodiazepines on GABAergic neurotransmission, but also explains why barbiturates at low concentrations have a pharmacological spectrum similar to that of the benzodiazepines, but when given at high (anesthetic) concentrations can produce a much more pronounced depression of the CNS due to their direct effects on the chloride conductance (hyperpolarization) of the neuronal membrane (see Sect. D.II).

The $GABA_A$ receptor complex also consists of an additional binding site for certain steriod anesthetic agents, e.g., the drug alfaxalone and the progesterone metabolite 3α-hydroxy-5α-dihydroprogesterone (GEE et al. 1988; HARRISON et al. 1987). These steroids produce barbiturate-like effects including enhancement of the binding of benzodiazepines and GABAergic agonists, enhancement of GABA-induced chloride currents at lower concentrations, and direct, GABA-independent activation of chloride currents at higher concentrations. Today, it can be assumed that a variety of other sedative or hypnotic drugs such as etomidate, methaqualone, and clomethiazole also interact with one or several of the different parts of the $GABA_A$ receptor complex (TICKU and MAKSAY 1983; OLSEN 1987; HAEFELY et al. 1985). However, the relevance of these effects for their pharmacological properties needs to be demonstrated.

II. Structural Aspects

Final proof for the existence of the $GABA_A$ receptor complex as functionally described above comes from molecular cloning experiments indicating the presence of mainly three distinct subunits (α, β, γ) finally forming the $GABA_A$ receptor complex (SCHOFIELD et al. 1987; PRITCHETT et al. 1989a; LÜDDENS and WISDEN 1991). Each of these subunits exists in several variants, which opens the possibility of several hundred $GABA_A$ receptor subclasses (LÜDDENS and WISDEN 1991). However, only few seem to be normally expressed in the CNS (see Sect. C.V). Up to now, it seems that none of the functional aspects of the $GABA_A$ receptor complex as summarized in Sect. C.I can be exclusively related to one of the structural subunits (LÜDDENS and WISDEN 1991).

III. Chain of Events from Benzodiazepine-Receptor Occupation to Pharmacological Response

Benzodiazepines modulate the $GABA_A$ receptor operated chloride channel complex (Fig. 4). This complex is part of the postsynaptic neuron (Fig. 5) GABAergic inhibitory synapse. Similar GABAergic synapses can be found in many areas of the CNS, because many if not all neurons receive inhibitory GABAergic input (Fig. 5). In a given brain region one type of neuron might be more sensitive than others for the enhanced GABAergic inhibition, which probably explains the specificity of the benzodiazepine effects. The best evidence available for this assumption is that for the anxiolytic activity of the benzodiazepines, where enhanced inhibition on serotoninergic target neurons in the limbic system might be relatively important (Fig. 5). The brain regions and target neurons involved in the other pharmacological properties of the benzodiazepines are also speculated upon in Fig. 5. Because benzodiazepines suppress all levels of the neuronal axis, it is not possible to localize clearly the specific structure relevant for their hypnotic activity. GABAergic interneurons seem to inhibit most neuronal systems determining the sleep-wakefulness cycle, such as the monoaminergic structures of the ascending mesencephalic activating system of the brain stem and the hypnogenic structures in the lower brain stem. Accordingly, it is reasonable to assume that enhanced GABAergic inhibition of these neuronal structures is relevant for the hypnotic properties of benzodiazepines (GAILLARD 1989).

The question which remains is, why some neurons or neuronal systems of a given brain region are significantly affected by benzodiazepines and

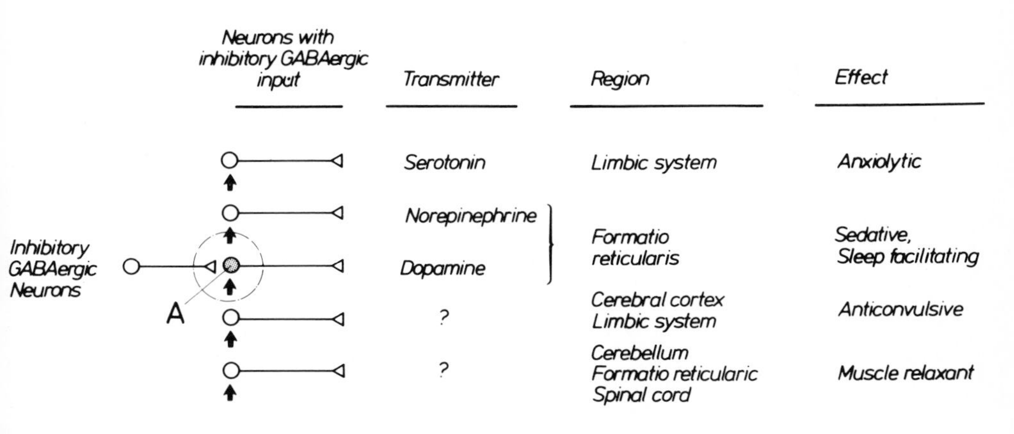

Fig. 5. Hypothetical model to explain the postsynaptic enhancement of GABAergic inhibition by benzodiazepines in several brain regions and its possible relationship to pharmacological effects. *Part A*, the GABAergic synapse with the benzodiazepine receptor, is shown in Fig. 3 on larger scale. (From MÜLLER 1987)

others not. It seems less likely that this can be explained by the presence or absence of benzodiazepine receptors or GABAergic synapses, because both are widely distributed in most brain areas. A possible explanation is the relatively small overall effect of benzodiazepines on neuronal activity, because the intensity of the potentiating action of benzodiazepines on GABAergic inhibition is rather small and only results in a small shift of the GABA dose-response curve without altering the intensity of the maximum effect of GABA (HAEFELY 1984) (see Fig. 6). Obviously, benzodiazepines are most active around the middle part of the dose-response curve of GABA alone, while a much reduced effect can be expected when the GABAergic activity alone is very high or very low (see arrows in Fig. 6). In other words, the maximum effect of benzodiazepines on GABAergic inhibition is not only rather small, but is also only seen within a narrow range of GABAergic activity. Both effects might contribute to the highly specific pharmacological properties of the benzodiazepines, because even in the same brain region the relatively small enhancement of GABAergic

BENZODIAZEPINES AND GABAergic SYNAPSES

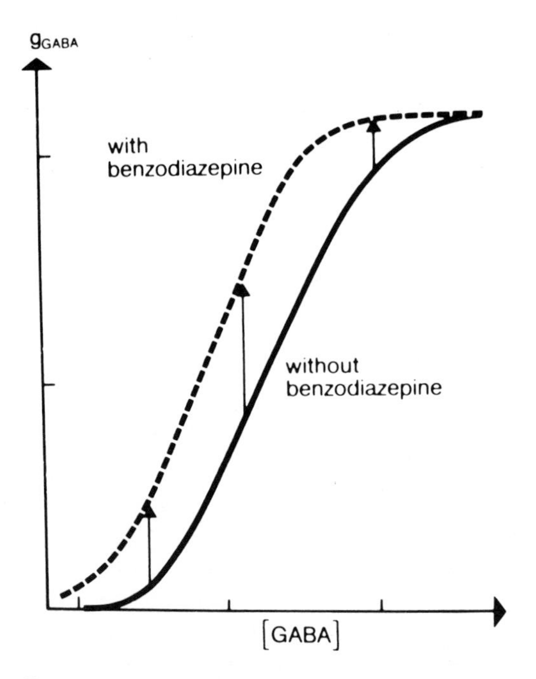

Fig. 6. Schematic diagram to show the possible effect of a benzodiazepine on the dose-response curve of GABA. The relative enhancement of GABAergic activity depends on the synaptic GABA concentration and accordingly on the activity of the GABAergic neuron, as indicated by the *arrows*. g_{GABA} indicates the conductance change induced by GABA. (From HAEFELY 1984)

inhibition by benzodiazepines might be sufficient to alter only the activity of one type of neuron but not that of another. Thus, the assumption that benzodiazepines act specifically in our brain by enhancing the synaptic function of the rather nonspecific inhibitory neurotransmitter GABA is not contradictory, but represents a very plausible concept to explain the mechanism of action of this important class of psychotropic drugs.

IV. Benzodiazepine-Receptor Ligands with Different Intrinsic Activities: Concept of Agonists, Antagonists, and Inverse Agonists

Initial studies on the pharmacological properties of some β-carboline derivatives indicated benzodiazepine-opposite (convulsive, anxiogenic) rather than benzodiazepine-like properties (NUTT 1983). Most of these β-carbolines bind with very high affinity and selectivity to the benzodiazepine receptor. Thus, their unusual pharmacological properties were explained by antagonistic properties at the benzodiazepine receptor. However, this assumption had to be revised after the discovery of benzodiazepine-receptor antagonists such as flumazenil. These compounds not only were nearly without any pharmacological effects themselves, but were also able to antagonize the effects of the classicical benzodiazepines as well as the opposite effects of some β-carboline derivatives (NUTT 1983). Due to the high selectivity of flumazenil for the benzodiazepine receptor, these observations supported the assumption that the pharmacological properties of these β-carbolines (see above) are mediated via the benzodiazepine receptor. Moreover, these findings suggested that two types of agonists exist for the benzodiazepine receptor, one with benzodiazepine-like and one with benzodiazepine-opposite properties, and that the effects of both types of agonist can be blocked by flumazenil, acting as a rather classical receptor antagonist. This concept, as unusual as it is in pharmacology, has received more and more experimental support over the years (BRAESTRUP et al. 1982; HAEFELY 1984, 1990; MÖHLER and RICHARDS 1983; TALLMAN and GALLAGHER 1985).

To conclude the review of all these data, Fig. 7 shows a scheme of the different types of benzodiazepine receptor ligands, which mainly indicates the existence of a continuum of agonistic properties, ranging from full agonists (benzodiazepines) on the one site to inverse agonists (dimethoxymethylcarboline carboxylate, DMCM) on the other site. The spectrum of benzodiazepine receptor agonists ranges from full agonists to partial agonists with reduced intrinsic activity (Fig. 7). The end point of the agonist spectrum are the antagonists such as flumazenil which are nearly, but not completely, devoid of agonistic properties (intrinsic activity), As we have seen for the agonists, the spectrum of inverse agonists ranges from antagonists with slight inverse intrinsic activity (or negative efficacy) over partial inverse agonists (methyl-β-carboline-3 carboxylate, MCC) to more or less full inverse agonists

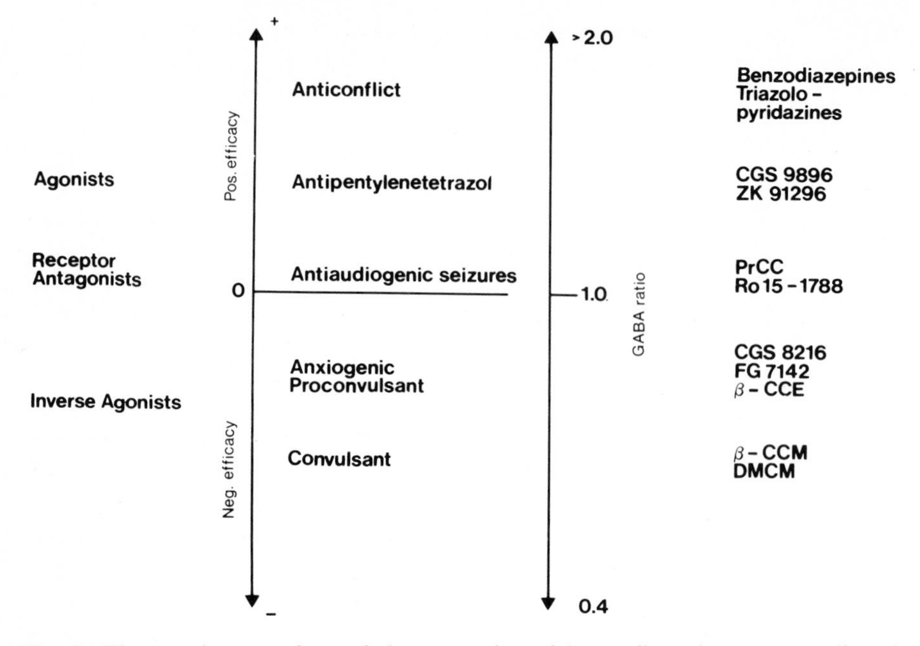

Fig. 7. The continuum of agonistic properties of benzodiazepine-receptor ligands ranging from positive to negative efficacy in relation to the pharmacological properties and to the GABA ratio as a biochemical in vitro test predictive of the different intrinsic activities. (From BRAESTRUP et al. 1983)

(DMCM), which are potent convulsants comparable to pentetrazol (BRAESTRUP et al. 1983).

Comparable to the benzodiazepine-receptor agonists, where the presence of GABA increases the benzodiazepine-receptor affinity with increasing agonistic properties (positive efficacy) (Fig. 7), the affinity of inverse agonists for the benzodiazepine receptor is decreased by GABA with increasing inverse agonistic properties (negative efficacy) (Fig. 7) (BRAESTRUP and NIELSEN 1981). Accordingly, the GABA ratio (IC_{50} in the absence of GABA over the IC_{50} in the presence of GABA) (see Sect. B.IV) can be used as an in vitro test for both positive and negative efficacy of benzodiazepine receptor ligands (BRAESTRUP et al. 1983). Again, the GABA effects on benzodiazepine-receptor affinity do not explain the inverse agonistic properties, but only seem to parallel these effects, because inverse agonists decrease GABAergic inhibition by interacting with the benzodiazepine receptor. Similar to the enhancement of GABAergic inhibition by benzodiazepine-receptor agonists, the biochemical mechanism of the decrease in GABAergic transmission by inverse agonists is not yet understood. However, the final effect of the inverse agonists is a decrease in the frequency of the chloride-channel opening events under GABA (BARKER et al. 1984), an effect exactly opposite to that of benzodiazepines or related agonists.

V. Receptor Subclasses and Hypnotic Benzodiazepines

All early binding experiments made with the classical anxiolytic or hypnotic benzodiazepine derivatives indicate the presence of only one class or population of benzodiazepine receptors with similar properties for all brain areas in all species investigated so far. However, further studies then indicated that a few non-benzodiazepine ligands of the benzodiazepine receptor were able to discriminate between two subclasses of the receptor. These two putative benzodiazepine receptor subclasses have tentatively been termed BZ_1 and BZ_2 receptors. Benzodiazepine receptor ligands binding with some degree of specificity to the BZ_1 subclass include the triazolopyridazine, CL 218 872, some β-carboline derivates such as propyl β-carboline-3-carboxylate (PrCC), and the hypnotics quazepam and zolpidem.

The concept of the presence of at least two benzodiazepine-receptor subclasses is also supported by biochemical data indicating the presence of different receptor proteins (for a review see SIEGHART 1985). On the other hand, the concept of the presence of two pharmacologically distinct subclasses of benzodiazepine receptors has been questioned by findings that the preferential affinity of some ligands for the BZ_1 subclass is mainly present at 4°C (at which temperature most binding experiments are carried out) but not at 37°C (for reviews see MARTIN et al. 1983; EHLERT et al. 1983; SIEGHART 1985). The existence of structurally distinct benzodiazepine-receptor subclasses has been finally confirmed by recombinant techniques where $GABA_A$ receptor subclasses with different subunit composition mimicking the BZ_1 and BZ_2 properties were coexpressed (PRITCHETT et al. 1989; LÜDDENS and WISDEN 1991). Again, the specificity of several benzodiazepine-receptor ligands for the recombinant BZ_1-like $GABA_A$ receptors in binding experiments was mainly present at 4°C and mostly disappeared at 37°C (PRITCHETT et al. 1989a,b). Using immunoprecipitation techniques, it was recently shown that $GABA_A$ receptors of the composition $\alpha_1\beta_2\gamma_2$ (resembling the BZ_1 subtype) represent a major proportion (60%–70%) of central $GABA_A$ receptors. The subunit composition $\alpha_1\beta_3\gamma_2$ (resembling the BZ_2 type) was present in approximately 20%–25% of $GABA_A$ receptors (MÖHLER et al. 1993).

Regardless of the rather sophisticated data about two biochemically distinct benzodiazepine receptor subclasses, the physiological, pharmacological, or even therapeutic significance of these putative subclasses is still a matter of dispute. The hypothesis that BZ_1-selective compounds in binding experiments at 4°C (e.g., CL 2318 872) possess a pharmacological spectrum different from that of the classical benzodiazepines (anxiolytic and anticonvulsive but much less sedative and muscle relaxant properties), as originally pointed out by LIPPA et al. (1979), was not substantiated in subsequent investigations (OAKLEY et al. 1983). Moreover, some ligands specific for the BZ_1 subclass, e.g., the anxiolytics CL 218 872 and alpidem and the hypnotic zolpidem, exhibit distinct differences in certain pharmacological models, which must be explained by different intrinsic activities rather than by subclass specificity (HAEFELY 1990; ZIRKOVIC et al. 1992). Similarly, the BZ_1-selective (binding experiments) hypnotic, zolpidem, and the unselective hypnotic, triazolam, showed nearly identical effects in suppressing neuronal activity within a large range of rat brain areas (PIERCEY et al. 1991). Very interestingly, both compounds were most active in brain regions rich in the BZ_2 subclass. This would support the old assumption of a specific relevance of the BZ_2 subclass for the hypnotic-sedative effects of benzodiazepines. This is further supported by the findings of CASUCCI et al. (1991) on healthy volunteers characterizing the BZ_1-selective benzodiazepine quazepam as a "nonsedative" agent, but the unselective triazolam as a "sedative." Accordingly, attempts to promote some newer hypnotics such as quazepam or zolpidem as drugs with relevant clinical selectivity due to their in vitro preference for the BZ_1 subclass must presently be considered cautiously (see Sect. D.III).

VI. Relationship Between Benzodiazepine-Receptor Affinity and Biological Activity

Convincing relationships between benzodiazepine receptor binding and pharmacological activity were noted in the first reports on the presence of specific benzodiazepine binding sites in the brain, inasmuch as pharmacologically inactive benzodiazepine derivatives or the inactive enantiomers of some benzodiazepine compounds exhibited very weak affinities for the central benzodiazepine receptor. This approach has been extended by several authors to a variety of pharmacologically active benzodiazepine derivatives, all exhibiting different receptor affinities and different potencies in various pharmacological tests indicative of anticonvulsive, anxiolytic, muscle-relaxing, and sedative properties (MÖHLER and OKADA 1977; MACKERER et al. 1978; BRAESTRUP and SQUIRES 1978; SPETH et al. 1980). In most of these cases, very good correlations were found between in vitro receptor affinity and in vivo activity in animals. This can be seen from the relationship between receptor affinity and pharmacological activity as muscle relaxant in the cat (Fig. 8), which is taken from the first report of MÖHLER and OKADA (1977) on the presence of benzodiazepine-specific binding sites in rat brain. Thus, these data provide strong evidence that binding to the benzodiazepine receptor is directly correlated to pharmacological activity in animals. Similar attempts have been made to correlate in vitro receptor affinity with therapeutic potency of benzodiazepines in man (SPETH et al. 1980). Figure 9 shows such a relationship between receptor affinity and the average hypnotic dose for adults, again indicating a very close correlation between affinity and potency. All these data clearly indicate a very good

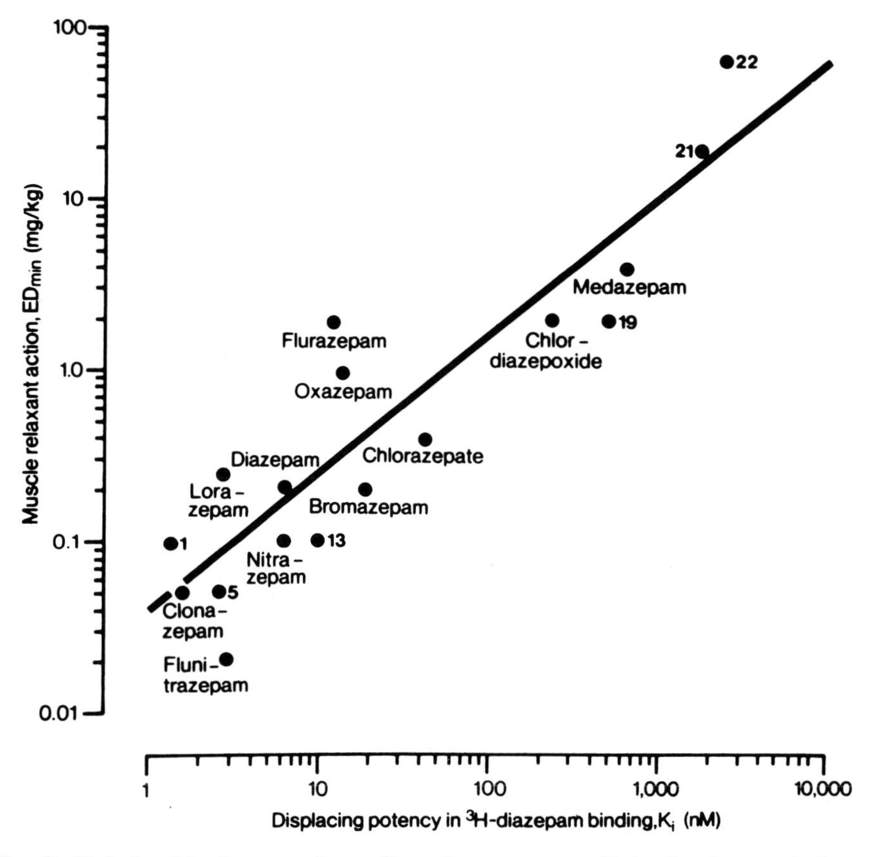

Fig. 8. Relationship between benzodiazepine receptor affinity in vitro as indicated by the inhibition constants (*K, abscissa*) and pharmacological potency as muscle relaxants in cats as indicated by the minimum effective dose (Ed_{min}, *ordinate*) of 17 different benzodiazepine derivates. (From MÖHLER and OKADA 1977)

relationship between the therapeutic potency of benzodiazepines and their affinity for the benzodiazepine receptor, strongly suggesting that affinity to the receptor is the most important determinant for their activity in man. Thus, in a very simplified view, receptor affinity represents the biochemical correlate to the therapeutic dose in man. It should be mentioned that there is no correlation between receptor affinity and the plasma half-life of the benzodiazepines in man (Table 2). Accordingly, the duration of the therapeutic effect of benzodiazepine derivatives shows no correlation at all to receptor affinity, because association to and dissociation from the benzodiazepine receptor take place in a few seconds (SPETH et al. 1978). Thus, the amount of a given benzodiazepine derivative bound to the receptor can change very rapidly in response to the drug concentration in the brain. Because the latter is directly related to the overall elimination

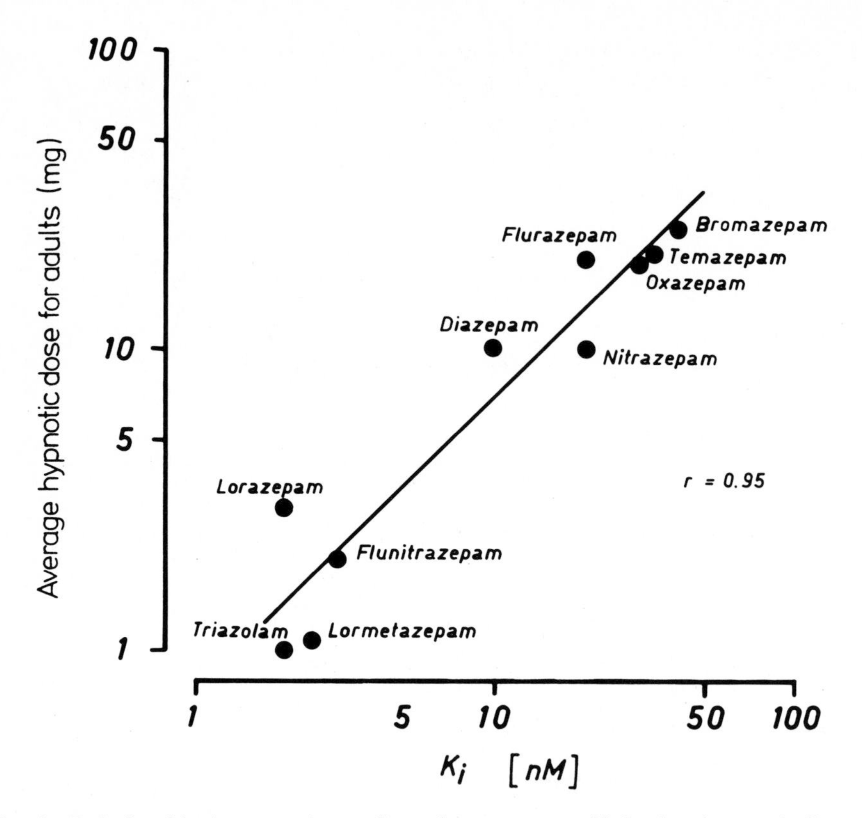

Fig. 9. Relationship between benzodiazepine-receptor affinity in vitro as indicated by the inhibition constants (*K, abscissa*) and the therapeutic potency in man as indicated by the average hypnotic dose for adults (*ordinate*) of ten different benzodiazepine derivates used clinically as hypnotics. (From Müller 1987)

rate of the benzodiazepine derivative in the body (usually determined by the plasma half-life), receptor occupancy and thereby the duration of the pharmacological effect are exclusively determined by the different pharmacokinetic parameters of the various benzodiazepine derivatives (Müller and Stillbauer 1983).

For single dosing, however, as is the case for the use of benzodiazepines as hypnotics, the duration of action of benzodiazepines seems to depend on three parameters (Greenblatt et al. 1990):

1. Regardless of individual pharmacokinetic properties, all benzodiazepines seem to produce acute tolerance to some CNS effects, which are diminished before the drug is eliminated from the CNS (Greenblatt et al. 1987, 1989; Smith et al. 1984).
2. The benzodiazepine can be distributed from the CNS to other tissues. This mechanism seems to be most relevant for the pronounced initial effects of most benzodiazepine hypnotics, whose duration is more dependent on this distribution mechanism (as measured by the volume of

Table 2. Comparison between benzodiazepine-receptor affinity and elimination half-life in man

Benzodiazepine	IC_{50} (nmol/l)	$t_{1/2}$ (h)
Alprazolam	20	10–18
Bromazepam	18	12–24
Brotiazolam	1	4–8
Camazepam	900	10–24
Chlordiazepoxide	350	10–18*
Clobazam	130	10–30*
Clonazepam	2	24–56
Clorazepate	59	2–3*
Clotiazepam	2	3–15
Diazepam	8	30–45*
Estazolam	9	11–20
Flunitrazepam	4	10–25
Flurazepam	15	2*
Ketazolam	1300	1.5*
Loprazolam	6	7–8
Lorazepam	4	10–18
Lormetazepam	4	9–15
Medazepam	870	2*
Metaclazepam	930	18–20*
Midazolam	5	1–3
Nitrazepam	10	20–50
Oxazepam	18	5–18
Oxazolam	10000	–
Desmethyldiazepam	9	50–80
Prazepam	110	1–3*
Quazepam	30	25–40*
Temazepam	16	6–16
Tetrazepam	34	12
Triazolam	4	2–4
Zolpidem	50	2–3
Zopiclone	30	3–6

Data for the inhibitory concentrations 50% against specific ^{3}H-diazepam binding (IC_{50}) were mostly taken from HAEFELY et al. (1985) and the data for the elimination half-lives ($t_{1/2}$) from SCHÜTZ (1982, 1989). * indicates that the elimination half-life of the active drug is much longer due to the presence of active, slowly eliminated metabolites. Oxazolam is a prodrug. As active component only desmethyldiazepam is present in human plasma after oral administration (SCHÜTZ 1982).

distribution) than on the terminal elimination half-life (ARENDT et al. 1983; GREENBLATT et al. 1989).

3. Rapid drug elimination by biotransformation also leads to a short pharmacodynamic duration of action. Slow elimination tends to increase the likelihood of unwanted residual sedation.

All three parameters must be taken into consideration if the duration of hypnotic activity needs to be explained. Especially the distribution phenomenon (from the CNS to other tissues) is often underestimated. For example, the duration of acute hypnotic effects (increase in beta activity in the EEG) was shorter for diazepam (long half-life) than for triazolam (short half-life) (ARENDT et al. 1983).

As already mentioned, most of the correlations between receptor affinity in vitro and biological activity as summarized above neglected differences in the pharmacokinetic properties of the drugs. To overcome these problems, receptor affinity can also be determined in vivo. In this case it is usually given as ED_{50}, e.g., that in vivo dose which inhibits specific ligand binding in vivo by 50% by occupying 50% of the receptors present.

If the same species is used for assaying in vivo receptor binding and biological activity, excellent correlations are found. One example is given in Fig. 10, which not only includes pharmacokinetically different but pharmacodynamically similar benzodiazepines such as midazolam and chlordiazepoxide, but also some nonbenzodiazepine agonists of the benzodiazepine receptor such as zopiclone and CL 218.872. These observations

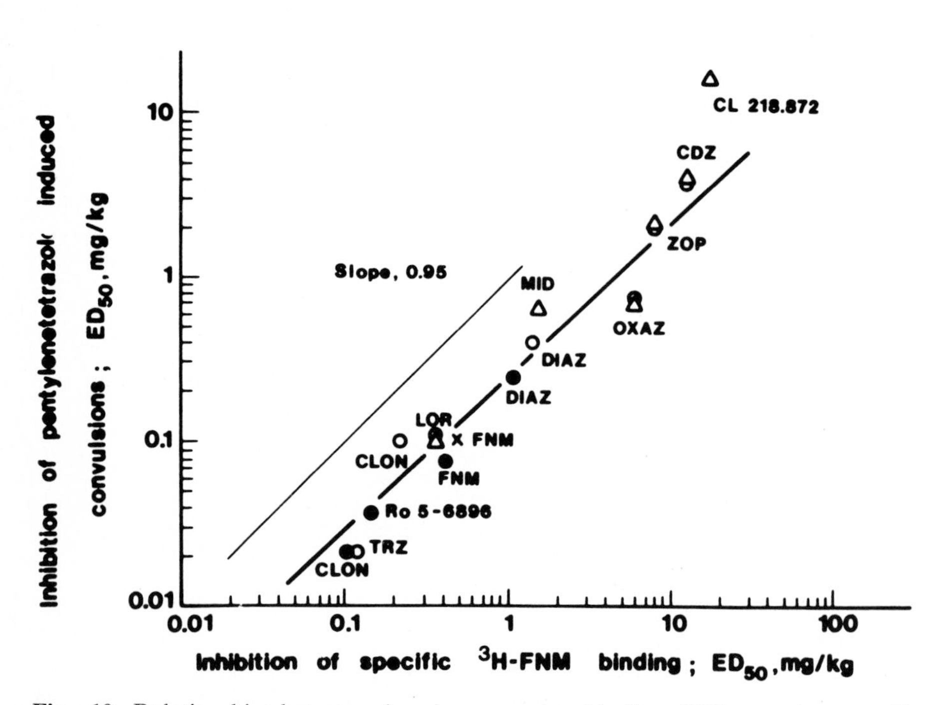

Fig. 10. Relationship between in vivo receptor binding (*ED_{50} against specific [^{3}H]flunitrazepam binding, abscissa*) and anticonvulsive potency (*ED_{50} against pentetrazol-induced convulsion, ordinate*) for several benzodiazepines and related drugs in mice. (From BRAESTRUP et al. 1983)

complete the conclusions made on the basis of correlations using in vitro receptor affinity and clearly indicate that for a given species a specific pharmacological effect is always observed following doses of benzodiazepine receptor agonists which occupy the same fraction of the total receptor population in a given brain area. These data again strongly suggest that binding to the benzodiazepine receptor represents the primary event within the molecular mechanism of action of benzodiazepines and related compounds.

The correlation between receptor occupancy and pharmacological activity has been investigated in some detail by PETERSEN et al. (1986). These authors demonstrated that in a given species receptor occupancy is relatively low for pharmacological tests predictive of anticonvulsive activity, somewhat higher for tests predictive of anxiolytic activity, and considerably higher for tests predictive of muscle relaxant activity (Fig. 11). Based on these and several other observations, the scheme in Table 3 can be given, indicating different levels of receptor occupancy for each of the different pharmacological effects of the benzodiazepines. Conversely, with an increasing level of receptor occupancy needed to elicit a given pharmacological response, the number of spare receptors (the receptors which have not been

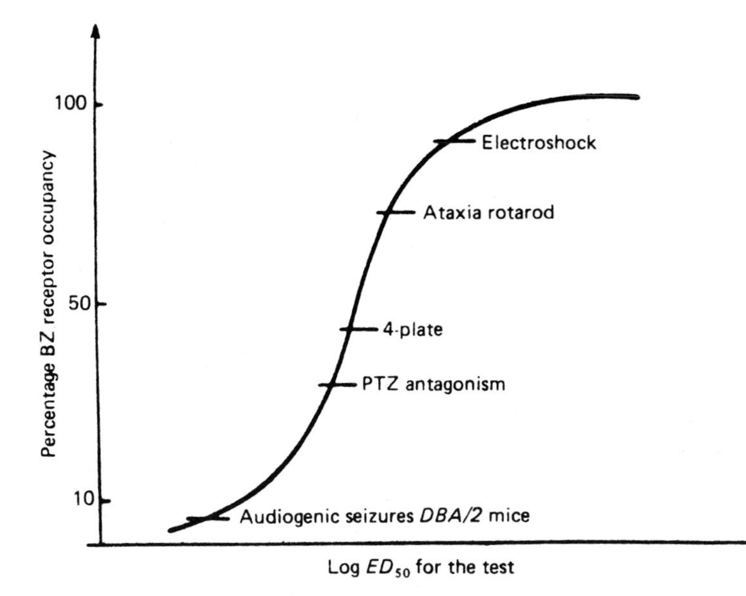

Fig. 11. Relationship between ED_{50} and benzodiazepine-receptor occupancy in various tests in mice. The *abscissa* is arbitrary (from PETERSEN et al. 1986). The effectiveness of benzodiazepines against audiogenic and pentetrazol-induced convulsions is predictive for anticonvulsive properties, that in the four-plate test for anxiolytic properties, and those in the ataxia rotarod and electrochock tests for muscle relaxant properties

Table 3. Relationship between benzodiazepine receptor occupancy and the pharmacological (and clinical?) effects of benzodiazepine derivates. Several of the effects may overlap

Benzodiazepine-receptor occupancy (receptor-mediated stimulus)	Loss of consciousness Muscle relaxation Motor impairment Amnesia Hypnotic effect Antiepileptic effect Anxiolytic effect	Number of spare receptors

occupied) decreases (Table 3). Thus, a major conclusion of all these findings is that the different doses of a given benzodiazepine derivative needed to elicit different pharmacological responses in experimental animals (for review see HAEFELY et al. 1981) can be explained by the different levels of receptor occupancy needed for these effects.

Sometimes the very impressive correlations between in vivo benzodiazepine receptor occupancy and pharmacological response (see above) do not hold true for all benzodiazepine receptor agonists and all pharmacological properties of the benzodiazepines. This is shown by some of our laboratory data in Fig. 12. In agreement with the conclusions drawn above, all five benzodiazepine derivatives are more potent (having lower pharmacological ED_{50} values) in the pentetrazol test (anticonvulsive activity) than in the fighting mice test (anxiolytic activity). Both pharmacological properties correlate linearly with the ED_{50} for in vivo receptor binding. The two straight lines obtained from both correlations suggest that both pharmacological properties are obtained at a different level of in vivo receptor occupation. This level, however, is similar for each of the five drugs in one single test. Similarly, muscle relaxant properties (horizontal grid test) are observed for diazepam, clobazam, chlordiazepoxide, and lofendazam at an even higher level of receptor occupation, which again is similar for the four drugs investigated. So far the data have been very compatible with the model presented in Fig. 11, indicating increasing but, for each agonist, similar fractional receptor occupation in relation to the different pharmacological properties. This model (Fig. 11), however, does not hold true for the 1,5-benzodiazepine arfendazam (MÜLLER 1985), which exhibits muscle relaxant properties only at much higher doses that one would expect from the correlation obtained for the other four benzodiazepine derivatives (Fig. 12), indicating that muscle relaxant properties of arfendazam are only observed at a considerably higher fractional receptor occupancy. Because studies with benzodiazepine-receptor antagonists have clearly indicated that all three pharmacological properties of arfendazam are mediated by interaction with the benzodiazepine receptor (MÜLLER et al. 1986), the classical explanation for the different properties of arfendazam is the assumption of a

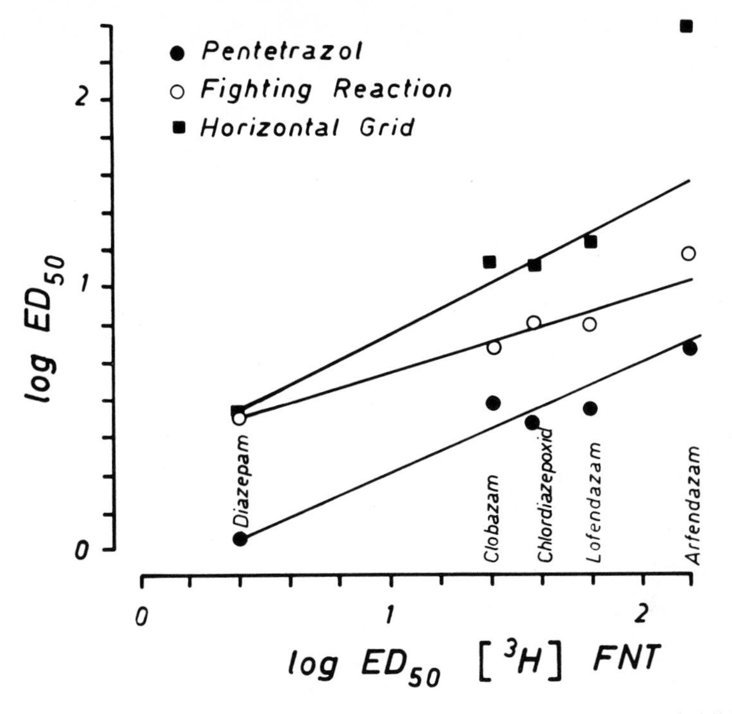

Fig. 12. Relationship between the potencies of five benzodiazepines as inhibitors of specific [^{3}H]flunitrazepam in vivo (whole mouse brain) as indicated by the ED_{50} [^{3}H]FNT (*mg/kg, abscissa*) and their potencies in three pharmacological tests in the mouse as indicated by the ED_{50} (*mg/kg, ordinate*). (From MÜLLER 1985)

lower intrinsic activity of arfendazam relative to other benzodiazepines. In other words, arfendazam only represents a partial agonist of the benzodiazepine receptor. Actually, arfendazam represents a pro-drug. The active metabolite, which is the true partial agonist of the benzodiazepine receptor, is *N*-desmethylclobazam (MÜLLER et al. 1986).

Similar pharmacological observations have been made for several other putative partial benzodiazepine-receptor agonists (STEPHENS et al. 1985; HAEFELY 1984), with the general impression that the span between the doses needed for anticonvulsive and anxiolytic effects is much higher for partial agonists than for the classical benzodiazepine receptor agonists. Evidence for the presence of only partial agonistic properties at the benzodiazepine receptor can also be obtained from some biochemical in vitro assays (Fig. 7). Whether the advantageous pharmacological profile of some of the newer partial agonists (alpidem, bretazenil, abecarnil) also holds true for their clinical use as anxiolytics in man is still a matter of dispute. A recent meta-analysis of clinical studies with bretazenil (DELINI-STULA 1992) showed a still high percentage of sedative side effects, although at a lower frequency than found for the classical benzodiazepine, diazepam.

D. Specific Hypnotic Drugs

I. Benzodiazepines

All benzodiazepine derivatives in current clinical use enhance GABAergic neurotransmission as outlined above. For the therapeutically available compounds (Table 2), we have very limited evidence for relevant pharmacodynamic differences, with the exceptions of clonazepam (RALL 1992), desmethylclobazam (MÜLLER et al. 1986), and desmethyldiazepam (BORLAND and NICHELSON 1977), which are probably somewhat less sedative than the other benzodiazepines at therapeutically comparable doses. The choice of only some of the many benzodiazepines as hypnotic drugs therefore is based less on significant pharmacodynamic advantages than on their pharmacokinetic properties (see Table 2) and on marketing considerations. A detailed description of the effects of benzodiazepines on sleep in man is given in Chap. 13 and 19, this volume.

Like many other sedative-hypnotic drugs, benzodiazepines when given in low doses and during the daytime decrease activity, moderate excitement, and calm the recipient. When the dose of a benzodiazepine is increased, sedation progresses to hypnosis and then to stupor. The latter state refers to the so-called anesthetic effect of benzodiazepines. However, one should keep in mind that benzodiazepines do not cause a true general anesthesia, because awareness usually persists. However, at high (pre-anesthetic) doses, all benzodiazepines induce amnesia for a certain period subsequent to their administration, which varies depending on the pharmacokinetics of the drugs. While this might be helpful for use of the drugs in anesthesiology, it creates rare, but disturbing side effects when the drugs are used as hypnotics (ROTH et al. 1990).

Benzodiazepines are notorious for their capacity to produce adaptation or tolerance to a number of their central actions. Tolerance for benzodiazepines is always a pharmacodynamic and not a pharmacokinetic phenomenon (GREENBLATT and SHADER 1978; FILE 1985). Tolerance may vary for the different pharmacological properties and may develop within hours even after a single dose or may develop over days or weeks of continuous use. In the first case, tolerance complicates the relationship between benzodiazepine levels to pharmacodynamic or clinical effects (GREENBLATT et al. 1987). In the second case, it terminates therapeutic efficacy. The latter case is very often seen with the use of benzodiazepines as anticonvulsants, and is sometimes but not always seen with the use of benzodiazepines as hypnotics (KALES et al. 1985). The mechanisms relevant for the tolerance to benzodiazepines are not known. Previous findings did not suggest benzodiazepine-receptor downregulation as a critical factor (for review MÜLLER 1987). However, more recent findings suggest a possible role of $GABA_A$-receptor downregulation with changes of sensitivity and/or density (MILLER et al. 1988, 1989). This is also supported by more recent findings indicating a decrease in $GABA_A$-receptor mRNA after chronic benzodiazepine treatment (KANG and MILLER 1991).

All benzodiazepines may cause paradoxical effects such as anxiety symptoms, hallucinations, and hypomanic and aggressive behavior. This phenomenon is usually related to disinhibition (depression of inhibitory centers) and seems to be mediated by the same GABAergic mechanisms as the therapeutic effects (HAEFELY et al. 1975; HALL and ZISOOK 1981; DIETCH and JENNINGS 1988). This is also supported by similar recent reports for the non-benzodiazepines zopiclone and zolpidem (see also Chaps. 13 and 20, this volume).

II. Barbiturates

Barbiturates have been extensively used as sedative-hypnotic drugs for decades, but, except for some specific uses, have been largely replaced by benzodiazepines. In contrast to the benzodiazepines, barbiturates depress the activity of all excitable tissues, although the CNS is specifically sensitive. Accordingly, relevant suppression in peripheral functions (e.g., cardiovascular) may occur in barbiturate intoxication (RALL 1992).

However, at therapeutic doses barbiturates act mainly on the CNS. At concentrations around 50 μmol/l, barbiturates enhance GABAergic inhibitory responses but also reduce glutamate-induced depolarizations. The first mechanism resembles that of the benzodiazepines. Both effects are similar for most hypnotic-anesthetic barbiturates (e.g., pentobarbital) and the anticonvulsant, phenobarbital. At higher concentrations, barbiturates can directly activate the chloride channel associated with the $GABA_A$ receptor and can enhance both GABA and benzodiazepine binding. These effects are largely mediated by barbiturate binding to the so-called picrotoxin site of the $GABA_A$-receptor complex (where the convulsant picrotoxin blocks chloride conductance). Anesthetic (e.g., pentobarbital) and anticonvulsant (e.g., phenobarbital) barbiturates differ in their potencies to produce such effects and to bind to the picrotoxin site. Accordingly, this site of action of the barbiturates is usually considered as being relevant for their ability to generally depress the CNS (TICKU and MAKSAY 1983; OLSEN 1987).

In contrast to benzodiazepines, which do not induce general anesthesia, the effects of barbiturates on the CNS range from mild sedation and tranquilizing effects to general anesthesia and even coma. This is the reason for the much higher toxicity of the barbiturates when compared with benzodiazepines. Similar to benzodiazepines, barbiturates might induce excitement on rare occasions. Again, these paradoxical reactions might result from disinhibition.

Both pharmacokinetic (within days or 1 week) and pharmacodynamic (within weeks) tolerance to barbiturates can occur to such an extent that the therapeutic dose is increased up to six times. Usually, tolerance is more pronounced to the therapeutic effects of the barbiturates than to the toxic effects. Thus, with increasing tolerance, the therapeutic index may decrease.

III. Zopiclone and Zolpidem

Both zopiclone and zolpidem are chemically non-benzodiazepines, which, however, act by activating the benzodiazepine receptor subsite of the $GABA_A$-receptor complex. This is best demonstrated by the possibility of blocking the pharmacological effects of both compounds with benzodiazepine-receptor antagonists. Both have relatively short elimination half-lives and have been introduced into therapy as short-acting hypnotics.

Zopiclone has a rather similar pharmacological profile when compared with the benzodiazepines, but seems to interact with a slightly different recognition site of the $GABA_A$-receptor complex. It also binds to the picrotoxin site of the $GABA_A$-receptor complex at lower concentrations (relative to its binding to the benzodiazepine receptor) than do benzodiazepines (Trifiletti and Snyder 1984; Trifiletti et al. 1984). Whether this is of functional relevance is not yet clear.

Zolpidem binds with higher affinity to the BZ_1 receptor (at 4°C) than to the BZ_2 receptor (Langer et al. 1988). As outlined above (Sect. B.V), the functional relevance of this effect is questionable. The use of the ω_1 and ω_2 nomenclature instead of the BZ_1 and BZ_2 terminology has not yet been generally accepted, especially since it completely neglects the fact that both are at best only subclasses of the benzodiazepine recognition site (Haefely 1990). On the other hand, zolpidem does show differences to the classical benzodiazepines in some pharmacological tests (Zirkovic et al. 1992). However, their relevance for its use as a hypnotic in man is not very obvious.

Probably the most important difference between zopiclone and zolpidem and the classical benzodiazepine hypnotics is their reduced potential to induce physical dependence after treatment with high doses in animal models (Piot et al. 1990; Perrault et al. 1992). Again, it is not yet clear at all whether this is of relevance for their use in man, because drug dependence has been reported for both compounds (Cavallaro 1993; Aranko et al. 1991).

E. Summary

Activation of central GABAergic inhibitory neurotransmission represents the cellular mechanism of action of hypnotics of the benzodiazepine class as well as of the newer compounds, zopiclone and zolpidem. Moreover, GABAergic effects are also involved in the pharmacological mechanism of action of other hypnotics such as the barbiturates, etomidate, and methaqualone.

Benzodiazepines act by activating specifically an allosterically modulating binding site of the $GABA_A$-receptor complex, usually called the benzodiazepine receptor. Activation of the benzodiazepine receptor by benzodiazepines or other agonists enhances the inhibitory synaptic response of central neurons to GABA (hyperpolarization), by increasing the frequency of chloride channel openings. Benzodiazepine-induced enhancement of GABAergic inhibition can be found in nearly all areas of the CNS, because many if not all central neurons receive inhibitory GABAergic input. However, not all central neurons are equally sensitive to benzodiazepines. In spite of the broad activity of benzodiazepines at nearly all levels of the neuronal axis, probably only some specific target neurons in different brain regions are relevant for the pharmacological properties of the benzodiazepines. Accordingly, it is not possible to localize clearly the specific neuronal structures relevant for their hypnotic activity. However, because GABAergic interneurons seem to inhibit most neuronal systems determining the sleep-wakefulness cycle, such as monoaminergic structures of the brain stem and the hypnogenic structures in the lower brain stem, it seems reasonable to assume that enhanced GABAergic inhibition by benzodiazepines in these structures is relevant for their hypnotic properties.

The benzodiazepine receptor exists in several subclasses, the molecular structures of which are known. Classical benzodiazepine hypnotics do not differentiate between receptor subclasses. However, some newer compounds such as the benzodiazepine derivative quazepam and the non-benzodiazepine, zolpidem exhibit some subclass specificity especially in vitro. The relevance of the subclass specificity for the hypnotic properties of these compounds has still not yet been finally clarified.

References

Aranko K, Henriksson M, Hublin C, Seppälainen AM (1991) Misuse of zopiclone and convulsions during withdrawal. Pharmacopsychiatry 24:138–140

Arendt RM, Greenblatt DJ, deJong RM et al. (1983) In vitro correlates of benzodiazepine cerebrospinal fluid uptake, pharmacodynamic action, and peripheral distribution. J Pharmacol Exp Ther 227:95–106

Barker JL, Gratz E, Owen DG, Study RE (1984) Pharmacological effects of clinically important drugs on the excitability of cultured mouse spinal neurons. In: Bovery NG (ed) Actions and interactions of GABA and benzodiazepines Raven, New York, pp 203–216

Borland RG, Nicholson AN (1977) Residual effects of potassium clorazepate, a precursor of nordiazepam. Br J Clin Pharmacol 4:86–89

Bowery NG, Hill DR, Hudson AL, Price GW, Turnbull MJ, Wilkin GP (1984) Heterogeneity of mammalian GABA receptors. In: Bovery NG (ed) Actions and interactions of GABA and benzodiazepines. Raven, New York, pp 81–108

Bowery NG, Knott C, Moratalla R, Pratt GD (1990) $GABA_B$ receptors and their heterogeneity. In: Biggio G, Costa E (eds) GABA and benzodiazepine receptor subtypes, vol 46. Raven, New York, pp 127–139

Braestrup C, Nielsen M (1981) GABA reduces binding of ^{3}H-methyl β-carboline-3-carboxylate to brain benzodiazepines receptors. Nature 294:472–474

Braestrup C, Nielsen M (1983) Benzodiazepine receptors. In: Iversen LL, Iversen SD, Snyder SH (eds) Handbook of psychopharmacology, vol 17. Plenum, New York, pp 285–384

Braestrup C, Squires RF (1977) Specific benzodiazepine receptors in rat brain characterized by high-affinity [^{3}H] diazepam binding. Proc Natl Acad Sci USA 74:3805–3809

Braestrup C, Schmiechen R, Neef G, Nielsen M, Petersen EN (1982) Interaction of convulsive ligands with benzodiazepine receptors. Science 216:1241–1243
Braestrup C, Schmiechen R, Neef G, Nielsen M, Petersen EN (1983) Benzodiazepine receptor ligands, receptor occupancy, pharmacological effect and GABA receptor coupling. In: Usdin E, Skolnik P, Tallman JF, Greenblatt D, Paul S (eds) Pharmacology of benzodiazepines. Vch, Weinheim, pp 71–85
Carlen PL, Gurevich N, Polc P (1983) The excitatory effects of the specific benzodiazepine antagonist Ro 14-7437 measured intracellularly in hippocampal CA 1 cells. Brain Res 271:115–119
Casucci G, di Costanzo A, Riva R, Alloca S, Tedeschi G (1991) A pharmacodynamic study of quazepam and triazolam. Hum Psychopharmacol 6:293–299
Cavallaro R (1993) Tolerance and withdrawal with zolpidem. Lancet 342:374–375
Costa E, Guidotti A, Mao CC (1975) Evidence for involvement of GABA in the action of benzodiazepines: studies on rat cerebellum. In: Costa E, Greengard P (eds) Mechanism of action of benzodiazepines. Raven, New York, pp 113–130
Delini-Stula A (1992) Bretazenil, clinical experience. Neurosci Facts 3:72
Dietch JT, Jennings RK (1988) Aggressive dyscontrol in patients treated with benzodiazepines. J Clin Psychiatry 52[Suppl 7]:16–23
Ehlert FJ, Roeske WR, Gee KW, Yamamura HI (1983) An allosteric model for benzodiazepine receptor function. Biochem Pharmacol 32:2375–2383
File SE (1985) Tolerance to the behavioral actions of benzodiazepines. Neurosci Behav Rev 9:113–121
Fuxe K, Agnati LF, Bolme P, Höfkelt T, Lidbrink P, Ljungdahl A, Perez de la Mora M, Ögren SO (1975) The possible involvement of GABA mechanisms in the action of benzodiazepines on central catecholamine neurons. In: Costa E, Greengard P (eds) Mechanism of action of benzodiazepine. Raven, New York, pp 45–62
Gaillard JM (1989) Benzodiazepines and GABA-ergic transmission. In: Kryger MH, Roth T Dement WC (eds) Principles and practice of sleep medicine. Saunder, Philadelphia, pp 213–218
Gavish M, Katz Y, Bar-Ami S, Weizman R (1992) Biochemical, physiological, and pathological aspects of the peripheral benzodiazepine receptor. J Neurochem 58:1589–1601
Gee KW, Bolger MB, Brinton RE, Coirini H, McEwen BS (1988) Steroid modulation of the chloride ionophore in rat brain: structure-activity requirements, regional dependence, and mechanism of action. J Pharmacol Exp Ther 246:803–812
Greenblatt DJ, Shader RI (1978) Dependence, tolerance, and addiction to benzodiazepines: clinical and pharmacological considerations. Drug Metab Rev 8:13–28
Greenblatt DJ, Allen MD, Noel BJ, Shader RI (1977) Acute overdosage with benzodiazepine derivatives. Clin Pharmacol Ther 4:497–514
Greenblatt DJ, Friedman HL, Shader RI (1987) Correlating pharmacokinetics and pharmacodynamics of benzodiazepines: problems and assumpions. In: Dahl R, Gram B, Paul S, Potter W (eds) Clinical pharmacology in psychiatry. Springer, Berlin Heidelberg New York, pp 62–71
Greenblatt DJ, Harmatz JS, Engelhardt N, Shader RI (1989) Pharmacokinetic determinants of dynamic differences among three benzodiazepine hypnotics. Arch Gen Psychiatry 46:326–332
Greenblatt DJ, Miller GL, Shader RI (1990) Neurochemical and pharmacokinetic correlates of the clinical action of benzodiazepine hypnotic drugs. Am J Med 88:3A18S–3A24S
Haefely W (1984) Actions and interactions of benzodiazepine agonists and antagonists at GABAergic synapses. In: Bowery NG (ed) Actions and interactions of GABA and benzodiazepines. Raven, New York, pp 263–285
Haefely W (1985) Tranquilizers. In: Grahame-Smith DG (ed) Preclinical psychopharmacology. Elsevier, Amsterdam, pp 92–182 (Psychopharmacology, vol 2/1)

Haefely W (1990) Concluding remarks. In: Biggion G, Costa E (eds) GABA and benzodiazepine receptor subtypes, vol 46. Raven, New York, pp 231–234
Haefely W, Kulcsar A, Möhler H, Pieri L, Polc P, Schaffner R (1975) Possible involvement of GABA in the central actions of benzodiazepines. In: Costa E, Greengard F (eds) Mechanism of action of benzodiazepines. Raven, New York, pp 131–152
Haefely W, Pieri L, Polc P, Schaffner R (1981) General pharmacology and neuropharmacology of benzodiazepine derivates. In: Hoffmeister F, Stille G (eds) Handbook of experimental pharmacology, vol 55/2. Springer, Berlin Heidelberg New York, pp 13–262
Haefely W, Kyburz E, Gerecke M, Möhler H (1985) Recent advnaces in the molecular pharmacology of benzodiazepine receptors and in the structure-activity relationships of their agonists and antagonists. Adv Drug Res 14:165–322
Hall RC, Zisook S (1981) Paradoxical reactions to benzodiazepines. Br J Clin Pharmacol 11[Suppl 1]:99S–104S
Häring P, Stähli C, Schoch P, Takacs B, Staehelin T, Möhler H (1985) Monoclonal antibodies reveal structural homogeneity of γ-aminobutyric acid/benzodiazepine receptors in different brain areas. proc Natl Acad Sci USA 82:4837–4841
Harrison NL, Majewska MD, Harrington JW, Barker JL (1987) Structure activity relationships for steriod interaction with the γ-aminobutyric acid A receptor complex. J Pharmacol Exp Ther 241:346–353
Johnston GAR, Skerritt JH (1984) GABARINS and the nexus between GABA and benzodiazepine receptors. In: Bowery NG (ed) Actions and interactions of GABA and benzodiazepines. Raven, New York, pp 179–189
Kales A, Soldatos CR, Vela-Bueno A (1985) Clinical comparison of benzodiazepine hypnotics with short and long elimination half lives. In: Smith DE, Wesson DR (eds) The benzodiapines current standard for medical practice. MTP Press, Lancaster
Kang I, Miller LG (1991) Decreased $GABA_A$ receptor subunit mRNA concentrations following chronic lorazepam administration. Br J Pharmacol 103:1285–1287
Krueger KE (1991) Peripheral-type benzodiazepine receptors: a second site of action for benzodiazepines. Neuropsychopharmacology 4:237–244
Kuhar MJ (1983) Radiohistochemichal localization of benzodiazepine receptors. In: Usdin E, Skolnik P, Tallmann JF Jr, Greenblatt D, Paul SM (eds) Pharmacology of benzodiazepines. Vch, Weinheim, pp 149–154
Langer SZ, Arbilla S, Scatton B, Niddam R, Dubois A (1988) Receptors involved in the mechanism of action of zolpidem. In: Sauvanet JP, Langer SZ, Moselli P (eds) Imidazopyridines in sleep disorders. Raven, New York
Lippa AS, Coupet J, Greenblatt EN, Klepner CA, Beer B (1979) A synthetic non-benzodiazepine ligand for benzodiazepine receptors: a probe for investigating neuronal substrates of anxiety. Pharmacol Biochem Behav 11:99–106
Lüddens H, Wisden W (1991) Function and pharmacology of multiple $GABA_A$ receptor subunits. Trends Pharmacol Sci 12:49–51
Lüddens H, Pritchett DB, Köhler M, Killisch I, Keinänen K, Monyer H, Sprengel R, Seeburg PH (1990) Cerebellar $GABA_A$ receptor selective for a behavioural alcohol antagonist. Nature 346:648–561
Mackerer CR, Kochman RL, Bierschenk BA, Bremner SS (1978) The binding of [^{3}H]diazepam to rat brain homogenates. J Pharmacol Exp Ther 206:405–413
Martin IL, Brown CL, Doble A (1983) Multiple benzodiazepine receptors: structures in the brain or structures in the mind? A critical review. Life Sci 32:1925–1933
Miller LG, Greenblatt DJ, Barnhill JG, Shader RI (1988) Chronic benzodiazepine administration. I. Tolerance is associated with benzodiazepine receptor down-regulation and decreased aminobutyric acid receptor function. J Pharmacol Exp Ther 246:170–176

Miller LG, Greenblatt DJ, Roy RB, Gaver A, Lopez F, Shader RI (1989) Chronic benzodiazepine administration. III. Upregulation of γ-aminobutyric $acid_A$ receptor binding and function associated with chronic treatment. J Pharmacol Exp Ther 248:1096–1101
Möler H, Okada T (1977) Benzodiazepine receptor: demonstration in the central nervous system. Science 198:849–851
Möhler H, Richards JG (1983) Benzodiazepine receptors in the central nervous system. In: Costa E (ed) The benzodiazepines from molecular biology to clinical practice. Raven, New York, pp 93–116
Möhler H, Benke D, Knoflach F, Fritschy JM (1993) Cellular and subcellular location of $GABA_A$ receptor subtypes and their pharmacological significance. Eur Neuropsychopharmacol 3:263–267
Müller WE (1985) Benzodiazepine receptor interactions of arfendazam, a novel 1,5-benzodiazepine. Pharmacopsychiatry 18:10–11
Müller WE (1987) The benzodiazepine receptor: drug accetor only or a physiologically relevant part of our central nervous system? Cambridge University Press, Cambridge
Müller WE, Stillbauer AE (1983) Benzodiazepine hypnotics: time course and potency of benzodiazepine receptor occupation after oral application. Pharmacol Biochem Behav 18:545–549
Müller WE, Groh B, Bub O, Hofmann HP, Kreiskott H (1986) In vitro and in vivo studies of the mechanism of action of arfendazam, a novel 1,5-benzodiazepine. Pharmacopsychiatry 19:314–315
Nutt D (1983) Pharmacological and behavioural studies of benzodiazepine antagonists and contragonists. In: Biggo G, Costa E (eds) Benzodiazepine recognition site ligands: biochemistry and pharmacology. Raven, New York, pp 153–173
Oakley NR, Jones BJ, Straughan DW (1983) The benzodiazepine receptor ligand CL 218.872 has both anxiolytic and sedative properties in rodents. Neuropharmacology 23:797–802
Olsen RW (1987) GABA-drug interactions. Prog Drug Res 31:224–238
Perrault G, Morel E, Sanger DJ, Zivkovic B (1992) Lack of tolerance and physical dependence upon repeated treatment with the novel hypnotic zolpidem. J Pharmacol Exp Ther 263:298–303
Petersen EN, Jensen LH, Drejer LH, Honoré T (1986) New perspectives in benzodiazepine receptor pharmacology. Pharmacopsychiatry 19:4–6
Piercey MF, Hoffmann WE, Cooper M (1991) The hypnotics triazolam and zolpidem have identical metabolic effects throughout the brain: implications for benzodiazepine receptor subtypes. Brain Res 554:244–252
Piot O, Betschard J, Stutzman JM, Blanchard JC (1990) Cyclopyrrolones, unlike some benzodiazepines, do not induce physical dependence in mice. Neurosci Lett 117:140–143
Placheta P, Karobath M (1979) Regional distribution of Na^+-independent GABA and benzodiazepine binding sites in rat CNS. Brain Res 178:580–583
Polc P (1988) Electrophysiology of benzodiazepine receptor ligands: multiple mechanisms and sites of action. Prog Neurobiol 31:349–423
Polc P, Laurent JP, Scherschlicht R, Haefely W (1981) Electrophysiological studies on the specific benzodiazepine antagonist Ro 15-1788. Naunyn Schmiedebergs Arch Pharmacol 316:317–325
Pritchett DB, Lüddens H, Seeburg PH (1989a) Type I and type II $GABA_A$-benzodiazepine receptors produced in transfected cells. Science 245:1389–1392
Pritchett DB, Sontheimer H, Shivers BD, Ymer S, Kettenmann H, Schofield PR, Seeburg PH (1989b) Importance of a novel $GABA_a$ receptor subunit for benzodiazepine pharmacology. Nature 338:582–585

Rall TW (1992) Hypnotics and sedatives: ethanol. In: Gilman AG, Rall TW, Nies AS, Taylor P (eds) Goodman and Gilman's the pharmacological basis of therapeutics. Pergamon, New York, pp 345–382
Roberts E (1984) γ-Aminobutryc acid (GABA): from discovery to visualization of GABAergic neurons in the vertebrate nervous system. In: Bowery NG (ed) Actions and interactions of GABA and benzodiazepines. Raven, New York, pp 1–25
Roth T, Roehrs TA, Stepanski EJ, Rosenthal LD (1990) Hypnotics and behavior. Am J Med 88:3A43S–3A46S
Saano V, Rägo L, Räty M (1989) Peripheral benzodiazepine binding sites. Pharmacol Ther 41:503–514
Schallek W, Horst WD, Schlosser W (1979) Mechanisms of action of benzodiazepines. Adv Pharmacol Chemother 16:45–87
Schoch P, Richards JG, Häring P, Takacs B, Stähli C, Staehlin T, Haefely W, Möhler H (1985) Co-localization of $GABA_A$ receptors and benzodiazepine receptors in the brain shown by monoclonal antibodies. Nature 314:168–171
Schofield PR, Darlison MG, Fujita N, Burt DR, Stephenson FA, Rodriquez H, Rhee LM, Ramachandran J, Reale V, Glencorse TA, Seeburg PH, Barnard EA (1987) Sequence and functional expression of the $GABA_A$ receptor shows a ligand-gated receptor super-family. Nature 328:221–227
Schütz H (1982) Benzodiazepines. A handbook. Springer, Berlin Heidelberg New York
Schütz H (1989) Benzodiazepines. II. A handbook. Springer, Berlin Heidelberg New York
Sieghart W (1985) Benzodiazepine receptors: multiple receptors or multiple conformations? J Neuronal Transm 63:191–208
Simmonds MA (1984) Physiological and pharmacological characterization of the actions of GABA. In: Bowery NG (ed) Actions and interactions of GABA and benzodiazepines. Raven, New York, pp 27–43
Smith RB, Kroboth PD, Vanderlugt JT, Philips JP, Juhl RP (1984) Pharmacokinetics and pharmacodynamics of alprazolam after oral and i.v. administration. Psychopharmacology (Berl) 84:452–456
Snodgrass SR (1983) Receptors for amino acid transmitters. In: Iversen LL, Iversen SS, Snyder SH (eds) Biochemical studies of CNS receptors. Plenum, New York, pp 167–219 (Handbook of psychopharmacology, vol 17)
Speth RC, Wastek GJ, Yamamura HI (1978) Benzodiazepine receptors: temperature dependence of [^{3}H]flunitrazepam binding. Life Sci 24:351–358
Speth RC, Johnson RW, Regan J, Reisine T, Kobayashi RM, Bresolin N, Roeske WR, Yamamura HI (1980) The benzodiazepine receptor of mammalian brain. Fed Proc 39:3032–3038
Squires RF, Braestrup C (1977) Benzodiazepine receptors in rat brain. Nature 266:732–734
Stephens DN, Kehr W, Wachtel H, Schmiechen R (1985) The anxiolytic activity of β-carboline derivates in mice, and its separation from ataxia properties. Pharmacopsychiatry 18:167–170
Tallman JF, Gallager DW (1985) The GABA-ergic system: a locus of benzodiazepine action. Annu Rev Neurosci 8:21–44
Ticku MK, Maksay G (1983) Minireview. Convulsant/depressant site of action at the allosteric benzodiazepine-GABA receptor-ionophore complex. Life Sci 33:2363–2375
Trifiletti RR, Snyder SH (1984) Anxiolytic cyclopyrrolones zopiclone and suriclone bind to a novel site linked allosterically to benzodiazepine receptors. Mol Pharmacol 26:458–469
Trifiletti RR, Snowman AM, Snyder SH (1984) Anxiolytic cyclopyrrolone drugs allosterically modulate the binding of [^{35}S]t-butylbicyclophosphorothionate to

the benzodiazepine/γ-aminobutryc acid-A receptor/chloride anionophore complex. Mol Pharmacol 26:470–476
Zirkovic B, Perrault G, Sanger DJ (1992) Receptor subtype-selective drugs: a new generation of anxiolytics and hypnotics. In: Mendlewicz J, Racagni G (eds) Target receptors for anxiolytics and hypnotics: from molecular pharmacology to therapeutics. Karger, Basel, pp 55–73 (International academy of biomedicine and drug research, vol 3)

CHAPTER 9

Pharmacology of the CNS Peptides

S. Inoué

A. Introduction

The experimental search for an endogenous sleep-promoting substance originates from the classical studies published by Ishimori in 1909 and Legendre and Piéron in 1913, although such pioneer endeavors were too premature to be evaluated and accepted by the contemporary sleep scientists (Inoué 1989a). However, the development of modern ideas and techniques between the late 1970s and the early 1980s finally yielded success in isolating and identifying a few candidate substances directly from body tissues or fluids of sleeping or sleep-deprived animals and humans. Eventually it was established that these and other substances may play an important role in the physiological and pathophysiological regulation of sleep. During the subsequent decade, a large number of endogenous substances were described as sleep modulators (Borbély and Tobler 1989; Inoué 1989a). The chemical nature of these hypnogenic compounds, i.e., *sleep substances*, is so diverse that it covers a wide variety of body constituents. Among them are several CNS peptides, i.e., *sleep peptides* (Inoué et al. 1988).

Delta-sleep-inducing peptide (DSIP), originally found in the venous blood of sleeping rabbits (Monnier and Schoenenberger 1973), is a typical example of the sleep peptides. DSIP along with factor S, muramyl peptides (MPs) (Pappenheimer et al. 1975), sleep-promoting substance (SPS-B) and oxidized glutathione (GSSG) (Uchizono et al. 1982) were all directly derived from extracts of body tissues or fluids. Many other CNS peptides were known for their specific nonsleep biological activities but were never previously regarded as sleep modulators.

Because studies on sleep peptides up to 1988 were extensively reviewed in several monographs (Inoué 1989a; Inoué and Borbély 1985; Inoué and Krueger 1990; Inoué and Schneider-Helmert 1988; Wauquier et al. 1985) and review articles (Borbély and Tobler 1989; Graf and Kastin 1984, 1986; Inoué 1990; Inoué and Kovalzon 1990; Krueger et al. 1989), the present author intends only to outline briefly the background already described in these publications and focus more attention on subsequent findings and new topics.

B. Delta-Sleep-Inducing Peptide: A Circadian Programmer?

I. Research Background

The possible existence of DSIP was first predicted by Monnier et al. (1963) through a crossed circulation experiment in rabbits. Thalamic electrical stimulation of the donor rabbit resuted in a significant rise in electroencephalographic (EEG) delta-wave activity in both the donor and the recipient in crossed circulation, suggesting that a blood-borne delta-sleep-inducing factor liberated by the donor acted on the brain of its partner. DSIP was then extracted from hemodialysates of sleeping rabbits (Monnier and Schoenenberger 1973) and finally identified as a novel nonapeptide (Schoenenberger et al. 1977). A phosphorylated analog of DSIP at Ser in position 7, P-DSIP, is known to occur endogenously (Schoenenberger et al. 1981) and its ratio to DSIP seems to be crucial in regulating the circadian time course of sleep and wakefulness in humans (Ernst and Schoenenberger 1988). The sleep-promoting potency of P-DSIP is fivefold stronger than that of DSIP in rats (Inoué et al. 1988; Kimura and Inoué 1989).

DSIP and P-DSIP enhance not only delta sleep, i.e., a deep stage of non-rapid eye movement (non-REM) sleep or slow-wave sleep (SWS), but also REM sleep or paradoxical sleep (PS) in many cases. However, their somnogenic effects are largely variable and depend on dosage (as evidenced by a typical bell-shaped, dose-response relation), route and timing of administration, species and physiological or pathological situations of recipients, and even sources of synthetic products (Inoué 1989a; Inoué et al. 1988). Such nonreproducible aspects sometimes give rise to difficulties in understanding the regulatory activities of these peptides and lead to controversies over their properties. Taking such aspects into consideration, Schoenenberger and associates (Ernst and Schoenenberger 1988; Schneider-Helmert 1988; Schoenenberger 1984; Schoenenberger et al. 1981) developed a concept of "programming effects" of DSIP, implying that DSIP never produces an effect over *normality*.

Apart from the modulation of sleep, DSIP exerts multivariate physiological functions, so-called "extra-sleep" activities, such as thermoregulation (see below), antinociception (see below), antistressor effect (for a recent review, see Sudakov 1991) and modulation of pituitary hormone secretion (Iyer and McCann 1987a–c; Iyer et al. 1988). Numerous literature on various activities of DSIP was comprehensively reviewed up to the late 1980s (Ernst and Schoenenberger 1988; Graf and Kastin 1984, 1986; Inoué 1989a; Schoenenberger 1984). DSIP has extensively been put to clinical use for insomnia and psychosis (Ernst et al. 1992; Schneider-Helmert 1988; Hermann-Maurer et al. 1992) and for obstructive sleep apnea (Becker et al. 1992).

II. Distribution and Biosynthesis in the Brain

Although DSIP was originally isolated from cerebral venous blood, DSIP-like immunoreactivity (DSIP-LI) was found throughout the body, making it difficult to specify the exact site of production (see the above-mentioned reviews). However, it is highly probable that certain neurons in the brain produce DSIP and release it neurohumorally, because cell bodies in a wide region of the rabbit basal forebrain display DSIP-immunoreactivity and their fibers terminate in circumventricular organs and the pituitary (CHARNAY et al. 1988). There have been similar observations with other species (cat, CHARNAY et al. 1990; humans, VALLET et al. 1990; rat, BJARTELL et al. 1991; SKAGERBERG et al. 1991). Various psychoactive drugs, such as haloperidol, imipramine, pentobarbital, zimeldine and zopiclone, modulate concentrations of DSIP-LI in the rat brain (TSUNASHIMA et al. 1992; WAHLESTEDT et al. 1989). It is suggested that DSIP stimulates morphogenesis of axosomatic synapses of cortical neurons (KURAEV et al. 1991; MENDEZHERITSKII et al. 1992).

Recently, it was reported that DSIP was colocalized with luteinizing hormone-releasing hormone (LHRH) in the same secretory granules of the same neuronal axons and terminals in the median eminence of rodents and that these peptides were possibly cosecreted (VALLET et al. 1991; PU et al. 1991). Colocalization of DSIP is also known with corticotropin (ACTH)-like intermediate lobe peptide (CLIP = $ACTH_{18-39}$) in the human pituitary (VALLET et al. 1988) and brain (ZAPHIROPOULOS et al. 1991). Interestingly, an antagonistic relationship was shown between secretory activities of DSIP and ACTH (BJARTELL et al. 1989). ZLOKOVIC et al. (1989) suggested that a high-affinity saturable mechanism exists for the transport of DSIP across the blood-brain barrier and that brain sites highly sensitive to L-tryptophan are located especially in the hippocampus for the subsequent uptake of DSIP.

Although DSIP and its numerous analogs have been artificially synthesized in several laboratories, its biosynthetic processes are still unknown. Recently, however, DSIP-like precursors were demonstrated in primary cultures of mouse anterior pituitary cells (BJARTELL et al. 1990). NAKAMURA and SHIOMI (1991) demonstrated that casein kinase II can phosphorylate DSIP in vitro, suggesting a possible involvement of this enzyme in the biosynthesis of P-DSIP, another endogenous DSIP analog. Furthermore, aminopeptidase in the cell membrane appears to metabolize liberated DSIP (NAKAMURA et al. 1993).

III. Circadian Sleep-Waking Rhythm

ERNST and SCHOENENBERGER (1988) proposed that the DSIP/P-DSIP ratio dynamically changes in accordance with circadian rest-activity rhythm in humans. KATO et al. (1990, 1991), using a highly sensitive enzyme immunoassay, largely confirmed the circadian rhythm of plasma and urine levels of

DSIP-LI in humans, i.e., high during daytime and low during the dark-night period. Interestingly, plasma concentrations of DSIP exhibit a reciprocal relationship to those of growth hormone (GH) in humans (KATO et al. 1991). Typically, stage 4 sleep, which is characterized by a GH surge, accompanies lowered levels of DSIP in normal and shifted sleep.

Recently, ERNST et al. (1992) analyzed time-course patterns of plasma levels of DSIP and other neuroendocrine activities by measuring DSIP, P-DSIP, cortisol, melatonin, thyrotropin (TSH) and prolactin (PRL) along with body temperature and cardiac pulse in humans. They found a significant cross correlation of DSIP to these factors and thus regarded DSIP as a "circadian organizer."

IV. Sleep-Promoting Activity of Structural Analogs

Nonreproducible somnogenicity of DSIP is sometimes ascribed to the structural instability of DSIP (GRAF and KASTIN 1986). Consequently, a number of DSIP analogs have been synthesized for structural stability, especially against cerebral aminopeptidases. INOUÉ et al. (1992, 1994) and KOVALZON et al. (1992, 1994) extensively examined sleep-modulatory action of several metabolically stable DSIP analogs in animals. Using two different administration methods of a 10-h nocturnal intracerebroventricular (i.c.v.) infusion at a dose of 2.5 nmol in rats and a single intravenous (i.v.) or i.c.v. injection at 7 nmol/kg in rabbits shortly before noon, they found that synthetic analogs such as [D-Tyr^1]DSIP and [D-Ala^2]DSIP were both non-REM and REM sleep-enhancing whereas [D-Ala^2]DSIP$_{1-6}$ was REM sleep-enhancing, indicating that there is structural dependency. It is suggested that the sleep-promoting action of DSIP analogs is mediated through interactions with other CNS peptides such as growth hormone-releasing hormone, GRH, α-melanophore-stimulating hormone (α-MSH), CLIP and others.

V. Aspects of "Extra-Sleep" Activities and Neurotransmitter Systems

1. Antinociceptive Activity and Sleep

NAKAMURA et al. (1988), NAKAMURA and SHIOMI (1990) and SHIOMI and NAKAMURA (1990) found antinociceptive activities of DSIP in rodents. These investigators demonstrated that such effects are mediated by the DSIP-induced release of Met-enkephalin (NAKAMURA and SHIOMI 1990; NAKAMURA et al. 1989a) in neurons of the restricted regions of the bulbospinal descending inhibitory noradrenergic system (SHIOMI and NAKAMURA 1990; NAKAMURA et al. 1989b). They further suggested that the antinociceptive and sleep-inducing effects of DSIP are closely related to the DSIP-induced release of Met-enkephalin (NAKAMURA et al. 1991).

2. Thermoregulation

TSUNASHIMA et al. (1990, 1991) demonstrated that DSIP and P-DSIP have a close relationship to the dopaminergic system with respect to apomorphine-induced hypothermia in rats. They further suggested that the primary thermoregulatory action of DSIP also involves the serotonergic system, especially a 5-HT_{1A} mechanism (TSUNASHIMA et al. 1991, 1993).

VI. Concluding Remarks

Recent findings as mentioned above suggest that, regardless of its original name, DSIP acts as a multipurpose neuromodulator in concert with several other neuromodulators and neurotransmitters in the CNS. In particular, colocalization and cosecretion of DSIP with hypothalamic hypophysiotropic hormones indicate a possible involvement of this peptide in the neuroendocrine regulation of reproduction and antistress, which are closely related changes in the state of vigilance.

C. Glutathione: A Neuronal Detoxification Factor?

I. Research Background

The claim that glutathione is a sleep peptide arose for the first time in 1990 from the studies of sleep-promoting substance (SPS) by the present author's research team (KOMODA et al. 1990a,b). SPS was originally extracted from the brain stems of 24-h sleep-deprived rats (NAGASAKI et al. 1974), which included at least four active components including SPS-A1, SPS-A2, SPS-B and SPS-X (INOUÉ et al. 1985). Purified SPS-A1 and SPS-B were finally identified as uridine (KOMODA et al. 1983) and GSSG (KOMODA et al. 1990a), respectively.

GSSG is a hexapeptide and the oxidized form of glutathione. GSSG is enzymatically convertible to a tripeptide, reduced glutathione (GSH) (ZIEGLER 1985). Both forms of glutathione are widely distributed in mammalian tissues including the CNS (MEISTER and ANDERSON 1983). In the brain, GSSG and GSH are detectable in the cerebral cortex, cerebellum, brain stem and cerebrospinal fluid (CSF) (FOLBERGROVA et al. 1979; SLIVAKA et al. 1987a). GSH levels in the rat cerebrum and cerebellum depend on aging, being high in adults and low in infants (KUDO et al. 1990). Concentrations of GSH but not GSSG are decreased in the substantia nigra of patients with Parkinson's disease (SOFIC et al. 1992). Glutathione is localized in glial cells along with neuronal axons and terminals (SLIVAKA et al. 1987b). In cultured cells, glutathione is present in high concentrations in astrocytes but not in neurons (RAPS et al. 1989) and biosynthesized in astrocytes (YUDKOFF et al. 1990).

The biological functions of glutathione are very diverse and cover the research subjects of "enzyme mechanisms, biosynthesis of macromolecules, intermediary metabolism, drug metabolism, radiation, cancer, oxygen toxicity, transport, immune phenomena, endocrinology, environmental toxins and aging" (Meister and Anderson 1983). It is well known that the outstanding property of glutathione is protection against oxidative stress at the cellular level. It should be noted that GSH and GSSG have a potent antinociceptive activity (Kubo et al. 1992) comparable to that of DSIP (see above).

II. Sleep-Modulatory Activity of Oxidized Glutathione

1. Bioassay

In order to quantify the somnogenic activity of sleep substances in unrestrained rats, the present author's research team originally developed a bioassay system called the long-term nocturnal intracerebroventricular infusion technique, which enables a small amount of sample solutions to be administered into the third cerebral ventricle of freely behaving male rats at a rate of 10 μl/h (Inoué et al. 1985; Honda and Inoué 1978). Cortical EEG, nuchal electromyogram (EMG), brain temperature (Tbr), locomotor activity and drinking behavior are continuously monitored for several weeks and processed by a computer-aided device system for statistical analyses (Honda and Inoué 1981). As a rule, a 4-day assay is conducted under continuous recordings of EEG, EMG and Tbr along with continuous i.c.v. infusion: day 1 as the baseline starting at the onset of the light period; day 2 as the experiment when the saline infusion is replaced by an infusion of a sample solution during a nocturnal 10-h period; and days 3 and 4 as the recovery when saline is i.c.v. infused as before. This technique was used throughout the investigations cited below.

2. Effects on Sleep

As summarized in Table 1, a 10-h nocturnal i.c.v. infusion of GSSG significantly enhances non-REM sleep at the dose range of 20–50 nmol and REM sleep at a dose of 25 nmol (Honda et al. 1990, 1991a,b, 1993, 1994; Inoué et al. 1990). The administration of 25 nmol results in the most dramatic increase in both non-REM sleep and REM sleep. Doses lower and higher than 25 nmol exert less somnogenic activity. Thus, the dose-response relationship in the range from 10 to 100 nmol exhibits a bell-shaped curve for the total time and duration of non-REM sleep and for the total time and frequency of REM sleep.

The time course changes in hourly amounts of non-REM sleep and REM sleep before and after the 25-nmol GSSG administration are illustrated in Fig. 1. Although the total time of both non-REM sleep and REM sleep

Table 1. Effects of nocturnally i.c.v. infused glutathione on percentage changes[a] in sleep parameters in male rats (means ± SEM)

Dose	n	Non-REM sleep (%)			REM sleep (%)		
(nmol)		Total time	Frequency	Duration	Total time	Frequency	Duration
Oxidized glutathione (GSSG)							
10	4	108 ± 7	94 ± 4	117 ± 10	112 ± 16	125 ± 15	91 ± 14
20	5	118 ± 4**	96 ± 7	125 ± 6	106 ± 10	117 ± 7	89 ± 4
25	5	135 ± 8***	101 ± 6	137 ± 12*	186 ± 27*	163 ± 18*	115 ± 12
50	5	113 ± 5*	94 ± 5	123 ± 9	109 ± 10	93 ± 5	118 ± 8
100	4	108 ± 6	113 ± 10	95 ± 3	113 ± 28	112 ± 26	97 ± 7
Reduced glutathione (GSH)							
10	5	102 ± 2	103 ± 7	103 ± 11	113 ± 21	118 ± 26	100 ± 8
25	7	116 ± 3*	112 ± 10	109 ± 10	121 ± 13	118 ± 9	103 ± 8
50	6	121 ± 4**	100 ± 7	120 ± 7	120 ± 6	110 ± 7	113 ± 10
100	7	111 ± 8	99 ± 4	114 ± 5	118 ± 15	106 ± 11	108 ± 6

[a] Data from the 12-h dark period were compared with those of the previous nighttime.
* $P < 0.05$; ** $P < 0.01$; *** $P < 0.001$: significantly different from the corresponding baseline values (Student's t-test).

are significantly increased, responses of non-REM sleep and REM sleep to 25 nmol GSSG are different: the enhancement of non-REM sleep is mainly due to an increase in a prolongation of episode duration whereas that of REM sleep is caused by an elevated frequency of episode number (Table 1). The prolonged duration of non-REM sleep episodes in the dark period is statistically not different from the duration observable in the light period (data not shown). This fact indicates that nocturnally administered GSSG brings about natural daytime non-REM sleep, i.e., the resting phase of the rat. The nocturnal amount of non-REM sleep is still higher than the baseline level in the first recovery night but it returns to the baseline level in the second recovery night.

3. Effects on Brain Temperature

Circadian pattern of Tbr is little affected by the administration of GSSG at 25 nmol (Honda et al. 1991b, 1993). Tbr exhibits almost no change from the baseline level in the first half of the GSSG infusion period, but it increases significantly by 0.38°C ($P < 0.001$) in the latter half of the infusion period and returns to the baseline level thereafter. Since no pyrogenic activity of GSSG per se is known, Tbr modulation is ascribed to a secondary reaction of the thermoregulatory system in the brain.

III. Sleep-Modulatory Activity of Reduced Glutathione

Since glutathione is known to exist in mammalian brain mostly in the form of GSH (Folbergrova et al. 1979; Slivaka et al. 1987a; Meister and

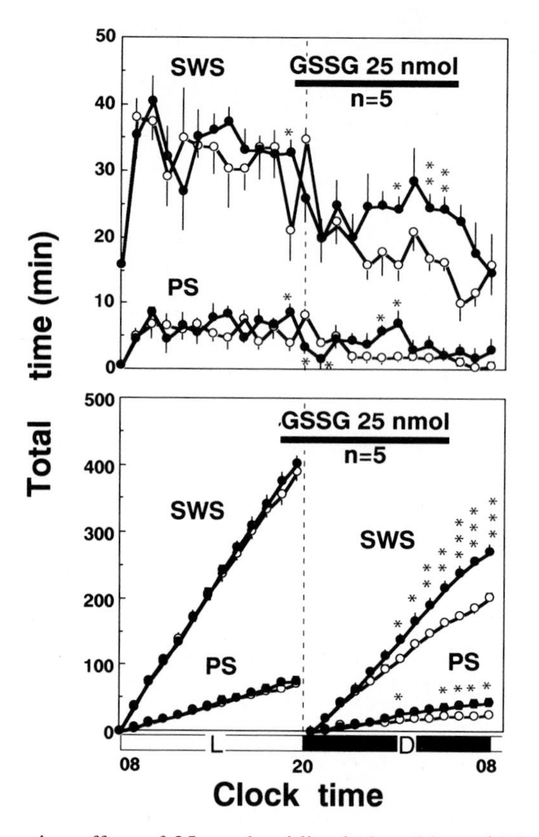

Fig. 1. Sleep-enhancing effect of 25 nmol oxidized glutathione (*GSSG*) i.c.v. infused for a nocturnal 10-h period in otherwise continuously i.c.v. saline-infused freely moving rats. *Top*: hourly amounts of non-REM sleep (*SWS*) and REM sleep (*PS*). *Bottom*: cumulative values in the environmental light (*L*) and dark (*D*) period. *Closed and open circles* indicate the baseline day and the experimental day, respectively. * $P < 0.05$; ** $P < 0.01$; *** $P < 0.001$: significantly different from the corresponding baseline values (Studient's *t*-test)

Anderson 1983) and GSH is easily converted into GSSG by a peroxidase (Ziegler 1985), the sleep-modulatory activity of GSH was examined by the same experimental procedures (Honda et al. 1991a,b; Inoué et al. 1991). It was found that GSH also exhibits non-REM sleep enhancements in the dose range from 25 to 50 nmol and an insignificant increase in REM sleep (Table 1). The dose-response relationship is thus bell shaped for the non-REM sleep-enhancing activity, in which the maximal effect is induced by 50 nmol GSH. Since GSH is structurally a half part of GSSG, it is reasonable that a twofold dose is required to give a somnogenicity of GSH equivalent to that of GSSG.

IV. Possible Mechanisms

Recently, there have been an increasing number of reports on the central action of glutathione. OGITA et al. (1986), OGITA and YONEDA (1987) and YONEDA and OGITA (1991) suggested that GSH and possibly GSSG have a binding site in the excitatory synaptic membrance of rat glutamatergic neurons and exert an inhibitory action on glutamatergic neurotransmission. GILBERT et al. (1991) demonstrated that GSSG but not GSH inhibits the redox regulation of the *N*-methyl-D-aspartate (NMDA) receptor in cultured rat forebrain neurons.

Taking these facts into consideration, the present author's research team proposed that the sleep-enhancing activity of GSSG and GSH is caused by their negative modulation on glutamatergic neurotransmission in the brain (HONDA et al. 1991a,b; INOUÉ 1992a; INOUÉ et al. 1991) (Fig. 2). This speculation is substantiated by the fact that the administration of NMDA and non-NMDA receptor antagonists such as DL-2-amino-5-phosphono-pentanoic acid (APV), 6-cyano-7-nitroquinoxaline-2,3-dione (CNQX), ketamine and riluzole, increases sleep in cats (JUHÁSZ et al. 1990) and rats (ARMSTRONG-JAMES and FOX 1988; STUTZMANN et al. 1988; FEINBERG and CAMPBELL 1993). It is also consistent with reports: that the release of glutamic acid in the cat brain changes in association with arousal and sleep (JASPER et

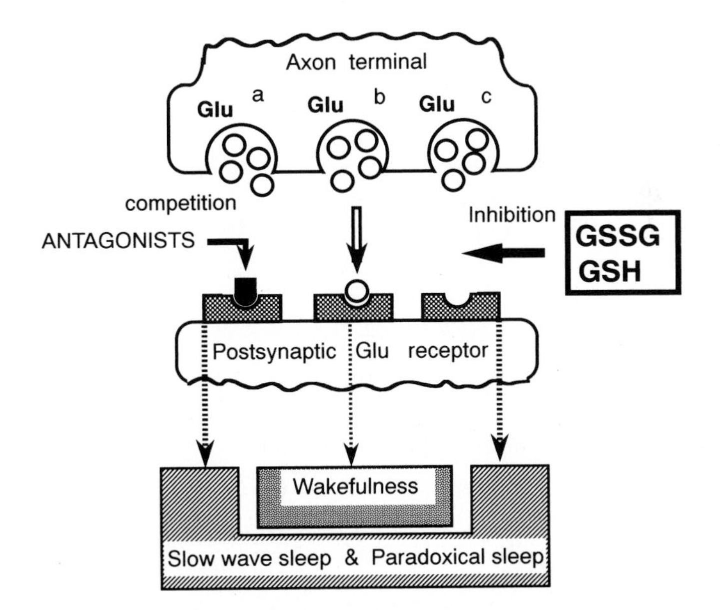

Fig. 2. Schematic illustration of the glutamatergic mechanism underlying the regulation of sleep and wakefulness modulated by two forms of glutathione (*GSSH* and *GSH*) and NMDA and non-NMDA receptor antagonists (see text)

al. 1965); that a larger quantity of glutamate exists in short-sleeper mice than in long-sleeper mice (Disbrow and Ruth 1984); and that the number of glutamate receptors in the rat forebrain is higher in wakefulness than in sleep (Tononi and Pompeiano 1990).

The present author (Inoué 1992b, 1993) further proposes that glutathione contributes to the detoxification of neurons in synchrony with sleep. Hence sleep at the behavioral level is a process of neuronal detoxification at the cellular level. Indeed, it was suggested that GSSG as well as GSH appear to play an important role in protecting against glutamate toxicity in neurons (Miyamoto et al. 1988). In addition, if cultured rat cortical neurons are exposed to a neurotoxic non-NMDA agonist, 2,4,5-trihydroxyphenylalanine (TOPA), GSH abolishes its agonist properties and blocks its excitotoxicity (Aizenman et al. 1992). It was also demonstrated that glutathione is essential for repair processes in neurons in slices of guinea pig hippocampus exposed to oxidative damage (Pellmar et al. 1992). If the present author's hypothesis is accepted, the classical hypnotoxicin theory of Ishimori-Piéron (Ishimori 1909; Legendre and Piéron 1913) is given a modern revival (Inoué 1992b).

It should be noted here that another SPS component, uridine, has an affinity to γ-aminobutyric acid (GABA) receptor binding sites in rat and bovine synaptic membranes in vitro (Guarneri et al. 1983; Yamamoto et al. 1990) and possibly to its proper receptor in the GABA-benzodiazepine-chloride channel complex (Yamamoto et al. 1993). On the basis of these findings, the present author's research team proposed that the sleep-promoting activity of uridine is caused through the positive modulation on GABAergic neurotransmission in the brain (Honda et al. 1993; Inoué 1992a; Inoué et al. 1991).

Thus, the two SPS components promote sleep by exerting a complementary action on two major CNS neurotransmitter systems, GABAergic and glutamatergic, that have mutually reciprocal functions, i.e., respectively inhibitory and excitatory (see also Chap. 8, this volume). Hence the two SPS components may regulate physiological sleep through a differential modulation of neuronal activity in the brain (for comprehensive reviews, see Honda et al. 1993; Inoué 1993).

D. Muramyl Peptides and Cytokines: Somnopyrogenic Immunomodulators?

I. Research Background

Pappenheimer et al. (1975) and Krueger et al. (1980) extracted a sleep-promoting factor S, first from the CSF of sleep-deprived goats, then from bovine and rabbit cerebral tissues, and finally from human urine. The active components of factor S were identified with a muramyl tetrapeptide, *N*-acetylglucosaminyl-*N*-1,6-anhydro-*N*-acetylmuramyl-alanyl-glutamyl-diamino-pimelyl-alanine, and a tripeptide, its derivative (MARTIN et al. 1984; KRUEGER et al. 1984a). Interestingly, these Muramyl peptides (MPs) enhance non-REM sleep by increasing EEG delta-wave amplitudes but suppress REM sleep in rabbits, simultaneously bringing about a striking fever (KRUEGER et al. 1984a). Somnogenicity and pyrogenicity of MPs are dissociable under certain experimental conditions (KRUEGER et al. 1982; SHOHAM and KRUEGER 1988). These MP-induced somnogenic and pyrogenic activities characterize a special role played by MPs in the regulation of pathophysiological sleep, because normal sleep is accompanied by a certain ratio of non-REM sleep and REM sleep and by a decrease in the core temperature during non-REM sleep.

Some of the many MP analogs, including muramyl dipeptide (MDP) as one of the simplest synthetic compounds, are also somnogenic and pyrogenic in several mammalian species (INOUÉ et al. 1984a; JOHANNSEN et al. 1989; KRUEGER et al. 1984b). Sleep assays for MPs are mainly performed by i.c.v. or i.v. administration in rabbits, because rabbits can react to MPs within a 6-h recording period during the light period (KRUEGER et al. 1980). However, similar to other sleep substances (INOUÉ et al. 1984b), MDP does not induce excess sleep in rats if it is administered during the light period, i.e., the resting phase of this animal species (FORNAL et al. 1984; MELTZER et al. 1989).

Muramyl peptides are constituents of peptidoglycans which form the bacterial cell walls and are not biosynthesized in the mammalian tissues. Because they are known to exert profound pyrogenic as well as immunomodulatory activities in the mammalian body (BLATTEIS 1990), the finding of MPs as a putative sleep peptide created a new area of attention in sleep science, i.e., the relationship between sleep and immunomodulation, which has eventually contributed to understanding further the function and regulation of sleep (INOUÉ 1989a; KRUEGER et al. 1988b).

II. Muramyl Peptides and Infectious Diseases

1. Bacterial and Viral Infections and Sleep

KRUEGER and his research team demonstrated that, in addition to MPs, a number of immune response modifiers such as a bacterial endotoxin (LANCEL et al. 1994; POLLMÄCHER et al. 1993; TRACHSEL et al. 1994) and lipid A (CADY et al. 1989a; KRUEGER et al. 1986), and a viral double-stranded RNA analog such as polyriboinosinic:polyribocytidylic acid (poly[rI:rC]) (KRUEGER et al. 1988a), are also somnogenic and pyrogenic. They further suggested that the biological activities of the microbial components are mediated by cytokines such as interleukin 1 (IL-1), interferon (IFN) and tumor necrosis factor (TNF), which respond acutely to microbial infections (KRUEGER 1990; KRUEGER et al. 1990a,c).

Indeed, bacterial cell walls or peptidoglycans isolated from certain bacteria induce the same reaction as that occurring after the administration of MPs (Johannsen and Cady 1990; Johannsen et al. 1990). Inoculation of rabbits with several microorganisms induces pathophysiological changes typically associated with infectious disease (Toth and Krueger 1989a,b). For example, in rabbits with bacterial infection, non-REM sleep and EEG delta-wave amplitude show an initial enhancement, followed by either subsequent recovery or inhibition and final recovery, depending on the time course and severity of disease (Toth and Krueger 1988, 1990). In contrast, REM sleep is suppressed for a long period. These changes indicate a similarity to MP- or IL-1-induced sleep and also to the recovery sleep after sleep deprivation. In this connection, it was reported that sleep quantity is closely correlated with intestinal bacterial flora in human insomniacs (Rhee and Kim 1987) and in developing rats (Brown et al. 1988). It was also reported that murine macrophages produce and release somnogenic and pyrogenic MPs through phagocytic digestion of bacteria (Johannsen et al. 1991). On the basis of these findings, Krueger and his associates (Krueger and Johannsen 1989; Krueger and Majde 1990) reported the new insight that sleep may serve as a host defense response based on a special case of microbial challenge.

This assumption is substantiated by the recent studies by Kimura-Takeuchi et al. (1992a,b), who inoculated rabbits i.v. with influenza virus or poly[rI:rC] twice at a 24-h interval. The first inoculation elicited several "acute-phase responses" (for a recent review, see Kushner 1991) including an enhancement of non-REM sleep, Tbr and serum antiviral activity. However, the second inoculation resulted in the diminishment or absence of all these acute-phase parameters, indicating the development of tolerance or hyporesponsiveness to influenza virus. Cytokines induced by the inoculations seem to be involved in the process of the host defense mechanisms.

2. Cerebral Site of Action

According to the early microinjection experiment, urinary factor S induces excess sleep if it is administered in brain sites including the preoptic area (POA) (García-Arrarás and Pappenheimer 1983). Recent lesion experiments showed that MDP induces a similar enhancement of sleep and fever even after bilateral lesions of the POA in rabbits (Shoham et al. 1989) and rats (Inoué et al. 1993; Kimura-Takeuchi and Inoué 1993a). These findings suggest that extra-POA sites are also involved in the induction of sleep and fever caused by MPs.

III. Interleukin-1

1. Somnopyrogenic Activities

Interleukin-1 appears to regulate physiological sleep by enhancing non-REM sleep in rabbits, rats and cats (Covelli et al. 1991; Opp et al. 1992a).

IL-1 is produced as an immune response modifier during microbial infections (KRUEGER 1990; KRUEGER et al. 1990a,c). As for MPs, the somnogenicity of IL-1 is differentially separable from its pyrogenicity (CADY et al. 1989b; SUSIC and TOTIC 1989; WALTER et al. 1989). There are two distinct gene products, i.e., IL-1α and IL-1β, which have approximately 26% structural homology and are produced by macrophages and astrocytes (FONTANA et al. 1984). IL-1β is the dominant form. OBÁL et al. (1990) extensively compared biological activities of different doses of human IL-1α and rat IL-1β and their various fragment peptides in rabbits, finding that IL-1α and IL-1β dose dependently enhance non-REM sleep, EEG delta-wave amplitudes and colonic temperature and inhibit REM sleep, and that only one fragment, IL-1β-(208–240), is somnogenic and pyrogenic. OPP et al. (1992b) further demonstrated that rabbit IL-1β-(208–240) and rat IL-1β-(208–240) exhibit species-specific amino acid sequences along with non-REM sleep- and Tbr-enhancing activities in their respective species.

The somnogenic activity of IL-1β in rats depends on the circadian timing of administration, effective in the dark period and noneffective in the light period (OPP et al. 1991), as is well known for other endogenous sleep substances in this animal species (INOUÉ et al. 1984b).

2. Transport

BLATTEIS et al. (1989) suggested that circulating cytokines such as IL-1, TNF and IFN do not enter the brain because of the blood-brain barrier, but they are detected by the receptors in the organum vasculosum laminae terminalis (OVLT), which in turn transduce its original messages into neuronal singals. However, BANKS and KASTIN (1991) recently demonstrated a possible existence of a saturable carrier-mediated system in mice that directly transports IL-1α from the blood into the CNS. LUE et al. (1988) detected an increase and a decrease in endogenous IL-1-like immunoreactivity in the cat CSF in phase with sleep and wakefulness, respectively.

3. IL-1 Receptors

The existence of IL-1 receptors in the mammalian brain is well known (DINARELLO et al. 1990). However, microinjection studies in rabbits (WALTER et al. 1989) revealed that IL-1 administered into various sites of the basal forebrain and the brain stem induces fever but no enhancement of sleep, and that it induces neither fever nor excess sleep after administration into the posterior hypothalamus. Thus the distribution of IL-1 receptors remains unknown with respect to the modulation of sleep.

Interestingly, IL-1-induced sleep and fever are blocked by an IL-1 receptor antagonist in rabbits (OPP and KRUEGER 1991). In addition, according to OPP et al. (1992b), the biological activities of rabbit and rat IL-1β-(208–240) fragments are completely blocked or significantly attenuated by pretreatment with an IL-1 receptor antagonist. On the basis of these findings they suggested the possible involvement of IL-1β receptors in the regulation of IL-1-mediated activities.

Miller et al. (1991) demonstrated that IL-1 acting through its own receptor sites augments $GABA_A$ receptor functions: a GABA-dependent increase in chloride uptake in mouse cortical synaptoneurosomes in vitro; a GABA-mediated increase in membrane permeability in chick cortical neurons in vitro; a suppression of open-field activity in mice; and an increase in the threshold for pentylenetetrazol-induced seizures in mice.

4. Interactions with Other Sleep Substances

Opp et al. (1989) reported that IL-1-induced enhancement of non-REM sleep and fever is blocked by combined administration of corticotropin-releasing hormone (CRH). REM sleep suppressed by IL-1 is partially restored by the CRH administration. These authors suggested that a negative feedback mechanism is operative in the regulation of IL-1 activities via IL-1-induced CRH release. Similar but less profound negative feedback action is noted for α-MSH by the same researchers (Opp et al. 1988). The possible existence of such a mechanism is substantiated by the fact that neonatal rats treated with monosodium glutamate exhibit a reduction of CRH and α-MSH in the hypothalamus and reduced sensitivity to IL-1 administration (Opp et al. 1990).

Kapás et al. (1991b) investigated whether the somnogenic activity of IL-1β is mediated by insulin, another sleep substance (see below), in normal and diabetic rats, because IL-1β is known to stimulate insulin secretion. However, they failed to show a positive interrelationship between the two substances.

IV. Tumor Necrosis Factor

Following the first report on the somnogenic and pyrogenic activities of human recombinant TNF (TNF-α, cachectin) (Shoham et al. 1987), which are largely similar to those of IL-1, Kapás and Krueger (1992) demonstrated that TNF-β (lymphotoxin) also has these activities in rabbits and suppresses food intake in rats. These two TNFs are known to share about 30% homology of amino acid sequences and the same receptors. Kapás et al. (1992) further examined the biological activities of TFN-α and its various fragments. The investigators found that some fragments elicit similar non-REM sleep- and/or fever-enhancing activities in rabbits, and that somnogenic fragments share the amino acid sequence 31–36. However, anorectic activity in rats is not manifested in these fragments but by another fragment. It seems likely that there are biologically active regions localized in the TNF-α molecule. Although TNF-α and TNF-β are known to be peripherally produced by macrophages and lymphocytes, respectively, and centrally by astrocytes, little information is available for the sleep-regulatory mechanisms of TNF. Shibata and Blatteis (1989, 1991) suggested that peripherally circulating TNF is detected by the receptors in the OVLT which in turn transduce its original messages into neuronal signals via serotonin.

V. Interferon

Although a number of reports have been published as to the antiviral and immune activity of INF, there are only a few papers on its sleep modulation. Two forms of IFN, α and β, are known to be produced in mammalian nucleated cells including lymphyocytes and glial cells. Subsequent to the first report on the somnogenic and pyrogenic activities of human recombinant IFNα-2 (KRUEGER et al. 1987), which are largely similar to those of IL-1, it was reported that human recombinant IFN-α and IFN-β along with rat IFN cause an increase in EEG slow-wave activity in rats, which is antagonized by naloxone (DE SARRO et al. 1990; BIRMANNS et al. 1990). More recently, KIMURA-TAKEUCHI et al. (1993a, 1994) compared the somnogenic and pyrogenic activities of human recombinant INF-β with those of INF-α in rabbits. According to these authors, INF-β induces excessive non-REM sleep and fever and a suppression of REM sleep in a dose-dependent manner, similar to those of IFN-α.

VI. Integrative Remarks

Muramyl peptide-mediated and cytokine-mediated sleep seem to be an extraordinary case of natural sleep, because they are characterized by exaggerated EEG slow-wave activity, a prolonged suppression or attenuation of REM sleep and even fever that never occurs during daily normal alternations in the state of vigilance. Furthermore, it is typically induced shortly after microbial challenge. Thus, the somnopyrogenic mechanisms reported by the recent studies of KRUEGER and his coworkers beautifully elucidate the immunomodulatory process as one special aspect of the multiple functions of sleep, especially of non-REM sleep (KRUEGER 1990; KRUEGER et al. 1990c; OPP et al. 1992a). Once the mammalian body requires the host defense response, this special type of physiological and/or pathological sleep is induced as an acute-phase response. However, the role that cytokines play in the regulation of normal healthy sleep appears to contribute rather little, although cytokines are linked with the sleep-modulatory neuroendocrine activity of the hypothalamo-hypophyseal axis (see below).

E. Hypothalamo-hypophyseal Hormones: Neuroendocrine Transducers?

I. Research Background

Secretion of hypophyseal hormones such as GH and PRL is well known to occur in phase with sleep in humans (see also Chap. 10, this volume), whereas that of other hormones such as ACTH and TSH is not sleep dependent but circadian rhythm dependent (Follenius et al. 1988; BrandenberGer 1993). Up to 1988, the effect of GH, PRL and hypothalamic GRH and somatostatin (SRIF) on sleep had not been clearly elucidated (Borbély and Tobler 1989; Inoué 1989a). Experimental results from different laboratories have resulted in controversial conclusions on the causal relationship as to whether GH enhances non-REM sleep and/or REM sleep or vice versa in humans and animals. However, a close interrelationship, although sometimes dissociable, has been well established to exist between a secretory surge of GH and deep non-REM sleep stages and between PRL secretion and shallower sleep stages. It should be noted that the enhancement of REM-sleep by GH is sometimes attributed to the secondary effects of SRIF via a feedback control mechanism for GH secretion, and that GH secretion is stimulated with the coexistence of DSIP.

In contrast, CRH and another CNS peptide, thyrotropin-releasing hormone (TRH), are known to be sleep-suppressive peptides (Borbély and Tobler 1989; Inoué 1989a). For the sleep-modulatory activities of proopiomelanocortin components, a convincing experimental result was demonstrated by Chastrette et al. (1988, 1990a,b; see also Jouvet 1988): Acth, desacetyl-α-MSH ($ACTH_{1-13}$) and CLIP ($ACTH_{18-39}$) differentially enhance wakefulness, non-REM sleep and REM sleep, respectively, in rats. Recently, Léger et al. (1994) reported that CLIP-containing neurons are synaptically connected with serotonergic neurons in the brain stem and may regulate REM sleep via modulation of the activity of these neurons.

II. Growth Hormone-Related Hormones

According to the recent studies by Obál et al. (1988, 1991), i.c.v. administration of GRH enhances deep non-REM sleep and GH secretion in rats and rabbits, while administration of its competitive antagonistic peptide blocks the effects of GRH in rats. Kupfer et al. (1991) showed an enhancement of EEG slow-wave activity after i.v. administration of GRH in humans. Steiger et al. (1992) reported that repetitive i.v. administration of GRH induces a long-lasting increase in plasma concentrations of GH and a suppression of cortisol along with a dramatic increase in deep non-REM sleep in healthy humans. REM sleep is little affected. They also demonstrated that repetitive i.v. administration of SRIF fails to modulate GH and cortisol, exerting no definite effect on sleep. A close correlation between EEG delta-wave activity and GH levels is also demonstrated in patients with GH deficiency (Aström and Jochumsen 1989) and acromegaly (Aström and Trojaborg 1991).

Inconsistent with the above results in humans, a positive effect of SRIF on REM sleep is known in rats (Danguir 1988, 1990). Hypothalamic SRIF levels, as measured by in vivo voltammetry in unrestrained rats, exhibit a decrease in sleep and an increase in wakefulness (DANGUIR and DE SAINT-HILAIRE KAFI 1989). With respect to the relations of SRIF with neurotransmitter systems, a possible involvement of serotonin (IMERI et al. 1990) and GABA (NAGAKI et al. 1990) is suggested. It should be noted that a definite circadian rhythmicity is known for SRIF-like immunoreactivity in the suprachiasmatic nucleus (SCN) (FUKUHARA et al. 1993), which is known to serve as a circadian clock in the mammalian species.

III. Corticotropin-Related Hormones

In addition to the established non-REM sleep-suppressive activity of CRH, MARROSU et al. (1990) reported that REM sleep rebound is enhanced by administration of murine and ovine CRH (1 μg i.c.v.) in REM sleep-deprived rats. As mentioned above, CRH antagonizes the somnogenic and pyrogenic activities of IL-1, exerting a negative feedback mechanism in the regulation of IL-1 activities via IL-1-induced CRH release (OPP et al. 1989). According to FEHM and BORN (1990), ACTH reduces REM sleep in humans. GULDNER et al. (1990) observed that an $ACTH_{4-9}$ fragment suppresses deep stages of non-REM sleep and REM sleep, and increases wakefulness in humans. Another ACTH fragment, CLIP, is anatomically and physiologically correlated with DSIP (see above).

IV. Thyrotropin-Related Hormones

In contrast to the established sleep-suppressive activity of TRH, SIDNEVA et al. (1988) reported that deep non-REM sleep is enhanced in rats by intraperitoneal (i.p.) administration of a TRH analog, pyroglutamyl-seryl-phenylalaninamide (0.5–5 mg). Recently, GOICHOT et al. (1992) suggested a correlation between TSH and the deep stages of non-REM sleep in humans, demonstrating that stage 3 and 4 sleep is closely in phase with declining plasma TSH levels.

V. Prolactin

OBÁL et al. (1989) reported that subcutaneous administration of PRL selectively increases REM sleep in rabbits. They further demonstrated that antiserum to PRL selectively suppresses REM sleep in rats (OBÁL et al. 1992). VALATX et al. (1990) suggested a correlation between extrapituitary (i.e., hypothalamic) PRL and REM sleep through estimating the amount of sleep in hypoprolactinemic mutant rats, hypophysectomized estradiol-induced hyperprolactinemic rats and normal rats. Recently, BRANDENBERGER (1993) demonstrated a close relationship between PRL and REM sleep in humans. However, PRL does not affect the quality of human sleep (SPIEGEL et al. 1994).

VI. Concluding Remarks

The hypothalamo-hypophyseal system plays an essential role in the maintenance or survival of the individual and the reproduction of the species. Its regulatory activities cover almost all basic biological functions in vertebrates. The system also serves as an interface between the neural, humoral and immune systems. Hence it is not surprising, but natural, that all these activities are closely harmonized with ultradian and circadian alternations of the resting phase, in which sleep may be involved. For example, growth and restitution of the body, including the brain, can be most efficiently achieved during the quiet sleeping phase. Thus, GRH antagonizes the hyperexcitability-inducing activity of the CRH-ACTH-cortisol axis on the one hand and augments sleep-inducing activities of the neural and humoral systems on the other hand. A number of CNS peptides are thus closely but not always completely associated with the extra-sleep functions of the hypothalamo-hypophyseal system.

F. Satiety-Related Substances: Ischymetric Factors or Postprandial Somnogens?

I. Research Background

A close relationship of sleep with energy metabolism and conservation has long been a topic in behavioral sleep science and the effects of nutrients on sleep have been discussed repeatedly (Inoué 1989a). Starvation or food deprivation easily modify sleep in humans and animals. In this respect, Danguir and Nicolaidis proposed in 1980 an ischymetric sleep regulation hypothesis (*ischys* = power), assuming that sleep is related to both the nature and the degree of utilization of circulating metabolites. Food-dependent modulation of sleep is evidenced by their extensive studies in rats (Danguir 1988). It was also demonstrated that pancreatic and possibly cerebral insulin regulates physiological sleep (Danguir 1988, 1990).

Postprandial satiety has also long been known to induce sleepiness. Indeed, satiety-inducing peptides such as cholecystokinin (CCK) induce sleep (see Borbély and Tobler 1989; Drucker-Colín et al. 1988; Inoué 1989a). Moreover CCK stimulates the pancreatic secretion of insulin. Thus there is the possibility of interactions between insulin and CCK. Below, recent studies on this and other aspects are dealt with briefly.

II. Cholecystokinin

Cholecystokinin is distributed in the brain and the gut in mammals as a 33-amino-acid-residue peptide (CCK-33). Because the CCK octapeptide fragment (CCK-8) is biologically active, CCK-8 is usually used in pharmacological experiments and referred to as CCK. CCK receptors seem to exist in the brain. Subsequent to previous studies from different laboratories, DE SAINT HILAIRE-KAFI et al. (1989a) demonstrated that anorexigenic doses of CCK-8 (16 mg/kg i.p.) induce a transient increase in non-REM sleep as well as unusual EEG spikes and slow waves even in quiet wakefulness in mildly food-deprived rats. A similar observation was also noted by the same research group (DEPOORTERE et al. 1990). The researchers concluded that CCK-induced sleep seems to be not physiological. They further demonstrated sleep-modulatory effects of a CCK receptor agonist in rats (DE SAINT HILAIRE et al. 1991). More recently, KAPÁS et al. (1991a) reported that CCK (10 or 50 mg/kg i.p.) induces a transient increase in non-REM sleep accompanied by simultaneous transient suppression of food intake in both normal and streptozotocin-induced diabetic rats. However, CCK-induced enhancement of insulin secretion is observable only in normal rats but not in diabetic rats, implying that pancreatic insulin plays no significant role in somnogenic and anorectic activity of CCK. KAPÁS et al. (1991c) also reported a similar observation in rabbits.

III. Bombesin

A tetradecapeptide, bombesin, and mammalian bombesin-like peptides are also satiety-inducing substances. According to DE SAINT HILAIRE-KAFI et al. (1989b), bombesin (8–32 mg/kg i.p.) dependently increases total sleep, total non-REM sleep, latency to REM sleep and duration of sleep cycles. Subsequent to these reports, DE SAINT HILAIRE and NICOLAÏDIS (1991) confirmed the findings and also reported that an intrahypothalamic injection of a hunger-inducing peptide, PYY, decreased both non-REM sleep and REM sleep in food-deprived rats.

IV. Acidic Fibroblast Growth Factor

OOMURA et al. (1992) found that CSF concentrations of acidic fibroblast growth factor (aFGF) change dependent on feeding in rats, and that this CNS peptide is the most potent satiety-inducing substance yet found. It is known that aFGF is localized in the ependymal cells and glial cells around the third cerebral ventricle walls and that after food intake aFGF is transferred to neurons in the broad regions of the basal forebrain and the hippocampus via an internalization of its receptor (OOMURA 1991). Recently, DE SAINT HILAIRE and NICOLAÏDIS (1992) demonstrated that i.c.v. administration of aFGF (40–80 ng) induced a dramatic increase in non-REM sleep but not in REM sleep along with a reduction of feeding.

V. Concluding Remarks

As a rule, feeding occurs during the active phase of circadian cycles and sleep occurs during the resting phase. Consequently, it seems likely that satiety-related CNS peptides are mainly responsible for the transient promotion of the postprandial sleep, but add little to the physiological sleep promotion during the circadian rest period.

G. Vasoactive Intestinal Polypeptide and Vasopressin: Circadian Timekeepers or REM Sleep Factors?

Vasoactive intestinal polypeptide (VIP) is a 28-amino-acid residue peptide originally known as a gastrointestinal hormone and today also as a CNS peptide. Up to 1988, several authors reported a sleep-enhancing activity for non-REM sleep and REM sleep in rats and for only REM sleep in cats (see Borbély and tobler 1989; Drucker-Colín et al. 1988; Inoué 1989a).

Mirmiran et al. (1988) observed that continuous i.c.v. infusion of a competitive VIP receptor antagonist selectively suppresses diurnal REM sleep in rats. On the basis of this finding, they concluded that VIP participates in the generation and maintenance of REM sleep. Obál et al. (1989) reported that VIP and its related peptide histidine methionine (0.01–1.0 nmol/kg i.c.v.) selectively enhanced REM sleep (and wakefulness at a high dose) in rabbits. They suggested that these effects are mediated via release of PRL. Pacheco-cano et al. (1990) demonstrated that VIP restores both non-REM sleep and REM sleep in basal forebrain-lesioned cats. The investigators emphasized that VIP exerts a powerful modulatory influence on REM sleep. This notion was substantiated by recent studies in cats (Jiménez-Anguiano et al. 1993) and rats (Bredow et al. 1994), but not in humans (Steiger et al. 1994).

Certain CNS peptides such as SRIF, VIP and a nonapeptide vasopressin (AVP) are known to exist in the SCN, the mammalian biological clock. Because the content of VIP-mRNA (Gozes et al. 1989) and SRIF-mRNA (Takeuchi et al. 1992) exhibits a circadian variation in the SCN of rats, the gene expression of the sleep-modulatory peptides raises an interesting topic in the field of chronobiology and sleep science today. As to AVP, Arnauld et al. (1989) reported that chronic i.c.v. infusion of AVP (1 ng/h) and an AVP agonist increases wakefulness in rats, whereas infusion of an AVP antagonist and anti-AVP antibodies decreases wakefulness. The investigators concluded that AVP increases the duration of waking via AVP receptors. These investigators further suggested that AVP participates in the generation or the expression of the circadian rhythm of sleep and wakefulness. In this connection, Born et al. (1992) recently demonstrated that AVP regulates human sleep by reducing REM sleep.

H. Vasotocin: A Brain Maturation Modulator?

The early report of Pavel et al. (1977) that arginine vasotocin (AVT), a putative pineal hormone, i.c.v. injected at the extremely small dose of

10^{-6} pg, induces an enhancement of non-REM sleep and a suppression of REM sleep in cats, was followed by several investigations on the sleep-modulatory activity of AVT in humans and animals resulting in controversial conclusions (see BORBÉLY and TOBLER 1989; INOUÉ 1989a). Recently, PAVEL and ADRIEN (1989) demonstrated that AVT (0.1–1 pg i.c.v.) dependently increases quiet sleep (the precursor of non-REM sleep) and suppresses active sleep (the precursor of REM sleep) in newborn cats. Interestingly, the i.c.v. administration of AVT antiserum in contrast increases active sleep. On the other hand, GOLDSTEIN (1984) speculated that AVT plays a role in inhibiting the maturation of the brain by enhancing REM sleep in developing cats in contrast to the assumption of ROFFWARG et al. (1966). Thus the possible physiological involvement of AVT remains unsettled to date.

I. General Concluding Remarks

Sleep has multidimensional functions. It modulates: metabolic processes at the molecular level; synaptic neurotransmission at the subcellular level; detoxicification, restitution and proliferation at the cellular level; thermoregulation and neuroimmunoendocrine information processing at the physiological level; antistress reactions and emotional fluctuations at the psychological level; immune and host defense responses at the pathological level; growth and time-keeping at the whole body level; pregnancy and lactation at the reproductive level; and strategy for survival at the evolutionary level (INOUÉ 1993). At each level, sleep seems to be supported by a number of corresponding highly but not strictly specific sleep substances (INOUÉ 1985, 1986, 1989a,b).

With respect to regulation, sleep is controlled by the dynamic interplay of two major systems in the brain: the neural system and the humoral system. As crucial regions for the former system, the evolutionally and ontogenetically older parts of the brain, *sleep-generating brain*, including the diencephalon, mesencephalon and rhombencephalon, actively participate in the alternations of the vigilance state in the younger part of the brain, *sleeping brain*, i.e., the telencephalon through generating two different kinds of sleep such as non-REM sleep and REM sleep (or possibly three kinds of sleep discriminating "deep non-REM sleep or EEG slow-wave sleep" from "light non-REM sleep or EEG spindle sleep") (INOUÉ 1989b). The diencephalon including the basal forebrain, in which the hypothalamus is located, seems to be mainly required for the regulation of non-REM sleep, evolutionally and ontogenetically new sleep, and the mesencephalon and the rhombencephalon (the brain stem) for REM sleep, the evolutionally and ontogenetically old sleep. In addition, circadian variations of sleep and wakefulness largely depend on the regulatory output of the SCN, which is localized in the ventromedial region of the hypothalamus.

Endogenous sleep substances, which are regarded as highly specific messengers of the humoral system, appear to modulate neuronal activities of the neural system, although a possible interrelationship between these systems has not been fully elucidated through experimental approaches. However, it is noteworthy that there is a general circulatory system throughout the sleep-generating brain and the sleeping brain, i.e., the CSF system. The location of such a cerebroventricular circulatory system has a special benefit for the harmonized control over the whole brain, because the third cerebral ventricle is so located in the center of the hypothalamus, the headquarters for the central regulation of sleep, biological time-keeping, feeding, immune response, reproduction, temperature and so forth that bioactive substances in the CSF can easily diffuse into the hypothalamus via ependymal cells and affect neurons regulating the vigilance state. Moreover, substances produced in the hypothalamus can easily be transported to the ventricular walls and liberated into the CSF. This location is also beneficial, since the liberated substances can be transported, via the aqueduct of the cerebrum, to the lower brain stem including the midbrain, pons and medulla, which plays an important role in the regulation of REM sleep, and affects their neural activities. Furthermore, the presence of a circumventricular organ like the OVLT in this region greatly contributes to the neuroendocrine communications across the blood-brain barrier. Thus a number of state-regulatory substances of both endogenous and exogenous origins can transfer their information to the neural system and vice versa on such a structural and functional basis.

Indeed, most CNS peptides produced in the hypothalamo-hypophyseal region have a potency more or less to affect sleep directly or indirectly at various dimensional levels. Their activities are so diverse and complicated that it is too premature to understand integratively the complex dynamic process of the neuroendocrine-immune aspects of sleep regulation. However, the phenomenon must be one of the most attractive targets that occur as a genuine biological regulation, because only highly evolved organisms with a greatly developed telencephalon manifest the essential necessity of differentiated sleep states. Thus the mechanisms for the regulation of sleep are complicated by nature. Experimental approaches to analyze the dynamic interactions of sleep-regulatory factors are then required to understand these mechanisms. In this respect, is should be noted here that some preliminary but promising attempts have already been made by several researchers to unveil the possible interactions among multiple sleep substances (Inoué 1989a,b, 1992c, 1993; Inoué et al. 1988, 1990; Kimura-Takeuchi and Inoué 1993b; Krueger et al. 1990b).

J. Summary

Since monographs and reviews are available on the studies of sleep substances up to 1988, the present author aims to review the literature mainly published between 1988 and 1992, which selectively deals with the sleep-modulatory activities of several CNS peptides, *sleep peptides*. Attention is focused on new findings as well as new topics.

DSIP as a "programming substance" exerts multivariate functions which cover the regulation of physiological sleep and the extra-sleep functions such as thermoregulation and antinociception. In hypothalamic neurons, DSIP is known to be colocalized and cosecreted with LHRH.

New data are presented on the sleep-enhancing activities of two forms of glutathione, GSH and GSSG, in rats. Glutathione seems to induce natural sleep by inhibiting the excitatory synaptic neurotransmission of glutamatergic neurons and serve as a neuronal detoxicification factor.

MPs and dsRNA released after microbial challenge provoke excessive non-REM sleep, fever and a suppression of REM sleep as an acute phase of the host defense response. These substances are capable of inducing production of cytokines such as IL-1, TNF and INF. All these peptides are somnogenic and pyrogenic. Cytokine-induced sleep seems to serve as a pathophysiological adaptation to infectious diseases.

Certain hypothalamic and pituitary hormones such as GRH, GH, CRH, ACTH and its derivatives, TRH, TSH and PRL are nonspecifically but correlatively involved in the modulation of sleep and wakefulness, serving as a possible neuroendocrine transducer.

Satiety-related factors such as CCK, bombesin and aFGF exert a transient increase in sleep shortly after feeding. These substances seem to manifest ischymetric and postprandial functions. VIP and AVP are possibly involved in the modulation of the circadian rhythm of sleep and wakefulness and also serve as a REM sleep factor. The role of AVT in the regulation of sleep is controversial although its inhibitory action on brain maturation through a suppression of active sleep is suggested.

The participation of multiple CNS peptides in the regulation of sleep requires further investigations for the dynamic interactions among these and other substances with special reference to their relations to the neural system.

Acknowledgments. This study was supported by grants-in-aid for scientific research (Nos. 02808046, 04807015 and 62108002) by the Ministry of Education, Science and Culture, Japan.

References

Aizenman E, Boeckman FA, Rosenberg PA (1992) Glutathione prevents 2,4,5-trihydroxyphenylalanine excitotoxicity by maintaining it in a reduced, non-active form. Neurosci Lett 144:233–236

Armstrong-James M, Fox K (1988) Evidence for a specific role for cortical NMDA receptors in slow-wave sleep. Brain Res 451:189–196

Arnauld E, Bibene V, Meynard J, Rodriguez F, Vincent JD (1989) Effects of chronic icv infusion of vasopressin on sleep-waking cycle of rats. Am J Physiol 256:R674–R684

Aström C, Jochumsen PL (1989) Decrease in delta sleep in growth hormone deficiency assessed by a new power spectrum analysis. Sleep 12:508–515
Astöm C, Trojaborg P (1991) Growth hormone and sleep energy. Sleep Res 20A:185
Banks WA, Kastin AJ (1991) Blood to brain transport of interleukin links the immune and central nervous systems. Life Sci 48:PL117–PL121
Becker PM, Fuchs IE, Jamieson AO, Brown WD, Roffwarg HP (1992) Delta sleep-inducing peptide in obstructive sleep apnea: changes with nasal CPAP tratment–a pilot study. Sleep Res 18:85
Birmanns B, Saphier D, Abramsky O (1990) α-Interferon modifies cortical EEG activity: dose-dependence and antagonism by naloxone. J Neurol Sci 100:22–26
Bjartell A, Ekman R, Bergquist S, Widerlöv E (1989) Reduction of immunoreactive ACTH in plasma following intravenous injection of delta sleep-inducing peptide in man. Psychoneuroendocrinology 14:347–355
Bjartell A, Ekman R, Loh YP (1990) Biosynthesis and processing of delta sleep-inducing peptide-like precursors in primary cultures of mouse anterior pituitary cells. Eur J Biochem 190:131–137
Bjartell A, Sundler F, Ekman R (1991) Immunoreactive delta sleep-inducing peptide in the rat hypothalamus, pituitary and adrenal gland: effects of adrenalectomy. Horm Res 36:52–62
Blatteis CM (1990) Neuromodulative actions of cytokines. Yale J Biol Med 63: 133–146
Blatteis CM, Dinarello CA, Shibata M, LLanos QJ, Quan N, Busija DW (1989) Does circulating interleukin-1 enter the brain? In: Mercer JB (ed) Thermal physiology. Elsevier, Amsterdam, pp 385–390
Borbély AA, Tobler I (1989) Endogenous sleep-promoting substances and sleep regulation. Physiol Rev 69:605–670
Born J, Kellner C, Uthgenannt D, Kern W, Fehm HL (1992) Vasopressin regulates human sleep by reducing rapid-eye-movement sleep. Am J Physiol 262:E295–E300
Born J, Späth-Schwalbe E, Schwakenhofer H, Kern W, Fehm HL (1989) Influences of corticotropin-releasing hormone, adrenocorticotropin, and cortisol on sleep in normal man. J Clin Endocrinol Metab 68:904–911
Brandenberger G (1993) Episodic hormone release in relation to REM sleep. J Sleep Res 2:193–198
Bredow S, Kacsóh B, Obál F Jr, Krueger JM (1994) VIP and PACAP induce PRL mRNA expression in rat hypothalamus. J Sleep Res 3 [Suppl 1]:32
Brown R, Price RJ, King MG, Husband AJ (1988) Autochthonous intestinal bacteria and coprophagy: a possible contribution to the ontogeny and rhythmicity of slow wave sleep in mammals. Med Hypotheses 26:171–175
Cady AB, Kotani S, Shiba T, Kusumoto S, Krueger JM (1989a) Somnogenic activities of synthetic lipid A. Infect Immun 57:396–403
Cady AB, Riveau G, Chedid L, Dinarello CA, Johannsen L, Krueger JM (1989b) Interleukin 1-induced sleep and febrile responses differentially altered by a muramyl dipeptide derivative. Int J Immunopharmacol 11:887–983
Charnay Y, Vallet PG, Guntern R, Bouras C, Constantinidis J, Tissot R (1988) Distribution du ≪delta sleep inducing peptide≫ dans le cerveau de lapin: étude par immunofluorescence. C R Soc Sci (Paris) 306:529–535
Charnay Y, Léger L, Golaz J, Sallanon M, Vallet PG, Guntern R, Bouras C, Constantinidis J, Jouvet M, Tissot R (1990) Immunohistochemical mapping of delta sleep-inducing peptide in the cat brain and hypophysis. Relationship with the LHRH system and corticotopes. J Chem Neuroanat 3:397–412
Chastrette N, Clement HW, Prevautel H, Cespuglio R (1988) Proopiomelanocortin components: differential sleep-waking regulation? In: Inoué S, Schneider-Helmert D (eds) Sleep peptides: basic and clinical approaches. Japan Scientific Societies Press, Tokyo/Springer Berlin Heidelberg New York, pp 27–52

Chastrette N, Cespuglio R, Jouvet M (1990a) Proopiomelanocortin (POMC)-derived peptides and sleep in the rat. I. Hypnogenic properties of ACTH derivatives. Neuropeptides 15:61–74
Chastrette N, Cespuglio R, Lin YL, Jouvet M (1990b) Proopiomelanocortin (POMC)-derived peptides and sleep in the rat. II. Aminergic regulatory processes. Neuropeptides 15:75–88
Covelli V, Cannuscio B, Munno I, Altamura M, Decandia P, Pellegrino NM, Maffione AB, Savastano S, Lombardi G (1991) Somnogenic cytokines with special reference to interleukin-1. Acta Neurol (Napoli) 13:520–526
Danguir J (1988) Internal milieu and sleep homeostasis. In: Inoué S, Schneider-Helmert D (eds) Sleep peptides: basic and clinical approaches. Japan Scientific Societies Press, Tokyo/Springer, Berlin Heidelberg New York, pp 53–72
Danguir J (1990) Insulin and somatostatin as sleep-inducing factors. In: Inoué S, Krueger JM (eds) Endogenous sleep factors. SPB Academic, Hague, pp 99–107
Danguir J, De Saint-Hilaire Kafi S (1989) Reversal of desipramine-induced suppression of paradoxical sleep by a long-acting somatostatin analogue (octreotide) in rats. Neurosci Lett 98:154–158
Danguir J, Nicolaidis S (1980) Intravenous infusions of nutrients and sleep in the rat: an ischymetric sleep regulation hypothesis. Am J Physiol 238:E307–E312
De Saint Hilaire Z, Nicolaïdis S (1991) Metabolic determinants of sleep and wakefulness: effect of metabolically active substances. Sleep Res 20A:82
De Saint Hilaire Z, Nicolaïdis S (1992) Enhancement of slow wave sleep parallel to the satiating effect of acidic fibroblast growth factor in rats. Brain Res Bull 29:525–528
De Saint Hilaire-Kafi Z, Depoortere H, Nicolaïdis S (1989a) Does cholecystokinin induce physiological satiety and sleep? Brain Res 488:304–310
De Saint Hilaire-Kafi Z, Gibbs J, Nicolaïdis S (1989b) Satiety and sleep: the effect of bombesin. Brain Res 478:152–155
De Saint Hilaire Z, Roques BP, Nicolaïdis S (1991) Effect of a highly selective central CCK-B receptor agonist: BC-264 on rat sleep. Pharmacol Biochem Behav 38:545–548
De Sarro GB, Masuda Y, Ascioti C, Audino MG, Nistico G (1990) Behavioural and ECoG spectrum changes induced by intracerebral infusion of interferons and interleukin 2 in rats are antagonized by naloxone. Neuropharamacology 29: 167–179
Depoortere H, de Saint Hilaire Z, Nicolaïdis S (1990) Intraperitoneal cholecystokinin-8 induces atypical somnolence and EEG changes. Physiol Behav 48:873–877
Dinarello CA, Clark BD, Ikejima T, Puren AJ, Savage N, Rosoff PM (1990) Interleukin 1 receptors and biological responses. Yale J Biol Med 63:87–93
Disbrow JK, Ruth JK (1984) Differential glutamate release in brain regions of long and short sleep mice. Alcohol 1:201–203
Drucker-Colín R, Prospéro-García O, Arankowsky-Sandoval G, Perez-Montfort R (1988) Gastropancreatic peptides and sensory stimuli as REM sleep factors. In: Inoué S, Schneider-Helmert D (eds) Sleep peptides: basic and clinical approaches. Japan Scientific Societies Press, Tokyo/Springer, Berlin Heidelberg New York, pp 73–94
Ernst A, Schoenenberger GA (1988) DSIP: basic findings in human beings. In: Inoué S, Schneider-Helmert D (eds) Sleep peptides: basic and clinical approaches. Japan Scientific Societies Press, Tokyo/Springer Berlin Heidelberg New York, pp 131–173
Ernst A, Schoenenberger GA, Schulz P (1992) DSIP – a circadian organizer? J Sleep Res 1 [Suppl 1]:69
Fehm HL, Born J (1990) Interactions between the hypothalamus-pituitary-adrenal (HPA) system and sleep in humans. In: Horne J (ed) Sleep '90. Pontenagel Bochum, pp 379–383

Feinberg I, Campbell IG (1993) Ketamine administration during waking increases delta EEG intensity in rat sleep. Neusopsychopharmacology 9:41–48

Folbergrova J, Rehncrona S, Siesjo BK (1979) Oxidized and reduced glutathione in the rat brain under normoxic and hypoxic conditions. J Neurochem 32: 1621–1627

Follenius M, Brandenberger G, Simon C, Schlienger JL (1988) REM sleep in human beigns during decreased secretory activity of the anterior pituitary. Sleep 11: 546–555

Fontana AE, Weber E, Dayer JM (1984) Synthesis of interleukin-1/endogenous pyrogen in the brain of endotoxin-treated mice: a step in fever induction? J Immunol 133:1696–1698

Fornal C, Markus R, Radulovacki M (1984) Muramyl dipeptide does not induce slow-wave sleep or fever in rats. Peptides 5:91–95

Fukuhara C, Shinohara K, Tominaga K, Otori Y, Inouye ST (1993) Endogenous circadian rhythmicity of somatostatin like-immunoreactivity in the rat suprachiasmatic nucleus. Brain Res 606:28–35

García-Arrarás J, Pappenheimer JR (1983) Site of actionof sleep-inducing muramyl peptide isolated from human urine: microinjection studies in rabbit brains. J Neurophysiol 49:528–533

Gilbert KR, Aizenman E, Reynolds IJ (1991) Oxidized glutathione modulates N-methyl-D-aspartate- and depolarization-induced increases in intracellular CA^{2+} in cultured rat forebrain neurons. Neurosci Lett 133:11–14

Goichot B, Brandenberger G, Saini J, Wittersheim G, Follenius M (1992) Nocturnal plasma thyrotropin variations are related to slow-wave sleep. J Sleep Res 1:186–190

Goldstein R (1984) The involvement of arginine vasotocin in the maturation of the kitten brain. Peptides 5;25–28

Gozes I, Shani Y, Lin B, Burbach JPH (1989) Diurnal variation in vasoactive intestinal peptide messenger RNA in the suprachiasmatic nucleus of the rat. Neurosci Res Commun 5:83–86

Graf MV, Kastin AJ (1984) Delta-sleep-inducing peptide (DSIP): a review. Neurosci Biobehav Rev 8:83–93

Graf MV, Kastin AJ (1986) Delta-sleep-inducing peptide (DSIP): an update. Peptides 7:1165–1187

Guarneri P, Guarneri R, Mocciaro C, Piccoli F (1983) Interaction of uridine with GABA binding sites in cerebellar membranes of the rat. Neurochem RES 8:1537–1545

Guldner J, Rothe B, Knisatschek H, Steiger A, Holsboer F (1990) Influence of the ACTH-(4-9)-fragment HOE 427 on sleep-EEG and sleep associated secretion of cortisol and growth hormone. In: Horne J (ed) Sleep '90. Pontenagel Bochum, pp 48–50

Hermann-Maurer EK, Ernst A, Schneider-Helmert D, Zimmermann A, Schoenenberger GA (1992) Effects of DSIP-therapy on memory in chronic insomnia. Sleep Res 18:53

Honda K, Inoué S (1978) Establishment of a bioassay method for the sleep-promoting substance. Rep Inst Med Dent Eng 12:81–85

Honda K, Inoué S (1981) Effects of sleep-promoting substance on sleep-waking patterns of male rats. Rep Inst Med Dent Eng 15:115–123

Honda K, Komoda Y, Inoué S (1990) Oxidized glutathione as another active component of sleep-promoting substance (SPS). II. Somnogenic activity in unrestrained rats. Sleep Res 16:61

Honda K, Komoda Y, Inoué S (1991a) Sleep-promoting activity of oxidized and reduced glutathione in unrestrained rats. Jpn J Psychiatry Neurol 45:958–959

Honda K, Komoda Y, Inoué S (1991b) Oxidized and reduced glutathione enhances sleep in unrestrained rats. Sleep Res 20A:139

Honda K, Komoda Y, Inoué S (1993) Differential sleep regulation by two SPS components: uridine and glutathione. In: Mohan Kumar V, Mallick HN, Nayar U (eds) Sleep-wakefulness. Wiley Eastern, New Delhi, pp 3–7
Honda K, Komoda Y, Inoué S (1994) Oxidized glutathione regulates physiological sleep in unrestrained rats. Brain Res 636:253–258
Imeri L, De Simoni MG, Giglio R, Clavenna, Mancia M (1990) Hypothalamic somatostatin during the sleep-waking cycle. In: Horne J (ed) Sleep '90. Pontenagel Bochum, pp 106–108
Inoué S (1985) Sleep substances: their roles and evolution. In: Inoué S, Borbély AA (eds) Endogenous sleep substances and sleep regulation. Japan Scientific Societies Press, Tokyo/VNU Science Press, Utrecht, pp 3–12
Inoué S (1986) Multifactorial humoral regulation of sleep. Clin Neuropharmacol 9 [Suppl 4]:470–472
Inoué S (1989a) Biology of sleep substances. CRC, Boca Raton
Inoué S (1989b) Brain and sleep (in Japanese). Kyoritsu-Shuppan, Tokyo
Inoué S (1990) Sleep peptides (in Japanese). Nippon Rinsho 48:1061–1065
Inoué S (1992a) Linkage between the neural and the humoral mechanisms in sleep regulation. J Sleep Res 1 [Suppl 1]:103
Inoué S (1992b) Sleep regulation and sleep substances (in Japanese). Seishin Igaku Rev 4:17–23
Inoué S (1992c) The humoral aspects of sleep regulation. In: Manchanda SK, Selvamurthy W, Mohan Kumar V (eds) Advances in physiological sciences. Macmillan India, New Delhi, pp 518–524
Inoué S (1993) Sleep-promoting substance (SPS) and physiological sleep regulation. Zool Sci 10:557–576
Inoué S, Borbély AA (eds) (1985) Endogenous sleep substances and sleep regulation. Japan Scientific Societies Press, Tokyo/VNU Science, Utrecht
Inoué S, Kovalzon VM (1990) Progress in studies on sleep substances (in Russian). In: Etingof EB (ed) Hypotheses and forecasts (the future of science), international annual. Znanie, Moscow, pp 210–225
Inoué S, Krueger JM (eds) (1990) Endogenous sleep factors. SPB Academic, Hague
Inoué S, Schneider-Helmert D (eds) (1988) Sleep peptides: basic and clinical approaches. Japan Scientific Societies Press, Tokyo/Springer, Berlin Heidelberg New York
Inoué S, Honda K, Komoda Y, Uchizono K, Ueno R, Hayaishi O (1984a) Differential sleep-promoting effects of five sleep substances nocturnally infused in unrestrained rats. Proc Natl Acad Sci USA 81:6240–6244
Inoué S, Honda K, Komoda Y, Uchizono K, Ueno R, Hayaishi O (1984b) Little sleep-promoting effect of three sleep substances diurnally infused in unrestrained rats. Neurosci Lett 49:207–211
Inoué S, Honda K, Komoda Y (1985) Sleep-promoting substances. In: Wauquier A, Gaillard JM, Monti JM, Radulovacki M (eds) Sleep: neurotransmitters and neuromodulators. Raven New York, pp 305–318
Inoué S, Kimura M, Honda K, Komoda Y (1988) Sleep peptides: general and comparative aspects. In: Inoué S, Schneider-Helmert D (eds) Sleep peptides: basic and clinical approaches. Japan Scientific Societies Press, Todyo/Springer, Berlin Heidelberg New York, pp 1–26
Inoué S, Kimura-Takeuchi M, Honda K (1990) Co-circulating sleep substances interactingly modulate sleep and wakefulness in rats. Endocrinol Exp 24:69–76
Inoué S, Honda K, Komoda Y (1991) Sleep-promoting substances (SPS). Sleep Res 20A:141
Inoué S, Kimura-Takeuchi M, Kovalzon VM, Mikhavela II, Prudchenko IA, Sviryaev VI, Kailikhevich VN, Chrukina (1992) Somnogenic effects of DSIP structural analogs (in Russian). Zh Vyssh Nerv Deiat 42:600–603
Inoué S, Kimura-Takeuchi M, Asala SA, Okano Y, Honda K (1993) The preoptic area as an interface of circadian and humaoral information of sleep and wakeful-

ness. In: Mohan Kumar V, Mallick HN, Nayar U (eds) Sleep-wakefulness. Wiley Eastern, New Delhi, pp 35–40

Inoué S, Kimura-Takeuchi M, Kovalzon VM, Kalikhevich VN, Churkina SI (1994) The influence of some DSIP analogs upon rat sleep after icv infusion. Bull Exp Biol Med 117:56–58

Ishimori K (1909) True cause of sleep: a hypnogenic substance as evidenced in the brain of sleep-deprived animals (in Japanese). Tokyo Igakkai Zasshi 23:429–457

Iyer KS, McCann SM (1987a) Dalta sleep-inducing peptide (DSIP) stimulates growth hormone (GH) release in the rat by hypothalamic and pituitary actions. Peptides 8:45–48

Iyer KS, McCann SM (1987b) Delta sleep inducing peptide inhibits somatostatin release via a dopaminergic mechanism. Neuroendocrinology 46:93–95

Iyer KS, McCann SM (1987c) Delta sleep-inducing peptide (DSIP) stimulates the release of LH but not FSH via a hypothalamic site of action in the rat. Brain Res Bull 19:535–538

Iyer KS, Marks GA, Kastin AJ, McCann SM (1988) Evidence for a role of delta sleep-inducing peptide in slow-wave sleep and sleep-related growth hormone release in the rat. Proc Natl Acad Sci USA 85:3653–3656

Jasper HH, Khan RT, Elliott KAC (1965) Amino acids released from the cerebral cortex in relation to its state of activation. Science 147:1448–1449

Jiménez-Anguiano A, Báez-Saldana A, Drucker-Colín R (1993) Cerebrospinal fluid (CSF) extracted immediately after REM sleep deprivation prevents REM rebound and contains vasoactive intestinal pepide (VIP). Brain Res 631:345–348

Johannsen L, Cady AB (1990) Sleep substances of microbial origin. In: Inoué S, Krueger JM (eds) Endogenous sleep factors. SPB Academic, Hague, pp 53–60

Johannsen L, Rosenthal RS, Martin SA, Cady AB, Obál F Jr, Guinand M, Krueger JM (1989) Somnogenic activity of O-acetylated and dimeric muramyl peptides. Infect Immun 57:2726–2732

Johannsen L, Toth LA, Rosenthal RS, Opp MR, Obál F Jr, Cady AB, Krueger JM (1990) Somnogenic, pyrogenic, and hematologic effects of bacterial peptidoglycan. Am J Physiol 259:R182–R186

Johannsen L, Wecke J. Obál F Jr, Krueger JM (1991) Macrophages produce somnogenic, muramyl peptides during digestion of staphylococci. Am J Physiol 260:R126–R133

Jouvet M (1988) The regulation of paradoxical sleep by the hypothalamo-hypophysis. Arch Ital Biol 126:259–274

Juhász G, Kékesi K, Emri Z, Soltesz I, Crunelli V (1990) Sleep-promoting action of excitatory amino acid antagonist: a different role for thalamic NMDA and non-NMDA receptors. Neurosci Lett 114:333–338

Kapás L, Krueger JM (1992) Tumor necrosis factor-β induces sleep, fever, and anorexia. Am J Physiol 263:R703–R707

Kapás L, Obál F Jr, Farkas I, Payne LC, Sary G, Rubiscek C, Krueger JM (1991a) Cholecystokinin promotes sleep and reduces food intake in diabetic rats. Physiol Behav 50:417–420

Kapás L, Payne L, Obál F Jr, Opp M, Johannsen L, Krueger JM (1991b) Sleep in diabetic rats: effects of interleukin 1. Am J Physiol 260:R995–R999

Kapás L, Obál F Jr, Opp MR, Johannsen L, Krueger JM (1991c) Intraperitoneal injection of cholecystokinin elicits sleep in rabbits. Physiol Behav 50:1241–1244

Kapás L, Hong L, Cady AB, Opp MR, Postlethwaite AE, Seyer JM, Krueger JM (1992) Somnogenic, pyrogenic, and anorectic activities of tumor necrosis factor-α TNF-α fragments. Am J Physiol 263:R708–R715

Kato N, Nagaki S, Someya T, Tsunashima K, Kawata E, Masui A, Sadamatsu M, Iida H (1990) DSIP and the circadian sleep-wake rhythm. In: Inoué S, Krueger JM (eds) Endogenous sleep factors. SPB Academic, Hague, pp 175–184

Kato N, Sadamatsu M, Iida H, Masui A, Takahashi S (1991) Sleep and growth hormone (GH): sleep dependency of GH secretion and 24-h secretion of delta sleep-inducing peptide (DSIP) (in Japanese). Horumon Rinsho 39:585–590

Kimura M, Inoué S (1989) The phosphorylated analogue of DSIP enhances slow wave sleep and paradoxical sleep in unrestrained rats. Psychopharamacology (Berl) 97:35–39

Kimura-Takeuchi M, Inoué S (1993a) Lateral preoptic area lesions void slow-wave sleep enhanced by uridine but not by muramyl dipeptide in rats. Neurosci Lett 157:17–20

Kimura-Takeuchi M, Majde JA, Toth LA, Krueger JM (1992a) Influenza virus-induced changes in rabbit sleep and acute phase responses. Am J Physiol 263:R1115–R1121

Kimura-Takeuchi M, Majde JA, Toth LA, Krueger JM (1992b) The role of double-stranded RNA in induction of the acute-phase response in an abortive influenza virus infection model. J Infect Dis 166:1266–1275

Kimura-Takeuchi M, Majde JA, Toth LA, Krueger JM (1993a) Effects of recombinant human interferons on sleep in rabbits. Sleep Res 22:36

Kimura-Takeuchi M, Inoué S (1993b) Differential sleep modulation by sequentially administered muramyl dipeptide and uridine. Brain Res Bull 31:33–37

Kimura M, Majde JA, Toth L, Opp MR, Krueger JM (1994) Somnogenic effects of rabbit and human recombinant interferons in rabbits. Am J Physiol 267: R53–R61

Komoda Y, Honda K, Inoué S (1990a) SPS-B, a physiological sleep regulaor, from the brainstems of sleep-deprived rats, identified as oxidized glutathione. Chem Pharm Bull (Tokyo) 28:2057–2059

Komoda Y, Honda K, Inoué S (1990b) Oxidized glutathione as another active component of sleep-promoting substance (SPS). I. Isolation from brainstem extract of sleep-deprived rats. Sleep Res 16:69

Komoda Y, Ishikawa M, Nagasaki H, Iriki M, Honda K, Inoué S, Higashi A, Uchizono K (1983) Uridine, a sleep-promoting substance from brainstems of sleep-deprived rats. Biomed Res [Suppl 4]:223–228

Kovalzon VM, Obál F Jr, Alfoldi P, Inoué S, Kimura-Takeuchi M, Mikhavels II, Prudchenko IA, Kalikhevich VN (1992) Hypnogenic effects of the delta sleep-inducing peptide (DSIP) analogs: somparative investigation in rabbits and rats (in Russian). J Evol Biochem Physiol 28:467–470

Kovalzon VM, Kuntsevich M, Gershovich YG, Mikhavela II, Prudchenko IA (1994) The study of synthetic DSIP analogs: from molecular structure to sleep profile. J Sleep Res [Suppl 1]:130

Krueger JM (1990) Somnogenic activity of immune response modifiers. Trends Pharmacol Sci 11:122–126

Krueger JM, Johannsen L (1989) Bacterial products, cytokines and sleep. J Rheumatol 116 [Suppl 19]:52–57

Krueger JM, Majde JA (1990) Sleep as a host defense: its regulation by microbial products and cytokines. Clin Immunol Immunopathol 57:188–199

Krueger JM, Bascik J, García-Arrarás J (1980) Sleep-promoting material from human urine and its relation to factor S from brain. Am J Physiol 238:E116–E124

Krueger JM, Pappenheimer JR, Karnovsky ML (1982) Sleep-promoting effects of muramyl peptides. Proc Natl Acad Sci USA 79:6102–6106

Krueger JM, Karnovsky ML, Martin SA, Pappenheimer JR, Walter J, Biemann K (1984a) Peptideglycans as promoters of slow-wave sleep. II. Somnogenic and pyrogenic activities of some naturally occurring muramyl peptides; correlation with mass spectrometric structure determination. J Biol Chem 259:12659–12662

Krueger JM, Walter J, Karnovsky ML (1984b) Muramyl peptides. Variation of somnogenic activity with structure. J Exp Med 159:68–76

Krueger JM, Kubillus S, Shoham S, Davenne D (1986) Enhancement of slow-wave sleep by endotoxin and lipid A. Am J Physiol 251:R591–R597

Krueger JM, Dinarello CA, Shoham S, Davenne D, Walter J, Kubillus S (1987) Interferon alpha-2 enhances slow-wave sleep in rabbits. Int J Immunopharmacol 9:23–30

Krueger JM, Majde JA, Blatteis CM, Endsley J, Ahokas A, Cady AB (1988a) Polyriboinosinic: polyribocytidylic acid enhances rabbit slow-wave sleep. Am J Physiol 255:R748–R755

Krueger JM, Toth LA, Cady AB, Johannsen L, Obal F Jr (1988b) Immunomodulation and sleep. In: Inoué S, Schneider-Helmert D (eds) Sleep peptides: basic and clinical approaches. Japan Scientific Societies Press, Tokyo/Springer, Berlin Heidelberg New York, pp 95–129

Krueger JM, Obál F Jr, Johannsen L, Cady AB, Toth LA (1989) Endogenous slow wave sleep substances: a review. In: Wauquier A, Dugovic C, Radulovacki M (eds) Slow wave sleep: physiological pathophysiological and functional aspects. Raven, New York, pp 75–90

Krueger JM, Obál F Jr, Opp Mr, Toth LA, Johannsen L, Cady AB (1990a) Somnogenic cytokines and models concerning their effects on sleep. Yale J Biol Med 63:157–172

Krueger JM, Obál F Jr, Toth L, Johannsen L, Cady AB, Opp M (1990b) Sleep factors: models concerning their multiple interactions and specificity for sleep. In: Inoué S, Krueger JM (eds) Endogenous sleep factors. SPB Academic, Hague, pp 19–29

Krueger JM, Opp MR, Toth LA, Johannsen L, Kapás L (1990c) Cytokines and sleep. Curr Top Neuroendocrinol 10:243–261

Kubo E, Horiuchi T, Nakamura A, Shiomi H (1992) Potent antinociceptive effects of centrally administered glutathione and glutathione-disulfide in mice. Jpn J Pharmacol 58 [Suppl 1]:311

Kudo H, Kokunai T, Kondoh T, Tamaki N, Matsumoto S (1990) Quantitative analysis of glutathione in rat central nervous system: comparison of GSH in infant brain with that in adult brain. Brain Res 511:326–328

Kupfer DJ, Jarrett DB, Ehlers CL (1991) The effect of GRF on the EEG sleep of normal males. Sleep 14:87–88

Kuraev GA, Mendezheritskii AM, Povilaitite PE (1991) Effect of delta-sleep peptide on the ultrastructural features of the rat sensorimotor cortex (in Russian). Tsitol Genet 25:13–16

Kushner I (1991) The acute phase response: from Hippocrates to cytokine biology. Eur Cytokine Network 2:75–80

Lancel M, Crönlein J, Müller-Preuß P, Holsboer F (1994) The effects of endotoxin on EEG activity within non REMS resemble the changes induced by sleep derivation in the rat. J Sleep Res 3 [Suppl 1]:136

Legendre R, Piéron H (1913) Recherches sur le besoin de sommeil consecutif à une veille prolongée. Z Allg Physiol 14:235–262

Léger L, Bonnet C, Cespuglio R, Jouvet M (1994) Immunocytochemical study of the release and metabolism of delta sleep-inducing peptide (DSIP) in the rat brain. Neuropeptides 24:131–138

Lue FA, Bail M, Jephtheh-Ochola J, Carayanniotis K, Gorczynski R, Moldofsky H (1988) Sleep and cerebrospinal fluid interleukin-1-like activity in the cat. Int J Neurosci 42:179–183

Marrosu F, Gessa GL, Giagheddu M, Fratta W (1990) Corticotropin-releasing factor (CRF) increases paradoxical sleep (PS) rebound in PS-deprived rats. Brain Res 515:315–318

Martin SA, Karnovsky ML, Krueger JM, Pappenheimer JR, Biemann K (1984) Peptideglycans as promoters of slow-wave sleep. I. Structure of the sleep-promoting factor isolated from human urine. J Biol Chem 259:12652–12658

Meister A, Anderson ME (1983) Glutathione. Annu Rev Biochem 52:711–760

Meltzer LT, Serpa KA, Moos WH (1989) Evaluation in rats of the somnogenic, pyrogenic, and central nervous system depressant effects of muramyl dipeptide. Psychopharmacology (Berl) 99:103–108

Mendezheritskii AM, Kuraev GA, Mikhaleva II, Povilaitite PE (1992) Morphometric evidence of axosomatic synapse activity upon introduction of delta-sleep-inducing peptide (in Russian). Biull Eksp Biol Med 113:202–203

Miller LG, Galpern WR, Dunlap K, Dinerello CA, Turner TJ (1991) Interleukin-1 augments γ-aminobutyric acid$_A$ receptor function in brain. Mol Pharmacol 39:105–108
Mirmiran M, Kruisbrink J, Bos NPA, Van der Werf D, Boer GJ (1988) Decrease of rapid-eye-movement sleep in the light by intraventricular application of a VIP-antagonist in the rat. Brain Res 458:192–194
Miyamoto M, Murphy TH, Schnaar RL, Coyle JT (1988) Antioxidants protect against glutamate cytotoxicity in a neuronal cell line. Soc Neurosci Abstr 14: 420
Monnier M, Schoenenberger GA (1973) Erzeugung, Isolierung und Characterisierung eines physiologischen Schlaffaktor "delta". Schweiz Med Wochenschr 103: 1733–1743
Monnier M, Koller T, Graver S (1963) Humoral influences of induced sleep and arousal upon electrical brain activity of animals with crossed circulation. Exp Neurol 8:264–277
Nagaki S, Kato N, Minatogawa Y, Higuchi T (1990) Effects of anticonvulsants and gamma-aminobutyric acid (GABA)-mimetic drugs on immunoreactive somatostatin and GABA contents in the rat brain. Life Sci 46:1587–1595
Nagasaki H, Iriki M, Inoué S, Uchizono K (1974) The presence of a sleep-promoting material in the brain of sleep-deprived rats. Proc Jpn Acad 50:241–246
Nakamura A, Shiomi H (1990) The mechanism of anticociceptive effects of DSIP. I. Involvement of endogenous opioid systems. In: Inoué S, Krueger JM (eds) Endogenous sleep factors. SPB Academic, Hague, pp 185–192
Nakamura A, Shiomi H (1991) Phosphorylation of delta sleep-inducing peptide (DSIP) by casein kinase II in vitro. Peptides 12:1375–1377
Nakamura A, Nakashima M, Sugao T, Kanemoto H, Fukumura Y, Shiomi H (1988) Potent antinociceptive effect of centrally administered delta-sleep-inducing peptide (DSIP). Eur J Pharmacol 155:247–253
Nakamura A, Nakashima M, Sakai K, Niwa M, Nozaki M, Shiomi H (1989a) Delta-sleep-inducing peptide (DSIP) stimulates the release of immunoreactive Met-enkephalin from rat lower brainstem slices in vitro. Brain Res 481:165–168
Nakamura A, Sugao T, Yamaue K, Kobatake M, Shiomi H (1989b) Involvement of spinal noradrenergic system in the mechanism of an antinociceptive effect of delta-sleep-inducing peptide (DSIP). Brain Res 480:82–86
Nakamura A, Sakai K, Takahashi Y, Shiomi H (1991) Characterization of delta-sleep-inducing peptide-evoked release of Met-enkephalin from brain synaptosomes in rats. J Neurochem 57:1013–1018
Nakamura A, Nakanishi H, Shiomi H (1993) Characterization of the release and metabolism of delta sleep-inducing peptide (DSIP) in the rat brain. Neuropeptides 24:131–138
Obál F Jr, Alföldi P, Cady AB, Johannsen L, Sary G, Krueger JM (1988) Growth hormone-releasing factor enhances sleep in the rat and rabbits. Am J Physiol 255:R310–R316
Obál F Jr, Opp M, Cady AB, Johannsen L, Krueger JM (1989) Prolactin, vasoactive intestinal peptide, and peptide histidine methionine elicit selective increases in REM sleep in rabbits. Brain Res 490:292–300
Obal F Jr, Opp MR, Cady AB, Johannsen L, Postlethwaite AE, Poppleton HM, Seyer JM, Krueger JM (1990) Interleukin 1α and an interleukin 1ß fragment are somnogenic. Am J Physiol 259:R439–R446
Obál F Jr, Kacsóh B, Alföldi P, Payne L, Markovic O, Grovenor C, Krueger JM (1992) Antiserum to prolactin decreases rapid eye movement sleep (REM sleep) in the male rat. Physiol Behav 52:1063–1068
Obál F Jr, Payne L, Kapás L, Opp M, Krueger JM (1991) Inhibition of growth hormone-releasing factor suppresses both sleep and growth hormone secretion in the rat. Brain Res 557:149–153
Ogita K, Yoneda Y (1987) Possible presence of [^{3}H]glutathione (GSH) binding site in synaptic membranes from rat brain. Neurosci Res 4:486–496

Ogita K, Kitago T, Nakamuta H, Fukuda Y, Koida M, Ogawa Y, Yoneda Y (1986) Glutathione-induced inhibition of Na^+-independent and dependent binding of L-[^{3}H]glutamate in rat brain. Life Sci 39:2411–2418
Okajima T, Hertting G (1986) Delta-sleep-inducing peptide (DSIP) inhibited CRF-induced ACTH secretion from rat anterior pituitary gland in vitro. Horm Metab Res 18:497–499
Oomura Y (1991) Neuronal and chemical factors on food intake regulation (in Japanese). Diabetes J 19:123–132
Oomura Y, Sasaki K, Suzuki K, Muto T, Li A, Ogita Z, Hanai K, Tooyama I, Kimura H, Yanaihara N (1992) A new brain glucosensor and its physiological significance. Am J Clin Nutr 55:278S–282S
Opp MR, Krueger JM (1991) Interleukin 1-receptor antagonist blocks interleukin 1-induced sleep and fever. Am J Physiol 260:R453–R457
Opp M, Obál F Jr, Krueger JM (1988) Effects of alpha-MSH on sleep, behavior, and brain temperature: interaction with IL 1. Am J Physiol 255:R914–R922
Opp M, Obál F Jr, Krueger JM (1989) Corticotropin-releasing factor attenuates interleukin 1-induced sleep and fever in rabbits. Am J Physiol 257:R528–R535
Opp MR, Obál F Jr, Krueger JM (1991) Interleukin 1 receptor alters rat sleep: temporal and dose-related effects. Am J Physiol 260:R52–R58
Opp MR, Obál F Jr, Payne L, Krueger JM (1990) Responsiveness of rats to interleukin-1: effects of monosodium glutamate treatment of neonates. Physiol Behav 48:451–457
Opp MR, Kapás L, Toth LA (1992a) Cytokine involvement in the regulation of sleep. Proc Soc Exp Biol Med 201:16–27
Opp MR, Postlethwaite AE, Seyer JM, Krueger JM (1992b) Interleukin 1 receptor antagonist blocks somnogenic and pyrogenic responses to an interleukin 1 fragment. Proc Natl Acad Sci USA 89:3726–3730
Pacheco-Cano MT, García-Hernández F, Prospéro-García O, Drucker-Colín R (1990) Vasoactive intestinal polypeptide induces REM recovery in insomniac forebrain lesioned cats. Sleep 13:297–303
Pappenheimer JR, Koski G, Fencl V, Karnovsky ML, Krueger J (1975) Extraction of sleep-promoting factor S from cerebrospinal fluid and from brains of sleep-deprived animals. J Neurophysiol 38:1299–1311
Pavel S, Adrien J (1989) Vasotocin increases quiet sleep and suppresses active sleep in newborn cats. Opposite effects after vasotocin immunoneutralization. Brain Res Bull 23:463–466
Pavel S, Psatta D, Goldstein R (1977) Slow-wave sleep induced in cats by extremely small amounts of synthetic and pineal vasotocin injected into the third ventricule of the brain. Brain Res Bull 2:251–254
Pellmar TC, Roney D, Lepinski DL (1992) Role of glutathione in repair of free radical damage in hippocampus in vitro. Brain Res 583:194–200
Pollmächer T, Schreiber W, Gudewill S, Vesser K, Fassbender K, Wiedermann K, Trachsel L, Galanos C, Holsboer F (1993) Influence of endotoxin on nocturnal sleep in humans. Am J Physiol 264:R1077–R1083
Pu LP, Charnay Y, Leduque P, Morel G, Dubois PM (1991) Light and electron microscopic immunohistochemical evidence that delta sleep-inducing peptide and gonadotropin-releasing hormone are coexpressed in the same nerve structures in the guinea pig median eminence. Neuroendocrinology 53:332–338
Raps SP, Lai JCK, Hertz L, Cooper AJL (1989) Glutathione is present in high concentrations in cultured astrocytes but not in cultured neurons. Brain Res 493:398–401
Rhee YH, Kim HI (1987) The correlation between sleeping-time and numerical change of intestinal normal flora in psychiatric insomnia patients. Bull Nat Sci Chungbuk Natl Univ 1:159–172
Roffwarg HP, Muzio JN, Dement WC (1966) Ontogenic development of the human sleep-dream cycle. Science 152:604–619

Schneider-Helmert D (1988) DSIP: clinical application of the programming effect. In: Inoué S, Schneider-Helmert D (eds) Sleep peptides: basic and clinical approaches. Japan Scientific Societies Press/Springer, Berlin Heidelberg New York, pp 175–198
Schoenenberger GA (1984) Characterization, properties and multivariate functions of delta-sleep-inducing peptide (DSIP). Eur Neurol 23:321–345
Schoenenberger GA, Maier PF, Tobler HJ, Monnier M (1977) A naturally occurring delta-EEG enhancing nonapeptide in rabbits. X. Final isolation, characterization and activity test. Pflugers Arch 369:99–109
Schoenenberger GA, Monnier M, Graf M, Schneider-Helmert D, Tobler HJ (1981) Biochemical aspects of sleep regulations and the involvement of delta-sleep-inducing peptide (=DSIP). In: Kamphuisen HAC, Bruyn GW, Visser P (eds) Sleep: normal and deranged function. Mefar, Leiden, pp 25–47
Shibata M, Blatteis CM (1989) Neurons in the organum vasculosum laminae terminalis respond to tumor necrosis factor and serotonin in slice preparations. In: Mercer JB (ed) Thermal physiology. Elsevier, Amsterdam, pp 413–414
Shibata M, Blatteis CM (1991) Recombinant human tumor necrosis factor and interferon affect the neuronal activity of the organum vasculosum laminae terminalis (OVLT). Brain Res 562:323–326
Shiomi H, Nakamura A (1990) The mechanism of anticociceptive effects of DSIP. II. Involvement of bulbospinal inhibitory system. In: Inoué S, Krueger JM (eds) Endogenous sleep factors. SPB Academic, Hague, pp 193–200
Shoham S, Krueger JM (1988) Muramyl dipeptide-induced sleep and fever: effects of ambient temperature and time of injection. Am J Physiol 255:R157–R165
Shoham S, Davenne D, Cady AB, Dinarello CA, Krueger JM (1987) Recombinant tumor necrosis factor and interleukin 1 enhance slow-wave sleep. Am J Physiol 253:R142–R149
Shoham S, Blatteis CM, Krueger JM (1989) Effects of preoptic area lesions on muramyl dipeptide-induced sleep and fever. Brain Res 376:396–399
Sidneva LN, Babichev VN, Airapetyants MG, Kolomeitseva IA, Shvachkin YP, Smirnova AP (1988) Effect of the thyrotropin-releasing hormone analogue on an electrographic picture of sleep. Neurosci Behav Physiol 18:282–286
Skagerberg G, Bjartell A, Vallet PG, Charnay Y (1991) Immunohistochemical demonstration of DSIP-like immunoreactivity in the hypothalamus of the rat. Peptides 12:1155–1159
Slivaka A, Spina MB, Cohen G (1987a) Reduced and oxidized glutathione in human and monkey brain. Neurosci Lett 74:112–118
Slivaka A, Mytilineou C, Cohen G (1987b) Histochemical evaluation of glutathione in brain. Brain Res 409:275–284
Sofic E, Lange KW, Jellinger K, Riederer P (1992) Reduced and oxidized glutathione in the substantia nigra of patients with Parkinson's disease. Neurosci Lett 142: 128–130
Spiegel K, Follenius M, Simon C, Saini J, Erhart J, Brandenberger G (1994) Prolactin secretion and sleep. Sleep 17:20–27
Steiger A, Guldner J, Hemmeter U, Rothe B, Wiedermann K, Holsboer F (1992) Effects of growth hormone-releasing hormone and somatostatin on sleep EEG and nocturnal hormone secretion in male controls. Neuroendocrinology 56: 566–573
Steiger A, Schier T, Colla-Müller M, Frieboes R, Guldner J, Wiedermann K, Holsboer F (1994) Dose-dependent effects of vasoactive intestinal polypeptide (VIP) on sleep EEG and prolactin release in man. J Sleep Res 3 [Suppl 1]:240
Stutzmann JM, Lucas M, Blanchard JC, Laduron PM (1988) Riluzole, a glutamate antagonist, enhances slow wave and REM sleep in rats. Neurosci Lett 88: 195–200
Susic V, Totic S (1989) "Recovery" function of sleep: effects of purified human interleukin-1 on the sleep and febrile response of cats. Metab Brain Dis 4:73–80

Sudakov KV (1991) Stress coping effects of delta-sleep inducing peptide (in Russian). Sechenov Physiol J SSSR 77:1–13

Takeuchi J, Nagasaki H, Shinohara K, Inouye ST (1992) A circadian rhythm of somatostatin messenger RNA levels, but not of vasoactive intestinal polypeptide/peptide histidine isoleucine messenger RNA levels in rat suprachiasmatic nucleus. Mol Cell Neurosci 3:29–35

Tononi G, Pompeiano M (1990) Changes in glutamate receptor binding during sleep-waking states in the rat. Sleep Res 19:64

Toth LA, Krueger JM (1988) Alteration of sleep in rabbits by Staphylococcus aureus infection. Infect Immun 56:1785–1791

Toth LA, Krueger JM (1989a) Effects of microbial challenge on sleep in rabbits. FASEB J 3:2062–2066

Toth LA, Krueger JM (1989b) Hematologic effects of exposures to three infective agents in rabbits. J Am Vet Med Assoc 195:981–986

Toth LA, Krueger JM (1990) Somnogenic, pyrogenic, and hematologic effects of experimental pasteurellosis in rabbits. Am J Physiol 258:R538–R542

Trachsel L, Schreiber W, Holsboer F, Pollmächer T (1994) Endotoxin enhances EEG alpha and beta power in human sleep. Sleep 17:132–139

Tsunashima K, Masui A, Kato N (1990) The effect of delta sleep-inducing peptide (DSIP) and phosphorylated DSIP (P-DSIP) on the apomorphine-induced hypothermia in rats. Brain Res 510:171–174

Tsunashima K, Kato N, Masui A, Takahashi K (1991) Delta sleep-inducing peptide (DSIP) has a potent and specific effect on thermoregulation via dopaminergic and serotonergic mechanisms. Sleep Res 20A: 179

Tsunashima K, Kato N, Masui A, Takahashi K (1993) The effect of delta sleep-inducing peptide (DSIP) on the change of body (core) temperature induced by serotonergic agonists in the rats. Peptides 15:61–65

Tsunashima K, Yamadera H, Kato N, Mitsushio H, Takahashi K (1992) Changes in delta sleep-inducing peptide-like immunoreactivity (DSIP-LI) in the rat brain and blood by zopiclone administration (in Japanese). Brain Sci Ment Disord 3:341–345

Uchizono K, Ishikawa M, Iriki M, Inoué S, Komoda Y, Nagasaki H, Higashi A, Honda K, McRae-Degeurce A (1982) Purification of sleep-promoting substances (SPS). In: Yoshida H, Hagihara Y, Ebashi S (eds) CNS pharmacology neuropeptides. Pergamon, Oxford, pp 217–226 (Advances in pharmacology and therapeutics II, vol 1)

Valatx JL, Roky R, Paut-Pagano L (1990) Prolactin and sleep regulation. In: Horne J (ed) Sleep '90. Pontenagel, Bochum, pp 346–348

Vallet PG, Charnay Y, Bouras C, Constantinidis J (1988) Distribution and colocalization of delta sleep inducing peptide (DSIP) with corticotropin-like intermediate lobe peptide (CLIP) in the human hypophysis. Neurosci Lett 90:78–82

Vallet PG, Charnay Y, Bouras C (1990) Distribution and colocalization of delta sleep-inducing peptide and luteinizing hormone-releasing hormone in the aged human brain: an immunohistological study. J Chem Neuroanat 3:207–214

Vallet PG, Charnay Y, Boura C, Kiss JZ (1991) Colocalization of delta sleep inducing peptide and luteinizing hormone releasing hormone in neurosecretory vesicles in rat median eminence. Neuroendocrinology 53:103–106

Wahlestedt C, Ekman R, Heilig M, Widerlöv E (1989) Effects of psychoactive drugs on delta sleep-inducing peptide concentrations in rat brain. Eur J Pharmacol 159:285–289

Walter JS, Meyers P, Krueger JM (1989) Microinjection of interleukin-1 into brain: separation of sleep and fever responses. Physiol Behav 45:169–176

Wauquier A, Gaillard JM, Monti JM, Radulovacki M (eds) (1985) Sleep: neurotransmitters and neuromodulators. Raven, New York

Yamamoto I, Kimura T, Watanabe K, Tateoka Y, Ho IK (1990) Action mechanism for hypnotic activity of N^3-benzyluridine and related compounds. In: Inoué S,

Krueger JM (eds) Endogenous sleep factors. SPB Academic, Hague, pp 133–142
Yamamoto I, Kimura T, Watanabe K, Kondo S, Ho IK (1993) Receptors of uridine and its derivatives. Sleep Res 22:512
Yoneda Y, Ogita K (1991) Neurochemical aspects of the N-methyl-D-aspartate recpetor complex. Neurosci Res 10:1–33
Yudkoff M, Pleasure D, Cregar L, Lin ZP, Nissim I, Stern J, Nissim I (1990) Glutathione turnover in cultured astrocytes: studies with [^{15}N]glutamate. J Neurochem 55:137–145
Zaphiropoulos A, Charnay Y, Vallet P, Constantinidis J, Bouras C (1991) Immunohistochemical distribution of corticotropin-like intermediate lobe peptide (CLIP) immunoreactivity in the human brain. Brain Res Bull 26:99–111
Ziegler DM (1985) Role of reversible oxidation-reduction of enzyme thiols-disulfides in metabolic regulation. Annu Rev Biochem 54:305–329
Zlokovic BV, Susic VT, Davson H, Begley DJ, Jankov RM, Mitrovic DM, Lipovac MN (1989) Saturable mechanism for delta sleep-inducing peptide (DSIP) at the blood-brain barrier of the vascularly perfused guinea pig brain. Peptides 10: 249–254

CHAPTER 10

Hormones and Sleep

E. VAN CAUTER

A. Introduction

There are several features of the interaction between sleep and circadian rhythmicity which appear to be unique to the human species. First, human sleep is generally consolidated in a single 6–9-h period, whereas fragmentation of the sleep period in several bouts is the rule in other mammals. Possibly as a result of this consolidation of the sleep period, the wake-sleep and sleep-wake transitions in man are associated with physiological changes which are usually more marked than those observed in animals. For example, the secretion of growth hormone (GH) in normal adults is tightly associated with the beginning of the sleep period whereas the relationship between GH secretory pulses and sleep stages is much less evident in rodents, primates and dogs. Secondly, humans are also unique in their capacity to ignore circadian signals and to maintain wakefulness despite an increased pressure to go to sleep. Finally, human subjects maintained for prolonged periods of time in so-called temporal isolation, i.e., in natural or artificial environments devoid of any periodicity (e.g., light-dark cycle) or temporal cues (i.e., clocks or watches), may show behavioral modifications which have not been observed in laboratory animals under constant conditions. These modifications consist of a desynchronization between the sleep-wake cycle and other rhythms, such as those of body temperature and cortisol secretion, which persist with a period of around 24 h (ASCHOFF 1979; WEVER 1979). Under conditions of so-called internal desynchronization, the sleep-wake cycle may be lengthened to 30 h and more, or shortened to less than 22 h. In view of these characteristics, animal studies of the interactions between sleep and hormonal release may be of limited clinical value.

B. Interactions Between Sleep, Circadian Rhythmicity and Hormonal Release

Figure 1 shows a schematic representation of the interaction between sleep and circadian rhythmicity and of the effects of these two central nervous system processes on hormonal release in the hypothalamopituitary axes (VAN CAUTER 1990). In mammalian species, a self-sustained oscillation with a period of approximately 24 h is generated in the suprachiasmatic nuclei of

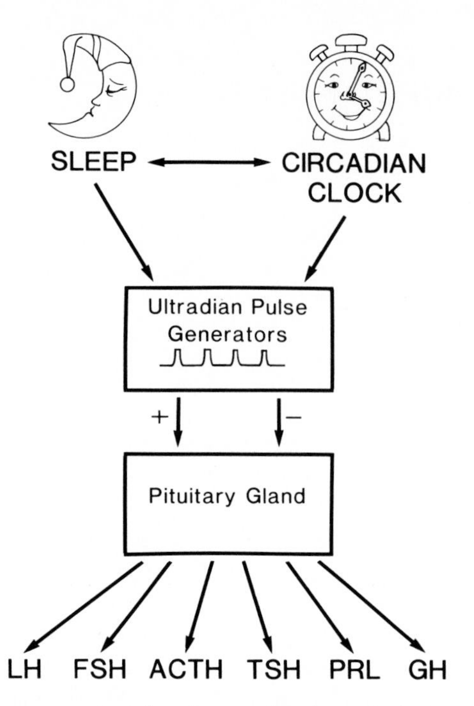

Fig. 1. Schematic representation for the modulation of pituitary secretion by sleep and circadian rhythmicity. (From VAN CAUTER 1990)

the hypothalamus. Circadian rhythmicity is thus endogenous and is not a response to the periodic changes in the environment, such as the alternation of light and dark. A series of investigations have shown that the timing, duration and characteristics of sleep are modulated by this circadian signal (CZEISLER et al. 1980; AKERSTEDT and GILLBERG 1981). Indeed, in subjects maintained under conditions of temporal isolation for prolonged periods of time, the duration of sleep episodes is correlated with the phase of the circadian rhythm in body temperature and not with the length of prior wakefulness. Similarly, the propensity for REM sleep is also dependent on circadian phase. In contrast, slow-wave (SW) sleep appears to be primarily regulated by a homeostatic process (BORBELY 1982). These interactions are further described and conceptualized in Chap. 2 of this volume.

The interaction between sleep and circadian rhythmicity is reflected in the modulation of hormonal release. Early studies suggested that the 24-h rhythms of certain hormones, such as GH and prolactin (PRL), were entirely "sleep-related," without any input from circadian rhythmicity, and that the 24-h profiles of other hormones, such as cortisol, were entirely "circadian dependent," without any sleep dependence. However, current evidence clearly indicates that there is no such dichotomy and that both circadian inputs and sleep inputs can be recognized in the 24-h profiles of all pituitary and pituitary-dependent hormones.

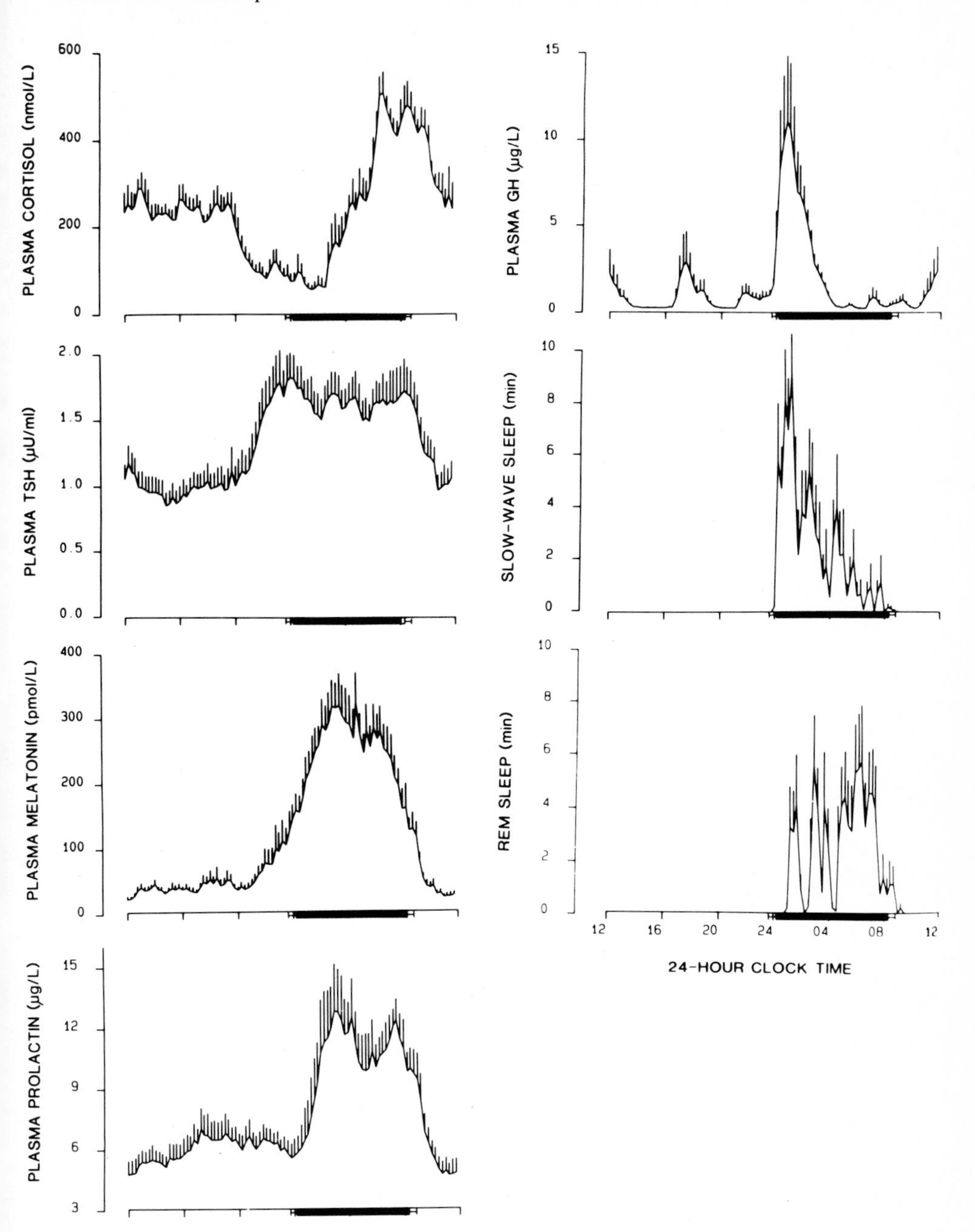

Fig. 2. *From top to bottom*, mean 24-h profiles of plasma cortisol, TSH, melatonin, PRL, GH levels and amounts of SW sleep and REM sleep in a group of young men (20–27 years of age). Data were sampled at 15-min intervals. At each time point, the *vertical line* represents the standard error for the group ($n = 8$). *Black bars* represent the mean sleep period. (Adapted from VAN COEVORDEN et al. 1991)

Figure 2 illustrates the mean 24-h profiles of plasma cortisol, thyrotropin-releasing hormone (TSH), PRL and GH observed simultaneously in a group of young healthy men (VAN COEVORDEN et al. 1991). The two lower panels show the concomitant distribution of SW stages and REM stages, respectively, during the sleep period. As is the case for the majority of hormones, the plasma concentrations of the four hormones shown in Fig. 2 follow a pattern which repeats itself day after day. The nocturnal rise of TSH starts at a time when cortisol secretion is quiescent and ends at the beginning of sleep, when GH and PRL concentrations surge. The early morning period is associated with low TSH, PRL and GH concentrations but high cortisol levels. Thus, the release of these four pituitary hormones follows a highly coordinated temporal program which results from the interaction of circadian rhythmicity (i.e., intrinsic effects of time of day, irrespective of the sleep or wake state), sleep (i.e., intrinsic effects of the sleep state, irrespective of the time of day when it occurs) and pulsatile release. This intricate temporal organization provides the endocrine system with remarkable flexibility.

Theoretically, the modulation of hormonal release by sleep and circadian rhythmicity could be achieved by two distinct types of mechanisms, represented schematically in Fig. 3 (VAN CAUTER et al. 1990). Circadian modulation could be achieved by modulation of pulse amplitude, with larger or smaller pulses occurring around the daily maximum or minimum, respectively. Alternatively, circadian modulation could be achieved by modulation of pulse frequency, with more pulses occurring around the time of the daily

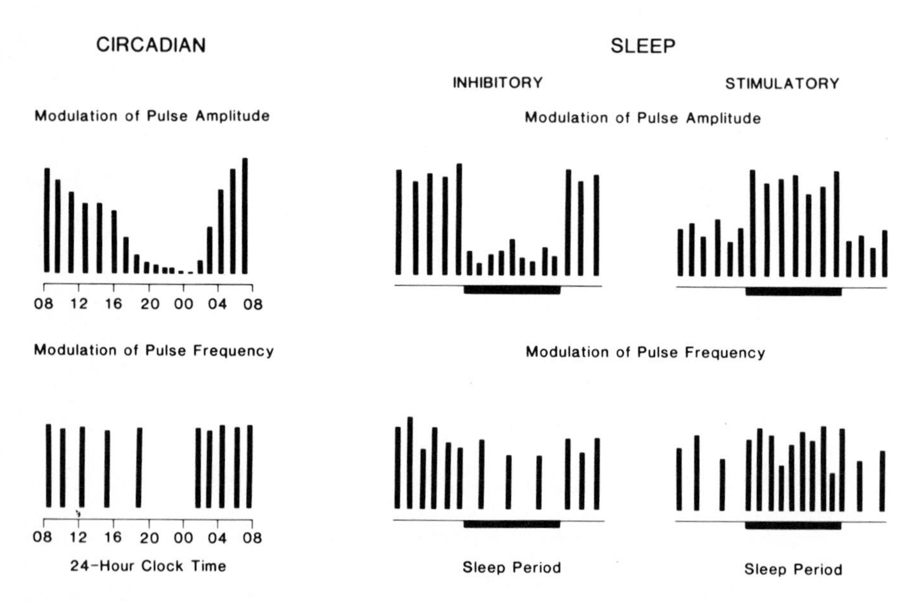

Fig. 3. Schematic representation of possible mechanisms for circadian modulation of pulsatility (*left*) and inhibitory and stimulatory effects of sleep on pulsatility (*right*). (From VAN CAUTER et al. 1990)

maximum and fewer pulses occurring around the time of the daily minimum. Similarly, modulatory effects of sleep on pulsatile release, whether stimulatory or inhibitory, could be exerted by either amplitude or frequency modulation. Importantly, to distinguish circadian effects from sleep effects, it is necessary to study circadian modulation in the absence of the confounding effects of sleep (i.e., during sleep deprivation), and sleep modulation at abnormal times of day, to avoid confounding circadian effects. These issues have only been examined in a handful of studies. For example, in the case of the corticotropic axis, there is good evidence to indicate that the pronounced circadian variation of adrenocorticotropic hormone (ACTH) and cortisol secretion is achieved by modulation of pulse amplitude without changes in pulse frequency (VELDHUIS et al. 1989a,b; VAN CAUTER et al. 1990). In the case of GH secretion, available data suggest that sleep exerts some control on pulse frequency since a pulse of GH release is consistently associated with sleep onset, irrespective of the time of day (VAN CAUTER 1990; VAN CAUTER et al. 1992a).

C. Methodology of Studies of Sleep and Hormonal Secretion

I. Design of Experimental Protocols

Studies on the roles of sleep on hormonal release need to be performed in specially designed units, allowing for the simultaneous recording of sleep and hormonal levels. Typically such units include a room where the light-dark cycle can be accurately controlled, which is equipped for recording of the EEG, EOG and EMG signals and which is adjacent to a room from which blood sampling can be performed without disturbing the subject's sleep using plastic tubing attached to the catheter and extending to this adjacent room through a hole in the wall. The subjects should spend 2–3 nights of habituation in the study unit prior to the beginning of the investigation. Indeed, the duration of sleep and the total duration of REM stages may be significantly reduced by the presence of the catheter (KERKHOFS et al. 1989). These alterations of sleep subside as the subject adapts to the experimental procedure. The catheter should be inserted at least 2 h before the collection of the first sample in order to avoid possible artifactual effects related to the venipuncture stress.

Because of the pulsatile nature of hormonal release, sampling needs to be performed at intervals not exceeding 20 min. To demonstrate the existence of intrinsic effects of sleep, protocols which involve shifts of the sleep-wake cycle are often used. These experimental designs indeed allow for the effects of sleep to be observed at an abnormal circadian time and for the effects of circadian time to be observed in the absence of sleep. Protocols which do not involve such shifts of sleep times do not permit the distinction between intrinsic effects of time of day and intrinsic effects of sleep because both effects are usually superimposed in normal conditions.

II. Analysis of Hormonal Data

The secretion of most, if not all, hormones is pulsatile and therefore the analysis of hormonal changes during sleep involves the need to define and characterize the variations in peripheral levels which are significant as compared with those that could arise from random changes reflecting measurement error (i.e., primarily assay error).

A number of computer algorithms for identification of pulses of hormonal concentration have been proposed. A detailed presentation of the operating principles of each of these procedures is beyond the scope of this chapter. Recent review articles (URBAN et al. 1988; ROYSTON 1989; EVANS et al. 1992) have provided comparisons of performance of several pulse detection algorithms. A comparative study has indicated that Ultra (VAN CAUTER 1981, 1988), Cluster (VELDHUIS and JOHNSON 1986) and Detect (OERTER et al. 1986) perform similarly when used with appropriate choices of parameters. Ultra, which was first developed, has the following characteristics: (1) it takes into account both the increasing limb and the declining limb in determining the significance of a pulse; (2) it allows for variable precision of the assay in various concentration ranges; (3) the significance of a pulse is evaluated independently of its width; and (4) the performance of the algorithm is not affected by the existence of a fluctuating baseline, as may be caused by sleep stimulation or inhibition or circadian variation. In summary, a pulse is considered significant if both its increment and its decline exceed, in relative terms, a threshold expressed in multiples of the intra-assay coefficent of variation in the relevant range of concentration. All changes in concentration which do not meet the criteria for significance are eliminated by an iterative process, providing a "clean" profile as an output. To test Ultra for false-positive and false-negative errors (VAN CAUTER 1988), the program has been applied to a large set of computer-generated series including both signal (i.e., pulses) and noise (i.e., measurement error). These simulation studies indicated that for series with a medium to high signal-to-noise ratio (i.e., large and frequent pulses), a threshold of 2 local intra-assay coefficient of variations (CV) minimizes both false-positive and false-negative errors. For series with low signal-to-noise ratios (i.e., low pulse amplitude and/or frequency), a threshold of 3 CV is preferable.

When the clearance kinetics of the hormone under study are known with reasonable accuracy, secretory rates can be derived from plasma levels by calculation using a mathematical model to remove the effects of hormonal distribution and degradation. This procedure, commonly referred to as "deconvolution," may be based on a one-compartment model, i.e., single exponential decay, or a two-compartment model, i.e., double exponential decay. Examining secretory rates rather than peripheral concentrations has obvious advantages. Indeed, the dynamic nature of the secretory process, e.g., entirely pulsatile or involving both tonic and intermittent secretion, is revealed and the mechanisms underlying modulatory influences such as those of sleep or circadian rhythmicity, can be analyzed. Moreover, as will be shown below, deconvolution provides a more accurate estimation of the temporal limits of the secretory process than peripheral concentrations and allows therefore for more precise evaluation of the temporal concordance between different endocrine and electroencephalographic events.

III. Analysis of the Interaction Between Sleep and Hormonal Release

Polygraphic sleep recordings are usually scored in 20- or 30-s epochs as stages Wake, 1 2, 3, 4 and REM. Thus, for a typical 8-h night scored at 20-s intervals, a series of 1440 scores is obtained. However, hormonal values are usually obtained at 10- to 20-min intervals, yielding 24–48 data points over an 8-h sleep period. Thus, some form of reduction of sleep data is necessary to correlate the hormonal profiles with changes in sleep stages. One of the procedures which has been used is to calculate the percentages of time spent in stages Wake, 1 + 2, 3 + 4 (SW) and REM during each time period between blood samplings. In this way, the sleep recordings are summarized in four series of data each containing the same number of points as the hormonal profile. This process of data reduction allows for straightforward calculations of correlations between hormonal and sleep changes. An example is given in Fig. 4, where the association between pulsatile GH secretion and sleep stages is studied in a single subject. The profile shown on the top represents the plasma levels of GH measured at 15-min intervals. Three pulses of plasma concentration were found to be significant using a computerized pulse detection algorithm. The corresponding profile of GH secretory rates is shown as the second profile from the top. These secretory rates were derived mathematically from the plasma levels by deconvolution using a single-compartment model for GH disappearance with a half-life of 19 min and a volume of distribution of 7% of the body weight. Pulse analysis of the secretory rates now reveals the occurrence of three additional pulses of GH secretion. The three lower profiles illustrate the percentages of each 15-min interval between blood samplings spent in stages Wake, slow-wave (SW: 3 + 4) and REM, respectively. When the profile of plasma concentrations is compared with the SW profile, it appears that, subsequent to its initiation concomitant with the beginning of the first SW period, the sleep-onset GH pulse spanned the first 3 h of sleep, without apparent modulation by the succession of non-REM and REM stages. However, the profile of secretory rates clearly reveals that GH was preferentially secreted during the SW stage, with interruptions of secretory activity coinciding with the intervening REM or Wake stages. Thus, deconvolution indicated the existence of a closer association between SW stages and active GH secretion than the

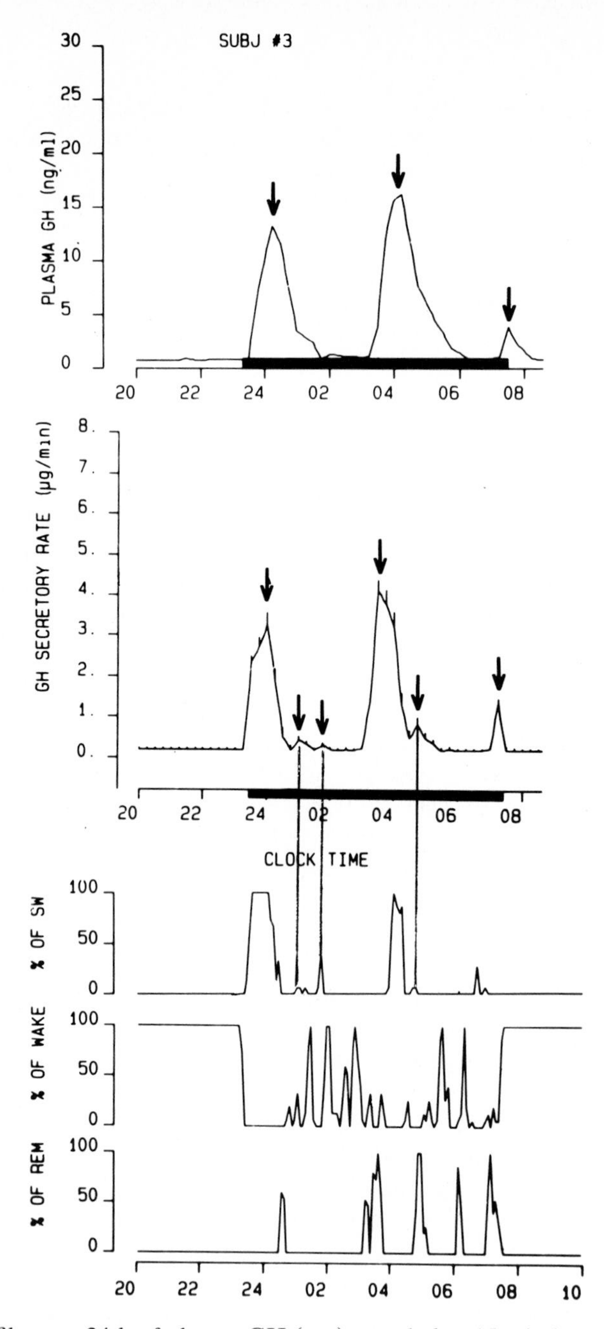

Fig. 4. Profile over 24-h of plasma GH (*top*) sampled at 15-min intervals in a normal man. *Black bar* indicates the sleep period. *Second panel from the top* shows the profile of GH secretory rates calculated by deconvolution from the profile of plasma levels. Significant pulses of plasma levels and of secretory rates are indicated by *arrows*. *The three lower panels* represent the temporal distribution of SW stages, Wake and REM during sleep. *Vertical lines* indicate the association between pulses of GH secretion and SW stages. (From VAN CAUTER 1990)

analysis of plasma concentrations because the temporal limits of each pulse were more accurately defined and additional pulses were detected (VAN CAUTER et al. 1992a).

D. Modulation of Hormonal Release by Sleep and Wake

I. Corticotropic Axis

For many years, the 24-h profile of cortisol secretion has been considered to be the epitomy of circadian rhythmicity and to remain unaffected by manipulations of the sleep-wake cycle. However, a careful analysis of 24-h cortisol profiles observed in subjects studied before and after sleep deprivation has indicated that the first few hours of sleep exert an inhibitory effect on cortisol secretion (WEITZMAN et al. 1983). Since cortisol secretion is already quiescent in the late evening, this inhibitory effect of sleep results in normal conditions in a prolongation of the quiescent period. Conversely, under conditions of sleep deprivation, the nadir of cortisol secretion is less pronounced and occurs earlier than under normal conditions of nocturnal sleep. This is illustrated in the upper panel of Fig. 5, which shows mean profiles of plasma cortisol observed in eight normal young men studied during a 53-h period including 8 h of nocturnal sleep (i.e., from 2300 hours until 0700 hours), followed by 28 h of sleep deprivation in the recumbent position, and 8 h of daytime sleep (i.e., from 1100 hours until 1900 hours). This experimental design thus involved a 12-h delay shift of the sleep period. During sleep deprivation, the nadir of cortisol levels was higher than during nocturnal sleep, because of the absence of the inhibitory effects of the first hours of sleep, and the morning acrophase was lower, reflecting the absence of the stimulating effects of morning awakening. Overall, the amplitude of the rhythm was reduced by approximately 15% during sleep deprivation as compared with normal conditions. The onset of daytime recovery sleep was associated with an immediate, but short-term, inhibition of cortisol concentrations. These effects of nocturnal sleep onset and daytime sleep onset on cortisol levels are further illustrated in Fig. 6. A preliminary analysis of a large data set on changes in cortisol secretory rate in relation to shifts in sleep stages in normal individuals studied both during nocturnal sleep and during daytime sleep has indicated that pulses of cortisol secretion are triggered in a consistent fashion by awakenings during the sleep period (VAN CAUTER et al. 1990). These observations are consistent with the findings that exogenous administration of glucocorticoids tends to reduce REM sleep and to increase the time spent awake or in stage 1 sleep (BORN et al. 1987, 1989). Somewhat paradoxically, cortisol, but not synthetic glucocorticoids, appears to stimulate SW sleep (BORN et al. 1989).

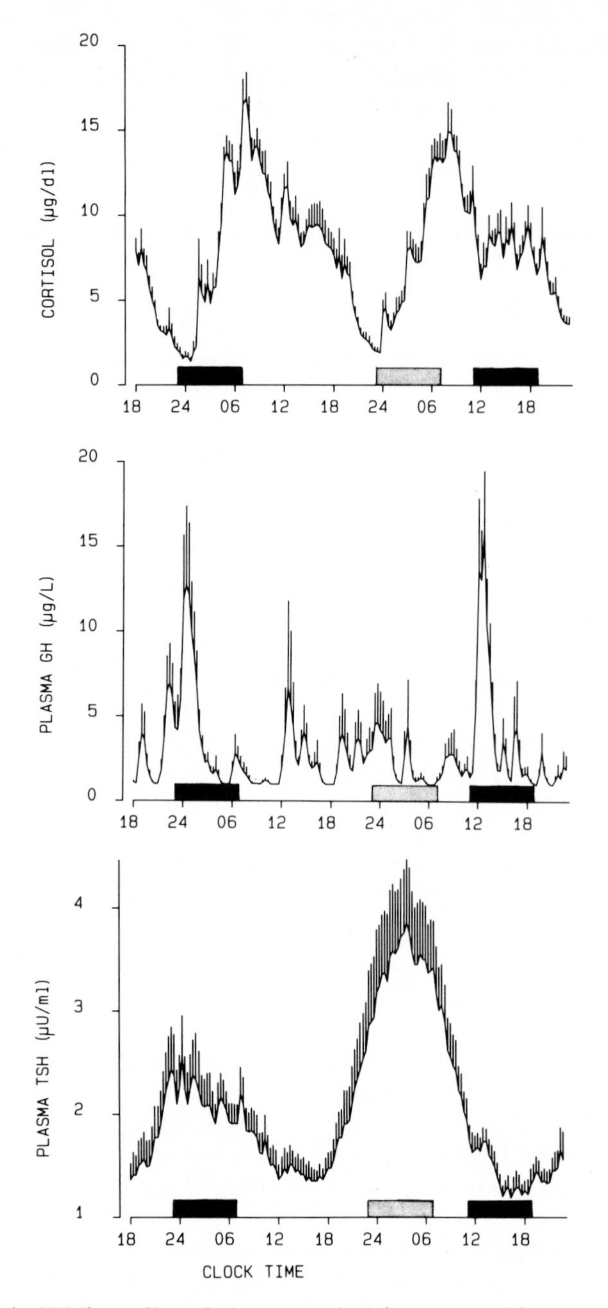

Fig. 5. Mean (+SEM) profiles of plasma cortisol (*upper panels*), plasma GH (*middle panels*) and plasma TSH (*lower panels*) in eight normal young men studied during a 53-h period including 8 h of nocturnal sleep, 28 h of sleep deprivation and 8 h of daytime sleep. *Vertical bars* at each time point represent the SEM. *Black bars* represent the sleep periods. *Open bar* represents the period of nocturnal sleep deprivation

II. Somatotropic Axis

In normal adult subjects, the 24-h profile of plasma GH levels consists of stable low levels abruptly interrupted by bursts of secretion. The most reproducible secretory pulse occurs shortly after sleep onset, in association with the first phase of SW sleep. Other pulses may occur in later sleep and during wakefulness, in the absence of any identifiable stimulus. In men, the sleep-onset GH pulse is generally the largest, and often the only, secretory episode observed over the 24-h span. The mean profiles of plasma GH and SW sleep shown in Fig. 2 are typical of normal young men. In women, daytime GH pulses are more frequent and the sleep-associated pulse, while present, does not generally account for the majority of the daily secretory output of GH. Sleep onset will elicit a pulse in GH secretion whether sleep is advanced, delayed, interrupted or fragmented. The center panel of Fig. 5 shows that GH is secreted following nocturnal sleep onset as well as diurnal sleep onset but that, during sleep deprivation, GH pulses do not occur at consistent times. The relationship between sleep onset and GH release is further illustrated in Fig. 6. Delta wave electroencephalographic activity consistently precedes the elevation in plasma GH levels. A study using blood sampling at 30-s intervals during sleep has shown that maximal GH release occurs within minutes of the onset of SW sleep (HOLL et al. 1991).

Approximately two-thirds (70%) of GH pulses during sleep occur during SW sleep and there is a quantitative correlation between the amount of GH secreted during these pulses and the duration of the SW episode (VAN CAUTER et al. 1992a). Furthermore, the longer the SW episode, the more likely it is to be associated with a GH pulse. These relationships remain significant even when sleep-onset pulses are not included in the calculations. The relationship is facilitatory rather than obligatory, since nocturnal GH secretion can occur in the absence of SW sleep and approximately one-third of the SW periods are not associated with significant GH secretion.

Finally, effects of sleep interruptions opposite to those observed for the corticotropic axis were found for GH release. Indeed, in a study where GH secretion was stimulated by the injection of the hypothalamic factor growth hormone-releasing hormone (GHRH) at the beginning of the sleep period, it was found that, whenever sleep was interrupted by a spontaneous awakening, the ongoing GH secretion was abruptly suppressed (VAN CAUTER et al. 1992b). This finding suggests that sleep fragmentation will generally decrease nocturnal GH secretion and is particularly interesting in view of the well-documented age-related decreases in GH secretion which occur in both men and women. A detailed study has shown that, in healthy elderly men, aged 67–84 years, the total amount of GH secreted over the 24-h span averaged one-third of the daily output of young men, and that the amount of SW sleep in the elderly was also reduced threefold (VAN COEVORDEN et al. 1991). Since in adult subjects, the sleep-onset GH pulse often constitutes the major secretory output, age-related decrements in GH secretion are particularly

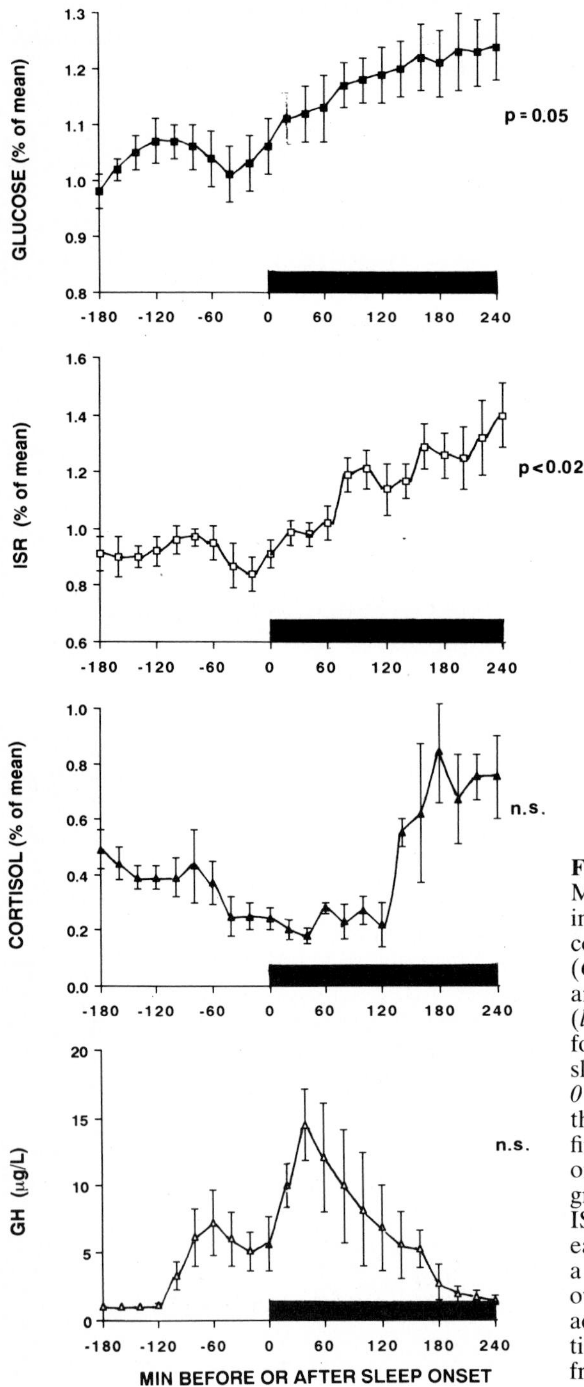

Fig. 6. Effects of sleep onset. Mean (±SEM) glucose levels, insulin secretion rates (*ISR*), cortisol and growth hormone (*GH*) concentrations before and after nocturnal sleep onset (*left*) and daytime sleep onset following a period of 28 h of sleep deprivation (*right*). *Time 0 on the abscissa* corresponds to the time of collection of the first sample following sleep onset as defined by the polygraphic recordings. For glucose, ISR and cortisol, the data for each subject were expressed as a percentage of the subject's overall mean level to take into account interindividual variations in mean levels. (Adapted from VAN CAUTER et al. 1991)

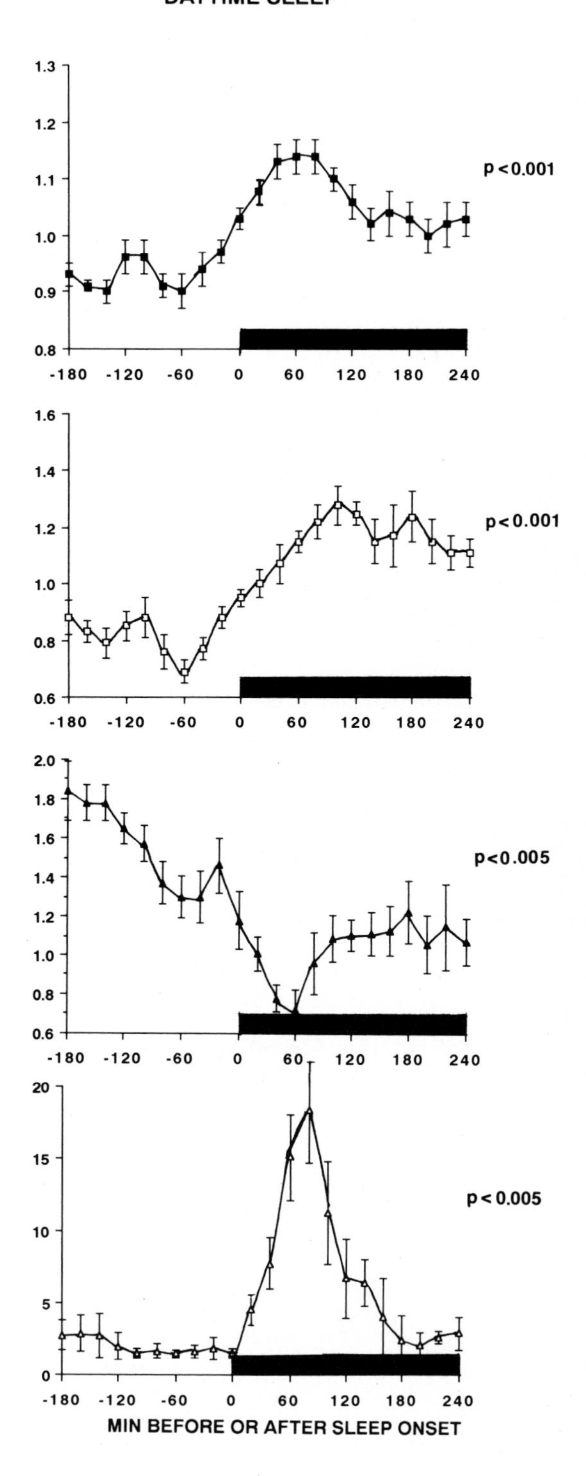
DAYTIME SLEEP
1.3
1.2
1.1
1.0
0.9
0.8
p<0.001
-180 -120 -60 0 60 120 180 240
1.6
1.4
1.2
1.0
0.8
0.6
p<0.001
-180 -120 -60 0 60 120 180 240
2.0
1.8
1.6
1.4
1.2
1.0
0.8
0.6
p<0.005
-180 -120 -60 0 60 120 180 240
20
15
10
5
0
p<0.005
-180 -120 -60 0 60 120 180 240
MIN BEFORE OR AFTER SLEEP ONSET

prominent at that time of day. Although early studies had generally concluded that sleep-related GH pulses were absent in the elderly, the findings of more recent studies are concordant in showing persistent, but reduced, GH secretion during sleep (Ho et al. 1987; Vermeulen 1987; van Coevorden et al. 1991). The parallelism between decreased depth of sleep and diminished somatotropic activity raises the interesting possibility that some of the peripheral effects of the hyposomatotropism of the elderly, such as the reduction in lean body mass, may partially reflect a central alteration in sleep control.

III. Thyrotropic Axis

The 24-h pattern of TSH secretion consists of relatively low daytime levels, followed by a rise starting in the early evening and culminating in a nocturnal maximum occurring around the beginning of the sleep period. The later part of sleep is associated with a progressive decline in TSH levels and daytime values resume shortly after morning awakening. A typical profile is shown in Fig. 2. Because the nocturnal rise of TSH occurs well before the time of sleep onset, it is believed to reflect a circadian effect. However, a marked effect of sleep on TSH secretion may be evidenced during sleep deprivation, when nocturnal TSH secretion is increased by as much as 300% over the levels observed during nocturnal sleep. Thus, sleep exerts an inhibitory influence on TSH secretion and this inhibition is relieved by sleep deprivation (Parker et al. 1976; Brabant et al. 1987, 1990). This is shown in the lower panels of Fig. 5, where the mean profiles of plasma TSH observed during the "12-h shift" study described in the preceding section are plotted. Interestingly, when sleep occurred during daytime hours, TSH secretion was not suppressed significantly below normal daytime levels. Thus, the inhibitory effect of sleep on TSH secretion appears to be operative when the nighttime elevation has taken place, indicating once again the interaction of effects of circadian time and effects of sleep. When the depth of sleep at the habitual time is increased by prior sleep deprivation, the nocturnal TSH rise is markedly decreased, suggesting that SW sleep is probably the primary determinant of the sleep-associated fall (Brabant et al. 1990). Indeed, a pulse-by-pulse analysis of TSH profiles and sleep stages has revealed a consistent association between descending slopes of TSH concentrations and SW stages (Goichot et al. 1992).

Because the thyroid hormones are largely bound to serum protein, the existence of a 24-h variation of thyroid hormone secretion, independent of the diurnal variation in hemodilution caused by postural changes, has been difficult to establish. However, under conditions of sleep deprivation, the increased amplitude of the TSH rhythm results in an increased amplitude of the rhythm of plasma triiodothyronine levels, which becomes clearly apparent, and parallels the nocturnal rise of TSH. If sleep deprivation is prolonged for a second night, the nocturnal rise of TSH is markedly diminished as compared with that occurring during the first night. It is likely that, following the first night of sleep deprivation, the elevated thyroid hormone levels, which persist during the daytime period because of the prolonged half-life of these hormones, limit the subsequent TSH rise at the beginning of the next nighttime period.

IV. Lactotropic Axis

Sleep onset, irrespective of the time of day, has a stimulatory effect on PRL release. In adults of both sexes, at least 50% of the total daily output of this hormone occurs during sleep in normal conditions. However, sleep is not the sole factor responsible for the nocturnal elevation of PRL concentrations. Indeed, a number of experiments involving abrupt advances or delays of sleep times have shown that the sleep-related rise of PRL is less pronounced when sleep does not occur at the normal nocturnal time and that maximal stimulation is observed only when sleep and circadian effects are superimposed (VAN CAUTER and REFETOFF 1985; VAN CAUTER 1990). As an example demonstrating the presence of the circadian component, the profiles of plasma PRL observed during adaptation to a 7-h delay of the sleep-wake and light-dark cycles (DESIR et al. 1982) are shown in Fig. 7. On the 1st day after the shift, and elevation of PRL occurred immediately after sleep onset but was not maintained throughout the night and a marked "anamnestic" elevation was observed during wakefulness, around the time of usual sleep onset before the shift (i.e., 2300 hours).

Because of the marked effect that sleep onset has on PRL release, several studies have examined the possible relationship between pulsatile PRL release during sleep and the alternation of REM and non-REM stages. An early report (PARKER et al. 1974) identified an association between nadirs of PRL levels and REM stages, on the one hand, and peaks of PRL levels and non-REM stages, on the other hand. This association was not confirmed in a later, more detailed, study (VAN CAUTER et al. 1982). However, a recent report by FOLLENIUS et al. (1988) suggested that the concept of a relationship between pulsatile PRL activity and the REM – non-REM cycle should be revised rather than abandoned. Indeed, these investigators noticed that REM sleep begins preferentially at a time of decreased PRL secretory activity. Prolonged awakenings interrupting sleep are consistently associated with decreasing prolactin concentrations. This is illustrated in Fig. 8, where the temporal profile of sleep stages and of PRL levels are examined in three normal men. The shaded areas indicate the correspondence between awakening and declining portions of secretory pulses. These data would indicate that fragmented sleep will generally be associated with lower nocturnal prolactin levels. As shown in Fig. 2, this is indeed what is observed in elderly subjects, who have an increased number of awakenings and decreased amounts of non-REM stages, and in whom a dampening of the nocturnal rise is evident. This diminished nocturnal rise in aging is associated

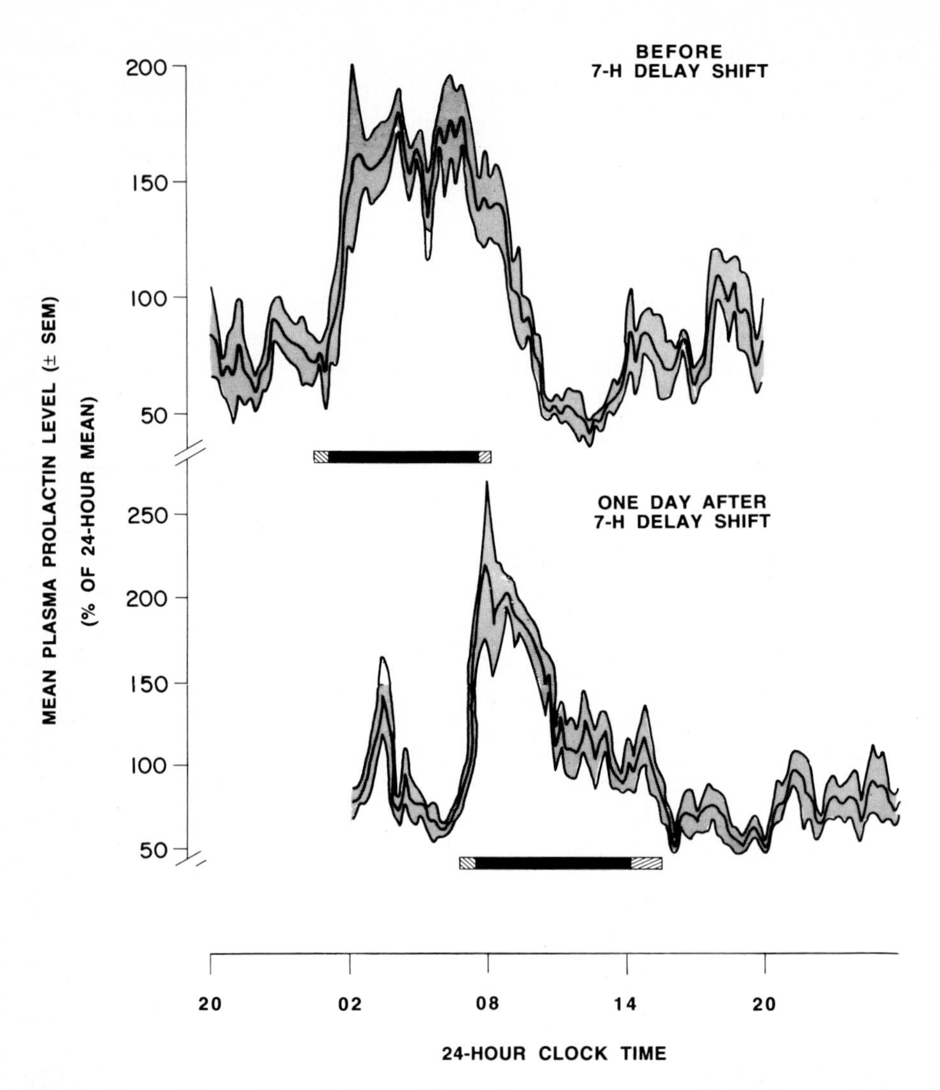

Fig. 7. Mean 24-h profiles of plasma PRL in five young men studied before and after a 7-h delay shift of the sleep-wake cycle associated with westward transmeridian travel (Désir et al. 1982). *Shaded area* represents one SEM above and below the mean. *Black bars* indicate the periods during which all subjects were asleep, with the *hatched ends* on both sides showing the range of individual bedtimes. For each subject, PRL levels were expressed as a percentage of the 24-h mean level to take into account interindividual variations in mean levels

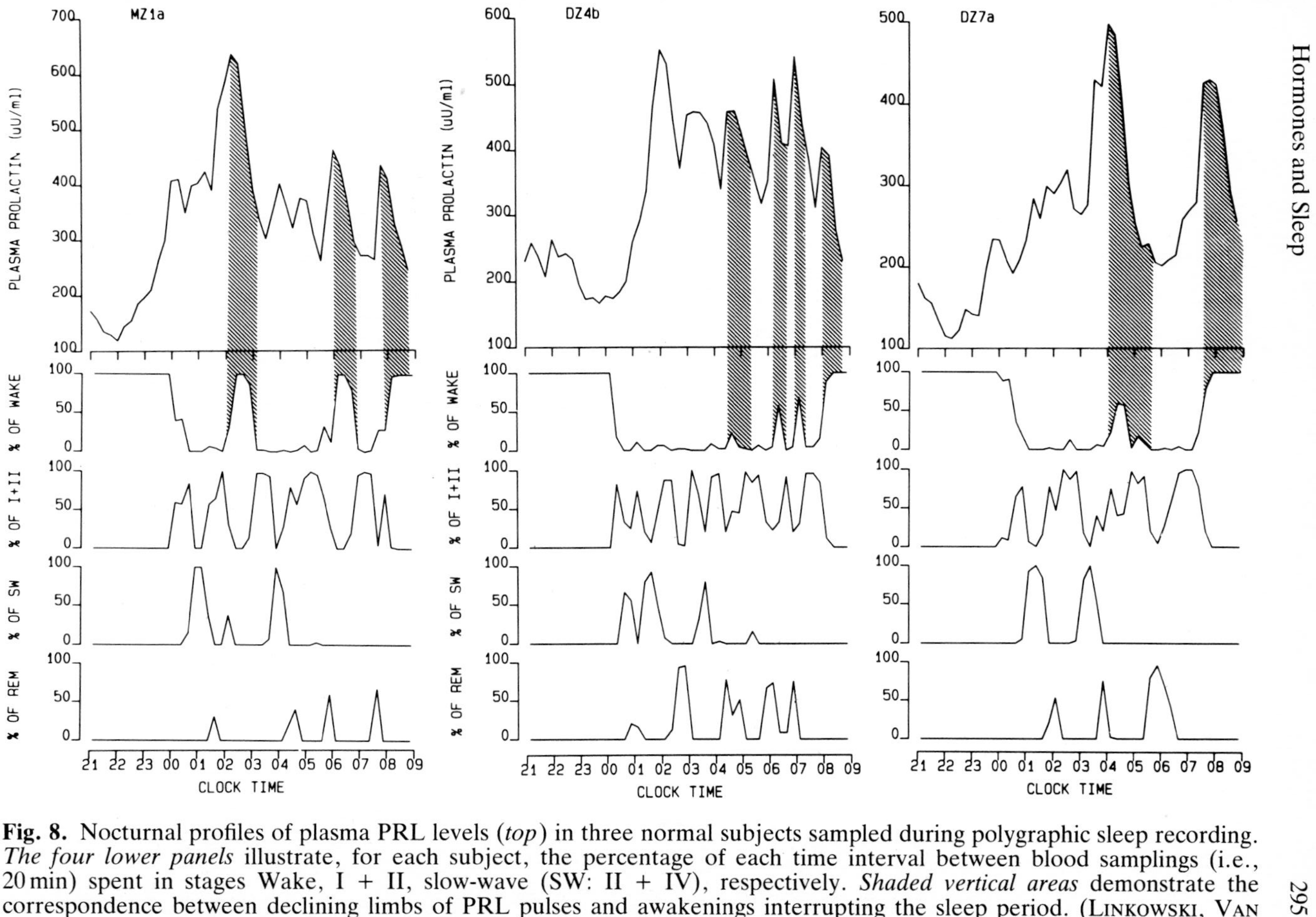

Fig. 8. Nocturnal profiles of plasma PRL levels (*top*) in three normal subjects sampled during polygraphic sleep recording. *The four lower panels* illustrate, for each subject, the percentage of each time interval between blood samplings (i.e., 20 min) spent in stages Wake, I + II, slow-wave (SW: II + IV), respectively. *Shaded vertical areas* demonstrate the correspondence between declining limbs of PRL pulses and awakenings interrupting the sleep period. (LINKOWSKI, VAN CAUTER, L'HERMITE-BALÉRIAUX et al., unpublished observations)

with a decrease in the amplitude of the secretory pulses (GREENSPAN et al. 1990). Benzodiazepine hypnotics taken at bedtime may cause an increase in the nocturnal PRL rise, resulting in concentrations in the pathological range for part of the night (COPINSCHI et al. 1990).

V. Gonadotropic Axis

Sleep effects on the gonadotropic axis were first demonstrated in the early 1970s. Prior to puberty, luteinizing hormone (LH) levels are very low and findings regarding the temporal pattern of release have been somewhat divergent, depending on the nature of the assay procedure. Recent studies generally concord in indicating that both LH and follicle-stimulating hormone (FSH) are secreted in a pulsatile pattern and that an augmentation of pulsatile activity is associated with sleep onset in a majority of both girls and boys (JAKACKI et al. 1982; WU et al. 1991; DUNKEL et al. 1992; APTER et al. 1993). In pubertal children, the magnitude of the nocturnal pulses of LH and FSH is consistently increased during sleep. As the pubescent child enters adulthood, the daytime pulse amplitude increases as well, eliminating or diminishing the diurnal rhythm. Based on an early study showing that, during reversal of the sleep-wake cycle of pubertal boys, LH augmentation occurred during the daytime sleep period, the nocturnal increase in overall LH levels during puberty has been first attributed solely to sleep (KAPEN et al. 1974). However, later analyses indicated that there was also an elevation of LH concentrations at the time when sleep occurred under basal conditions, indicating the presence of an inherent circadian component. Patterns of LH release in adult men exhibit large interindividual variability (SPRATT et al. 1988). The diurnal variation is dampened and may become undetectable. Some studies have, however, observed modest elevations of nocturnal LH, and possibly FSH, levels in young adult men (SPRATT et al. 1988; TENOVER et al. 1988) and a recent report noted a sleep-associated augmentation with a coupling of LH pulses to the REM – non-REM cycle (FEHM et al. 1991). In normal women, inhibitory, rather than stimulatory, effects of sleep have been consistently observed during the early part of the follicular phase of the menstrual cycle (WEITZMAN 1973; SOULES et al. 1985; FILICORI et al. 1986; ROSSMANITH and YEN 1987). Nighttime sleep is then associated with a slowing of LH pulsatility. In fact, it appears that the hypothalamic ultradian pacemaker responsible for LH pulse generation may be totally inhibited during sleep and that the only large nocturnal pulses that may be observed correspond to sleep interruptions.

VI. Glucose Tolerance and Insulin Secretion

Sleep and circadian effects have been most investigated for hormonal secretions that are directly dependent on the hypothalamopituitary axis. However, these effects are also present in other endocrine systems, which may be thought of as less dependent on central nervous system control. One of the more remarkable examples is glucose regulation. In the early 1980s, attention was given to the so-called dawn phenomemon, defined as an early morning (i.e., 4–6 a.m.) rise in glucose levels and/or insulin requirements which may be observed in a majority of diabetic patients (BOLLI et al. 1984; BOLLI and GERICH 1984). Detailed studies in diabetic patients indicated that this early morning rise may be causally related to GH secretion in the beginning of the sleep period (CAMPBELL et al. 1985a,b).

Effects of sleep on glucose regulation in normal subjects have been demonstrated more recently. Glucose tolerance is normally lower in the evening than in the morning and further deteriorates as the evening progresses, reaching a minimum around the middle of the night (JARRETT 1979; SHAPIRO et al. 1988; VAN CAUTER et al. 1989). This diurnal variation is not caused by changes in activity level because it persists during continuous bedrest (VAN CAUTER et al. 1989). Since glucose concentrations begin to increase in the late afternoon or early evening, well before bedtime, and continue to rise until approximately the middle of the night, both sleep-dependent effects and sleep-independent effects, reflecting circadian rhythmicity, could be involved in producing this overall 24-h pattern. To differentiate between the effects of circadian rhythmicity and those of sleep, an experimental protocol involving a 12-h delay of the sleep period has been used (VAN CAUTER et al. 1991). Glucose was infused at a constant rate throughout the study. Figure 9 shows the mean profiles of plasma glucose and insulin secretion rate during the 53-h study period. During nocturnal sleep, levels of glucose and insulin secretion increased by approximately 30% and 60%, respectively, and returned to baseline in the morning. During sleep deprivation, glucose levels and insulin secretion rose again to reach a maximum at a time corresponding to the beginning of the habitual sleep period, indicating the existence of an intrinsic circadian modulation. The magnitude of the rise above morning levels was less than during nocturnal sleep. Daytime sleep was associated with marked elevations of glucose levels and insulin secretion, indicating that the sleep condition per se, irrespective of the time of day when sleep does occur, exerts modulatory influences on glucose regulation. The effects of nocturnal sleep onset and daytime sleep onset on glucose levels and insulin secretion are further illustrated in Fig. 6. Examination of correlations with the temporal variations of GH, a counter-regulatory hormone, indicated that sleep-associated rises in glucose correlated with the amount of concomitant GH secreted (VAN CAUTER et al. 1991). These studies show that glucose regulation is markedly influenced by sleep and suggest that these effects could be partially mediated by GH.

VII. Hormones Regulating Water Balance

Water and salt homeostasis is under the combined control of the posterior pituitary, the renin-angiotensin system and the atrial natriuretic peptide.

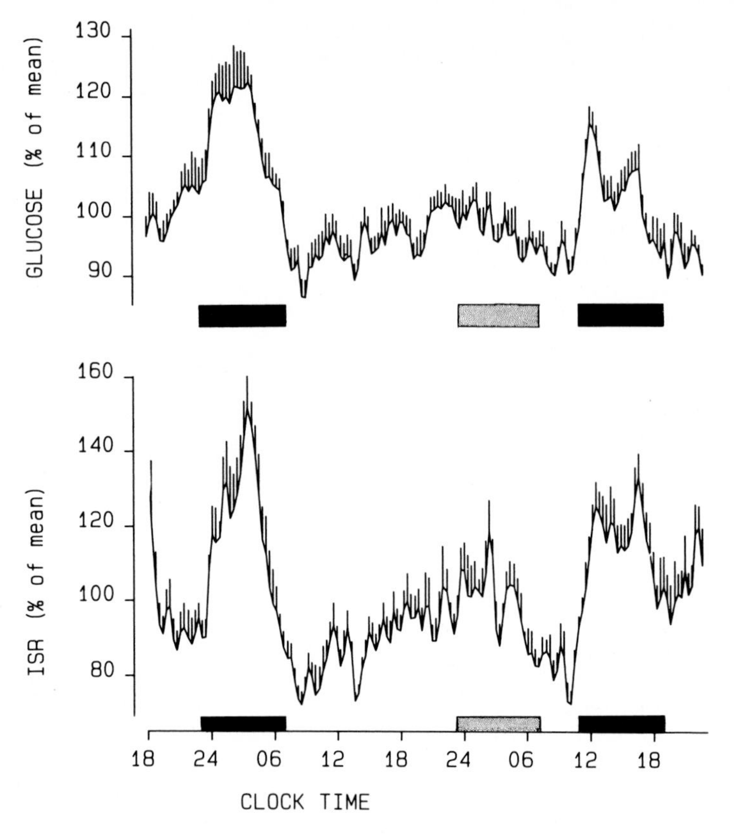

Fig. 9. Mean profiles of plasma glucose and insulin secretion rates (*ISR*) observed in a group of eight subjects studied during a 53-h period, including 8 h of nocturnal sleep, 28 h of sleep deprivation and 8 h of daytime sleep. *Vertical bars* at each time point represent the SEM. To eliminate the effects of interindividual variations in mean glucose, insulin, ISR and cortisol level on the group pattern, the individual values are expressed as percentages of the mean. *Black bars* represent the sleep periods. *Shaded bar* represents the period of nocturnal sleep deprivation. (Adapted from VAN CAUTER et al. 1991)

Early reports have demonstrated the existence of a 90- to 100-min rhythm in urine volume and osmolality, synchronized with the REM–non-REM cycle (MANDELL et al. 1966). Levels of atrial natriuretic peptide are relatively stable and do not show fluctuations related to the sleep-wake or REM–non-REM cycles. Vasopressin release is pulastile, but without apparent relationship to sleep stages. A close relationship between the beginning of REM episodes and decreased activity has been consistently observed for plasma renin activity (BRANDENBERGER et al. 1985, 1990). Figure 10 illustrates the 24-h rhythm of plasma renin activity in a subject studied during a normal

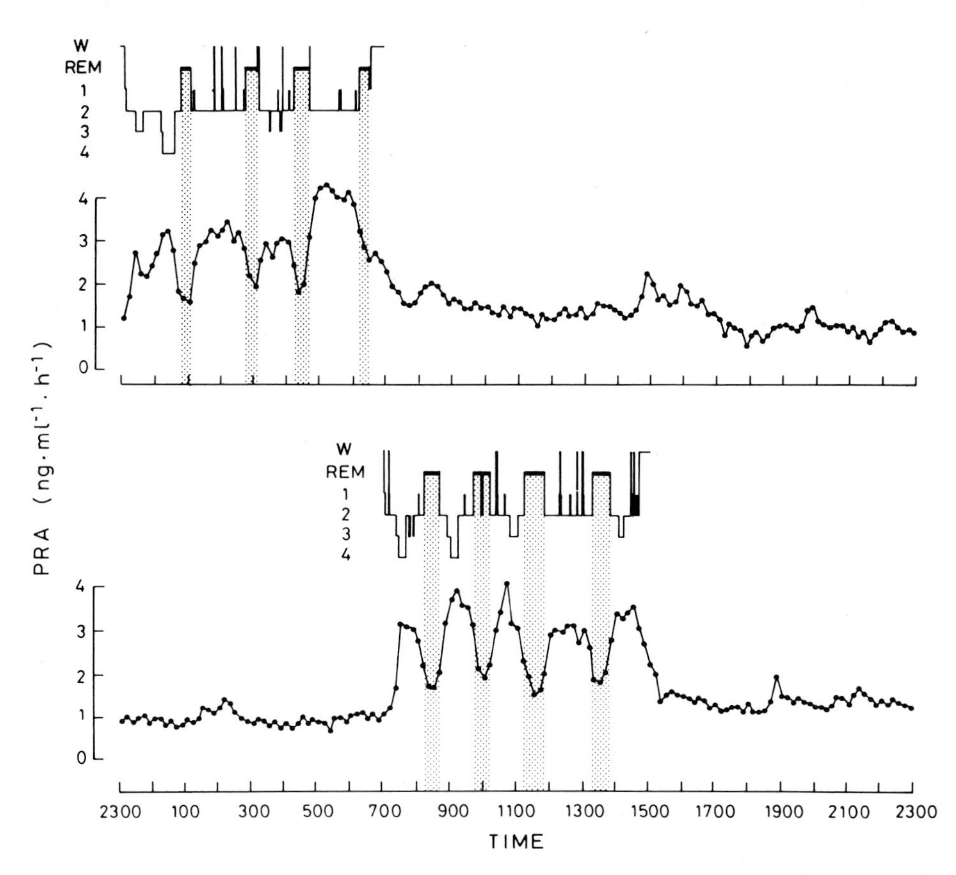

Fig. 10. Profiles over 24 h of plasma renin activity (*PRA*) in two individual subjects who had regular sleep-wake cycles. In the subject shown *in the lower panel*, sleep times had been shifted from 2300–0700 hours to 0700–1500 hours. The temporal distribution of stages wake (*W*), rapid eye movement (*REM*), 1, 2, 3 and 4 are shown above the PRA profiles. *Shaded vertical areas* show the temporal association of REM stages and declining PRA. (Unpublished data from BRANDENBERGER et al., with permission)

sleep-wake cycle (top) and in a subject studied following a shift of the sleep period (bottom). A remarkable synchronization between decreased plasma renin activity and REM stages is apparent during both sleep periods. This relationship was confirmed in studies with selective REM sleep deprivation in normal subjects (BRANDENBERGER et al. 1988). In patients with obstructive sleep apnea who have increased urine and sodium excretion and decreased activity of the renin-angiotensin system, treatment with continuous nasal positive airways restores regular REM–non-REM cyclicity as well as its relationship to plasma renin activity, which in turn contributes to the normalization of urine and sodium outputs (BRANDENBERGER 1991).

E. Sleep-Promoting Effects of Hormones

I. Growth Hormone-Releasing Hormone and Growth Hormone

Sleep-promoting effects of GHRH have been demonstrated in various mammalian species, including man. Intracerebroventricular injections of GHRH in rats and rabbits increase both REM and non-REM sleep (EHLERS et al. 1986; OBÁL et al. 1988). Inhibition of endogenous GHRH, either by administration of a competitive antagonist or by immunoneutralization, inhibits sleep (OBÁL et al. 1991). While intravenous administration of synthetic GHRH during the daytime does not modify sleep in normal young men (GARRY et al. 1985), recent studies have shown that administration of GHRH during sleep may decrease the amount of wake and increase the amount of SW (STEIGER et al. 1992; KERKHOFS et al. 1993). In a study examining the effects of a single injection of a near physiological dose of GHRH (0.3 μg/kg body weight) at the beginning of the third REM period during normal nocturnal sleep, the amount of wake was markedly reduced, and the amount of SW sleep was enhanced almost threefold (KERKHOFS et al. 1993). Such sleep-inducing effects of the peptide could not be evidenced when the injection was given at the beginning of the sleep period, probably because, in these young men who were good sleepers, the amounts of wake during the first 3 h of sleep were already minimal and the amounts of SW stages were normal. These findings raise the interesting possibility that pharmacologically induced elevation of GHRH before bedtime may be beneficial in elderly adults in whom sleep-related GH secretion is drastically decreased, SW sleep is markedly diminished and sleep is fragmented and of shorter duration than in young adults.

Whether or not GHRH can cross the blood-brain barrier is a matter of controversy. Theoretically, the somnogenic effects of GHRH could be mediated by GH. This hypothesis is not supported by the finding of a significant decrease in stage 3 sleep following GH therapy in GH-deficient children (WU and THORPY 1988). Furthermore, injections of pharmacological doses of GH in normal subjects have been reported to either increase REM, rather than SW, sleep (MENDELSON et al. 1980) or to have no effects on sleep (KERN et al. 1993).

II. Melatonin

Exogenous administration of the pineal hormone, melatonin, has been shown to exert effects both on circadian rhythmicity and on sleep. In particular, melatonin administration is capable of re-synchronizing certain overt rhythms, and in particular, the sleep-wake cyle, in a variety of conditions, including real and simulated "jet lag" (PETRIE et al. 1989; SAMEL et al. 1991) and delayed sleep phase syndrome (DAHLITZ et al. 1991). Lewy and associates have recently demonstrated that a low dosage of oral melatonin

(0.5 mg) on four consecutive days causes phase advances when administered in the afternoon and evening and phase delays when given in the morning (LEWY et al. 1992). Nocturnal administration does not result in any phase-shift. It is possible that melatonin exerts its synchronizing effects indirectly, via an effect on the sleep-wake cycle, rather than directly on the central circadian pacemaker. Indeed, in several human studies which have examined the effects of pharmacological doses of melatonin, sedative effects were demonstrated (BORBÉLY 1986; ARENDT 1988; CLAUSTRAT et al. 1990). Sleepiness has been reported to increase even for oral doses as low as 0.3 mg. Nighttime administration can prolong REM latency in young subjects (JAMES et al. 1987). Daytime administration causes decreases in alertness and increases in fatigue ratings (LIEBERMAN 1986). There are also data showing correlations between endogenous melatonin secretion and sleep regulation. In a study involving 64 h of sleep deprivation under constant conditions, a high correlation was found between performance ratings and urinary melatonin levels (AKERSTEDT et al. 1982). In elderly subjects, there appears to be a significant correlation between daily endogenous melatonin production, total REM sleep and REM latency (SINGER et al. 1988).

F. Conclusions

From this brief review of the current state of knowledge on the interactions between hormonal secretions and sleep, we may conclude that effects of sleep are ubiquitous in the endocrine system and that these effects are of sufficient magnitude to be of functional significance. Conversely, hormonal release may affect sleep and be of functional significance to the maintenance and quality of sleep. Therefore, the maintenance of normal sleep-associated endocrine events should be considered as an important criterion in the pharmacology of sleep.

G. Summary

Sleep and circadian rhythmicity, two processes controlled by the central nervous system, both exert marked modulatory effects on hormonal release. Moreover, the timing, duration and characteristics of sleep are partially controlled by circadian rhythmicity. Early studies suggested that the 24-h rhythms of certain hormones, such as growth hormone and prolactin, were entirely "sleep-related," without any input from circadian rhythmicity, and that the 24-h profiles of other hormones, such as cortisol, were entirely "circadian-dependent," and were not influenced by the sleep or wake condition. Current evidence clearly indicates that there is no such dichotomy and that both circadian inputs and sleep inputs can be recognized in the 24-h profiles of all pituitary and pituitary-dependent hormones. Moreover, during the past decade, a series of studies have demonstrated that hormones which are not directly controlled by the hypothalamopituitary axes, such as insulin and the hormones regulating water balance, also undergo consistent variations during sleep and wakefulness.

Effects of sleep on endocrine secretions may be stimulatory or inhibitory. Sleep stimulates the production of growth hormone prolactin and is associated with an increase in plasma renin activity. Sleep inhibits the secretion of cortisol and interrupts the nocturnal elevation of thyroid-stimulating hormone, which is normally initiated during the early evening. Conversely, sleep deprivation results in an approximate twofold increase in nighttime TSH levels. In the reproductive axis, the effects of sleep vary according to the stage of sexual maturation. In pubertal children, sleep enhances the release of the gonadotropins. In adult men, this sleep-associated increase in gonadotropin secretion becomes undetectable. In contrast, in adult women, the pulsatile release of the gonadotropins is inhibited during the follicular phase of the menstrual cycle, whereas no discernible effects of sleep are apparent during the luteal phase. Studies in subjects maintained under constant glucose infusion have shown that glucose tolerance is decreased during sleep, because of a marked reduction in glucose utilization by the brain and peripheral tissues. Thus, effects of sleep appear to be ubiquitous in the endocrine system and these effects are of sufficient magnitude to be of functional significance. Conversely, hormones such as growth hormone-releasing hormone and melatonin may affect sleep and be of functional significance to the maintenance and quality of sleep. Therefore, the maintenance of normal sleep-associated endocrine events should be considered as an important criterion in the pharmacology of sleep.

Acknowledgements. This work was supported in part by grant DK-41814 from the National Institutes of Health, Bethesda, Md., USA, and by grants 90-0222 and 93-0188 from the US Air Force Office of Scientific Research. Studies shown in Figs. 5, 6 and 9 were performed in the University of Chicago General Clinical Research Center, which is supported by grant RR-00055 from the National Institutes of Health, Bethesda, USA. Studies shown in Figs. 2, 4, 7 and 8 were performed in the Sleep Laboratory of Hôpital Erasme, Université Libre de Bruxelles, Brussels, Belgium, and were partially supported by a grant from the Belgian Foundation for Medical Scientific Research (FRSM).

References

Akerstedt A, Gillberg M (1981) The circadian variation of experimentally displaced sleep. Sleep 4:159–169

Akerstedt T, Gillberg M, Wetterberg L (1982) The circadian covariation of fatigue and urinary melatonin. Biol Psychiatry 17:547–554

Apter D, Bützow TL, Laughlin GA, Yen SSC (1993) Gonadotropin-releasing hormone pulse generator activity during pubertal transition in girls: pulsatile and diurnal patterns of circulating gonadotropins. J Clin Endocrinol Metab 76:940–949

Arendt J (1988) Melatonin. Clin Endocrinol 29:205–229

Aschoff J (1979) Circadian rhythms: general features and endocrinological aspects. In: Krieger DT (ed) Endocrine rhythms. Raven, New York, pp 1–61

Bolli G, Gerich J (1984) The "dawn phenomenon" – a common occurrence in both non-insulin-dependent and insulin-dependent diabetes mellitus. N Engl J Med 310:746–50

Bolli G, De Feo P, De Cosmo S, Perriello G, Ventura M, Calcinaro F, Lolli C, Campbell P, Brunetti P, Gerich J (1984) Demonstration of a dawn phenomenon in normal human volunteers. Diabetes 33:1150–53

Borbély AA (1982) A two process model of sleep regulation. Hum Neurobiol 1:195–204

Borbély AA (1986) Endogenous sleep substances and sleep regulation. J Neural Transm [Suppl] 21:243–254

Born J, Zwick A, Roth G, Fehm-Wolfsdorf G, Fehm HL (1987) Differential effects of hydrocortisone, flucortolone, and aldosterone on nocturnal sleep in humans. Acta Endocrinol (Copenh) 116:129–137

Born J, Späth-Schwalbe E, Schwakenhofer H, Kern W, Fehm HL (1989) Influences of corticotropin-releasing hormone, adrenocorticotropin, and cortisol on sleep in normal man. J Clin Endocrinol Metab 68:904–911

Brabant G, Brabant A, Ranft U, Ocran K, Köhrle J, Hesch RD, von zur Mühlen A (1987) Circadian and pulsatile thyrotropin secretion in euthyroid man under influence of thyroid hormone and glucocorticoid administration. J Clin Endocrinol Metab 65:83–88

Brabant G, Prank K, Ranft U, Schuermeyer T, Wagner TOF, Hauser H, Kummer B, Feistner H, Hesch RD, von zur Mühlen A (1990) Physiological regulation of circadian and pulsatile thyrotropin secretion in normal man and woman. J Clin Endocrinol Metab 70:403–409

Brandenberger G (1991) Hydromineral hormones during sleep and wakefulness. Sleep Res 20A:186

Brandenberger G, Follenius M, Muzet A, Ehrhart H, Schieber JP (1985) Ultradian oscillations in plasma renin activity: their relationship to meals and sleep stages. J Clin Endocrinol Metab 61:280–284

Brandenberger G, Follenius M, Simon C, Ehrhart J, Libert JP (1988) Nocturnal oscillations in plasma renin activity and REM – NREM sleep cycles in man: a common regulatory mechanism? Sleep 11:242–250

Brandenberger G, Krauth MO, Ehrhart J, Libert JP, Simon C, Follenius M (1990) Modulation of episodic renin release during sleep in humans. Hypertension 15:370–375

Campbell P, Bolli G, Cryer P, Gerich J (1985a) Pathogenesis of the dawn phenomenon in patients with insulin-dependent diabetes mellitus. N Engl J Med 312:1473–1479

Campbell P, Bolli G, Cryer P, Gerich J (1985b) Sequence of events during development of the dawn phenomenon in insulin-dependent diabetes mellitus. Metabolism 34:1100–1104

Claustrat B, Brun J, Chazot G (1990) Melatonin in man, neuroendocrinological and pharmacological aspects. Nucl Med Biol 17:625–632

Copinschi G, Van Onderbergen A, L'Hermite-Balériaux M, Szyper M, Caufriez A, Bosson D, L'Hermite M, Robyn C, Turek FW, Van Cauter E (1990) Effects of the short-acting benzodiazepine triazolam, taken at bedtime, on circadian and sleep-related hormonal profiles in normal men. Sleep 13:232–244

Czeisler CA, Weitzman ED, Moore-Ede MC, Zimmerman JC, Knauer RS (1980) Human sleep: its duration and organization depends on its circadian phase. Science 210:1264–1267

Dahlitz M, Alvarez B, Vignau J, English J, Arendt J, Parkes JD (1991) Delayed sleep phase syndrome response. Lancet 337:1121–1124

Desir D, Van Cauter E, L'Hermite M, Refetoff S, Jadot C, Caufriez A, Copinschi G, Robyn C (1982) Effects of "jet lag" on hormonal patterns. III. Demonstration of an intrinsic circadian rhythmicity in plasma prolactin. J Clin Endocrinol Metab 55:849–857

Dunkel L, Alfthan H, Stenman UH, Selstam G, Rosberg S, Albertsson-Wikland K (1992) Developmental changes in 24-hour profiles of luteinizing hormone and follicle-stimulating hormone from prepuberty to midstages of puberty. J Clin Endocrinol Metab 74:890–897

Ehlers CL, Reed TK, Henriksen SJ (1986) Effects of corticotropin-releasing factor and growth hormone-releasing factor on sleep and activity in rats. Neuroendocrinology 42:467–474

Evans WS, Sollenberger MJ, Booth RAJ, Rogol AD, Urban RJ, Carlsen EC, Johnson ML, Veldhuis JD (1992) Contemporary aspects of discrete peak-detection algorithms. II. The paradigm of the luteinizing hormone signal in women. Endocr Rev 13:81–104

Fehm HL, Clausing J, Kern W, Pietrowsky R, Born J (1991) Sleep-associated augmentation and synchronization of luteinizing hormone pulses in adult men. Neuroendocrinology 54:192–195

Filicori M, Santoro N, Merriam GR, Crowley WF Jr (1986) Characterization of the physiological pattern of episodic gonadotropin secretion throughout the menstrual cycle. J Clin Endocrinol Metab 62:1136–1144

Follenius M, Brandenberger G, Simon C, Schlienger JL (1988) REM sleep in humans begins during decreased secretory activity of the anteiror pituitary. Sleep 11:546–555

Garry P, Roussel B, Cohen R, Biot-Laporte S, Elm Charfi A, Jouvet M, Sassolas G (1985) Diurnal administration of human growth hormone-releasing factor does not modify sleep and sleep-related growth hormone secretion in normal young men. Acta Endocrinol (Copenh) 110:158–163

Goichot N, Brandenberger G, Sainio J, Wittersheim G, Follenius M (1992) Nocturnal plasma thyrotropin variations are related to slow-wave sleep. J Sleep Res 1:186–190

Greenspan SL, Klibanski A, Rowe JW, Elahi D (1990) Age alters pulsatile prolactin release: influence of dopaminergic inhibition. Am J Physiol 258:E799–E804

Ho KY, Evans WS, Blizzard RM, Veldhuis JD, Merriam GR, Samojlik E, Furlanetto R, Rogol AD, Kaiser DL, Thorner MO (1987) Effects of sex and age on the 24-hour profile of growth hormone secretion in man: importance of endogenous estradiol concentrations. J Clin Endocrinol Metab 64:51–58

Holl RW, Hartmann ML, Veldhuis JD, Taylor WM, Thorner MO (1991) Thirty-second sampling of plasma growth hormone in man: correlation with sleep stages. J Clin Endocrinol Metab 72:854–861

Jakacki RI, Kelch RP, Sauder SE, Lloyd JS, Hopwood NJ, Marshall JC (1982) Pulsatile secretion of luteinizing hormone in children. J Clin Endocrinol Metab 55:453–458

James S, Mendelson W, Sack R, Rosenthal N, Wehr T (1987) The effects of melatonin on normal sleep. Neuropsychopharmacology 1:41–44

Jarrett RJ (1979) Rhythms in insulin and glucose. In: Krieger DT (ed) Endocrine rhythms. Raven, New York, pp 247–258

Kapen S, Boyar RM, Finkelstein JW, Hellman L, Weitzman ED (1974) Effect of sleep-wake cycle reversal on luteinizing hormone pattern in puberty. J Clin Endocrinol Metab 39:293–299

Kerkhofs M, Linkowski P, Mendlewicz J (1989) Effects of intravenous catheter on sleep in healthy men and in depressed patients. Sleep 12:113–119

Kerkhofs M, Van Cauter E, Van Onderbergen A, Caufriez A, Thorner MO, Copinschi G (1993) Sleep-promoting effects of growth hormone-releasing hormone in norml men. Am J Physiol 264:E594–E598

Kern W, Halder R, Al-Reda S, Späth-Schwalbe E, Fehm HL, Born J (1993) Systemic growth hormone does not affect human sleep. J Clin Endocrinol Metab 76:1428–1432

Lewy AL, Ahmed S, Jackson JML, Sack RL (1992) Melatonin shifts human circadian rhythms according to a phase-response curve. Chronobiol Int 9:380–392
Lieberman HR (1986) Behavior, sleep and melatonin. J Neural Transm 21:233–241
Mandell AJ, Chaffey B, Brill P, Mandell MP, Rodnick J, Rubin RT, Sheff R (1966) Dreaming sleep in man: changes in urine volume and osmolality. Science 151: 1558–1560
Mendelson WB, Slater S, Gold P, Gillin JC (1980) The effect of growth hormone administration on human sleep: a dose-response study. Biol Psychiatry 15:613–618
Obál FJ, Alfödi P, Cady AP, Johannsen L, Sary G, Krueger JM (1988) Growth hormone-releasing factor enhances sleep in rats and rabbits. Am J Physiol 255:R310–R316
Obál FJ, Payne L, Kapás L, Opp M, Alföldi P, Krueger JM (1991) Growth hormone releasing hormone (GHRH) in sleep regulation. Sleep Res 20A:192
Oerter KE, Guardabasso V, Rodbard D (1986) Detection and characterization of peaks and estimation of instantaneous secretory rate for episodic pulsatile hormone secretion. Comput Biomed Res 19:170–191
Parker DC, Rossman LG, Vanderlaan EF (1974) Relation of sleep-entrained human prolactin release to REM–non REM cycles. J Clin Endocrinol Metab 38:646–651
Parker DC, Pekary AE, Hershman JM (1976) Effect of normal and reversed sleep-wake cycles upon nyctohemeral rhythmicity of thyrotropin (TSH): evidence suggestive of an inhibitory influence in sleep. J Clin Endocrinol Metab 43:315–22
Petrie K, Conaglen JV, Thompson L, Chamberlain K (1989) Effect of melatonin on jet lag after long haul flights. Br Med J 298:705–707
Rossmanith WG, Yen SSC (1987) Sleep-associated decrease in lutẹinizing hormone pulse frequency during the early follicular phase: evidence for an opioidergic mechanism. J Clin Endocrinol Metab 65:715–718
Royston JP (1989) The statistical analysis of pulsatile hormone secretion data. Clin Endocrinol (Copenh) 30:201–210
Samel A, Wegmann H-M, Vejvoda M, Maab H, Gundel A, Schütz M (1991) Influence of melatonin treatment on human circadian rhythmicity before and after a simulated 9-Hr time shift. J Biol Rhythms 6:235–248
Shapiro ET, Tillil H, Polonsky KS, Fang VS, Rubenstein AH, Van Cauter E (1988) Oscillations in insulin secretion during constant glucose infusion in normal man: relationship to changes in plasma glucose. J Clin Endocrinol Metab 67:307–314
Singer CM, Sack RL, Denney D, Byerly LR, Blood ML, Vandiver RF, Lewy AJ (1988) Melatonin production and sleep in the elderly. Sleep Res 399
Soules M, Steiner R, Cohen N, Bremner WJ, Clifton DK (1985) Nocturnal slowing of pulsatile luteinizing hormone secretion in women during the follicular phase of the menstrual cycle. J Clin Endocrin Metab 61:43–49
Spratt DI, O'Dea LL, Schoenfeld D, Butler JP, Rao TN, Crowley WF Jr (1988) Neuroendocrine-gonadal axis in men: frequent sampling of LH, FSH and testosterone. Am J Physiol 254:E658–E666
Steiger A, Guldner J, Hemmeter U, Rothe B, Wiedemann K, Holsboer F (1992) Effects of growth hormone-releasing hormone and somatostatin on sleep EEG and nocturnal hormone secretion in male controls. Neuroendocrinology 56:566–573
Tenover JS, Matsumoto AM, Clifton DK, Bremner WJ (1988) Age-related alterations in the circadian rhythms of pulsatile luteinizing hormone and testosterone secretion in healthy men. J Gerontol 43:M163–M169
Urban RJ, Evans WS, Rogol AD, Kaiser DL, Johnson ML, Veldhuis JD (1988) Contemporary aspects of discrete peak-detection algorithms. I. The paradigm of the luteinizing hormone pulse signal in men. Endocr Rev 9:3–37
Van Cauter E (1981) Quantitative methods for the analysis of circadian and episodic hormone fluctuations. In: Copinschi G, Van Cauter E (eds) Human pituitary hormones: circadian and episodic variations. Nyhoff, the Hague, pp 1–25

Van Cauter E (1988) Estimating false-positive and false-negative errors in analyses of hormonal pulsatility. Am J Physiol 254:E786–E794

Van Cauter E (1990) Diurnal and ultradian rhythms in human endocrine function: a mini-review. Horm Res 34:45–53

Van Cauter E, Refetoff S (1985) Multifactorial control of the 24-hour secretory profiles of pituitary hormones. J Endocrinol Invest 8:381–391

Van Cauter E, Desir D, Refetoff S, Spire J-P, Noel P, L'Hermite M, Robyn C, Copinschi G (1982) The relationship between episodic variations of plasma prolactin and REM-non-REM cyclicity is an artifact. J Clin Endocrinol Metab 54:70–75

Van Cauter E, Desir D, Decoster C, Féry F, Balasse EO (1989) Nocturnal decrease of glucose tolerance during constant glucose infusion. J Clin Endocrinol Metab 69:604–611

Van Cauter E, van Coevorden A, Blackman JD (1990) Modulation of neuroendocrine release by sleep and circadian rhythmicity. In: Yen SSC, Vale WW (eds) Advances in neuroendocrine regulation of reproduction. Serono Symposia USA, Norwell, pp 113–122

Van Cauter E, Blackman JD, Roland D, Spire JP, Refetoff S, Polonsky KS (1991) Modulation of glucose regulation and insulin secretion by circadian rhythmicity and sleep. J Clin Invest 88:934–942

Van Cauter E, Kerkhofs M, Caufriez A, Van Onderbergen A, Thorner MO, Copinschi G (1992a) A quantitative estimation of GH secretion in normal man: reproducibility and relation to sleep and time of day. J Clin Endocrinol Metab 74:1441–1450

Van Cauter E, Caufriez A, Kerkhofs M, Van Onderbergen A, Thorner MO, Copinschi G (1992b) Sleep, awakenings and insulin-like growth factor I modulate the growth hormone secretory response to growth hormone-releasing hormone. J Clin Endocrinol Metab 74:1451–1459

Van Coevorden A, Mockel J, Laurent E, Kerkhofs M, L'Hermite-Balériaux M, Decoster C, Nève P, Van Cauter E (1991) Neuroendocrine rhythms and sleep in aging. Am J Physiol 260:E651–E661

Veldhuis JD, Johnson ML (1986) Cluster analysis: a simple, versatile, and robust algorithm for endocrine pulse detection. Am J Physiol 250:E486–E493

Veldhuis JD, Iranmanesh A, Lizarralde G, Johnson ML (1989a) Amplitude modulation of a burstlike mode of cortisol secretion subserves the circadian glucocorticoid rhythm. Am J Physiol 257:E6–E14

Veldhuis JD, Iranmanesh A, Johnson ML, Lizarralde G (1989b) Amplitude, but not frequency, modulation of adrenocorticotropin secretory bursts gives rise to the nyctohemeral rhythm of the corticotropic axis in man. J Clin Endocrinol Metab 71:452–463

Vermeulen A (1987) Nyctohemeral growth hormone profiles in young and aged men: correlation with somatomedin-C levels. J Clin Endocrinol Metab 64:884–888

Weitzman ED (1973) Effect of sleep-wake cycle shifts on sleep and neuroendocrine function. In: Burch N, Altschuler HL (eds) Behavior and brain electrical activity. Plenum, New York, pp 93–111

Weitzman ED, Zimmerman JC, Czeisler CA, Ronda JM (1983) Cortisol secretion is inhibited during sleep in normal man. J Clin Endocrinol Metab 56:352–358

Wever RA (1979) The circadian system of man: results of experiments under temporal isolation. Springer, Berlin Heidelberg New York

Wu FCW, Butler GE, Kelnar CJH, Stirling HF, Huhtaniemi L (1991) Patterns of pulsatile luteinizing hormone and follicle-stimulating hormone secretion in prepubertal (midchildhood) boys and girls and patients with idiopathic hypogonadotropic hypogonadism (Kallman's syndrome): a study using an ultrasensitive time-resolved immunofluorometric assay. J Clin Endocrinol Metab 72: 1229–1237

Wu RH, Thorpy MJ (1988) Effect of growth hormone treatment on sleep EEGs in growth hormone-deficient children. Sleep 11:425–429

CHAPTER 11

Pharmacology of the Adenosine System

M. RADULOVACKI

A. Introduction

Adenosine is a nucleoside which consists of the purine base adenine linked to ribose (Fig. 1). It is produced by two main enzymatic reactions involving dephosphorylation of 5′-nucleotidase (EC 3.1.5.5) and alkaline phosphatase (EC 3.1.3.1) as well as hydrolysis of *S*-adenosyl-L-homocysteine (SAH) by SAH-hydrolase (3.3.1.1). Adenosine may be formed intracellularly, as a result of a breakdown of cytosolic adenosine 5′-triphosphate (ATP), or extracellularly from ATP. In nerve tissue, an increase in nerve firing releases both adenosine and ATP, which is then extracellularly degraded to adenosine (STONE et al. 1990).

Substantial experimental evidence suggesting that adenosine acts as a neurotransmitter or neuromodulator in the mammalian central nervous system (CNS) has been extensively reviewed (PHILLIS and WU 1981; STONE 1981). At least two distinct subtypes of extracellular adenosine receptors have been identified in mammalian CNS: A_1 receptors, whose stimulation inhibits adenylate cyclase, and A_2 receptors, whose stimulation activates adenylate cyclase in vitro (VAN CALKER et al. 1979; LONDOS et al. 1980). In addition, A_1 receptors have high affinity (nanomolar) for adenosine and are present in many areas of the brain, including hippocampus, cerebral cortex, thalamus, cerebellum, spinal cord and striatum. In contrast, A_2 receptors have low affinity (micromolar) for adenosine and are present only in striatum, nucleus accumbens and olfactory tubercle (YEUNG and GREEN 1984: REDDINGTON et al. 1986).

Because adenosine is rapidly metabolized by adenosine deaminase, compounds that are resistant to the action of the enzyme are available. These adenosine receptor ligands bind either selectively to A_1 receptors [i.e., N^6-L-(phenylisopropyl)-adenosine, L-PIA] or nonselectively to both A_1 and A_2 receptors (i.e., adenosine-5′-*N*-ethylcarboxamide, NECA). Recently, adenosine receptor ligands with greater in vivo selectivity between A_1 and A_2 receptors than previously available compounds have been developed and/or characterized. To date, the ligands for A_1 adenosine receptors, both agonists and antagonists, display substantially greater selectivity than do A_2 receptor ligands. For example, the most selective A_1 adenosine receptor agonists are N^6-cyclopentyladenosine (CPA) and N^6-cyclohexyladenosine

(CHA), which exhibit 780- and 390-fold selectivity, respectively, and the most selective A_1 receptor antagonists are 8-cyclopentyl-1,3-dipropylxanthine (CPX) and 8-cyclopentyltheophylline (CPT), which exhibit 740- and 130-fold selectivity, respectively. In contrast, the most selective known A_2 adenosine receptor agonists, 2-(phenylamino) adenosine (CV-1808) and 2-(4-methoxyphenyl) adenosine (CV-1674), were only 4.8- and 2.2-fold, respectively, more potent at A_2 than at A_1 adenosine receptors (BRUNS et al. 1986, 1987a,b; MARTINSON et al. 1987). Antagonist ligands for A_2 adenosine receptors also exhibit considerably less selectivity than do A_1 antagonists. Among the most selective A_2 antagonist ligands in vivo are 1-propargyl-3,7-dimethylxanthine (PDX), the triasoloquinazoline CGS1594A, and alloxazine, which exhibit 11.0-, 6.2-, and 1.98-fold selectivity, respectively, for A_2 receptors (BRUNS et al. 1987a; UKENA et al. 1986; WILLIAMS et al. 1987).

Our interest in the possible hypnotic role of adenosine was stimulated by findings that behavioral stimulant effects of caffeine and theophylline, methylated dioxypurines or methylxanthines, which naturally occur in plants, involve a blockade of central adenosine receptors (SNYDER et al. 1981). The structural formulae of adenosine and caffeine (Fig. 1) show that both of these compounds, although different, have a purine base in common. Thus, caffeine acts on adenosine receptors not as an agonist, but as an antagonist.

In addition, experiments with iontophoretic application of adenosine showed that adenosine had depressant effects on the responses of neurons in several brain regions (PHILLIS and WU 1981), and general neurophysiologic effects of adenosine were shown to be inhibitory (STONE 1981; PHILLIS et al. 1979a). Thus, it was conceivable that stimulation of adenosine receptors

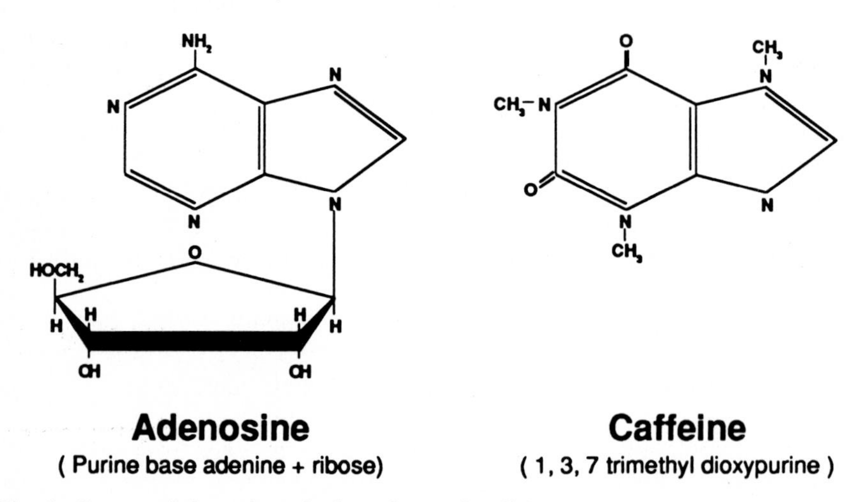

Fig. 1. Structural formulae of adenosine and caffeine

may produce sedation or sleep. Moreover, preliminary data in dogs (HAULICA et al. 1973) indicated a possible hypnotic role for adenosine, whereas administration of adenosine into the brains of rats, cats and fowls produced behavioral sleep (BUDAY et al. 1961; FELDBERG and SHERWOOD 1954; MARLEY and NISTICO 1972). Administration of relatively low doses of adenosine analogs to mice and rats produced marked sedation and hypothermia (SNYDER et al. 1981; DUNWIDDIE and WORTH 1982). In accordance, administration of adenosine triphosphate to rabbits or mice caused sedation (BHATTACHARYA et al. 1970; MATHIEU-LEVY 1968).

These behavioral inhibitory actions of adenosine may be related to its inhibition of the release of excitatory amino acid glutamate (DOLPHIN et al. 1985) as well as acetylcholine (SAWYNOK and JHAMANDAS 1976), dopamine (MICHAELIS et al. 1979), norepinephrine (HEDQUIST and FREDHOLM 1976) and serotonin (HARMS et al. 1979) into the synaptic cleft. Recent findings by HAAS and GREENE (1988) showed that endogenous adenosine reduces neuronal excitability.

There is no direct evidence whether or not endogenous adenosine is also a behavioral depressant, but our work with deoxycoformycin (RADULOVACKI et al. 1983), a potent inhibitor of adenosine deaminase, and caffeine (RADULOVACKI 1982), an adenosine receptor antagonist (SNYDER et al. 1981), leads us to believe that it may be.

B. Effects of Adenosine and Adenosine-Related Compounds on Sleep

In order to test the effects of adenosine on sleep, we administered adenosine intracerebroventricularly to rats because of its short plasma half-life (KLABUNDE 1983) and because it crosses the blood-brain barrier slowly (BERNE et al. 1974). We also administered deoxycoformycin to rats, an adenosine deaminase inhibitor, which prevents the conversion of adenosine to inosine and thus extends the plasma half-life of adenosine. By administering deoxycoformycin we assumed that the effects on sleep produced by the drug were those of endogenous adenosine. Administration of adenosine analogs L-PIA, CHA and NECA, which are resistant to the action of adenosine deaminase, was also of interest because these compounds are agonists of either adenosine A_1 receptors (L-PIA, CHA) or both adenosine A_1 and A_2 receptors (NECA).

I. Intracerebroventricular Administration of Adenosine and Sleep in Rats

We examined and compared the dose-response effects of intracerebroventricular (i.c.v.) infusion of the pyrimidine ribonucleosides, cytidine and uridine, and the purine ribonucleoside, adenosine, on sleep and waking in rats, because iontophoretic administration of cytidine and uridine was shown to be devoid of depressant effects on brain neurons (PHILLIS et al. 1974) while iontophoretic application of adenosine depressed neuronal responses in several brain areas (PHILLIS and WU 1981). We administered all three drugs at doses of 1, 10 and 100 nmol and electroencephalographically monitored rats for 6 h following administration of the drugs. The results showed that i.c.v. infusion of 1 and 10 nmol cytidine significantly suppressed sleep and that administration of uridine did not affect sleep. In contrast, adenosine exhibited significant hypnotic effects at all doses examined (see Table 1). All three doses of adenosine significantly reduced waking and increased total sleep (TS). Both the 1- and 100-nmol doses of adenosine also significantly reduced the latencies to the onset of rapid eye movement (REM) sleep (RADULOVACKI et al. 1985).

II. Microinjections of Adenosine to Preoptic Area and Sleep in Rats

For almost half of this century researchers have sought to find discrete groups of neurons which controlled and regulated the occurrence of mammalian sleep or its component stages. Although several brain regions that modulate sleep are now recognized, there is little consensus among current workers concerning the specific contribution of any neuronal group to the normal sleep cycle. One of the brain areas that was investigated as a

Table 1. Dose-response effects of intracerebroventricular injection of adenosine on sleep and wakefulness in rats (from RADULOVACKI et al. 1985)

Sleep state	Hours	Saline	Adenosine (nmol)		
			1	10	100
W	0–3	61.5 ± 10.6	28.7 ± 5.2[d]	35.7 ± 5.9[b]	25.2 ± 3.4[d]
	3–6	51.2 ± 17.0	21.8 ± 7.3	29.8 ± 8.2	9.5 ± 1.6[c]
	0–6	112.7 ± 26.6	50.5 ± 10.5[b]	65.5 ± 11.6[a]	34.7 ± 3.0[d]
SWS1	0–3	33.5 ± 1.1	33.3 ± 6.0	38.8 ± 4.9	34.8 ± 2.0
	3–6	42.7 ± 3.7	41.7 ± 5.6	44.7 ± 3.9	34.7 ± 5.6
	0–6	76.2 ± 3.7	75.0 ± 10.6	83.5 ± 7.6	69.5 ± 6.5
SWS2	0–3	75.2 ± 9.2	105.0 ± 7.0[b]	87.8 ± 9.8	105.0 ± 3.1
	3–6	71.3 ± 13.7	95.0 ± 7.5	89.5 ± 9.8	111.2 ± 4.6[c]
	0–6	146.2 ± 22.1	200.0 ± 13.6	177.3 ± 18.1	216.2 ± 7.0[c]
REM	0–3	9.0–3.2	13.0 ± 2.2	17.7 ± 1.0	13.3 ± 3.3
	3–6	15.5 ± 5.0	21.5 ± 3.0	16.0 ± 2.0	18.7 ± 2.6
	0–6	24.5 ± 7.6	34.5 ± 4.3	33.7 ± 1.5	32.0 ± 5.2
TS	0–3	118.5 ± 10.6	151.3 ± 5.2[d]	144.3 ± 5.9[b]	154.8 ± 3.4[d]
	3–6	128.8 ± 17.0	158.2 ± 7.3	150.2 ± 8.2	170.5 ± 1.6[c]
	0–6	247.3 ± 26.6	309.5 ± 10.5[b]	294.5 ± 16.6[a]	325.3 ± 3.0[d]

Note: All values reported are means ± SEM in a minimum of six rats per group. Significantly different from saline group: [a] $P < 0.050$; [b] $P < 0.025$; [c] $P < 0.010$; [d] $P < 0.005$.

potential sleep "center" was the preoptic area. This was because STERMAN and CLEMENTE (1962) showed that both low- and high-frequency electrical stimulation of the preoptic area produced sleep. We have examined the effects on sleep of bilateral microinjections of adenosine in the preoptic area of the rat. The results showed that administration of 12.5 nmol adenosine increased deep slow-wave sleep (SWS2), REM sleep and TS during the first 3 h of polygraphic recording. The dose of 25 nmol adenosine did not affect sleep (TICHO and RADULOVACKI 1991). This finding is in accordance with our data with adenosine analogs where high doses of L-PIA and CHA were devoid of hypnotic effects, possibly due to their action on body temperature.

III. Inhibition of Adenosine Deaminase and Sleep

We administered to rats deoxycoformycin, a potent inhibitor of adenosine deaminase (AGARWAL et al. 1977), which would be expected to elevate the levels of adenosine in the CNS, to determine whether an increase in endogenous adenosine would promote sleep. Deoxycoformycin (0.5 or 2.0 mg/kg) was administered intraperitoneally and animals were polygraphically recorded for 6 h. The 0.5-mg/kg dose of the drug was shown to increase REM sleep and reduce REM sleep latency, while the dose of 2 mg/kg increased SWS2 (RADULOVACKI et al. 1983). These results were consistent with the results that we had previously reported for the adenosine analog L-PIA (RADULOVACKI et al. 1982), indicating a hypnotic role for endogenous adenosine.

IV. Adenosine Analogs and Sleep

Following our studies with L-PIA, an adenosine A_1 receptor agonist, and deoxycoformycin, we examined the effects on sleep in rats of another adenosine A_1 receptor stimulant, CHA, and the adenosine A_1 and A_2 receptor stimulant, NECA, in a dose-related manner (RADULOVACKI et al. 1983). This was of interest since adenosine is rapidly metabolized by adenosine deaminase to inosine (SKOLNICK et al. 1978) and the effects on sleep of metabolically stable adenosine analogs, L-PIA, CHA and NECA, were expected to be of equal or of longer duration than those of adenosine (see Fig. 2). The effects of L-PIA, CHA and NECA on sleep consisted of increased deep SWS2 from 6.6% to 45.7% in all doses used for CHA and NECA and 0.1 and 0.3 μmol/ kg L-PIA, respectively. All three agents reduced REM sleep at 0.9 μmol/kg dose, whereas L-PIA at this dose increased waking as well. The results showed that the effect on sleep was obtained with nanomolar doses of the drugs and that it diminished or disappeared when the drug dose increased (0.9 μmol/kg). The only exception was administration of 0.9 μmol/kg NECA, which increased SWS2 and

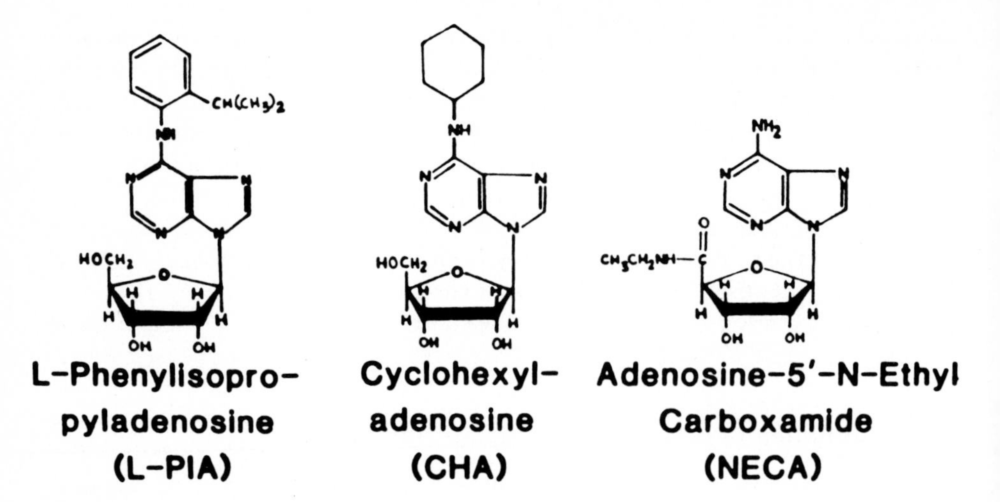

Fig. 2. Structural formulae of three adenosine receptor agonists: L-PIA, CHA and NECA

TS for the first 3h of polygraphic recording. This suggests that stimulation of adenosine A_1 receptors by NECA is also relevant to behavioral inhibition and sleep.

C. Adenosine A_1 Receptors in Sleep and Waking

I. Central A_1 Receptor Variation During the Circadian Period

In view of the possible involvement of adenosine in the modulation of sleep and demonstrated circadian variations in the receptors for other putative neurotransmitter substances (KAFKA et al. 1983), we examined the possible circadian variations in adenosine receptors by examining [^{3}H]L-PIA binding to whole rat brain membranes. Our data indicated that the number (B_{max}) – but not the dissociation constant (K_d)–of adenosine receptors in rat brain exhibited a statistically significant circadian rhythm, with a major peak at 2300h and a minor peak at 1100h, 3h after the onset of dark and light phases, respectively, of the diurnal cycle (VIRUS et al. 1984a). This biphasic circadian variation in the number of adenosine receptors, with the highest number during the rats' more active period and the next highest number when rats are less active, complicates the correlation of the B_{max} of adenosine receptors with the reported hypnogenic effects of adenosine and its congeners in rats. This apparent paradox may, however, be explained in part by measurements of the circadian variation of adenosine concentrations in the rat brain (CHAGOYA-DESANCHEZ et al. 1983). During the light period (when rats spend most time asleep), rat brain adenosine concentrations were high, as were B_{max} values, suggesting significant stimulation of adenosine receptors, while brain adenosine concentrations were low during the dark phase of the diurnal cycle (during which rats exhibit most activity), suggesting very low levels of receptor stimulation despite the presence of the greatest number of receptors. Therefore, these results were not inconsistent with the hypothesis that adenosine acts as a positive modulator of sleep in rats.

II. Central A_1 Receptors in Young and Old Rats

Aged rats are known to exhibit less sleep and altered patterns of sleep as compared to young rats (ZEPELIN et al. 1972; ROSENBERG et al. 1979). Because adenosine appeared to mediate sleep in rats, we investigated possible age-dependent changes in the binding of [^{3}H]L-PIA to membrane preparations of whole brains from normal male Sprague-Dawley rats 12 and 84 weeks of age (VIRUS et al. 1984b). Two populations of binding sites, probably corresponding to adenosine A_1 and A_2 receptors, are reduced in old rats. Because treatment with both A_1 and A_2 agonists enhanced SWS2 or TS in young rats (RADULOVACKI et al. 1984), the reduced affinities of both A_1 and A_2 receptors observed in old rats may explain the reduced amounts and altered patterns of sleep reported in old rats by Rechtschaffen's group (ZEPELIN et al. 1972; ROSENBERG et al. 1979).

III. Central A_1 Receptors in REM Sleep Deprivation

If adenosine or adenosine receptors have a role in sleep, then what would be the effect of REM sleep deprivation on both these parameters? We decided to investigate to what extent CNS adenosine concentration and adenosine A_1 receptors interact in the modulation of sleep and arousal using the "flowerpot" method of REM sleep deprivation (YANIK and RADULOVACKI 1987). This technique had been shown to selectively deprive rats of REM sleep and result in REM rebound following REM sleep deprivation (MENDELSON et al. 1974). If adenosine indeed has a role in sleep, then during REM sleep rebound, when the amount of REM sleep almost doubles the normal values, either adenosine concentration or adenosine receptors, or both, could be expected to be elevated. We assayed endogenous adenosine in microwave-fixed brain tissue of rats deprived of REM sleep for 48 h and found no significant changes in adenosine concentration when compared to controls. However, these data did not reflect the role that intrasynaptic adenosine concentration might play in the modulation of adenosine receptors in the CNS. We determined adenosine receptor binding in the brains of rats deprived of REM sleep for 48 and 96 h using [^{3}H]L-PIA. The data showed that adenosine A_1 receptors (B_{max}) were significantly increased in the cortex and corpus striatum and that the increase was sleep specific. We concluded that the effect of endogenous adenosine on sleep after REM sleep deprivation may not have been the result of its accumulation in specific brain areas, but could rather be a consequence of changes that occurred at the level of the adenosine A_1 receptor.

D. Methylxanthines and Sleep: Effects of Blockade of Adenosine A_1 and/or Adenosine A_2 Receptors

Methylxanthines are best known as CNS stimulants. Extensive studies have documented the ability of caffeine to decrease drowsiness and fatigue and to facilitate clarity of thought. Although lower doses of xanthines improve overall mental functioning, higher doses elicit tremor, insomnia and hyperalgesia. In addition, both caffeine and theophylline readily produce grand mal convulsions. As we have mentioned above, these stimulant actions of methlyxanthines are now believed to be due to the blockade of central adenosine receptors (Snyder et al. 1981). We discuss here the results of our studies dealing with the relationship between methylxanthines, adenosine receptors and sleep.

I. Caffeine and Adenosine Agonists: Behavioral Interaction

If behavioral stimulant effects of methylxanthines involve a blockade of central adenosine receptors (Snyder et al. 1981), then the effects on sleep of adenosine or an adenosine analog would be abolished by caffeine. We tested this possibility and determined the effect of L-PIA on sleep in the presence of caffeine. The rats were implanted with electrodes for EEG recording and the effects of L-PIA (0.115 mg/kg), caffeine (15 mg/kg) and L-PIA + caffeine were recorded for 6 h. The results showed that administration of L-PIA increased SWS2 by 54 min, suggesting that stimulation of adenosine receptors promoted deep sleep. However, administration of L-PIA failed to produce the same effect in the presence of caffeine, a finding consistent with the hypothesis that the CNS stimulant effect of caffeine is due to its ability to antagonize depressant effects of endogenous adenosine (Radulovacki et al. 1982).

II. Dose-Response Effects of Caffeine on Sleep

In humans, caffeine produces insomnia, a reduction in total sleep time and an increase in wakefulness, and causes a suppression of REM sleep rebound in REM sleep-deprived rats (Radulovacki et al. 1980). Additionally, caffeine produced a biphasic effect on locomotor behavior in mice, with low doses resulting in profound depression (Snyder and Sklar 1984).

We investigated the possibility of whether or not low doses of caffeine, in addition to depressing locomotor activity, would also affect sleep (Yanik et al. 1987) as has been shown with low doses of apomorphine (Mereu et al. 1979) or bromocriptine (Loew and Spiegel 1976). The rats were implanted with electrodes for polygraphic recording, and the effects of caffeine (0.125, 1.25, 12.5 and 25 mg/kg) were monitored by the EEG for 6 h. The results showed that the 12.5- and 25-mg/kg doses of caffeine increased wakefulness and decreased sleep. The 0.125- and 1.25-mg/kg doses of caffeine increased light sleep at the expense of deep sleep and did not affect total sleep time. This finding was of interest since adenosine or adenosine agonists had been shown to increase deep sleep at the expense of waking or light sleep without an increase in total sleep time. Thus, the obtained effects of low doses of caffeine on sleep suggested that caffeine administration antagonizes the effects of adenosine not only at the receptor level, but also at the behavioral level.

III. Caffeine Blocks Both Adenosine A_1 and A_2 Receptors: Is the Blockade of A_1 Receptors More Important for CNS Stimulation?

Although the methylxanthine, caffeine, is a potent CNS stimulant, the molecular and cellular basis for this CNS stimulation is only incompletely understood. Considerable experimental evidence suggests that the most important of the proposed mechanisms of methylxanthine action for the production of CNS stimulation is adenosine receptor blockade. Caffeine blocks both adenosine A_1 and A_2 receptors (DALY et al. 1983), but in vitro radioligand binding studies have shown that caffeine has a slightly greater affinity at A_1 adenosine receptors than at A_2 receptors as indicated by the A_2K_i/A_1K_i ratio of 1.65 (BRUNS et al. 1986a). Therefore, it was of interest for us to determine the role of A_1 and A_2 adenosine receptors in the stimulant effects of caffeine and to document these findings by the EEG (VIRUS et al. 1990). We implanted rats for EEG recording and polygraphically monitored the effects on wakefulness for 6h after administration of: caffeine (12.5 mg/kg i.p.); 8-cyclopentyl-theophylline (CPT), a selective A_1 receptor antagonist (BRUNS et al. 1987a) (10, 20 and 40 mg/kg i.p.); alloxazine, an A_2 receptor antagonist (BRUNS et al. 1987b) (12.5, 25 and 50 mg/kg i.p.); and CPT (20 mg/kg) + alloxazine (50 mg/kg). The results showed that both CPT and alloxazine injected individually produced sleep suppression qualitatively similar to that produced by caffeine, but of a lower magnitude. However, when 20 mg/kg CPT and 50 mg/kg alloxazine were injected together, their sleep suppressant effect was of the same magnitude as that of 12.5 mg/kg caffeine.

The results supported the hypothesized involvement of adenosine receptor blockade in the effects of caffeine on sleep in rats. They further suggested that A_1 adenosine receptor blockade may be more important than A_2 receptor blockade, because behavioral effects of the selective in vitro antagonist, CPT, were generally similar to those of nonselective in vitro adenosine receptor antagonists, caffeine and alloxazine.

E. Benzodiazepines and Central Adenosine Receptors

Considerable attention has been focused on the possibility that many actions of the benzodiazepines (see also Chap. 8, this volume) may be mediated by adenosine (PHILLIS and WU 1982). This was initially supported by the observation that diazepam potentiated the depressant actions of adenosine on the spontaneous firing of cerebral cortical neurons (PHILLIS 1979) and further experiments revealed that the depressant action of the benzodiazepine, flurazepam, on neuronal firing rates was blocked by the methylxanthine adenosine antagonist, theophylline (PHILLIS et al. 1979b). Because benzodiazepines do not act as agonists at the adenosine receptor (WILLIAMS and RISLEY 1981), the demonstration of benzodiazepine interactions with the purinergic system suggested that benzodiazepines might exert their actions by potentiating the effects of locally released adenosine. This possibility was strengthened by the observation that benzodiazepines inhibit adenosine uptake by rat brain cerebral cortical synaptosomes (PHILLIS et al. 1981). Indeed the potencies of benzodiazepines on the inhibition of adenosine uptake showed a good correlation with their clinical anxiolytic and anticonflict potencies (PHILLIS and WU 1982), suggesting that inhibition of adenosine uptake plays an important role in the central actions of the benzodiazepines. We were interested in the interaction of benzodiazepines and the adenosinergic system at the adenosine receptor level. Accordingly, we chronically administered diazepam and triazolam to rats and determined adenosine A_1 and A_2 receptors in specific brain areas.

I. Diazepam and Adenosine Receptors in Specific Brain Areas in Rats

Because some of the actions of benzodiazepines may involve adenosine uptake inhibition and increase adenosine concentrations at synapses, we tested the hypothesis that chronic administration of diazepam may decrease adenosine receptor binding. Following chronic administration for 10–20 days of diazepam (5 mg/kg per day, subcutaneous pellets), adenosine receptors in different rat brain areas were assessed by radioligand-binding studies using [^{3}H]R-PIA for A_1 and the [^{3}H]NECA + [^{3}H]R-PIA assay for A_2 receptors. The results showed that chronic administration of diazepam for 10 days decreased A_1 receptors in the striatum by 46% and A_1 receptors in the hippocampus by 13% ($P < 0.05$) (HAWKINS et al. 1988a). The results were also in accordance with the postulate of PHILLIS and WU (1982), who proposed that the sedative effect of benzodiazepines could be related to their inhibition of adenosine uptake. Thus, the data gave new evidence for a role of adenosine in the central actions of benzodiazepines by showing that diazepam modifies adenosine receptor binding.

II. Triazolam and Central Adenosine Receptors

Triazolam is a triazolobenzodiazepine derivative with a short plasma half-life which differs in several aspects from diazepam, a benzodiazepine with a long plasma half-life. O'REGAN and PHILLIS (1988) showed that triazolam, like diazepam (PHILLIS 1979), potentiated adenosine-evoked depression of cerebral cortical neuronal firing in the rat and that those effects were blocked by the adenosine antagonist, caffeine. Because our previous work (HAWKINS et al. 1988a) showed that chronic administration of diazepam decreased central adenosine receptors, we examined whether or not central adenosine receptors would be affected by prolonged administration of triazolam as well. The results showed that, following continuous subcutaneous administration of triazolam (0.5, 1 and 2 mg/day, pellets) for 10 days, radioligand binding to adenosine A_2 receptors in the rat striatum either decreased (31%, 2 mg/day) or increased (15%, 0.5 mg/day) (HAWKINS et al. 1988b). The data indicated that, although triazolam has different pharmacokinetic properties from diazepam, its administration affected central adenosine receptors in a similar manner as for diazepam.

III. Inhibition of Adenosine Transport and Central Adenosine Receptors

If the downregulation of central adenosine receptors by prolonged administration of diazepam (HAWKINS et al. 1988b) was accomplished by the inhibition of adenosine uptake, then, we assumed, prolonged administration of adenosine transport inhibitors to rats should produce similar effects. We tested this postulate by administering to rats for 14 days a potent adenosine transport inhibitor, soluflazine (VAN BELLE 1985) (1 μmol, 0.5 μ/h via Alzet miniosmotic pumps), and examined its effects on central adenosine receptors in specific brain areas (O'CONNOR et al. 1991). Soluflazine decreased adenosine A_1 radioligand binding in the hippocampus as measured by [^{3}H]R-PIA and lowered A_2 binding sites in the striatum, as estimated by the "NECA minus R-PIA" assay. The data showed that a specific adenosine transport inhibitor produced the same effect on adenosine receptors as benzodiazepines and further suggested a role for adenosine in CNS effects of benzodiazepines, as originally proposed by PHILLIS and WU (1982).

F. Hypothesis for Adenosine Hypnotic Action

What do we propose as the mechanism for hypnotic action of adenosine? Because there are two types of adenosine receptors, i.e., A_1 and A_2, whose role in sleep is not yet clear, we start from evidence that stimulation of adenosine A_1 receptors by adenosine or adenosine-related compounds leads to suppression of calcium influx into presynaptic nerve terminals (TEN BRUGGENCATE et al. 1977), possibly as a consequence of an inhibition of adenylate cyclase (VAN CALKER et al. 1979; LONDOS et al. 1980), and this inhibits the release of brain neurotransmitters (DOLPHIN et al. 1985; SAWYNOK and JHAMANDAS 1976; MICHAELIS et al. 1979; HEDQUIST and FREDHOLM 1976; HARMS et al. 1979). The end result of this process is the reduced amount of neurotransmitter at synapses in brain regions critical for sleep generation, which may lead to the induction of sleep. This should not be interpreted to mean that sleep is a passive process; sleep may be actively brought up by stimulation of adenosine A_1 receptors which would initiate the chain of events as described above.

G. Conclusions

Adenosine and adenosine analogs were shown to produce hypnotic effects in several animal species. Although their effects on sleep architecture differ from the hypnotic effect of benzodiazepines, our studies showed that benzodiazepines interact with central adenosine receptors. We also reported that deprivation of REM sleep upregulates adenosine receptors in the brains of rats in a manner similar to that of long-term administration of caffeine (MARANGOS et al. 1985). This suggests the existence of an "endocaffeine," whose normal role would be to block adenosine receptors during prolonged sleep deprivation–a mechanism that could be responsible for the increased number of adenosine receptors (RADULOVACKI 1987).

H. Summary

Adenosine is a nucleoside which consists of the purine base adenine linked to ribose. Substantial experimental evidence suggests that adenosine acts as a neurotransmitter or neuromodulator in the mammalian central nervous system. At least two distinct subtypes of adenosine receptors have been identified: A_1 receptors, whose stimulation inhibits adenylate cyclase, and A_2 receptors, whose stimulation activates adenylate cyclase in vitro. Adenosine A_1 receptors, which have high affinity (n*M*) for adenosine, are located in many areas of the brain, whereas adenosine A_2 receptors, which have low affinity (n*M*) for adenosine, are present only in striatum, nucleus accumbens and olfactory tubercle.

Adenosine and adenosine analogs have been shown to produce a hypnotic effect in several animal species. This hypnotic effect may occur as a result of stimulation of adenosine A_1 receptors, which leads to suppression of calcium influx into presynaptic nerve terminals, inhibiting the release of brain neurotransmitters. The end result of this process is the reduced amount of neurotransmitter at synapses in brain regions critical for sleep generation, which may lead to the induction of sleep.

Adenosine and adenosine analogs affect sleep by increasing slow-wave sleep and REM sleep. Although their effects on sleep architecture differ from the hypnotic effect of benzodiazepines, our studies showed that benzodiazepines interact with central adenosine receptors. We also reported that prolonged deprivation of REM sleep increases the number of adenosine receptors (B_{max}) in the brains of rats in a manner similar to that of long-term administration of caffeine. This suggests the existence of an "endocaffeine," whose normal role would be to block adenosine receptors during prolonged sleep deprivation–a mechanism that could be responsible for the increased number of adenosine receptors.

References

Agarwal RP, Spector T, Parks RE (1977) Tight-binding inhibitors. IV. Inhibition of adenosine deaminase by various inhibitors. Biochem Pharmacol 26:359–367

Berne RM, Rubio R, Curnish RR (1974) Release of adenosine from ischaemic brain: effect on cerebral vascular resistance and incorporation into cerebral adenine nucleotides. Circ Res 35:262–271

Bhattacharya IC, Goldstein L, Pfeiffer CC (1970) Influence of acute and chronic nicotine administration on EEG reactivity to drugs in rabbits. I. Nucleosides and nucleotides. Res Commun Chem Pathol Pharmacol 1:99–108

Bruns RF, Lu GH, Pugsley TA (1986) Characterization of the A_2 adenosine receptor labeled by [^{3}H]NECA in rat striatal membranes. J Pharmacol Exp Ther 29:331–346

Bruns RF, Fergus JH, Badger EW, Bristol JA, Santay LA, Hartman JD, Hays SJ, Huang CC (1987a) Binding of the A_1-selective adenosine antagonist 8 cyclopentyl-1,3-dipropylxanthine to rat brain membranes. Naunyn Schmiedebergs Arch Pharmacol 335:59–63

Bruns RF, Fergus JH, Badger EW, Bristol JA, Santay LA, Hays SJ (1987b) PD115-199: an antagonist ligand for adenosine A_2 receptors. Naunyn Schmiedebergs Arch Pharmacol 335:64–69

Buday PV, Carr CJ, Miya TS (1961) A pharmacologic study of some nucleosides and nucleotides. J Pharm Pharmacol 13:290–299

Chagoya DeSanchez V, Hernandez-Munoz R, Diaz-Munoz M, Suarez J, Vidrio S, Yanez L (1983) Circadian variations of adenosine and its physiological meaning in the energetic homeostasis of the cell and the sleep-wake cycle of the rat. 4th International Congress on Sleep Research, Washington, p 255

Daly JW, Butts-Lamb P, Padgett W (1983) Subclasses of adenosine receptors in the central nervous system: interactions with caffeine and related methylxanthines. Cell Mol Pharmacol 3:69–80

Dolphin AC, Prestwich SA, Forda SR (1985) Presynaptic modulation by adenosine analogues: relationship to adenylate cyclase. In: Stefanovich V, Rudolphi E, Schubert P (eds) Adenosine modulation of cell function. IRL, Oxford, p 107

Dunwiddie TV, Worth T (1982) Sedative and anti-convulsant effects of adenosine analogs in mouse and rat. J Pharmacol Exp Ther 220:70–76

Feldberg W, Sherwood SL (1954) Injections of drugs into the lateral ventricle of the cat. J Physiol (Lond) 123:148–167

Haas HH, Greene RW (1988) Endogenous adenosine inhibits hippocampal CAI neurons: further evidence from extra- and intra-cellular recordings. Naunyn Schmiedebergs Arch Pharmacol 337:561–565

Harms HH, Warden G, Mulder AH (1979) Effects of adenosine on depolarization-induced release of various radiolabeled neurotransmitters from slices of rat corpus striatum. Neuropharmacology 18:577–580

Haulica I, Ababei L, Branisteanu D, Topoliceanu F (1973) Preliminary data on the possible hypnogenic role of adenosine. J Neurochem 21:1019–1020

Hawkins M, Pravica M, Radulovacki M (1988a) Chronic administration of diazepam downregulates adenosine receptors in the rat brain. Pharmacol Physiol Behav 21:479–482

Hawkins M, Hajduk P, O'Connor S, Radulovacki M, Starz KE (1988b) Effects of prolonged administration of triazolam on adenosine A_1 and A_2 receptors in the brain of rats. Brain Res 505:141–144

Hedquist P, Fredholm BB (1976) Effects of adenosine on adrenergic neurotransmission: prejunctional inhibition and postjunctional enhancement. Naunyn Schmiedebergs Arch Pharmacol 293:217–224
Kafka MS, Wirz-Justice A, Naber D, Moore RY, Benedito MA (1983) Circadian rhythms in rat brain neurotransmitter receptors. Fed Proc 42:2796
Klabunde RE (1983) Dipyridamole inhibition of adenosine metabolism in human blood. Eur J Pharmacol 93:21–26
Loew DM, Spiegel R (1976) Polygraphic sleep studies in rats and humans. Their use in psychopharmacologic research. Arzneimittelforschung 26:1032–1035
Londos C, Cooper MF, Wolff J (1980) Subclasses of external adenosine receptors. Proc Natl Acad Sci USA 77:2551–2554
Marangos PJ, Boulenger JP, Patel J (1985) Effects of chronic caffeine on brain adenosine receptors: regional and ontogenic studies. Life Sci 34:899–907
Marley E, Nistico G (1972) Effects of catecholamines and adenosine derivatives given into the brain of fowls. Br J Pharmacol 46:619–636
Martinson EA, Johnson PA, Wells JN (1987) Potent adenosine receptor antagonists that are selective for the A_1 receptor subtype. Mol Pharmacol 31:247–252
Mathieu-Levy N (1968) Contribution à l'étude du mechanisme de la potentialisation du sommeil experimental par l'acide adenosine triphosphorique (ATP). Sur quelques actions d'ATP au niveu du système nerveaux central. Therapie 23: 1157–1173
Mendelson WB, Guthrie RD, Frederick G, Wyatt RJ (1974) The flowerpot technique of rapid eye movement (REM) sleep deprivation. Pharmacol Biochem Behav 2:553–556
Mereu GP, Scarnatti E, Paglietti E, Chessa P, Chicara G, Gessa GI (1979) Sleep induced by low doses of apomorphine in rats. Electroencephalogr Clin Neurophysiol 46:214–219
Michaelis ML, Michaelis EK, Myers SL (1979) Adenosine modulation of synaptosomal dopamine release. Life Sci 24:2083–2092
O'Connor SD, Hawkins M, Radulovacki M (1991) The effect of soluflazine on adenosine receptors in the rat brain. Neuropsychopharmacology 30:93–95
O'Regan MH, Phillis JW (1988) Potentiation of adenosine-evoked depression on rat cerebral cortical neurons by triazolam. Brain Res 445:376–379
Phillis JW (1979) Diazepam potentiation of purinergic depression of central neurons. Can J Physiol Pharmacol 57:432–435
Phillis JW, Wu PH (1981) The role of adenosine and its nucleotides in central synaptic transmission. Prog Neurobiol 16:187–193
Phillis JW, Wu PH (1982) Adenosine and benzodiazepine action. In: Usdin E, Skolnick P, Tallman JF, Greenblatt D, Paul SM (eds) Adenosine and benzodiazepine action. Macmillan, London, p 497
Phillis JW, Kostopoulos GK, Limacher JJ (1974) Depression of corticospinal cells by various purines and pyrimidines. Can J Physiol Pharmacol 52:1226–1230
Phillis JW, Edstrom JP, Kostopoulos GK, Kirkpatrick JR (1979a) Effects of adenosine and adenosine nucleotides on synaptic transmission in the cerebral cortex. Can J Physiol Pharmacol 57:1289–1312
Phillis JW, Edstrom JP, Ellis SW, Kirkpatrick JR (1979b) Theophylline antagonizes flurazepam-induced depression of cerebral cortical neurons. Can J Physiol Pharmacol 57:917–920
Phillis JW, Wu PH, Bender AS (1981) Inhibition of adenosine uptake into rat brain synaptosomes by the benzodiazepines. Gen Pharmacol 12:67–70
Radulovacki M (1987) Progress in sleep. N Engl J Med 316:1275
Radulovacki M, Walowitch P, Yanik G (1980) Caffeine produces REM sleep rebound in rats. Brain Res 201:497–500
Radulovacki M, Miletich RS, Green RD (1982) N^6(L-Phenylisopropyl) adenosine (L-PIA) increases slow wave sleep (S2) and decreases wakefulness in rats. Brain Res 246:178–180

Radulovacki M, Virus RM, Djuricic-Nedelson M, Green RD (1983) Hypnotic effects of deoxycoformycin in rats. Brain Res 271:392–395
Radulovacki M, Virus RM, Djuricic-Nedelson M, Green RD (1984) Adenosine analogs and sleep in rats. J Pharmacol Exp Ther 228:268–274
Radulovacki M, Virus RM, Rapoza D, Crane R (1985) A comparison of the dose response effects of pyrimidine ribonucleosides and adenosine on sleep in rats. Psychopharmcology (Berl) 87:136–140
Reddington M, Erfurth A, Lee KS (1986) Heterogeneity of binding sites of N-ethylcarboxamido-[^{3}H] adenosine in rat brain: effects of N-ethylmaleimide. Brain Res 399:232–239
Rosenberg RS, Zepelin H, Rechtschaffen A (1979) Sleep in young and old rats. J Gerontol 34:525–532
Sawynok J, Jhamandas KH (1976) Inhibition of acetylcholine release from cholinergic nerves by adenosine, adenosine nucleotides, and morphine: antagonism by theophylline. J Pharmacol Exp Ther 197:379–390
Skolnick P, Nimilkitpaisan Y, Stalvey I, Daley JW (1978) Inhibition of brain adenosine deaminase by 2′-deoxycoformycin and erythro-9-(2-hydroxy-3-nonyl) adenine. J Neurochem 30:1579–1583
Snyder SH, Sklar P (1984) Psychiatric progress. Behavioral and molecular actions of caffeine: focus on adenosine. J Psychiatry Res 18:91–106
Snyder SH, Katims JJ, Annau Z, Bruns RF, Daly JW (1981) Adenosine receptors and behavioral actions of methylxanthines. Proc Natl Acad Sci USA 78:3260–3264
Sterman MB, Clemente CD (1962) Forebrain inhibitory mechanisms: critical synchronization induced by basal forebrain stimulation. Exp Neurol 6:91–102
Stone TW (1981) Physiological roles of adenosine and adenosine 5′-triphosphate in the nervous system. Neuroscience 6:523–552
Stone TW, Newby AC, Lloyd HGA (1990) Adenosine release. In: Williams M (ed) The adenosine receptors. Humana, Clifton, p 173
Ten Bruggencate D, Steinberg R, Stockle H, Nicholson C (1977) Modulation of extracellular CA^{++} and K^+-levels in the mammalian cerebellar cortex. In: Ryall RW, Kelly JS (eds) Iontophoresis and transmitter mechanisms in the mammalian central nervous system. Elsevier/North-Holland, Amsterdam, p 442
Ticho SR, Radulovacki M (1991) Role of adenosine in sleep and temperature regulation in the preoptic area of rats. Pharmacol Physiol Behav 40:33–40
Ukena D, Shamin MT, Padgett W, Daly JW (1986) Analogs of caffeine: antagonists with selectivity for A_2 adenosine receptors. Life Sci 39:743–750
Van Belle H (1985) Myocardial purines during ischemia, reperfusion and pharmacological protection. Mol Physiol 8:615–630
Van Calker D, Muller M, Hambrecht V (1979) Adenosine regulates, via two different types of receptors, the accumulation of cyclic AMP in cultured brain cells. J Neurochem 33:999–1005
Virus RM, Baglajewski T, Radulovacki M (1984a) Circadian variation of [^{3}H]N^6-(L-phenylisopropyl) adenosine binding in rat brain. Neurosci Lett 46:219–222
Virus RM, Baglajewski T, Radulovacki M (1984b) [^{3}H]N^6-(L-Phenylisopropyl) adenosine binding in brains from young and old rats. Neurobiol Aging 5:61–62
Virus RM, Ticho BS, Pilditch M, Radulovacki M (1990) A comparison of the effects of caffeine, 8-cyclopentyl theophylline, and alloxazine on sleep in rats; possible roles of central nervous system adenosine receptors. Neuropsychopharmacology 3:243–249
Williams M, Risley EA (1981) Interaction of putative anxiolyts agentic agents with central adenosine receptors. Can J Physiol Pharmacol 59:897–900
Williams M, Francis J, Ghai G, Braunwalder A, Psychoyos S, Stone GA, Cash WD (1987) Biochemical characterization of the triazoloquinazoline, CGS15843A, a novel non-xanthine adenosine antagonist. J Pharmacol Exp Ther 241:415–420

Yanik G, Radulovacki M (1987) REM sleep deprivation upregulates adenosine A_1 receptors. Brain Res 402:362–364
Yanik G, Glaum S, Radulovacki M (1987) The dose response effects of caffeine on sleep in rats. Brain Res 403:177–180
Yeung SH, Green RD (1984) [^{3}H]5′-N-ethyl-carboxamide adenosine binds to both R_a and R_i adenosine receptors in rat striatum. Naunyn Schmiedebergs Arch Pharmacol 325:218–225
Zepelin H, Whitehead WE, Rechtschaffen A (1972) Aging and sleep in the albino rat. Behav Biol 7:65–74

CHAPTER 12

Methodological Issues in Pharmacological Studies of Sleep

E.O. BIXLER, A.N. VGONTZAS, and A. KALES

A. Introduction

There are four phases of investigation for a new hypnotic drug. The overall goal of these four phases is to identify an effective and safe dose of the drug. The specific objectives vary for each phase of the evaluation of a new hypnotic drug. The initial phases of evaluation are quite conservative in terms of risk exposure to the patient. As more confidence is gained in the safety of the drug with each successive phase of investigation, the drug is assessed in broader, more heterogeneous study samples ultimately including patients with medical illness and the elderly. Measures of both safety and efficacy as well as the design of a particular study are driven necessarily by the objectives of the specific phase of evaluation for a drug.

In this chapter we first describe the objectives for each of the four phases of investigation. Next, methods of measurement or assessment are outlined for pharmacokinetics, efficacy and safety. Finally, design and analysis issues are described for each of the four phases of investigation.

B. Objectives of the Phases of Investigation of a New Hypnotic

This review of the four phases of investigation for a new hypnotic drug is necessarily oversimplified for the purposes of this chapter. It is based on the Guidelines for the Clinical Evaluation of Hypnotic Drugs developed by the U.S. Food and Drug Administration (FDA 1977). The objectives of the first phase of evaluation of a new hypnotic focus primarily on safety and secondarily on efficacy. Specifically, the objectives are to identify toxicity levels and a range of doses which may be effective. In addition, the pharmacokinetic and pharmacodynamic characteristics of the new drug are determined because these are relevant to the evaluation of both safety and efficacy.

The objectives of the second phase of evaluation are to define more specifically the effective doses and broaden the assessment of safety with longer periods of drug administration and use of the drug beyond normal controls (FDA 1977). Specifically studied in this phase are initial and short-term efficacy including identifying the optimal dose. The safety objectives include assessment of effects on daytime performance and identification of frequent and severe adverse drug reactions (ADRs) as well as evaluating withdrawal effects following drug discontinuation after short-term administration.

The objectives of the third phase are focused primarily on the study of long-term efficacy and safety (FDA 1977). Specifically, these objectives include the evaluation of long-term efficacy including the rapidity of development of tolerance. These objectives also include the optimization of the appropriate clinical dose and identification of less frequent, but possibly severe ADRs associated with drug administration as compared to the more frequent and severe ADRs identified in the second phase.

The objective of the fourth phase of evalution of a hypnotic drug is to continue to monitor the drug after it has been approved and marketed (FDA 1977). This is done through post-marketing surveillance studies which attempt to identify severe but rare ADRs and/or ADRs previously not identified with use of the drug. In addition, the appropriateness of the recommended clinical dose is continually monitored. Further, in this phase, the efficacy and toxicity of the recommended clinical dosage is monitored for appropriateness in patients usually excluded from pre-marketing evaluation such as children, pregnant women and the elderly.

C. Methods of Measurement

I. Pharmacokinetic

Data derived from pharmacokinetic studies are useful in terms of understanding the unique properties and therapeutic actions of hypnotic drugs. This is in spite of the fact that a close relationship between clinical sedative and/or anxiolytic effects and plasma concentration of benzodiazepines has not been demonstrated (Breimer 1977; Greenblatt et al. 1982; Greenblatt and Shader 1985). Many findings observed during clinical and sleep laboratory studies of various benzodiazepine hypnotics can be understood from the drugs' pharmacokinetic properties. For example: rapid rates of absorption and distribution are associated with a shorter onset of sleep (Greenblatt et al. 1982); slow rates of elimination are associated with a greater degree of morning sedation (Kales et al. 1976a); and rapid rates of elimination are associated with hyperexcitability states such as rebound insomnia following abrupt withdrawal (Kales et al. 1978) while the most rapidly eliminated drugs are also associated with daytime anxiety and early morning insomnia during drug use (Morgan and Oswald 1982; Kales et al. 1983a; ADAM and Oswald 1989).

II. Efficacy

1. Objective Measures

a) Sleep Laboratory Studies

In assessing hypnotic drugs, sleep laboratory studies are especially valuable in determining a range of efficacy as well as defining an optimal clinical dose. Sleep laboratory measurements are unique in that they provide valuable objective and precise information on both initial effectiveness and effectiveness with continued use as well as potential withdrawal effects of a hypnotic (Kales et al. 1975, 1979b; Kales and Bixler 1975; Soldatos and Kales 1979; Bixler and Kales 1985). The objective measurement of sleep is accomplished through continuous, second-by-second electrophysiologic monitoring throughout the night. Because of a standardized system for scoring sleep, interlaboratory reliability is high (Rechtschaffen and Kales 1968). The sleep laboratory provides a sound-attenuated, temperature-controlled environment that is free of noise or interruption. The environment also insures strict compliance for administration of the drug or placebo as well as completing various questionnaires pre- and post-sleep. It also provides a regular time for going to bed and arising and thus the total time spent in bed is precisely quantified. In addition, the ascertainment of side effects, including paradoxical and unexpected ones, is enhanced because of the control over many external, potentially confounding variables and because of the very close daily monitoring by sleep laboratory staff.

Prior to the advent of the sleep laboratory, many critical properties and effects of hypnotic drugs went unrecognized in clinical trials conducted during this period. Sleep laboratory studies with a sample size of only six to eight subjects allowed for the identification and precise determination of the following with nonbenzodiazepine hypnotics (glutethimide, methyprylon and barbiturates) used prior to the benzodiazepines: (a) rapid loss of initial efficacy (tolerance) within 1–2 weeks of administration of most nonbenzodiazepine hypnotics (Kales et al. 1970, 1977); (b) degrees of nocturnal wakefulness associated with chronic use of nonbenzodiazepine hypnotics that are similar to those noted in insomniacs not taking medication (Kales et al. 1974); and (c) disturbed sleep including more frequent and intense dreams and occasionally nightmares following abrupt withdrawal of chronically used nonbenzodiazepine hypnotics (Kales et al. 1974). Similarly, sleep laboratory studies provided a number of important and original findings on benzodiazepine drugs including: (a) rapid development of tolerance to the initial efficacy of rapidly eliminated benzodiazepines with continued use over only short- and intermediate-term periods (Kales et al. 1983a); (b) carryover effectiveness into the withdrawal period after discontinuation of slowly eliminated benzodiazepines (Kales et al. 1976a); and (c) rebound insomnia following abrupt withdrawal of rapidly eliminated benzodiazepines (Kales et al. 1978, 1979a, 1983b).

b) Non-Polygraphic Devices

Several methods of measurement have been developed which objectively monitor multiple sleep and breathing parameters and do not rely on polygraphic quantification of sleep (ANCOLI-ISRAEL et al. 1981; STOOHS and GUILLEMINAULT 1990, 1992; REDLINE et al. 1991; MAN 1994; WHITE et al. 1994). Most of these devices monitor body movement as a means of differentiating sleep from wakefulness. In addition, a system for discriminating REM from non-REM sleep (WHITE et al. 1994) has evolved from the recent emphasis on developing screening devices for the detection of sleep apnea both in Europe and in the United States. Thus, there is a potential for using non-polygraphic measures of sleep and wakefulness in large-scale drug evaluation studies that needs further investigation.

2. Subjective Estimates

Patients' subjective estimates have been the traditional method of measurement used in clinical trials of hypnotic drugs (KALES et al. 1979b; BIXLER and KALES 1985). Until the 1960s, when investigators began to use the sleep laboratory for the assessment of the effectiveness and withdrawal effects of hypnotic drugs, patients' subjective estimates were the only means of evaluating their safety and efficacy (KALES and KALES 1984). Ever since, however, both methods have been used to establish a drug's profile of efficacy and overall side effects. Experience accumulated over many years indicates that the most thorough and clinically relevant approach to hypnotic drug evaluation is one that balances the strengths and weaknesses of both clinical trials and sleep laboratory evaluations (KALES et al. 1979b).

Patients' subjective estimates have a large degree of variability even with short-term use while with long-term use this variability is even greater because subjective estimates of baseline sleep difficulty may be influenced by recent as well as past experiences. Thus, in order to obtain adequate power in these studies large sample sizes need to be utilized. Such a large sample size is the primary advantage of clinical trials in that it potentially allows for a thorough assessment of the severity and frequency of side effects. In addition, clinical trials can be conducted with special target groups (KALES et al. 1979b). This is important because a hypnotic drug may be used in patients with virtually any medical condition. Because clinical trials based on patients' subjective estimates can be conducted in almost any setting, they permit evaluation of the drug's efficacy and side effects in such varied populations as geriatric patients or individuals suffering from chronic cardiovascular or respiratory conditions (KALES et al. 1979b).

III. Safety

1. Objective Measures

a) Clinical Laboratory Tests

Safety and/or toxicity are primarily assessed by repeated physical examinations including monitoring of vital signs and laboratory tests to assess the hematopoietic, hepatic, renal and cardiovascular systems.

b) Studies of Effects on Daytime Performance

An important safety question is how a hypnotic drug affects daytime performance. Impairment of daytime performance is associated with some degree of risk particularly when subjects drive or operate machinery. The primary consideration for this risk is whether there is a drug carryover or hangover effect on subsequent daytime functioning (KALES et al. 1971, 1976a; BIXLER et al. 1975). This area of research utilizes various objective measurements that can be made at multiple time points for a given subject to assess the time course of any changes in daytime functioning (BIXLER et al. 1975; JOHNSON and CHERNIK 1982). In addition, many of these measurements can be applied to relatively large samples of subjects, thus increasing the statistical power of a study.

2. Patient Reports

The majority of safety data in the evaluation of a new hypnotic are obtained through patient reports. These complaints are made to study personnel, either directly or by means of questionnaires, or to the patient's physician during phase IV monitoring.

D. Design and Analysis Considerations

Design considerations vary for each phase of the evaluation of a new hypnotic drug. In general, as experience is accumulated in terms of a drug's safety and efficacy by progressing through the phases of evaluation, the drug's use is broadened to more heterogeneous groups of patients including special target groups.

I. Phase I Studies

The major objectives of phase I evaluation of a new hypnotic are evaluating safety or toxicity of the drug at various doses (FDA 1977). Such an assessment requires the design to minimize the risk placed on subjects. This is done primarily by using young, healthy, noninsomniac subjects. In addition, administration of the drug is done during the day when subjects can be monitored conveniently and closely. Also, only single drug exposures are used initially in an open label fashion.

The major advantage of early phase I studies is that subjects are monitored closely and frequently, thus minimizing risk to the subject. This monitoring allows for assessment of sedation and side effects or toxicity when drug levels are at peak concentration. However, such studies do not provide a complete picture of hypnotic efficacy and safety under the usual clinical conditions, i.e., drug administration at bedtime (SOLDATOS and KALES 1979; BIXLER and KALES 1985).

Pharmacokinetic studies provide useful information about the onset and duration of action of a drug. Although these studies are limited in their scope and often do not mimic the usual conditions of clinical use, their findings are most helpful in understanding general differences among various hypnotics (SOLDATOS and KALES 1979; BIXLER and KALES 1985). Pharmacokinetic studies are accomplished by measuring blood levels of the parent drug and its active metabolites in a very precise and objective manner (BREIMER 1977; GREENBLATT et al. 1982; GREENBLATT and SHADER 1985). Most protocols obtain data from a single subject across several time points in order to describe changes in blood levels of the drug over time. Data are typically obtained from only a small number of healthy subjects. A disadvantage of this type of study is that the techniques are invasive and, therefore, blood levels of the hypnotic and its metabolites cannot be continuously monitored without potentially interfering with other types of assessment. In addition, these studies usually are carried out during the day and involve only a limited number of drug administrations (most often only one).

II. Phase II Studies

The major objective of phase II is to define more specifically the efficacy of a drug (FDA 1977). Because it is still early in the development of the drug, minimizing risk to the patient is still a primary concern. Thus, small samples of subjects are used who are carefully screened to eliminate concomitant use of medications and various medical conditions. In this phase of evaluation, insomniac patients are used instead of young, healthy, noninsomniac subjects. In addition, the drug is administered at night before sleep, drug exposure is for only a short period (up to a week) and is usually done in a double-blind fashion.

1. Design of Phase II Studies

Sleep laboratory studies have had a major role in phase II evaluations of hypnotic drugs. There are two basic designs that have been employed in these sleep laboratory studies: the placebo-drug-placebo (PDP) design within a single group or within parallel groups; and the crossover design. Other types of studies (e.g., performance studies and clinical trials relying on subjective estimates) have employed similar designs. Each of these designs is reviewed in the following sections and contrasted in terms of their advantages and disadvantages.

a) Placebo-Drug-Placebo Design in a Single Group (Within Comparisons)

This design evaluates a single drug and dose, employing three successive study conditions for each subject: placebo baseline, drug administration, and placebo withdrawal. Because of the amount of night-to-night variability in terms of sleep efficiency each condition of a PDP design should include at least 3 nights in order to establish stable values. All PDP designs used in the sleep laboratory include a 1st night for adaptation to the laboratory environment (RECHTSCHAFFEN and VERDONE 1964; AGNEW et al. 1966) followed by nights 2–4 for placebo-baseline measurements (Table 1). The length, however, of the drug administration period may vary from 3, 7, 14, 28, or 35 nights. Depending on the length of drug administration, initial, short-, intermediate-, and even long-term effectiveness and overall effects of the drug can be assessed. The placebo-withdrawal period allows for determining if sleep disturbances occur or effectiveness persists following the discontinuation of the drug. At times, as is the case with studies of slowly eliminated hypnotics, the length of the withdrawal period may extend up to 15 consecutive nights. Thus, by using the PDP design in the sleep laboratory, a complete profile of a drug's efficacy and overall effects can be quantified throughout its administration and following its withdrawal (SOLDATOS and KALES 1979; BIXLER and KALES 1985).

Table 1. Placebo-drug-placebo design in a single group (within comparisons)

Condition	PDP protocols of varying duration					
	Placebo	Drug	11-night protocol	14-night protocol	22-night protocol	47-night protocol
Adaptation	X		1	1	1	1
Baseline	X		2–4	2–4	2–4	2–4
Short-term effects		X	5–7	5–7	5–7	5–7
Continued effects		X		9–11		
Readaptation		X			15	15
Intermediate-term effects		X			16–18	16–18
Readaptation		X				29
Long-term effects	X					30–32
Initial withdrawal	X		9–11	12–14	19–22	33–35
Readaptation						46
Continued withdrawal	X					45–47

The numbers listed under each of the four protocol designs commonly used in the sleep laboratory indicate nights spent in the sleep laboratory. Exceptions are the 11- and 14-night protocols where night 8 is spent in the laboratory but its values are not included in the comparisons of the 3-night conditions.

Because this design requires that all subjects be recorded across the three conditions (placebo-baseline, drug, and placebo-withdrawal), each subject becomes his own control rather than being compared to other subjects. Consequently, a primary advantage of this design is that the precision of the experiment is increased by eliminating between-subject variability (LINDQUIST 1956; HAYS 1985; WINER et al. 1992). One can argue that the PDP design violates the statistical assumption that drug and placebo conditions are independent because of the carryover of pharmacologic effects. However, in sleep laboratory studies this design does not weaken the precision of the experiment but rather results in valuable clinical information. Specifically, the observed changes in sleep disturbances during the transition from the condition of drug administration to the condition of placebo withdrawal will define whether values for sleep and wakefulness gradually return to baseline, as is the case with slowly eliminated benzodiazepine hypnotics, or there is an abrupt worsening of sleep to levels significantly greater than baseline, as is the case with most rapidly eliminated benzodiazepine hypnotics (see KALES et al. 1983b for a review).

The primary disadvantage of the PDP design is that measurements are repeated for the same subject on a daily basis. A statistical analysis requiring repeated measures can potentially cause a false-positive (type I) error if the several successive conditions are unequally correlated with one another (LANA and LUBIN 1963; HAYS 1965; WINER et al. 1992). Use of techniques to correct for this potential problem tends to reduce the power of the statistical analysis and, thus, increase the likelihood of a false-negative (type II) error (BOX 1954; GEISSER and GREENHOUSE 1958). However, at an early stage of the investigation of a drug (late phase I or early phase II) when sleep laboratory studies are usually employed, one is more willing to commit a false-positive than a false-negative error (LINDQUIST 1956). Generally, more is to be gained scientifically if an investigator incorrectly assumes that a drug is effective and continues investigating it than if the investigation of a potentially effective drug is prematurely terminated. In retrospect, the fact that neither type of error is common is demonstrated by the high degree of replicability of sleep laboratory findings across many sleep laboratories (KALES et al. 1975, 1979b; KALES and BIXLER 1975; SOLDATOS and KALES 1979; BIXLER and KALES 1985).

b) Placebo-Drug-Placebo Design With Parallel Groups (Between and Within Comparisons)

In this design several groups of subjects are evaluated in parallel during a single experiment. In sleep laboratory studies using this design in phase II, each group is usually studied in a standard placebo-drug-placebo protocol, but during the drug period each group is administered either a different dose of the same hypnotic, a different hypnotic, or a placebo. In addition to the advantages described for the PDP design in a single group, this design allows for controlling variables that are related to the time frame of the experiment itself (LINDQUIST 1956; WINER et al. 1992). Because all of the groups are recorded during a single experiment, each group is exposed to similar laboratory conditions.

A potential disadvantage of the PDP design with parallel groups is that due to the small sample size, simply randomizing patients will not always balance the groups in terms of baseline severity. However, this is actually not a significant problem because the primary analysis for this type of study should be completed by statistical comparisons within each group. Comparisons made between groups should be made by comparing statistical findings in the within group analyses or by evaluating change from baseline between the groups.

It can be argued that a parallel placebo group is necessary in order to control adequately a study with a PDP design (LADER and LAWSON 1987; ROEHRS et al. 1990). However, in sleep laboratory studies, we believe that this additional costly control group is not necessary because the objective measurements and rigorous control of multiple variables in the sleep laboratory environment result in stable values for sleep and wakefulness. We have reported on a sleep laboratory study which included three parallel groups, each evaluated across four conditions (KALES et al. 1971). One of the groups was studied for a period of 4 nights when nothing was administered, followed by two, successive 5-night periods during which placebo was administered, and ending with a final period of 4 nights when again nothing was administered. Across these four conditions there was little difference observed in terms of the values of wakefulness measures, demonstrating the general stability of such sleep laboratory data. In contrast, clinical trials rely on rather imprecise subjective estimates and provide relatively little control over potentially disturbing variables, thus resulting in much greater night-to-night variability as well as a potentially changing frame of reference over time in values for sleep and wakefulness. A final argument against the necessity of a parallel placebo group in sleep laboratory studies using the PDP design is the stability of the original baseline values; in this regard the sleep disturbance observed during the placebo-withdrawal period usually returns to approximately baseline levels at some point depending primarily on the pharmacokinetics of the drug.

c) Crossover Design

The primary advantage of the crossover design over the PDP design is that because each condition is presented to each subject, a smaller number of subjects is required to evaluate the same number of conditions (LINDQUIST 1956). This design in its simplest form is a balanced two-way classification method (COCHRAN and COX 1957). When used in the sleep laboratory it is often employed to evaluate several drug conditions. Thus, all subjects complete all of the drug conditions as well as a placebo-control condition. In order to control for the order of administration of each drug or placebo condition, subjects are assigned the conditions in a counterbalanced order utilizing a latin square design (COCHRAN and COX 1957). This ordering insures that each condition is represented an equal number of times in each ordinal position. For the crossover design to be balanced, the number of subjects employed must be equal to or a multiple of the number of conditions. This requirement is at times ignored (ROTH et al. 1980; NICHOLSON et al. 1982; NICHOLSON and STONE 1983; ROEHRS et al. 1983), introducing potentially biased order effects among conditions.

A balanced design can be achieved by using a single latin square if the number of treatments is even (WILLIAMS 1949). When the number is odd, however, a sample size that is a multiple of two times the number of conditions is required. Further, only first-degree order effects are controlled in the typical latin square type of counterbalancing (COCHRAN and COX 1957). For example, if drug A is one of five drug conditions, counterbalancing will require that drug A is administered first through fifth in terms of order an equal number of times. Because only first-degree order effects are controlled, drug A will be administered after another drug which is always the same, with one exception, when drug A is first in the series. Therefore, a systematic error can be introduced due to a second-degree order effect, i.e., an effect due to the condition preceding the last experimental condition. This is of special concern in any pharmacological evaluation where carryover may occur. However, there do appear to be special designs that control for second-degree order effects, in addition to the immediately preceding treatment (COCHRAN and COX 1957). Such a design requires a completely orthogonal set of latin squares (for n treatments, $n - 1$ squares are used). Thus, if four treatments are employed, three latin squares must be included, requiring multiples of 12 subjects, eliminating the crossover design's primary advantage of requiring a small sample size.

The crossover design, however, has a number of disadvantages. Most important, it assumes that no effects carry over from condition to condition (WILLIAMS 1949; LINDQUIST 1956; HAYS 1965; WINER et al. 1992), which is seldom possible, especially in a pharmacologic study. To counteract carryover problems, investigators attempt at times to strike a balance between maintaining an adequate washout period to control completely for any drug carryover, but not so long that the baseline sleep patterns of the subjects change, making a new baseline necessary. However, it is difficult to know with certainty when all effects of a drug have been eliminated. For example, OSWALD and PRIEST (1965) have shown that the effects of hypnotic administration can be detected for several weeks following withdrawal. Another method of counteracting carryover from one condition to the next is to identify the amount of carryover statistically and extract this effect from the treatment mean values (COCHRAN and COX 1957; NICHOLSON and STONE 1983). However, BROWN (1980) has shown that adjusting for residual effects using this technique requires a larger sample than that needed for a parallel comparison. Thus, again the major advantage of the crossover design (small sample size) is lost.

Finally, the crossover design may be limited from a practical standpoint. The duration of a given condition is usually only 1 or 2 nights (NICHOLSON et al. 1982; NICHOLSON and STONE 1983), which precludes the evaluation of continued and long-term effects as well as withdrawal effects.

2. General Issues: Sleep Laboratory Studies

The advantages of sleep laboratory studies, which are described previously in this chapter (see CII.1.a), far outweigh any of their limitations (KALES et al. 1975; SOLDATOS and KALES 1979; BIXLER and KALES 1985). Moreover, some of the apparent limitations are actually strengths when the total context is taken into proper perspective. For example, because of space constraints, only a small number of subjects can be studied together at the same time in the sleep laboratory. While at first this may appear to constitute a limitation, a very small number of subjects studied in the sleep laboratory can provide statistically significant data that have high clinical relevance. This is because of the highly objective, precise and sensitive measurements made in the sleep laboratory. In our own studies, a high degree of statistical power has been demonstrated for evaluations of hypnotic drugs in the sleep laboratory. For these analyses, a 40% change in wakefulness was considered to be the minimum change to be clinically significant. We found that when using a small number of subjects (six to eight) and the clinically significant change of 40% or more, we were able to achieve a median statistical power which was more than 80%.

Another apparent limitaton of the sleep laboratory is difficulty in evaluating special target groups such as medical and surgical patients. As a consequence, these studies at first do not appear to be well suited to provide an adequate evaluation of the frequency and severity of drug-related adverse events (SOLDATOS and KALES 1979; BIXLER and KALES 1985). However, because of the much more frequent contact between the patient and the investigator each evening and morning of the study, there is a greater opportunity and likelihood, than in most clinical trials, for patients to report ADRs both on the written forms as well as spontaneously.

Finally, the setting of the sleep laboratory itself and the encumbrance of the recording electrodes provide an unnatural environment for sleep. It is well established that there is an adaptation effect to the sleep laboratory (RECHTSCHAFFEN and VERDONE 1964; AGNEW et al. 1966) as well as a readaptation effect (SCHARF et al. 1975), both effects consisting of increased values of wakefulness measures. However, this limitation is readily compensated for with an adequate adaptation period (Table 1).

Protocols with ad-lib or fixed recording times clearly evaluate opposite conditions (BIXLER et al. 1984). In the ad-lib protocol, the emphasis is on the evaluation of sleep per se, whereas in the fixed protocol the emphasis is on the evaluation of wakefulness during the sleep period. Because in the ad-lib design the amount of wake time after sleep onset is confounded with the varying amount of total laboratory time from patient to patient and from night to night, it is impossible to compare values for total wake time or wake time after sleep onset obtained from fixed and ad-lib protocols. With an ad-lib design, it is not possible to control for factors which may affect the amount of time a subject spends in bed. For example, time constraints involving work or social situations can influence the length of subjects' recording times. Also, a variation in total laboratory time may be due to the increased wakefulness caused by the rapid elimination of a benzodiazepine hypnotic with a relatively short elimination half-life, i.e., the subject may shorten the time spent in bed. Such changes induced by certain benzodiazepines disrupt sleep not only following drug withdrawal (rebound insomnia (see KALES et al. 1978, 1979a, 1983b for reviews) but even during actual drug administration (early morning insomnia) (KALES et al. 1983a). With a fixed protocol, the effects of such drug-induced changes are entirely measurable and accounted for, providing findings which are more clinically meaningful.

Another issue is that there has been at times a failure to concentrate on the clinically relevant aspects of testing for hypnotic efficacy and an undue focus on drugs' effects on sleep stages. Although this approach is interesting and may be promising in terms of basic research, at the present time no definite clinical significance has been ascribed to sleep stage alterations (KALES and KALES 1970; LASAGNA 1972). Also, see Chap. 13, this volume, for a discussion of this issue.

3. General Issues: Clinical Trials

Patients' estimates of the effects of hypnotic drugs are quite variable because they are subjective. They are based on subjects' perceptions of the quantitative and qualitative aspects of their night's sleep as reported in questionnaires or interviews. Although a positive relationship has been found between a clinical complaint of sleep difficulty and the presence of objective sleep disturbance in the sleep laboratory, it is also well established that insomniacs tend to overestimate the degree of their sleep difficulty (MONROE 1967; KALES and BIXLER 1975; CARSKADON et al. 1976).

During the first few nights of drug administration, subjective estimates are relatively reliable because they are based on comparisons to immediately preceding placebo or no-drug, baseline conditions. As drug administration is extended, however, subjects may compare their sleep to immediately preceding drug nights rather than to the original baseline (HELSON 1964). This continually changing frame of reference may affect the reliability and interpretation of subjective estimates in clinical trials. Also, because most clinical trials of hypnotic drugs do not include an evaluation of the drug-withdrawal period, carryover effectiveness of a drug or worsening of sleep following a drug's withdrawal cannot be assessed (KALES and KALES 1984).

4. General Issues: Studies of Effects on Daytime Performance

The major difficulties of evaluating effects of hypnotic drugs on daytime performance tasks are due primarily to the effects of learning (BIXLER et al. 1975). In most evaluations of the effects of hypnotic drugs on daytime performance, learning per se is not considered. Instead, the assumption is made that no learning effects are present, but this is seldom the case. In addition, most of the tasks commonly employed in this type of research have little relation to "real world" tasks. Thus, when a change in performance is observed, it is difficult to relate these findings in a meaningful manner to a patient population.

Another issue in the measurement of performance is the general acceptance that an individual subject seldom, if ever, performs at his maximum level (HELSON 1964). One manner of viewing this phenomenon has been proposed by Helson in his "hypothesis of par" or "tolerance." This hypothesis states that in most tasks, individuals set for themselves a level of performance which is usually below their maximum capability.

The design of any study assessing effects on daytime performance must take into account that the level of performance of any individual fluctuates throughout the day (BIXLER et al. 1975; FOLKARD et al. 1976; HARRIS 1977). Kleitman in 1938 reviewed a set of studies that demonstrated that performance efficiency and body temperature paralleled each other closely (KELITMAN 1963). Further, this need to evaluate multiple time points throughout the day necessarily restricts the potential pool of subjects to those available during the day (also see Chap. 3, this volume, for a review of circadian effects).

In 1971, physicians were first alerted to the possible disadvantages of the carryover effectiveness of slowly eliminated hypnotics (KALES et al. 1971). Further, a subsequent review of the available literature evaluating the effects of hypnotics on performance showed a general paucity of studies in which 8h of sleep were allowed between the administration of the hypnotic and performance testing (BIXLER et al. 1975). In 1976, physicians were again alerted to both the potential advantages and disadvantages of a benzodiazepine hypnotic that accumulates due to its long elimination half-life (KALES et al. 1976a). Finally, in 1979, the Institute of Medicine dealt extensively with the problem of carryover effects on daytime performance (Institute of Medicine 1979).

In a review of 52 studies of the effects of hypnotics on performance, four major conclusions were drawn (JOHNSON and CHERNIK 1982). First, in only eight of these studies were insomniac patients used. Second, the majority of the performance studies focused on psychomotor measures; thus, little consistent data were collected regarding effects of hypnotic drugs on cognitive functioning and more complex behaviors, i.e., few "real world" tasks had been assessed. A third, rather surprising finding was that, although long-acting drugs generally showed more decrements on daytime performance, the relationship to half-life was inconsistent; all hypnotics, at some dose, produced daytime decrements. Thus, the single strongest factor affecting daytime performance appears to be the dose level of an hypnotic drug.

III. Phase III Studies

The major objective for phase III of the evaluation of a new hypnotic is to assess long-term efficacy and safety. The concerns regarding risk to the patient have been minimized by the time this phase is reached. Thus, larger samples of subjects are employed and there are many less exclusionary criteria. The double-blind drug exposure can be for long periods of time, i.e., up to several months.

More recently there has been a trend to employ sleep laboratory studies in phase III evaluation. Thus, these sleep laboratory studies record much larger samples, at times in multicenter studies approximating 100 subjects, compared to the traditional small sample size of 6–8 subjects of phase II studies in the sleep laboratory (KRIPKE et al. 1990; BONNET et al. 1988; LEE 1992; SCHARF et al. 1994). Due to their larger sample size, these types of studies are more representative of the general population taking hypnotics. In addition, these studies should be able to detect a much smaller change from baseline as being significant due to increased statistical power and, also, to detect ADRs that occur less frequently. However, there is a major pitfall for these studies which has been ignored. This serious shortcoming relates to confusing statistical significance for efficacy with clinical relevance (WADE and WATERHOUSE 1977). Specifically, relatively small and clinically meaningless differences in sleep and wakefulness measurements can become statistically significant because of a large sample size. However, for a difference to be a meaningful difference, it must make a difference clinically (WADE and WATERHOUSE 1977). Hopefully, manufacturers, investigators and regulatory agency staff alike will not be misled, nor will they mislead, by allowing a larger sample size to mask a clinical lack of relevance with statistical significance.

IV. Phase IV Studies

The function of post-marketing surveillance in its broadest sense is to obtain information about the effects of a drug after the approval of the product for marketing (FASSIHI and ROBERTSON 1990). Post-marketing surveillance can also include the monitoring of defects in the manufacturing process such as an incorrect strength of product, contamination and mislabeling. However, for the purposes of this discussion, this latter area is not included.

In general, phase I through phase III testing of a new hypnotic has several limitations: restricted patient populations; limited duration of patient exposure; and limited patient numbers (FASSIHI and ROBERTSON 1990). Thus, phase IV studies are important in order to ascertain: less common ADRs; delayed adverse events; efficacy and toxicity in patients usually excluded from pre-marketing evaluation (for example, children, pregnant women and the elderly); efficacy and toxicity in patients with other illnesses; and toxic effects of overdose.

INMAN (1987) has described three types of post-marketing surveillance. Promotional post-marketing surveillance is carried out by individual drug companies in order to introduce doctors to their new product and at times to satisfy questions raised by regulating authorities. Regulatory post-marketing surveillance is conducted by governmental agencies. This monitoring is usually restricted to spontaneous reporting systems and published or unpublished reports. Independent post-marketing surveillance is conducted by universities and free-standing research institutions.

Post-marketing surveillance studies sponsored by companies have made only a limited contribution to the assessment of drug safety usually due to methodological weaknesses (WALLER et al. 1992; STEPHENS 1993). Further, INMAN (1987) suggests that drug companies should not undertake these types of studies. First, companies often pay doctors to use their drug rather than the older products they customarily prescribe. Secondly, a rival company is unlikely to agree to their product being prescribed as a comparison. Third, because companies are anxious to avoid ADRs, they may apply special precautions in use which make their drug appear safer. Finally, companies have great difficulty recruiting sufficient numbers of patients; thus, their studies may take 5 or more years to complete.

There are several study designs for conducting post-marketing surveillance (JICK 1977; ROSSI et al. 1983; FASSIHI and ROBERTSON 1990): clinical trials, cohort studies, case-controlled studies, and spontaneous reporting. In clinical trials, patients are randomly assigned for the duration of the study to one of the treatments under comparison. These patients are then prospectively followed for a predefined period of time to ascertain the possible development of any reaction (JICK 1977). The advantage of this method is that carefully selected patients are randomized to various drug conditions. The disadvantage of this method is that it is costly, and sample sizes are usually limited to a few thousand users (FASSIHI and ROBERTSON 1990).

Cohort studies are follow-up studies in which the choice of drug regimen is dictated by ordinary clinical practices rather than the interest of scientific comparison (JICK 1977; FASSIHI and ROBERTSON 1990). These methods allow relative risk to be established. These methods also lend themselves to large-scale surveillance which may be greater than 10000. This design has the disadvantage of potential biases in patient selection inherent in normal clinical practice.

A case control study contrasts patients with a given illness with patients who do not have that illness. The proportions of patients within these two series who have used the drug of interest are then compared (JICK 1977;

Fassihi and Robertson 1990). The method is comparatively inexpensive. However, the quality of a comparison of this type depends upon how precise and objective the target illness is defined.

The Spontaneous Reporting System (SRS) relies primarily on the active participation of astute and conscientious physicians in reporting their observations to a regulatory agency such as the FDA (Rossi et al. 1983; Fisher 1987). This method is the oldest, most general and the most productive method of gathering data on ADRs (Fassihi and Robertson 1990). Using data from the FDA's Spontaneous Reporting System, we evaluated the rates of adverse reactions associated with the three benzodiazepine hypnotics that were then commercially available (Bixler et al. 1987). The number of ADRs was controlled for by the number of new prescriptions. For each drug, only the 1st year on the market was considered because this is when ADRs are more likely to be reported. The rate for all CNS ADRs for flurazepam and temazepam was approximately the same. Triazolam, however, had an ADR rate that was more than seven times greater than either flurazepam or temazepam. We found an inverse relationship between the incidence of CNS hyperexcitability and withdrawal symptoms on the one hand and elimination half-lives of the three benzodiazepine hypnotics on the other. This is consistent with previous findings of an increased frequency and severity of rebound phenomena with rapidly eliminated benzodiazepine hypnotics, and much lower frequency and much milder rebound reactions following withdrawal of drugs with longer halflives (for reviews, see Kales et al. 1978, 1979a, 1983a,b). Another finding in our SRS study of an increased tolerance liability and dependency potential with the more rapidly eliminated drugs, triazolam and temazepam, compared to flurazepam, was suggested by previous sleep laboratory findings (Kales et al. 1976a,b; Bixler et al. 1978). Finally, the SRS study also showed extremely high rates of amnesia for triazolam versus both temazepam (47.5:1) and flurazepam (47.5:0). This finding is also in agreement with the results of controlled studies and clinical case reports noting frequent next-day amnesia as well as other cognitive impairments with triazolam (See Kales 1991; Public Citizen 1992 for reviews).

Wysowski and Barash (1991), of the FDA's Division of Epidemiology and Surveillance, also evaluated the SRS ADR reports for hypnotic drugs and found that the reporting rates for triazolam for CNS and psychiatric adverse reactions were 22–99 times those for temazepam. Further, when they corrected for a number of possible sources of bias, including differential manufacturer reporting, concomitant use of alcohol, narcotics, or psychoactive drugs, psychiatric or neurologic disorders and dose, reporting rates for triazolam versus temazepam were still 4–26 times greater.

Reports in clinical journals provide another means of communicating data about ADRs especially unexpected ones (Gelenberg 1993) particularly if the ADRs are paradoxical, e.g., hyperexcitability states induced by a drug with the primary effects of CNS sedation.

Another type of post-marketing surveillance study which is increasingly gaining use is record-linkage evaluation (Fassihi and Robertson 1990). Record-linkage involves integrating different records regarding a single person. One example of this method is "prescription-event monitoring" (Inman 1987). In this method, large cohorts of patients are identified by prescriptions written by general practitioners. Linkage by patient is completed between the general practitioner and hospital records and the Office of Population Census and Surveys which supplies data from death certificates. Many other examples exist of linked data being used in post-marketing surveillance (e.g., Ray and Griffin 1989; Shapiro 1989). In general, this method has the advantage of assembling large cohorts based on pre-existing data. This method suffers from several difficulties which include: confounding and a lack of adequate definition of exposure and outcome (Shapiro 1989).

E. Summary

The assessment of a hypnotic, both in terms of design and methods used, is derived from the objectives of the specific phase of evaluation. Phase I is focused primarily on safety. The design of this phase minimizes the risk placed on the subject by daytime administration of a single dose of the drug to young, nondiseased subjects under conditions of close monitoring.

Phase II studies are focused more on efficacy because initial concerns regarding safety have been satisfied. During this phase, insomniac subjects are used who have been carefully screened to eliminate concomitant use of medications and various medical conditions. The most useful design in this phase has proven to be the placebo-drug-placebo (PDP) design used in the sleep laboratory setting, where subjects act as their own controls, thus increasing the precision of the experiment. Clinical trials which primarily utilize subjective estimates were employed in the period prior to the development of the sleep laboratory for the evaluation of hypnotic efficacy. During this period, many important clinical properties of hypnotics went unrecognized. The sleep laboratory is an ideal environment for assessing hypnotic drugs because of precise and objective, second-by-second measurements obtained in a well-controlled setting. Sleep laboratory studies have provided the major findings related to hypnotic drugs including: rapid development of tolerance with continued use to the initial efficacy of non-benzodiazepine and rapidly eliminated benzodiazepine drugs; carryover effectiveness into the withdrawal period with slowly eliminated benzodiazepine drugs; and withdrawal sleep disturbances (rebound insomnia) following abrupt discontinuation of rapidly eliminated benzodiazepine drugs.

During phase III studies the major objective is to assess efficacy and safety more thoroughly, for example, with long-term use and in special target populations. Risk to the patient is greatly reduced because prior experience with the drug has provided considerable assurance regarding the safety of the hypnotic. Thus, these studies employ much large samples of insomniac patients using less rigorous exclusion criteria.

Finally, phase IV studies begin following approval and the initial marketing of a hypnotic. In general, pre-marketing studies are limited in terms of restricted patient populations, limited duration of patient exposure and limited patient numbers. Thus, the designs of post-marketing studies are based on large samples of patients (more than 10000) monitored in normal clinical practice. This allows for addressing less common or delayed side effects, efficacy and toxicity in patients excluded from pre-marketing evaluation, and drug interactions with other illnesses and medications.

References

Adam K, Oswald I (1989) Can a rapidly eliminated hypnotic cause daytime anxiety? Pharmacopsychiatry 22:115–119

Agnew HW, Webb WB, Williams RL (1966) The first night effect: an EEG study of sleep. Psychophysiology 2:263–266

Ancoli-Israel S, Kripke DF, Mason W, Messin S (1981) Comparisons of home sleep recordings and polysomnograms in older adults with sleep disorders. Sleep 4:283–291

Bixler EO, Kales A (1985) Clinical laboratory evaluation of hypnotic drugs. In: McMahon FG (ed) Principles and techniques of human research and therapeutics: selected topics, vol 2. Futura, Mount Kisco, pp 153–189

Bixler EO, Scharf MB, Leo LA, Kales A (1975) Hypnotic drugs and performance: a review of theoretical and methodological considerations. In: Kagan F, harwood T, Rickels K, Rudzik A, Sorer H (eds) Hypnotics: methods of development and evaluation. Spectrum, New York, pp 175–194

Bixler EO, Kales A, Soldatos CR, Scharf MB, Kales JD (1978) Effectiveness of temazepam with short-, intermediate-, and long-term use: sleep laboratory evaluation. J Clin Pharmacol 18:110–118

Bixler EO, Kales A, Jacoby JA, Soldatos CR, Vela-Bueno A (1984) Nocturnal sleep and wakefulness: effects of age and sex in normal sleepers. Int J Neurosci 23:33–42

Bixler EO, Kales A, Brubaker BH, Kales JD (1987) Adverse reactions to benzodiazepine hypnotics: spontaneous reporting system. Pharmacology 35:286–300

Bonnet MH, Dexter JR, Gillin JC, James SP, Kripke D, Mendelson W, Mitler M (1988) The use of triazolam in phase-advance sleep. Neuropsychopharmacology 1:225–234

Box GEP (1954) Some theorems on quadratic forms applied in the study of analysis of variance problems. II. Effects of inequality of variance and correlation between errors in the two-way classification. Ann Math Stat 25:484–498

Breimer DD (1977) Clinical pharmacokinetics of hypnotics. Clin Pharmacokinet 2:93–109

Brown BW Jr (1980) The crossover experiment for clinical trials. Biometrics 36:69–79

Carskadon MA, Dement WC, Mitler MM, Guilleminault C, Zarcone VP, Spiegel R (1976) Self reports versus sleep laboratory findings in 122 drug-free subjects with complaints of chronic insomnia. Am J Psychiatry 133:1382–1388

Cochran WG, Cox GM (1957) Experimental designs, 2nd edn. Wiley, New York

Fassihi AR, Robertson SSD (1990) Post-marketing drug surveillance – concepts, insights and applications. S Afr Med J 77:577–580
Fisher S (1987) Post-marketing surveillance of adverse drug reactions In: Meltzer H (ed) Psychopharmacology: the third generation of progress. Reven, New York, pp 1667–1673
Folkard S, Monk T, Knauth P, Rutenfranz J (1976) The effect of memory load on circadian variation in performance efficiency under rapidly rotating shift system. Ergonomics 19:479–488
FDA (1977) Guidelines for the clinical evaluation of hypnotic drugs. US Government Printing Office, Washington
Geisser S, Greenhouse SW (1958) An extension of Box's results on the use of the F distribution in multivariate analysis. Ann Math Stat 29:885–891
Gelenberg AJ (1993) Post-marketing surveillance: perspectives of a journal editor. Psychopharmacol Bull 29:135–137
Greenblatt DJ, Shader RI (1985) Clinical pharmacokinetics of the benzodiazepines. In: Smith DG, Wesson DR (eds) The benzodiazepines current standards for medical practice. MTP Press, Boston
Greenblatt DJ, Divoll M, Abernethy DR, Shader RI (1982) Benzodiazepine hypnotics: kinetic and therapeutic options. Sleep 5:18S–27S
Harris W (1977) Fatigue, circadian rhythm, and truck accidents. In: Macke R (ed) Vigilance theory, operational performance, and physiological correlates. Plenum, New York, pp 133–146
Hays WL (1985) Statistics for psychologists. Rinehart and Winston, New York
Helson H (1964) Adaptation-level theory: an experimental and systematic approach to behavior. Harper and Row, New York
Inman WHW (1987) Prescription-event monitoring: its strategic role in post-marketing surveillance for drug safety. In: Specialized strategies in benefit and risk monitoring. Ministry of Health and Social Assistance, Caracas, Venezuela
Institute of Medicine (1979) Sleeping pills, insomnia, and medical practice. National Academy of Sciences, Washington
Jick H (1977) The discovery of drug-induced illness. N Engl J Med 296:481–485
Johnson LC, Chernik DA (1982) Sedative-hypnotics and human performance. Psychopharmacology (Berl) 76:101–113
Kales A (1991) An overview of safety problems of triazolam. Int Drug Ther Newslett 26:25–28
Kales A, Bixler EO (1975) Sleep profiles of insomnia and hypnotic drug effectiveness. In: Burch N, Altshuler HL (eds) Behavior and brain electrical activity. Plenum, New York, pp 81–91
Kales A, Kales JD (1970) Sleep laboratory evaluation of psychoactive drugs. Pharmacol Physicians 4:1–5
Kales A, Kales JD (1983) Sleep laboratory studies of hypnotic drugs: efficacy and withdrawal effects. J Clin Psychopharmacol 3:140–150
Kales A, Kales JD (1984) Evaluation and treatment of insomnia. Oxford University Press, New York
Kales A, Scharf MB, Kales JD (1970) Hypnotic drugs and their effectiveness: all-night EEG studies of insomniac patients. Arch Gen Psychiatry 23:226–232
Kales J, Kales A, Bixler EO, Slye ES (1971) Effects of placebo and flurazepam on sleep patterns in insomniac subjects. Clin Pharmacol Ther 12:691–697
Kales A, Bixler EO, Tan T-L, Scharf MB, Kales JD (1974) Chronic hypnotic-drug use: ineffectiveness, drug-withdrawal insomnia and dependence. JAMA 227: 513–517
Kales A, Kales JD, Bixler EO, Scharf MB (1975) Methodology of sleep laboratory drug evaluations: further considerations. In: Kagan F, Harwood T, Rickels K, Rudzik A, Sorer H (eds) Hypnotics: methods of development and evaluation. Spectrum, New York, pp 109–126

Kales A, Bixler EO, Scharf MB, Kales JD (1976a) Sleep laboratory studies of flurazepam: a model for evaluating hypnotic drugs. Clin Pharmacol Ther 19: 576–583
Kales A, Kales JD, Bixler EO, Scharf MB, Russek E (1976b) Hypnotic efficacy of triazolam: sleep laboratory evaluation of intermediate-term effectiveness. J Clin Pharmacol 16:399–406
Kales A, Bixler EO, Kales JD, Scharf MB (1977) Comparative effectiveness of nine hypnotic drugs: sleep laboratory studies. J Clin Pharmacol 17:207–213
Kales A, Scharf MB, Kales JD (1978) Rebound insomnia: a new clinical syndrome. Science 201:1039–1041
Kales A, Scharf MB, Kales JD, Soldatos CR (1979a) Rebound insomnia: a potential hazard following withdrawal of certain benzodiazepines. JAMA 241:1692–1985
Kales A, Scharf MB, Soldatos CR, Bixler EO (1979b) Clinical evaluation of hypnotic drugs: contributions from sleep laboratory studies. J Clin Pharmacol 19:329–336
Kales A, Soldatos CR, Bixler EO, Kales JD (1983a) Early morning insomnia with rapidly eliminated benzodiaepines. Science 220:95–97
Kales A, Soldatos CR, Bixler EO, Kales JD (1983b) Rebound insomnia and rebound anxiety: a review. Pharmacology 26:121–137
Kleitman N (1963) Sleep and wakefulness. University of Chicago Press, Chicago
Kripke DF, Hauri P, Ancoli-Israel S, Roth T (1990) Sleep evaluation in chronic insomniacs during 14-day use of flurazepam and midazolam. J Clin Psychopharmacol 10:32S
Lader M, Lawson C (1987) Sleep studies and rebound insomnia: methodological problems, laboratory findings, and clinical implications. Clin Neuropharmacol 10:291–312
Lana RE, Lubin A (1963) The effect of correlation on the repeated measures design. Educ Psychol Measurem 23:729–739
Lasagna L (1972) Drug therapy: hypnotic drugs. N Engl J Med 287:1182–1184
Lee T (1992) Technical report 9159-92-001 of protocol M/2100/0232. Psychopharmacologic Drugs Advisory Committee 5/18/92, Food and Drug Administration, Washington
Lindquist EF (1956) Design and analysis of experiments in psychology and education. Hougton Mifflin, Boston
Man GCW (1994) Validation of a portable sleep apnea monitoring device (PolyG) (Abstr). American Lung Association and the Thoracic Society Annual Conference, Boston, MA
Monroe LJ (1967) Psychological and physiological differences between good and poor sleepers. J Abnorm Psychol 72:255–264
Morgan K, Oswald I (1982) Anxiety caused by a short-life hypnotic. Br Med J 284:942–944
Nicholson AN, Stone BM (1983) Imidazobenzodiazepines: sleep and performance studies in humans. J Clin Psychopharmacol 3:72–75
Nicholson AN, Stone BM, Pascoe PA (1982) Hynotic efficacy in middle age. J Clin Psychopharmacol 2:118–121
Oswald I, Priest RG (1965) Five weeks to escape the sleeping-pill habit. Br Med J 11:1093–1095
Public Citizen (1992) Citizen's petition to withdraw approval of triazolam (Halcion). Public Citizen, Washington
Ray WA, Griffin MR (1989) Use of Medicaid data for pharmacoepidemiology. Am J Epidemiol 129:837–849
Rechtschaffen A, Kales A (eds) (1968) A manual of standardized terminology, techniques, and scoring system for sleep stages of human subjects. National Institutes of Health, Bethesda (NIH no 204)
Rechtschaffen A, Verdone P (1964) Amount of dreaming: effect of incentive, adaptation to laboratory, and individual differences. Percept Mot Skills 19:947–958

Redline S. Tosteson T, Boucher MA, Millman RP (1991) Measurement of sleep-related breathing disturbances in epidemiologic studies: assessment of the validity and reproducibility of a portable monitoring device. Chest 22:1281–1286
Roehrs T, Zorick F, Sicklesteel J, Wittig R, Hartse K, Roth T (1983) Effects of hypnotics on memory. J Clin Psychopharmacol 3:310–313
Roehrs T, Vogel G, Roth T (1990) Rebound insomnia: its determinants and significance Am J Med 88 Suppl [3A]:39S–42S
Rossi AC, Knapp DE, Anello C et al. (1983) Discovery of adverse drug reactions: a comparison of selected phase IV studies with spontaneous reporting methods. JAMA 249:2226–2228
Roth T, Hartse K, Saab P, Piccone P, Kramer M (1980) The effects of flurazepam, lorazepam, and triazolam on sleep and memory. Psychopharmacology (Berl) 70:231–237
Scharf MB, Kales A, Bixler EO (1975) Readaptation to the sleep laboratory in insomniac subjects. Psychophysiology 12:412–415
Scharf MB, Roth T, Vogel GW (1994) A multicenter, placebo-controlled study evaluating zolpidem in the treatment of chronic insomnia. J Clin Psychiatr 55:192–199
Shapiro S (1989) The role of automated record linkage in the post-marketing surveillance of drug safety: a critique. Clin Pharmacol Ther 46:371–386
Soldatos CR, Kales A (1979) Role of the sleep laboratory in the evaluation of hypnotic drugs. In: Priest RG, Pletscher A, Ward J (eds) Sleep research. MTP Press, Lancaster, pp 181–195
Stephens MDB (1993) Marketing aspects of company-sponsored post-marketing surveillance studies. Drug Safety 8:1–8
Stoohs R, Guilleminault C (1990) Investigations of an automatic screening device (MESAM) for obstructive sleep apnoea. Eur Respir J 3:823–829
Stoohs R, Guilleminault C (1992) MESAM 4: an ambulatory device for the detection of patients at risk for obstructive sleep apnea syndrome (OSAS). Chest 101: 1221–1227
Wade OL, Waterhouse JAH (1977) Significant or important? Br Med J 4:411–412
Waller PC, Wood SM, Langman MJS, Breckenridge AM, Rawlins MD (1992) Review of company post-marketing surveillance studies. Br Med J 304:1470–1472
White DM, Gibb TJ, Wall JM, Westbrook PR (1994) Assessment of accuracy and analysis time of a novel device to monitor sleep and breathing in the home. (in press)
Williams EJ (1949) Experimental designs balanced for the estimation of residual effects of treatments. Aust J Sci Res 2:149–168
Winer BJ, Brown DR, Michels KM (1992) Statistical principles in experimental design 3rd edn. McGraw-Hill, New York
Wysowski DK, Barash D (1991) Adverse behavioral reactives attributed to triazolam in the Food and Drug Administration's Spontaneous Reporting System. Arch Intern Med 151:2003–2008

CHAPTER 13

Hypnotic Drugs

A. KALES, A.N. VGONTZAS, and E.O. BIXLER

A. Introduction

Over the years, a variety of drugs have been used in the management of anxiety, tension and insomnia. In the early part of this century, the pharmacotherapy of anxiety and insomnia relied on two classes of substances: the bromides and the barbiturates (HOLLISTER and CSERNANSKY 1990). The bromides soon were shown to have major limitations, primarily because of their tendency to accumulate which often resulted in toxic delirium. They were also abused, as they were sold over-the-counter. Similarly, by the 1950s, the barbiturates were shown to result in a number of clinical dilemmas. They were found to produce tolerance, withdrawal reactions and physical dependence and to be highly lethal with only moderate overdose (ISBELL 1950; BARLEY and JATLON 1975). These difficulties underlined the need for safer alternatives for the pharmacotherapy of anxiety disorders and insomnia.

B. The Condition of Insomnia

Insomnia is a term referring to a relative deficiency in the amount or quality of sleep (KALES and KALES 1984). Insomnia is a highly prevalent disorder with about one-third of the general population having some type of complaint of sleep difficulty (KARACAN et al. 1976; BIXLER et al. 1979a,b). It is more prevalent with increasing age, among women, in association with psychiatric disturbances and among individuals of lower socioeconomic status. Regional and nationwide surveys show that about one-third of the adult population have a current complaint of some difficulty in sleeping: difficulty falling asleep, difficulty staying asleep or early final awakening (BIXLER et al. 1979b). Similar high percentages of persons with insomnia were found in two other surveys: a national survey of physicians showed that 17% of adult patients had a complaint of insomnia; and, a survey of the general population in San Marino, Italy, reported that 20% had chronic insomnia.

Sleep difficulty most often is the result of the complex interaction of a variety of factors. Psychologic factors, the aging process, situational factors, medical conditions, the hospital environment and administration and withdrawal of certain drugs may adversely affect sleep (KALES and KALES 1984). As individuals age, there is an increase in complaints of insomnia. Older persons obtain less total sleep and have very little of the "'deeper" stages 3 and 4 sleep (KALES et al. 1967; FEINBERG and CARLSON 1968). With aging, sleep often returns to a more polyphasic pattern, with periods of sleepiness during the day and frequent nap taking. Another factor contributing to the sleep difficulties experienced by elderly persons is an increase in general medical illnesses, along with emotional factors related to declining function and concern regarding approaching death (SOLDATOS et al. 1979).

When insomnia is chronic, psychopathology (psychologic disturbances and behaviorally conditioned patterns) is by far the most predominant etiological factor, although medical conditions and aging often contribute (KALES and KALES 1984; TAN et al. 1984). Most often chronic insomnia develops at a time when life-stress factors are prevalent in individuals who are predisposed by having inadequate methods of coping (HEALEY et al. 1981). The internalization of emotions typically experienced by patients with chronic insomnia leads to emotional arousal that, in turn, result in physiologic activation and insomnia (KALES et al. 1976b, 1983a; KALES and KALES 1984). A fear of sleeplessness and its consequences soon develops, establishing a vicious circle of physiologic activation, sleeplessness, more fear of sleeplessness, further emotional arousal and physiologic activation and still further sleeplessness.

I. Treatment of Insomnia

Knowing that a multiplicity of factors may underlie the condition of chronic insomnia, the physician must conduct a thorough evaluation of patients including taking histories of sleep difficulty, drug use and psychiatric problems (KALES et al. 1980b; SOLDATOS et al. 1979; KALES and KALES 1984). Taking a thorough sleep history: provides definition of the specific sleep problem and any coexisting sleep disorder; assesses the clinical course of the specific sleep problem; evaluates sleep/wakefulness patterns; includes reports from the bed partner; and examines the impact of the sleep problem on the patient's life.

To treat these patients successfully, a multidimensional approach is required (SOLDATOS et al. 1979; KALES and KALES 1984). This approach selectively combines several elements from the following modalities: general measures for improving sleep hygiene and lifestyle; supportive, insight-oriented and behavioral psychotherapeutic techniques; and adjunctive use of hypnotic drugs or use of antidepressant agents.

C. Development of Benzodiazepines and Other Hypnotic Drugs

The most frequently prescribed hypnotic drugs by far are the benzodiazepines. These drugs act by facilitating GABAergic transmission (for a detailed discussion see Chap. 8, this volume). The first benzodiazepines introduced were 1,4-benzodiazepines such as diazepam and flurazepam that were similar in pharmacokinetics (relatively slow elimination rate) and pharmacodynamics (low to moderate receptor-binding affinity/potency) (HOLLISTER and CSERNANSKY 1990). Accordingly, the most frequent side effect of these benzodiazepines was excessive daytime sleepiness.

Late in the 1970s, additional 1,4-benzodiazepines were introduced such as lorazepam and flunitrazepam, which were more rapidly eliminated and had higher receptor-binding affinities/potencies. The elimination half-lives and binding affinities of various benzodiazepines are illustrated in Table 1 (GREENBLATT and SHADER 1985; ARENDT et al. 1987; MULLER 1987; GUSTAVSON and CARRIGAN 1990; LANGTRY and BENFIELD 1990; MILLER et al. 1992; FERNANDEZ et al. 1993; BAILEY et al. 1994). The use of these benzodiazepines was not only associated with more potent anxiolytic and hypnotic effects, but also with more rapid development of tolerance, significant withdrawal difficulties such as rebound insomnia and anxiety (KALES et al. 1978, 1983d) following termination of treatment and other "unexpected" side effects such as memory impairment during drug administration (SCHARF et al. 1982; BIXLER et al. 1991).

In the late 1970s and early 1980s, a new group of benzodiazepines with greater potency and more rapid elimination was introduced, the triazolobenzodiazepines. This new class also differed chemically from the classical 1,4-benzodiazepines in that it included a triazolo ring attached to the basic diazepine structure. In fact, the triazolobenzodiazepines have been promoted for their unique clinical applications based on this structure; for example, alprazolam has been marketed as an antipanic drug with, in addition, potential antidepressant effects. It appears that special properties associated with the triazolo ring also account in part for adverse reactions associated with the triazolobenzodiazepines. Clinical case reports, spontaneous reporting system data, controlled clinical trials and sleep laboratory studies have

Table 1. Pharmacokinetics of hypnotics

	Elimination half-life (h)	Binding affinity (K_i)
Flurazepam	40–120	17.2
Quazepam	40–200	–
Nitrazepam	20–50	11.5
Temazepam	6–16	23.0
Flunitrazepam	10–25	3.8
Lormetazepam	9–15	4.8
Midazolam	1–3	0.4
Estazolam	9–27	17.0
Brotizolam	4–8	0.9
Triazolam	2–4	0.4
Zopiclone	3–7	–
Zolpidem	1–2	–

shown that the use of triazolobenzodiazepines is associated with frequent and severe CNS and psychiatric adverse reactions in three major categories (see KALES 1991; PUBLIC CITIZEN 1992 for reviews): amnesia and other cognitive impairments (KALES et al. 1976c; VAN DER KROEF 1979); daytime anxiety, tension or panic and early morning insomnia (MORGAN and OSWALD 1982; KALES et al. 1983c; GARDNER and COWDRY 1985; PECKNOLD and FLEURY 1986; ADAM and OSWALD 1989); and withdrawal difficulties such as rebound insomnia, anxiety and seizures (KALES et al. 1978); PECKNOLD et al. 1988).

Recently, two new non-benzodiazepine hypnotics, zopiclone and zolpidem, have been introduced. Both of these drugs share certain pharmacokinetic and pharmacodynamic properties, such as short elimination half-life (HINDMARCH and MUSCH 1990; SAUVANET et al. 1988) and binding to the GABA-benzodiazepine receptor complex. Although these drugs were purported to have a better side effect profile than the benzodiazepines, cognitive impairments (WHO 1990; DREWES et al. 1991; ANSSEAU et al. 1992; BALKIN et al. 1992; MAAREK et al. 1992; IRUELA et al. 1993) and rebound insomnia following withdrawal (DORIAN et al. 1983; LADER and FRCKA 1987; KRYGER et al. 1991; SCHARF et al. 1994) have been reported.

I. Evaluation of Hypnotic Drugs

This chapter focuses primarily on sleep laboratory studies of hypnotic drugs rather than clinical trials. The reason for this is the precise, objective and carefully controlled measurements of the sleep laboratory as opposed to the imprecise subjective estimates of clinical trials in which different study variables are usually not controlled rigorously. This, in turn, allows sleep laboratory studies to have strong statistical power, with only a small number of subjects, in assessing efficacy, development of tolerance and withdrawal effects. In turn, clinical trials have statistical power only through the use of a large number of subjects and often may show statistically significant changes that are relatively small in magnitude and, thus, of little clinical relevance (KALES and KALES 1983). Also, see Chap. 12, this volume, for a detailed discussion of the advantages of sleep laboratory studies and of the placebo-drug-placebo design used in the sleep laboratory.

Within this context, we do not include a number of sleep laboratory studies that have the shortcomings of: an inadequate period allowed for adaptation and baseline; an inadequate washout period with carryover of drug effects from one night to the other; assessment of only one night of drug administration for each drug; or lack of a period to assess withdrawal effects. Because there are considerably fewer sleep laboratory assessments of benzodiazepines on a long-term basis, some clinical trials are included that assessed benzodiazepine hypnotics over lengthy periods. Finally, to assess properly the hypnotic efficacy of drugs, we rely primarily on those studies utilizing insomniac patients rather than healthy volunteers. Studies of healthy volunteers or normal sleepers or studies with some of the other shortcomings noted above are cited only when there are no other or few sleep laboratory studies for a given drug or dose.

Individual benzodiazepines are grouped in this chapter according to their elimination half-life, receptor-binding affinity and chemical structure, as we have found that pharmacokinetic and pharmacodynamic properties are closely related to the drugs' clinical profiles, including efficacy, the development of tolerance, adverse events and withdrawal effects.

D. 1,4-Benzodiazepines with Slow Elimination and Low Binding Affinity

I. Flurazepam

1. Efficacy

With short-term administration of 30 mg flurazepam a marked improvement in sleep was found both in terms of sleep induction and sleep maintenance (Table 2) (KALES et al. 1970a,b, 1971, 1975, 1976a, 1982b). This efficacy was maintained with intermediate-term use (Fig. 1), i.e., consecutive, nightly administration of the drug over a 2-week period of time (KALES et al. 1975, 1976a, 1982b). With long-term use of 30 mg flurazepam (28 consecutive nights) sleep continued to be markedly and significantly improved with only a slight loss in efficacy during this lengthy period of administration (Table 2) (KALES et al. 1975, 1976a, 1982b).

Peak effectiveness of flurazepam was found to occur on the second and third consecutive drug nights, although sleep was markedly improved on the first night of administration (KALES et al. 1976a). These data suggest that the short elimination half-life components of the drug contribute significantly to both the drug's sleep induction and sleep maintenance properties on the first night while the long half-life metabolite, desalkylflurazepam, is unavailable on the first night of use to affect sleep induction and available to only a slight degree to affect sleep maintenance (KAPLAN et al. 1973; GREENBLATT et al. 1981). From these findings, we predicted that the carryover effect demonstrated for flurazepam would prove to be an advantage in facilitating withdrawal and a disadvantage in that it would increase the possibility of daytime sedation during drug administration (KALES et al. 1976a).

Other groups have conducted additional sleep laboratory evaluations of 30 mg flurazepam (VOGEL et al. 1976; DEMENT et al. 1978; PETRE-QUADENS et al. 1983; ADAM and OSWALD 1984; BLIWISE et al. 1984; KRIPKE et al. 1990). Data from these studies are in strong agreement with our original findings on the short-term effectiveness of flurazepam, as well as on its maintaining intermediate- and long-term efficacy with little development of tolerance.

Table 2. Efficacy of benzodiazepine hypnotics: percentage change of total wake time from baseline

	Dose	Drug administration nights			
		(1–3)	(5–7)	(12–14)	(26–28)
Flurazepam	15	−65.4%			
	30	−51.2%**		−59.8%**	
	30	−48.6%**		−56.3%**	−42.6%**
	30	−51.1%**		−42.7%**	−35.1%*
Quazepam	7.5	−39.7%*			
	15	−46.4%**			
	15	−54.8%**		−49.0%**	
	15	−37.5%**		−41.8%**	−29.6%*
	30	−57.4%**		−30.3%	−22.6%
Nitrazepam	10	−36.0%*	−43.4%*		
Temazepam	7.5	−30.8%*	−14.0%		
	15	−28.9%*		−14.5%	
	30	−11.7%		1.8%	2.9%
Flunitrazepam	1	−5.1%	1.6%		
	2	−61.3%**	−52.3%**		
	2	−39.4%**		−16.7%	−9.0%
Lormetazepam	0.5	−33.5%*	−35.3%*		
	1.0	−26.5%	−12.4%		
	1.5	−49.5%*	−39.0%		
	2.0	−47.1%*	−27.9%		
Midazolam	10	−27.7%*	−21.2%		
	20	−31.1%	−2.4%		
	30	−27.8%	−7.9%		
Brotizolam	0.25	−32.5%*			
Triazolam	0.25	−24.3%		−6.1%	
	0.5	−45.1%**		−17.0%	

All of the data in this table and in Figs. 1 and 2 are from studies conducted in the Sleep Laboratory at Pennsylvania State University College of Medicine using standardized methods and techniques.
* $P < 0.05$; ** $P < 0.01$.

A number of sleep laboratory studies have been conducted using 15 mg flurazepam (KALES and SCHARF 1973; DEMENT et al. 1973; FROST and DELUCCHI 1979; ROEHRS et al. 1982; KRIPKE et al. 1990). This lower dose was found to be quite effective over a three-night period of administration for both sleep induction and maintenance, with a marked shortening of sleep latency and reduction in wake time after sleep onset and total wake time (KALES and SCHARF 1973). In other studies, 15 mg flurazepam was reported to be effective for both inducing and maintaining sleep (DEMENT et al. 1973; FROST and DELUCCHI 1979; ROEHRS et al. 1982; KRIPKE et al. 1990). In a meta-analysis of three of these studies which had a 1-week period of drug administration, there was no evidence for the development of tolerance both in relation to efficacy being maintained over the 1-week period of

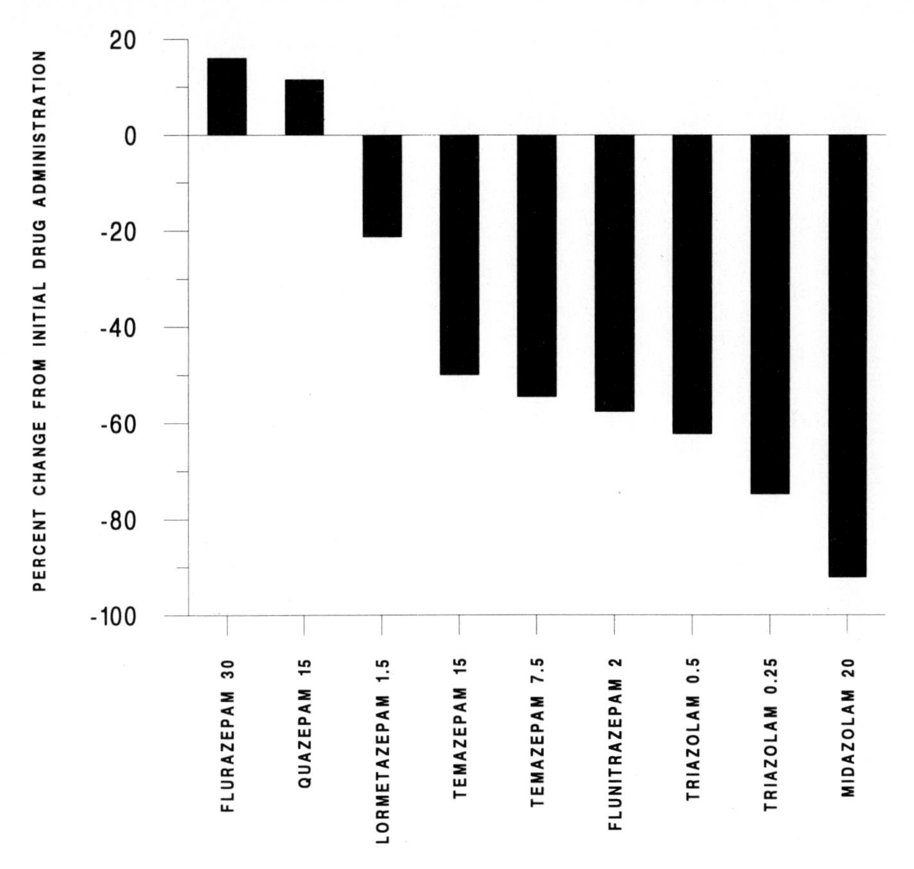

Fig. 1. Loss of efficacy with continued use of benzodiazepine hypnotics. Percent change from first three nights of drug administration of mean values of total wake time is indicated for continued use at 1 week or 2 weeks of drug administration based upon mean differences from baseline for both conditions. Midazolam and lormetazepam were evaluated for continued use following 1 week of drug administration while all others were evaluated for continued use following 2 weeks of administration. Flurazepam and quazepam demonstrated no tolerance after 2 weeks of use while strong tolerance was observed for midazolam after 1 week and triazolam (both doses) after 2 weeks of continued use. Somewhat less tolerance was observed with flunitrazepam and temazepam and even less tolerance with lormetazepam

continued use and there being a carryover effectiveness demonstrated following drug withdrawal (SNYDER and THOMAS 1972).

2. Withdrawal Effects

Carryover effectiveness for flurazepam has been demonstrated across the initial withdrawal nights, i.e., levels of total wake time remained slightly to moderately below baseline values (KALES et al. 1970a,b, 1971, 1975, 1976a, 1982b). Accordingly, none of these studies demonstrated any rebound

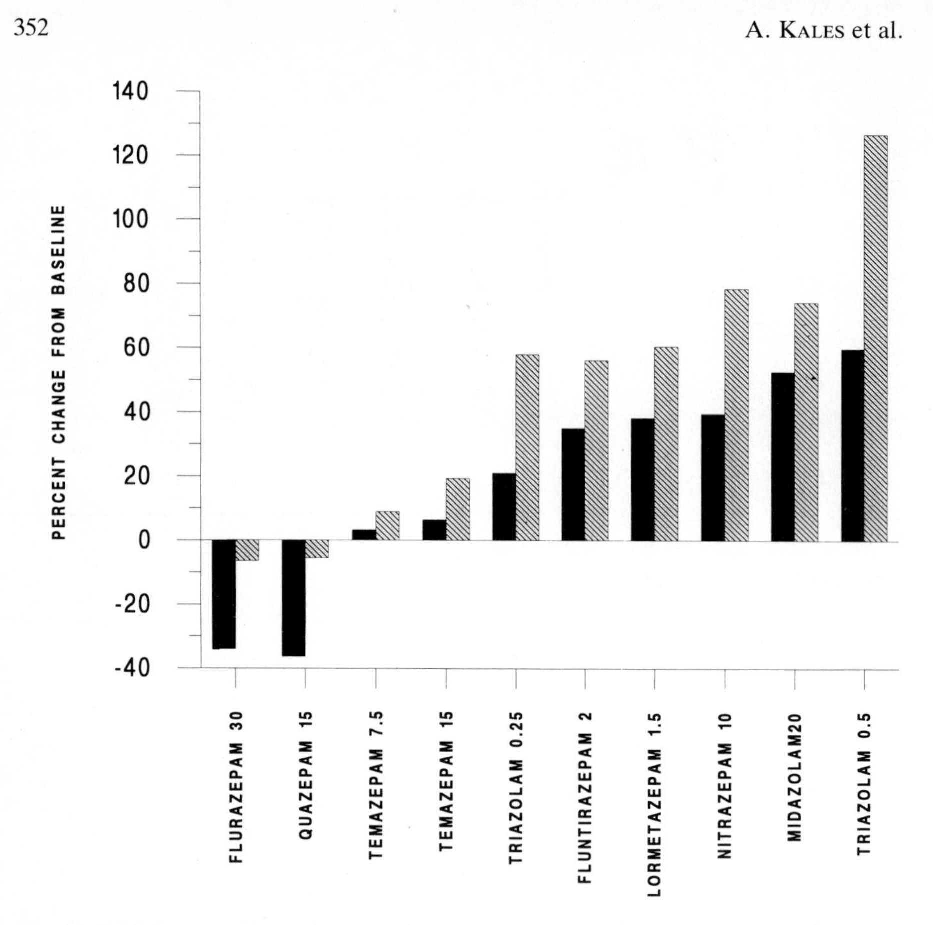

Fig. 2. Withdrawal effects of hypnotic benzodiazepines. Percent change from baseline of mean values of total wake time is indicated for the initial withdrawal condition (nights 1–3) (*solid bar*) and the single highest withdrawal night (*hashed bar*). Both flurazepam and quazepam showed some carryover efficacy on withdrawal while triazolam withdrawal resulted in the strongest degree of rebound insomnia followed by midazolam and nitrazepam

insomnia as originally defined (KALES et al. 1978, 1979), i.e., mean significant increases or increases of 40% or greater in total wake time (Fig. 2). In the most comprehensive study (KALES et al. 1982b), which assessed 15 consecutive withdrawal nights for each subject, the initial period of carryover effectiveness was followed by a period when total wake time generally approximated the baseline value. A mild degree of sleep disturbance was observed on only two nights of the 15, but this level of increase of total wake time was only 20%–25% above the baseline value. In agreement with our studies, carryover effectiveness was noted in several other studies, and none have demonstrated the presence of rebound insomnia (VOGEL et al. 1976; DEMENT et al. 1978; OSWALD et al. 1979). Finally, flurazepam 15 mg has also been shown to produce minimal withdrawal phenomena (SNYDER and THOMAS 1972; DEMENT et al. 1973; FROST and DELUCCHI 1979; ROEHRS et al. 1982).

3. Side Effects

The most frequent side effect of this group of benzodiazepines is daytime sedation. Daytime sedation is an "augmented," i.e., expected side effect with any hypnotic drug because it represents a direct extension of the drug's therapeutic effect. Manifestations of CNS depressant effects, such as daytime drowsiness, hangover and dizziness, are in general considered to be more prevalent with benzodiazepines with long elimination half-lives such as flurazepam (KALES et al. 1976a; OSWALD et al. 1979). This contention is supported by findings from specially designed studies of performance. These studies have shown that impairment of daytime psychomotor functioning is usually greater with slowly eliminated benzodiazepines such as flurazepam which accumulate with successive nightly administration (KALES et al. 1976a; OSWALD et al. 1979; CHURCH and JOHNSON 1979; JOHNSON and CHERNIK 1982).

II. Quazepam

1. Efficacy

Our group has conducted five sleep laboratory studies of quazepam. In the first study both induction and maintenance of sleep were improved with short-term use of 15 mg quazepam while with intermediate-term use of the drug (14 consecutive nights of drug administration), total wake time continued to be significantly reduced from baseline (Table 2) (KALES et al. 1980a). In the second sleep laboratory study, a dose-response study of 7.5, 15 and 30 mg quazepam (KALES et al. 1981), all three doses were effective in inducing and maintaining sleep with short-term administration. In the third study, 15- and 30-mg doses of quazepam were assessed in separate 47-night sleep laboratory evaluations of insomniac patients that included 28 consecutive nights of drug administration (KALES et al. 1982b). Both doses were effective for sleep induction and maintenance with short-, intermediate- and long-term use; during long-term use, there was some loss of efficacy noted with both doses of quazepam but 15 mg still produced a statistically significant decrease in total wake time (Table 2). In the last two studies (KALES et al. 1986a,b), quazepam improved sleep significantly during both short- and intermediate-term use, showing minimal development of tolerance with continued use (KALES et al. 1986b). Similar findings have been reported by other investigators for the efficacy of 15 mg (MAMELAK et al. 1981) and 30 mg (MAMELAK et al. 1984) quazepam. In both studies, each of the two doses of quazepam were effective both with short- and intermediate-term use.

2. Withdrawal Effects

In all five sleep laboratory studies of quazepam conducted by our group, the potential for rebound insomnia following withdrawal of quazepam was assessed (KALES et al. 1980a, 1981, 1982b, 1986a,b). None of the studies demonstrated rebound insomnia or any sleep disturbances following withdrawal of quazepam (Fig. 2). Furthermore, in each of the studies, there was a carryover effectiveness for quazepam on the first night of withdrawal. Results of studies conducted by other groups of investigators are consistent with this finding of carryover effectiveness and thus an absence of withdrawal sleep difficulties with quazepam (MAMELAK et al. 1981, 1984; KALES 1990).

3. Side Effects

In separate long-term evaluations (28-consecutive nights of drug use) of 30 mg flurazepam and 30 mg and 15 mg quazepam, we found that 30 mg quazepam produced more frequent and severe daytime sedation than 30 mg flurazepam (KALES et al. 1982b). This is not surprising when one considers that quazepam's three active components (the parent compound and two active metabolites) all have long elimination half-lives and, thus, the drug would be expected to produce a greater carryover effect. Evidence for such a greater degree of carryover effect was the fact that quazepam produced a continued improvement in sleep compared to baseline for a longer period following drug withdrawal than flurazepam. Quazepam in a 15-mg dose produced less frequent and severe daytime sedation than the 30-mg dose of quazepam (KALES et al. 1981).

E. 1,4-Benzodiazepines with Intermediate to Slow Elimination and Moderate Binding Affinity

I. Nitrazepam

1. Efficacy

Nitrazepam in a dose of 10 mg significantly reduced both sleep latency and wake time after sleep onset over a 1-week period of drug administration (KALES et al. 1978). In several other studies assessing the hypnotic efficacy of 5 mg nitrazepam (ADAM et al. 1976; ADAM and OSWALD 1982; JOVANOVIC and DREYFUS 1983), nitrazepam was effective with initial and short-term use (Table 2). However, two studies which assessed the drug's hypnotic efficacy over more than 2 weeks of usage reported conflicting results regarding the development of tolerance. In the first study ADAM et al. (1976) reported that nitrazepam remained effective even after 10 weeks of use, whereas in the second study by the same authors (ADAM and OSWALD 1982) significant tolerance was found by the 3rd week of administration.

2. Withdrawal Effects

Following administration of 10 mg nitrazepam at bedtime over a 7-night period, we evaluated the effects of withdrawal over a 3-night period (KALES et al. 1978). Following drug withdrawal, rebound insomnia was demonstrated with wake time after sleep onset and total wake time increased significantly over baseline levels (Fig. 2). Our observations for rebound insomnia with nitrazepam agreed with those of ADAM et al. (1976), who first reported that sleep worsened following withdrawal of nitrazepam.

3. Side Effects

Increased morning drowsiness or daytime sedation is the most reported side effect for nitrazepam.

F. 1,4-Benzodiazepines with Rapid to Intermediate Elimination and Low Binding Affinity

I. Temazepam

1. Efficacy

To explain more clearly findings related to temazepam, it is important to consider different formulations of this drug. First, temazepam, as it was marketed originally in a hard gelatin capsule formulation in the United States, had a very slow rate of absorption from the gastrointestinal tract. Peak plasma concentrations were reached on average 1.8–4.7 h after oral ingestion (DIVOLL et al. 1981). Therefore, not surprisingly, little if any efficacy for inducing sleep was noted for the drug either in our earlier study (BIXLER et al. 1978) or in some other early sleep laboratory studies of this drug with this formulation (FDA 1978/1980; MITLER et al. 1979).

Our own sleep laboratory evaluation of 30 mg temazepam in the hard gelatin capsule formulation employed a protocol that included 28 consecutive nights of drug administration (BIXLER et al. 1978). The drug had little effect on sleep induction and, although there was more effect on sleep maintenance, wake time after sleep onset was not significantly decreased during any of the drug conditions (Table 2). Total wake time was decreased only slightly with short-term drug administration and was similar to baseline with intermediate- and long-term use.

The effects of 30 mg temazepam in the hard capsule formulation have been evaluated in two other sleep laboratory studies (MITLER et al. 1979; ROEHRS et al. 1984). One study found the drug effective for inducing and maintaining sleep in insomniac patients, but the drug was administered 30 min before bedtime (ROEHRS et al. 1984). In the other study, temazepam was effective for maintaining sleep but not for inducing sleep (MITLER et al.

1979), with the improved sleep maintenance continuing throughout the whole 35-night period of drug administration. ROEHRS et al. (1990) also assessed the dose-response effects of 7.5, 15 and 30 mg temazepam using a hard gelatin capsule and administering the drug 30 min before bedtime. They found that total sleep time and sleep efficiency increased significantly in a linear fashion with increasing doses of temazepam.

In contrast, more recent studies have shown that 30 mg temazepam in the soft gelatin form or reformulated tablet is more effective in improving sleep, both sleep induction and sleep maintenance (ROEHRS et al. 1986; MITLER et al. 1986; SCHARF et al. 1990c; KALES et al. 1991), while one of these studies showed considerable loss of efficacy or tolerance at the end of a 2-week period of administration (SCHARF et al. 1990b). The hypnotic efficacy of 20 mg temazepam has also been evaluated in the soft gelatin capsule formulation (WITTIG et al. 1985). During both initial and continued administration, temazepam decreased sleep latency and increased total sleep time.

Several sleep laboratory studies have demonstrated significant improvement in sleep with administration of 15 mg temazepam (MITLER et al. 1979, 1986; KALES et al. 1986a; SCHARF et al. 1990b,c). In one of these studies (KALES et al. 1986a), however, it was found that tolerance for temazepam's efficacy developed rapidly with intermediate use, with the drug losing virtually all of its efficacy with short-term use after 2 weeks of continued drug administration (Fig. 1). More recently, temazepam 7.5 mg was found to be effective in elderly insomniacs with short-term use (VGONTZAS et al. 1994). However, some tolerance developed to the drug's effectiveness at the end of 1 week of drug administration (Table 2).

2. Withdrawal Effects

The findings for rebound insomnia following withdrawal of temazepam have been inconsistent with only a few studies reporting this degree of sleep disturbance. Four studies of 30 mg temazepam given at bedtime did not demonstrate rebound insomnia following withdrawal of temazepam (BIXLER et al. 1978; ROEHRS et al. 1984, 1986; SCHARF et al. 1990b). Rebound insomnia was observed in three other studies following withdrawal of temazepam. Although MITLER et al. (1979) stated that in their study of 30 mg temazepam there was "no evidence of rebound insomnia," total wake time increased almost 50% above baseline for the entire three-night withdrawal period. Another study of 30 mg temazepam reported that total wake time rose sharply on the first night after withdrawal (FDA 1978/1980). Finally, in a more recent sleep study of brief and intermittent administration and withdrawal of 30 mg temazepam (KALES et al. 1991), there was a significant increase in total wake time over baseline during one of the two withdrawal periods.

The only sleep laboratory evaluation of 20 mg temazepam that included a withdrawal period did not find evidence of rebound manifestations upon drug discontinuation (WITTIG et al. 1985). In one study of 15 mg temazepam, drug withdrawal was associated with some sleep and mood disturbance on the first withdrawal night (KALES et al. 1986a). However, in three other studies with 15 mg temazepam, rebound insomnia was not found (MITLER et al. 1979; ROEHRS et al. 1986; SCHARF et al. 1990b). Finally, no rebound insomnia was found following withdrawal of 7.5 mg temazepam in elderly insomniacs (Fig. 2) (VGONTZAS et al. 1994).

3. Side Effects

For temazepam, the most commonly reported side effect is daytime sedation (HEEL et al. 1981), associated with performance decrements with the higher dose and the slower absorption rate of the previously used hard gelatin capsule preparation (GREENBLATT et al. 1982). However, the frequency and severity of daytime sedation is significantly less than that of flurazepam (BIXLER et al. 1987) or quazepam (KALES et al. 1986a).

Cognitive impairments (including memory impairment/amnesia) have been rarely noted with temazepam use. In two controlled studies assessing next-day memory impairment/amnesia, there was no impairment with temazepam, compared to a significant number of episodes of memory impairment/amnesia with triazolam (SCHARF et al. 1988; BIXLER et al. 1991). There is only one clinical case report of amnesia with temazepam use (REGESTEIN and REICH 1985). In the Food and Drug Administration (FDA) Spontaneous Reporting System (SRS) data, the rate ratio for amnesia for triazolam versus temazepam was 47.5:1 (BIXLER et al. 1987). Finally, WYSOWSKI and BARASH (1991), in a later report of SRS data for triazolam and temazepam, noted that while there were 29 episodes of complex behavioral actions with triazolam, no such episodes were reported for temazepam.

Use of temazepam has been rarely associated with the occurrence of hyperexcitability phenomena such as daytime anxiety. There have been only several clinical case reports of paradoxical and unexpected reactions such as daytime anxiety, tenseness and anger (REGESTEIN and REICH 1985; HEEL et al. 1981). Furthermore, in the FDA SRS data, the frequency of hyperexcitability phenomena with temazepam is much lower than that with triazolam, but, as would be expected from temazepam's pharmacokinetics, it is much higher than that for flurazepam (BIXLER et al. 1987).

G. 1,4-Benzodiazepines with Rapid to Intermediate Elimination and Moderate to High Binding Affinity

I. Flunitrazepam

1. Efficacy

We found that 1 mg flunitrazepam administered nightly for 1 week did not result in statistically significant improvement in sleep induction or sleep maintenance (Table 2) (BIXLER et al. 1977). In separate studies, 2 mg flunitrazepam was assessed for short- and long-term efficacy for periods of three and seven (BIXLER et al. 1977) and 28 consecutive nights (SCHARF et al. 1979) of drug administration. Sleep induction and maintenance was improved in all three studies with short-term drug administration. However, in the study where drug administration extended over 28 consecutive nights, tolerance developed for sleep maintenance during the intermediate- (Fig. 1) and long-term drug administration periods (SCHARF et al. 1979).

A number of studies by other investigators have confirmed the short-term efficacy of 2 mg flunitrazepam in inducing and maintaining sleep (CERONE et al. 1974; MONTI et al. 1974; HAURI and SILBERFARB 1976; JOVANOVIC 1977), while one of these studies also demonstrated continued efficacy over a long-term period (28 nights) of drug administration (ROSEKIND et al. 1979). Another study included a 16-day period of drug administration; total wake time was nonsignificantly decreased with short-term drug administration, with tolerance to this effect developing with continued use (MONTI et al. 1974).

2. Withdrawal Effects

Withdrawal of flunitrazepam has been shown to produce rebound insomnia on the first withdrawal night both after short-term use of 1 mg (BIXLER et al. 1977) and after long-term use of 2 mg (Fig. 2) (SCHARF et al. 1979). Surprisingly, rebound insomnia was not present following withdrawal of flunitrazepam in two separate short-term studies with the 2-mg dose (BIXLER et al. 1977).

Five sleep laboratory studies were conducted by other investigators in which there was an assessment for rebound insomnia following the withdrawal of 2 mg flunitrazepam (HAURI and SILBERFARB 1976; CERONE et al. 1974; JOVANOVIC 1977; MONTI et al. 1974; ROSEKIND et al. 1979). In all five studies, the withdrawal values cited were for the entire withdrawal period rather than individual nights. Thus, it is not surprising that rebound insomnia was not reported for any of these studies, as the presence of rebound insomnia on any given individual night could be diluted by averaging values over several nights. Nevertheless, even under these conditions, values for wakefulness appeared to approximate those for rebound insomnia in one study (ROSEKIND et al. 1979).

3. Side Effects

Several sleep laboratory studies have reported mild degrees of daytime sedation with 1 and 2 mg flunitrazepam (BIXLER et al. 1977; SCHARF et al. 1979).

Anterograde amnesia has also been reported to occur with flunitrazepam in the experimental setting, i.e., specially designed studies where information is presented during the peak drug activity shortly after administration of certain benzodiazepines, and the subjects are asked to recall it in the morning (GEORGE and DUNDEE 1977; KORTTILA et al. 1978; BIXLER et al. 1979c; KLEIN et al. 1986; SMIRNE et al. 1989). In one study of 2 mg flunitrazepam, one subject had total amnesia in the morning for being awakened and tested during the previous night (BIXLER et al. 1979c).

II. Loprazolam

1. Efficacy

Few sleep laboratory studies have assessed the hypnotic efficacy of loprazolam. Only one sleep laboratory study employed an adequate design and used insomniac subjects for the evaluation of loprazolam's efficacy (ADAM et al. 1984). In this study of 3 weeks of drug administration, 1.0 mg loprazolam was effective in significantly improving sleep with initial and continued use. However, there was evidence of some tolerance by the 3rd week of drug administration. In the same report, the 0.5-mg dose was not found to be an effective hypnotic dose.

2. Withdrawal Effects

Following withdrawal of 1 mg loprazolam after 3 weeks of drug administration, rebound insomnia was observed during the 3rd night of the 7-night withdrawal period (ADAM et al. 1984). In another sleep laboratory study in which the drug was administered to healthy volunteers at a dose of 1 mg for a 1-week period, rebound manifestations were not observed. However, the withdrawal period in this study consisted of only one night (SCHITTECATTE et al. 1988).

3. Side Effects

Daytime sedation has been the most frequently reported side effect with loprazolam (CLARK et al. 1986). However, several studies have found no difference in daytime sedation between 1 mg loprazolam and placebo (ELIE et al. 1983; KESSON et al. 1984; MCALPINE et al. 1984). Psychomotor decrements with loprazolam appear to be mild. In one study, the single nighttime dose of loprazolam 2 mg had a much higher incidence of dizziness and ataxia the following morning than 0.5 or 1 mg loprazolam, placebo, or 15 mg flurazepam (ELIE et al. 1983). However, in another single-dose study of loprazolam 0.5, 1 or 2 mg, and nitrazepam 5 mg, these differences were not found (FISHER et al. 1984). Long-term studies are associated with fewer unwanted effects than short-term studies, suggesting a tolerance to side effects despite maintenance of clinical efficacy (COSTA E SILVA and BORGES 1982). In a comparative study of loprazolam administered at bedtime at a 1-mg dose, and triazolam at a 0.5-mg dose, there was little subjective awareness of impairment in alertness, whereas in performance tests there were small decreases during the morning testing (MORGAN et al. 1984). Finally, in a clinical trial, daytime anxiety and early morning insomnia were reported for loprazolam but significantly more frequently with triazolam (MOON et al. 1985).

III. Lormetazepam

1. Efficacy

Lormetazepam was evaluated by our group in doses of 0.5, 1.0, 1.5 and 2.0 mg in four separate sleep laboratory protocols of 7 nights of drug administration followed by 3 withdrawal nights (KALES et al. 1982a). Although a moderate degree of efficacy was shown across all four doses, there was considerable inconsistency; no dose-response effect was observed for efficacy on either the first 3 or last 3 nights of this short-term drug administration period (Table 2). In general, there was less efficacy on later drug nights, indicating a potential for the development of tolerance over a relatively short period of drug use.

In another study, ADAM and OSWALD (1984) found that lormetazepam was effective in improving sleep in doses of 1.0 and 2.5 mg. Whereas our findings suggested some development of tolerance for the drug's effects even with short-term drug administration, they reported that both doses of the drug were effective, even after a 3-week period of drug administration. One possible explanation for the differences between studies is that their subjects were somewhat older than ours and consequently more sensitive to the drug's effects.

2. Withdrawal Effects

In our study (KALES et al. 1982a), although there was no dose-response effect for efficacy, there was a dose-related worsening of sleep above baseline levels following drug withdrawal with rebound insomnia occurring following withdrawal of the 2.0-, 1.5- and 1.0-mg doses. The degree of rebound insomnia noted with the 1.5-mg dose is illustrated in Fig. 2. An even greater dose-response effect was noted when making comparisons between each group for the single withdrawal night that showed the greatest increase in total wake time over the respective baseline period. These data are in agreement with those of Adam and Oswald (1984), who reported rebound insomnia for the 2.5-mg dose for the first three nights after withdrawal and sleep disturbance of a nonsignificant degree for the 1-mg dose.

3. Side Effects

Reports of daytime sedation were present in almost half of the subjects in our dose-response study with lormetazepam (Kales et al. 1982a). These side effects did not appear to be persistent, severe or dose related.

H. 1,4-Benzodiazepines with Rapid Elimination and High Binding Affinity

I. Midazolam

1. Efficacy

Doses of 10, 20 and 30 mg midazolam were assessed in three separate protocols, each of which included four placebo-baseline nights, seven drug nights and three placebo-withdrawal nights (Kales et al. 1983b). Only a slight to moderate degree of effectiveness was shown across the three doses with no dose-response effect for efficacy with either initial or continued drug administration (Table 2).

Additional studies of 30 mg midazolam showed significant decreases in total wake time and wake time after sleep onset (Monti et al. 1982, 1983); neither of these studies reported tolerance following 2 weeks of drug use. In another study, sleep latency was significantly decreased and sleep duration significantly increased with administration of midazolam 30 mg (Monti et al. 1987).

Two sleep laboratory studies with 20 mg midazolam showed statistically significant improvement in sleep (Hauri et al. 1983; Vogel and Vogel 1983). Both studies demonstrated short-term efficacy, and one also showed intermediate-term efficacy (Vogel and Vogel 1983). Results regarding the efficacy of the 15-mg dose have been fairly consistent; eight studies reported that this dose was effective in improving sleep (Scollo-Lavizzari 1983; Ziegler et al. 1983; Vogel and Vogel 1983; Borbély et al. 1985; Roth et al. 1985; Monti et al. 1987; Kripke et al. 1990; Lamphere et al. 1990), whereas two studies (Gath et al. 1983; Hauri et al. 1983) reported little efficacy with this dose.

Across all of the midazolam studies with 10, 15, 20 and 30 mg, the results are inconsistent in regard to rapid development of tolerance with continued use. While several studies showed efficacy with continued drug administration (MONTI et al. 1982, 1983, 1987; LAMPHERE et al. 1990), some showed rapid development of tolerance (KALES et al. 1983b; KRIPKE et al. 1990).

2. Withdrawal Effects

In our dose-response study of 10, 20 and 30 mg midazolam, there was a statistically significant and dose-related occurrence of rebound insomnia following withdrawal of midazolam (KALES et al. 1983b). The degree of rebound insomnia that occurred with 20 mg midazolam is illustrated in Fig. 2. Rebound insomnia was shown to occur both with the 20-mg and 30-mg doses of midazolam in four other studies (MONTI et al. 1982, 1983; VOGEL and VOGEL 1983; GATH et al. 1983). With the 15-mg dose, one study found rebound insomnia (MONTI et al. 1982), while four did not (VOGEL and VOGEL 1983; GATH et al. 1983; MONTI et al. 1987; LAMPHERE et al. 1990).

3. Side Effects

Sleep laboratory studies with midazolam have produced sporadic daytime sedation effects that seemed to be unrelated to dose. In one study (KALES et al. 1983b), three subjects out of six who were given 10 mg midazolam reported manifestations of hangover, while only one subject in each group of six receiving 20 and 30 mg, respectively, reported daytime sedation.

Six studies have demonstrated significant memory impairment with midazolam when subjects were tested several hours after drug administration, i.e., at the time of peak drug concentration and sedation (MONTI et al. 1982; HAURI et al. 1983; SUBHAN and HINDMARCH 1983; VOGEL and VOGEL 1983; LANGLOIS et al. 1987; BORBELY et al. 1988). While next-day amnesia has not been reported in clinical case reports, it has been noted as a side effect associated with midazolam use in the sleep laboratory (KALES et al. 1983b).

While there are no controlled studies reporting daytime anxiety with midazolam, this drug was reported to produce early morning insomnia under the conditions of continued drug administration and a rapid development of tolerance to the drug's initial efficacy (KALES et al. 1983c).

I. Triazolobenzodiazepines

I. Brotizolam

1. Efficacy

There are only a few sleep laboratory studies of brotizolam's efficacy that are adequately designed (VELA-BUENO et al. 1983; MAMELAK et al. 1983a; ROEHRS et al. 1983a). Brotizolam in a 0.25-mg dose was evaluated in a 10-

night sleep laboratory study of normal subjects and demonstrated significant efficacy (VELA-BUENO et al. 1983). In another sleep laboratory study, 0.5 mg brotizolam was administered to insomniac patients for 14 nights (MAMELAK et al. 1983a). While efficacy was initially observed, a considerable degree of tolerance developed at the end of the 2-week period of drug administration. The 0.25- and 0.5-mg doses of brotizolam were assessed in another sleep laboratory study (ROEHRS et al. 1983a). The 0.5-mg dose was found to be more consistently effective; it reduced sleep latency, number of awakenings and wake time after sleep onset.

2. Withdrawal Effects

Rebound insomnia was observed following withdrawal of 0.25 mg brotizolam in normal subjects (VELA-BUENO et al. 1983). One study (ROEHRS et al. 1983a) found some indication of rebound insomnia following withdrawal after three nights of administration of either 0.25 or 0.5 mg. In other studies, rebound insomnia was reported to occur in insomniac patients on the first withdrawal night following 2 weeks of administration of 0.5 mg brotizolam (MAMELAK et al. 1983a) and after 3 weeks of administration of the 0.25-mg and 0.5-mg doses of the drug in separate clinical trials (RICKELS et al. 1986).

3. Side Effects

Brotizolam was associated in one study with an increase in reports of daytime anxiety/tension as compared to baseline (VELA-BUENO et al. 1983).

II. Estazolam

1. Efficacy

Estazolam's hypnotic efficacy has been assessed in only a few sleep laboratory studies (ISOZAKI et al. 1976; ROTH et al. 1983; ROEHRS et al. 1983b; LAMPHERE et al. 1986), with one of these studies having a design making interpretation of the results difficult (ISOZAKI et al. 1976). Estazolam 2 mg (ISOZAKI et al. 1976; ROTH et al. 1983; ROEHRS et al. 1983b) and 4 mg (ISOZAKI et al. 1976) were found to be effective with initial use. In a dose-response study in the sleep laboratory, the 0.25-, 0.5-, 1.0-mg and 2.0-mg doses were reported to be effective in significantly improving wake time after sleep onset in a dose-response manner but did not reduce sleep latency significantly (ROEHRS et al. 1983b). Only one study (LAMPHERE et al. 1986) assessed hypnotic efficacy of 2.0 mg estazolam beyond 1 week of drug administration. No tolerance was reported in this study with long-term use over a 6-week period.

A recent sleep laboratory study evaluated the hypnotic effects of 1 mg estazolam in a sample of elderly insomniacs (VOGEL and MORRIS 1992). Patients received placebo for 2 weeks (baseline), estazolam for the next 4 weeks and placebo for 2 weeks (withdrawal). Sleep recordings were obtained only on the first two nights of each week. Efficacy was demonstrated on the first two nights of treatment. However, by the 2nd week of treatment, there was clear evidence of development of tolerance.

2. Withdrawal Effects

Although several studies (ISOZAKI et al. 1976; ROEHRS et al. 1983b; ROTH et al. 1983; LAMPHERE et al. 1986; VOGEL and MORRIS 1992) have been conducted on estazolam's hypnotic efficacy, data on estazolam's withdrawal effects are very limited. Several of the sleep laboratory studies either did not include a withdrawal period (ROEHRS et al. 1983b), included a single withdrawal night following only a single night of drug administration (ISOZAKI et al. 1976), or did not report wakefulness measurements during the withdrawal period (ROTH et al. 1983). One study evaluating 1.0 mg over a 4-week period of administration in elderly insomniacs reported rebound insomnia on the 1st withdrawal night (VOGEL and MORRIS 1992). Another sleep laboratory study assessed withdrawal following 6 weeks of nightly drug use and reported a significant increase in wake time after sleep onset and a nonsignificant decrease of sleep efficiency during the second withdrawal night (LAMPHERE et al. 1986). On an individual basis, 7 out of the 12 subjects exhibited worsened sleep during the first and second withdrawal nights. In addition, two multicenter, double-blind clinical trials of the drug failed to include a withdrawal period in their design (WALSH et al. 1984; SCHARF et al. 1990a). The failure to study withdrawal effects is a major design flaw, particularly for a hypnotic drug that is relatively rapidly eliminated and structurally belongs to the group of the triazolobenzodiazepines, a class of drugs that is associated with major difficulties following withdrawal.

3. Side Effects

For its 1st year on the market, as reported in the FDA's Spontaneous Reporting System (SRS), there were high rates reported for a number of psychiatric adverse events for estazolam including: amnesia and other cognitive impairments; hyperexcitability rates during drug administration; and withdrawal difficulties (TSONG 1992).

III. Triazolam

1. Efficacy

Triazolam, in a dose of 0.5 mg, has consistently been shown to be effective across a number of sleep laboratory studies. Our initial sleep laboratory study of triazolam 0.5 mg extended for 22 nights and included four placebo-baseline nights, a 2-week period of nightly drug administration and four placebo-withdrawal nights (KALES et al. 1976c). With short-term use, both sleep induction and sleep maintenance improved (Table 2). However, by the end of 2 weeks of triazolam use, none of the efficacy parameters was significantly decreased from baseline, as there was a rapid development of tolerance (Fig. 1).

In a number of additional sleep laboratory studies of 0.5 mg triazolam, drug administration periods extended for four (VOGEL et al. 1976), seven (VOGEL et al. 1975), 14 (ROTH et al. 1976; MAMELAK et al. 1984), 21 (PEGRAM et al. 1980; ADAM et al. 1984) and 35 (MITLER et al. 1984) consecutive nights, while one study assessed intermittent use (KALES et al. 1991). In general, 0.5 mg triazolam was found to be effective both for inducing and maintaining sleep during short-term administration. Five of these studies extended beyond short-term administration; however, in two of these studies, total recording time was either not held constant or not specified (ROTH et al. 1976; PEGRAM et al. 1980). In another study, development of tolerance was noted within 3 weeks of continued use (ADAM et al. 1984).

Triazolam, in a dose of 0.25 mg, has demonstrated only limited or marginal efficacy in non-elderly insomniacs, for whom it has been the recommended starting dose in the United States since 1988. In our group's second study of triazolam (KALES et al. 1986b), this at a 0.25-mg dose, hypnotic efficacy was limited and nonsignificant with initial and short-term use. Continued use of triazolam for a 2-week period resulted in a rapid development of tolerance to even this limited efficacy; total sleep time at the end of 2 weeks of drug administration was only 7 min greater than on baseline. Of a total of 13 sleep laboratory studies in non-elderly insomniacs or normal volunteers with induced insomnia (ROTH et al. 1977; FERNANDEZ-GUARDIOLA and JURADO 1981; KALES et al. 1986b; SEIDEL et al. 1986; JOHNSON et al. 1987; BONNET et al. 1988; O'DONNELL et al. 1988; BALKIN et al. 1989; SCHARF et al. 1990b,c; SCHWEITZER et al. 1991; LEE 1992; VOGEL 1992), six studies did not show significant improvement in sleep (FERNANDEZ-GUARDIOLA and JURADO 1981; KALES et al. 1986b; SEIDEL et al. 1986; O'DONNELL et al. 1988; BALKIN et al. 1989; SCHARF et al. 1990c). However, because of the highly precise and objective measurements in the sleep laboratory and their strong statistical power, such studies have a strong probability of demonstrating a hypnotic drug's efficacy, if present, with only a small sample size. Additionally a number of the seven studies showing significant improvement in sleep with 0.25 mg triazolam did so by utilizing sample sizes much larger than those usually employed in the sleep laboratory ($N = 6 - 8$) and able to demonstrate significant efficacy. Thus, of the 13 sleep laboratory evaluations of 0.25 mg triazolam, only three of the eight studies which had a sample size of less than 12 subjects showed significant efficacy while four out of five studies which had a sample size of 12 or more subjects demonstrated efficacy. In the discussion of their study, Seidel and his colleagues raise the possibility that the lack of efficacy for 0.25 mg triazolam may be due to the occurrence of early morning insomnia (SEIDEL et al. 1986). Finally, there is even one study in which 0.25 mg triazolam when administered to elderly insomniacs did not show significant improvement in sleep (MOURET et al. 1990).

There are two sleep laboratory studies assessing the efficacy of 0.125 mg triazolam in both elderly and non-elderly insomniacs (ROEHRS et al. 1985; SCHARF et al. 1990c). One study in elderly insomniacs showed that 0.125 mg triazolam induced and maintained sleep and increased total sleep time over the two nights of assessment (ROEHRS et al. 1985). Another study in non-elderly patients with chronic insomnia showed that triazolam 0.125 mg was initially effective in inducing and maintaining sleep, but lost its effectiveness with intermediate-term administration over a 2-week period (SCHARF et al. 1990c).

2. Withdrawal Effects

Our first study of 0.5 mg triazolam showed the presence of rebound insomnia upon withdrawal of the drug following 2 weeks of nightly administration (KALES et al. 1976c). Among the three withdrawal nights, rebound insomnia was most severe on the first night when total wake time was 130% greater than on baseline (Fig. 2). There is considerable agreement that withdrawal of triazolam consistently produces rebound insomnia. There are an additional 10 controlled studies that demonstrate rebound insomnia following withdrawal of 0.5 mg triazolam (VOGEL et al. 1975, 1976; ROTH et al. 1976; MITLER et al. 1984; ADAM et al. 1984; MAMELAK et al. 1984; SOLDATOS et al. 1986; GREENBLATT et al. 1987; MAMELAK et al. 1990; KALES et al. 1991). In one of these studies, although ROTH et al. (1976) did not report rebound insomnia in their study of 0.5 mg triazolam, there was clear evidence for rebound insomnia when we subsequently reanalyzed their data (KALES et al. 1978). Of special interest are two recent studies assessing withdrawal effects of 0.5 mg triazolam after only one night of drug administration (MAMELAK et al. 1990; KALES et al. 1991). Both studies showed significant increases in wakefulness above baseline following the withdrawal of triazolam, i.e., the occurrence of rebound insomnia, even after use of the drug for only one night. The consistency of frequent and severe rebound insomnia occurring following the withdrawal of triazolam is emphasized by the fact that only one sleep laboratory study did not report rebound insomnia following withdrawal of 0.5 mg triazolam (PEGRAM et al. 1980). This study, however, had a number of methodologic limitations.

Rebound insomnia has also been demonstrated with the 0.25-mg dose even though this dose has marginal efficacy (KALES et al. 1986b). In fact, in this study, the degree of sleep difficulty following withdrawal was much greater than the nonsignificant improvement in sleep during drug administration. Of the remaining 12 sleep laboratory studies evaluating 0.25 mg in non-elderly insomniacs or volunteers with induced insomnia, only three evaluated withdrawal (ROTH et al. 1977; SCHARF et al. 1990c; LEE 1992). Of these three, one reported rebound insomnia (SCHARF et al. 1990c). In another study in geriatric insomniacs that compared the efficacy and withdrawal effects of 0.25 mg triazolam and 7.5 mg zopiclone, rebound insomnia was not found following withdrawal of triazolam (MOURET et al. 1990). However, in this study, triazolam was not effective in improving sleep with drug administration.

In addition, there are 10 clinical case reports describing sleep disturbances following withdrawal of triazolam (See KALES 1991; PUBLIC CITIZEN 1992 for reviews). Furthermore, in the FDA's SRS data, the rates for withdrawal and dependency problems associated with triazolam were much higher than those with temazepam or flurazepam (BIXLER et al. 1987; ANELLO 1989).

3. Side Effects

A controversy over the safety of Halcion began in 1979 (VAN DER KROEF 1979). Now, however, many major independent sources of data and information including controlled studies, clinical case reports and post-marketing surveillance data from drug regulatory agencies collectively document the high frequency and severity of safety problems of Halcion. These studies (See KALES 1991; PUBLIC CITIZEN 1992; O'DONOVAN and MCGUFFIN 1993 for reviews) show that frequent and serious CNS and psychiatric adverse reactions are associated with triazolam in doses ranging from 0.125 to 1.0 mg. These psychiatric adverse reactions include: memory impairment/amnesia and other cognitive impairments such as confusion, disorientation, paranoia and other delusions and hallucinations which are indicative of organic mental disorders (organic brain syndromes); hyperexcitability states during drug administration including early morning insomnia and daytime anxiety, tension or panic attacks, disinhibition and agitated episodes; and hyperexcitability states following drug withdrawal including rebound insomnia, rebound anxiety and seizures. As a consequence of these cognitive impairments and hyperexcitability states, unusual and aberrant behavior including bizarre and violent actions has been reported frequently to be associated with triazolam use and has led to numerous litigations.

In six controlled studies assessing memory several hours after drug administration, triazolam produced significantly more impairment than other benzodiazepine drugs (flurazepam, lorazepam, secobarbital) or placebo (ROTH et al. 1980; SPINWEBER and JOHNSON 1982; ROEHRS et al. 1983c; GREENBLATT et al. 1989, 1991; PENETAR et al. 1989). Two other controlled studies recorded frequent and significant next-day memory impairment/amnesia with triazolam but not with temazepam (SCHARF et al. 1988; BIXLER et al. 1991). Furthermore, 28 case reports (11 with 0.25 mg and/or 0.125 mg triazolam) cite amnesia and other cognitive impairments (see KALES 1991; PUBLIC CITIZEN 1992 for reviews). Finally, in the FDA's SRS data, amnesia is reported at very high ratios for triazolam; in one analysis, 47.5/1 and 47.5/0 versus temazepam and flurazepam, respectively (BIXLER et al. 1987), and in another, 40.5/1 versus temazepam (WYSOWSKI and BARASH 1991).

Increased daytime anxiety, tension or panic and early morning insomnia were demonstrated in eight controlled studies in which triazolam was administered for at least 5 days (MORGAN and OSWALD 1982; CARSKADON et al. 1982; KALES et al. 1983c; MOON et al. 1985; BAYER et al. 1986; KALES et al. 1986b; BLIWISE et al. 1988; DETULLIO et al. 1989; ADAM and OSWALD 1989), while one controlled study showed a nonsignificant increase in morning tension (BLIWISE et al. 1988). Furthermore, anxiety, nervousness, tremulousness, agitation, mania or early morning insomnia were noted in 10 clinical case reports, three with 0.25 mg and/or 0.125 mg (see KALES 1991; PUBLIC CITIZEN 1992 for reviews). Finally, SRS data showed much higher rates for hyperexcitability for triazolam than for temazepam or flurazepam (BIXLER et al. 1987; ANELLO 1989; WYSOWSKI and BARASH 1991).

J. New Non-Benzodiazepine Hypnotics

I. Zopiclone

1. Efficacy

The sleep-inducing and -maintaining properties of 7.5 mg zopiclone have been evaluated in a number of sleep laboratory studies (JOVANOVIC and DREYFUS 1983; MAMELAK et al. 1983b; FLEMING et al. 1988; PECKNOLD et al. 1990; MOURET et al. 1990; PETRE-QUADENS et al. 1983). In one study with a 3-week drug administration period, efficacy was demonstrated with short-term administration (MAMELAK et al. 1983b). However, at the end of the drug administration period, none of these variables was significantly different from baseline, indicating the rapid development of tolerance. A second study had a number of methodologic limitations (PETRE-QUADENS et al. 1983). In a third study, 7.5 mg zopiclone was administered for 2 weeks, preceded and followed by four and ten nights of placebo (JOVANOVIC and DREYFUS 1983). The data showed a significant decrease in sleep latency, number of awakenings and total wake time, and a significant increase in total sleep time. The changes were present throughout the entire drug period. However, in this study, the data were difficult to interpret because the duration of the actual washout period was unclear as the baseline nights were defined additionally as being part of the washout. In a fourth study, 7.5 mg zopiclone was assessed in the sleep laboratory with elderly insomniac patients (MOURET et al. 1990). The drug was found to be significantly effective for increasing total sleep time both with short- and intermediate-term administration.

Two studies that evaluated the long-term efficacy of 7.5 mg zopiclone provided conflicting findings. Both of these studies, however, had methodologic limitations. In one study, in which the drug was given for 17 weeks (FLEMING et al. 1988), significant efficacy was demonstrated, with tolerance developing at the 5th week. In another sleep laboratory evaluation, long-term efficacy of zopiclone was reported with no tolerance occurring throughout an 8-week administration period (PECKNOLD et al. 1990).

2. Withdrawal Effects

Several studies have evaluated the withdrawal effects of 7.5 mg zopiclone in the sleep laboratory (PETRE-QUADENS et al. 1983; JOVANOVIC and DREYFUS 1983; MAMELAK et al. 1983b; MOURET et al. 1990; PECKNOLD et al. 1990). One study that assessed three nights of withdrawal following 3 weeks of drug administration did not demonstrate rebound insomnia both on the first withdrawal night or throughout the entire withdrawal period (MAMELAK et al. 1983b). Another study that used a protocol that included 10 withdrawal nights after 2 weeks of drug use also showed no rebound insomnia during the early and late conditions of withdrawal (JOVANOVIC and DREYFUS 1983). In this study, however, no analysis of individual nights was reported. Another study that did not report withdrawal manifestations after discontinuation of zopiclone 7.5 mg also used inadequate methods to assess rebound insomnia (PETRE-QUADENS et al. 1983). In addition to not stating clearly if patients were given placebo during the withdrawal period, the recorded nights were the 11th to 13th after drug discontinuation. Because zopiclone has a short elimination half-life, rebound phenomena should only be observed in the first few nights of withdrawal. Another study of the effects of 7.5 mg zopiclone in geriatric insomniacs did not report rebound insomnia during the three nights of withdrawal following 15 nights of administration (MOURET et al. 1990). However, the mean increase of total wake time (34%) over the three-night period and the more than 1 h decrease of total sleep time during the first withdrawal night suggested strong sleep disturbance.

Of the two studies evaluating the long-term effects of 7.5 mg zopiclone in the sleep laboratory (FLEMING et al. 1988; PECKNOLD et al. 1990), only one reported data for the withdrawal period (PECKNOLD et al. 1990). Further, the assessment of rebound insomnia in this study was methodologically limited; sleep recordings were obtained only on the 1st, 7th, 8th, 13th and 14th nights of the 2-week withdrawal period. Thus, although no rebound insomnia was found, there was a lack of thorough assessment of the first several nights following drug withdrawal. Also, it should be noted that one subject dropped out of the study due to "intense rebound insomnia."

In summary, only two of the five studies that evaluated withdrawal of zopiclone did so adequately and, in one of these, reassessment of the findings suggested the presence of rebound insomnia.

Finally, clinical studies have shown that rebound insomnia may occur after discontinuation of zopiclone. Rebound insomnia was noted in two studies following withdrawal after 2 (Dorian et al. 1983) or 3 (Lader and Frcka 1987) weeks of drug administration to normal subjects.

3. Side Effects

There are few reports of excessive daytime sedation with zopiclone in either clinical or sleep laboratory studies. In contrast, there are several published reports regarding unexpected side effects. Thus, memory impairment has been reported, as well as other cognitive impairments (Fleming et al. 1990; Who 1990; Drewes et al. 1991). Also, there have been some reports of hyperexcitability from various spontaneous reporting systems for adverse events (Who 1990).

II. Zolpidem

1. Efficacy

There have been eight studies with adequate methodology that have evaluated the hypnotic properties of zolpidem in the sleep laboratory in insomniac patients (Herrmann et al. 1988; Oswald and Adam 1988; Monti 1989; Kryger et al. 1991; Declerck et al. 1992; Scharf et al. 1994; Kummer et al. 1993). Another study consisted of only one night of administration to normal subjects with induced transient insomnia (Walsh et al. 1990).

Five studies have assessed efficacy using 10 mg zolpidem (Oswald and Adam 1988; Monti 1989; Kryger et al. 1991; Declerck et al. 1992; Scharf et al. 1994); in all five studies initial efficacy was noted. Two of these studies evaluated continued efficacy and reported no development of tolerance (Monti 1989; Scharf et al. 1994). One of these studies also evaluated 15 mg zolpidem and found both initial and continued efficacy (Scharf et al. 1994). Finally, two studies evaluated 20 mg zolpidem and found it to be efficacious both initially as well as following 4 weeks (Herrmann et al. 1988) and 6 months of drug administration in elderly psychiatric patients (Kummer et al. 1993), the latter study in elderly psychiatric patients.

2. Withdrawal Effects

Rebound insomnia during initial withdrawal was assessed in five of the eight sleep laboratory studies evaluating efficacy. Rebound insomnia was not found in two sleep laboratory studies of 10 mg zolpidem (Monti 1989; Scharf et al. 1994). However, in another study of 10 mg zolpidem, although the authors stated that there was no rebound insomnia, a significant decrease in total sleep time compared to baseline was noted (Kryger et al. 1991). In the higher dosages, Scharf et al. (1994) reported rebound insomnia following withdrawal of 15 mg zolpidem, while Herrmann et al. (1988) did not report rebound insomnia following withdrawal of 20 mg.

3. Side Effects

In some studies, daytime sedation has been reported to be the most frequent side effect associated with zolpidem (HERRMANN et al. 1988). However, there are several reports of cognitive impairments (amnesia, confusion, hallucinations) associated with zolpidem use (ANSSEAU et al. 1992; MAAREK et al. 1992; BALKIN et al. 1992; IRUELA et al. 1993; SCHARF et al. 1994). Also, hyperexcitability was found in a controlled study (KRYGER et al. 1991), and there have been some isolated case reports of daytime anxiety (MONTI 1989). Further, in geriatric patients, a dose of 20 mg was associated with frequent problems of unsteadiness, faintness and giddiness and infrequently with unclear speech, nausea and vertigio (KUMMER et al. 1993). Finally, in a controlled study, two subjects were withdrawn because of dizziness, nausea and visual disturbance associated with the use of 15 mg zolpidem (SCHARF et al. 1994).

K. Conclusions on Efficacy of Hypnotic Drugs

As this chapter details, the vast majority of benzodiazepine and non-benzodiazepine hypnotic drugs are effective for inducing and maintaining sleep with initial and short-term use, i.e., for the first several nights or so. A different picture emerges with continued use over 1- (short-term use), 2- (intermediate-term use) and 4-week (long-term use) periods. Benzodiazepine hypnotic drugs that are slowly eliminated (and also have low receptor-binding affinity) have been found to have a relatively slow development of tolerance. Thus, a number of sleep laboratory studies have demonstrated significant efficacy for flurazepam and quazepam for periods up to 1 month of consecutive nightly administration with relatively little loss of the drugs' initial effectiveness. In contrast, hypnotic drugs which are relatively rapidly eliminated such as triazolam, temazepam and midazolam rapidly lose much of their initial effectiveness, in some cases after only 1 week of consecutive, nightly administration. These drugs, with the exception of temazepam, also have high or relatively high receptor-binding affinities.

A final comment on the efficacy of hypnotic drugs relates to the more recent trend toward using lower recommended starting doses, both for existing drugs on the market and for recently approved drugs. This trend has occurred because of an attempt by the manufacturers and regulatory agencies to avoid the relatively high frequency and severity of psychiatric adverse events noted for a number of hypnotic agents which are rapidly eliminated and have relatively high receptor-binding affinities. This has resulted in recommended doses for adult insomniacs (for example, 0.25 mg triazolam, 2 mg estazolam and 10 mg zolpidem) that have not been clearly demonstrated to be effective especially with continued use; investigators have shown efficacy in some sleep laboratory studies by using much larger samples than are usually needed in the sleep laboratory to demonstrate statistically significant hypnotic efficacy or by administering the hypnotic 30 min before bedtime thereby exaggerating the drug's effects on sleep latency. The larger samples used in this strategy may have resulted in statistically significant improvements in sleep when the actual changes are modest in magnitude and are thus of limited clinical relevance (see also Chap. 12, this volume).

L. Conclusions on Underlying Mechanisms for Adverse Events

For a number of years, the only benzodiazepines on the market were 1,4-benzodiazepines, which were slowly eliminated and had low receptor-binding affinities. It was well recognized that daytime sedation, an expected side effect, could occur frequently and, therefore, physicians should appropriately warn their patients and advise various precautions, as well as adjust the dose when needed. With the introduction of newer benzodiazepines with more rapid elimination and higher receptor-binding affinities, and especially triazolobenzodiazepines, which have these pharmacokinetic and pharmacodynamic properties as well as a unique triazolo chemical structure, many paradoxical or unexpected psychiatric adverse effects were reported.

A number of investigators have clouded this issue by claiming that all benzodiazepines are alike and differ in their adverse events primarily on the basis of dose. However, if one carefully reviews the scientific and clinical literature as detailed in this chapter, it is clear that the frequency and severity of psychiatric adverse effects among benzodiazepine drugs vary considerably. Further, this variance is accounted for by considering the pharmacokinetic and pharmacodynamic characteristics of each drug and any special properties related to its chemical structure as well as dose level.

I. Pharmacokinetic Factors

The importance of pharmacokinetic factors in the appearance of adverse events with hypnotic drugs is best exemplified by daytime sedation and withdrawal difficulties (rebound insomnia and rebound anxiety). Manifestations of CNS depressant effects of hypnotics generally are considered to be more prevalent with drugs that have long elimination half-lives; daytime sedation is more frequent, intense and prolonged with high doses and in the elderly. There is also strong agreement that following withdrawal of benzodiazepines, hyperexcitability states such as rebound insomnia are frequently present and to a strong degree with rapidly eliminated drugs and infrequently present and, if so, on a delayed basis and to a milder degree with slowly eliminated drugs. In fact, no investigator has demonstrated clear-cut rebound insomnia following the withdrawal of flurazepam or quazepam; only mild degrees of sleep disturbance have occasionally been reported (see KALES et al. 1983d for a review).

Daytime anxiety, tenseness or panic and early morning insomnia occurring during the use of benzodiazepine hypnotic drugs is another hyperexcitability state, but one that has not been studied as extensively as rebound insomnia. Nevertheless, when such adverse events have been observed in controlled studies or reported in clinical case studies, they have been invariably associated with the use of very rapidly eliminated drugs (see KALES 1991 and PUBLIC CITIZEN 1992 for reviews).

II. Pharmacodynamic Factors

Receptor-binding affinity is a contributing factor to the hyperexcitability states seen with benzodiazepines including those observed both during drug administration and following drug withdrawal. Thus, benzodiazepines with rapid elimination and relatively high binding affinities produce more frequent and severe adverse events related to hyperexcitability (see KALES and VGONTZAS 1990; KALES et al. 1994 for reviews). Thus, in the case of triazolam, which is very rapidly eliminated and has an extremely high receptor-binding affinity, there are numerous controlled studies and clinical case studies reporting daytime anxiety during drug administration and remarkable consistency among 12 sleep laboratory studies in demonstrating rebound insomnia following withdrawal (see KALES 1991 and PUBLIC CITIZEN 1992 for reviews).

In the case of memory impairment/amnesia and other cognitive impairments such as confusion, delusions and hallucinations, receptor-binding affinity appears to have a major role in determining their frequency and severity (see VGONTZAS et al. in press for a review). Thus, triazolam, which has an extremely high receptor-binding affinity, in contrast to temazepam which has a low receptor-binding affinity, produces frequent and severe cognitive impairments. It also appears that pharmacokinetics (rapid elimination) are a secondary factor in a drug's potential for producing cognitive impairments.

III. Special Properties Related to Chemical Structure

Triazolam and alprazolam were initially marketed as having unique therapeutic benefits and applications based on their triazolo structure. The accumulation of adverse events associated with the use of these medications has provided evidence that the special properties previously claimed to underlie their unique clinical properties can equally well account for their adverse event profiles (KALES and VGONTZAS 1990; KALES et al. 1994). However, there has been reluctance on the part of certain regulators and investigators to recognize the role of these special properties in regard to the occurrence of ADRs.

Based on experimental data, we have proposed that the more frequent and severe side effects with the newer triazolobenzodiazepines are related to an interaction of several factors including rapid elimination, high binding affinity and special chemical properties (see KALES and VGONTZAS 1990; KALES et al. 1994 for reviews). Triazolam, in addition to its ultra-short elimination half-life and very high binding affinity, has the unique chemical properties of the triazolobenzodiazepines. For example, triazolobenzodiazepines activate α_2-adrenoreceptors, inhibit platelet-activating factor (PAF), and suppress corticotropin-releasing hormone (CRH). Therefore, triazolobenzodiazepines, by activating α_2-adrenoreceptors, inhibiting PAF and suppressing CRH, seem to have a direct suppressing effect on the LC and the hypothalamic-pituitary-adrenal (HPA) axis. The initial dampening or suppression of the LC-NE system and HPA axis is followed by a rebound and activation of the LC-NE system and HPA axis, when the drug is rapidly eliminated from the system. The repetitive daily pattern of suppression followed by activation can additionally explain neurochemically (excessive NE and CRH) and neurophysiologically (kindling phenomenon) many of the behavioral side effects during administration and withdrawal of triazolam, such as daytime anxiety, panic, mania, hostility and withdrawal difficulties. Also, it can additionally explain the frequent memory impairment caused by triazolam.

Further, in recently analyzed data from the FDA's SRS, alprazolam compared to lorazepam was associated with significantly greater rates of panic, anxiety and withdrawal changes (KALES et al. 1994). These differences exist in spite of the fact that lorazepam and alprazolam share similar kinetics and dynamics. Finally, in the FDA's SRS data, of 329 drugs associated with at least one report of hostility, the largest number by far was reported for triazolam, with the second highest for alprazolam (WISE 1990). Thus, both sets of data support the significant role of the triazolo structure in the high rate of adverse events associated with the use of triazolobenzodiazepines.

M. Comments on Drug-Induced Sleep Stage Alterations

In this chapter and Chap. 14, we do not discuss the effects of benzodiazepines on sleep stages for a number of reasons: (1) sleep stage effects are fairly consistent across all benzodiazepine hypnotic and anxiolytic drugs; (2) a number of the sleep stage effects are quite nonspecific, for example, most psychotropic drugs will cause a delay in REM sleep onset, (3) The physiologic or clinical significance of drug-induced sleep stage alterations has not been determined (KALES and KALES 1975). Nevertheless, with these caveats in mind, we discuss here the general effects of benzodiazepines on sleep stages both during drug administration and following drug withdrawal.

The major effects of benzodiazepines on sleep architecture are on slow-wave sleep and REM sleep. Specifically, there is a progressive and marked decrease in the percentage of stages 3 and 4 sleep combined (slow-wave sleep) during drug administration and a very gradual return of this sleep stage to baseline following withdrawal. These changes have been demonstrated in many studies with numerous benzodiazepines independent of their pharmacokinetic and pharmacodynamic profiles. For example, it has been shown that administration of flurazepam (KALES et al. 1976a), lormetazepam (ADAM and OSWALD 1984), nitrazepam (ADAM and OSWALD 1982), temazepam (BIXLER et al. 1978; MITLER et al. 1979) and triazolam (KALES et al. 1976) are all associated with a marked to significant suppression of slow-wave sleep during drug administration with a very gradual return to baseline levels following drug withdrawal.

The administration of benzodiazepines is also associated with a mild suppression of REM sleep, which can be marked especially during the first third of the night, and manifested by a decrease in the percentage of REM sleep and/or a lengthening of REM latency. Following drug withdrawal, there is a rather rapid return of REM sleep to baseline. Again, these alterations appear to be independent of the pharmacokinetic and pharmacodynamic profiles of the drugs. For example, administration of flurazepam (KALES et al. 1976a), lormetazepam (ADAM and OSWALD 1984), nitrazepam (ADAM and OSWALD 1982), triazolam (KALES et al. 1976c) and, to a lesser degree, temazepam (BIXLER et al. 1978; MITLER et al. 1979) have been associated with a slight decrease of REM sleep and/or increase of REM latency while on withdrawal REM sleep levels returned quickly to baseline values.

Finally, the newer non-benzodiazepine hypnotics such as zopiclone and zolpidem are reported in some studies not to have a suppressing effect on slow-wave sleep (OSWALD and ADAM 1988; MONTI 1989; HINDMARCH and MUSCH 1990; SCHARF et al. 1992). However, as detailed in Chap. 2, this volume, computerized EEG analysis showed no differences in terms of spectral changes between benzodiazepines and zolpidem.

N. Summary

The use of hypnotic drugs is an important part of an overall, multidimensional approach to the treatment of insomnia. Over the years, the hypnotic drugs used have evolved from the bromides and barbiturates to benzodiazepines and newer compounds acting through the GABAergic system.

Within the benzodiazepine class of hypnotic drugs, 1,4-benzodiazepines (flurazepam) were first introduced; these drugs have slow rates of elimination and low receptor-binding affinities. Next, more rapidly eliminated and more potent 1,4-benzodiazepines (midazolam) were marketed along with the introduction of triazolobenzodiazepines (triazolam), which are rapidly eliminated and have high receptor-binding affinities in addition to their unique triazolo ring. More recently, non-benzodiazepine hypnotics acting through the GABAergic system (zopiclone, zolpidem) have been marketed.

Efficacy, development of tolerance for efficacy, adverse events and withdrawal effects of benzodiazepine hypnotics are highly related to:

pharmacokinetics; pharmacodynamics; and special properties related to chemical structure. With continued use, slowly eliminated benzodiazepines maintain significant improvement in sleep with little development of tolerance while a number of rapidly eliminated benzodiazepines lose most of their initial efficacy even after only 1 week of consecutive nightly use.

Rapid elimination is a primary factor in the potential for a benzodiazepine drug to produce hyperexcitability states including both those during drug administration (daytime anxiety, tension, or panic, and/or early morning insomnia) and those following drug withdrawal (rebound insomnia and rebound anxiety). In the case of memory impairment/amnesia and other cognitive impairments, receptor-binding affinity has a major role. Rapid elimination is a contributing factor to the frequency and severity of cognitive impairments, while binding affinity is a contributing factor to the frequency and severity of hyperexcitability states. Finally, special properties related to the triazolo ring of triazolobenzodiazepines contribute further to the frequency and severity of hyperexcitability states and cognitive impairments.

Flurazepam and quazepam are eliminated very slowly and have low receptor/binding affinities. These characteristics account for the following: a high degree of efficacy with initial and short-term use with little development of tolerance even after 1 month of consecutive nightly use; a relatively high rate of daytime sedation; extremely low rates of paradoxical or unexpected adverse events, including hyperexcitability states and cognitive impairments during drug administration; and no reports from controlled studies of rebound insomnia following withdrawal. At the other end of the pharmacokinetic and pharmacodynamic spectrum are benzodiazepines with rapid elimination and relatively high binding affinities. When a drug has extremely rapid elimination and high binding affinity and also the special properties of a triazolobenzodiazepine (e.g., triazolam), it produces frequent and severe psychiatric adverse events related to hyperexcitability and cognitive impairments.

Finally, the newer non-benzodiazepines, zopiclone and zolpidem, are claimed to be as efficacious as benzodiazepines and to have better profiles for adverse events. A number of cautionary comments are in order. These drugs have not been studied as thoroughly as others in the sleep laboratory and, in addition in the case of zolpidem, its withdrawal period has often not been assessed. Nevertheless, two of the five studies that assessed withdrawal of zolpidem reported rebound insomnia. With zopiclone, of the two studies that adequately assessed withdrawal, reassessment of the findings of one study suggested the presence of strong sleep disturbance during withdrawal. Also, with both drugs, there are a number of clinical reports of serious paradoxical or unexpected adverse events, i.e., psychotic reactions associated with their use.

References

Adam KI, Oswald I (1982) A comparison of the effects of chlormezanone and nitrazepam on sleep. Br J Clin Pharmacol 14:57–65

Adam KI, Oswald I (1984) Effects of lormetazepam and of flurazepam on sleep. Br J Clin Pharmacol 17:531–538

Adam KI, Oswald I (1989) Can a rapidly-eliminated hypnotic cause daytime anxiety? Pharmacopsychiatry 22:115–119

Adam KI, Adamason L, Brezinova V, Hunter W (1976) Nitrazepam: lastingly effective but trouble on withdrawal. Br Med J 1:1558–1560

Adam KI, Oswald I, Shapiro C (1984) Effects of loprazolam and of triazolam on sleep and overnight urinary cortisol. Psychopharmacology (Berl) 82:389–394

Anello C (1989) Adverse behavior reactions attributed to triazolam in the FDA's Spontaneous Reporting System. Pharmacologic Drugs Advisory Committee Meeting, 22 Sept, Washington

Ansseau M, Pitchot W, Hansenne M, Moreno AG (1992) Psychotic reactions to zolpidem. Lancet 339:809

Arendt RM, Greenblatt DJ, Liebisch DC, Luu MD, Paul SM (1987) Determinants of benzodiazepine uptake: lipophilicity versus binding affinity. Psychopharmacology (Berl) 93:72–76

Bailey L, Ward M, Musa MN (1994) Clinical pharmacokinetics of benzodiazepines. J Clin Pharmacol 34:804–811

Balkin TJ, O'Donnell VM, Kamimori GH, Redmond DP, Belenky G (1989) Administration of triazolam prior to recovery sleep: effects on sleep architecture, subsequent alertness and performance. Psychopharmacology (Berl) 99:526–531

Balkin TJ, O'Donnell VM, Wesensten N, McCann U, Belenky G (1992) Comparison of the daytime sleep and performance effects of zolpidem versus triazolam. Psychopharmacology (Berl) 107:83–88

Barley DN, Jatlon PI (1975) Barbital overdose and abuse. Am J Clin Pathol 64:291–296

Bayer AJ, Bayer EM, Pathy MSJ, Stoker MJ (1986) A double-blind controlled study of chlormethiazole and triazolam and hypnotics in the elderly. Acta Psychiatr Scand 73:104–111

Bixler EO, Kales A, Soldatos CR, Kales JD (1977) Flunitrazepam, an investigational hypnotic drug: sleep laboratory evaluations. J Clin Pharmacol 17:569–578

Bixler EO, Kales A, Soldatos CR, Scharf MB, Kales JD (1978) Effectiveness of temazepam with short-, intermediate-, and long-term use: sleep laboratory evaluation. J Clin Pharmacol 18:110–118

Bixler EO, Kales A, Soldatos CR (1979a) Sleep disorders encountered in medical practice: a national survey of physicians. Behav Med 6:1–6

Bixler EO, Kales A, Soldatos CR, Kales JD, Healey S (1979b) Prevalence of sleep disorders in the Los Angeles metropolitan area. Am J Psychiatry 136:1257–1262

Bixler EO, Scharf MB, Soldatos CR, Mitsky DJ, Kales A (1979c) Effects of hypnotic drugs on memory. Life Sci 25:1379–1388

Bixler EO, Kales A, Brubaker BH, Kales JD (1987) Adverse reactions to benzodiazepine hypnotics: spontaneous reporting system. Pharmacology 35:286–300

Bixler EO, Kales A, Manfredi RL, Vgontzas AN, Tyson KL, Kales JD (1991) Next-day memory impairment with triazolam use. Lancet 337:827–831

Bliwise DL, Seidel WF, Greenblatt DJ, Dement WC (1984) Nighttime and daytime efficacy of flurazepam and oxazepam in chronic insomnia. Am J Psychiatry 141:191–195

Bliwise DL, Seidel WF, Cohen SA, Bliwise NG, Dement WC (1988) Profile of mood states changes during and afte 5 weeks of nightly triazolam administration. J Clin Psychiatry 49:349–355

Bonnet MH, Dexter JR, Gillin JC, James SP, Kripke D, Mendelson W, Mitler M (1988) The use of triazolam in phase-advanced sleep. Neuropsychopharmacology 1:225–234

Borbély AA, Balderer G, Trachsel L, Tobler I (1985) Effects of midazolam and sleep deprivation on day-time sleep propensity. Arzneimittelforschung 35:1696–1699

Borbély AA, Schlapfer B, Trachsel L (1988) Effect of midazolam on memory. Arzneimittelforschung 38:824–827

Carskadon MA, Seidel WF, Greenblatt DJ, Dement WC (1982) Daytime carry-over of triazolam and flurazepam in elderly insomniacs. Sleep 5:361–371

Cerone G, Cirignotta R, Coccagna G, Milone F, Lion P, Lorizio A, Lugaresi E, Mantovani M, Muratorio A, Murri L, Mutani R, Riccio A (1974) All-night polygraphic recordings on the hypnotic effects of a new benzodiazepine: flunitrazepam (Ro 5-4200, Rohypnol). Eur Neurol 11:172–179

Church MW, Johnson LC (1979) Mood and performance of poor sleepers during repeated use with flurazepam. Psychopharmacology (Berl) 61:309–316

Clark BG, Jue SG, Dawson W, Ward W (1986) Loprazolam: a preliminary review of its pharmacodynamic and pharmacokinetic properties and therapeutic efficacy in insomnia. Drugs 31:500–516

Costa e Silva JA, Borges RS (1982) Long term loprazolam (Ru 31158) tolerance in psychiatric patients with insomnia. In: Koella W (ed) Sleep 1982. 6th European congress on sleep research, Zurich. Karger, Basel, pp 418–419

Declerck AC, Ruwe F, O'Hanlon JF, Wauquier A (1992) Effects of zolpidem and flunitrazepam on nocturnal sleep of women subjectively complaining of insomnia. Psychopharmacology (Berl) 106:497–501

Dement WC, Zarcone VP, Hoddes E, Smythe H, Carskadon M (1973) Sleep laboratory and clinical studies with flurazepam. In: Garrattini S, Mussini E, Randall LO (eds) The benzodiazepines. Raven, New York, pp 599–611

Dement WC, Carskadon MA, Mitler MM, Phillips RL, Zarcone VP (1978) Prolonged use of flurazepam: a sleep laboratory study. Behav Med 5:25–31

DeTullio PL, Kirking DM, Zacardelli DK, Kwee P (1989) Evaluation of long-term triazolam use in an ambulatory Veterans Administration medical center population. DICP Ann Pharmacother 23:290–293

Divoll M, Greenblatt DJ, Harmatz JS, Shader RI (1981) Effect of age and gender on disposition of temazepam. J Pharmacol Sci 70:1104–1107

Dorian P, Sellers EM, Kaplan H, Hamilton C (1983) Evaluation of zopiclone physical dependence liability in normal volunteers. Pharmacology 27:228–234

Drewes AM, Andreasen A, Jennum P, Nielsen KD (1991) Zopiclone in the treatment of sleep abnormalities in fibromyalgia. Scand J Rheumatol 20:288–293

Elie R, Caille G, Levasseur FA, Gareau J (1983) Comparative hypnotic activity of single doses of loprazolam, flurazepam and placebo. J Clin Pharmacol 23: 32–36

FDA (1978, 1980) Psychopharmacologic drugs advisory meeting minutes on evaluation of temazepam. Food and Drug Administration, Washington

Feinberg I, Carlson V (1968) Sleep variable as a function of age in man. Arch Gen Psychiatry 18:239–250

Fernandez C, Maradeix V, Ginenez F, Thuillier A, Farinotti R (1993) Pharmacokinetics of zopiclone and its enantiomers in caucasian young health volunteers. Drug Metab Dispos 21:1125–1128

Fernandez-Guardiola A, Jurado JL (1981) The effect of triazolam on insomniac patients using a laboratory sleep evaluation. Curr Ther Res 29:950–958

Fisher HBJ, Middleton RSW, Mahmoud OMM, Pines A, Ankier SI (1984) Loprazolam – a multicentre dose-ranging study in general medical and preoperative patients. Clin Trials J 21:109–120

Fleming JAE, Bourgouin J, Hamilton P (1988) A sleep laboratory evaluation of the long-term efficacy of zopiclone. Can J Psychiat 33:103–107

Fleming JAE, McClure DJ, Mayes C, Phillips R, Bourgouin J (1990) A comparison of the efficacy, safety and withdrawal effects of zopiclone and triazolam in the treatment of insomnia. Int Clin Psychopharmacol 6 [Suppl] 2:29–37

Frost JD, DeLucchi MR (1979) Insomnia in the elderly: treatment with flurazepam hydrochloride. J Am Geriatr Soc 27:541–546
Gardner DL, Cowdry RW (1985) Alprazolam-induced dyscontrol in borderline personality disorder. Am J Psychiatry 142:98–100
Gath I, Bar-On E, Rogowski Z, Bental E (1983) Automativ scoring of polygraphic sleep recordings: midazolam in insomniacs. Br J Clin Pharmacol 16:89S–96S
George KA, Dundee JW (1977) The relative amnesic actions of diazepam, flunitrazepam and lorazepam in man. Br J Clin Pharmacol 4:45–50
Greenblatt DJ, Shader RI (1985) Clinical pharmacokinetics of the benzodiazepines. In: Smith DE, Wesson DR (eds) The benzodiazepines: current standards for medical practice. MTP Press, Lancaster, pp 43–58
Greenblatt DJ, Divoll M, Harmatz JS, MacLaughlin DS, Shader RI (1981) Kinetics and clinical effects of flurazepam in young and elderly noninsomniacs. Clin Pharmacol Ther 30:475–486
Greenblatt DJ, Shader RI, Abernethy DR, Ochs HR, Divoll M, Sellers EM (1982) Benzodiazepines and the challenge of pharmacokinetic taxonomy. In: Usdin E, Skolnick P, Tallman JF Jr, Greenblatt D, Paul SM (eds) Pharmacology of benzodiazepines. National Institute of Health, Bethesda, pp 257–270
Greenblatt DJ, Harmatz JS, Zinny MA, Shader RI (1987) Effect of gradual withdrawal on the rebound sleep disorder after discontinuation of triazolam. N Engl J Med 317:722–728
Greenblatt DJ, Harmatz JS, Engelhardt N, Shader RI (1989) Pharmacokinetic determinants of dynamic differences among three benzodiazepine hypnotics. Flurazepam, temazepam, and triazolam. Arch Gen Psychiatry 46:326–322
Greenblatt DJ, Harmatz JS, Shapiro L, Engelhardt N, Gouthro TA, Shader RI (1991) Sensitivity to triazolam in the elderly. N Engl J Med 324:1691–1698
Gustavson LE, Carrigan PS (1990) The clinical pharmacokinetics of single doses of estazolam. Am J Med 88:2S–5S
Hauri P, Roth T, Sateia M, Zorick F (1983) Sleep laboratory and performance evaluaion of midazolam in insomniacs. Br J Clin Pharmacol 16:109S–114S
Hauri PJ, Silberfarb PM (1976) Effects of RO5-4200 on sleep. Sleep Res 5:65
Healey ES, Kales A, Monroe LJ, Bixler EO, Chamberlin K, Soldatos CR (1981) Onset of insomnia: role of life-stress events. Psychosom Med 43:439–451
Heel RC, Brogden RN, Speight TM, Avery GS (1981) Temazepam: a review of its pharmacological properties and therapeutic efficacy as an hypnotic. Drugs 21:323–337
Herrmann WM, Kubicki S, Wober W (1988) Zolpidem, a four-week pilot polysomnographic study in patients with chronic sleep disturbances In: Sauvanet JP, Langer SZ, Morselli PC (eds) Imidazopyridines in sleep disorders. Raven, New York, pp 261–278
Hindmarch I, Musch B (1990) Zopiclone in clinical practice. Int Clin Psychopharmacol 5 [Suppl 2]:1–158
Hollister LE, Csernansky JG (1990) Clinical pharmacology of psychotherapeutic drugs, 3rd edn. Churchill Livingstone, New York
Iruela LM, Ibañez-Rojo V, Baca E (1993) Zolpidem-induced macropsia in anorexic woman. Lancet 342:443
Isbell H (1950) Addiction to barbiturates and the barbiturate abstinence syndrome. Ann Intern Med 33:108–121
Isozaki H, Tanaka M, Inanaga K (1976) Effect of triazolobenzodiazepine derivative, estazolam, on all-night sleep pattern. Curr Ther Res 20:493–509
Johnson LC, Chernick DA (1982) Sedative-hypnotics and human performance. Psychopharmacology (Berl) 76:101–113
Johnson LC, Spinweber CL, Webb SC, Muzet AG (1987) Dose level effects of triazolam on sleep and response to a smoke detector alarm. Psychopharmacology (Berl) 91:397–402

Jovanovic UJ (1977) Polygraphic sleep recordings before and after the administration of flunitrazepam. J Int Med Res 5:77–84
Jovanovic UJ, Dreyfus JF (1983) Polygraphic sleep recordings in insomniac patients under zopiclone or nitrazepam. Pharmacology 27 [Suppl 2]:136–145
Kales A (1990) Quazepam: hypnotic efficacy and side effects. Pharmacotherapy 10:1–12
Kales A (1991) An overview of safety problems of triazolam. Int Drug Ther Newslett 26:25–28
Kales A, Kales JD (1983) Sleep laboratory studies of hypnotic drugs: efficacy and withdrawal effects. J Clin Psychopharmacol 3:140–150
Kales A, Kales JD (1984) Evaluation and treatment of insomnia. Oxford University Press, New York
Kales A, Kales JD (1975) Shortcomings in the evaluation and promotion of hypnotic drugs. N Engl J Med 293:826–827
Kales A, Scharf MB (1973) Sleep laboratory and clinical studies of the effects of benzodiazepines on sleep: flurazepam, diazepam, chlordiazepoxide, and RO 5-4200. In: Garrattini S, Mussini E, Randall LO (eds) The benzodiazepines. Raven, New York, pp 577–598
Kales A, Vgontzas A (1990) Not all benzodiazepines are alike. In: Stefanis CN, Rabavilas AD, Soldatos CR (eds) Psychiatry: a world perspective, vol 3. Elsevier, Amsterdam, pp 379–384
Kales A, Wilson T, Kales JD, Jacobson A, Paulson MJ, Kollar E, Walter RD (1967) Measurements of all-night sleep in normal elderly persons: effects of aging. J Am Geriatr Soc 15:405–414
Kales A, Allen C, Scharf MB, Kales JD (1970a) Hypnotic drugs and effectiveness: all-night EEG studies of insomniac subjects. Arch Gen Psychiatry 23:226–232
Kales A, Kales JD, Scharf MB, Tan T-L (1970b) Hypnotics and altered sleep-dream patterrns. II. All-night EEG strudies of chloral hydrate, flurazepam, and methalqualone. Arch Gen Psychiatry 23:219–225
Kales A, Kales JD, Bixler EO, Scharf MB (1975) Effectiveness of hypnotic drugs with prolonged use: flurazepam and pentobarbital. Clin Pharmacol Ther 18:356–363
Kales A, Bixler EO, Scharf MB, Kales JD (1976a) Sleep laboratory studies of flurazepam: a model for evaluating hypnotic drugs. Clin Pharmacol Ther 19:576–583
Kales A, Caldwell AB, Preston TA, Healey S, Kales JD (1976b) Personality patterns in insomnia. Arch Gen Psychiatry 33:1128–1134
Kales A, Kales JD, Bixler EO, Scharf MB, Russek E (1976c) Hypnotic efficacy of triazolam: sleep laboratory evaluation of intermediate-term effectiveness. J Clin Pharmacol 16:399–406
Kales A, Scharf MB, Kales JD (1978) Rebound insomnia: a new clinical syndrome. Science 201:1039–1041
Kales A, Scharf MB, Kales JD, Soldatos CR (1979) Rebound insomnia: a potential hazard following withdrawal of certain benzodiazepines. JAMA 241:1692–1695
Kales A, Scharf MB, Soldatos CR, Bixler E, Bianchi S, Schweitzer PK (1980a) Quazepam, a new benzodiazepine hypnotic: intermediate-term sleep laboratory evaluation. J Clin Pharmacol 20:184–192
Kales A, Soldatos CR, Kales JD (1980b) Taking a sleep history. Am Fam Physician 22:101–108
Kales A, Scharf MB, Bixler EO, Schweitzer PK, Jacoby JA, Soldatos CR (1981) Dose-response studies of quazepam. Clin Pharmacol Ther 30:194–200
Kales A, Bixler EO, Soldatos CR, Mitsky DJ, Kales JD (1982a) Dose-response studies of lormetazepam: efficacy, side effects, and rebound insomnia. J Clin Pharmacol 22:520–530
Kales A, Bixler EO, Soldatos CR, Vela-Bueno A, Jacoby J, Kales JD (1982b) Quazepam and flurazepam: long-term use and extended withdrawal. Clin Pharmacol Ther 32:781–788

Kales A, Caldwell AB, Soldatos CR, Bixler EO, Kales JD (1983a) Biopsychobehavioral correlates of insomnia II: MMPI pattern specificity and consistency. Psychosom Med 45:341–356
Kales A, Soldatos CR, Bixler EO, Goff PJ, Vela-Bueno A (1983b) Midazolam: dose-response studies of effectiveness and rebound insomnia. Pharmacology 26:138–149
Kales A, Soldatos CR, Bixler EO, Kales JD (1983c) Early morning insomnia with rapidly eliminated benzodiazepines. Science 220:95–97
Kales A, Soldatos CR, Bixler EO, Kales JD (1983d) Rebound insomnia and rebound anxiety: a review. Pharmacology 26:121–137
Kales A, Bixler EO, Soldatos CR, Vela-Bueno A, Jacoby JA, Kales JD (1986a) Quazepam and temazepam: effects of short- and intermediate-term use and withdrawal. Clin Pharmacol Ther 39:345–352
Kales A, Bixler EO, Vela-Bueno A, Soldatos CR, Niklaus DE, Manfredi RL (1986b) Comparison of short and long half-life benzodiazepine hypnotics: triazolam and quazepam. Clin Pharmacol Ther 40:378–386
Kales A, Manfredi RL, Vgontzas AN, Bixler EO, Vela-Bueno A, Fee EC (1991) Rebound insomnia after only brief and intermittent use of rapidly eliminated benzodiazepines. Clin Pharmacol Ther 49:468–476
Kales A, Vgontzas A, Bixler EO (1994) Not all benzodiazepines are alike: update 1993. In: Beigel A, Lopez Ibor JJ, Costa e Silva JA (eds) Past, present and future of psychiatry. IX World Congress of Psychiatry, vol II World Scientific, New Jersey, pp 942–946
Kales JD, Kales A, Bixler EO, Slye ES (1971) Effects of placebo and flurazepam on sleep patterns in insomniac subjects. Clin Pharmacol Ther 12:691–697
Kaplan SA, de Silva AF, Jack ML, Alexander K, Strojny N, Weinfeld RE, Puglisi CV, Weisman L (1973) Blood level profile in man following chronic oral administration of flurazepam hydrochloride. J Pharm Sci 62:1932–1935
Karacan I, Thornby JI, Anch M, Holzer CE, Warheit GJ, Schwab JL, Williams RL (1976) Prevalence of sleep disturbance in a primarily urban Florida County. Soc Sci Med 10:239–244
Kesson CM, Gray JMB, Lawson DH, Ankier SI (1984) Long-term efficacy and tolerability of a new hypnotic – loprazolam. Br J Clin Pract 38:306–312
Klein MJ, Patat A, Manuel C (1986) Etude des effets sur la mémoire et les performances psychomotrices induits chez le sjet sain par 3 benzodiazepines hypnotiques (triazolam, flunitrazepam, loprazolam). Therapie 41:299–304
Korttila K, Saarnivaara L, Tarkkanen J, Himberg J, Hytonen M (1978) The comparison of diazepam and flunitrazepam for sedation during local anesthesia for bronchoscopy. Br J Clin Anaesth 50:281–287
Kripke DF, Hauri P, Ancoli-Israel S, Roth T (1990) Sleep evaluation in chronic insomniacs during 14-day use of flurazepam and midazolam. J Clin Psychopharmacol 10:32S–43S
Kryger MH, Steljes D, Pouliot Z, Neufeld H, Odynski T (1991) Subjective versus objective evaluation of hypnotic efficacy: experience with zolpidem. Sleep 14: 399–407
Kummer J, Guendel L, Linden J, Eich FX, Ahali P, Coquelin JP, Kyrein HJ (1993) Long-term polysomnographic study of the efficacy and safety of zolpidem in elderly psychiatric in-patients with insomnia. J Int Med Res 21:171–184
Lader M, Frcka G (1987) Subjective effects during and on discontinuation of zopiclone and temazepam in normal subjects. Pharmacopsychiatry 20:67–71
Lamphere J, Roehrs T, Zorick F, Koshorek G, Roth T (1986) Chronic hypnotic efficacy of estazolam. Drugs Exp Clin Res 12:687–691
Lamphere J, Roehrs T, Vogel G, Koshorek G, Fortier J, Roth T (1990) The chronic efficacy of midazolam. Int Clin Psychopharmacol 5:31–39
Langlois S, Kreeft JH, Chouinard G, Ross-Chouinard A, East S, Ogilvie RI (1987) Midazolam: kinetics and effects on memory, sensorium, and haemodynamics. Br J Clin Pharmacol 23:273–278

Langtry HD, Benfield P (1990) Zolpidem, a review of its pharmacodynamic and pharmacokinetic properties and therapeutic potential. Drugs 40:291–313

Lee T (1992) Technical Report 9159-92-001 of protocol M/2100/0232, submitted on February 21, 1992; additional analysis submitted on March 24, 1992. Psychopharmacologic Drug Advisor Committee Meeting, May 18, Washington

Lugaresi E, Zucconi M, Bixler EO (1987) Epidemiology of sleep disorders. Psychiatr Ann 17:446–453

Maarek L, Cramer P, Ahali P, Coquelin JP, Morselli PL (1992) The safety and efficacy of zolpidem in insomniac patients: a long-term open study in general practice. J Int Med Res 162–170

Mamelak M, Csima A, Price V (1981) The effects of quazepam on the sleep of chronic insomniacs. Curr Ther Res 29:135–147

Mamelak M, Csima A, Price V (1983a) Effects of brotizolam on the sleep of chronic insomniacs. Br J Clin Pharmacol 16:377S–382S

Mamelak M, Csima A, Price V (1983b) Effects of zopiclone on the sleep of chronic insomniacs. Pharmacology 27:156–164

Mamelak M, Csima A, Price V (1984) A comparative 25-night sleep laboratory study on the effects of quazepam and triazolam on chronic insomniacs. J Clin Pharmacol 24:67–77

Mamelak M, Csima A, Price V (1990) The effects of a single night's dosing with triazolam on sleep the following night. J Clin Pharmacol 30:549–555

McAlpine CJ, Ankier SI, Elliott CSC (1984) A multicentre hospital study to compare the hypnotic efficacy of loprazolam and nitrazepam. J Int Med Res 12:229–237

Miller LG, Galpern WR, Byrnes JJ, Greenblatt DJ (1992) Benzodiazepine receptor binding of benzodiazepine hypnotics: receptor and ligand specificity. Pharmacol Biochem Behav 43:413–416

Mitler MM, Carskadon MA, Phillips RL, Sterling WR, Zarcone VP, Spiegel R, Guilleminault C, Dement WC (1979) Hypnotic efficacy of temazepam: a long-term sleep laboratory evaluation. Br J Clin Pharmacol 8:63S–68S

Mitler MM, Seidel BA, van den Hoed J, Greenblatt DJ, Dement WC (1984) Comparative hypnotic effects of flurazepam, triazolam, and placebo: a long-term simultaneous nighttime and daytime study. J Clin Psychopharmacol 4:2–13

Mitler MM, Browman CP, Menn SJ, Gujavarty K, Timms RM (1986) Nocturnal myoclonus: treatment efficacy of clonazepam and temazepam. Sleep 31:385–392

Monti JM (1989) Effect of zolpidem on sleep in insomnia patients. Eur J Clin Pharmacol 36:461–466

Monti J, Trenchi HM, Morales F, Monti L (1974) Flunitrazepam (Ro 5-4200) and sleep cycle in insomniac patients. Psychopharmacologia (Berl) 35:371–379

Monti JM, Debellis J, Gratadoux E, Alterwain P, Altier H, D'Angelo L (1982) Sleep laboratory study of the effects of midazolam in insomniac patients. Eur J Clin Pharmacol 21:479–484

Monti JM, Alterwain P, Debellis J, Altier H, Pellejero T, Monti D (1983) Short-term sleep laboratory evaluation of midazolam in chronic insomniacs. Arznheim Forsch/Drug Res 37:54–57

Monti JM, DeBellis J, Alterwain P, D'Angelo L (1987) Midazolam and sleep in insomniac patients. Br J Clin Pharmacol 16:87S–88S

Moon CAL, Ankier SI, Hayes G (1985) Early morning insomnia and daytime anxiety: a multicenter general practice study comparing loprazolam and triazolam. Br J Clin Pract 39:352–358

Morgan K, Oswald I (1982) Anxiety caused by a short half-life hypnotic. Br Med J 284:942

Morgan K, Adam K, Oswald I (1984) Effects of loprazolam and of triazolam on psychological functions. Psychopharmacology 82:386–388

Mouret J, Ruel D, Maillard F, Bianchi M (1990) Zopiclone versus triazolam in insomniac geriatric patients: a specific increase in delta sleep with zopiclone. Int Clin Psychopharmacol 5:47–55
Muller WE (1987) The benzodiazepine receptor. Cambridge University Press, New York
O'Donnell VM, Balkin TJ, Andrade JR, Simon LM, Kamimori G, Redmond DP, Belenky G (1988) Effects of triazolam on performance and sleep in a model of transient insomnia. Hum Perform 1:145–160
O'Donovan MC, McGuffin P (1993) Short-acting benzodiazepines. Br Med J 306: 945–946
Oswald I, Adam K (1988) A new look at short-acting hypnotics. In: Sauvanet JP, Langer Z, Morselli PL (eds) Imidazopyridines in sleep disorders. Raven, New York, pp 253–259
Oswald I, Adam K, Borrow S, Idzikowski C (1979) The effects of two hypnotics on sleep, subjective feelings and skilled performance. In: Passouant J, Oswald I (eds) Pharmacology of the states of awareness. Pergamon, New York, pp 51–63
Pecknold JC, Fleury D (1986) Alprazolam-induced manic episode in two patients with panic disorder. Am J Psychiatry 143:652–653
Pecknold JC, Swinson RP, Kuch K, Lewsi CP (1988) Alprazolam in panic disorder: results from a multicenter trial. Arch Gen Psychiatry 45:429–436
Pecknold J, Wilson R, LeMorvan L (1990) Long term efficacy and withdrawal of zopiclone: a sleep laboratory study. Int Clin Psychopharmacol 2:57–67
Pegram V, Hyde P, Linton P (1980) Chronic use of triazolam: the effects on the sleep patterns of insomniacs. J Int Med Res 8:224–231
Penetar DM, Belenky G, Garrigan JJ et al. (1989) Triazolam impairs learning and fails to improve sleep in a long-range aerial deployment. Aviat Space Environ Med 60:594–598
Petre-Quadens OP, Hoffman G, Buytaert G (1983) Effects of zopiclone as compared to flurazepam on sleep in women over 40 years of age. Pharmacology Suppl 27:146–155
Public Citizen (1992) Citizen's petition to withdraw approval of triazolam (Halcion). Public Citizen, Washington
Regestein QR, Reich P (1985) Agitation observed during treatment with newer hypnotic drugs. J Clin Pharmacol 46:280–283
Rickels K, Morris RJ, Mauriello R, Rosenfled H, Chung HR, Newman HM, Case WG (1986) Brotizolam, a triazolothienodiazepine, in insomnia. Clin Pharmacol Ther 40:293–299
Roehrs T, Zorick F, Kaffeman M, Sicklesteel J, Roth T (1982) Flurazepam for short-term treatment of complaints of insomnia. J Clin Pharmacol 22:290–296
Roehrs T, Zorick F, Koshorek GL, Wittig R, Roth T (1983a) The effect of acute administration of brotizolam in subjects with disturbed sleep. Br J Clin Pharmacol 16:371S–376S
Roehrs T, Zorick F, Lord N, Koshorek GL, Roth T (1983b) Dose-related effects of estazolam on sleep of patients with insomnia. J Clin Pharmacol 3:152–156
Roehrs T, Zorick FJ, Sicklesteel JM, Wittig RM, Hartse KM, Roth T (1983c) Effects of hypnoticcs on memory. J Clin Psychopharmacol 3:310–313
Roehrs T, Lamphere J, Paxton C, Wittig R, Zorick F, Roth T (1984) Temazepam's efficacy in patients with sleep onset insomnia. Br J Clin Pharmacol 17:691–696
Roehrs T, Zorick F, Wittig R, Roth T (1985) Efficacy of reduced triazolam dose in elderly insomniacs. Neurobiol Aging 6:293–296
Roehrs T, Vogel G, Vogel F, Wittig R, Zorich F, Paxton C, Lamphere J, Roth T (1986) Dose effects of temazepam tablets in sleep. Drugs Exp Clin Res 12:693–2699
Roehrs T, Vogel G, Sterling W, Roth T (1990) Dose effects of temazepam in transient insomnia. Drug Res 40:859–862

Rosekind MR, Seidel WF, Brown E, Davison H, van den Hoed J, Dement WC (1979) 28-night sleep laboratory evaluation of flunitrazepam. Sleep Res 8:104
Roth T, Kramer M, Lutz T (1976) Intermediate use of triazolam: a sleep laboratory study. J Int Med Res 4:59–62
Roth T, Kramer M, Lutz T (1977) The effects of triazolam (0.25 mg) on the sleep of insomniac subjects. Drugs Exp Clin Res 1:271–277
Roth T, Hartse KM, Saab PG, Piccione PM, Kramer M (1980) The effects of flurazepam, lorazepam, triazolam on sleep and memory. Psychopharmacology (Berl) 70:231–237
Roth T, Zorick F, Sicklesteel J, Stepanski E (1983) Effects of benzodiazepines on sleep and wakefulness. Br J Clin Pharmacol 11:31S–35S
Roth T, Hauri P, Zorick F, Sateia M, Roehrs T, Kipp J (1985) The effects of midazolam and temazepam on sleep and performance when administered in the middle of the night. J Clin Psychopharmacol 5:66–69
Sauvanet JP, Langer SZ, Morselli PL (eds) (1988) Imidazopyridines in sleep disorders: a novel experimental and therapeutic approach. Raven, New York
Scharf MB, Bixler EO, Kales A, Soldatos CR (1979) Long-term sleep laboratory evaluation of flunitrazepam. Pharmacology 19:173–181
Scharf MB, Kales A, Bixler EO, Jacoby JA, Schweitzer PK (1982) Lorazepam: efficacy, side effects and rebound phenomena. Clin Pharmacol Ther 31:175–179
Scharf MB, Fletcher K, Graham JP (1988) Comparative amnesic effects of benzodiazepine hypnotic agents. J Clin Psychiatry 49:134–137
Scharf MB, Roth PB, Dominguez RA, Catesby Ware J (1990a) Estazolam and flurazepam: a multicenter, placebo-controlled comparative study in outpatients with insomnia. J Clin Pharmacol 30:461–467
Scharf MB, Sachais BA, Mayleben DW, Fletcher K, Jennings SW (1990b) A polysomnographic comparison of temazepam 15 and 30 mg with triazolam 0.125 and 0.25 mg in chronic insomnia. Curr Ther Res 48:555–567
Scharf MB Sachais BA, Mayleben DW, Jennings SW (1990c) The effects of a calcium channel blocker on the effects of temazepam and triazolam. Curr Ther Res 48:516–523
Scharf MB, Roth T, Vogel GW, Walsh JK (1994) A multicenter, placebo-controlled study evaluating zolpidem in the treatment of chronic insomnia. J Clin Psychiatry 55:192–199
Schittecatte M, Crine A, Wilmotte J, Kuperberg E (1988) The effects of two benzodiazepines with rapid elimination on sleep, subjective feelings, and skilled performance. Curr Ther Res 44:397–401
Schweitzer PK, Koshorek G, Muehlbach MJ, Morris DD, Roehrs T, Walsh JK, Roth T (1991) Effects of estazolam and triazolam on transient insomnia associated with phase-shifted sleep. Hum Psychopharmacol 6:99–107
Scollo-Lavizzari G (1983) Hypnotic efficacy and clinical safety of midazolam in shift-workers. Br J Clin Pharmacol 16:73S–78S
Seidel WF, Cohen SA, Bliwise NG, Roth T, Dement WC (1986) Dose-related effects of triazolam and flurazepam on a circadian rhythm insomnia. Clin Pharmacol Ther 40:314–320
Smirne S, Ferini-Strambi L, Pirola R, Tancredi O, Franceschi M, Pinto P, Bareggi SR (1989) Effects of flunitrazepam on cognitive functions. Psychopharmacology (Berl) 98:251–256
Snyder J, Thomas GB (1972) Interlaboratory reliability in testing a new hypnotic. Sleep Res 1:73
Soldatos CR, Kales A, Kales JD (1979) Management of insomnia. Annu Rev Med 30:301–312
Soldatos CR, Sakkas PN, Bergiannaki JD, Stefanis CN (1986) Behavioral side effects of triazolam in psychiatric inpatients: report of five cases. Drug Intell Clin Pharmacol 20:294–297

Spinweber CL, Johnson LC (1982) Effects of triazolam (0.5 mg) on sleep, performance, memory, and arousal threshold. Psychopharmacology (Berl) 76:5–12
Subhan Z, Hindmarch I (1983) The effects of midazolam in conjunction with alcohol on iconic memory and free-recall. Neuropsychobiology 9:230–234
Tan T-L, Kales JD, Kales A, Soldatos CR, Bixler EO (1984) Biopsychobehavioral correlates of insomnia, IV: diagnosis based on DSM-II. Am J Psychiatry 141: 357–362
Tsong Y (1992) Statistical comparison of ADE reporting rates between triazolam and temazepam. FDA Center for Drug Evaluation and Research, Psychopharmacologic Drug Advisor Committee Meeting, May 18, Washington
Van der Kroef C (1979) Reactions to triazolam. Lancet 2:526
Vela-Bueno A, Oliveros JC, Dobladez-Blanco B, Arrigain-Ijurra S, Soldatos CR, Kales A (1983) Brotizolam: a sleep laboratory evaluation. Eur J Clin Pharmacol 25:53–56
Vgontzas AN, Kales A, Bixler EO (in press) Benzodiazepine side effects: Role of pharmacologic properties. Pharmacology
Vgontzas A, Kales A, Bixler EO, Myers DC (1994) Temazepam 7.5 mg: effects on sleep in elderly insomniacs. Eur J Clin Pharmacol 46:209–213
Vogel GW (1992) Technical report 7251-79-047. The effects of triazolam, flurazepam and placebo on "middle of the night" insomnia. Psychopharmacologic Drug Advisor Committee Meeting, May 18, Washington
Vogel GW, Morris D (1992) The effects of estazolam on sleep, performance, and memory: a long-term sleep laboratory study of elderly insomniacs. J Clin Pharmacol 32:647–651
Vogel GW, Vogel F (1983) Effect of midazolam on sleep of insomniacs. Br J Clin Pharmacol 16 [Suppl] 1:103S–108S
Vogel GW, Thurmond A, Gibbons P, Edwards K, Sloan KB, Sexton K (1975) The effect of triazolam on the sleep of insomniacs. Psychopharmacology (Berl) 41:65–69
Vogel GW, Barker K, Gibbons P, Thurmond A (1976) A comparison of the effects of flurazepam 30 mg and triazolam 0.5 mg on the sleep of insomniacs. Psychopharmacology (Berl) 47:81–86
Walsh JK, Targum SD, Pegram V, Allen RP, Fillingim JM, Parwatikar S, Schweitzer PK (1984) A multi-center clinical investigation of estazolam: short-term efficacy. Curr Ther Res 36:866–874
Walsh JK, Schweitzer PK, Sugerman JL, Muelbach MJ (1990) Transient insomnia associated with a 3-h phase advance of sleep time and treatment with zolpidem in the treatment of chronic insomnia. J Clin Psychiatry 55:5
WHO (1990) Zopiclone and neuropsychiatric reactions. WHO Drug Info 4:179
Wise RP (1990) Alert to increased frequency report (IFR) for triazolam. Division of Epidemiology and Surveillance, FDA, Washington
Wittig R, Zorick F, Roehrs T, Paxton C, Lamphere J, Roth T (1985) Effects of temazepam soft gelatin capsules on the sleep of subjects with insomnia. Curr Ther Res 38:15–22
Wysowski DK, Barash D (1991) Adverse behavioral reactions attributed to triazolam in the Food and Drug Administration's Spontaneous Reporting System. Arch Inter Med 151:2003–2008
Ziegler G, Ludwig L, Klotz U (1983) Effect of midazolam on sleep. Br J Clin Pharmacol 16:81S–86S

CHAPTER 14
Anxiolytic Drugs

A. VELA-BUENO

A. Introduction

Anxiety and the states of vigilance are closely interrelated. Thus, any anxious state, normal or pathologic, is associated with hyperarousal. Also, clinical experience shows that sleep disturbances are common among patients with diagnoseable anxiety disorders. For example, difficulties in falling and staying asleep are among the symptoms of generalized anxiety disorder (AMERICAN PSYCHIATRIC ASSOCIATION 1994). Also, a considerable proportion of patients with panic disorder suffer from panic attacks during sleep and may secondarily develop a sleep disorder (MELLMAN and UHDE 1989). Further, insomnia and nightmares are among the main manifestations of post-traumatic stress disorder (AMERICAN PSYCHIATRIC ASSOCIATION 1994). Finally, patients with obsessive compulsive disorder often present with sleep disturbances (INSEL et al. 1982).

Anxiety can be a common finding among patients who complain of sleep disorders. Thus, it is present in a large proportion of insomniac patients, as a trait or as the clinical picture fulfilling the diagnostic criteria for an anxiety disorder (KALES and KALES 1984). Also, patients complaining of other sleep disorders such as nightmares or night terrors may have anxious symptoms or traits (A. KALES et al. 1980; J.D. KALES et al. 1980).

Interest in studying the effects on sleep of the pharmacological agents used in the treatment of the anxiety disorders is manifold. First, drugs marketed as anxiolytics are often prescribed as hypnotics. Second, some anxiolytic drugs have side effects involving sleep, e.g., insomnia and hypersomnia. Third, most drugs used for the treatment of anxiety disorders may induce changes in the structure of sleep, i.e., sleep stages. Finally, the mechanisms of action of drugs with anxiolytic properties seem to involve neurotransmitter systems that also play a role in the regulation of the sleep-wakefulness cycle.

This chapter overviews the effects on sleep of the currently available anxiolytic drugs as assessed in the sleep laboratory. It also includes studies on the investigational anxiolytics.

B. Anxiolytic Drugs and Sleep: Overview

Drugs with anxiolytic properties are encountered in virtually all categories of psychotropic drugs. For example, psychotic patients with anxiety may benefit from the use of sedative neuroleptics. Also, antidepressants of different types are efficacious for the treatment of a broad spectrum of anxiety disorders. Because the effects on sleep of these two types of drugs are covered in different chapters of this volume (see also Chaps. 16 and 17, this volume), the focus in this chapter is on: drugs primarily marketed for the treatment of anxiety; those drugs with a major indication as anxiolytics; and those drugs being investigated for their anxiolytic effects.

The main categories of anxiolytics currently in clinical use and those under investigation are known to act on one or more of the neurotransmitter systems considered to be involved in the production of anxiety. Anatomical studies, behavioral experiments, manipulations of neurotransmitter turnover and studies at the receptor level point to GABA, norepinephrine and serotonin as the three main neurotransmitters contributing to the modulation of anxious states (for reviews, see Gershon and Eison 1983; Charney et al. 1983; Norman and Burrows 1986; Oakley and Tyers 1992; Kahn and Moore 1993; Hamon 1994).

Benzodiazepines, the most prescribed anxiolytics, act by facilitating GABA-ergic transmission. They bind to a site in the $GABA_A$-receptor complex and lead to an increase in the affinity of this receptor for GABA (for a detailed discussion, see Chap. 8, this volume). The clinical implication of this mechanism is evidenced by the fact that the potency of benzodiazepines as anxiolytics relates to their binding affinity.

In the last few years, two types of non-benzodiazepine anxiolytics have appeared that bind to the $GABA_A$-receptor complex in a site different from that for benzodiazepines. They belong to two chemical varieties: the imidazopyridines and the cyclopyrrolones. Alpidem, an imidazopyridine, has been shown to possess anxiolytic efficacy in patients with generalized anxiety disorder and other conditions (Pancheri et al. 1993; Dufour et al. 1993). Suriclone, a cyclopyrrolone derivative, has demonstrated anxiolytic effects in the treatment of patients with generalized anxiety disorder (Ansseau et al. 1991).

The monoamines, norepinephrine, dopamine and serotonin, appear to be involved in the effects of benzodiazepines (see review in Norman and Burrows 1986). Whereas most of this effect seems to be mediated by their GABA-ergic activity, a subclass of benzodiazepines, the triazolobenzodiazepines, have in addition a direct effect on the noradrenergic system due to their action as agonists at α_2-adrenergic receptors. They also inhibit platelet-activating factor, which activates cortical releasing hormone (CRH). In turn, CRH activates the locus coeruleus noradrenergic system (see review in Kales et al. 1994).

Some drugs with anxiolytic effects exert their actions directly on adrenergic receptors. Although marketed for other purposes, the β-adrenergic antagonists have been in use for years in the treatment of different anxiety conditions (NOYES 1982). Also, the α_2-agonist clonidine has been shown to possess short-term anxiolytic effects (CHARNEY and HENINGER 1986; UHDE et al. 1989). The use of both types of adrenergic agents in anxiety disorders has been recently reviewed (DAVIDSON 1992; LADER 1988).

A decrease in the level of serotonin may result in anxiolytic effects. Thus, a 5-HT_{1A} agonist marketed for the treatment of anxiety, such as buspirone, has shown efficacy in patients with generalized anxiety disorder, but not in those with panic disorder (see review in KAHN and MOORE 1993). A 5-$HT_{2/1C}$ antagonist, ritanserin, has also shown anxiolytic properties in patients with generalized anxiety disorder (see review in KAHN and MOORE 1993).

The basic methodology of sleep laboratory studies evaluating the hypnotic effects of drugs is described in Chap. 12 in this volume. The main focus of such studies on the effects of anxiolytics on sleep varies; thus for some drugs it is not hypnotic efficacy but effects on sleep stages. In this chapter, studies with major shortcomings are generally excluded including those with: lack of constant dosage; use of nonstandard dosage; lack of adequate recording techniques; inadequate duration of recording; use of nonstandard scoring methods; lack of statistical analysis; and unclear therapeutic profile of the drug.

C. Benzodiazepines

In this chapter benzodiazepines are classified according to their pharmacokinetic (elimination half-lives) and pharmacodynamic (receptor-binding affinity) properties as well as their chemical structure. These properties relate to their efficacy, to their potential for the development of tolerance as well as to their capacity to cause withdrawal difficulties and other adverse reactions.

Benzodiazepines are marketed as hypnotics or anxiolytics based on a set of pharmacokinetic and pharmacodynamic properties that are not always consistent. For instance, a desired quality for a hypnotic is that of being quickly absorbed, because the main difficulty insomniac patients complain of, is difficulty falling asleep. However, when comparing the absorption time of different benzodiazepines, that of some anxiolytics is shorter than that of some hypnotics.

I. 1,4-Benzodiazepines with Slow Elimination and Low Binding Affinity

1. Ketazolam

a) Efficacy

The administration of 30 mg ketazolam at bedtime for seven nights to anxious subjects increased total sleep time and sleep efficiency and decreased sleep latency, percentage of wake time and number of awakenings. Only total sleep time was not significantly increased at the end of the drug period (BONNET et al. 1980). In another sleep laboratory evaluation, the drug was given for seven nights at doses of 15, 30 or 45 mg to normal subjects (MOORE et al. 1981). The 30-mg dose, the only one for which statistical analysis was reported, significantly decreased the percentage of wakefulness and the number of awakenings; this value was decreased only at the end of the drug period.

b) Withdrawal Effects

The withdrawal effects of ketazolam were evaluated in two studies with the 30-mg dose. In the one study with anxious patients, the withdrawal period lasted 4 days (BONNET et al. 1980). Whereas most variables changed by drug use tended to return to baseline, sleep latency remained decreased, indicating some carryover effectiveness. In the other study, half of the subjects were studied for four withdrawal nights and the other half for seven (MOORE et al. 1981). Wakefulness variables (percentage of wakefulness and number of awakenings) returned to baseline. No statistical analysis was reported for the continued withdrawal period.

c) Side Effects

The only significant change found with bedtime administration of ketazolam 30 mg, in a checklist for side effects, was impaired coordination (BONNET et al. 1980). The other study did not report on side effects.

2. Chlordiazepoxide

a) Efficacy

The single administration of 100 mg chlordiazepoxide to normal subjects increased total sleep time (HARTMANN 1968). This study was part of a larger one comparing different drugs. Chlordiazepoxide was administered at a dose of 50 mg for 28 consecutive nights to normal subjects who were recorded for the first five nights on drug and then weekly, for the next 3 weeks. The data of the drug period were compared to those of a group of subjects on placebo using a similar protocol. With initial administration, there was an increase in total sleep time and a decrease in waking (HARTMANN and CRAVENS 1973). These measures did not differ, after three nights of administration, from those of the placebo group. The interpretation of these results is difficult due to the methodological flaws of the protocol, such as the lack of a placebo-baseline condition.

b) Withdrawal Effects

There are no studies of the withdrawal of chlordiazepoxide that utilized placebo administration on withdrawal nights. In the only study assessing the effects of discontinuation, the first six nights of the 32 nights of that period and three additional nights, with 1 week of interval between them, were recorded (HARTMANN and CRAVENS 1973). No rebound insomnia was reported with the 50-mg dose.

c) Side Effects

No paradoxical or unexpected side effects (for definition see Chap. 13, this volume) were found with the bedtime administration of chlordiazepoxide 50 mg (HARTMANN and CRAVENS 1973). Only fatigue, as measured by a scale, showed a trend to be increased both with drug use and withdrawal. Side effects were not assessed in the 100-mg study (HARTMANN 1968).

3. Clorazepate

a) Efficacy

The single-night administration of 15 mg to normal subjects resulted in an increase of total sleep time and a decrease of sleep onset latency (NICHOLSON et al. 1976a). When the drug was administered to normal subjects in three doses of 7.5 mg through the 24 h (total daily dose, 22.5 mg), for eight consecutive days, a significant increase in total sleep time and sleep efficiency and a significant decrease in sleep latency, number of awakenings and percentage of wakefulness were noted with drug administration (KARACAN et al. 1973). Tolerance was not assessed.

b) Withdrawal Effects

No rebound insomnia was found in the three placebo-withdrawal night period (KARACAN et al. 1973). Variables such as sleep efficiency and sleep latency returned to predrug levels. Total sleep time remained at drug levels throughout the withdrawal period, indicating some carryover effectiveness.

c) Side Effects

A decrease in alertness was noted with the administration to normal subjects of clorazepate 7.5 mg t.i.d. and an increase in alertness following discontinuation of the drug (KARACAN et al. 1973). The fact that the drug was given both at bedtime and during the daytime makes it difficult to exclusively relate this side effect to the nighttime dose.

4. Fosazepam

a) Efficacy

The hypnotic effects of 60 mg fosazepam given to poor sleepers at bedtime were assessed in the sleep laboratory for 3 weeks after 1 week of placebo. An increase in total sleep time was noted, which was significant with continued drug use compared with baseline (ALLEN and OSWALD 1976). In another study, 100 mg fosazepam was given to normal subjects at bedtime for ten nights after two placebo-baseline nights (RISBERG et al. 1977). No significant changes in sleep latency, total sleep time and wakefulness were noted.

Finally, one study assessed in normal subjects the acute effects of the 80-mg and 60-mg doses. Both doses resulted in a significant increase in total sleep time (NICHOLSON et al. 1976b).

b) Withdrawal Effects

No rebound insomnia has been shown to occur after withdrawal of fosazepam. With the 60-mg dose, the withdrawal period lasted 3 weeks (ALLEN and OSWALD 1976), whereas with the 100-mg dose it was four nights (RISBERG et al. 1977). Rebound anxiety occurred with the 60-mg dose.

c) Side Effects

A decrease in "morning vitality" was the only side effect reported with the 60-mg dose (ALLEN and OSWALD 1976). In contrast with the 100-mg dose, drowsiness reports were not different than with placebo (RISBERG et al. 1977).

II. 1,4-Benzodiazepines with Intermediate to Slow Elimination and Moderate to High Binding Affinity

1. Clonazepam

a) Efficacy

The administration of 0.5 mg clonazepam to insomniac subjects for seven consecutive nights, following four placebo-baseline nights, significantly reduced the number of wakes and total wake time (Fig. 1) and increased the percentage of sleep time with both short-term and continued administration (KALES et al. 1991). In another study, 1 mg clonazepam or 30 mg temazepam

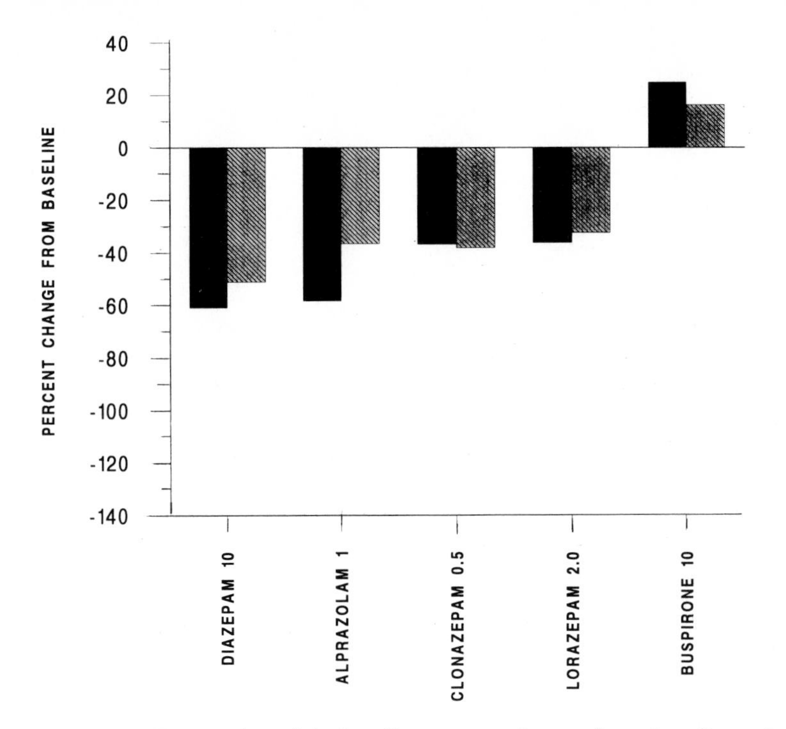

Fig. 1. Hypnotic efficacy of anxiolytics. Percentage change from baseline of mean values of total wake time for short-term (nights 1–3) drug administration (*solid bar*) and continued (1 week) drug administration (*hashed bar*). Alprazolam resulted in the largest amount of tolerance while buspirone resulted in mild sleep disturbance. All of the data in Figs. 1 and 2 are from studies conducted in the Sleep Laboratory at Pennsylvania State University College of Medicine using standardized methods and techniques

was administered for seven consecutive nights to insomniacs with nocturnal myoclonus according to a crossover design. No placebo was used on the two baseline nights. Sleep recordings were obtained only on the last two nights of drug use. A significant increase in sleep duration and efficiency and a decrease in wakefulness occurred with clonazepam (MITLER et al. 1986).

b) Withdrawal Effects

The withdrawal effects of 0.5 mg clonazepam (KALES et al. 1991), but not those of the 1-mg dose (MITLER et al. 1986), were evaluated in the sleep laboratory. Rebound insomnia was absent during the short-term withdrawal condition (first three nights) with the 0.5-mg dose (KALES et al. 1991). However, in a night-by-night analysis, significant worsening of sleep (rebound insomnia) was found on the third withdrawal night (Fig. 2). Through the extended withdrawal period (fourth and fifth withdrawal nights), no significant increase in wakefulness variables were evident.

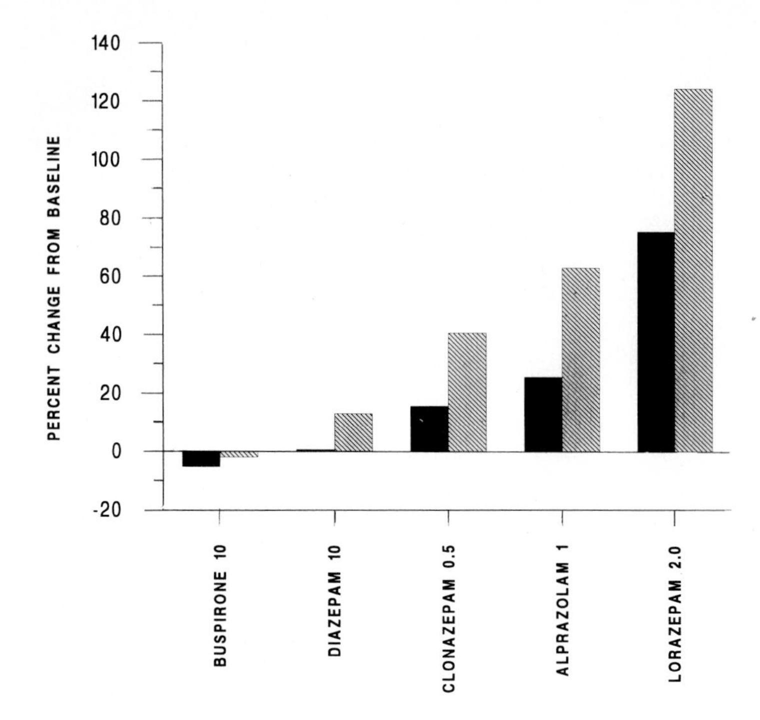

Fig. 2. Hypnotic withdrawal effects of anxiolytics. Percentage change from baseline of mean values of total wake time is indicated for initial (nights 1–3) withdrawal (*solid bar*) and single highest night (*hashed bar*). The anxiolytics with the least withdrawal effects were buspirone and diazepam while those with the strongest withdrawal effects were alprazolam and lorazepam

c) Side Effects

No changes in the estimates of daytime or evening sleepiness, anxiety, tension or panic with initial and continued administration of 0.5 mg clonazepam at bedtime were found (KALES et al. 1991). With the 1-mg dose, daytime sleepiness did not change, but a decrease in tiredness was reported (MITLER et al. 1986).

2. Diazepam

a) Efficacy

There are several sleep laboratory evaluations of the hypnotic efficacy of diazepam. One study assessed the 10-mg dose given to insomniacs for seven consecutive nights following four placebo nights (KALES et al. 1988b). With initial administration, there was a significant reduction of sleep latency, wake time after sleep onset and total wake time (Fig. 1). At the end of the drug period, sleep latency and total wake time remained significantly decreased.

Another study assessed the 2-, 5- or 10-mg doses of diazepam, given for two nights each to normal subjects. No significant changes were found in various sleep efficiency measures (BONNET et al. 1981). In two studies on the single-night effect of 5 mg diazepam a significant decrease in sleep latency was noted (NICHOLSON et al. 1976b; NICHOLSON and STONE 1979). In another study, in which only the last night of a 1-week administration to insomniacs at the 10-mg dose level was recorded, a significant decrease in total wake time and a significant increase in sleep efficiency were evident (SCHARF et al. 1985). Finally, the effects on sleep of 5 mg diazepam bid were evaluated on the night preceding the last day of a 1-week period of administration (SEIDEL et al. 1985b); total sleep time was increased and nocturnal sleep latency decreased.

b) Withdrawal Effects

No rebound insomnia or rebound anxiety were present in the two studies that assessed the withdrawal effects of 10 mg diazepam administered at bedtime. One study included a seven-night withdrawal period following seven nights of drug use (KALES et al. 1988b); the findings in this study included a nonsignificant increase of 33.7% in total wake time on the 6th withdrawal night. The other included ten placebo-withdrawal nights (SCHARF et al. 1985). Actually, in this study a carryover hypnotic effectiveness was found during the early withdrawal nights.

c) Side Effects

Increased daytime sleepiness was observed with the initial and continued administration of 10 mg diazepam at bedtime (KALES et al. 1988b). The lack of tolerance to the sedative effect of diazepam was also found when it was administered at a dose of 5 mg b.i.d. (SEIDEL et al. 1985b). Rebound anxiety was not reported in either study.

III. 1,4-Benzodiazepines with Rapid to Intermediate Elimination and Low Binding Affinity

1. Oxazepam

a) Efficacy

The hypnotic efficacy of 30 mg oxazepam has been assessed in the sleep laboratory (BLIWISE et al. 1984). The drug was given nightly to seven insomniacs for a period of 1 week. No significant changes were observed in the sleep efficiency measures. The administration of 15 mg oxazepam for 1 week to insomniacs showed no hypnotic efficacy; in addition, in the same study, the same dose given for 2 days to normal subjects increased total wake time (ZIEGLER et al. 1983). The fact that insomniacs were recorded the last two nights of the drug period does not rule out the development of tolerance. The 10-mg dose showed some hypnotic efficacy in that it significantly increased total sleep time during the daytime in shift workers (EHRENSTEIN et al. 1972).

b) Withdrawal Effects

Rebound insomnia was observed after withdrawing 30 mg oxazepam (BLIWISE et al. 1984). Total sleep time (1 h less compared to the drug period) and sleep efficiency were markely reduced.

c) Side Effects

One study that assessed daytime CNS depression with the 30-mg dose of oxazepam administered at bedtime did not find daytime sleepiness using the Multiple Sleep Latency Test (BLIWISE et al. 1984). However, an increase in peripheral reaction time was found, whereas performance on mental arithmetic tasks improved.

IV. 1,4-Benzodiazepines with Rapid to Intermediate Elimination and Moderate to High Binding Affinity

1. Lorazepam

a) Efficacy

The hypnotic efficacy of lorazepam given at bedtime to insomniacs has been demonstrated in several studies in the sleep laboratory. In one study, 4 mg lorazepam was administered for seven nights and its effects compared to those of four placebo-baseline nights (SCHARF et al. 1982). Total wake time (as well as sleep latency and wake time after sleep onset) was significantly reduced with drug administration. The decrease in total wake time persisted throughout the entire period of drug administration. Another study, using a similar protocol to assess the effects of the 3-mg dose of lorazepam, also showed that the drug was effective in inducing (decreasing sleep latency) and maintaining (decreasing wake time after sleep onset) sleep (WALSH et al. 1983). No tolerance to these effects developed through the week of drug administration. Two different studies evaluated the efficacy of 2 mg lorazepam (KALES et al. 1986; MCCLURE et al. 1988). The first study used a 16-night protocol including seven nights of drug use (KALES et al. 1986). This dose was shown to induce a moderate but nonsignificant decrease in total wake time with both initial and continued drug administration (Fig. 1). The other study used a 14-week crossover protocol to compare 2 mg lorazepam and 30 mg flurazepam (McCLURE et al. 1988). Subjects slept in the sleep laboratory about twice weekly as follows: three nights in the 2-week baseline-placebo period; six nights in the first 3-week drug period; four nights in the 3-week placebo period between drug periods; six nights in the second, 3-week drug period; and six nights in the final 3-week placebo period. A significant increase in sleep efficiency and a significant decrease in total wake time after sleep onset were shown with lorazepam throughout the 3 weeks of drug administration.

b) Withdrawal Effects

The withdrawal effects of three doses of lorazepam were assessed in the sleep laboratory (SCHARF et al. 1982; WALSH et al. 1983; KALES et al. 1986; McCLURE et al. 1988). With the exception of one study (McCLURE et al. 1988), different degrees of rebound insomnia have been noted with the three dose levels. The withdrawal period in the study evaluating the 4-mg dose consisted of seven nights of placebo administration (SCHARF et al. 1982). Significant increases in total wake time (rebound insomnia) and sleep latency occurred on the third and fifth withdrawal nights. Rebound anxiety was also present after the second withdrawal night. Continued efficacy on the first withdrawal night was found with the 3-mg dose (sleep latency and total wake time remained decreased). However, wake time after sleep onset and number of awakenings were significantly increased on the second night, whereas wake time after sleep onset, total wake time and number of awakenings rose significantly on the third withdrawal night (WALSH et al. 1983). There also was an increase of tension after the first and third withdrawal nights. In one study assessing the effects of the 2-mg dose (KALES et al. 1986), the withdrawal period included five placebo nights. An intense degree of rebound insomnia (Fig. 2) was noted on the third withdrawal night (total wake time rose 125% above the baseline values). Also, an increase in tension and anxiety (rebound anxiety) was evident the next day. Finally, the only study (McCLURE et al. 1988) that did not find rebound insomnia used a design (crossover, long-term administration) that makes the interpretation of its results difficult.

c) Side Effects

Lorazepam studied as a hypnotic in the sleep laboratory has been shown to produce daytime sedation in a dose-related manner. Thus, with the 2-mg dose no significant daytime sedation was reported (KALES et al. 1986; McCLURE et al. 1988). With 3 mg (WALSH et al. 1983), there were feelings of sleepiness and fatigue throughout the day during the drug administration period with sleepiness persisting into the initial withdrawal period; four subjects (of six) reported being "dizzy" or "groggy." There was some worsening of performance during drug administration. The 4-mg dose (SCHARF et al. 1982) caused severe hangover persisting for the first 3 days of drug administration. By the third drug night, daytime sedation manifestations diminished, with some of them of mild degree persisting through the seventh night on drug. Anterograde amnesia was reported to occur in three subjects with 4 mg. With 2 mg lorazepam there were some complaints of memory impairment and confusional state in two patients (Kales et al. 1986). Daytime anxiety was found in the same study with continued use of lorazepam. After 1 week of drug administration, measurements of anxiety in the evening were significantly higher than baseline levels.

V. Triazolobenzodiazepines

1. Alprazolam

a) Efficacy

The hypnotic efficacy of 1 mg alprazolam was evaluated in the sleep laboratory (Kales et al. 1987). The drug was administered to insomniac subjects for seven consecutive nights following four placebo-baseline nights. A marked and significant decrease in total wake time was evident on the first three nights of administration. However, there was a loss of about 40% of this efficacy at the end of the 1 week of drug administration (Fig. 1).

The 0.5-mg dose administered to elderly insomniacs (Kramer et al. 1984) was reported to decrease sleep latency, wake time during sleep and percentage wake time and to increase sleep efficiency; some tolerance was also reported in this study at the end of the 1-week period of administration. Unfortunately it did not include any statistical analysis. Two dose-response studies on the hypnotic effectiveness of alprazolam were done on normal subjects. In the first one, the dose of 0.25, 0.5 and 1.0 mg were administered for two nights each on consecutive weeks (Bonnet et al. 1981). No significant effects on sleep efficiency variables were evident. In the other study, 1-, 2- and 3-mg doses were assessed (Kramer 1982). The subjects were studied in the sleep laboratory for five consecutive weeks for two nights each week. The first two nights were for adaptation; the following three pairs of nights the subjects received alprazolam 1 or 2 mg or placebo; during the 5th week they received 3 mg alprazolam on the first night and placebo on the second. Sleep efficiency was significantly increased with all three doses, whereas sleep latency was significantly shorter only with the 3-mg dose; the number of awakenings decreased with the 1- and 2-mg doses but not with the 3-mg one. The effects of 0.5 mg alprazolam b.i.d. on the nocturnal sleep of healthy volunteers were assessed on the night preceding the last day of a 1-week treatment period (Seidel et al. 1985b). Total sleep time was significantly increased. Two studies evaluated the effect of alprazolam on the sleep of patients with major depression. The daily doses were 4–9 mg (Hubain et al. 1990) and 6–10 mg (Zarcone et al. 1994). In each study, the subjects were monitored in the sleep laboratory on two occasions; the first time baseline recordings were obtained and the second time was after they had completed 6 weeks of drug use. In the first study there was a significant increase in total sleep time, whereas in the second, sleep latency was significantly shortened.

b) Withdrawal Effects

The occurrence of rebound insomnia (Fig. 2) after the withdrawal of alprazolam has been documented in a sleep laboratory evaluation of the 1-mg dose (KALES et al. 1987). The withdrawal period consisted of five nights following seven nights of drug use. On the third withdrawal night, total wake time increased by 63% over the baseline level. In another study of the 0.5-mg dose administered to elderly insomniacs, there was a worsening of sleep (increased sleep latency, wake time and percentage wake time) upon withdrawal (KRAMER et al. 1984). However, this study did not present statistical analysis of the data.

c) Side Effects

The presence of symptoms of CNS depression has been reported with the administration of 3 mg (KRAMER 1982). Tolerance to the daytime sedation caused by the administration for 1 week of 0.5 mg alprazolam given b.i.d. was reported (SEIDEL et al. 1985b). This is in agreement with the rapid development of tolerance to the hypnotic efficacy of alprazolam (KALES et al. 1987). No effects on performance or other side effects were reported with the bedtime administration of 0.5 mg to elderly insomniacs (KRAMER et al. 1984). Some cognitive side effects were reported with the 3-mg dose (KRAMER 1982); one subject showed general disorientation. Also, hyperexcitability symptoms with alprazolam have been described in the sleep laboratory. With the 1-mg dose, one subject felt "extremely pressured," irritable, abrupt and had difficulty in controlling inappropriate emotions (KALES et al. 1987), whereas with the 3-mg dose increased restlessness was found (KRAMER 1982).

2. Adinazolam

a) Efficacy

The effects of adinazolam on the sleep of normal subjects were assessed in the sleep laboratory (KRAMER and SCHOEN 1986). In a crossover study, subjects were given 15 mg, 30 mg or placebo for three consecutive nights. The drug significantly decreased the number of awakenings and the percentage of wakefulness with the 30-mg dose. In another study the long-term administration of higher doses (80–120 mg daily) of adinazolam to depressed patients resulted in a significant reduction in sleep latency, which was maintained through 4 weeks of administration (VOGEL et al. 1987).

b) Withdrawal Effects

Rebound insomnia, as manifested by a significant increase in the percentage of wakefulness, was observed with the 15-mg dose but not with the 30-mg one (KRAMER and SCHOEN 1986). For both doses, the withdrawal period lasted three nights following three nights of drug use. The data on withdrawal of the 80- to 120-mg doses were not reported, although the protocol included 11 placebo-withdrawal nights (VOGEL et al. 1987).

c) Side Effects

Only one study reported on side effects (KRAMER and SCHOEN 1986). Performance was unaffected by the administration of adinazolam both in the morning and in the evening. Two subjects showed incoordination, ataxia and auditory hallucinations with the 45-mg dose.

D. Non-Benzodiazepines, $GABA_A$ Agonists

I. Cyclopyrrolones

1. Suriclone

Suriclone is a cyclopyrrolone derivative with an elimination half-life of about 14 h (2 for the parent compound and 12 for the active metabolites) (cf. SALETU et al. 1990). The only published study in the sleep laboratory was done on 16 healthy volunteers (SALETU et al. 1990). The study consisted of a double-blind, placebo-controlled, crossover design. The single night effects of 0.2 and 0.4 mg suriclone were tested and compared to one baseline night. One "washout" night following each drug condition was recorded (it is unclear if subjects were given placebo on these nights). Each drug condition (the two doses of suriclone, lorazepam and placebo) were separated by 1-week intervals.

a) Efficacy

Some dose-dependent changes occurred as shown by the significant decrease in wake time and increase in sleep efficiency with the 0.4-mg dose but not with the 0.2-mg dose (SALETU et al. 1990). Also, there were less nocturnal arousals with suriclone 0.4 mg as compared to placebo.

b) Withdrawal Effects

Nocturnal awakenings were more frequent on the "washout" night after 0.4 mg suriclone than in the night after placebo (SALETU et al. 1990). This increase was small and nonsignificant as compared to baseline.

c) Side Effects

No significant effects on mood and cognitive functions appeared after the administration of either dose (SALETU et al. 1990). Similarly the psychomotor functions were unaffected by suriclone.

II. Imidazopyridines

1. Alpidem

Alpidem is an imidazopyridine compound with an elimination half-life of about 19h in healthy adults (BIANCHETTI et al. 1993). In two studies, the effects of different doses of alpidem on the sleep of healthy volunteers were assessed in the sleep laboratory. In one study, the 75-mg dose was evaluated using a double-blind, placebo-controlled, parallel-group design. Subjects spent nine nights in the laboratory: one for adaptation; two for baseline; three for drug administration; and three for withdrawal. Ten subjects were administered alpidem and ten subjects, placebo (SALETU et al. 1986). In the other study published as a summary (see DANJOU et al. 1993), the doses of 125 and 250mg were administered to four healthy controls. The protocol was not specified.

a) Efficacy

No significant changes in objective sleep variables concerning sleep induction and maintenance occurred with the administration of 75mg alpidem (SALETU et al. 1986). The 125- and 250-mg doses did not significantly change sleep latency (the only measure reported) (DANJOU et al. 1993).

b) Withdrawal Effects

The withdrawal of 75mg alpidem did not result in any significant changes when the whole condition was considered. However, in the night-by-night analysis, a significant decrease in total sleep time and sleep efficiency was noted on the last withdrawal night (SALETU et al. 1986). Withdrawal was not reported for the 125- and 250-mg doses (DANJOU et al. 1993).

c) Side Effects

No significant effects on alertness and cognitive functioning were reported during the days following the nighttime administration of 75mg alpidem (SALETU et al. 1986) or 125 and 250mg (DANJOU et al. 1993). Also, psychomotor functions were basically unaffected by the 75-mg dose (SALETU et al. 1986).

E. β-Adrenoceptor Antagonists

β-Adrenoceptor antagonists (β-blockers) have a high benefit-to-risk ratio, with a low incidence of central nervous system (CNS) side effects (KOELLA 1985; DAHLÖF and DIMENÄS 1990; MCAINSH and CRUICKSHANK 1990). In clinical practice, sleep-related side effects (insomnia, vivid dreams, nightmares and drowsiness) are among the most disturbing complaints. Therefore, the sleep laboratory evaluation of the different β-blockers is clinically relevant.

The available sleep laboratory studies included normal or hypertensive subjects but did not focus on anxiety-disorder patients. The experimental protocols vary widely in terms of design, length of drug use and time of the day when the drug was administered. In this section, information is considered regarding sleep efficiency, withdrawal effects and other side effects. Because sleep disorders as side effects with the clinical use of β-blockers are considered in Chap. 20, this volume, in this chapter the side effects reviewed are those reported in sleep laboratory studies.

Several mechanisms could be involved in the appearance of CNS side effects with the use of β-blockers (KOELLA 1985). A recent comprehensive review pointed to lipophilicity as the major determinant (MCAINSH and CRUICKSHANK 1990), whereas the plasma concentration has been postulated by other authors, who also pointed out that nonselective β-blockers seem to cause more CNS side effects than β_1-selective blockers (DAHLÖF and DIMENÄS 1990). In this section the effects of β-blockers on sleep are reviewed according to their lipophilicity.

I. β-Blockers with High Lipophilicity

1. Propranolol

a) Efficacy

The effects on sleep of propranolol given at bedtime to normal volunteers have been assessed in the sleep laboratory. In one study of 120 mg propranolol, sleep was studied for seven nonconsecutive nights (DUNLEAVY et al. 1971); no changes in total sleep time and number of wakes occurred. In another study (a placebo-controlled trial), propranolol was given for six consecutive nights (two nights, 80 mg, and four nights, 120 mg), with no significant changes in the "waking scores" obtained with automated EEG analysis (BETTS and ALFORD 1985). The administration of a total daily dose of 160 mg propranolol (80 mg b.i.d.) to healthy subjects increased the number of awakenings and the time of wakefulness after sleep onset. The study consisted of a placebo-controlled, Latin-square design comparing the effects of four β-blockers (atenolol, pindolol, metoprolol and propranolol). The single recording night was at the end of a 7-day period of administration (KOSTIS and ROSEN 1987).

Propranolol 20–80 mg given b.i.d. was administered for 3 months to middle-aged hypertensives in a crossover study that included placebo and clonidine (Kostis et al. 1990). Polygraphic sleep recordings were obtained before and after each treatment period; no significant changes in any sleep variable occurred with propranolol. In a similar study of elderly hypertensives comparing propranolol (20–40 mg given b.i.d.) with placebo, no significant changes in sleep were evident (Kostis et al. 1990).

b) Withdrawal Effects

No changes in sleep and wakefulness measures were reported after withdrawing the 120- and 80-mg doses of propranolol (Dunleavy et al. 1971; Betts and Alford 1985). Withdrawal was not assessed in the other studies (Kostis and Rosen 1987; Kostis et al. 1990).

c) Side Effects

There are very few reports of side effects in sleep laboratory assessments of propranolol. In one study with normal volunteers no true side effects were reported; only an increase in the number of remembered awakenings and recalled dreams was observed during the first few days on the drug (Betts and Alford 1985).

Another study did not find significant effects on mood and performance (Kostis and Rosen 1987). An increase in sleepiness and drowsiness in middle-aged subjects and a decrease in cognitive performance in elderly hypertensives have been reported with propranolol administration (Kostis et al. 1990). However, the drug was administered b.i.d., and patients were not screened for sleep apnea. It is known that propranolol worsens sleep apnea (Boudoulas et al. 1983), and hypertensives have a high prevalence of sleep apnea (Kales et al. 1984). On the other hand, daytime sleepiness has been shown to decrease in narcoleptics with the administration of propranolol (Kales et al. 1979). This effect was lost after 2 years of use (Meier-Ewert et al. 1985).

The effects of propranolol on nocturnal penile tumescence (NPT) have been assessed. Whereas in one study with normal subjects the daily dose of 160 mg did not change tumescence (Kostis and Rosen 1987), another study by the same authors found a decrease in the amplitude of erection (Kostis et al. 1990). This finding was observed in elderly hypertensives given 20–40 mg b.i.d.

II. β-Blockers with Moderate Lipophilicity

1. Metoprolol

a) Efficacy

Metoprolol has been assessed in the sleep laboratory in several studies of normal subjects using different doses. No effects on wakefulness were found when metoprolol was given for six consecutive nights (50 mg for two nights and 100 mg on the following four nights) (Betts and Alford 1985). However, with 400 mg administered for two nights, a few changes in sleep efficiency measures were found on the first night (a decrease in sleep efficiency index and an increase in sleep latency, number of awakenings and percentage of wakefulness during the sleep period), but no changes were found on the second night (Kayed and Godtlibsen 1977).

In another study in normal subjects, 100 mg metoprolol given b.i.d. caused an increase in number of awakenings and time of wakefulness after sleep onset on the last night of 1 week of drug administration (Kostis and Rosen 1987). Metoprolol in a 100-mg dose was administered for 8 days (sleep was recorded the last 2) to hypertensive patients with sleep apnea. It is unclear what time of day the drug was administered or whether placebo was given on the two baseline nights (Weichiler et al. 1991). A significant reduction in total sleep time, as compared to baseline, was found.

b) Withdrawal Effects

No significant changes were found after withdrawing 100 mg metoprolol (Betts and Alford 1985). The variables studied were number of remembered awakenings and automated EEG analysis of waking scores. Other studies did not include a withdrawal condition.

c) Side Effects

No side effects or worsening of performance were reported with any of the dose levels studied in the sleep laboratory. Contrary to what was expected, 100 mg metoprolol significantly decreased the apnea index in a hypertensive sample (Weichler et al. 1991). Apneas were equally diminished in REM and non-REM sleep. In contrast, in another study with hypertensive patients who snored, an increase of obstructive apnea was observed when they were administered 50–100 mg metoprolol (Kantola et al. 1991). One study that assessed sexual function (NPT) found no significant changes as compared to placebo after 1 week of administration of 100 mg metoprolol given b.i.d. (Kostis and Rosen 1987).

2. Pindolol

a) Efficacy

An increase in the "waking scores" (the only objective measurement assessed) has been reported with the administration of pindolol for six consecutive nights (10 mg for two nights and 15 mg for the following four nights) to normal subjects. An increase in remembered awakenings was also found (Betts and Alford 1985). Pindolol at the dose of 10 mg b.i.d. increased number of awakenings and time of wakefulness after sleep onset in normal subjects (Kostis and Rosen 1987).

b) Withdrawal Effects

In one study no significant changes were present upon withdrawal of pindolol (Betts and Alford 1985). In the other study, no withdrawal period was included (Kostis and Rosen 1987).

c) Side Effects

In the sleep laboratory, very few side effects of pindolol have been reported. Only an increase in remembered dreams has been found (Betts and Alford 1985). NPT was not affected by the use of pindolol 10 mg b.i.d. during 1 week as compared to placebo (Kostis and Rosen 1987). In this study mood and performance were also unaffected.

III. β-Blockers with Low Lipophilicity

1. Acebutolol

a) Efficacy

In the only published sleep laboratory evaluation of acebutolol, no effects on different sleep measurements were found (Kayed and Godtlibsen 1977). The drug was administered for two consecutive nights to normal subjects at the dose of 200 mg, following two placebo-baseline nights.

b) Withdrawal Effects

Withdrawal was not assessed in the only sleep laboratory study available with acebutolol (Kayed and Godtlibsen 1977).

c) Side Effects

Few side effects were reported with the administration of 200 mg acebutolol at bedtime (Kayed and Godtlibsen 1977). A certain degree of drowsiness was reported after the first night of drug administration but less so on the second.

2. Atenolol

a) Efficacy

No significant changes in "waking scores" were found when atenolol was administered for six consecutive nights (two nights at the 50-mg dose and the following four nights at the 100-mg dose in the sleep laboratory). Subjects were healthy volunteers and were studied in a placebo-controlled trial (Betts and Alford 1985). Also, no change in the number of awakenings and time of wakefulness was found when 100 mg atenolol was administered in the morning to normal subjects (KOSTIS and ROSEN 1987). In this study, sleep was recorded at the end of 1 week of drug administration.

b) *Withdrawal Effects*

In one study, two nights of withdrawal were included (BETTS and ALFORD 1985); no significant differences were evident in either the subjective or objective data. In the other study, no withdrawal nights were included (KOSTIS and ROSEN 1987).

c) *Side Effects*

No side effects with attenolol were reported in the sleep laboratory (BETTS and ALFORD 1985). In the other study, no effects on mood and psychomotor performance were reported; sexual function as assessed by NPT was also unaffected (KOSTIS and ROSEN 1987).

3. Nadolol

a) *Efficacy*

The effects on sleep of 20 and 80 mg nadolol have been evaluated in the sleep laboratory (KALES et al. 1988a). Subjects were patients with mild hypertension. The study protocol consisted of 32 experimental nights following a 1-week period of placebo administration at home. All subjects ($N = 6$) were given placebo each morning for the first 4 days in the sleep laboratory and nadolol 20 mg for the following 14 days; half the subjects took this dose through the 32nd day, whereas the other half were administered 80 mg each morning until the end of the study. No significant changes in sleep induction and maintenance were observed with either dose. With the 80-mg dose, a slight sedative effect was indicated by a decrease in nocturnal wakefulness; the small number of subjects ($N = 3$) might account for the lack of statistical significance.

b) *Withdrawal Effects*

There are no available data on withdrawal for nadolol. Since the only study on this drug in the sleep laboratory was on hypertensive patients (KALES et al. 1988a), a withdrawal period was not included.

c) *Side Effects*

There is little information regarding side effects of nadolol on sleep-wakefulness patterns. In the sleep laboratory, no side effects were reported (KALES et al. 1988a).

4. Sotalol

a) Efficacy

One study assessed the effects on sleep of 320 mg and 960 mg sotalol (BENDER et al. 1979). Subjects were normal volunteers who slept in the laboratory for 11 days and took the drug at one or the other dose for four nights at nighttime, following five placebo nights. No significant changes were reported for sleep latency (the only sleep efficiency measurement reported).

b) Withdrawal Effects

The withdrawal condition in the study previously mentioned consisted of two final placebo nights (BENDER et al. 1979). No significant changes were reported.

c) Side Effects

No side effects were reported with the nighttime administration of 320 mg or 960 mg sotalol (BENDER et al. 1979). The effects of both doses on performance the next day were equivocal.

F. α_2-Receptor Agonists

Studies on the effects of α_2-receptor agonists on human sleep had several objectives: understanding the involvement of norepinephrine in the regulation of REM sleep; explaining the drugs' sedative effects; and assessing the impact of the drugs on breathing during sleep. The available studies included normal subjects or depressive, hypertensive or sleep apneic patients. Therefore, there are no sleep laboratory evaluations of clonidine or guanfacine used as anxiolytics. The studies differ widely in terms of doses used and design. In this section, sleep laboratory studies on clonidine and guanfacine are summarized; excluded were those studies in which the drug was administered only during the daytime, during sleep-recorded naps, or after sleep onset at night.

I. Clonidine

1. Efficacy

The administration of 0.3 mg for two nights to normal volunteers was followed by a significant increase in total sleep time (AUTRET et al. 1977). Other studies using different dose regimes did not find any change in the sleep and wakefulness measures (either in normal subjects or in depressive or sleep apnea patients) (GAILLARD and KAFI 1979; ISSA 1992; JIMERSON et al. 1980; KANNO and CLARENBACH 1985; MALING et al. 1979; NICHOLSON and PASCOE 1991).

Two additional studies showed some worsening of sleep maintenance with clonidine. In one study using 0.15 mg and 0.30 mg clonidine given to normal subjects for one night each, the 0.15-mg dose (but not the 0.30-mg dose) caused an increase in the number of shifts to wakefulness and stage 1 (SPIEGEL and DEVOS 1980). Finally, the administration of clonidine (initially 0.1 mg b.i.d. and then 0.3 mg b.i.d.) for 3 months was followed on the only recording night (at the end of the drug administration period) by a decrease in total sleep time and sleep maintenance (KOSTIS et al. 1990).

2. Withdrawal Effects

Two studies on clonidine included withdrawal nights (GAILLARD et al. 1983; JIMERSON et al. 1980). However, data on wakefulness variables were not reported for those nights.

3. Side Effects

Sedation is one of the main side effects reported when clonidine has been studied for anxiolytic purposes (CHARNEY and HENINGER 1986; UHDE et al. 1989). Few sleep laboratory studies on clonidine have focused on the drug's side effects. The only study reporting on performance the day after clonidine administration at bedtime reported no changes 9 and 11 h after drug intake (NICHOLSON and PASCOE 1991), which could be somewhat expected because of the elimination half-life (6–12 h) of clonidine (REID 1981).

A decrease in duration of nocturnal penile tumescence has been reported with long-term administration of clonidine to middle-aged hypertensives (KOSTIS et al. 1990). Another study including obstructive sleep apneic patients showed that clonidine induced an improvement in REM-related breathing abnormalities in a subgroup of patients, whereas it worsened those of another subgroup (ISSA 1992). In this study, subjects did not report unpleasant side effects.

II. Guanfacine

1. Efficacy

The single-night administration of 1 and 2 mg guanfacine to normal subjects did not change total sleep time (SPIEGEL and DEVOS 1980). This was the only sleep efficiency measure reported.

2. Withdrawal Effects

There is no information about sleep changes following withdrawal of guanfacine in the sleep laboratory.

3. Side Effects

No information is available regarding the side effects of guanfacine when administered at bedtime in the sleep laboratory.

G. Serotonin Receptor–Agonists

I. Buspirone

Buspirone is an azapirone derivative that possesses a strong affinity for the 5-HT_{1A} receptors. It acts primarily as a presynaptic 5-HT_{1A} agonist, but it also acts as a 5-HT_{1A} postsynaptic partial agonist (for a review, see EISON and TEMPLE 1986). Its elimination half-life is about 2–3 h in healthy volunteers (GAMMANS et al. 1986).

There are a few sleep laboratory evaluations of buspirone; they differ in terms of protocol design (length and timing of administration and withdrawal period), dose used and type of subjects studied.

1. Efficacy

The acute administration of buspirone t.i.d. (5 mg at 9 and 14 h and 10 mg at 21 h) to chronic insomniacs did not induce any significant change in sleep efficiency measures (SEIDEL et al. 1985a). The administration for more than 3 weeks to patients with generalized anxiety disorder (DEROECK et al. 1989) resulted in a significant decrease in sleep latency after 3 weeks of use. It also produced a decrease in percentage of wakefulness that reached significance on the second and 22nd drug nights as compared to the two baseline nights. The drug was administered at a dose of 5 mg at 8 a.m. and 8 p.m. for 2 days and 10 mg at the same hours for the next 21 days. Since the anxiolytic effect of buspirone is said to be delayed, these changes in sleep could be related to the improvement of anxiety. A more recent study evaluated the hypnotic effects of 10 mg buspirone administered at bedtime to insomniac patients (MANFREDI et al. 1991). The study protocol included seven drug nights, preceded by four placebo-baseline nights. There was a significant increase in wake time after sleep onset on the first night of drug administration, whereas the increase was nonsignificant for the remaining six drug nights.

Two additional studies evaluated the effect of buspirone on the sleep of patients with mild to moderate obstructive sleep apnea. In one, the single night of administration of 20 mg buspirone did not induce any significant change in total sleep time, sleep latency or wakefulness after sleep onset (MENDELSON et al. 1991); in the other, 10 mg buspirone administered for three consecutive nights after three baseline nights had no effect on sleep latency, wake time, sleep efficiency or total sleep time (SCHARF et al. 1993).

2. Withdrawal Effects

The two studies in the sleep laboratory that evaluated withdrawal did not find any significant change. The first one included only one withdrawal night (DEROECK et al. 1989), whereas in the second the withdrawal period consisted of five nights (MANFREDI et al. 1991).

3. Side Effects

Studies on the effects of buspirone on breathing add evidence to the stimulant effect observed in the sleep laboratory (MANFREDI et al. 1991). It has been suggested that buspirone, due to its lack of respiratory depressant effects, might be a safe anxiolytic for patients with mild to moderate obstructive sleep apnea (MENDELSON et al. 1991; SCHARF et al. 1993). Finally, studies with the Multiple Sleep Latency Test and performance tasks have demonstrated the lack of sedative effects of buspirone, both with initial (SEIDEL et al. 1985a; DEROECK et al. 1989) and continued use (DEROECK et al. 1989).

H. Serotonin Receptor–Antagonists

I. Ritanserin

The effects on sleep of different doses of ritanserin have been assessed in a series of sleep laboratory studies. These evaluations show wide variations in terms of their methodology. Thus most of them consist of one or two nights of drug administration, whereas only a few have a sufficient number of nights to assess the development of tolerance. Most studies lack an adequate baseline, and a considerable proportion of them consist of crossover designs. The majority of the studies do not include a withdrawal period, and some of them that have, do not report the data for this condition. In addition the studies were done on healthy volunteers or in patients with heterogeneous psychiatric disorders. Finally, the time of administration of ritanserin varies across the studies. Thus, whereas in some studies the drug was administered in the morning, other studies included evening times of administration. The relatively long elimination half-life of ritanserin, which is about 40 h (VAN PEER et al. 1985), might allow for some effects on sleep even when administered in the morning.

The virtual totality of sleep laboratory studies with ritanserin have focused on its effects on sleep stages, especially slow-wave sleep. Nevertheless, many of them have included information on the hypnotic efficacy of the drug and a few of them on withdrawal.

1. Efficacy

A number of studies have evaluated the hypnotic effects of ritanserin in patients with different psychiatric disorders. Some of them used an adequate methodology, whereas other had shortcomings such as those mentioned in the introduction to this section.

Ritanserin in a dose of 5 mg administered after the evening meal showed some hypnotic efficacy when administered to middle-aged poor sleepers (ADAM and OSWALD 1989). The period of drug administration consisted of 20 consecutive nights that followed 7 days of placebo. A significant decrease in the number of awakenings was noted.

Another study evaluated the effects of 10 mg ritanserin administered at 9 a.m. to abstinent alcoholics with dysthymia and a personality disorder (MONTI et al. 1993) during 28 nights following ten nights of placebo. A significant decrease in total wake time occurred after 10 days of ritanserin administration and no tolerance to this effect took place. Total sleep time and sleep efficiency were significantly increased with similar time course. Wake time after sleep onset was significantly decreased with ritanserin in the early and late phases of treatment.

Ritanserin in a dose of 10 mg showed some hypnotic effectiveness when given to schizophrenic patients (BENSON et al. 1991). They were given the drug in the morning for two consecutive days that followed two baseline days. Ritanserin decreased the number of awakenings and wake time after sleep onset but did not change sleep latency and total sleep time.

Two other studies on psychiatric patients have documented the hypnotic efficacy of 5 mg ritanserin. Both studies included only one night of drug administration. In one study the drug was given 90 min before bedtime to patients with generalized anxiety disorder (DAROZA DAVIS et al. 1992a). Sleep efficiency increased and wake time after sleep onset decreased. In the other, 5 mg ritanserin was given in the morning to major depressive patients (STANER et al. 1992). An increase in total sleep time and in sleep efficiency were evident.

Negative findings regarding the hypnotic efficacy of ritanserin have been reported in other studies. Thus, the administration in the morning of 10 mg for 5 days to a group of insomniac patients (together with asymptomatic subjects) did not show any significant change in sleep latency, total sleep time and sleep efficiency. There were no differences in sleep parameters between insomniacs and controls (RUIZ-PRIMO et al. 1989). The study protocol consisted of a crossover double-blind design. Only the last two nights of drug administration were recorded and there was only one baseline night. In another study, 10 mg of the drug was administered with breakfast to dysthymic patients throughout 4 weeks following a 2-week placebo period (PAIVA et al. 1988). No changes in sleep efficiency, total sleep time and sleep latency were found. However, sleep was recorded in the laboratory the last three nights of drug administration. Thus, the negative findings do not rule out some initial efficacy and the development of tolerance. The other negative study consisted of a single night of administration of 5 mg ritanserin 90 min before bedtime to recovered depressed patients (DAROZA DAVIS et al. 1992b).

Studies with normal volunteers have yielded inconsistent results. Thus the 5-mg dose has been shown to have some hypnotic efficacy either when given in the morning (STANER et al. 1992) or 90 min before bedtime (DAROZA DAVIS et al. 1992a). In contrast, another study did not show any efficacy of this dose given 90 min before bedtime (DAROZA DAVIS et al. 1992b). Finally, seven nights after daily morning administration of the same dose only a trend toward decrease in wake time was reported (DECLERCK et al. 1987).

A dose-response study (1, 5 and 10 mg) found no efficacy when the drug was given once in the evening (SHARPLEY et al. 1990). Studies on the 10-mg dose did not find any change in various sleep efficiency measures, regardless of the time of the day when the drug was administered or whether it was initial or continued administration (IDZIKOWSKI et al. 1986, 1987). In a dose-response study the 10-mg dose, along with the 1-mg and 30-mg doses, were efficacious in that they reduced sleep latency (IDZIKOWSKI et al. 1991). Because the 3-mg dose did not reduce sleep latency, a clear dose response effect was not evident.

Ritanserin has been administered to pateints with sleep disorders. There is one case report showing the efficacy of ritanserin in insomnia (DAHLITZ et al. 1990). Also, a reduction in wakefulness after sleep onset and a subjective improvement in the morning with a reduction in daytime sleepiness was observed when 5 mg ritanserin daily for 4 weeks was added to the usual medication in narcoleptics (LAMMERS et al. 1991).

2. Withdrawal Effects

The effects of withdrawing ritanserin at various doses studied in psychiatric patients have been assessed in only a few sleep laboratory evaluations (ADAM and OSWALD 1989; BENSON et al. 1991; MONTI et al. 1993). Two studies reported the data for the withdrawal period (ADAM and OSWALD 1989; MONTI et al. 1993). In one study, the withdrawal of 5 mg ritanserin administered for 20 days caused sleep deterioration. This was particularly marked on the third (last) withdrawal night when a significant decrease in total sleep time occurred (ADAM and OSWALD, 1989). In the other study, the withdrawal of 10 mg after 28 days of administration was not followed by any rebound phenomena (MONTI et al. 1993). Actually, wake time after sleep onset remained significantly decreased upon withdrawal.

Only one sleep laboratory study with normal subjects assessed withdrawal. No clear effects were seen on the only night of withdrawal after the administration of 5 mg through seven days (DECLERCK et al. 1987).

3. Side Effects

Side effects after the administration of ritanserin were assessed in some of the existing sleep laboratory studies including those with normal subjects (DECLERCK et al. 1987; IDZIKOWSKI et al. 1987; DAROZA DAVIS et al. 1992a,b) or patients (ADAM and OSWALD 1989; DAROZA DAVIS 1992a,b; MONTI et al. 1993). The reports on side effects were based on subjective ratings and/or symptoms. The only side effects reported indicate some CNS depressant action of the drug and appear to be dose related. Thus, with the 5-mg dose, vigilance was unaffected (ADAM and OSWALD 1989; DAROZA DAVIS et al. 1992a,b; DECLERCK et al. 1987). In one study with the 10-mg dose (the time of administration is unclear), four of the seven normal subjects who participated complained of increased morning sleepiness during the first 3 or 4 days of drug use (BIRMANNS and CLARENBACH 1989). Dizziness, numbness to touch and impaired coordination appeared in one patient each in a study with alcoholics (MONTI et al. 1993). In contrast, no side effects were found in another study with the same dose (IDZIKOWSKI et al. 1987).

Finally, dizziness and lack of concentration were reported by three of four subjects after one initial dose of 25 mg (DECLERCK et al. 1987). Because the drug was administered in the morning and the authors did not specify when these side effects occurred, it is difficult to know if they were related to peak concentration or were delayed or next-day effects.

I. Summary

Anxiolytic drugs are often prescribed as hypnotics, especially for patients with anxiety disorders. They can be administered as single bedtime doses or in several doses with the greater amount at night. The most prescribed anxiolytics are the benzodiazepines. Although they are referred to as a single class of drugs, they are a heterogeneous group in terms of pharmacokinetic, pharmacodynamic and chemical structure properties. Differences in these properties account for the varying effects among these drugs in terms of efficacy, development of tolerance, withdrawal manifestations and adverse events.

Studies in the sleep laboratory have shown that the majority of benzodiazepine anxiolytics have some degree of hypnotic efficacy. However, only those benzodiazepines with slow elimination half-lives maintain their efficacy after several weeks of consecutive administration. Also drugs with more rapid rates of elimination and higher binding affinities show more potential for withdrawal manifestations such as rebound insomnia. As for the capacity to induce side effects, those with slower elimination half-lives are more likely to cause daytime sedation, whereas those more rapidly eliminated and more potent, especially those with a triazolo structure, have greater capacity to induce hyperexcitability states and other unexpected side effects.

Two new non-benzodiazepines, which are $GABA_A$ receptor-complex agonists, have been investigated for their anxiolytic effects. These drugs belong to the groups of cyclopyrrolones (suriclone) and imidazopyridines (alpidem). There are only a few sleep laboratory assessments with these drugs. Only minor or no changes in sleep and wakefulness were found with them; the fact that only normal subjects were studied makes these findings of little clinical relevance.

Drugs acting on adrenergic receptors such as the β-blockers and the α_2-agonists, although not marketed as psychotropic drugs, appear to have some anxiolytic effects. The interest for studying the β-blockers in the sleep laboratory stemmed from the clinical observation of sleep-related adverse events with these drugs. Generally speaking, sleep laboratory studies point to more sleep changes (increased wakefulness) with the more lipophilic drugs, although other pharmacological factors may play a role.

Drugs acting on the serotoninergic receptors have been marketed or are being investigated for their anxiolytic properties. Buspirone, a 5-HT_{1a} agonist, has shown more stimulant than depressant effects while ritanserin, a 5-$HT_{2/1c}$ antagonist, has shown some hypnotic efficacy, although not consistently across studies.

To conclude, sleep laboratory studies provide an important and objective means for assessing anxiolytic drugs. Along with clinical trials, they contribute significantly in establishing the benefit-to-risk ratio of an anxiolytic drug.

References

Adam K, Oswald I (1989) Effects of repeated ritanserin on middle-aged poor sleepers. Psychopharmacology (Berl) 99:219–221

Allen S, Oswald I (1976) Anxiety and sleep after fosazepam. Br J Clin Pharmacol 3:165–168

American Psychiatric Association (1994) Diagnostic statistical manual – IV (DSM-IV). Anceutan Psychiatric Association, Washington DC

Ansseau M, Olie JP, von Frenckell R, Jourdain G, Stehle B, Guillet P (1991) Controlled comparison of the efficacy and safety of four doses of suriclone, diazepam and placebo in generalized anxiety disorder. Psychopharmacology (Berl) 104:439–443

Autret A, Minz M, Beillevaire T, Cathala HP, Schmitt H (1977) Effect of clonidine on sleep patterns in man. Eur J Clin Pharmacol 12:319–322

Bender W, Greil W, Ruther E, Schnelle K (1979) Effects of the β-adrenoceptor blocking agent sotalol on CNS: sleep, EEG, and psychophysiological parameters. J Clin Pharmacol 19:505–512

Benson KL, Csernansky JG, Zarcone VP (1991) The effect of ritanserin on slow wave sleep deficits and sleep continuity in schizophrenia. Sleep Res 20:170

Betts TA, Alford C (1985) β-blockers and sleep: a controlled trial. Eur J Clin Pharmacol 28(suppl):65–68

Bianchetti G, Dubruc C, Ascalone V, Thénot JPG (1993) Pharmacokinetic profile of alpidem: effect of age and pathological conditions. In: Bartholini G, Garreau M, Morselli PL, Zivkovic B (eds) Imidazopyridines in anxiety disorders: a novel experimental and therapeutic approach. Raven, New York, pp 121–131

Birmanns B, Clarenback P (1989) Effects of serotonin-receptor antagonists on sleep and sleep related release of HGH, prolactin and cortisol. In: Wauquier A, Dugovic C, Radulovacki M (eds) Slow wave sleep: physiological, pathophysiological and functional aspects. Raven, New York, pp 235–242

Bliwise D, Seidel W, Greenblatt DJ, Dement W (1984) Nighttime and daytime efficacy of flurazepam and oxazepam in chronic insomnia. Am J Psychiatry 141:191–195

Bonnet M, Kramer M, Rosa R (1980) The hypnotic effectiveness of ketazolam in anxious subjects. Curr Ther Res 28:284–293.

Bonnet M, Kramer M, Roth T (1981) A dose response study of the hypnotic effectiveness of alprazolam and diazepam in normal subjects. Psychopharmacology (Berl) 75:258–261

Boudoulas H, Schmidt H, Geleris P, Clark PW, Lewis RP (1983) Case reports on deterioration of sleep apnea during therapy with propranolol. Preliminary studies. Res Commun Chem Pathol Pharmacol 39:3–10

Charney DS, Heninger GR (1986) Abnormal regulation of noradrenergic function in panic disorders. Arch Gen Psychiatry 43:1042–1054

Charney DS, Heninger GR, Hafstad KM, Capelli S, Redmond E (1983) Neurobiological mechanisms in human anxiety: recent clinical studies. Psychopharmacol Bull 19:470–475

Da Roza Davis JM, Sharpley AL, Cowen PJ (1992a) Slow wave sleep and 5-HT_2 receptor sensitivity in generalized anxiety disorder: a pilot study with ritanserin. Psychopharmacology (Berl) 108:387–389

Da Roza Davis JM, Sharpley AL, Solomon RA, Cowen PJ (1992b) Sleep and 5-HT_2 receptor sensitivity in recovered depressed patients. J Affect Disord 24:177–182

Dahlitz M, Wells, James R, Idzikowski C, Parkes JD (1990) Treatment of insomnia with ritanserin. Lancet 336:379

Dahlöf C, Dimenas E (1990) Side effects of β-blocker treatments as related to the central nervous system. Am J Med Sci 299:236–244

Danjou P, Court LA, Feuerstein C, Vandel B, Rosenzweig P (1993) Anxiolytics and electrogenesis: alpidem effects on waking EEG and sleep architecture. In: Bartholini G, Garreau M, Morselli PL, Zivkovic B (eds) Imidazopyridines in anxiety disorders: a novel experimental and therapeutic approach. Raven, New York, pp 133–142

Davidson J (1992) Drug therapy of post-traumatic stress disorder. Br J Psychiatry 160:309–314

De Roeck J, Cluydts R, Schotte C, Rouckhout D, Cosyns P (1989) Explorative single-blind study on the sedative and hypnotic effects of buspirone in anxiety patients. Acta Psychiatr Scand 79:129–135

Declerck AC, Wauquier A, Van der Ham-Veltman PHM, Celders Y (1987) Increase in slow-wave sleep in humans with the serotonin-S_2 antagonist ritanserin. Curr Ther Res 41:427–432

Dufour H, Wade A, Legris P, L'Heritièr C, Priore P, Garreau M (1993) Efficacy of alpidem in anxiety disorders. In: Bartholini G, Garreau M, Morselli PL, Zivkovic B (eds) Imidazopyridines in anxiety disorders: a novel experimental and therapeutic approach. Raven, New York, pp 165–173

Dunleavy DLF, McLean AW, Owald I (1971) Desibroquine, guanethidine, propranolol and human sleep. Psychopharmacology (Berl) 21:101–110

Ehrenstein W, Schaffler K, Müller-Limmroth W (1972) Die Wirkung von Oxazepam auf den gestörten Tagschlaf nach Nachtschichtarbeit. Arzneimittel forschung 22: 421–427

Eison MS, Temple DL (1986) Buspirone: review of its pharmacology and current perspectives on its mechanism of action. Am J Med 80 [Suppl 3B]:1–9

Gaillard JM, Kafi S (1979) Involvement of pre and postsynaptic receptors in catecholaminergic control of paradoxical sleep in man. Eur J Clin Pharmacol 15:83–89

Gaillard JM, Iorio G, Kafi S, Blois R (1983) Paradoxical sleep rebound without previous debt: the effect of minute doses of clonidine in man. Sleep 6:60–66
Gammans RE, Mayol RF, Labudde JA (1986) Metabolism and disposition of buspirone. Am J Med 80(suppl 3b):41–51
Gershon S, Eison AS (1983) Anxiolytic profiles. J Clin Psychiatry 44:45–56
Hamon M (1994) Neuropharmacology of anxiety: perspectives and prospects. TIPS 15:36–39
Hartmann E (1968) The effect of four drugs on sleep pattern in man. Psychopharmacologia (Berl) 12:346–353
Hartmann E, Cravens J (1973) The effects of long-term administration of psychotropic drugs on human sleep: VI. The effects of chlordiazepoxide. Psychopharmacologia (Berl) 33:233–245
Hubain PP, Castro P, Mesters P, De Maertelaer V, Mendlewicz J (1990) Alprazolam and amitriptyline in the treatment of major depressive disorder: a double-blind clinical and sleep EEG study. J Affect Disord 18:67–73
Idzikowski C, Mills FJ, Glennard R (1986) 5-Hydroxytryptamine-2 antagonist increases human slow wave sleep. Brain Res 378:164–168
Idzikowski C, Cowen PJ, Nutt D, Mills FJ (1987) The effects of chronic ritanserin treatment on sleep and the neuroendocrine response to L-tryptophan. Psychopharmacology (Berl) 93:416–420
Idzikowski C, Mills FJ, James RJ (1991) A dose-response study examining the effects of ritanserin on human slow wave sleep. Br J Clin Pharmacol 31:193–196
Insel TR, Gillin JC, Moore A, Mendelson WB, Loewenstein RJ, Murphy DL (1982) The sleep of patients with obsessive-compulsive disorder. Arch Gen Psychiatry 39:1372–1377
Issa FG (1992) Effect of clonidine in obstructive sleep apnea. Am Rev Respir Dis 145:435–439
Jimerson DC, Post RM, Stoddard FJ, Gillin JC, Bunney WE (1980) Preliminary trial of the noradrenergic agonist clonidine in psychiatric patients. Biol Psychiatry 15:45–57
Kahn RS, Moore C (1993) Serotonin in the pathogenesis of anxiety. In: Hoehn-Saric R, McLeod DR (eds) Biology of anxiety disorders. American Psychiatric Press, Washington, pp 61–102
Kales A, Kales JD (1984) Evaluation and treatment of insomnia. Oxford University Press, New York
Kales A, Soldatos CR, Cadieux RJ, Bixler EO, Tan T-L, Scharf MB (1979) Propranolol in the treatment of narcolepsy. Ann Intern Med 91:742–743
Kales A, Soldatos CR, Caldwell AB, Charney DS, Kales JD, Markel D, Cadieux R (1980) Nightmares: clinical characteristics and personality patterns. Am J Psychiatry 137:1197–1201
Kales A, Bixler EO, Cadieux RJ, Schneck DW, Shaw LC III, Locke TW, Vela-Bueno A, Soldatos CR (1984) Sleep apnoea in a hypertensive population. Lancet 2:1005–1008
Kales A, Bixler EO, Soldatos CR, Jacoby JA, Kales JD (1986) Lorazepam: effects on sleep and withdrawal phenomena. Pharmacology 32:121–130
Kales A, Bixler EO, Vela-Bueno A, Soldatos CR, Manfredi RL (1987) Alprazolam: effects on sleep and withdrawal phenomena. J Clin Pharmacol 27:508–515
Kales A, Bixler EO, Vela-Bueno A, Cadieux RJ, Manfredi RL, Bitzer S, Kantner T (1988a) Effects of nadolol on blood pressure, sleep efficiency and sleep stages. Clin Pharmacol Ther 43:655–662
Kales A, Soldatos CR, Bixler EO, Kales JD, Vela-Bueno A (1988b) Diazepam: effects on sleep and withdrawal phenomena. J Clin Psychopharmacol 8:340–345
Kales A, Manfredi RL, Vgontzas AN, Baldassano CF, Kostakos K, Kales JD (1991) Clonazepam: sleep laboratory study of efficacy and withdrawal. J Clin Psychopharmacol 11:189–193

Kales A, Vgontzas AN, Bixler EO (1994) Not all benzodiazepines are alike: update 1993. Proceedings of the 9th Congress of Psychiatry, Rio de Janeiro, 1993
Kales JD, Kales A, Soldatos CR, Caldwell AB, Charney DS, Martin ED (1980) Night terrors: clinical characteristics and personality patterns. Arch Gen Psychiatry 37:1413–1417
Kanno O, Clarenbach P (1985) Effect of clonidine and yohimbine on sleep in man: polygraphic study and EEG analysis by normalized slope descriptors. Electroencephalogr Clin Neurophysiol 60:478–484
Kantola I, Rauhala E, Erkinjunti M, Mansury L (1991) Sleep disturbances in hypertension: a double-blind study between isradipine and metoprolol. J Cardiovasc Pharmacol 18 [Suppl 3]:S41–S45
Karacan I, O'Brien GS, Williams RL, Salis PJ, Thornby JI (1973) Methodology for electroencephalographic sleep evaluation of drugs. In: Koella WP, Levin P (eds) Sleep: physiology, biochemistry, psychology, pharmacology, clinical implications. Karger, Basel, pp 463–476
Kayed K, Godtlibsen OB (1977) Effects of the β-adrenoceptor antagonists acebutolol and metoprolol on sleep pattern in normal subjects. Eur J Clin Pharmacol 12:323–326
Koella WP (1985) CNS-related (side) effects of β-blockers with special reference to mechanisms of action. Eur J Clin Pharmacol 28 Suppl:55–63
Kostis JB, Rosen RC (1987) Central nervous system effects of β-adrenergic-blocking drugs: the role of ancillary properties. Circulation 75:204–212
Kostis JB, Rosen RC, Holzer BC, Randolph C, Taska L, Miller MH (1990) CNS side effects of centrally-active antihypertensive agents: a prospective, placebo-controlled study of sleep, mood state, and cognitive and sexual function in hypertensive males. Psychopharmacology (Berl) 102:163–170
Kramer M (1982) Dose-response effects of alprazolam on sleep architecture in normal subjects. Curr Ther Res 31:960–968
Kramer M, Schoen L (1986) Short-term effects of adinazolam on sleep and postsleep behavior. Curr Ther Res 40:924–932
Kramer M, Schoen LW, Scharf M (1984) Effects of alprazolam on sleep and performance in geriatric insomniacs. Curr Ther Res 35:67–76
Lader M (1988) β-adrenoceptor antagonists in neuropsychiatry: an update. J Clin Psychiatry 49:213–223
Lammers GJ, Arends J, Declerck AC, Kamphuisen HAC, Schouwink G, Troost J (1991) Ritanserin, a 5-HT_2 receptor blocker, as add-on treatment in narcolepsy. Sleep 14:130–132
Maling TJB, Dollery CT, Hamilton CA (1979) Clonidine and sympathetic activity during sleep. Clin Sci 57:509–514
Manfredi RL, Kales A, Vgontzas AN, Bixler EO, Isaac MA, Falcone CM (1991) Buspirone: sedative or stimulant effect? Am J Psychiatry 148:1213–1217
McAinsh J, Cruickshank JM (1990) Beta-blockers and central nervous system side effects. Pharmacol Ther 46:163–197
McClure DJ, Walsh J, Chang H, Olah A, Wilson R, Pecknold JC (1988) Comparison of lorazepam and flurazepam as hypnotic agents in chronic insomniacs. J Clin Pharmacol 28:52–63
Meier-Ewert K, Matsubayashi K, Benter L (1985) Propranolol: long-term treatment in narcolepsy-cataplexy. Sleep 8:95–104
Mellman TA, Uhde TW (1989) Electroencephalographic sleep in panic disorder. Arch Gen Psychiatry 46:178–184
Mendelson WB, Maczaj M, Holt J (1991) Effects of buspirone on sleep-related respiration in obstructive sleep apnea. Sleep Res 20:69
Mitler MM, Browman CP, Menn SJ, Gujavarty K, Timms RM (1986) Nocturnal myoclonus: treatment efficacy of clonazepam and temazepam. Sleep 9:385–392
Monti JM, Alterwain P, Estevez F, Alvariño F, Giusti M, Olivera S, Labraga P (1993) The effects of ritanserin on mood and sleep in abstinent alcoholic patients. Sleep 16:647–654

Moore S, Bonnet M, Kramer M, Roth T (1981) A dose-response study of the hypnotic effectiveness of ketazolam in normal subjects. Curr Ther Res 29:704–713
Nicholson AN, Pascoe PA (1991) Presynaptic alpha$_2$-adrenoceptor function and sleep in man: studies with clonidine and idazoxan. Neuropharmacology 30:367–372
Nicholson AN, Stone BM (1979) Diazepam and 3-hydroxydiazepam (temazepam) and sleep of middle age. Br J Clin Pharmacol 7:463–468
Nicholson AN, Stone BM, Clarke CH, Ferres HM (1976a) Effect of N-desmethyldiazepam (nordiazepam) and a precursor, potassium clorazepate on sleep in man. Br J Clin Pharmacol 3:429–438
Nicholson AN, Stone BM, Clarke CH (1976b) The effect of diazepam and fosazepam (a soluble derivative of diazepam) on sleep in man. Br J Clin Pharmacol 3:533–541
Norman TR, Burrows GD (1986) Anxiety and the benzodiazepine receptor. Prog Brain Res 65:73–90
Noyes R (1982) Beta-blocking drugs and anxiety. Psychosomatics 23:155–170
Oakley NR, Tyers MB (1992) 5-HT receptor types and anxiety. In: Mendlewicz J, Racagni G (eds) Target receptors for anxiolytics and hypnostics: from molecular pharmacology to therapeutics. Karger, Basel, pp 11–23 (International academy of biomedicine and drug research, vol 3)
Paiva T, Arriaga F, Wauquier A, Lara E, Largo R, Leitao JN (1988) Effects of ritanserin on sleep disturbances of dysthymic patients. Psychopharmacology (Berl) 96:395–399
Pancheri P, Bressa GM, Borghi C (1993) Double-bind randomized studies on the therapeutic action of alpidem in generalized anxiety disorders. In: Bartholini G, Garreau M, Morselli PL, Zivkovic B (eds) Imidazopyridines in anxiety disorders: a novel experimental and therapeutic approach. Raven, New York, pp 155–164
Reid JL (1981) The clinical pharmacology of clonidine and related central antihypertensive agents. Br J Clin Pharmacol 12:295–302
Risberg AM, Henricsson S, Ingvar DH (1977) Evaluation of the effect of fosazepam (a new benzodiazepine), nitrazepam and placebo on sleep patterns of normal subjects. Eur J Clin Pharmacol 12:105–109
Ruiz-Primo E, Haro R, Valencia M (1989) Polysomnographic effects of ritanserin in insomniacs: a crossed double-blind controlled study. Sleep Res 18:72
Saletu B, Schultes M, Grünberger J, Chaudry HR (1986) Sleep laboratory study of a new antianxiety drug, alpidem: short-term trial. Curr Ther Res 40:769–779
Saletu B, Frey R, Grünberger J, Krupka M, Anderer P, Musch B (1990) Sleep laboratory studies on single dose effects of suriclone. Br J Clin Pharmacol 30:703–710
Scharf MB, Kales A, Bixler EO, Jacoby JA, Schweitzer PK (1982) Lorazepam efficacy, side effects and rebound phenomena. Clin Pharmacol Ther 31:175–179
Scharf MB, Denson DD, Thompson GA, Goff PJ (1985) Carryover hypnotic effectiveness of diazepam. Curr Ther Res 37:309–317
Scharf MB, Brannen DE, Berkowitz DV, McDannold M (1993) Comparative effects of buspirone and alprazolam on snoring and sleep apnea in geriatric patients. Sleep Res 22:47
Seidel WF, Cohen SA, Bliwise NG, Dement WC (1985a) Buspirone: an anxiolytic without sedative effects. Psychopharmacology (Berl) 87:371–373
Seidel WF, Cohen SA, Wilson L, Dement WC (1985b) Effects of alprazolam and diazepam on the daytime sleepiness of nonanxious subjects. Psychopharmacology (Berl) 87:194–197
Sharpley AL, Solomon RA, Fernando AI, Da Roza Davis JM, Cowen PJ (1990) Dose-related effects of selective 5-HT$_2$ receptor antagonists on slow wave sleep in humans. Psychopharmacology (Berl) 101:568–569
Spiegel R, Devos JE (1980) Central effects of guanfacine and clonidine during wakefulness and sleep in healthy subjects. Br J Clin Pharmacol 10:165S–168S

Staner L, Kempenaers C, Simonnet MP, Fransolet L, Mendlewicz J (1992) 5-HT_2 receptor antagonism and slow-wave sleep in major depression. Acta Psychiatr Scand 86:133–137

Uhde TW, Stein MB, Vittone BJ, Siever LJ, Boulenger JF, Klein E, Mellman TA (1989) Behavioral and physiologic affects of short-term and long-term administration of clonidine in panic disorder. Arch Gen Psychiatry 46:170–177

Van Peer A, Gasparini R, Woestenborghs R, Heykants J, Gelders Y (1985) Intravenous pharmacokinetics and effect of food on the bioavailability of ritanserin in healthy volunteers. Naunyn Schmiedebergs Arch Pharmacol [Suppl 330]:R15

Vogel G, Minter K, Clifton T (1987) The effect of adinazolam on sleep variables in patients with melancholia. Sleep Res 16:155

Walsh JK, Schweitzer PK, Parwatikar P (1983) Effects of lorazepam and its withdrawal on sleep, performance and subjective state. Clin Pharmacol Ther 34: 496–500

Weichler U, Herres-Mayer B, Mayer J, Weber K, Hoffmann R, Peter JH (1991) Influence on antihypertensive drug therapy on sleep pattern and sleep apnea activity. Cardiology 78:124–130

Zarcone VP, Benson KL, Greene KA, Csernansky JG, Faull KF (1994) The effect of chronic alprazolam on sleep and bioamine metabolities in depression. J Clin Psychopharmacol 14:36–40

Ziegler G, Ludwig L, Klotz U (1983) Effect of midazolam on sleep. Br J Clin Pharmacol 16:81S–86S

CHAPTER 15
Stimulant Drugs

Y. Hishikawa

A. Early Use of Plant Preparations Containing Central Stimulants

Plants containing central stimulants were identified by man several thousand years ago. Most of these plants were used as ancient drugs, examples of which are tea leaves in China and India, coffee beans and cola nuts in South America, ma huang in China, khat leaves in Africa, and coca leaves in South America. The early histories of the central stimulant drugs are described in detail by Angrist and Sudilovsky (1978).

Various aqueous extracts of dried tea leaves, coffee beans, and cola nuts have long been used as beverages with the action of mild psychomotor excitation and suppression of both fatigue and sleepiness. The main active constituent of tea, coffee, and cola is caffeine, and this has been used as a central stimulant also in modern Western medicine.

The Chinese herb ma huang has long been used as a drug with circulatory stimulant, diaphoretic, antipyretic, and antitussive action in traditional Chinese medicine. Only recently was its main active ingredient found to be the sympathomimetic, ephedrine, (Chen and Schmidt 1925), which was introduced into clinical use in modern Western medicine for the treatment of asthma and similar conditions, and also for the treatment of narcolepsy as described later in this chapter.

The use of khat as a "food stimulant" originated many centuries ago. Chewing of fresh khat leaves has long been a socially accepted habit of people living in the eastern regions of Africa. The chewing of fresh khat leaves produces effects of mild mental stimulation characterized by feelings of contentment, loquacity, mild psychomotor excitation, suppression of both fatigue and sleepiness, and loss of appetite. In recent years, the main psychoactive substance contained in the fresh khat leaf was identified to be an alkaloid called cathionine with amphetamine-like action in man and animals (Kalix 1992). As with amphetamine, when overindulgence of chewing fresh khat leaves occurs, a sense of elation and euphoria are induced, and this may give way to an episode characterized by mania, auditory hallucinations, and other schizophrenia-like symptoms, or to a confused delusional syndrome (Laurent 1962). Modern medical use of cathionine has not yet been reported.

Since times prior to the advent of the Incas, chewing of fresh coca leaves has been widespread among the inhabitants in South America. It was known to induce exhilaration and euphoria. Cocaine isolated from coca leaf was used for local anesthesia and for other medical purposes (Angrist and Sudilovsky 1978). But, because the dangers of developing cocaine dependence and cocaine psychosis soon became apparent, cocaine came to be regarded with alarm and its use is now controlled by legislative measures in many countries.

B. Indications for Medical Use of Central Stimulant Drugs

Among the various drugs available for treatment of sleep disorders, central stimulant drugs are pharmacological agents used for maintaining vigilance and alertness by means of dissipating intense daytime sleepiness and suppressing irresistible episodes of sleep, which often occur unexpectedly, impeding appropriate and adaptive, social activities of patients under different circumstances.

Different pathological conditions and sleep disorders cause intense sleepiness and irresistible episodes of sleep in the daytime during work and other social activities. For example, brain tumor, hypothyroidism, post-traumatic, and other pathological conditions may accompany intense sleepiness and intolerable sleep episodes. Overdosage of hypnotic, anxiolytic, and anticonvulsive drugs also often induces excessive sleepiness and irresistible sleep episodes in the daytime. These symptoms are also found in patients with circadian rhythm sleep disorders, sleep apnea syndrome, or periodic-limb movements in sleep. Even healthy people with very short nocturnal sleep due to their being engaged in a second job at night or idling away their time until late in the night often experience excessive sleepiness and short irresistible or unexpected naps in the daytime.

Therapeutic measures for excessive daytime sleepiness and unexpected, irresistible episodes of sleep under the conditions described above must be directed primarily to remove the etiological factors, to correct the underlying pathological conditions and to reestablish a regular circadian sleep-wake rhythm suitable for normal daily life of patients. Central stimulant drugs are used only rarely as a subsidiary or adjunctive measure for the treatment of excessive daytime sleepiness and unwanted sleep episodes found in patients with the different pathological conditions described above or circadian rhythm sleep-wake disorders.

Central stimulant drugs are primarily indicated for, and really used for, dispelling or suppressing intense daytime sleepiness and irresistible sleep episodes in hypersomniac conditions due to unknown etiologies, such as narcolepsy, idiopathic hypersomnia, and recurrent hypersomnia (periodic hypersomnia). Many central stimulant drugs are effective for relieving hypersomniac symptoms in narcoleptic patients. However, most of the central stimulants available at present for medical use are only partially effective or are ineffective for idiopathic hypersomnia and recurrent hypersomnia. The clinical features of these hypersomniac conditions are explained in the next section. Their diagnostic criteria are not described in this chapter; they are detailed in the Intenational Classification of Sleep Disorders (DIAGNOSTIC CLASSIFICATION STEERING COMMITTEE 1990).

Pharmacologically, stimulant agents encompass a wide variety of drugs, including phenylethylamines (ephedrine, amphetamines), cocaine, thyroid hormones, and various xanthine derivatives (caffeine, theophylline). Stimulant drugs with the central effects of maintaining wakefulness and alertness are used for the treatment of excessive daytime sleepiness in narcoleptic patients and other hypersomniac conditions. Central stimulant drugs are also employed for the treatment of attention deficit disorder and hyperkinetic states in children and for reducing appetite and body weight in obese patients. Many other stimulants are used for their peripheral sympathomimetic effects as antihypotensives, bronchodilators, etc.

Descriptions in this chapter are limited to central stimulant drugs used mainly for the treatment of narcoleptic patients and only secondarily for other sleep disorders. Before the descriptions of such stimulant drugs are presented, the clinical features of narcolepsy and a few other hypersomniac conditions are briefly explained. Also in this chapter are outlined the actual, therapeutic measures for hypersomniac conditions using central stimulants and other adjunctive drugs.

C. Clinical Features of Narcolepsy and Other Hypersomniac Conditions

I. Narcolepsy

Narcolepsy is characterized by a set of clinical symptoms consisting of abnormal sleep characteristics: excessive daytime sleepiness, irresistible short episodes of sleep, hypnagogic hallucinations, disturbed nocturnal sleep, and manifestations of suddenly occurring and short-lasting muscular weakness, cataplexy, and sleep paralysis (DALY and YOSS 1974; KALES et al. 1982a).

Irresistible episodes of sleep recur several times a day, not only under favorable circumstances for sleep, such as during monotonous work in the sitting position or after a heavy meal, but also in different situations in which normal subjects rarely fall asleep, such as during very important work or while discussing business with other persons. The duration of the sleep episode may vary from a few minutes to over 1 h. Narcoleptic patients usually wake up refreshed from such brief sleep episodes. Apart from sleep episodes, patients often feel very sleepy. This excessive daytime sleepiness often causes-memory lapses, poor performance at work, and misadaption in social life (KALES et al. 1982b; BROUGHTON et al. 1983).

Cataplexy is an abrupt and reversible, marked decrease or loss of muscle tone, most frequently triggered by emotional experience, such as laughter, anger, and surprise. It involves voluntary muscles in some of the extremities or in the entire body. Complete loss of muscle tone resulting in a collapse may be noted during a cataplectic attack. An attack may be only a slight buckling of the knees, and patients may simply stop walking or stand against a wall. The muscle weakness may also involve the facial and speech muscles. Cataplexy usually lasts from several seconds up to a few minutes. During cataplectic attacks, consciousness is not disturbed and the patient is well aware of his or her surroundings, and of him- or herself.

Sleep paralysis is a terrifying experience that frequently occurs in narcoleptic patients on falling asleep, or on awakening, in the relaxed position, such as lying recumbent on a sofa or a bed. The episode of sleep paralysis usually lasts several minutes. The patient suddenly finds him- or herself unable to move the limbs and to speak. This state is frequently accompanied by terrifying dream-like hallucinations. During the episode of sleep paralysis and hypnagogic hallucinations, the patient is usually very anxious often with the fear of dying. Nocturnal sleep of narcoleptic patients is often disturbed with frequent interrupting awakenings and vivid dreams.

Cataplexy, sleep paralysis, and hypnagogic hallucinations are pathological, dissociated manifestations of REM sleep, which often occur at the onset of sleep in narcoleptic patients. Pronounced inhibition of muscle tone and vivid dreams in sleep-onset REM sleep probably induce these pathological phenomena (HISHIKAWA et al. 1968; ALDRICH 1990).

Narcolepsy is not a rare condition, occurring in 0.02%–0.05% of the general population. Age at onset varies from childhood to the 3rd decade, with a peak in the 2nd decade.

The etiology of narcolepsy is unknown. The genetic feature of this disease has been shown by KESSLER et al. (1974) and HONDA et al. (1983b). The recent discovery of a link between a certain type of human leukocyte antigen (HLA), known as DR2, and narcolepsy is probably a major advance in the understanding of the genetic basis of this disorder (HONDA et al. 1983a).

After the onset of narcolepsy, the condition usually lasts many years, in many patients throughout life. As the etiological therapeutic measures are not available at present, they must be treated by symptomatic therapeutic measures with the use of central stimulants and other adjunctive drugs (KALES et al. 1987).

II. Idiopathic Hypersomnia

Idiopathic hypersomnia is much less frequent than narcolepsy, and was only recently distinguished from narcolepsy on the basis of the absence of cataplexy and sleep-onset REM sleep episodes. This hypersomnia has been successively referred to as essential narcolepsy, non-REM narcolepsy, idiopathic CNS hypersomnia, and, finally, idiopathic hypersomnia (THORPY 1990).

Idiopathic hypersomnia is characterized by complaints of constant or recurrent excessive daytime sleepiness and lengthy nonrefreshing daytime sleep episodes lasting from one to several hours. The capacity to arouse the subject may be normal, but some patients report great difficulty in awakening and experience disorientation after awakening (sleep drunkenness). Age of onset is from the 1st to the 5th decade with a peak in the 2nd decade. Idiopathic hypersomnia tends to be chronic with no remission of the excessive daytime sleepiness over the life span. Because of this, the social life of patients is often significantly disturbed. Associated symptoms suggesting dysfunction of the autonomic nervous system are not uncommon in patients with idiopathic hypersomnia. They include headaches, fainting episodes, orthostatic hypotension, and, most commonly, peripheral vascular complaints (Raynaud-type phenomena with cold hands and feet). The etiology of idiopathic hypersomnia is unknown. Differential diagnosis from narcolepsy without cataplexy must be carefully considered, and is often difficult, since cataplexy may not be present or may never appear in the course of an otherwise typical narcolepsy (DALY and YOSS 1974; THORPY 1990).

III. Recurrent Hypersomnia

Recurrent hypersomnia is a rare condition, and was often called periodic hypersomnia or periodic somnolence. Hypersomniac episodes in this condition develop either abruptly or gradually, lasting several days to a few weeks, and recur at varying frequencies from once to several times a year. During a hypersomniac episode, the subject retires to his or her bed, and often refuses to leave it. Urinary incontinence does not occur. The subject is usually calm, but is sometimes agitated, when his or her sleep is disturbed. During hypersomniac episodes, disorientation, forgetfulness, depression, and depersonalization are often observed. Abnormal behavior characterized by compulsive overeating and sexual disinhibition sometimes appears. The condition characterized by both recurrent hypersomnia and compulsive overeating has been called the Kleine-Levin syndrome. In the interval between episodes, physical and mental functions are normal and there is no significant personality and sleep disorders. Febrile conditions, fatigue, and heavy drinking of alcohol often precipitate a hypersomniac episode. In many patients, the initial hypersomniac episode occurs in the 2nd decade of life and episodes gradually decrease in frequency, and eventually disappear completely in the 3rd decade of life. However, in some patients, the symptoms reoccur even after the 3rd decade. The etiology is unknown and some functional disorder in the diencephalon has been suggested (TAKAHASHI 1965; THORPY 1990).

D. Central Stimulants

Central stimulants described in this section are caffeine, ephedrine, amphetamine, methamphetamine, methylphenidate, pemoline, mazindol, and modafinil. These drugs were, and are, used for dispelling excessive sleepiness and maintaining wakefulness in patients with narcolepsy and other hypersomniac conditions due to different causes. Some of the central stimulants are also used for the treatment of depression, attention deficit disorder, and hyperactivity in children with minimal brain disorder, and for reducing appetite and body weight in obese patients.

I. Caffeine

Caffeine is contained in the beverages of coffee, tea, and cola, which have long been routinely used by many people in the world (Fig. 1). Therefore, caffeine is an extremely old and widely used central stimulant. Normal people often drink coffee and tea in order to dissipate sleepiness and to maintain wakefulness in ordinary daily life. Recent research has suggested that caffeine acts as an antagonist of benzodiazepine receptors (MARANGOS et al. 1979) or as an antagonist of adenosine receptors in the brain (PHILLIS et al. 1979). These mechanisms probably underlie the pharmacological action of caffeine as a central stimulant (see also Chaps. 8, 11, this volume).

Many narcoleptic patients, when their disorder is not timely diagnosed and not properly treated, often drink coffee or tea in order to suppress excessive daytime sleepiness and to maintain alertness. The caffeine content of six cups of strong black coffee has about the same stimulant effect as 5 mg amphetamine. But coffee and tea are usually ineffective in suppressing intense daytime sleepiness and irresistible sleep episodes in narcoleptic patients. Caffeine tablets and ephedrine as described below are inferior to amphetamines and methylphenidate in their effects on the hypersomniac symptoms of narcoleptic patients (DALY and YOSS 1974). The clinical efficacy of caffeine on hyperactivity in children has been studied with positive or negative results (SCHNACKENBERG 1973; KLEIN 1987).

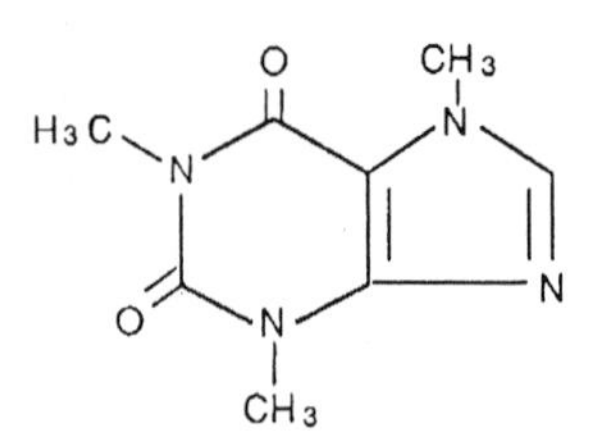

Fig. 1. Chemical structure of caffeine

II. Ephedrine

Ephedrine was isolated from the Chinese drug ma huang (Fig. 2), and was shown to have sympathomimetic actions by CHEN and SCHMIDT (1925). Ephedrine was generally considered a sympathomimetic agent acting on the peripheral visceral organs. However, it has been found that ephedrine also has an action as a dopaminergic agonist in the brain (ANGRIST et al. 1977).

Before amphetamine was found to be very effective for narcolepsy by PRINZMETAL and BLOOMBERG (1935), ephedrine was used as the standard drug for the treatment of narcolepsy (DANIELS 1930; JANOTA 1931). More recently, it was shown that sophisticated subjects could not distinguish between roughly equivalent doses of dextroamphetamine, methamphetamine, phenmetrazine, methylphenidate, and ephedrine under blind conditions (MARTIN et al. 1971). However, Hishikawa et al. (unpublished data) found that the effect of ephedrine on excessive daytime sleepiness in narcoleptic patients was inferior to that of methylphenidate. A similar view was also presented by DALY and YOSS (1974).

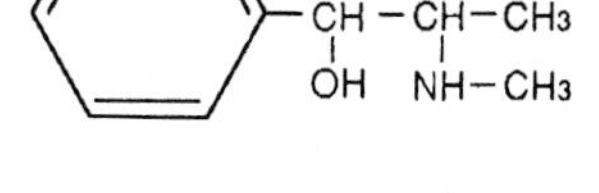

Fig. 2. Chemical structure of ephedrine

III. Amphetamine and Methamphetamine

Amphetamine was first synthesized by Edeleano (1887), but its central stimulant effects were not noted until it was independently resynthesized by Alles (1927) as a part of his work for synthesizing sympathomimetic agents that might substitute for ephedrine in the treatment of asthma (Fig. 3). Methamphetamine was first synthesized by Ogata (1919) in Japan (see ANGRIST and SUDILOVSKY 1978) (Fig. 4). Central stimulant effects of amphetamine and methamphetamine are considered to be due to their actions enhancing dopaminergic and noradrenergic nerve transmission in the brain (KUCZENSKI 1983) (see also Chap. 4, this volume).

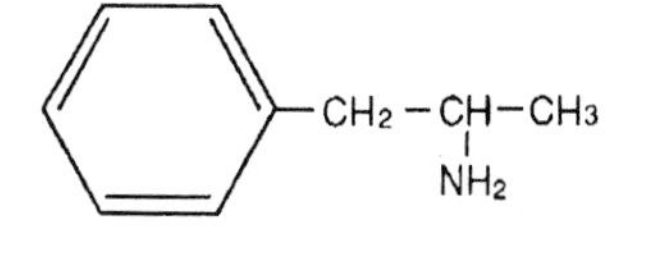

Fig. 3. Chemical structure of amphetamine

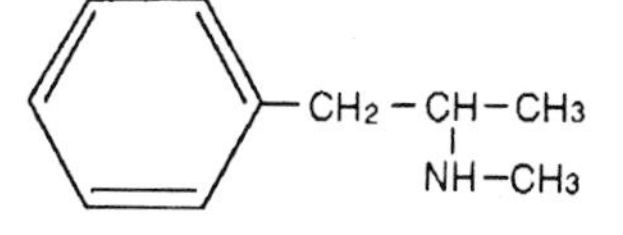

Fig. 4. Chemical structure of methamphetamine

Racemic amphetamine was found to have better effects than ephedrine on irresistible episodes of sleep in narcoleptic patients (PRINZMETAL and BLOOMBERG 1935). Later, the dextrorotatory isomer, dextroamphetamine, produced equally effective arousal with much less unpleasant autonomic side effects (PRINZMETAL and ALLES 1940). Several years later, methamphetamine was introduced as a therapeutic agent for narcolepsy (EATON 1943). Methamphetamine was found to be the best tolerated of the amphetamines, because it causes fewer sympathomimetic side effects (DALY and YOSS 1974).

The alerting effect of a single oral dose of amphetamine is at its maximum 2–4 h after oral administration and its clinical effects last about 6 h. Many narcoleptic patients require a single daily dose of 5–20 mg in the morning or twice daily doses. The daily dose should be varied, depending upon severity of daytime sleepiness and body weight of the patient. Amphetamines as well as other central stimulants should not be administered late in the evening or in the night, because they interfere with nocturnal sleep. Duration of action of methamphetamine is usually 5–6 h and sometimes its action lasts 10 h or longer. But, in any individual, the duration of drug action appears to be quite constant. So, a single daily dose in the morning or twice daily doses must be selected for different patients.

In some countries including the United States, methamphetamine and other amphetamines are available in slow release forms. The slow release forms of amphetamines are considered to be superior for the treatment of narcolepsy, because they produce a longer alerting response for 6–8 h without the sudden and short peak action with conventional release preparations (MITLER et al. 1993; PARKES and DAHLITZ 1993). However, before considering the use of these slow release forms, it is advisable to try the conventional preparations of amphetamine to determine their effectiveness and duration of action at the smallest effective doses (DALY and YOSS 1974). In practice, amphetamine and methamphetamine should first be given in a single dose in the morning in amounts selected on the basis of clinical assessment of the severity of hypersomniac symptoms and body weight of patients. When the symptoms are mild and the body weight is low, a small, single dose of 5–10 mg of either drug must first be tried. After the effect of the initial dose and its side effects are assessed, the single dose in the morning should be gradually increased until the appropriate dose for each patient is found.

Many narcoleptic patients will not continue using dextroamphetamine or methamphetamine for long periods, because tolerance to the alerting action of these central stimulants gradually develops. Because of this, many narcoleptic patients need increasing amounts of these drugs to suppress excessive daytime sleepiness and to maintain wakefulness in the daytime. However, patients using increased amounts of the drugs are often troubled by sympathomimetic side effects. These side effects include anorexia, tachycardia, palpitation, tremulousness, irritability, and anxiety (GUILLEMINAULT et al.

1974; PARKES et al. 1975). Sometimes, nocturnal sleep disturbance and drug dependence, and, rarely, paranoid psychosis may arise (GUILLEMINAULT 1993; PARKES and DAHLITZ 1993). Indeed, amphetamine is known to cause acute exacerbation of preexisting schizophrenic psychopathology (ANGRIST and SUDILOVSKY 1978). For these reasons, some physicians rarely use amphetamine and methamphetamine for the treatment of narcoleptic patients, although many physicians are likely continuing to prescribe them (DALY and YOSS 1974; GUILLEMINAULT 1989, 1993; MITLER et al. 1993). Especially in Japan and some European countries, the medical use of amphetamines has been stringently controlled by legislative measures. This is due to the epidemic abuse of the central stimulants which occurred soon after World War II. A large number of studies reported that both dextroamphetamine and methamphetamine successfully reduced hyperactivity in children (GITTELMAN 1983).

IV. Methylphenidate

Methylphenidate was developed by Ciba Laboratories in Switzerland (Fig. 5) and its central stimulant effects were reported by MEIER et al. (1954). The dextroisomer of methylphenidate has far stronger stimulant effects. The central stimulant effects of this drug are similar to those of amphetamines, and its stimulant action is probably due also to facilitation of catecholaminergic nerve transmission in the central nervous system, although the biological effects of methylphenidate on catecholamine metabolism in the brain have not yet been well documented (KUCZENSKI 1983).

Methylphenidate was initially marketed for the treatment of depressive states and not, as had been the case with other central stimulants, as an anorectic agent. The first clinical use of the drug for the treatment of narcolepsy showed its favorable effects (DALY and YOSS 1956; YOSS and DALY 1959). Its long-term effectiveness for narcoleptic patients was described by HONDA et al. (1979). This drug is also effective for hyperkinetic behavior disorder in children (ANGRIST and SUDILOVSKY 1978).

Methylphenidate is a frequently used drug for narcolepsy. This is probably because development of tolerance is less frequent and fewer sympathomimetic side effects occur with this drug than with other central stimulant drugs. Methylphenidate also has the advantage of prompt onset of drug action, but it has the disadvantages of short-lasting action and effective

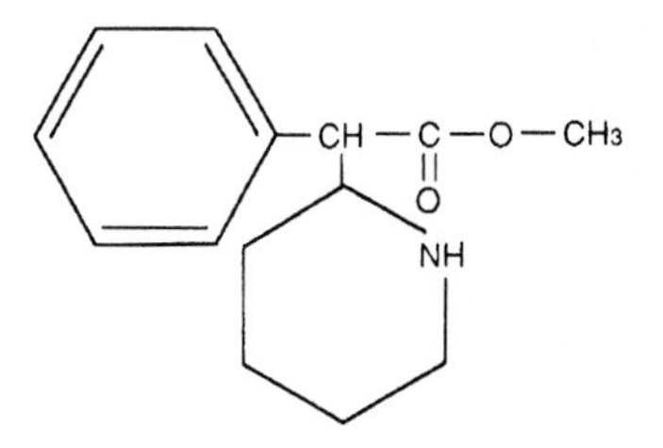

Fig. 5. Chemical structure of methylphenidate

absorption only from an empty stomach. Because of this latter problem, patients must be advised to take the drug at least 30 min before eating or not earlier than 1 h after eating (Daly and Yoss 1974). Among the narcoleptic symptoms, excessive daytime sleepiness and irresistible episodes of sleep are significantly reduced by this drug, resulting in increased alertness of patients and improvement of their disturbed social life in the daytime. But the drug effects upon cataplexy and other narcoleptic symptoms are often insufficient (Hishikawa et al. 1966, 1978; Honda et al. 1979). This is probably related to the polysomnographic finding in narcoleptic patients that the effect of methylphenidate in reducing the occurrence of REM sleep at sleep onset was milder than the drug effect in suppressing the initiation of sleep (Takahashi et al. 1977, cited by Honda et al. 1979).

In the practice of treating narcoleptic patients with methylphenidate, it is recommended to prescribe 10–20 mg twice daily, on awakening in the morning and at midday, depending upon the severity of excessive daytime sleepiness and body weight of patients. The drug action lasts from 3 to 5 h. This relatively short drug action is probably related to the short half-life (2–3 h) of drug concentration in the serum. Administration of the drug later than the evening may interfere with nocturnal sleep.

In some patients, transient development of increased sleepiness called "paradoxical sleepiness effect" may appear 15–30 min after drug ingestion in the morning, and it often lasts for 10–20 min, being replaced by alertness usually about 1 h after drug ingestion (Honda et al. 1979). Sometimes, during long-term use of the drug, tolerance to its alerting action gradually develops and larger doses of 30–50 mg are required. However, increased amounts of the drug often induce sympathomimetic side effects including reduced appetite, tachycardia, and palpitation. As in the cases of amphetamine and methamphetamine, this drug may provoke dramatic worsening of psychotic symptoms in patients with schizophrenia (Angrist and Sudilovsky 1978).

Recently, a slowly released form of methylphenidate has become available, and its 10-mg tablet administered twice daily was successfully used for the treatment of children with attention deficit disorder. Effects of the slowly released tablet last 8 h or longer (Pelham et al. 1987). But the efficacy of this drug form in narcoleptic patients has not yet been reported.

V. Pemoline

Pemoline is a mild stimulant of the central nervous system, but it produces minimal sympathomimetic side effects (Fig. 6). The drug has been used in doses of 20–50 mg after breakfast and lunch for the treatment of depressive states. It has also been advocated for the treatment of hyperkinetic states in children (Gittelman 1983). Favorable effects of pemoline in narcoleptic patients were preliminarily reported by Takahashi and Honda (1964). Its

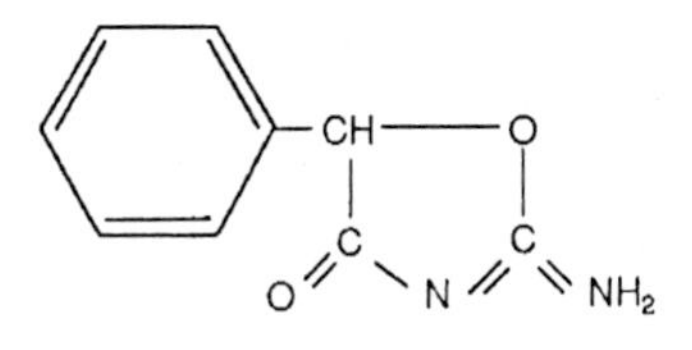

Fig. 6. Chemical structure of pemoline

usefulness for long-term treatment of narcolepsy was described by HONDA and HISHIKAWA (1980).

For the treatment of narcolepsy, usually doses of 50–100 mg administered twice daily, after breakfast and after lunch, significantly reduce excessive sleepiness and increase alertness in the daytime. But the drug effects upon cataplexy and othe narcoleptic symptoms are mild or nearly absent. As compared to amphetamine and methylphenidate, the advantages of pemoline for long-term treatment of narcolepsy are: milder and less frequent sympathomimetic side effects including stomach discomfort and decreased appetite; and no development of tolerance and drug dependence (BIEL and BOPP 1978; HONDA and HISHIKAWA 1980). The disadvantages of pemoline are milder alerting effects and slower appearance of the drug action after ingestion. The maximal serum concentration of this drug occurs 1–2 h after ingestion and the half-life of the serum concentration is about 10 h. This is probably related to the longer duration of the drug effects of pemoline than those of methylphenidate and the amphetamines.

VI. Mazindol

Mazindol was developed in 1967 by Sandoz Co. in the United States (Fig. 7). This drug with central stimulant action has been marketed as a drug for reducing appetite and body weight for the treatment of obese patients. Daily doses of 1–2 mg mazindol appear comparable to 5–10 mg dextroamphetamine in reducing the body weight of obese patients, with less profound peripheral side effects (SIRTORI et al. 1971; HADLER 1972). The maximal

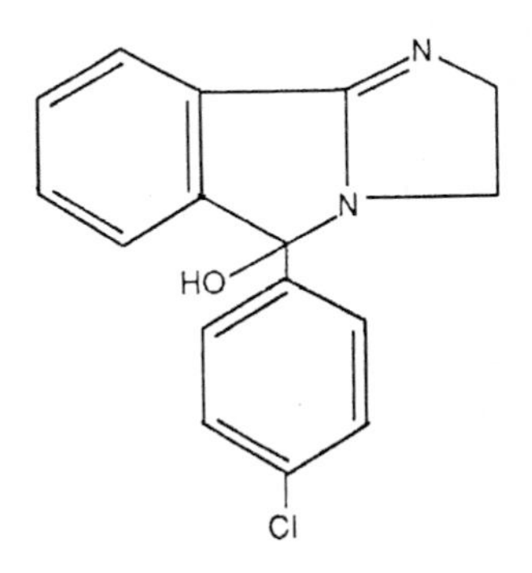

Fig. 7. Chemical structure of mazindol

serum concentration occurs 2–4 h after ingestion and the half-life of the serum concentration is about 9 h. The drug enhances noradrenergic and serotonergic nerve transmission (ENGSTROM et al. 1975; SHIMIZU et al. 1991).

Therapeutic effects of mazindol on narcolepsy were first reported by PARKES and SCHACHTER (1979). The drug is effective not only for excessive sleepiness and sleep attacks in the daytime, but also for cataplexy and other narcoleptic symptoms (IIJIMA et al. 1986). Usually, one or two daily doses of 1–2 mg are used for the treatment of narcolepsy.

During long-term treatment, tolerance to the stimulating drug action sometimes develops and administration of larger daily doses of 4–6 mg is required. As compared to dextroamphetamine and methylphenidate, the therapeutic effects of mazindol upon excessive daytime sleepiness are milder, but sympathomimetic side effects are also milder and less frequent.

VII. Modafinil

Modafinil is a novel stimulant drug developed by Lafon Laboratories in France (Fig. 8). It is a centrally active α_1-adrenergic agonist. Unlike amphetamines and methylphenidate, the drug has minimal peripheral sympathomimetic side effects at therapeutic doses (LYONS and FRENCH 1991).

A 200-mg nighttime dose of modafinil administered to sleep-deprived subjects significantly reduced subjective sleepiness, and improved vigilance and performance on search and memory tests (BENOIT et al. 1987). Daily doses of 200–500 mg had favorable therapeutic effects in patients with narcolepsy or idiopathic hypersomnia. Regular daytime treatment of narcoleptic patients with daily doses of at least 200 mg modafinil for as long as 3 years produced neither tolerance nor drug dependence (BATSUJI and JOUVET 1988, BILLARD et al. 1991). Good therapeutic effects of modafinil on excessive daytime sleepiness in patients with narcolepsy or idiopathic hypersomnia have also been reported by other investigators (BILLIARD et al. 1988). It is suggested that modafinil can be an ideal replacement for amphetamine in such uses as military operations, in which fatigue and excessive sleepiness might threaten successful completion of a mission (LYONS and FRENCH 1991).

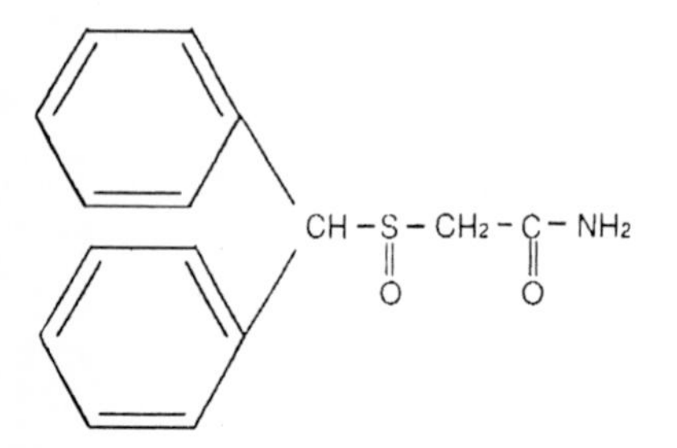

Fig. 8. Chemical structure of modafinil

E. Treatment of Narcolepsy and Other Hypersomniac Conditions with Central Stimulants and Other Adjunctive Drugs

I. Narcolepsy

Central stimulants are the most widely used drugs for the treatment of excessive daytime sleepiness and irresistible episodes of sleep in narcoleptic patients. Therapeutic daily doses of different central stimulants available at present are shown in Table 1. In clinical practice, at the beginning of treatment, one of the central stimulants should be administered as a single daily dose in the morning. The initial daily dose may be varied, depending upon the severity of daytime sleepiness and body weight of the patient, but it should be limited to one-sixth to one-fourth or less of the maximal therapeutic daily doses shown in Table 1. After assessment of the alerting effect and its duration attained by the initial dose, an increased single daily dose or twice daily doses, in the morning and at midday, may be administered. During long-term treatment with most of the central stimulants, tolerance to their alerting action may gradually develop and their therapeutic effects often become milder so that administration of larger daily doses is required. But increased amounts of the drugs may frequently cause a number of sympathomimetic side effects, including irritability, reduced appetite, tachycardia, palpitation, and nocturnal sleep disturbance. Usually, the use of methylphenidate is encouraged, because of the faster appearance of drug

Table 1. Drugs currently available for the treatment of narcolepsy

Drug	Therapeutic dose/day (mg/day) (all drugs administered p.o.)
Treatment of excessive daytime somnolence	
Stimulants	
Amphetamine	5 ~ 40
Methamphetamine	5 ~ 40
Methylphenidate	10 ~ 60
Mazindol	1 ~ 5
Pemoline	50 ~ 200
Modafinil	50 ~ 300
Treatment of cataplexy, sleep paralysis, and hypnagogic hallucinations	
Antidepressants	
Imipramine	25 ~ 200
Clomipramine	25 ~ 200
Desipramine	25 ~ 200
Protriptyline	5 ~ 20
Viloxazine	25 ~ 200
Fluoxetine	10 ~ 60

action and lower incidence of side effects than with amphetamines (YOSS and DALY 1959). Pemoline and mazindol are relatively well tolerated with less side effects, but they are less efficient. When continued use of some central stimulant drug is stopped, excessive daytime sleepiness returns, often being accompanied by severe rebound sleepiness for a few days, but no other significant withdrawal symptoms occur.

When an increase of the daily dose of central stimulant drugs is not desirable because of side effects, patients should be recommended to take regularly one to three short daytime naps of 10–20 min each, which will help them maintain a satisfactory level of vigilance for the rest of the daytime. Modafinil, available at present only in France and some other European countries, is better tolerated with few side effects, and brings a substantial improvement in excessive daytime sleepiness not only in narcoleptic patients but also in patients with idiopathic hypersomnia (BASTUJI and JOUVET 1988; BILLIARD et al. 1988). Phenelzine, a nonspecific monoamine oxidase (MAO) inhibitor, when used at daily doses of 60–90 mg, significantly diminished excessive daytime sleepiness and irresistible episodes of sleep as well as cataplexy and other narcoleptic symptoms related to REM sleep, and these effects continued for over 1 year (WYATT et al. 1971). However, bothersome side effects, including hypotension, edema, and impaired sexual function, have prevented the widespread investigation of MAO inhibitors in narcolepsy. Selegiline, a putative selective MAO-B inhibitor, a major metabolite of which is amphetamine, has been used as an adjuvant drug in the levodopa treatment of Parkinson's disease in some European countries for the past decade. When orally administered at the single daily doses of 20–30 mg in the morning, selegiline had alerting and mood-elevating effects in narcoleptic patients and these effects appeared to be equal to those of amphetamines in the same doses. It is suggested that selegiline is an effective alternative to other central stimulant drugs in the treatment of narcolepsy (ROSELAAR et al. 1987). Oral administration of *l*-tyrosine, a precursor of cathecholamine synthesis, was reported to be of benefit in narcoleptic patients (MOURET et al. 1988). However, a double-blind study showed that dietary supplementation with *l*-tyrosine 9 g daily had a mild stimulant action on the central nervous system but this effect was not clinically significant in the treatment of the narcoleptic syndrome (ELWES et al. 1989). γ-Hydroxybutyrate (GHB) also does not belong to the group of central stimulants, and it is a putative neurotransmitter in the human brain (MAMELAK 1989). After oral administration, GHB has a hypnotic action. GHB has a very short plasma half-life and it is undetectable after 3–4 h after oral administration. However, when given orally in the dose of 25–30 mg/kg twice, once at bedtime and again at the time of a night awakening, the drug is often of definite value for reducing excessive daytime sleepiness and cataplexy, even though its efficacy varies among patients and additional use of one of the central stimulants is often necessary (BROUGHTON and MAMELAK 1980; SCHARF et al. 1985; MAMELAK et al. 1986; SCRIMA et al. 1989; LAMMERS et al. 1993).

For the treatment of cataplexy, sleep paralysis, and hypnogogic hallucinations in narcoleptic patients, central stimulants are sometimes fairly effective, but their effects upon these symptoms are usually insufficient. For the treatment of these symptoms, tricyclic antidepressant drugs, clomipramine, imipramine, desipramine, and protriptyline, have been widely used, usually with good responses. However, these drugs are usually not effective for excessive daytime sleepiness. The good therapeutic effects of the tricyclic antidepressants upon cataplexy and other narcoleptic symptoms were found in Japan (AKIMOTO et al. 1960; HISHIKAWA et al. 1966). Potent REM sleep-suppressing action of these tricyclic antidepressants probably brings about their favorable effects upon the narcoleptic symptoms related to REM sleep occurring at the onset of sleep (HISHIKAWA et al. 1965, 1966). REM sleep-suppressing effects of the tricyclic antidepressants are probably due to their norepinephrine and/or serotonin re-uptake blocking action and anticholinergic action.

Initially, a single daily or twice daily doses of 25 mg of one of the antidepressant drugs in the morning and/or in the night is often very effective. But, during long-term treatment with the antidepressant drugs, tolerance to their REM sleep-suppressing action often develops gradually, and then higher doses of the drugs are required. The tricyclic antidepressants sometimes induce atropinic side effects, including nausea, constipation, tachycardia, palpitation, tremulousness, dried mouth, difficulty in micturition, and delayed ejaculation in men. These side effects have led to the search for other new compounds.

One group of the new compounds consists of specific serotonin re-uptake inhibitors including zimelidine, femoxetine, and fluoxetine. These drugs (femoxetine 300 mg, twice daily, fluoxetine 60 mg, once daily) significantly reduced the frequency and severity of cataplexy and sleep paralysis, but their effects upon excessive daytime sleepiness were insufficient or absent (MONTPLAISIR and GOBOUT 1983; SCHRADER et al. 1986; LANGDON et al. 1986). Viloxazine hydrochloride, a norepinephrine re-uptake blocker, belongs to another group of the new compounds with good effects upon cataplexy. Although viloxazine, 100 mg daily, had good effects on cataplexy and was well tolerated by patients including the elderly, it seemed to be ineffective for excessive daytime sleepiness (GUILLEMINAULT et al. 1986). Phenelzine, a nonspecific MAO inhibitor, has been successfully used for the treatment of patients with intractable narcolepsy-cataplexy (WYATT et al. 1971), but frequent, dangerous side effects of MAO inhibitors have severely limited the routine use of them. Selegiline, a selective MAO-B inhibitor, appears to be effective for cataplexy as well as excessive daytime sleepiness in narcoleptic patients (ROSELAAR et al. 1987). Tricyclic antidepressants and other recent drugs effective for cataplexy were, and have been, developed mainly as antidepressant drugs. Pharmacology of these drugs and their therapeutic use for sleep disorders are also described in more detail in Chap. 17 of this volume. Recent experimental studies in dogs with narcolepsy have revealed that several serotonin 1A agonists and an analog of TRH have marked suppressing effects upon cataplexy and central stimulating effects (Nishino et al. 1993; Arrigoni et al. 1993). Some of these chemical substances should be shown to be effective, and applicable for the treatment of cataplexy and excessive daytime sleepiness in narcolepsy of man.

Nocturnal sleep disturbance characterized by frequent interrupting awakenings and vivid, anxious dreams is often found in narcoleptic patients. Their nocturnal sleep disturbance can usually be well controlled by benzodiazepine hypnotics or by GHB (Mamelak et al. 1986; Lammers et al. 1993). Improving nocturnal sleep with these drugs has also been reported to have positive effects not only on alleviating excessive daytime sleepiness but also on reducing the frequency of cataplectic episodes in the daytime. However, benzodiazepines and GHB used for improving nocturnal sleep disturbance are usually insufficient for the narcoleptic symptoms in the daytime.

In general, central stimulants effective for excessive daytime sleepiness and irresistible episodes of sleep are not sufficiently effective or are ineffective for cataplexy and disturbed nocturnal sleep. Tricyclic antidepressants and other recent drugs that are effective for cataplexy are not sufficiently effective or are ineffective for excessive daytime sleepiness. The drugs effective for treating disturbed nocturnal sleep are not sufficiently effective or are ineffective for excessive sleepiness and cataplexy in the daytime. Therefore, for narcoleptic patients with typical symptoms, combined drug treatment, particularly with central stimulants and antidepressant medications, is often necessary in combination with a benzodiazepine hypnotic or GHB. This combined drug treatment significantly suppresses or reduces the narcoleptic symptoms, and improves the personal and social life of patients. However, following interruption of the drug treatment, narcoleptic symptoms return within several days, and sometimes very frequent cataplectic episodes occur for a few weeks or longer. Finally, some drugs used for the treatment of hypertension, reserpine and prazosin, an α_1-adrenergic blocking agent, markedly exacerbate cataplexy (Aldrich and Roger 1989; Hishikawa 1990).

II. Idiopathic Hypersomnia

The treatment of idiopathic hypersomnia is not yet well established. Central stimulant drugs including amphetamines, methylphenidate, and pemoline have been recommended. However, these drugs have often been found to be less effective in idiopathic hypersomnia than in narcolepsy and are not as well tolerated in patients with idiopathic hypersomnia. Patients often report more side effects, such as tachycardia and irritability, and such medications tend to exacerbate the associated autonomic symptoms including headache.

As already cited in the above section on narcolepsy, modafinil, an α_1-adrenergic agonist, was recently found to be very effective for the treatment of excessive daytime sleepiness in idiopathic hypersomnia, with considerably fewer or no side effects (Billiard et al. 1988; Batsuji and Jouvet 1988).

III. Recurrent Hypersomnia

Treatment of recurrent hypersomnia includes symptomatic and preventive measures. The symptomatic measures are oriented toward cessation or interruption of the hypersomniac episode. For this purpose, central stimulant drugs have been used, but their effect is usually insufficient. Excessive daytime sleep of patients may be reduced with these drugs but patients often become very irritable while on such stimulant medication.

Preventive measures have to be considered when hypersomniac episodes are very frequent. Avoidance of precipitating factors, overfatigue and overdrinking of alcohol, may reduce the frequency of hypersomniac episodes (Hishikawa 1987). Positive results in reducing the frequency of hypersomniac episodes and their prevention have been obtained with lithium carbonate in some patients (Ogura et al. 1976; Abe 1977; Hart 1985; Kayukawa et al. 1980). In a few patients with menstruation-linked recurrent hypersomnia, administration of an ovulation inhibitor, an estrogen preparation or other combined contraceptives, was very effective; the hypersomniac episode disappeared completely while the patients were on such medication (Billiard et al. 1975; Papy et al. 1982; Sachs et al. 1982).

F. Other Problems Related to the Use of Central Stimulant Drugs

It is well known that widespread and unlimited use of central stimulants is often accompanied by the undesirable problems of stimulant drug dependence, addiction, abuse, and drug-induced psychoses; these problems easily leading to socially harmful accidents, illegal violence, and homicide. Because of these, stimulant drug use is controlled by means of more or less stringent legislative measures in different countries. On the other hand, the medical use of central stimulant drugs for narcoleptic and other hypersomniac patients, hyperkinetic children, and patients with obesity significantly reduces or ameliorates their symptoms, especially irresistible daytime sleepiness of narcoleptic patients, and markedly improves their well-being and social adaptation. In addition, the medical use of central stimulants at limited doses, performed under regular consultation with physicians, very rarely causes the undesirable problems of stimulant drug dependence, abuse, or psychoses. Only one of 300 narcoleptic patients treated with amphetamine at the daily dose of 20 mg developed an amphetamine psychosis (Parkes and Dahlitz 1993).

Many central stimulant drugs including amphetamines, methylphenidate, and mazindol are transferred to the fetus, and are secreted in breast milk.

Thus, their administration to pregnant women and those breast-feeding should be avoided, although their teratogenicity in man is very unlikely (CONNEL 1972; GUILLEMINAULT 1993). Another possible side effect, as one of the sympathomimetic side effects due to central stimulant drugs, is hypertension of the systemic arterial blood pressure. Narcoleptic patients taking therapeutic daily doses of amphetamines usually have normal blood pressure and rarely hypertension. However, it is possible that some patients may have deceptive normal blood pressure, because they usually have low blood pressure as one of the mild signs of autonomic dysfunction before treatment (GUILLEMINAULT 1993). In general, when central stimulant drugs are used carefully within the therapeutic dose range cited in Table 1, their side effects are minor and not serious. Heavy sweating is probably the most common problem encountered by patients while on medication with amphetamines or methylphenidate.

G. Summary

Varied plants containing central stimulants, such as tea leaves, coca leaves, and coffee beans, have long been used by people in different areas in the world for their action of exciting psychomotor activity and dissipating both fatigue and sleepiness.

Modern central stimulant drugs were originally extracted from those plants, and later synthesized. Most of them have been used not only for the purposes of maintaining wakefulness and alertness in patients with excessive daytime sleepiness, but also for the treatment of hyperkinetic states in children, hypotension, obesity, and bronchial asthma.

In this chapter, descriptions are limited to central stimulant drugs used mainly for the treatment of excessive daytime sleepiness in narcolepsy and other hypersomniac conditions. The effects of central stimulant drugs as well as the clinical features of hypersomniac conditions and their actual treatment with these drugs are reviewed.

Central stimulant drugs available at present are more or less markedly, but not sufficiently, effective for excessive daytime sleepiness in narcolepsy and a subtype of idiopathic hypersomnia. They often significantly improve the well-being and social adaptation of patients suffering from these conditions.

Central stimulant drugs include caffeine, ephedrine, amphetamine, methamphetamine, methylphenidate, pemoline, mazindol, and modafinil. Caffeine and ephedrine are inferior to the other stimulants in the treatment of excessive daytime sleepiness. Amphetamine, methamphetamine, methylphenidate, and pemoline are the most commonly prescribed and considered the most efficacious medications for these disorders. Mazindol, which is less frequently prescribed, appears to be less effective. Modafinil is a novel stimulant which acts centrally as an α_1-adrenergic agonist with reportedly good therapeutic effects.

Widespread and unlimited use of certain central stimulant drugs is often accompanied by undesirable problems of creating stimulant dependence and abuse, and of inducing schizophrenia-like psychotic states. Because of these potential adverse reactions, some of the central stimulant drugs are under rigorous legislative control, and are not available for medical use in certain countries. However, medical use of these controlled central stimulant drugs for hypersomnia very rarely causes these undesirable problems if the doses are limited and regular consultation with a physician is maintained.

References

Abe K (1977) Lithium prophylaxis of periodic hypersomnia. Br J Psychiatry 130: 312–316

Akimoto H, Honda Y, Takahashi Y (1960) Pharmacotherapy in narcolepsy. Dis Nerv Syst 21:1–3

Aldrich MS (1990) Narcolepsy. N Engl J Med 323:389–394

Aldrich MS, Roger AE (1989) Exacerbation of human cataplexy by prazosin. Sleep 12:254–256

Angrist B, Sudilovsky A (1978) Central nervous system stimulants: historical aspects and clinical effects. In: Iversen LL, Iversen SD, Snyder SH (ed) Stimulants. Plenum, New York, pp 99–165 (Handbook of psychopharmacology, vol 11)

Angrist B, Rotorosen J, Kleinberg D, Merriam V, Gershon S (1977) Dopaminergic agonist properties of ephedrine: theoretical implications. Psychopharmacology (Berl) 55:115–120

Arrigoni J, Nishino S, Shelton J, Elizaga R, Mignot E (1993) CG3703, an analog of TRH, decreases cataplexy in canine narcolepsy. Sleep Res 22:488

Bastuji H, Jouvet M (1988) Successful treatment of idiopathic hypersomnia and narcolepsy with modafinil. Prog Neuropsychopharmacol Biol Psychiatry 12: 695–700

Benoit O, Clodore M, Touron N, Paihous E (1987) Effects of modafinil on sleepiness in normal sleep deprived and symptomatic subjects. 5th International Congress of Sleep Research, June 28–July 3, Copenhagen

Biel JH, Bopp BA (1978) Amphetamines: structure-activity relationships. In: Iversen LL, Iversen SD, Snyder SH (eds) Stimulants. Plenum, New York, pp 1–39 (Handbook of psychopharmacology, vol 11)

Billiard M, Guilleminault C, Dement WC (1975) A menstruation-linked periodic hypersomnia. Neurology (Minneap) 25:436–443

Billiard M, Dissoubray C, Lubin S, Touchon J, Besset A (1988) Effects of modafinil in patients with narcolepsy-cataplexy or with idiopathic hypersomnia. 9th European Congress of Sleep Research, Sept 4–9, Jerusalem

Billiard M, Laffont F, Goldenberg F, Weil J-S, Lubin S (1991) Placebo-controlled, cross-over study of modafinil therapeutic effect in narcolepsy. Sleep Res 20A: 289

Broughton R, Mamelak M (1980) Effects of nocturnal gamma-hydroxybutyrate on sleep/waking patterns in narcolepsy-cataplexy. Can J Neurol Sci 7:23–31

Broughton R, Ghanem Q, Hishikawa Y, Sugita Y, Nevsimalova S, Roth B (1983) Life effects of narcolepsy: relationships to geographic origin (North American, Asian or European) and to other patient and illness variables. Can J Neurol Sci 10:100–104

Chen KK, Schmidt CF (1925) The action of ephedrine, the active principle of the Chinese drug Ma Huang. J Pharmacol Exp Ther 24:339–357

Connel PH (1972) Central nervous system stimulants. In: Herxheimer ML (ed) Side effects of drugs, vol 7. Excerpta Medica, Amsterdam, pp 1–16

Daly DD, Yoss RE (1956) The treatment of narcolepsy with methyl-phenylpiperidylacetate: a preliminary report. Proc Mayo Clin 31:620–625
Daly DD, Yoss RT (1974) Narcolepsy. In: Vinken PJ, Bruyn GW (eds) The epilepsies. North-Holland, Amsterdam, pp 836–852 (Handbook of clinical neurology, vol 15)
Daniels LE (1930) A symptomatic treatment of narcolepsy. Proc Mayo Clin 5: 299–300
Diagnostic Classification Steering Committee (1990) International classification of sleep disorders: diagnostic and coding manual. American Sleep Disorders Association, Rochester
Eaton LM (1943) Treatment of narcolepsy with desoxy-ephedrine hydrochloride. Proc Mayo Clin 18:262–264
Elwes RDC, Crewes H, Chesterman LP, Summers B, Jenner P, Binnie CD (1989) Treatment of narcolepsy with L-tyrosine: double-blind placebo-controlled trial. Lancet 2:1067–1069
Engstrom RG, Kelly LA, Gogerty JH (1975) The effects of 5-hydroxy-5-(4,1-chlorophenyl)-2,3-dihydro-5H-imidazo[2,1-a] isonidole (mazindol, SaH 42-548) on the metabolism of brain norepinephrine. Arch Int Pharmacodyn Ther 214: 308–321
Gittelman R (1983) Experimental and clinical studies of stimulant use in hyperactive children and children with other behavioral disorders. In: Creese I (ed) Stimulants: neurochemical, behavioral, and clinical perspectives. Raven, New York, pp 205–226
Guilleminault C (1989) Narcolepsy syndrome. In: Kryger MH, Roth T, Dement WC (eds) Principles and practice of sleep medicine. Saunders, Philadelphia, pp 338–346
Guilleminault C (1993) Amphetamines and narcolepsy. Sleep 16:199–201
Guilleminault C, Carskadon M, Dement WC (1974) On the treatment of rapid eye movement narcolepsy. Arch Neurol 30:90–93
Guilleminault C, Mancuso J, Salva MA, Hayes B, Mitler M, Poirier G, Montplaisir J (1986) Viloxazine hydrochloride in narcolepsy: a preliminary report. Sleep 9:275–279
Hadler AJ (1972) Mazindol, a new non-amphetamine anorexigenic agent. J Clin Pharmacol 12:453–458
Hart EJ (1985) Kleine-Levine syndrome: normal CSF monoamines and response to lithium therapy. Neurology 35:1395–1396
Hishikawa Y (1987) Sleep and wakefulness disorders (in Japanese). In: Kakeda K, Shimazono Y, Okuma T, Takahashi R, Hosaki H (eds) Handbook of psychiatry, vol 87-B. Nakayama, Tokyo, pp 189–236
Hishikawa Y (1990) Narcolepsy (in Japanese). Clin Neurosci 8:1266–1268
Hishikawa Y, Nakai K, Ida H, Kaneko Z (1965) The effect of imipramine, desmethylimipramine and chlorpromazine on the sleep-wakefulness cycle of the cat. Electroencephalogr Clin Neurophysiol 19:518–521
Hishikawa Y, Ida H, Nakai K, Kaneko Z (1966) Treatment of narcolepsy with imipramine (Tofranil) and desmethylimipramine (Pertofran). J Neurol Sci 3:453–461
Hishikawa Y, Nan'no H, Tachibana M, Furuya E, Koida H, Kaneko Z (1968) The nature of sleep attack and other symptoms of narcolepsy. Electroencephalogr Clin Neurophysiol 24:1–10
Hishikawa Y, Sugita Y, Iijima S (1978) Therapeutic effects of methylphenidate in narcoleptic patients (in Japanese). Basic Pharmacol Ther 6:2166–2178
Honda Y, Hishikawa Y (1980) A long-term treatment of narcolepsy and excessive daytime sleepiness with pemoline (Betanamine). Curr Ther Res 27:429–441
Honda Y, Hishikawa Y, Takahashi Y (1979) Long-term treatment of narcolepsy with methylphenidate (Ritalin). Curr Ther Res 25:288–298

Honda Y, Asaka A, Tanaka Y, Juji T (1983a) Discrimination of narcoleptic patients by using genetic markers and HLA. Sleep Res 12:254
Honda Y, Asaka A, Tanimura M, Furusho T (1983b) A genetic study of narcolepsy and excessive daytime sleepiness in 308 families with a narcolepsy of hypersomnia proband. In: Guilleminault C, Lugaresi E (eds) Sleep/wake disorders: natural history, epidemiology and long term evolution. Raven, New York, pp 187–199
Iijima S, Sugita Y, Teshima Y, Hishikawa Y (1986) Therapeutic effects of mazindol on narcolepsy. Sleep 9:265–268
Janota O (1931) Symptomatische Behandlung der pathologischen Schlafsucht, besonders der Narkolepsie. Med Klin 27:278–281
Kales A, Cadieux RJ, Soldatos CR, Bixler EO, Schweitzer PK, Prey WT, Vela-Bueno A (1982a) Narcolepsy-cataplexy. I. Clinical and electrophysiologic characteristics. Arch Neurol 39:164–168
Kales A, Soldatos CR, Bixler EO, Caldwell A, Cadieux RJ, Verrichio JM, Kales JD (1982b) Narcolepsy-cataplexy. II. Psychosocial consequences and associated psychopathology. Arch Neurol 39:169–171
Kales A, Vela-Bueno A, Kales JD (1987) Sleep disorders: sleep apnea and narcolepsy. Ann Intern Med 106:434–443
Kalix P (1992) Cathionine, a natural amphetamine. Pharmacol Toxicol 70:77–86
Kayukawa Y, Yoshiga S, Okada T (1980) Therapeutic effects of lithium carbonate on periodic hypersomnia (in Japanese). Psychiatr Neurol Jpn 82:313–314
Kessler S, Guilleminault C, Dement WC (1974) A family of 50 REM narcolepsy. Arch Neurol Scand 50:503–512
Klein RG (1987) Pharmacotherapy of childhood hyperactivity: an update. In: Meltzer HY (ed) Psychopharmacology: the third generation of progress. Raven, New York, pp 1215–1224
Kuczenski R (1983) Biochemical actions of amphetamine and other stimulants. In: Creese I (ed) Stimulants: neurochemical, behavioral, and clinical perspectives. Raven, New York, pp 31–61
Lammers GJ, Arends J, Declerck AC, Ferrari MD, Schouwink G, Troost J (1993) Gamma-hydroxybutyrate and narcolepsy: a double-blind placebo-controlled study. Sleep 16:216–220
Langdon N, Bandak S, Shindler J, Parkes JD (1986) Fluoxetine in the treatment of cataplexy. Sleep 9:371–372
Laurent JM (1962) Toxique et toxicomanie peu connus, "le cath". Ann Med Psychol (Paris) 120:649–657
Lyons TJ, French J (1991) Modafinil: the unique properties of a new stimulant. Aviat Space Environ Med May: 432–435
Mamelak M (1989) Gamma-hydroxybutyrate: an endogenous regulator of energy metabolism. Neurosci Biobehav Rev 13:187–198
Mamelak M, Scharf MB, Woods M (1986) Treatment of narcolepsy with gamma-hydroxybutyrate: a review of clinical and sleep laboratory findings. Sleep 9: 285–289
Marangos PJ, Paul SM, Parma AM, Goodwin FK, Syapin P, Skolnick P (1979) Purinergic inhibition of diazepam binding to rat brain. Life Sci 24:851–857
Martin WR, Sloan JW, Sapria JD, Jasinski DR (1971) Physiologic, subjective and behavioral effects of amphetamine, methamphetamine, ephedrine, phenmetrazine and methylphenidate in man. Clin Pharmacol Ther 12:245–258
Meier R, Gross F, Tripod J (1954) Ritaline, eine neuartige synthetische Verbindung mit spezifischer zentralerregender Wirkungskomponente. Klin Wochenschr 32:445–450
Mitler MM, Erman M, Hajdukovic R (1993) Amphetamine and narcolepsy: the treatment of excessive somnolence with stimulant drugs. Sleep 16:203–206
Montplaisir J, Gobout R (1983) A new treatment for cataplexy. 4th International Congress of Sleep Research, Bologna

Mouret J, Lemoine P, Sanchez P, Robelin N, Taillard J, Canini F (1988) Treatment of narcolepsy with L-tyrosine. Lancet 2:1458–1459
Nishino S, Shelton J, Renaud A, Dement WC, Mignot E (1993) The role of 5HT-1A receptor mechanisms in canine narcolepsy. Sleep Res 22:42
Ogura C, Okuma T, Nakazawa K, Kishimoto A (1976) Treatment of periodic somnolence with lithium carbonate. Arch Neurol 33:143
Papy JJ, Conte-Devoix B, Sormani J, Porto R, Guillaume V (1982) Syndrome d'hypersomnie périodique avec mégaphagie chez une jeune femme, rhythmé par le cycle menstruel. Rev Electroencephalogr Clin Neurophysiol 12:54–61
Parkes JD, Dahlitz M (1993) Amphetamines and narcolepsy: amphetamine prescription. Sleep 16:201–203
Parkes JD, Schachter M (1979) Mazindol in the treatment of narcolepsy. Acta Neurol Scand 60:250–254
Parkes JD, Baraitser M, Marsden CD, Asselman P (1975) Natural history, symptoms and treatment of the narcoleptic syndrome. Acta Neurol Scand 52:337–353
Pelhan WE Jr, Sturges J, Hoza J, Schmidt C, Bijlsma JJ, Milich R, Moore S (1987) Sustained release and standard methylphenidate effects on cognitive and social behavior in children with attention deficit disorder. Pediatrics 80:491–501
Phillis JW, Edstrom JP, Kostopoulos GK, Kirkpatrick JR (1979) Effects of adenosine and adenine nucleotides on synaptic transmission in the cerebral cortex. Can J Physiol Pharmacol 57:1289–1312
Prinzmetal M, Alles GA (1940) The central nervous system stimulant effects of dextro-amphetamine sulphate. Am J Med Sci 200:665–673
Prinzmetal M, Bloomberg W (1935) Use of benzedrine for treatment of narcolepsy. JAMA 105:2051–2054
Roselaar SE, Langdon N, Lock CB, Jenner P, Parkes JD (1987) Selegiline in narcolepsy. Sleep 10:491–495
Sachs C, Persson HE, Hagenfeldt K (1982) Menstruation-related periodic hypersomnia: a case study with successful treatment. Neurology 32:1376–1379
Scharf M, Brown D, Woods M, Brown L, Hirschowitz J (1985) The effects and effectiveness of gamma-hydroxybutyrate in patients with narcolepsy. J Clin Psychiatry 46:222–225
Schnackenberg RC (1973) Caffeine as a substitute for schedule II stimulants in hyperactive children. Am J Psychiatry 130:796–800
Schrader H, Kayed K, Bendixen Markset A-C, Treiden HE (1986) The treatment of accessory symptoms in narcolepsy: a double-blind cross-over study of a selective serotonin re-uptake inhibitor (femoxetine) versus placebo. Acta Neurol Scand 74:297–303
Scrima L, Hartman PG, Johnson FH Jr, Hiller FC (1989) Efficacy of gamma-hydroxybutyrate versus placebo in treating narcolepsy-cataplexy: double-blind subjective measures. Biol Psychiatry 26:331–343
Shimizu N, Take S, Hori T, Oomura Y (1991) Hypothalamic microdialysis of mazindol causes anorexia with increase in synaptic serotonin in rats. Physiol Behav 49:131–134
Sirtori C, Hurwitz A, Azarnof DL (1971) Hyperinsulinemia seconday to chronic administration of mazindol and d-amphetamine. Am J Med Sci 261:341–349
Takahashi Y (1965) Clinical studies of periodic somnolence: analysis of 28 personal cases. Psychiatr Neurol Jpn 67:853–889
Takahashi Y, Honda Y (1964) Pharmacotherapy in narcolepsy. I. Effects of central nervous system stimulant (in Japanese). Clin Psychiatry 6:673–682
Thorpy MJ (1990) Handbook of sleep disorders. Dekker, New York
Wyatt R, Fram D, Buchbinder R, Snyder F (1971) Treatment of intractable narcolepsy with a monoamine oxidase inhibitor. N Engl J Med 285:987–991
Yoss RE, Daly DD (1959) Treatment of narcolepsy with Ritalin. Neurology (Minneap) 9:171–173

CHAPTER 16

Neuroleptics, Antihistamines and Antiparkinsonian Drugs: Effects on Sleep

C.R. SOLDATOS and D.G. DIKEOS

A. Introduction

Although neuroleptics, antihistamines and antiparkinsonians are among the most widely used drugs, their effects on sleep have been relatively poorly studied. There is a scarcity of sleep laboratory studies of practically all these drugs. Moreover, clinical trials usually provide little information relevant to their effects on sleep.

This review is based on selected studies mentioned in previous reviews (KAY et al. 1976; HARTMANN 1978) as well as those retrieved through a Medline search covering almost 3 decades. In the selection of sleep laboratory studies reviewed in this chapter, certain exclusion criteria were used, such as: nonconventional recording and scoring procedures, very small number of subjects and serious design deficiencies. Nonetheless, the studies which were excluded were relatively few. Further, whenever appropriate, data from clinical trials and animal studies were utilized.

B. Neuroleptics

I. Introduction

From a clinical standpoint, studying the effects of neuroleptics on sleep is very important (see also Chap. 4, this volume). Many psychotic patients complain of various degrees of sleeplessness (ZARCONE and DEMENT 1969; FEINBERG and HIATT 1978; GANGULI et al. 1987; SOLDATOS et al. 1987; TANDON et al. 1992; BENCA et al. 1992), while with others (mainly chronic schizophrenics) mobilization may be necessitated, which is incompatible with the sedative effects of medication. Thus, the clinician needs to know whether a neuroleptic has sedative potency and, if so, to what extent.

There is a lack of sleep laboratory studies for certain widely used neuroleptics such as promazine. Moreover (based on the previously mentioned criteria), all available studies pertaining to drugs such as thiothixene, fluphenazine, perphenazine and loxapine were excluded from this review. Thus, by necessity, the drugs reviewed are grouped into three major categories: phenothiazines (chlorpromazine, promethazine, thioridazine, mesoridazine and trifluoperazine), butytophenones (haloperidol and pimozide) and nonclassical neuroleptics (sulpiride, clozapine, remoxipride and ritanserin).

II. Phenothiazines

1. Chlorpromazine

This oldest and still widely used antipsychotic drug is the most extensively studied neuroleptic in the sleep laboratory. However, in only one of 12 sleep laboratory studies (with a total number of 99 subjects) were all 13 subjects schizophrenics (Kaplan et al. 1974); the remaining 11 studies included either only healthy volunteers (Lester and Guerrero-Figueroa 1966; Lewis and Evans 1969; Sagales et al. 1969; Lester et al. 1971; Fujii 1973; Hartmann and Cravens 1973; Hata 1975) or mixed patient and/or healthy volunteer samples (Toyoda 1964; Feinberg et al. 1969; Kupfer et al. 1971; Moldofsky and Lue 1980).

A major issue to consider when interpreting the findings of sleep laboratory studies of chlorpromazine is the wide methodological differences between studies, differences which pertain to: design characteristics [crossover (Lester and Guerrero-Figueroa 1966; Feinberg et al. 1969; Hartmann and Cravens 1973), drug administration versus baseline (Toyoda 1964; Lewis and Evans 1969; Sagales et al. 1969; Kupfer et al. 1971; Lester et al. 1971; Fujii 1973; Kaplan et al. 1974; Hata 1975) or parallel group (Moldofsky and Lue 1980)]; route of administration [intramuscular in one study (Sagales et al. 1969) and oral in the remaining 11 studies]; diversity of dose levels [25–50 mg (Toyoda 1964; Lewis and Evans 1969; Sagales et al. 1969; Fujii 1973; Hartmann and Cravens 1973; Hata 1975) or 100–400 mg (Lester and Guerrero-Figueroa 1966; Feinberg et al. 1969; Lewis and Evans 1969; Kupfer et al. 1971; Kaplan et al. 1974; Moldofsky and Lue 1980)]; and time of administration [divided daily doses (Kupfer et al. 1971; Kaplan et al. 1974) or once at bedtime (Toyoda 1964; Lester and Guerrero-Figueroa 1966; Feinberg et al. 1969; Lewis and Evans 1969; Sagales et al. 1969; Kupfer et al. 1971; Lester et al. 1971; Fujii 1973; Hartmann and Cravens 1973; Hata 1975; Moldofsky and Lue 1980)]. Further, in only three of the 12 studies did the duration of drug administration allow for the assessment of the effects on sleep beyond 1 week of continued use (Hartmann and Cravens 1973; Kaplan et al. 1974; Moldofsky and Lue 1980). Finally, the design of only one study provided for the evaluation of any withdrawal effects following drug discontinuation (Hartmann and Cravens 1973).

It has been confirmed that chlorpromazine at all dose levels evaluated in the sleep laboratory has a sleep-promoting effect upon initial administration. These initial sedative effects were documented through an increase in total sleep time and a decrease in nocturnal wakefulness (Kupfer et al. 1971;

HARTMANN and CRAVENS 1973; KAPLAN et al. 1974). In one study of healthy volunteers, chlorpromazine 50 mg lost most of its sleep-promoting effect in the 3rd and 4th weeks of nightly administration (HARTMANN and CRAVENS 1973). However, in another study of chronic schizophrenics taking chlorpromazine 100 mg q.i.d., total sleep time was significantly longer at 3 weeks or more following initial administration than on baseline (375 min vs. 307 min) (KAPLAN et al. 1974). In the same long-term study, sleep latency and wake time after sleep onset were also found to be significantly reduced (by 43 min and 25 min, respectively) (KAPLAN et al. 1974). The only study evaluating withdrawal effects following discontinuation of 50 mg chlorpromazine administered in healthy volunteers for 28 days showed an absence of rebound regarding percent sleep time (HARTMANN and CRAVENS 1973).

Based on data from both sleep laboratory studies (KUPFER et al. 1971; HARTMANN and CRAVENS 1973; KAPLAN et al. 1974) and clinical trials (PRIEN and COLE 1968; SHOPSIN et al. 1971; SWETT 1974; SINGH and KAY 1975; CHOUINARD 1990), it is quite certain that chlorpromazine possesses an initial sedative effect. This is obviously a pharmacological property of the drug, because it has been observed at various dose levels in both psychiatric patients and normal subjects. A degree of tolerance to this effect, however, may develop after a few weeks of drug administration (PRIEN and COLE 1968; HARTMANN and CRAVENS 1973); the apparent continued hypnotic efficacy of chlorpromazine in psychiatric patients (KAPLAN et al. 1974) could very well reflect the reduction in sleeplessness related to an overall clinical improvement which may have occurred in the meantime.

The effects of chlorpromazine on sleep stages were generally less consistent across studies. The enhancement of slow-wave sleep (SWS) by chlorpromazine was the most consistent finding; in many studies time spent in SWS and/or percentage of SWS were increased (LESTER and GUERRERO-FIGUEROA 1966; FEINBERG et al. 1969; SAGALES et al. 1969; LESTER et al. 1971; HARTMANN and CRAVENS 1973; KAPLAN et al. 1974; MOLDOFSKY and LUE 1980). Reports on the effects of chlorpromazine on stage 2 are conflicting; the majority of studies in healthy volunteers report a reduction of this stage (LEWIS and EVANS 1969; SAGALES et al. 1969; FUJII 1973), whereas studies in psychiatric patients show stage 2 to be increased (KAPLAN et al. 1974; KUPFER et al. 1971). However, the increase in time spent in stage 2 may reflect an increase in total sleep time (KAPLAN et al. 1974).

The effects of chlorpromazine on REM sleep seem to be dose dependent. Low doses of chlorpromazine (25–50 mg) usually increased the time in minutes and/or percentage of REM sleep (TOYODA 1964; LEWIS and EVANS 1969), while higher doses either decreased (FEINBERG et al. 1969; LEWIS and EVANS 1969) or did not change REM significantly (LESTER and GUERRERO-FIGUEROA 1966; SAGALES et al. 1969; KUPFER et al. 1971; LESTER et al. 1971; HARTMANN and CRAVENS 1973; KAPLAN et al. 1974; HATA 1975). In the majority of studies, the latency to stage REM either remained unchanged (FEINBERG et al. 1969; LEWIS and EVANS 1969; SAGALES et al. 1969; KUPFER et al. 1971; Hartmann and Cravens 1973) or was shortened (Lester and Guerrero-Figueroa 1966; Lester et al. 1971). In the study of chronic schizophrenics, however, REM latency was prolonged by chlorpromazine 100 mg q.i.d. (Kaplan et al. 1974). The latter finding may be considered to be due either to the effect of the higher drug dose studied, or to the treatment-induced "normalization" of the abnormally short REM latency which is frequently detected in untreated schizophrenic patients (Zarcone and Dement 1969; Feinberg and Hiatt 1978; Ganguli et al. 1987; Soldatos et al. 1987; Zarcone et al. 1987; Keshavan et al. 1990; Thaker et al. 1990; Tandon et al. 1992).

2. Other Phenothiazines

There has been a widespread clinical impression that, similarly to chlorpromazine, other aliphatic phenothiazines possess sedative effects, while the non-aliphatic ones do not. Unfortunately, the existing sleep laboratory studies of phenothiazines other than chlorpromazine are quite limited not only because of their small number, but also because of methodological insufficiencies.

Besides chlorpromazine, the only aliphatic phenothiazine studied in the sleep laboratory is promethazine, mostly at relatively low dose levels. There have been four studies of this drug: two on 14 healthy volunteers [25 mg for 3 days and 50 mg for 1 day at h.s. in one study (Kales et al. 1969), 50–200 mg up to 9 days at h.s. in the other (Risberg et al. 1975)], another on 12 poor sleepers (20 mg or 40 mg at h.s. for 1 day) (Adam and Oswald 1986a), and a fourth on four chronic schizophrenics (50 mg at h.s. for 1 day) (Brannon and Jewett 1969). As expected, promethazine did increase total sleep time in patients (Brannon and Jewett 1969; Adam and Oswald 1986a). In healthy volunteers, however, promethazine did not demonstrate this effect in the sleep laboratory (Kales et al. 1969; Risberg et al. 1975) but only in performance/clinical studies (Clarke and Nicholson 1978; Hindmarch and Parrott 1978). It also increased stage 2 (Risberg et al. 1975; Adam and Oswald 1986a) and decreased the REM stage (Brannon and Jewett 1969; Kales et al. 1969; Risberg et al. 1975; Adam and Oswald 1986a); upon withdrawal REM sleep rebounded (Kales et al. 1969; Risberg et al. 1975). Similarly to chlorpromazine, SWS was increased with promethazine (Risberg et al. 1975).

Two piperidinic phenothiazines have been studied in the sleep laboratory to a rather limited extent: thioridazine in five healthy volunteers (1 mg/kg on the first night and 3 mg/kg on the second, at h.s.) (Gaillard and Aubert 1975) and in 16 insomniacs [75 mg at h.s. for 14 nights (Kales et al. 1974) and 25–100 mg for three nights (Scarone et al. 1978)]; mesoridazine in seven healthy volunteers (10 mg at h.s. for 3 weeks) (Adam et al. 1976); and both thioridazine and mesoridazine in 12 schizophrenic patients (100–800 mg/day and 50–400 mg/day, respectively, for 12 weeks each, using a crossover design) (GARDOS et al. 1978). Both drugs were found to increase total sleep time and/or decrease nocturnal wakefulness (KALES et al. 1974; GAILLARD and AUBERT 1975; ADAM et al. 1976; GARDOS et al. 1978; SCARONE et al. 1978). Further, these drugs at relatively low dose levels increased REM sleep (KALES et al. 1974; ADAM et al. 1976; SCARONE et al. 1978) and decreased REM latency (GAILLARD and AUBERT 1975). No REM enhancement, however, was observed with the high dosage of either thioridazine or mesoridazine (GARDOS et al. 1978). It is noteworthy that the sedative effects of thioridazine were confirmed in a clinical trial of the 25-mg dose administered in 20 psychogeriatric patients for 2 weeks (LINNOILA and VIUKARI 1976).

There is only one sleep laboratory study of a piperazinic phenothiazine (trifluoperazine 5 mg administered to four chronic schizophrenics) (BRANNON and JEWETT 1969). In this study, no significant changes in sleep efficiency or sleep stages were observed. However, the sample size was small and the drug was administered for only one night at a relatively low dose. Thus, no firm conclusion can be drawn about the effects of piperazinic phenothiazines on sleep.

III. Butyrophenones

In spite of the widespread use of butyrophenones, there have been relatively few sleep laboratory studies of these drugs: four of haloperidol and three of pimozide. Among the studies of haloperidol, one reported on the effects of 2–20 mg administered for 3 weeks in 11 schizophrenic patients (KESHAVAN et al. 1989), another study on the effects of 1–5 mg administered for 7–10 days to six Tourette patients (MENDELSON et al. 1980) and two studies on the withdrawal effects following abrupt discontinuation of 5–40 mg in 28 chronic schizophrenics (THAKER et al. 1989; NEYLAN et al. 1992). One study of pimozide reported on the effects of 1 mg and 4 mg administered for 1 day in six healthy volunteers (SAGALES and ERILL 1975) and two others on the effects of 4–20 mg administered for about 1 month in a total of 17 psychiatric patients (GILLIN et al. 1977, 1978b).

When taking into account the evidence stemming from the above studies in conjunction with findings of clinical trials (COOKSON and WELLS 1973; SINGH and KAY 1975; FRUENSGAARD et al. 1977; GERLACH et al. 1985; LEWANDER et al. 1990; MENDLEWICZ et al. 1990) and experimental pharmacology studies (SAYERS and KLEINLOGEL 1974; MONTI 1979; WAUQUIER et al. 1980; TRAMPUS and ONGINI 1990; ONGINI et al. 1992), it appears that there may be a bimodal effect of butyrophenones on sleep. Thus, low doses of these drugs were shown to increase indices of nocturnal wakefulness (WAUQUIER et al. 1980; KESHAVAN et al. 1989), whereas higher doses had a sedative effect (COOKSON and WELLS 1973; SAYERS and KLEINLOGEL 1974; SINGH and KAY 1975; FRUENSGAARD et al. 1977; GILLIN et al. 1977, 1978b; MONTI 1979; MENDELSON et al. 1980; WAUQUIER et al. 1980; GERLACH et al.

1985; Keshavan et al. 1989; Lewander et al. 1990; Mendlewicz et al. 1990; Trampus and Ongini 1990; Ongini et al. 1992), especially when taken for prolonged periods (Sayers and Kleinlogel 1974; Gillin et al. 1977, 1978b; Mendelson et al. 1980; Gerlach et al. 1985; Keshavan et al. 1989; Mendlewicz et al. 1990). This sedative effect was dissipated following drug withdrawal (Gillin et al. 1977; Thaker et al. 1989; Neylan et al. 1992), especially in patients who relapsed (Neylan et al. 1992). Finally, the effects of butyrophenones on sleep stages were quite inconsistent (Sayers and Kleinlogel 1974; Gillin et al. 1977, 1978b; Monti 1979; Wauquier et al. 1980; Keshavan et al. 1989; Trampus and Ongini 1990; Ongini et al. 1992).

IV. Nonclassical Neuroleptics

1. Sulpiride

The benzamide derivative, sulpiride, possesses a selective D_2 receptor-blocking action (Kebabian and Calne 1979). This drug has been evaluated in three sleep laboratory studies. Two of them had an identical design (2 days of 200–400 mg i.m. followed by 3 weeks of 300–600 mg per os): one study included 11 psychiatric patients (schizophrenics, depressives, and neurotics) (Schneider et al. 1974) and the other 16 subjects (psychotic patients and controls with "autonomic symptoms") (Schneider et al. 1975). In these two studies, sulpiride was found to reduce sleep latency (Schneider et al. 1974) and wake time after sleep onset (Schneider et al. 1974, 1975). It also increased stage 2 (Schneider et al. 1974, 1975), decreased stage 3 (Schneider et al. 1975), and either increased (Schneider et al. 1974, 1975) or decreased (Schneider et al. 1975) REM sleep. A third study on schizophrenic patients confirmed the sedative action of sulpiride, showing a reduction in the number of wakes and a shortening of the subjective (but not of the objective) sleep latency (Scarone et al. 1976).

2. Clozapine

The dibenzodiazepine, clozapine, is a neuroleptic drug practically devoid of extrapyramidal side effects; however, the relatively increased risk of the occurrence of blood dyscrasias has limited its use (Kane 1990). Clozapine antagonizes D_1 more than D_2 receptors, while it is also a potent antagonist of D_4 receptors (Coward 1992). Further, many other neurotransmitter systems (adrenergic, serotoninergic, cholinergic, histaminergic) are also influenced by clozapine (Coward 1992).

There are only two early sleep laboratory studies assessing the effects of very low doses of clozapine: 12.5 mg (Touyz et al. 1978) and 25 mg (Touyz et al. 1977) at h.s. in 6 and 14 healthy volunteers, respectively. A sedative effect was reported on the first night of administration of the 25-mg dose (Toyuz et al. 1977), while slight sedation was suggested even with 12.5 mg at the beginning of a 2-week drug administration period (TOUYZ et al. 1978). These findings have been corroborated by data from animal studies (SAYERS and KLEINLOGEL 1974; SPIERINGS et al. 1977; COWARD 1992) as well as clinical trials (ACKENHEIL et al. 1976; FACTOR and BROWN 1992) assessing therapeutic dose levels, showing initial sedation (SAYERS and KLEINLOGEL 1974; ACKENHEIL et al. 1976; COWARD 1992; FACTOR and BROWN 1992) with subsequent development of tolerance (ACKENHEIL et al. 1976; COWARD 1992). The effects of clozapine on sleep stages were inconsistent (SAYERS and KLEINLOGEL 1974; SPIERINGS et al. 1977; TOUYZ et al. 1977, 1978), but REM sleep was suppressed both in humans (TOUYZ et al. 1978) and in animals (SAYERS and KLEINLOGEL 1974; SPIERINGS et al. 1977).

3. Remoxipride

Remoxipride is a selective D_2 dopamine-receptor antagonist with a preferential action on the mesolimbic and other dopaminergic structures besides the striatum (OGREN et al. 1984). Moreover, it has virtually no effect on other neurotransmitter receptor systems in the brain (OGREN et al. 1984). The selectivity of action and lack of striatal activity of remoxipride are considered to exempt it from major side effects, especially the extrapyramidal ones (LEWANDER et al. 1990).

No sleep laboratory studies of remoxipride are available. An animal study showed that, in contrast to haloperidol, remoxipride 1–10 mg/kg did not have sedative effects; it was found only to suppress REM stage (ONGINI et al. 1992). Further, based on comparative data stemming from clinical trials, it appears that remoxipride has significantly fewer sedative effects than chlorpromazine (CHOUINARD 1990) and haloperidol (LEWANDER et al. 1990; MENDLEWICZ et al. 1990).

4. Ritanserin

Ritanserin, considered to be effective for the treatment of negative symptoms of schizophrenia, is a selective 5-HT_2 antagonist with negligible action on other neurotransmitter systems (LEYSEN et al. 1985).

Many controlled sleep laboratory studies have evaluated the effects of ritanserin; none of them, however, was conducted on schizophrenic patients. In most cases, 5–10 mg was assessed in healthy volunteers (IDZIKOWSKI et al. 1986, 1987, 1991; IDZIKOWSKI and MILLS 1986; DECLERCK et al. 1987), poor sleepers (ADAM and OSWALD 1989), and dysthymic patients (PAIVA et al. 1988) for up to 4 weeks. None of these studies showed any effects on sleep efficiency parameters, which is corroborated by the results of a clinical trial in healthy volunteers (KAMALI et al. 1992). Nonetheless, SWS was consistently found to be increased in all seven sleep laboratory studies, while stage 2 was shown to be decreased in some of them (IDZIKOWSKI and MILLS 1986; IDZIKOWSKI et al. 1986, 1987; ADAM and OSWALD 1989).

V. Clinical Implications

As reviewed in this chapter, almost all neuroleptics except for ritanserin and possibly remoxipride possess a considerable sedative effect. This effect is probably more pronounced in aliphatic phenothiazines such as chlorpromazine and promethazine. Nonetheless, as shown above, there are as yet no adequate data to clearly differentiate among sedative and nonsedative phenothiazines and butyrophenones. As far as the effects of neuroleptics on sleep stages are concerned, the findings are quite inconsistent among studies not only across drugs, but, in most cases, even for an individual drug. The inconsistencies observed may be due to methodological differences across studies. Nonetheless, they may also reflect a genuine diversity of effects of neuroleptics on the internal architecture of sleep. The only exception is an increase of SWS with ritanserin which has also been observed with the aliphatic phenothiazines. This finding, however, has no obvious clinical relevance, since the exact physiological significance of sleep stages is still unknown.

Occasionally, the effects of certain neuroleptics on sleep efficiency or on sleep stages were found to be biphasic. Thus, haloperidol and pimozide in small doses may have a sleep-reducing effect while at higher doses they may have a sleep-promoting one. Similar biphasic effects on REM sleep were shown for chlorpromazine and thioridazine, which in small doses enhance REM and in higher doses either suppress or do not alter it. This phenomenon may reflect differential drug effects on either different receptor systems or various subgroups of dopamine receptors, i.e., low doses affecting mostly the presynaptic receptors and high doses affecting the postsynaptic receptors.

From a purely clinical standpoint, when psychotic patients suffer from insomnia, a more sedative neuroleptic should be prescribed with most or all of the daily dosage administered at bedtime, while when patients' mobilization is desirable a nonsedative neuroleptic is generally preferable. In any case, the overall treatment strategy should be individualized taking into account the multifaceted nature of schizophrenia (Stefanis 1990), which is reflected in patients' symptomatology (Kane 1990). Also, it should be emphasized that tolerance appears to develop rather quickly to all neuroleptics' sedative effects. Thus, generally, these effects should not be a deterrent for long-term treatment with an appropriate neuroleptic.

C. Antihistamines

I. Introduction

Antihistamines have been notorious for their sedative side effects (see also Chap. 6, this volume), which may inadvertently interfere with daytime performance (Reynolds 1989). Thus, sedation as an adverse effect of these drugs, which are usually taken as antiallergenics, may pose serious risks for patients who are otherwise expected to drive, operate machinery, compute, and generally preserve high levels of cognitive functioning. Sedation, however, may be desirable in allergic patients with sleep difficulties; antihistamines may be occasionally used in these patients as sleeping aids.

The prevailing trend in clinical therapeutics is to develop antihistamines with fewer sedative side effects (BRITISH MEDICAL ASSOCIATION 1993). Thus, over the years, a major effort has been to produce drugs which preferably do not cross the blood-brain barrier and/or have a greater affinity for peripheral than central histamine receptors. As a consequence, mainly based on clinical evidence, two major groups of antihistamines have been distinguished conventionally: the sedative and the nonsedative (BRITISH MEDICAL ASSOCIATION 1993).

In this chapter both groups of antihistamines are reviewed in terms of their effects on sleep. The reviewed drugs include only the H_1-blocking agents, which are utilized as antiallergenics and not the H_2 blockers (cimetidine and ranitidine), which are used to control gastric hypersecretion (REYNOLDS) 1989). Besides data stemming from clinical trials and sleep laboratory studies, findings of neuropsychological assessments are also reviewed, because they constitute measures of the effects of antihistamines on daytime alertness and performance.

II. Sedative Antihistamines

In this group, 13 drugs are included. Four of them (chlorpheniramine, diphenhydramine, hydroxyzine, and promethazine) were found to have quite sedative effects in sleep laboratory studies (VOGEL et al. 1953; NOELL et al. 1955; BRANNON and JEWETT 1969; ADAM and OSWALD 1986a; ALFORD et al. 1992) and/or clinical trials (JICK et al. 1969; TEUTSCH et al. 1974; HINDMARCH and PARROTT 1978; SUNSHINE et al. 1978; RICKELS et al. 1983; BORBELY and YOUMBI-BALDERER 1988; MONROE 1992). Moreover, neuropsychological studies demonstrated that they lower alertness and impair performance to a considerable degree (CARRUTHERS et al. 1978; CLARKE and NICHOLSON 1978; HINDMARCH and PARROTT 1978; LEVANDER et al. 1985; MATTILA et al. 1986; ROTH et al. 1987; SALETU et al. 1988; NICHOLSON et al. 1991; WALSH et al. 1992; WITEK et al. 1992); they were also found to shorten sleep latency more than placebo in the Multiple Sleep Latency Test (ROEHRS et al. 1984; ROTH et al. 1987; SEIDEL et al. 1990; NICHOLSON et al. 1991). However, it should be emphasized that tolerance was shown to develop after the 1st day of administration of diphenhydramine (MATTILA et al. 1986). Finally, diphenhydramine and promethazine were found to suppress REM sleep (VOGEL et al. 1953; BRANNON and JEWETT 1969; KALES et al. 1969; RISBERG et al. 1975; ADAM and OSWALD 1986a; SALETU et al. 1987).

A subgroup of nine antihistamines (azatadine, brompheniramine, carbinoxamine, clemastine, dimethindene, ketotifen, mequitazine, phenindmine, triprolidine) were found to have a relatively weaker sedative effect. Only three of them (brompheniramine, mequitazine, triprolidine) were evaluated in the sleep laboratory (Nicholson et al. 1985); none of the three had a large effect on sleep efficiency parameters (Nicholson et al. 1985), although they were found to interfere somewhat with performance and to induce a degree of sedation subjectively (Peck et al. 1975; Seppala et al. 1981; Betts et al. 1984; Nicholson and Stone 1984; Bradley and Nicholson 1987). REM sleep was suppressed by brompheniramine and triprolidine (Nicholson et al. 1985). For the other six (azatadine, carbinoxamine, clemastine, dimethindene, ketotifen, phenindamine), some evidence of drug-induced sedation was provided through clinical and/or neuropsychological studies (Peck et al. 1975; Clarke and Nicholson 1978; Hindmarch and Parrott 1978; Seppala et al. 1981; Levander et al. 1985; Nicholson et al. 1991; Hopes et al. 1992; Witek et al. 1992). Moreover, when assessed, tolerance to the sedative effects of brompheniramine, carbinoxamine, and clemastine was shown to develop after the 1st day of drug administration (Seppala et al. 1981). It is also noteworthy that some studies showed that the peripheral drug effects (antiallergic action) are dissociated from the central ones (sedation) (Peck et al. 1975; Levander et al. 1985).

III. Nonsedative Antihistamines

This group includes nine drugs (astemizole, cetirizine, ebastine, loratadine, mebhydroline, mepyramine, niaprazine, temelastine, terfenadine). Most of them have been assessed by neuropsychological studies, while five (loratadine, mebhydroline, mepyramine, niaprazine, temelastine) have also been evaluated in clinical trials and/or sleep laboratory studies. The latter showed that temelastine, mepyramine, and niaprazine did not increase sleep efficiency (Nicholson et al. 1985; Adam and Oswald 1986b; Zucconi et al. 1988). In terms of sleep stage effects, niaprazine increased SWS (Zucconi et al. 1988). Neuropsychological studies showed that none of these drugs in the usual dose range impaired performance (Clarke and Nicholson 1978; Hindmarch and Parrot 1978; Kulshrestha et al. 1978; Reinberg et al. 1978; Nicholson 1982; Nicholson and Stone 1982; Betts et al. 1984; Mattila et al. 1986; Bradley and Nicholson 1987; Roth et al. 1987; Feldman et al. 1992; Hopes et al. 1992; Murri et al. 1992; Ramaekers et al. 1992; Walsh et al. 1992; Witek et al. 1992) or shortened sleep latency in the Multiple Sleep Latency Test (Roehrs et al. 1984; Roth et al. 1987; Zucconi et al. 1988; Seidel et al. 1990; Murri et al. 1992). In the exceptional cases of loratadine and cetirizine, performance and/or the Multiple Sleep Latency Test were found to be affected either at the higher dosage tested (Bradley and Nicholson 1987; Roth et al. 1987) or in combination with alcohol (Ramaekers et al. 1992). Finally, subjective evaluations in clinical trials (Sorkin and Heel 1985; Herman et al. 1992; Monroe 1992) and performance studies (Hindmarch and Parrott 1978; Betts et al. 1984; Mattila et al.

1986; ROTH et al. 1987; HOPES et al. 1992; WALSH et al. 1992; WITEK et al. 1992) showed that none of the nine drugs produced a major impairment of alertness; however, in two studies, terfanadine was found to exert an alerting effect (CLARKE and NICHOLSON 1978; REINBERG et al. 1978).

IV. Clinical Implications

For the reasons detailed in the introduction of this section, it is generally desirable to use nonsedative antihistamines. However, because the choice of an antiallergic agent often depends on various clinical considerations, it is useful to the prescribing physician to be aware of the exact degree of sedation associated with the administration of a specific antihistamine. The above review actually distinguishes not only between sedative and nonsedative antihistamines, but also defines two subgroups among the sedative ones, depending on the degree of drug-induced sedation.

When high levels of daytime alertness are required, a nonsedative antihistamine should be prescribed (REYNOLDS 1989; BRITISH MEDICAL ASSOCIATION 1993). Even in this case, however, the patient should be cautioned regarding the occurrence of some sedation, especially when higher doses are used and/or alcohol or other sedatives are taken concomitantly (BRADLEY and NICHOLSON 1987; ROTH et al. 1987; RAMAEKERS et al. 1992). When considerations regarding unimpaired daytime alertness are not imperative, a sedative antihistamine can be prescribed. Indeed, this should be the case when a degree of sedation is desirable (RICKELS et al. 1983) or whenever the allergic condition is considered to have a neuropsychological substrate, e.g., central pruritus (BRITISH MEDICAL ASSOCIATION 1993). The individualization of antihistamine treatment in relation to drug-induced sedation is facilitated by the distinction of less sedative compounds among the sedative antihistamines. It should also be mentioned that drug reactions vary widely from case to case. Thus, even a drug which is expected to cause sedation may occasionally cause hyperexcitability instead (REYNOLDS 1989).

D. Antiparkinsonian Drugs

I. Introduction

There are two main groups of antiparkinsonian drugs: the dopaminergic and the anticholinergic (REYNOLDS 1989). Both are used for treating Parkinson's disease. In neuroleptic-induced parkinsonism, however, only anticholinergic drugs are indicated, because dopaminergic drugs are not generally considered to be effective (REYNOLDS 1989) and they may counteract the therapeutic dopamine-blocking effect of neuroleptics.

Based on clinical studies (COTZIAS et al. 1969; HORVATH and MEARES 1974; SHARF et al. 1978; CHOUZA et al. 1989), dopaminergic drugs have been shown to induce insomnia or somnolence as well as vivid dreaming/nightmares. The clinical effects of anticholinergics on sleep are less clear; existing information, however, suggests that sleeplessness may occasionally occur either during their administration (TANDON et al. 1991) or upon their withdrawal (BATEGAY 1966; JELLINEK et al. 1981; MCINNIS and PETURSSON 1985). Thus, there is a need to assess systematically the effects of antiparkinsonian drugs on sleep through sleep laboratory and neuropsychological studies. Unfortunately, existing studies of this kind are relatively few and show methodological weaknesses.

II. Dopaminergic Antiparkinsonian Drugs

Only two dopaminergic antiparkinsonians, L-dopa and bromocriptine, have been evaluated in the sleep laboratory. Sleep laboratory studies showed that L-dopa in healthy volunteers did not affect sleep efficiency (KALES et al. 1971; AZUMI et al. 1972), while in parkinsonian (WYATT et al. 1970), narcoleptic (BOIVIN et al. 1989), and depressed (FRAM et al. 1970) patients it had a sleep-reducing effect. The highest dosage of L-dopa tested in healthy volunteers, however, was lower than that taken by patients (5 g vs. 12.6 g). On the other hand, enhancement of sleep was noted in parkinsonian patients who improved under treatment with low doses of L-dopa (0.6–1.4 g) combined with carbidopa and/or other antiparkinsonians (ASKENASY and YAHR 1985).

The effects of L-dopa on sleep stages were inconsistent across studies, particularly regarding REM sleep; some studies showed REM suppression (FRAM et al. 1970; WYATT et al. 1970; CASTALDO et al. 1973; GILLIN et al. 1973; ASKENASY and YAHR 1985), while others showed REM enhancement (KALES et al. 1971; AZUMI et al. 1972; CASTALDO et al. 1973). It cannot be ascertained whether this is due to methodological differences. Nonetheless, it should be noted that animal study data suggest that there is a bimodal effect on sleep stages of the dopa agonists apomorphine, bromocriptine, and pergolide (MONTI 1982, 1983; MONTI et al. 1988). Finally, bromocriptine in daily doses of 1.25–40 mg administered to chronic schizophrenics (BRAMBILLA et al. 1983) and narcoleptics (BOIVIN et al. 1993) was not found to affect any of the sleep efficiency or sleep stage parameters.

III. Anticholinergic Antiparkinsonian Drugs

Only biperiden (2–10 mg at h.s.) has been studied in the sleep laboratory. This drug had no effect on sleep efficiency (GILLIN et al. 1991a; SALIN-PASCUAL et al. 1993). However, in one study two of 12 healthy volunteers had almost complete sleeplessness following 6–10 mg biperiden at h.s. (TANDON et al. 1991). In terms of its effects on sleep stages, biperiden was found to suppress REM sleep and prolong REM latency (GILLIN et al.

1991a; SALIN-PASCUAL et al. 1991a,b, 1992a, 1993; TANDON et al. 1991); tolerance to this effect was noted after the 2nd day of drug administration (SALIN-PASCUAL et al. 1993), while on withdrawal following four nights of administration of 6 mg biperiden a shortening of REM latency (obviously of a rebound nature) was observed (SALIN-PASCUAL et al. 1993). These effects of biperiden on REM sleep are in keeping with similar effects of scopolamine, another anticholinergic (SAGALES et al. 1969, 1975; SITARAM et al. 1978, 1979; GILLIN et al. 1979a,b, 1991b; SITARAM and GILLIN 1980; POLAND et al. 1989; RIEMANN et al. 1991), as well as with the opposite effects of cholinomimetics (physostigmine, arecoline, etc.) on human sleep (SITARAM et al. 1976, 1977, 1978, 1980, 1982; GILLIN et al. 1978a; SPIEGEL 1984; BERGER et al. 1985). Moreover, the REM-suppressing effect of biperiden and scopolamine was also observed in animal experiments (SHIROMANI and FISHBEIN 1986; SUTIN et al. 1986; SALIN-PASCUAL et al. 1992b).

IV. Clinical Implications

The issue of drug-induced sleep difficulty in parkinsonian patients is of special clinical importance. Complaints of disturbed sleep are quite common in parkinsonian patients (LEES et al. 1988; FACTOR et al. 1990; JANSEN and MEERWALDTT 1990; LEES 1990). Difficulty turning in bed, cramps, problems with urination, and various other nocturnal events frequently disrupt the sleep of these patients. L-dopa in high doses administered to patients with Parkinson's disease may worsen their sleep difficulty (WYATT et al. 1970). Fortunately, nowadays, preparations containing low doses of L-dopa combined with carbidopa are widely prescribed and have been shown to improve patients' sleep (ASKENASY and YAHR 1985). It is possible that this combined medication may affect sleep efficiency indirectly through the alleviation of the patients' nocturnal difficulties, because low doses of L-dopa were found not to change sleep efficiency on their own (KALES et al. 1971; AZUMI et al. 1972).

Anticholinergic antiparkinsonians are usually administered to control extrapyramidal side effects of neuroleptics, drugs which are prescribed to psychotic patients (KANE 1990). These patients, suffering from schizophrenia, organic brain syndromes, mania, or other affective illnesses, often present with sleep difficulties inherent in their psychiatric disorders (ZARCONE and DEMENT 1969; FEINBERG and HIATT 1978; GANGULI et al. 1987; SOLDATOS et al. 1987; BENCA et al. 1992; TANDON et al. 1992). Although neuroleptics do not directly disturb sleep by themselves (see Sect. II), they often cause parkinsonism (quite severe at times), which may lead to disturbed sleep. Anticholinergic antiparkinsonians do not seem to have a major effect on sleep efficiency on their own (GILLIN et al. 1991a; SALIN-PASCUAL et al. 1993). However, by improving the neuroleptic-induced parkinsonism, they should be expected to have an overall beneficial effect on patient's sleep.

E. Summary

In this chapter the effects of neuroleptics, antihistamines and antiparkinsonian drugs on sleep are reviewed. With the exception of chlorpromazine, the effects of classical neuroleptics on sleep have been studied to a rather limited extent. This holds true even for the newer, nonclassical, neuroleptics except for ritanserin. Almost all neuroleptics, except for ritanserin and possibly remoxipride, were found to possess a considerable sedative effect. However, even for the most sedative ones, such as chlorpromazine, tolerance to sedation may develop rather quickly. The effects of neuroleptics on sleep stages were rather inconsistent, with the exception of ritanserin and the aliphatic phenothiazines, which were shown to increase slow-wave sleep. Nonetheless, any drug effects on sleep stages lack clinical relevance because their exact physiological significance is still unknown.

In spite of the scarcity of sleep laboratory studies of antihistamines, their sedative effects have been adequately evaluated through clinical and/or neuropsychological studies. Two main categories are distinguished: the sedative antihistamines and the nonsedative ones; the latter including astemizole, cetirizine, ebastine, loratadine, mebhydroline, mepyramine, niaprazine, temelastine and terfenadine. Even with the sedative antihistamines, however, two subgroups can be differentiated; chlorpheniramine, diphenhydramine, hydroxyzine and promethazine are more sedative, while azatadine, brompheniramine, carbinoxamine, clemastine, dimethindene, ketotifen, mequitazine, phenindamine and triprolidine are less sedative. In terms of their effects on sleep stages, REM suppression has been reported with brompheniramine, diphenhydramine, promethazine and triprolidine, while niaprazine was found to increase slow-wave sleep.

Among the dopaminergic antiparkinsonian drugs, only L-dopa and bromocriptine have been studied in the sleep laboratory. High doses of L-dopa had an alerting effect, while low doses of this drug and bromocriptine were not found to affect sleep. Subjects with parkinsonism, who are known to suffer from sleep difficulties, appear to sleep better under treatment with low doses of L-dopa combined with carbidopa. Biperiden is the only anticholinergic antiparkinsonian drug evaluated through sleep laboratory studies. Generally, the drug had no appreciable effects on sleep efficiency, whereas it was found to suppress REM sleep.

Acknowledgment. The authors appreciate the assistance provided by Drs. O. Mouzas and S. Markoulaki in the collection of the bibliographic material.

References

Ackenheil M, Brau H, Burkhart A, Franke A, Pacha W (1976) Antipsychotic efficacy in relation to plasma levels of clozapine. Arzneimittelforschung 26:1156–1158

Adam K, Oswald I (1986a) The hypnotic effects of an antihisamine: promethazine, Br J Clin Pharmacol 22:715–717

Adam K, Oswald I (1986b) No effects on sleep of a histamine H_1-receptor antagonist: temelastine. Br J Clin Pharmacol 22:718–720
Adam K, Oswald I (1989) Effects of repeated ritanserin on middle-aged poor sleepers. Psychopharmacology (Berl) 99:219–221
Adam K, Allen S, Carruthers-Jones I, Oswald I, Spence M (1976) Mesoridazine and human sleep. Br J Clin Pharmacol 3:157–163
Alford O, Rombaur N, Jones J, Foley S, Idzikowski C, Hindmarch J (1992) Acute effects of hydroxyzine on nocturnal sleep and sleep tendency the following day. A C-EEG study. Hum Psychopharmacol 7:25–35
Askenasy JJM, Yahr MD (1985) Reversal of sleep disturbance in Parkinson's disease by antiparkinsonian therapy: a preliminary study. Neurology 35:527–532
Azumi K, Jinnai S, Takahashi (1972) The effects of l-dopa on sleep pattern and SPR in normal adults. Sleep Res 1:40
Bategay R (1966) Drug depenence as a criterion for differentiating psychotropic drugs. Compr Psychiatry 7:501–509
Benca RM, Obermeyer WH, Thisted RA, Gillin JC (1992) Sleep and psychiatric disorders. A meta-analysis. Arch Gen Psychiatry 49:651–668
Berger M, Hochli D, Zulley J, Lauer C, von Zerssen D (1985) Cholinomimetic drug RS 86, REM sleep, and depression. Lancet 1:1385–1386
Betts T, Markman D, Debenham S, Mortiboy D, McKevitt T (1984) Effects of two antihistamine drugs on actual driving performance. Br Med J 288:281–282
Boivin DB, Montplaisir J, Poirier G (1989) The effects of L-dopa on periodic leg movements and sleep organization in narcolepsy. Clin Neuropharmacol 12:339–345
Boivin DB, Montplaisir J, Lambert C (1993) Effects of bromocriptine in human narcolepsy. Clin Neuropharmacol 16:120–126
Borbély AA, Youmbi-Balderer G (1988) Effect of diphenhydramine on subjective sleep parameters and on motor activity during bedtime. Int J Clin Pharmacol Ther Toxicol 26:392–396
Bradley CM, Nicholson AN (1987) Studies on the central effects of the H_1-antagonist, loratadine. Eur J Pharmacol 32:419–421
Brambilla F, Scarone S, Pugnetti L, Massironi R, Penati G, Nobile P (1983) Bromocriptine therapy in chronic schizophrenia: effects on symptomatology, sleep patterns and prolactin response to stimulation. Psychiatry Res 8:159–169
Brannon JO, Jewett RE (1969) Effects of selected phenothiazines on REM sleep in schizophrenics. Arch Gen Psychiatry 21:284–290
British Medical Association (1993) British national formulary no 25. Royal Pharmaceutical Society, London
Carruthers SG, Shoeman DW, Hignite CE, Azarnoff DL (1978) Correlation between plasma diphenhydramine level and sedative and antihistamine effects. Clin Pharmacol Ther 23:375–382
Castaldo V, Krynicki VE, Crade M (1973) L-dopa and REM sleep in normal and mentally retarded subjects. Biol Psychiatry 6:295–299
Chouinard G (1990) A placebo-controlled clinical trial of remoxipride and chlorpromazine in newly admitted schizophrenic patients with acute exacerbation. Acta Psychiatr Scand 82 [Suppl 358]:111–119
Chouza C, Aljanati R, Scaramelli A, De Medina O, Caamano JL, Buzo R, Fernandez A, Romero S (1989) Combination of selegiline and controlled release levodopa in the treatment of fluctuations of clinical disability in parkinsonian patients. Acta Neurol Scand 126:127–137
Clarke CH, Nicholson AN (1978) Performance studies with antihistamines. Br J Clin Pharmacol 6:31–36
Cookson IB, Wells PG (1973) Haloperidol in the treatment of stutterers (Letter). Br J Psychiatry 123:491
Cotzias GC, Papavasiliou PS, Gellene R (1969) Modification of parkinson – chronic treatment with l-dopa. N Engl J Med 280:337–345

Coward DM (1992) General pharmacology of clozapine. Br J Psychiatry 168 [Suppl 17]:5–11
Declerck AC, Wauquier A, Van der Ham-Veltman PHM, Gelders Y (1987) Increase in slow-wave sleep in humans with the serotonin-S2 antagonist ritanserin. Curr Ther Res 41:427–432
Factor SA, Brown D (1992) Clozapine prevents recurrence of psychosis in Parkinson's disease. Mov Disord 7:125–131
Factor SA, McAlarney T, Senchez-Ramos JR, Weiner WJ (1990) Sleep disorders and sleep effect in Parkinson's disease. Mov Disord 5:280–285
Feinberg I, Hiatt JF (1978) Sleep patterns in schizoprenia: a selective review. In: Williams RL, Karacan I (eds) Sleep disorders. Diagnosis and treatment. Wiley, New York, pp 205–231
Feinberg I, Wender PH, Koresko RL, Gottlieb F, Piehuta JA (1969) Differential effects of chlorpromazine and phenobarbital on EEG sleep patterns. J Psychiatr Res 7:101–109
Feldman W, Shanon A, Leiken L, Ham-Pong A, Peterson R (1992) Central nervous system side-effects of antihistamines in school-children. Rhinol Suppl 13:13–19
Fram DH, Murphy DL, Goodwin FK et al. (1970) L-dopa and sleep in depressed patients. Psychophysiology 7:316–317
Fruensgaard K, Korsgaard S, Jorgensen H, Jensen K (1977) Loxapine versus haloperidol parenterally in acute psychosis. Acta Psychiatr Scand 56:256–264
Fujii S (1973) Effects of some psychotropic and hypnotic drugs on the human nocturnal sleep. Psychiatr Neurol Jpn 75:545–573
Gaillard JM, Aubert C (1975) Specificity of benodiazepine action on human sleep confirmed. Another contribution of automatic analysis of polygraph recordings. Biol Psychiatry 10:185–197
Ganguli R, Reynolds CF, Kupfer DJ (1987) Electroencephalographic sleep in young, never-medicated schizophrenics: a comparison with delusional and nondelusional depressives and with healthy controls. Arch Gen Psychiatry 44:36–44
Gardos G, Tecce JL, Hartmann E, Bowers P, Cole JO (1978) Treatment with mesoridazine and thioridazine in chronic schizophrenia. I. Assessment of clinical and electrophysiologic responses in refractory hallucinating schizophrenics. Comp Psychiatry 19:517–525
Gerlach J, Behnke K, Heltberg J, Munk-Andersen E, Nielsen H (1985) Sulpiride and haloperidol in schizophrenia: a double-blind cross-over study of therapeutic effect, side effects and plasma concentrations. Br J Psychiatry 147:283–288
Gillin JC, Post RM, Wyatt RJ et al. (1973) REM inhibitory effect of l-dopa infusion during human sleep. Electroencephalogr Clin Neurophysiol 35:181–186
Gillin JC, van Kammen DP, Post R, Bunney WE Jr (1977) Effects of prolonged administration of pimozide on sleep-EEG patterns in psychiatric patients. Commun Psychopharmacol 1:225–232
Gillin JC, Sitaram N, Mendelson WB, Wyatt RJ (1978a) Physostigmine alters onset but not duration of REM sleep in man. Psychopharmacology (Berl) 58:111–114
Gillin JC, van Kammen DP, Bunney WE Jr (1978b) Pimozide attenuates d-amphetamine-induced sleep changes in man. Life Sci 22:1805–1810
Gillin JC, Duncan W, Pettigrew KD, Frankel BL, Snyder F (1979a) Successful separation of depressed, normal, and insomniac subjects by EEG sleep data. Arch Gen Psychiatry 36:85–90
Gillin JC, Sitaram N, Duncan WC (1979b) Muscarinic supersensitivity: a possible model for the sleep disturbance of primary depression? Psychiatry Res 1:17–22
Gillin JC, Sutton L, Ruiz C et al. (1991a) Dose dependent inhibition of REM sleep in normal volunteers by biperiden, a muscarinic antagonist. Biol Psychiatry 30:151–156
Gillin JC, Sutton L, Ruiz C et al. (1991b) The effects of scopolamine on sleep and mood in depressed patients with a history of alcoholism and a normal comparison group. Biol Psychiatry 30:157–169

Hartmann E (1978) Effects of psychotropic drugs on sleep: the catecholamines and sleep. In: Lipton MA, DiMascio A, Killam KF (eds) Psychopharmacology: a generation of progress. Raven, New York, pp 711–728

Hartmann E, Cravens J (1973) The effects of long term administration of psychotrophic drugs on human sleep. IV. The effects of chlorpromazine. Psychopharmacology (Berl) 33:203–218

Hata H (1975) Effects of some neuro-active drugs on REM sleep and rapid eye movements during REM sleep in man. Psychiatr Neurol Jpn 77:29–52

Herman D, Arnaud A, Dry J et al. (1992) Efficacite clinique et tolerance de la loratadine versus cetirizine dans le traitement de la rhinite allergique saisonniere. Allerg Immunol (Paris) 24:270–274

Hindmarch I, Parrott AC (1978) A repeated dose comparison of the side effects of five antihistamines on objective assessments of psychomotor performance, central nervous system arousal and subjective appraisals of sleep and early morning behaviour. Arzneimittelforschung 28:483–486

Hopes H, Meuret GH, Ungethum W, Leopold G, Wiemann H (1992) Placebo controlled comparison of acute effects of ebastine and clemastine on performance and EEG. Eur J Clin Pharmacol 42:55–59

Horvath TB, Meares RA (1974) L-dopa and arousal. J Neurol Neurosurg Psychiatry 37:416–421

Idzikowski C, Mills FJ (1986) 5-HT_2 antagonist causes sustained increase in human slow wave sleep. Clin Sci 70:89

Idzikowski C, Mills FJ, Glennard R (1986) 5-Hydroxytryptamine-2 antagonist increases human slow wave sleep. Brain Res 378:164–168

Idzikowski C, Cowen PJ, Nutt D, Mills FJ (1987) The effects of chronic ritanserin treatment on sleep and the neuroendocrine response to 1-tryptophan. Psychopharmacology 93:416–420

Idzikowski C, Mills FJ, James RJ (1991) A dose-response study examining the effects of ritanserin on human slow wave sleep. Br J Clin Pharmacol 31:193–196

Jansen ENH, Meerwaldtt JD (1990) Madopar HBS in nocturnal symptoms of Parkinson's disease. Adv Neurol 53:527–531

Jellinek T, Gardos G, Cole JO (1981) Adverse effects of antiparkinson drug withdrawal. Am J Psychiatry 138:1567–1571

Jick H, Slone D, Shapiro S, Lewis GP (1969) Clinical effects of hypnotics. I. A controlled trial. JAMA 209:2013–2015

Kales A, Malmstrom EJ, Scharf MB, Rubin RT (1969) Psychophysiological and biochemical changes following use and withdrawal of hypnotics. In: Kales A (ed) Sleep, physiology and pathology. Lippincott, Philadelphia, pp 331–343

Kales A, Ansel RD, Markham CH, Scharf MB, Tan TL (1971) Sleep in patients with Parkinson's disease and normal subjects prior to and following levodopa administration. Clin Pharmacol Ther 12:397–405

Kales A, Scharf MB, Kales JD, Bixler EO, Djoko M (1974) Sleep laboratory drug evaluation: thioridazine (mellaril), a REM enhancing drug (Abstr). Sleep Res 3:55

Kamali F, Stansfield SC, Ashtin CH, Hammond GL, Emanuel MB, Rawlins MD (1992) Absence of withdrawal effects of ritanserin following chronic dosing in healthy volunteers. Psychopharmacology (Berl) 108:213–217

Kane JM (1990) Psychopharmacologic treatment of schizophrenia. In: Kales A, Stefanis CN, Talbott J (eds) Recent advances in schizophrenia. Springer, Berlin Heidelberg New York, pp 257–276

Kaplan J, Dawson S, Vaughan T, Green R, Wyatt RJ (1974) Effect of prolonged chlorpromazine administration on the sleep of chronic schizophrenics. Arch Gen Psychiatry 31:62–66

Kay DC, Blackburn AB, Buckingham JA, Karacan I (1976) Human pharmacology of sleep. In: Williams RL, Karacan I (eds) Psychopharmacology of sleep. Wiley, New York, pp 83–210

Kebabian JW, Calne DB (1979) Multiple receptors for dopamine. Nature 277:93–96
Keshavan MS, Reynolds C, Brar J et al. (1989) Sleep EEG changes in schizophrenia during haloperidol treatment. Biol Psychiatry 25:181A
Keshavan MS, Reynolds CF, Kupfer DJ (1990) Electroencephalographic sleep in schizophrenia: a critical review. Compr Psychiatry 30:34–47
Kulshrestha VK, Gupta PP, Turner P, Wadsworth J (1978) Some clinical pharmacological studies with terfenadine, a new antihistaminic drug. Br J Clin Pharmacol 6:25–30
Kupfer DJ, Wyatt RJ, Snyder F, Davis JM (1971) Chlorpromazine and sleep in psychiatric patients. Arch Gen Psychiatry 24:185–189
Lees AJ (1990) Madopar HBS (hydrodynamically balanced system) in the treatment of Parkinson's disease. Adv Neurol 53:475–482
Lees AJ, Blackburn MA, Campoell VL (1988) The nighttime problems of Parkinson's disease. Clin Neuropharmacol 11:512–519
Lester BK, Guerrero-Figueroa R (1966) Effects of some drugs on electroencephalographic fast activity and dream time. Psychophysiology 2:224–236
Lester BK, Coulter JD, Cowden LC, Williams HL (1971) Chlorpromazine and human sleep. Psychopharmacology (Berl) 20:280–287
Levander S, Hagermark O, Stahle M (1985) Peripheral antihistamine and central sedative effects of three H_2-receptor antagonists. Eur J Pharmacol 28:523–529
Lewander T, Westerbergh SE, Morrison D (1990) Clinical profile of remoxipride – a combined analysis of a comparative double-blind multicentre trial programme. Acta Psychiatr Scand 82 [Suppl 358]:92–98
Lewis SA, Evans JI (1969) Dose effects of chlorpromazine on human sleep. Psychopharmacologia (Berl) 14:342–348
Leysen J, Gommeren W, Van Gompel P, Wynants J, Janssen PFM, Laudron PM (1985) Receptor-binding properties in vitro and in vivo of ritanserin. Mol Pharmacol 27:600–611
Linnoila M, Viukari M (1976) Efficacy and side effects of nitrazepam and thioridazine as sleeping aids in psychogeriatric in-patients. Br J Psychiatry 128:566–569
Mattila MJ, Mattila M, Konno K (1986) Acute and subacute actions on human performance and interactions with diazepam of temelastine (SK & F93944) and diphenhydramine. Eur J Pharmacol 31:291–298
McInnis M, Petursson H (1985) Withdrawal of trihexyphenidyl. Acta Psychiatr Scand 71:297–303
Mendelson WB, Caine ED, Goyer P, Ebert M, Gillin JC (1980) Sleep in Gilles de la Tourette syndrome. Biol Psychiatry 15:339–343
Mendlewicz J, de Bleeker E, Cosyns P et al. (1990) A double-blind comparative study of remoxipride and haloperidol in schizophrenic and schizophreniform disorders. Acta Psychiatr Scand 82 [Suppl 358]:138–141
Moldofsky H, Lue FA (1980) The relationship of alpha and delta EEG frequencies to pain and mood in "fibrositis" patients treated with chlorpromazine and L-tryptophan. Electroencephalogr Clin Neurophysiol 50:71–80
Monroe EW (1992) Relative efficacy and safety of loratadine, hydroxyzine and placebo in chronic idiopathic urticaria and atopic dermatitis. Clin Ther 14:17–21
Monti JM (1979) The effects of neuroleptics with central dopamine and noradrenaline receptor blocking properties in the l-dopa and (+)-amphetamine-induced waking EEG in the rat. Br J Pharmacol 67:87–91
Monti JM (1982) Catecholamines and the sleep-wake cycle. I. EEG and behavioral arousal. Life Sci 30:1145–1157
Monti JM (1983) Catecholamines and the sleep-wake cycle. II. REM sleep. Life Sci 32:1401–1415
Monti JM, Hawkins M, Jantos H, D'Angelo L, Fernandez M (1988) Biphasic effects of dopamine D-2 receptor agonists on sleep and wakefulness in the rat. Psychopharmacology (Berl) 95:395–400

Murri L, Massetani R, Krause M, Dragonetti C, Iudice A (1992) Evaluation of antihistamine-related daytime sleepiness. A double-blind, placebo-controlled study with terfenadine. Allergy 47:532–534

Neylan TC, van Kammen DP, Kelley ME, Peters JL (1992) Sleep in schizophrenic patients on and off haloperidol therapy. Clinically stable vs. relapsed patients. Arch Gen Psychiatry 49:643–649

Nicholson AN (1982) Antihistaminic activity and central effects of terfenadine. A review of European studies. Arzneimittelforschung Res 32:1191–1193

Nicholson AN, Stone BM (1982) Performance studies with the H_1-antihistamine receptor antagonists, astemizole and terfenadine. Br J Clin Pharmacol 13: 199–202

Nicholson AN, Stone BM (1984) The H_2-antagonists, cimetidine and ranitidine: studies on performance. Eur J Pharmacol 26:579–582

Nicholson AN, Pascoe PA, Stone BM (1985) Histaminergic systems and sleep. Studies in man with H_1 and H_2 antagonists. Neuropharmacology 24:245–250

Nicholson AN, Pascoe PA, Turner C et al. (1991) Sedation and histamine H_1-receptor antagonism: studies in man with the enantiomers of chlorpheniramine and dimethindene. Br J Pharmacol 104:270–276

Noell WK, Chinn HI, Haberer CE (1955) Electroencephalographic evaluation of the sedative effects of antihistaminic drugs. United States Air Force, report no 55–35

Ogren SO, Hall H, Kohler C et al. (1984) Remoxipride, a new potential antipsychotic compound with selective antidopaminergic actions in the rat brain. Eur Pharmacol 102:459–474

Ongini E, Bo P, Dionisotti S, Trampus M, Savoldi F (1992) Effects of remoxipride, a dopamine D-2 antagonist antipsychotic, on sleep-waking patterns and EEG activity in rats and rabbits. Psychopharmacology (Berl) 107:236–242

Paiva T, Arriaga F, Wauquier A, Lara E, Largo R, Leitao JN (1988) Effects of ritanserin on sleep disturbances of dysthymic patients. Psychopharmacology (Berl) 96:395–399

Peck AW, Fowle ASE, Bye C (1975) A comparison of triprolidine and clemastine on histamine antagonism and performance tests in man: Implications for the mechanism of drug induced drowsiness. Eur J Clin Pharmacol 8:455–463

Poland RE, Tondo L, Rubin RT, Trelease RB, Lesser IM (1989) Differential effects of scopolamine on nocturnal cortisol secretion, sleep architecture, and REM latency in normal volunteers: relation to sleep and cortisol abnormalities in depression. Biol Psychiatry 25:403–412

Prien RF, Cole JO (1968) High dose chlorpromazine therapy in chronic schizophrenia. Arch Gen Psychiatry 18:482–495

Ramaekers JG, Uiterwijk MM, O'Hanlon JF (1992) Effects of loratadine and cetirizine on actual driving and psychometric test performance, and EEG during driving. Eur J Clin Pharmacol 42(4):363–369

Reinberg A, Levi F, Guillet P, Burke JT, Nicolai A (1978) Chronopharmacological study of antihistamines in man with special reference to terfenadine. Eur J Clin Pharmacol 14:245–252

Reynolds EF (ed) (1989) Martindale. The extra pharmacopoeia. Pharmaceutical Press, London

Rickels K, Morris R, Newman H, Rosenfeld H, Schiller H, Weinstock R (1983) Diphenhydramine in insomniac family practice patients: a double-blind study. J Clin Pharmacol 23:235–242

Riemann D, Hohagen F, Fleckenstein P, Schredi M, Berger M (1991) The cholinergic REM induction test with RS 86 after scopolamine pretreatment in healthy subjects. Psychiatry Res 38:247–260

Risberg AM, Risberg J, Ingvar DH (1975) Effects of promethazine on nocturnal sleep in normal man. Psychopharmacology (Berl) 43:279–284

Roehrs TA, Tietz EI, Zorick FJ, Roth T (1984) Daytime sleepiness and antihistamines. Sleep 7:137–141

Roth T, Roehrs T, Koshorek J, Sicklesteel BA, Zorick F (1987) Sedative effects of antihistamines. J Allergy Clin Immunol 80:94–98

Sagales T, Erill S (1975) Effects of central dopaminergic blockade with pimozide upon the EEG stages of sleep in man. Psychopharmacology (Berl) 41:53–56

Sagales T, Erill S, Domino EF (1969) Differential effects of scopolamine and chlorpromazine on REM and NREM sleep in normal male subjects. Clin Pharmacol Ther 10:522–529

Sagales T, Erill S, Domino EF (1975) Effects of repeated doses of scopolamine on the electroencephalographic stages of sleep in normal volunteers. Clin Pharmacol Ther 18:727–732

Saletu B, Grunberger J, Krupka M, Schuster P (1987) Comparative double-blind, placebo-controlled sleep laboratory studies of a combination of lorazepam and biphenhydramine (SM-1014) and its single components. Curr Ther Res 42: 1037–1058

Saletu B, Grunberger J, Anderer P, Barbanoj MJ (1988) Pharmacodynamic studies of a combination of lorazepam and diphenhydramine and its single components: electroencephalographic brain mapping and safety evaluation. Curr Ther Res 44:909–937

Salin-Pascual RJ, Granados-Fuentes D, Galicia-Polo L, Nieves E (1991a) Rapid eye movement (REM) sleep increases by auditory stimulation reverted with biperiden administration in normal volunteers. Neuropsychopharmacology 5:183–186

Salin-Pascual RJ, Granados-Fuentes D, Galicia-Polo L, Nieves E, Echeverry J (1991b) Biperiden administration in normal sleep and after rapid eye movement sleep deprivation in healthy volunteers. Neuropsychopharmacology 5:97–102

Salin-Pascual RJ, Grandos-Fuentes D, Galicia-Polo L, Nieves E, Roehrs TA, Roth T (1992a) Biperiden administration during REM sleep deprivation diminished the frequency of REM sleep attempts. Sleep 156:252–256

Salin-Pascual RJ, Jimenez-Anguiano A, Granados-Fuentes D, Drucker-Colin R (1992b) Effects of biperiden on sleep at baseline and after 72 h of REM sleep deprivation in the cat. Psychopharmacology (Berl) 106:540–542

Salin-Pascual RJ, Granados-Fuentes D, Galicia-Polo L, Nieves E, Gillin JC (1993) Development of tolerance after repeated administration of a sedative muscarinic M_1 anatagonist biperiden in healthy human volunteers. Biol Psychiatry 33: 188–193

Sayers AC, Kleinlogel H (1974) Neuropharmakologische Befunde unter chronischer Verabreichung von Haloperidol, Loxapin und Clozapin. Arzneimittelforschung 24:981–983

Scarone S, Spoto G, Fenati G, Canger R, Moja EA (1976) A study of the EEG sleep patterns and the sleep and dream experience of a group of schizophrenic patients treated with sulpiride. Arzneimittelforschung 26:1626–1628

Scarone S, Soldatos CR, Bixler EO, Scharf MB, Kales A (1978) Dose effects of thioridazine on phasic sleep events. Sleep Res 7:112

Schneider E, Ziegler B, Maxion H, Badawi M (1974) Kurz- und Langzeiteffekte von Sulpirid auf das Nachtschlaf-EEG des Menschen. Arzneimittelforschung 24: 990–993

Schneider E, Ziegler B, Maxion H, Badawi M (1975) Der Einfluß von Sulpirid auf den Schlaf. Ergebnisse kontinuierlicher EEG-Untersuchungen. Pharmakopsychiatr Neuropsychopharmakol 7:90–98

Seidel WF, Cohen S, Gourash-Bliwise N, Dement WC (1990) Direct measurement of daytime sleepiness after administration of cetirizine and hydroxyzine with a standarized electroencephalographic assessment. J Allergy Clin Immunol 86 [Suppl 2]:1029–1033

Seppala T, Nuotto E, Korttila K (1981) Single and repeated dose comparison of three antihistamines and phenylpropanolamine: psychomotor performance and subjective appraisals of sleep. Br J Clin Pharmacol 12:179–188

Sharf B, Moskovitz C, Lupton MD, Klawans HL (1978) Dream phenomena induced by chronic levodopa therapy. J Neural Transm 43:143–151
Shiromani PJ, Fishbein W (1986) Continuous pontine cholinergic microinfusion via mini-pump induces sustained alterations in rapid eye movement (REM) sleep. Pharmacol Biochem Behav 25:1253–1261
Shopsin B, Kim SS, Gershon S (1971) A controlled study of lithium vs. chlorpromazine in acute schizophrenics. Br J Psychiatry 119:435–440
Singh MM, Kay SR (1975) A comparative study of haloperidol and chlorpromazine in terms of clinical effects and therapeutic reversal with benztropine in schizophrenia. Theoretical implications for potency differences among neuroleptics. Psychopharmacology (Berl) 43:103–113
Sitaram NJ, Gillin JC (1980) Development and use of pharmacological probes of the CNS in man: evidence of cholinergic abnormality in primary affective illness. Biol Psychiatry 15:925–955
Sitaram NJ, Mendelson WB, Wyatt RJ, Gillin JC (1977) The time-dependent indution of REM sleep and arousal by physostigmine infusion during normal human sleep. Brain Res 122:562–567
Sitaram NJ, Wyatt RJ, Dawson S, Gillin JC (1976) REM sleep induction by physostigmine infusion during sleep. Science 191:1281–1283
Sitaram NJ Moore AM, Gillin JC (1978) Experimental acceleration and slowing of REM sleep ultradian rhythm by cholinergic agonist and antagonist. Nature 274:490–492
Sitaram NJ, Moore AM, Gillin JC (1979) Scopolamine-induced muscarinic supersensitivity in normal man: changes in sleep. Psychiatry Res 1:9–16
Sitaram NJ, Nurnberger JL Jr, Gershon ES et al. (1980) Faster cholinergic REM induction in euthymic patients with primary affective illness. Science 208:200–202
Sitaram NJ, Nurnberger JI, Gershon ES, Gillin JC (1982) Cholinergic regulation of mood and REM sleep: potential model and marker of vulnerability to affective disorder. Am J Psychiatry 139:571–576
Soldatos CR, Vela-Bueno A, Kales A (1987) Sleep in psychiatric disorders. Psychiatr Med 4:119–132
Sorkin EM, Heel RC (1985) Terfenadine. A review of its pharmacodynamic properties and therapeutic efficacy. Drugs 29:34–56
Spiegel R (1984) Effects of RS 86, an orally active cholinergic agonist, on sleep in man. Psychiatry Res 11:1–13
Spierings EL, Dzoljic MR, Godschalk M (1977) Effect of clozapine on the sleep pattern in the rat. Pharmacology 15:551–556
Stefanis CN (1990) On the concept of schizophrenia. In: Kales A, Stefanis CN, Talbott J (eds) Recent advances in schizophrenia. Springer, Berlin Heidelberg New York, pp 25–57
Sunshine A, Zighelboim I, Laska E (1978) Hypnotic activity of diphenhydramine, methapyrilene and placebo. J Clin Pharmacol 18:425–431
Sutin EL, Shiromani PJ, Kelsoe JR Jr, Storch F, Gillin JC (1986) Rapid-eye movement sleep and muscarinic receptor binding in rats are augmented during withdrawal from chronic scopolamine treatment. Life Sci 39:2419–2427
Swett C (1974) Drowsiness due to chlorpromazine in relation to cigarette smoking. Arch Gen Psychiatry 31:211–213
Tandon R, Shipley JE, Greden JF, Mann NA, Eisner WH, Goodson J (1991) Muscarinic cholinergic hyperactivity in schizophrenia. Relationship to positive and negative symptoms. Schizophr Res 4:23–30
Tandon R, Shipley JE, Taylor S et al. (1992) Electroencephalographic sleep abnormalities in schizophrenia: relationship to positive/negative symptoms and prior neuroleptic treatment. Arch Gen Psychiatry 49:185–194
Teutsch G, Mahler DL, Colin RB et al. (1974) Hypnotic efficacy of diphenhydramine, methapyrilene and pentobarbital. Clin Pharmacol Ther 17:195–201

Thaker GK, Wagman AM, Kirkpatrick B, Tamminga CA (1989) Alterations in sleep polygraphy after neuroleptic withdrawal: a putative supersensitive dopaminergic mechanism. Biol Psychiatry 25:75–86
Thaker GK, Wagman AMI, Tamminga CA (1990) Sleep polygraphy in schizophrenia: methodological issues. Biol Psychiatry 28:240–246
Touyz SW, Beumont PJV, Saayman GS, Zabow T (1977) A psychophysiological investigation of the short-term effects of clozapine upon sleep parameters of normal young adults. Biol Psychiatry 12:801–822
Touyz SW, Saayman GS, Zabow T (1978) A psychophysiological investigation of the long-term effects of clozapine upon sleep patterns of normal young adults. Psychopharmacology (Berl) 56:69–73
Toyoda J (1964) The effects of chlorpromazine and imipramine on the human nocturnal sleep electroencephalogram. Folia Psychiatr Neurol Jpn 18:198–221
Trampus M, Ongini E (1990) The D_1 dopamine receptor antagonist SCH 23390 enhances REM sleep in the rat. Neuropharmacology 29:889–893
Vogel GW, Schultz EN, Swenson GW (1953) Concerning the effectiveness of diphenhydramine hydrochloride as a sleep aid. Over the Counter, vol 050041
Walsh JK, Muehlbach MJ, Schweitzer PK (1992) Simulated assembly line performance following ingestion of cetirizine or hydroxyzine. Ann Allergy 69:195–200
Wauquier A, Van den Broeck WAE, Janssen PAJ (1980) Biphasic effects of pimozide on sleep-wakefulness in dogs. Life Sci 27:1469–1475
Witek TJ Jr, Canestrari DA, Miller RD, Yang JY, Riker DK (1992) The effects of phenindamine tartrate on sleepiness and psychomotor performance. J Allergy Clin Immunol 90:953–961
Wyatt RJ, Chase TN, Scott J, Snyder F, Engleman K (1970) Effect of L-dopa on the sleep of man. Nature 228:999–1001
Zarcone V, Dement W (1969) Sleep disturbances in schizophrenia. In: Kales A (ed) Sleep, physiology and pathology. Lippincott, Philadelphia, pp 192–199
Zarcone VP, Benson KL, Berger PA (1987) Abnormal rapid eye movement latencies in schizophrenia. Arch Gen Psychiatry 44:45–48
Zucconi M, Mondini S, Gerardi R et al. (1988) Niaprazine: a polysomnographic study of noctural sleep and daytime sleepiness in healthy volunteers. Curr Ther Res 44:118–132

CHAPTER 17
Antidepressant and Antimanic Drugs

R. Fritsch Montero, D. Riemann, and M. Berger

A. Introduction

Investigations on sleep have a long history. After the discovery of REM sleep (Aserinsky and Kleitmann 1953) interest in this phenomenon progressively increased. This interest was fostered when investigators noted that some CNS-active drugs changed the amount of REM sleep. An important group of such drugs is that of antidepressants. Subsequently the attention of many investigators has focused on this particular field. Studies testing the effects of antidepressants on sleep have been carried out in animals, normal humans, depressed patients and subjects suffering from other psychiatric disorders. The aims of these investigations have been heterogeneous and have included: evaluation of the effects of these drugs on sleep parameters; relationships between neurotransmitters and sleep regulation; drug classification based on their effects on sleep; and drug effects on sleep as indicators of treatment responsiveness.

Due to the heterogeneity of the subjects investigated, dosages of drugs, designs and statistical methods utilized unambiguous conclusions were very difficult to draw. Only few studies were carried out in a controlled manner, using large samples and standardized and therapeutic doses. Reviews of such studies (Kupfer 1981) based on the evaluation of adequacy of design, sufficient number of patients and choice of statistical procedure have indicated that not more than seven studies of tricyclic antidepressants have ever met such criteria. A review of these studies demonstrated that the available database did not constitute a body of evidence on which to base any conclusions on the possible interrelationships between EEG sleep and tricyclic administration (Kupfer 1981). Regarding this problem a systematic and critical description of the information accumulated so far on antidepressants and sleep is warranted to plan more systematic investigations.

The most commonly used antimanic drugs are lithium carbonate, carbamazepine and neuroleptics. In this chapter the effects of lithium carbonate and carbamazepine on sleep will be discussed. For a more detailed description of the effects of neuroleptics and carbamazepine on sleep see Chaps. 16 and 18, this volume.

B. The Antidepressant Drugs

I. Mechanisms of Action

The theoretical and clinical research of antidepressant drugs was initiated by the discovery of the antidepressant properties of imipramine and iproniazide by KUHN (1957) and KLINE (1958). Research on the mechanism of action of antidepressants has raised questions about early concepts of the pathogenesis of depression and the mechanism of action of antidepressant treatments. The monoamine deficiency hypothesis (SCHILDKRAUT 1965) was in part based on the premise that the therapeutic action of antidepressants was a consequence of an increased availability of synaptic norepinephrine or serotonin. This effect was conceptualized as a consequence of a high-affinity reuptake inhibition, by diminution of α_2-presynaptic receptors, or decreased catabolism as an effect of inhibition of monoaminoxidase. There is preclinical evidence that all antidepressants produce changes in monoamine receptor sensitivity, and the time course of these changes is similar to the time course of therapeutic actions of antidepressant drugs (GARVER and DAVIS 1979; MAAS 1979; SULSER and MOBLEY 1980; WALDMEIER 1981). Long-term antidepressant treatment has been found to reduce β-adrenergic sensitivity and the cyclic AMP response to norepinephrine (SUGRUE 1980). This type of drug has also been found to enhance responses to α-adrenergic, serotonergic and dopaminergic stimulation. On the other hand, clinical data indicate that postsynaptic α_1- and α_2-receptor systems may be subsensitive in depressed patients (GARCIA-SEVILLA 1989; CHARNEY et al. 1982). These abnormalities do not clearly improve with antidepressants. Some data suggest that depressed patients have an abnormality in the serotonergic system, which is normalized under antidepressant treatment. The data also suggest, however, that antidepressant drugs may act at the neurochemical level through different mechanisms.

II. Classification

Antidepressant drugs (ADs) can be classified from different points of view. One approach is a classification based on chemical structure, which leads to three types of drugs: tricyclic, tetracyclic and those with another structure. Another approach is to classify ADs on the basis of their neuronal mechanism of action at the synaptic level. A third approach is to take into account their affinity to different receptors. Different classifications permit an understanding of different collateral symptom profiles and the differential action of ADs on depressive symptomatology. A classification according to the action of ADs on depressive symptoms, for example, enables ADs to be subtyped according to their effects on psychomotor function. For our purposes two of these classifications are of special interest.

1. Sedative and Nonsedative Antidepressants

This classification is based on the differential effect of ADs on psychomotor function of depressed patients (Fig. 1).

a) Nonsedative and Activating Antidepressants

There are three types of antidepressants included in this grouping: nortriptyline, imipramine and tranylcypromine. The nortriptyline type includes nortriptyline, desimipramine and protriptyline. The imipramine type includes imipramine, clomipramine and dibenzepine. The third type includes tranylcypromine and other monoaminoxidase inhibitors (MAOIs). The secondary amines or amines of the nortriptyline type are the most activating antidepressants in comparison to the other tricyclics. Only the MAOIs are comparable in this aspect to this group (Kruse and Hoermann 1960). While the greater activating effect of imipramine in comparison to amitriptyline is well demonstrated (Angst et al. 1970), the differences between imipramine and desimipramine are less clear but have been postulated by many authors.

b) Sedative Antidepressants

This group includes two types of drugs: the amitriptyline and opipramol types. The first type is represented by amitriptyline, doxepine and trimipramine, the second by opipramol and trazodone. The sedative effects of the amitriptyline type are more potent than those of the other group.

2. Effects on Norepinephrine, Dopamine and Serotonin

Most tricyclic antidepressants inhibit norepinephrine reuptake to a variable degree (see also Chap. 4, this volume). The tertiary amines imipramine, clomipramine and amitriptyline are the most potent in this regard. It is important to mention that the tricyclic antidepressants, amineptine, butriptyline, iprindol, opipramol and trimipramine do not inhibit norepinephrine reuptake. Only a few antidepressants (i.e., bupropion and nomifensine) produce an inhibition of dopamine reuptake. No tricyclic agent produces this kind of inhibition. Only a few antidepressants do not affect

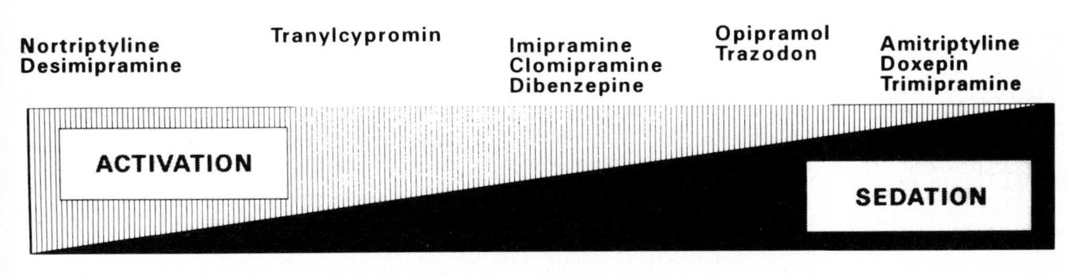

Fig. 1. Activating and sedative spectrum of antidepressants

serotonin reuptake (see also Chap. 5, this volume). The most potent tricyclic serotonin-reuptake inhibitor is clomipramine. Other compounds such as fluoxetine, fluvoxamine and zimelidine possess a comparable potency in inhibiting serotonin. (For an overview see Table 1.)

III. An Integrative Model for Depression

Three decades of biological-psychiatric research have confirmed the hypothesis that alterations of central nervous neurochemical processes play a prominent role in the induction and maintenance of depressive disorders. Although the influence of genetic factors is no longer controversial, until now no evidence has been found for predisposing alterations at a molecular level. However, a disturbance of the balance of neuronal transmission during the depressive episode appears to be certain. The extension of the classic amine depletion hypothesis to the theory of cholinergic-aminergic imbalance by taking into account transmitters, receptors and second-messenger mechanisms seems to describe most accurately the neurochemical basis of depressive disorders. Experimental animal studies on the interaction of stress response and neurochemical parameters support and extend a model proposed by Bohus and Berger (1992). Furthermore this model permits conclusions with respect to neurochemical correlates of psychosocial traumatas during childhood and adulthood (Fig. 2). This psychobiological model attempts to integrate the imbalance hypothesis of depression and the frequently observed REM sleep disinhibition by considering accepted psychological, social and biological concepts of depression. Central to this model is the cholinergic-aminergic imbalance, which can be triggered by somatic disorders, for example, puerperal conditions, depressiogenic drugs or psychological stressors. These conditions may produce an inhibition of the

Table 1. Effects of different antidepressants on histaminic receptors subtype 1 (H_1), muscarinic (M), α_1- and α_2-receptors, norepinephrine (NE) and serotonin reuptake

	NE reuptake	Serotonin reuptake	H_1	M	α_1	α_2
Clomipramine	↓	↓↓↓		↓↓	↓	
Zimelidine	↓	↓↓↓		↓		
Fluoxetine	=	↓↓↓↓				
Fluvoxamine	=	↓↓↓↓				
Desipramine	↓↓↓	↓↓		↓		
Imipramine	↓↓	↓		↓	↓	
Amitriptyline	↓↓	↓↓↓	↓	↓↓	↓↓	
Nortriptyline	↓↓↓	↓↓↓		↓	↓	
Trimipramine	(↓)	=	↓↓	↓↓	↓↓	
Trazodone	(↓)	↓		↓	↓	↓

Inhibition: ↓↓↓↓, extremely marked; ↓↓↓, very marked; ↓↓, marked to moderate; ↓, moderate; (↓), not known but supposed; =, no action.

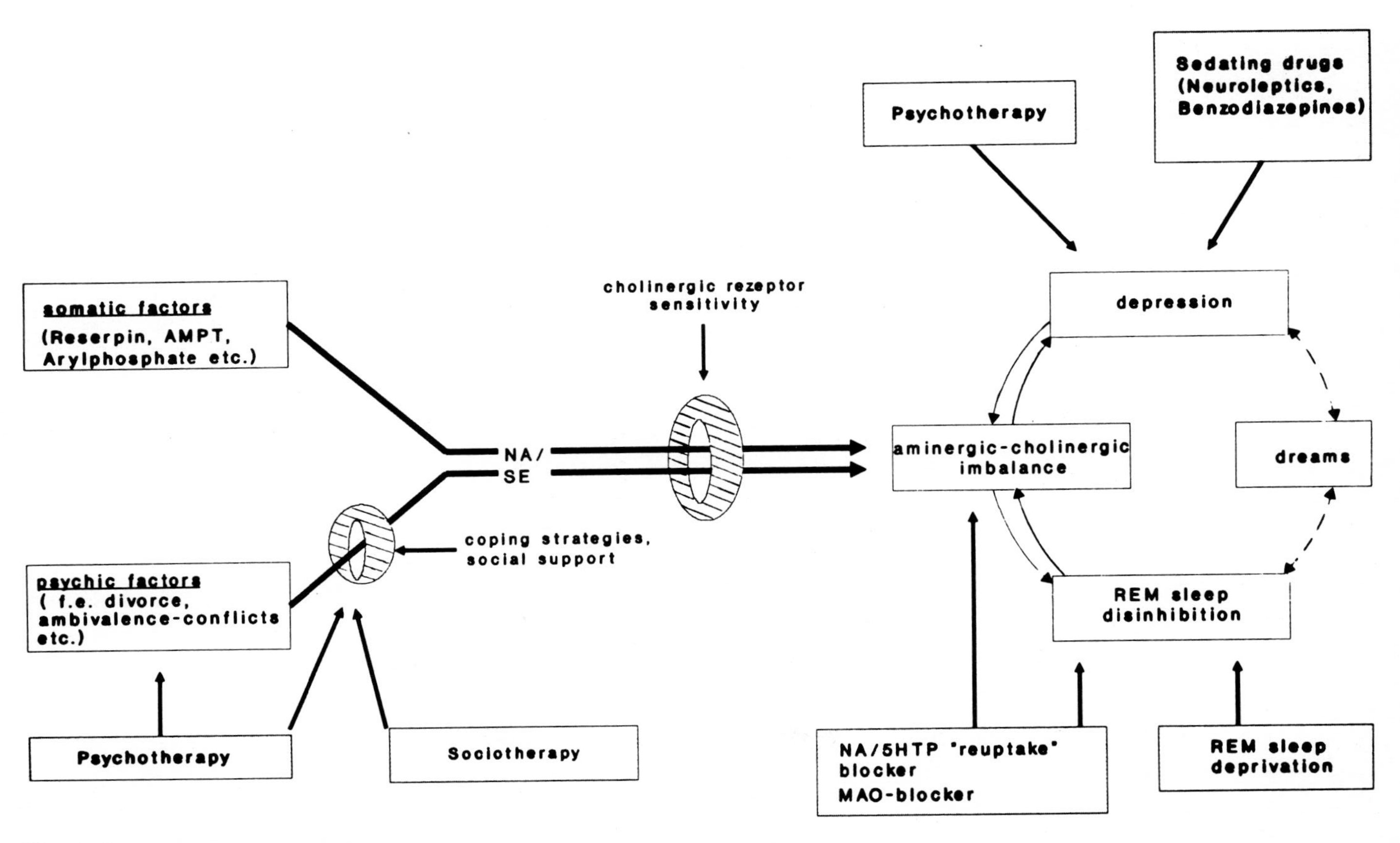

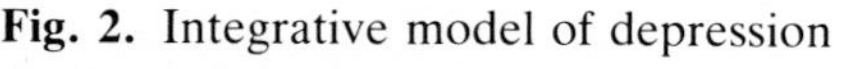
Fig. 2. Integrative model of depression

aminergic system and a stimulation of cholinergic neurotransmission. An important aspect of the model is the assumption of interindividual differences in the stability of the transmitter balance, e.g., a variable vulnerability. A supersensitivity of the cholinergic system is postulated to be a possible biological marker reflecting this vulnerability. The imbalance of the two systems can account for the REM sleep disinhibition encountered in depressed patients. According to this model, therapeutic interventions can be carried out on different levels. Classical antidepressants influence the aminergic/cholinergic imbalance by blocking the reuptake of amines or by an anticholinergic effect. This final common pathway may simultaneously lead to inhibition and suppression of REM sleep.

IV. Sleep and Depression

From a clinical point of view early morning awakening has long been advocated as a specific diagnostic sign for endogenous depression (KILOH and GARSIDE 1963) and for predicting a good response to electroconvulsive therapy. Polygraphic sleep studies by MENDELS and HAWKINS (1967) clearly demonstrated that early morning awakening could not be used to distinguish neurotic and psychotic depressives. Nevertheless, polygraphic sleep studies in depressed patients have consistently shown that, in comparison to normal controls, they have difficulty in falling asleep, frequent shifts of sleep stages, increased time spent awake, early morning awakening and a considerable reduction of stage 4 sleep (REYNOLDS and KUPFER 1987; GILLIN et al. 1984). Discrepancies exist in the literature with regard to other sleep variables such as total sleep time, REM sleep and REM latency. Although depressed patients have often been reported to have a reduced total sleep time, a proportion may in fact sleep normally at night or show hypersomnia (GARVER et al. 1984; HAWKINS et al. 1985).

REM sleep in depressed patients has been reported to be reduced (KUPFER 1976) or increased (HAURI and HAWKINS 1973). However, the increase in REM sleep in depressed patients has never been substantially or universally demonstrated. SNYDER (1972) suggested that depressed patients may suffer REM deprivation as a result of sleep curtailment in the initial waxing phase of the illness. This may lead to a gradual buildup of a hypothetical REM pressure that eventually leads to a compensatory REM sleep rebound during the waning phase of the illness. Thus the amount of REM sleep reported in any study of depressed patients may only reflect the condition of sleep at one point of their longitudinal clinical course. This hypothesis is contradicted by data of COBLE et al. (1979), who continuously investigated depressed patients for a period of 5 weeks, nightly. The authors concluded that at any time during the investigation the occurrence of REM sleep abnormalities was equal. As patients with sleep disorders who are not depressed do not show abnormalities of REM sleep (GILLIN et al. 1979), Snyder's hypothesis is not supported.

A shortened REM latency was frequently demonstrated in early investigations (SNYDER 1972; HARTMANN 1968a). A shortened REM latency was advocated forcefully as a psychobiological marker for depressive illness. Shortened REM latency was claimed to be present in virtually all drug-free patients suffering from unipolar and bipolar primary depressive illness. However, clinical conditions other than depression may also display a shortened REM latency, e.g., elderly psychotics (FEINBERG et al. 1965), healthy insomniacs (JONES and OSWALD 1968), patients with eating disorders (KATZ et al. 1984), obsessive-compulsive disorder (INSEL et al. 1982) and schizophrenia (ZARCONE et al. 1987; RIEMANN et al. 1991; TANDON et al. 1988) and healthy subjects following selective REM sleep deprivation.

KUPFER et al. (1976) described that, in a group of 18 unipolar depressive inpatients with shortened REM latency, the clinical response to amitriptyline could reliably be predicted by the degree of sleep changes in the first two nights during medication. Good responders showed more pronounced reduction of REM sleep, REM activity and a prolonged REM latency following medication. GILLIN et al. (1978) were able to replicate this finding in six depressed patients. HÖCHLI et al. (1986) found that the amount of REM sleep suppressed on the initial night of clomipramine treatment in comparison to baseline seems to be a good predictor of clinical response, a result which was, however, not confirmed in a larger sample by the same group (RIEMANN et al. 1990). The applicability of REM sleep suppression as a predictor for therapy response remains unclear for other antidepressants.

V. Effects of Antidepressants on Sleep

One of the most consistent findings in psychiatric sleep research is the suppression of REM sleep observed with many psychopharmacological agents (BERGER and RIEMANN 1993); of these drugs those that alleviate depression seem to be especially powerful. When placebo is substituted for an active drug, an excess of REM sleep associated with intense REM activity and shortened REM latency occurs (KALES et al. 1969; OSWALD 1969). This phenomenon is commonly referred to as REM sleep rebound, and can also be provoked by selective REM sleep deprivation. REM rebound has been put forward as evidence for hypothetical REM pressure, which increases linearly following prior REM sleep suppression (VOGEL 1975).

Several lines of gradually accumulating evidence suggested that REM sleep deprivation may be the biological process underlying the antidepressant properties of drugs which improve depression. VOGEL (1975) demonstrated that after three consecutive weeks of REM sleep deprivation by selective awakenings, patients were significantly less depressed than patients undergoing non-REM sleep deprivation. Another argument in favor of this hypothesis is that statistical comparisons across studies indicate that in terms of hospital discharge rates and reduction of Hamilton depression scores REM sleep deprivation and imipramine have similar efficacy (VOGEL et al. 1975).

Statistical comparisons of Hamilton depression score reductions in successive weeks of treatment showed no significant differences between REM sleep deprivation and imipramine treatment. Imipramine did not improve the mood of endogenous depressives who did not respond to REM sleep deprivation.

As a result of these findings it was hypothesized for many years that REM-sleep-suppressing properties were a prerequisite for an antidepressant drug, and, if a drug is able to suppress REM sleep, then it should have antidepressant properties (VOGEL 1983). This hypothesis, however, appears to be an oversimplification, because other REM-reducing drugs, such as barbiturates, amphetamines and meprobamate, do not possess antidepressant activity. On the other hand, some antidepressant drugs, such as trimipramine, trazodone and amineptine, do not change the amount of REM sleep (DUNLEAVY et al. 1972; WARE et al. 1985; WIEGAND and BERGER 1989; MOURET et al. 1988; WARE and PITTARD 1990; POIGNANT 1979).

1. Effects of Antidepressants on REM Sleep

As already mentioned, the most consistent effect of antidepressant drugs on sleep is their REM-sleep-suppressing property, which is common to almost all tricyclic, tetracyclic and atypical antidepressant agents, as well as to the MAOIs and the new serotonin-reuptake inhibitors. Suppression of REM sleep with these drugs has been reported for investigations carried out on animals, healthy individuals and depressed patients. There are three questions of importance concerning this point: the effects of acute and chronic administration and of discontinuation of antidepressants on REM sleep. The available data are not entirely consistent. Drugs that enhance catecholaminergic activity appear more likely to suppress REM sleep while those that reduce catecholaminergic activity are more likely to enhance it. WARE and PITTARD (1990) assumed that the two major factors in the suppression of REM sleep are a decrease in cholinergic activity and an increase in noradrenergic activity. They ranked antidepressants according to their relative ability to provoke these changes, but studies concerning this issue have not been carried out to date.

It has been suggested that the timing of REM sleep cycles is controlled by cholinergic mechanisms, so that a lengthening of REM latency would be considered an anticholinergic effect (SITARAM et al. 1977). The data presented below do not support the notion that only drugs with strong anticholinergic properties significantly change REM sleep timing. Even drugs with a relatively minor cholinergic effect lead to prominent effects on REM sleep suppression. Differential effects on REM sleep may, however, represent the varying degree of anticholinergic properties of different antidepressants. In a study by KUPFER et al. (1982) REM latency changed in a differential manner, according to the different anticholinergic properties of the drugs studied (amitriptyline and nortriptyline). This finding was not confirmed in a study involving fluvoxamine and desipramine (KUPFER et al. 1991a).

a) Effects of Acute Administration

Studies simultaneously comparing different antidepressants are sparse. The most studied antidepressants are tricyclic drugs. SCHERSCHLICHT et al. (1982) undertook one of the most comprehensive investigations in animals. A whole series of antidepressant drugs and their impact on REM sleep was studied in cats. The drugs studied were amitriptyline, clomipramine, imipramine, desipramine, nortriptyline, citalopram, iprindole, maprotiline, binodaline, diclofensine, nomifensine, zimelidine, mianserin, pargyline, femoxetine, fluvoxamine, paroxetine and viloxazine. For all antidepressants studied a dose-dependent REM sleep suppression was demonstrated. The highest dose used in most cases completely abolished REM sleep. The tertiary amines, e.g., clomipramine, amitryptiline and imipramine, proved to be more potent than secondary amines, e.g., desipramine or nortriptyline. Clomipramine is a tricyclic antidepressant with weak anticholinergic properties and is the most potent tricyclic serotonin-uptake inhibitor. Its major metabolite, desmethylchlorimipramine, is a potent inhibitor of noradrenergic uptake (HYTTEL 1982). The effect of clomipramine on REM sleep was confirmed in other animal species such as rats (KHAZAN and BROWN 1970; MIRMIRAN et al. 1981; HILAKIVI and SINCLAIR 1984). DIJK et al. (1991) replicated this property of clomipramine in rats. In this study, animals were sleep deprived for one night during several days of drug treatment; there was a tendency towards an increase in REM sleep during the sleep deprivation recovery night, despite administration of clomipramine. After sleep deprivation alone no effect on REM sleep could be demonstrated. The suppressing effect of clomipramine on REM sleep was abolished by sleep deprivation. The authors speculated that this unexpected effect was related to the enhanced pressure for REM sleep caused by the combined action of sleep deprivation and clomipramine. Due to the methodological problems of this study, conclusions are difficult to draw. The animals received the drug orally and the administered dose was only estimated. But the question is still open whether drugs, which under normal conditions suppress REM sleep, exert the same effect following sleep deprivation.

Depressed patients have alterations of REM sleep characterized by shortened REM latency and augmented amounts of REM sleep in the first half of the night. In these patients clomipramine produced a suppression of REM sleep as pronounced as in animals. KUPFER et al. (1989) studied the effect of 150 mg during the first night and 200 mg clomipramine during the second night in 19 depressed patients. During the first two nights after administration, REM time, REM activity and number of REM periods were significantly diminished, whereas REM latency was significantly prolonged. HÖCHLI et al. (1986), by studying the effects of clomipramine in ten depressed patients, demonstrated the potent REM-suppressing efficiency of even lower doses administered orally. A dose of 75 mg clomipramine led to total REM sleep suppression in four of the ten patients.

Very low i.v. doses of clomipramine produce similar effects. After 12.5 mg i.v., KUPFER et al. (1991b) found in eight depressed patients a significant reduction in REM time, REM density, REM activity and number of REM periods. REM latency was significantly prolonged. Desipramine, a tricyclic antidepressant, is a serotonin-reuptake inhibitor, possesses some anticholinergic activity and markedly inhibits norepinephrine reuptake. In healthy volunteers desipramine led to an immediate suppression of REM sleep (ZUNG 1969). Its effect is maximal during the first night and declines in later nights in depressive patients (KUPFER et al. 1980).

Imipramine, another tricyclic antidepressant, an important norepinephrine-reuptake inhibitor, which additionally possesses some serotonin reuptake and anticholinergic activity, produced a prolongation of REM latency and a decrease in REM sleep time and number of REM periods in animals (HISHIKAWA et al. 1965; HILL et al. 1979). In order to clarify how far anticholinergic activity of this compound plays a role in suppressing REM sleep, HILL et al. (1980) studied the effects of different single doses of imipramine and physostigmine in rats. Physostigmine is an inhibitor of cholinesterase activity and induces REM sleep in normal men (SITARAM et al. 1976). The investigators found no changes in REM sleep after administration of low doses of imipramine. However, when such doses were administered in combination with physostigmine, a decrease in REM sleep was observed. This reduction in REM was almost certainly a direct result of both the longer latency to sleep and the reduced total sleep time observed. Of principal interest was the fact that significantly longer REM latency was observed following administration of a higher dose of imipramine and, furthermore, that this effect could be antagonized by administration of physostigmine. This antagonism occurred even though physostigmine did not, by itself, alter REM latency. These findings indicate that anticholinergic activity may play an important role in suppressing REM sleep, but certainly not exclusively. But anticholinergic mechanisms may be involved in the effects of many antidepressant drugs. They can explain the different potency of drugs investigated by SCHERSCHLICHT et al. (1982). When anticholinergic activity was compared, clomipramine and amitriptyline were more potent than desipramine and nortriptyline. According to SCHERSCHLICHT et al. (1982), clomipramine and amitriptyline led to a more pronounced suppression of REM sleep than desipramine and nortriptyline.

The tricyclic antidepressant, amitriptyline, a tertiary amine, is an important serotonin-reuptake blocker and anticholinergic drug, with only weak noradrenergic-reuptake blockade. The acute impact on REM sleep is similar to that described for the other tricyclic antidepressants. The first study investigating its effects on REM sleep was carried out by HARTMANN (1968b). Paradoxical sleep was reduced by an average of 60%–70% in comparison to baseline levels when 75 mg or 50 mg was administered to healthy subjects. A decreased number of REM periods and a significant prolongation of REM latency was observed (HARTMANN and CRAVENS 1973). A dose of 35 mg did not suppress REM sleep (Saletu 1976). It is still an open question whether the failure to affect REM sleep was due to the small dose used or the time of administration. In depressed patients evening doses of 50 mg are capable of significantly suppressing REM sleep and REM activity, diminishing the number of REM periods and prolonging REM latency. REM density appeared to be not affected by the drug (Kupfer et al. 1978, 1981). A relatively new antidepressant drug, amitriptyline oxide, did not exhibit any REM-sleep-suppressing properties when given at a dose of 60 mg/day (3 × 20 mg) to endogenously depressed patients (Jovanovic and Schulte 1978). This drug itself has almost no noradrenergic reuptake-inhibiting properties and only weak anticholinergic and serotonin reuptake-inhibiting effects. But this result was surprising, because this drug is rapidly metabolized to amitriptyline, which strongly suppresses REM sleep, as mentioned above. A possible explanation of this apparent paradox is that, when equal doses of amitriptyline and amitriptyline oxide were given, the drugs in the CNS showed a ratio of 4:1, respectively (Melzacka and Danek 1983). Riemann et al. (1990) has demonstrated the ability of both drugs to suppress REM sleep. They compared the effects of comparable evening single doses of amitriptyline (75 mg) and amitriptyline oxide (150 mg) over two nights in healthy volunteers. Both drugs produced REM sleep suppression, but the effect of amitriptyline was more pronounced. Amitriptyline led to a relative decrease in REM sleep to 30% and amitriptyline oxide to 60% of placebo values. This is in accordance with the fact that after equal doses the CNS concentration ratio of amitriptyline to amitriptyline oxide is 4:1. REM latency was significantly lengthened after administration of both drugs (increment of 74% for amitriptyline oxide and 84% for amitriptyline). REM latency does not seem to show proportional changes to the different CNS drug concentrations, as do the amounts of REM sleep. The amount of REM sleep seems to be more representative of central nervous effects than REM latency, indicating that when variables such as administration time are controlled and plasma and CNS concentrations are considered for adequate dose selection, REM percentage will be a good indicator for CNS activity.

Nortriptyline blocks serotonin and norepinephrine reuptake and has some anticholinergic activity. This latter property is much less important than with amitriptyline. Kupfer et al. (1982) investigated the differential effects of both drugs on REM sleep of depressive patients. Twenty patients received increasing evening doses of nortriptyline and 49 patients received 50 mg amitriptyline in the evening, which were compared with the 25-mg nortriptyline period. This drug produced, during three dose periods (25, 50, 75 mg), a significant reduction of percentage REM and number of REM periods, and a significant lengthening of REM latency. REM activity was significantly decreased under the lowest dose and showed a rapid progressive tolerance, again reaching baseline levels at the highest dose. For other REM sleep parameters, e.g., percentage REM and REM latency, no tolerance during acute administration developed. Amitriptyline, compared with nortriptyline, significantly lengthened REM latency. However, none of the other REM sleep measures, such as REM activity or REM density, were differentially affected despite the different anticholinergic properties of the drugs.

Investigations indicate that some antidepressants, two of them tricyclics, produce no apparent acute changes in REM sleep (e.g., trimipramine trazodone, amineptine). Trimipramine has only weak effects on norepinephrine and serotonin reuptake. It is a strong adrenergic-receptor blocker and has marked antagonistic H_1 activity. Its action is centered on the α_1-receptors. DUNLEAVY et al. (1972) reported that in two healthy subjects trimipramine did not affect REM sleep. Further studies on depressive patients confirmed this observation. WARE et al. (1985) compared trimipramine and imipramine effects on REM sleep in 30 depressed insomniac patients. Under trimipramine, percentage REM and REM latency were not affected. WIEGAND and BERGER (1989) reported similar findings in ten depressive patients after 2 days of treatment. No alteration of REM latency was observed. Percentage REM even increased. The authors explained this phenomenon as a normalization of REM sleep, which happened to be abnormally low at baseline in the sample investigated. MOURET et al. (1988) studied the effect of 150 mg trimipramine in seven depressive patients. They found a lengthening of REM latency, but no alteration of percentage REM. The lengthening of REM latency was interpreted as a consequence of SWS increment. Effects of trimipramine on other sleep stages will be discussed later.

Amineptine, the other tricyclic antidepressant which does not affect REM sleep, only presynaptically enhances dopamine release and inhibits serotonin reuptake. When administered in rats an increase in REM sleep can be observed directly after administration (POIGNANT 1979). DI PERRI et al. (1986) confirmed these data in 12 depressive patients.

Trazodone, a triazolopyridine derivative, is a potent adrenergic-receptor blocker (α_1) as is trimipramine, but is not a strong norepinephrine-reuptake blocker. Compared with tricyclic antidepressants, it has much lower affinity to muscarinic receptors. WARE and PITTARD (1990) compared three different doses of trazodone (50, 100, 200 mg) with three doses of trimipramine (25, 50, 100 mg). Both drugs exerted no influence on REM sleep.

The new selective serotonin-reuptake inhibitors led to a suppression of REM sleep comparable to tricyclics. Fluoxetine, a potent serotonin-uptake inhibitor, in dosages between 30 and 80 mg, reduced REM sleep in normal subjects (NICHOLSON and PASCOE 1988; BARDELEBEN et al. 1989). SALETU et al. (1991) compared the effects of two serotonin-uptake inhibitors in 18 healthy subjects, i.e., three different morning doses (20, 30 and 40 mg), and a 30-mg evening dose of paroxetine and 40 mg fluoxetine in the morning. The half-life of paroxetine is around 16 h and its metabolites are much less potent than the parent compound and do not contribute to its activity. In animals and man it produces a pronounced suppression of REM sleep

(KLEINLOGEL and BURKI 1987; OSWALD and ADAM 1986). When 40 mg fluoxetine was given in the morning before recording, only a significant increase in REM latency occurred. On the other hand, 30 mg paroxetine administered in the evening produced a marked suppression of REM sleep and no effect on REM latency, in contrast to the morning dose, which produced a marked suppression of REM sleep and a prolongation of REM latency. During the withdrawal night after the evening dose REM latency was still significantly prolonged and REM sleep remained diminished. It is possible that this variability of effect on REM sleep with time depends on plasma concentrations. Because of the dynamic changes in plasma levels the action on the CNS may vary dependent on time. This variation may be independent from the sleep-wake schedule. These data underline the problem of using REM sleep, especially REM latency, as an indicator of the drugs' central action. REM sleep parameters can only be measured during a limited time, e.g., the "window" for REM sleep is narrow and "cannot be open" every time. Because of the strong time dependence of changes that will be observed, the "landscape" is very dynamic. Administration time then has a very important role in the studies that investigate single-dose effects, as demonstrated in this study. Unfortunately only few studies consider this point.

Fluvoxamine, a relatively new serotonergic reuptake-blocker and an atypical antidepressant, has the same capacity to suppress REM sleep as the other serotonin-uptake inhibitors described above (BERGER et al. 1986). KUPFER et al. (1991a) compared the effects of fluvoxamine and desipramine in 40 depressed patients. The effects of 100 mg desipramine and 200 mg fluvoxamine on REM sleep were immediate and comparable, and a reduction of REM time, percentage REM, REM activity and number of REM periods was observed. But the substances differed with regard to the latency to exert a maximal suppressing effect. The desipramine effect was maximal during the first drug night and declined during the second drug night, whereas fluvoxamine exerted its maximal effect during the second drug night. It is possible that active metabolites of fluvoxamine potentiate the effect on REM sleep not before the second night, or this effect may be due to a longer half-life of this drug, or a longer absorption time may produce a delay of its effects on REM sleep.

Zimelidine, another antidepressant, with a similar receptor specificity, e.g., strong serotonin-uptake inhibition and weak noradrenergic-uptake inhibition and anticholinergic activity, is capable of suppressing REM sleep in newborn rats (HILAKIVI et al. 1987) and in depressive patients (KUPFER 1982).

Only a few studies have been carried out using MAOIs. The drugs that have been studied in normal and depressed subjects include phenelzine (DUNLEAVY and OSWALD 1973; KUPFER and BOWERS 1972; WYATT et al. 1969, 1971), moclobemide (HOFF et al. 1986; MONTI et al. 1990), clorgyline (COHEN et al. 1982), pargyline (COHEN et al. 1982), nialamide (AKINDELE et al. 1970), brofaromine (Steiger et al. 1987; Hohagen et al. 1993) and tranylcypromine (Gassicke et al. 1965). All investigators reported a significant suppression of REM sleep and prolongation of REM latency after administration. This effect became maximal not initially but after many days. The acute effects were not systematically investigated and results varied substantially across different studies. For example, Wyatt et al. (1969) studied the effects of four different MAOIs in five subjects. From night 1 to night 5 an increase in REM sleep latency appeared with a consecutive progressive diminution of REM sleep. On the other hand, Hoff et al. (1986) studied the effects of two selective and reversible MAO-A inhibitors, cimoxaton and moclobemide, a selective inhibitor of MAO-A, in 20 depressive patients. After a single morning dose of cimoxaton they observed a significant decrease of REM sleep and prolongation of REM latency during the subsequent night. After 300 mg moclobemide (given in three portions, last dose at 5 p.m.) only a tendency toward suppression of REM sleep and prolongation of REM latency was observed. This difference can be explained by the different halflifes of the drugs used and the times of administration. As moclobemid has only a short half-life of about 2 h (compared with 12 h for cimoxaton) and the last dose was administered at 5 p.m. according to the study design, the pharmacological influence may already have disappeared by the time the sleep recording started. Similar findings with regard to moclobemide-suppressant properties in depressed patients were reported by Monti et al. (1990). Steiger et al. (1987) found a positive correlation between the amount of REM sleep suppression and plasma concentrations in healthy volunteers under treatment with brofaromine, another selective inhibitor of the MAO-A.

In this section we have reviewed the acute effects of different antidepressants on sleep. The majority of studies carried out so far indicate that almost all antidepressants produce a rapid suppression of REM sleep and prolongation of REM latency in animals, healthy subjects and patients suffering from depression. Only three drugs appear not to alter REM sleep, these being trimipramine, trazodone and amineptine. But many methodological problems reported in the studies may contribute to divergent results. The most important problem may be administration time. To assess the acute effect of drugs on sleep, it is desirable that sleep recording coincides with high plasma levels of the drug studied. Therefore, the pharmacokinetics must be studied and dosage and administration time should be decided accordingly.

b) Effects of Chronic Administration

The duration of studies varies from 1 week to many months. Very long studies are sparse and the majority of investigations have taken around 4 weeks, due to the habitual duration of response onset to antidepressant drugs. The effects of chronic administration are similar to those of acute administration, e.g., sustained REM sleep suppression and prolongation of REM latency can be observed with almost all antidepressants. KUPFER et al. (1991a), in their study comparing fluvoxamine and desipramine, observed that whereas after 3 weeks of drug administration the total REM counts for the whole night were restored to predrug levels in patients treated with desipramine; in the fluvoxamine-treated patients they were not restored to the predrug levels. The very potent and persistent effects of fluvoxamine on inhibition of serotonin uptake in the synaptosomal preparations may account for its more persistent effect on REM suppression. On the other hand, desipramine acts both noradrenergically and moderately anticholinergically, whereas the action of fluvoxamine is primarily serotonergic. Perhaps both neurochemical (noradrenergic and anticholinergic) effects enable total REM counts for the whole night to be restored to predrug levels of desipramine.

MAOIs seem to produce a similar sustained REM inhibition as described for tricyclic and serotonin-reuptake inhibitors. MONTI et al. (1990), however, observed in depressed patients a significant prolongation of REM latency after 1 month of treatment with moclobemide. The amount of REM sleep showed a tendency to increase, but not significantly so, in contrast to the marked REM suppressant properties of the selective MAO-A inhibitors, clorgyline and cimoxatone (COHEN et al. 1982; HOFF et al. 1986).

In the case of antidepressants that do not produce an initial suppression of REM sleep, no posterior inhibition of REM sleep occurs. DI PERRI et al. (1986) noted a significant increment of REM sleep after 14 days of treatment with amineptine in depressed patients. WIEGAND and BERGER (1989) noted an increment in the REM latency and amount of REM sleep after 11 and 21 days treatment with trimipramine in patients with major depression, contradicting the findings of WARE et al. (1985), who found no modification of REM sleep after 4 weeks treatment. This discrepancy can be explained as a normalization of REM sleep, in particular REM latency, in the patients studied by WIEGAND and BERGER (1989), as a consequence of the improvement of depression, and not as a direct pharmacological effect as demonstrated for most antidepressants in the acute phase of administration. The increment in the amount of REM sleep can be explained as a consequence of the decrease in wake time and probably an improvement of sleep continuity. The baseline sleep patterns were characterized by a pronounced disturbance of sleep, which resulted in a decrease in REM sleep prior to treatment.

c) *Effects of Discontinuation*

Many authors have mentioned an increase in REM sleep and a shortening of REM latency after discontinuation of drugs that suppress REM sleep, as an expression of a high REM pressure. This phenomenon is also known as REM rebound. Studies on this point are sparse and involve only few patients or controls. For example, HARTMANN and CRAVENS (1973) reported that after 28 days administration of amitriptyline to normals, a significant increase in REM sleep and shortening of REM latency can be observed in the three nights after discontinuation. KUPFER et al. (1991b) observed a persistence of the prolongation of REM latency and suppression of REM sleep on the first follow-up night after a single dose of clomipramine in depressive patients. After two nights of oral administration of clomipramine, a gradual restoration of REM sleep began to occur during the three nights after discontinuation. However, even at the end of 72 h, only 25%–30% of REM sleep time and REM activity was restored. Thus, it is clear that no acute rebound of REM sleep occurred during this period. It is not known whether the absence of REM rebound is specific for depressed individuals or whether it represents a general property of clomipramine withdrawal following a pulse-loading procedure.

In the case of the MAOIs, many authors reported that after treatment with irreversible MAOIs a total abolition of REM sleep may persist for several days after withdrawal and a REM rebound may occur only later (COHEN et al. 1982). When reversible and short-acting MAOIs were tested, a REM rebound was observed immediately after discontinuation of the drug, in a parallel manner to the decrease in plasma concentrations (STEIGER et al. 1987). Despite the mild REM suppression observed with moclobemide, a REM rebound occurs after discontinuation (MONTI et al. 1990). This exception to the other MAOIs can be explained as a downregulation of aminergic receptors during drug administration in a similar fashion to the tricyclic antidepressants.

2. Effects of Antidepressants on Other Sleep Stages

The sedative antidepressant amitriptyline, when given acutely to depressive patients, produces an immediate significant increase in sleep efficiency (SE), sleep period time (SPT) and sleep stage 2. After 4 days treatment a diminution of early morning awakening (EMA) and the amounts of stage 1 can be observed (KUPFER et al. 1978). This effect increases after 2 weeks treatment. A diminution of sleep onset latency (SOL) was evident not before 2 weeks, under 200 mg of the drug. In normal subjects this effect does not occur (HARTMANN and CRAVENS 1973; RIEMANN et al. 1990).

Trimipramine produces similar changes in depressed patients, i.e., an increase in SPT, total sleep time (TST) and sleep efficiency, and a decrease in SOL, EMA and amount of wake time after 3 days of treatment (MOURET et al. 1989; WIEGAND and BERGER 1989).

Doxepine leads to a restoration of sleep continuity, but the effect does not seem to be stronger than with imipramine. This compound produces a transient improvement of SOL, but a subsequent deterioration of SOL and SE can be demonstrated (WARE et al. 1985).

KUPFER et al. (1989) found that clomipramine modified slow-wave sleep (SWS). When the delta wave activity was analyzed across 60-min intervals, on the second drug night and the second withdrawal night a major shift of delta wave sleep to the 1st and 2nd h of sleep following the active drug was observed.

Amitriptyline and zimelidine have also been found to strengthen the rate of production of delta wave activity in the first non-REM period, thereby moving the distribution of delta wave activity during the night in a more normalized direction (KUPFER et al. 1987). It was found that the mean power of the three delta bands can be correlated significantly with ultimate clinical response, postulating it as an important predictor of eventual clinical response.

The activating antidepressant, nortriptyline, surprisingly does not produce a deterioration of sleep continuity, which remained unaltered under 25, 75 and 100 mg nortriptyline. But in comparision there was a marked difference in sleep continuity effects of amitriptyline compared with nortriptyline, with the former drug improving SE and TST (KUPFER et al. 1982).

Fluvoxamine produces an initial prolongation of SOL, an increase in the amount of wake time and a decrease in sleep efficiency. These alterations were restored to predrug levels after 7 days of treatment.

VI. Use of Antidepressants in Sleep-Related Disorders

1. Psychophysiological Insomnia

The widely used benzodiazepines present many problems in the treatment of patients suffering from insomnia. These include an altered sleep architecture, the development of treatment resistance after a few days of use, rebound insomnia after discontinuation and a high risk of dependence. These problems stimulated several investigators to search for alternative treatments for insomnia. As already mentioned, some antidepressants, principally the sedative antidepressants, improve sleep continuity. Based on this observation we carried out a study on 17 insomniacs (HOHAGEN et al. 1994). They were administered trimipramine for 4 weeks. Then the active drug was discontinued and replaced by placebo for the subsequent 2 weeks. To assess for the hypnotic effects, and short- and long-term rebound insomnia, seven polygraphic recordings were carried out during the study period. A significant improvement of TST and SE, a diminution of wake time and an increase in sleep stage 2 was observed after 4 weeks of treatment. REM sleep remained unaltered, i.e., it seemed that treatment with trimipramine restored a physiological sleep architecture. Subjective reports of patients agreed with the objective data. After drug discontinuation a progressive deterioration of sleep continuity was noted, but the values did not exceed baseline levels, so that no rebound insomnia occurred. These results are promising and indicate that sedative antidepressants can represent an alternative to benzodiazepines, without their inherent problems.

2. Sleep Apnea Syndrome

In the early 1970s, several case reports suggested that tricyclic antidepressants might be useful in the treatment of sleep apnea. Since protriptyline is less sedative than most tricyclics (SCHERSCHLICHT et al. 1982), it seemed an appropriate agent for clinical trials. Its usefulness has been confirmed, though the mechanism of action is not certain (CLARK et al. 1979). It seems to decrease the frequency of disordered breathing events, in non-REM sleep, and may be of help in patients whose apneas are largely confined to non-REM sleep. Side effects that should be considered include anticholinergic symptoms such as dry mouth or cardiac alterations, which is of particular concern, since patients with sleep apnea may already be more likely to experience nocturnal arrhythmias.

3. Narcolepsy

Narcolepsy is a disorder of excessive sleepiness in combination with other auxiliary symptoms (KALES et al. 1982). The sleepiness is often expressed as sleep attacks, episodes of a seemingly irresistible need to sleep, usually lasting about 15 min or less, from which the patient often awakens feeling at least briefly refreshed. Another very disturbing symptom is cataplexy, which is characterized by a brief loss of muscle tone, occurring frequently in association with the experience of strong emotions. In general, patients retain consciousness during these episodes, although during longer ones visual hallucinatory experiences may occur. Other symptoms include sleep paralysis and hypnagogic or hypnopompic hallucinations, vivid dream-like experiences that occur at these times. Patients may also display episodes of automatic behavior, associated with microsleep episodes. Paradoxically, some narcoleptics may complain of disturbed nocturnal sleep. REM sleep occurring in the first few minutes of sleep occurs in the majority of patients during nocturnal sleep.

The traditional pharmacotherapy for narcolepsy is determined by the target symptoms. Amphetamines are used for daytime sleepiness. To treat cataplexy, tricyclic antidepressants are often used. Perhaps the most widely administered compound has been imipramine, usually in doses of 50–100 mg. Protriptyline and clomipramine are employed by many clinicians with acceptable results. In recent years it has been demonstrated that the selective MAOIs, such as brofaromin or moclobemid, prevent somnolence and secondary symptoms in a considerable number of the patients treated (HOHAGEN et al. 1993).

C. Antimanic Drugs

Sleep EEG data suggest similarities of lithium with carbamazepine, and differences compared with the antidepressants. Both antimanic drugs produce an increase in SWS (KUPFER et al. 1974; FRISTON et al. 1989; YANG et al. 1989; RIEMANN et al. 1993), which is not observed with the tricyclics or the monoaminoxidase inhibitors. The SWS-enhancing effect is interpreted as the result of an inhibition of cGMP production.

I. Lithium

Lithium carbonate therapy is an accepted strategy for manic-depressive illness and the drug of choice in acute mania and chronic recurrent mania. Changes in sleep often signal the termination of a manic or depressive episode. Serum lithium carbonate levels of 0.7 mEq/l or greater are associated with REM suppression most clearly reflected in reduction of percentage REM sleep (CHERNIK and MENDELS 1974) in normals (ITOH et al. 1987), depressed patients (KUPFER et al. 1970; MENDELS and CHERNIK 1973) and subjects with mania (HUDSON et al. 1989).

Despite the considerable reduction in REM sleep during the drug period, abrupt discontinuation of lithium is associated with an immediate restoration of REM sleep, but not a compensatory REM rebound (KUPFER et al. 1974). This compensation may not occur within 1 month after discontinuation. This lack of compensatory rebound after a prolonged REM suppression is unusual, as most psychotropic drugs which suppress REM sleep show a REM rebound (KALES et al. 1969; OSWALD 1969). The REM sleep suppression observed with lithium is due to the decreased length of each REM period, rather than a lengthening of the inter-REM intervals. The latter effect is typical for antidepressant agents.

KUPFER et al. (1974) demonstrated an increase in delta sleep under lithium treatment. During the drug period there was a significant inverse correlation between percentage REM and percentage delta sleep. While delta sleep latency did not change significantly on lithium, REM latency changed significantly and also correlated with serum lithium levels, findings that suggest REM sleep changes are more responsive to varying dosages of lithium than delta sleep. But lithium by itself should not be considered a sedative drug. There are no significant changes in the time spent asleep during the medication periods. In normal subjects this increment in delta sleep was not observed by BERT et al. (1963). FRISTON et al. (1989) reported an increase in SWS in healthy subjects. This increment is not a hypnotic effect of lithium because wake after sleep onset was significantly higher under treatment. Authors postulated an inhibition of 5-HT_2 receptors as being responsible for this increase.

There is no evidence at present to suggest that lithium may act, through the modification of the sleep stages it induces, on the mood of subjects with affective disorders. It could be that lithium acts by correcting the abnormally phase-advanced temporal distribution of REM sleep relative to the rest of sleep (WEHR and WIRZ-JUSTICE 1982). In fact it has been shown that lithium alters the internal phase angles between different rhythm systems, either by delaying one oscillator differentially or in altering rhythm-coupling parameters (KRIPKE 1983).

In recent years it has been claimed that lithium might be able to prevent the recurrence of the hypersomniac episodes associated with overeating and mental disturbances typical of Kleine-Levin syndrome. Up to now successful results have been reported either in incomplete cases of the syndrome (OGURA et al. 1976; GOLDBERG 1983) or in typical cases (ROTH et al. 1980; HART 1985), with a disappearance of the symptoms while on medication and a relapse when the drug was discontinued.

II. Carbamazepine

Carbamazepine, a drug commonly used in several neurological and psychiatric illnesses, exhibits acute antimanic and long-term prophylactic effects in the treatment of mono- and bipolar affective disorders. It has been suggested that carbamazepine may have acute antidepressant properties (POST 1987), and a significant association between the response to sleep deprivation and the response to carbamazepine has been observed in depressed patients (ROY-BYRNE et al. 1984). The mechanism of action is so far unknown. In animals and normal subjects carbamazepine led to an improvement in parameters of sleep continuity (GIGLI et al. 1988). The impact on SWS is prominent (YANG et al. 1989), doubling baseline values after 5 days of treatment (RIEMANN et al. 1993). These effects resemble those observed under lithium treatment. Under higher doses of the drug, a reduction in REM sleep can be observed (YANG et al. 1989) but not under 400 mg daily. In the case of abnormal REM sleep parameters at baseline, as in the case in rapid-cyclers, a normalization of shortened REM latencies was observed (RIEMANN et al. 1993).

D. Summary

It is now widely accepted that almost all antidepressant and antimanic drugs influence sleep. The present chapter gives an overview of the effects of different types of antidepressant and antimanic drugs on polysomnographically recorded sleep. After briefly reviewing mechanisms of action and classifications of antidepressant drugs, an interactive psychobiological model for depressive disorders is presented, which focuses especially on the effects of antidepressant drugs on REM sleep. This model is based on the assumption of a cholinergic-aminergic neurotransmitter imbalance which is held responsible for depressive and manic states. When combining this model with data from animal experiments on REM sleep regulation, the REM sleep abnormalities (i.e., shortened REM latency and enhanced REM density) often observed in depressed patients may be conceived as the result of a cholinergic hyperfunction or a muscarinic supersensitivity in depressive disorders.

The effects of antidepressants on sleep and especially REM sleep are reviewed with respect to the duration of administration (acute versus chronic). Summarizing, it can be stated that most of the antidepressant durgs suppress REM sleep; however, there are a few exceptions to this rule, e.g., trimipramine.

Our chapter also gives a brief overview of the effects of antidepressants in sleep disorders which are not primarily due to depression, such as psychophysiological insomnia and narcolepsy. Whereas for psychophysiological insomnia promising results with sedative antidepressants (for example, doxepine or trimipramine) have been published, for narcolepsy substances with marked REM sleep suppressing and vigilance-activating properties (for example, clomipramine, fluvoxamine or fluoxetine) seem to be helpful.

Concerning antimanic drugs, much less is known in comparison to antidepressant drugs. Data for lithium and carbamazepine are summarized, which convincingly demonstrate that both types of drugs enhance slow-wave sleep and only slightly influence REM sleep.

Although to date many studies have been conducted on the effects of antidepressant and antimanic drugs on sleep in healthy subjects and depressed and manic patients, there are still many unexplored areas. It is still unknown how the long-lasting administration of antidepressant and antimanic drugs affects sleep. Furthermore, the effects of discontinuation of antidepressant and antimanic drugs on sleep have not been fully elucidated and studies utilizing sophisticated methods such as spectral analysis of the sleep EEG are still scarce.

References

Akindele M, Evans J, Oswald I (1970) Mono-amine oxidase inhibitors, sleep and mood. Electroenceph Clin Neurophysiol 29:47–56

Aserinsky E, Kleitmann N (1953) Regularly occurring periods of eye motility, and concomitant phenomena, during sleep. Science 188:273–274

Angst J, Theobold W, Bleuler M, Kuhn R (1970) Tofranil (imipramine). Stampfli, Bern

Bardeleben U, Steiger A, Gerken A, Holsboer F (1989) Effects of fluoxetine upon pharmacoendocrine and sleep EEG parameters in normal controls. Int Clin Psychopharmacol 4:1–5

Berger M, Riemann D (1993) REM sleep in depression: An overview. J Sleep Res 2:211–223

Berger M, Emrich H, Lund R, Riemann D, Lauer C, von Zerssen D (1986) Sleep-EEG variables as course criteria and predictors of antidepressant therapy with flauvxomine/oxoprotiline. In: Stille G, Wagner W, Herrmann W (eds) Advances in pharmacotherapy. Karger, Basel, pp 110–120

Bert J, Saier J, Dufour J, Scotto J, Julien R, Sutter J (1963) Modifications du sommeil provoquees par le lithium en administration aigue et en administration chronique. Electroencephalogr Clin Neurophysiol 15:599–609

Bohus M, Berger M (1992) Der Beitrag biologisch-psychiatrischer Befunde zum Verstaendnis depressiver Erkrankungen. Z Klin Psychol 2:156–171

Charney D, Heninger G, Sternberg D, Hafstad K, Giddings F, Landis H (1982) Adrenergic receptor sensitivity in depression: effects of clonidine in depressed patients and controls. Arch Gen Psychiatry 39:290–294

Chernik D, Mendels J (1974) Longitudinal study of the effects of lithium carbonate on the sleep of hospitalized depressed patients. Biol Psychiatry 9(2):117–123
Clark RW, Schmidt HS, Schaal SF, Boudoulas H, Schuller DE (1979) Sleep apnea treatment with protryptyline. Neurology 29:1287–1292
Coble P, Kupfer D, Spiker D, Neil J, McPartland R (1979) EEG sleep in primary depression: a longitudinal placebo study. J Affective Dis 1:131–138
Cohen R, Pickar D, Garnett D, Lipper S, Gillin C, Murphy D (1982) REM sleep suppression induced by selective monoamine oxidase inhibitors. Psychopharmacology 78:137–140
Di Perri R, Mailland F, Bramanti P (1986) The effects of amineptine and the mood and nocturnal sleep of depressed patients. Prog Neuropsychopharmacol Biol Psychiatry 11:65–70
Dijk D, Strijkstra A, Daan S, Beersma D, Van Den Hoofdakker R (1991) Effect of clomipramine on sleep and EEG power spectra in the diurnal rodent Eutamias sibiricus. Psychopharmacology 103:375–379
Dunleavy D, Oswald I (1973) Phenelzine, mood response, and sleep. Arch Gen Psychiatry 28:353–366
Dunleavy D, Brezinova V, Oswald I, Maclean A, Tinker M (1972) Changes during weeks in effects of tricyclic drugs on the human sleeping brain. Br J Psychiatry 120:663–672
Feinberg I, Koresko R, Heller N, Steinberg H (1965) Unusually high dream time in a hallucinating patient. Am J Psychiatry 121:1018–1020
Friston J, Sharpley A, Solomon R, Cowen P (1989) Lithium increases slow wave sleep: possible mediation by brain 5-HT_2 receptors? Psychopharmacology 98: 139–140
Garcia-Sevilla J (1989) The platelet alpha-2 adrenoceptor as a potential marker in depression. Br J Psychiatry 154 [Suppl 4]:67–72
Garver DL, Davis JM (1979) Biogenic amine hypothesis of affective disorders. Life Sci 24:303–311
Garvey M, Mungas D, Tollefson G (1984) Hypersomnia in major depressive disorders. J Affective Disord 6:283–286
Gassicke J, Ashcroft G, Eccleston D, Evans J, Oswald I, Ritson E (1965) The clinical state, sleep and amine metabolism of a tranylcypromine (Parnate) addict. Br J Psychiatry III:357–364
Gigli G, Gotman J, Thomas S (1988) Sleep alterations after acute administration of carbamazepine in cats. Epilepsia 29(6):748–752
Gillin J, Wyatt R, Fram D, Snyder F (1978) The relationship between changes in REM sleep and clinical improvement in depressed patients treated with amitriptyline. Psychopharmacology 59:267–272
Gillin J, Duncan W, Pettigrew K, Frankel B, Snyder F (1979) Successful separation of depressed, normal and insomniac subjects by EEG sleep data. Arch Gen Psychiatry 36:85–90
Gillin J, Sitaram N, Wehr T et al. (1984) Sleep and affective illness. In: Post R, Ballenger J (eds) Neurobiology of mood disorders. Williams and Wilkins, Baltimore, pp 157–189
Goldberg A (1983) The treatment of Kleine-Levin syndrome with lithium. Can J Psychiatry 28:491–493
Hart E (1985) Kleine-Levin syndrome: normal CSF monoamines and response to lithium therapy. Neurology 35:1395–1396
Hartmann E (1968a) Longitudinal studies of sleep and dream patterns in manic depressive patients. Arch Gen Psychiatry 19:312–329
Hartmann E (1968b) The effect of four drugs on sleep patterns in man. Psychopharmacologia 12:346–353
Hartmann E, Cravens J (1973) The effects of long term administration of psychotropic drugs on human sleep: III. The effects of amitriptyline. Psychopharmacologia 33:185–202

Hauri P, Hawkins D (1973) Individual differences in the sleep of depression. In: Jovanovic U (ed) The nature of sleep. Fischer, Stuttgart, pp 193–197

Hawkins D, Taub J, Van De Castle R (1985) Extended sleep (hypersomnia) in young depressed patients. Am J Psychiatry 142:905–910

Hilakivi L, Sinclair J (1984) Neonatal active sleep suppression by clomipramine and later alcohol related bahavior in rat. In: Koella W et al. (eds) Sleep 1984. Fischer, Stuttgart, pp 307–309

Hilakivi L, Stenberg D, Sinclair J, Kiianmaa K (1987) Neonatal desipramine or zimelidine treatment causes long-lasting changes in brain monaminergic systems and alcohol related behavior in rats. Psychpharmacology 91:403–409

Hill SY, Reyes RB, Kupfer DJ (1979) Physostigmine induction of REM sleep in imipramine treated rats. Community Psychopharmacol 3:261–266

Hill SY, Reyes RB, Kupfer DJ (1980) Imipramine and REM sleep. Cholinergic mediation in animals. Pyschopharmacology 69:5–9

Hishikawa Y, Nakai K, Idah, Kaneko Z (1965) The effect of imipramine, desmethylimipramine and chlorpromazine on the sleep-wakefulness cycle of the cat. Electroencephalogr Clin Neurophysiology 19:518–521

Höchli D, Riemann D, Zulley J, Berger M (1986) Initial REM sleep suppression by clomipramine: a prognostic tool for treatment response in patients with a major depressive disorder. Biol Psychiatry 21:1217–1220

Hoff P, Golling H, Kapfhammer R, Lund R, Pakesh E, Ruether E, Schmauss M (1986) Cimoxaton and moclobemid, two new MAO-inhibitors: influence on sleep parameters in patients with major depressive disorders. Pharmacopsychiatry 19:249–250

Hohagen F, Mayer G, Menche A, Riemann D, Volle S, Meier-Ewert KH, Berger M (1993) Treatment of narcolepsy-catoplexy-syndrome with brofaromine. J Sleep Res 2:250–256

Hohagen F, Fritsch-Montero R, Weiss E, Lis S, Schönbrunn E, Dressing H, Riemann D, Berger M (1994) Treatment of insomnia with trimipramine. Eur Arch Psychiatry Clin Neurosci 242:329–336

Hudson J, Lipinski J, Frankenburg F, Tohen M, Kupfer D (1989) Effects of lithium on sleep in mania. Biol Psychiatry 25:665–668

Hyttel J (1982) Citalopram-pharmacological profile of a specific serotonin uptake inhibitor with antidepressant activity. Prog Psychopharmacol Biol Psychiatry 6:277–295

Insel T, Gillin C, Moore A, Wallace B, Löwenstein R, Murphy D (1982) The sleep of patients with obsessive-compulsive disorder. Arch Gen Psychiatry 39:1372–1377

Itoh H, Kabashima T, Tamura M, Onda M, Takahashi T, Higuchi H, Sasaki M, Atsuyoshi M (1987) Lithium influence on nocturnal sleep and daytime sleepiness of normal subjects. Sleep Res 16:617

Jones H, Oswald I (1968) Two cases of healthy insomnia. Electroencephalogr Clin Neurophysiol 24:378–380

Jovanovic U, Schulte W (1978) Polygraphic sleep recordings in patients with endogenous depression before and after treatment with amitriptyline-n-oxide. Drug Res 28:1924–1925

Kales A, Malstrom EJ, Scharf MB, Rubin RT (1969) Psychophysiological and biochemical changes following use and withdrawal of hypnotics. In: Kales A (ed) Sleep physiology and pathology. Lippincott, Philadelphia, pp 331–343

Kales A, Cadieux RJ, Soldatos CR, Bixler EO, Schweitzer PK, Prey WT, Vela-Bueno A (1982) Narcolepsy-cataplexy, clinical and electrophysiologic characteristics. Arch Neurol 39:164–168

Katz J, Kuperberg A, Pollack C, Walsh B, Zumoff B, Weiner H (1984) Is there a relationship between eating disorder and affective disorder? New evidence from sleep recordings. Am J Psychiatry 141:753–759

Khazan H, Brown P (1970) Differential effects of three tricyclic antidepressants on sleep and REM sleep in the rat. Life Sci 9:279–284

Kiloh L, Garside R (1963) The independence of neurotic depression and endogenous depression. Br J Psychiatry 109:451–463
Kleinlogel H, Burki H (1987) Effects of the selective 5-hydroxytryptamine uptake inhibitors paroxetine and zimelidine on EEG sleep and waking stages in the rat. Neuropsychology 17:206–211
Kline NS (1958) Clinical experience with iproniacid (Marsilid). J Clin Exp Psychopathol 19 [Suppl]:72–78
Kripke D (1983) Phase advance theories for affective illnesses. In: Wehr TA, Goodwin FK (eds) Circadian rhythms in psychiatry. Boxwood, Pacific Grove, pp 41–69
Kruse W, Hoermann MG (1960) Clinical evaluation of four antidepressant drugs. Curr Ther Res 2:111–115
Kuhn R (1957) Ueber die Behandlung depressiver Zustaende mit einem Iminodibenzylderivat (G 22355). Schweiz Med Wochenschr 95:1135–1140
Kupfer D (1976) REM latency: a psychobiologic marker for primary depressive disease. Biol Psychiatry 11:159–174
Kupfer D (1981) EEG sleep and tricyclic antidepressants in affective disorders. In: Usdin E (ed) Clinical pharmacology in psychiatry. Elsevier, New York, pp 325–338
Kupfer D (1982) Interaction of EEG sleep, antidepressants, and affective disease. J Clin Psychiatry 43(11):30–35
Kupfer D, Bowers M (1972) REM sleep and central monoamine oxidase inhibition. Psychopharmacologia 27:183–190
Kupfer D, Wyatt R, Greenspan K, Snyder F (1970) Lithium carbonate and sleep in affective illness. Arch Gen Psychiatry 23:35–40
Kupfer D, Reynolds C, Weiss B, Foster G (1974) Lithium carbonate and sleep in affective disorders. Arch Gen Psychiatry 30:79–84
Kupfer D, Reich L, Thompson K, Weiss B (1976) EEG sleep changes as predictors in depression. Am J Psychiatry 133:622–626
Kupfer D, Spiker D, Coble P, McPartland R (1978) Amitriptyline and EEG sleep in depressed patients: I. Drug effect. Sleep 1(2):149–159
Kupfer D, Brondy D, Coble P, Spiker D (1980) EEG sleep and affective psychosis. J Affective Disord 2:17–25
Kupfer D, Spiker D, Coble P, Neil J, Ulrich R, Shaw D (1981) Sleep and treatment prediction in endogenous depression. Am J Psychiatry 138(4):429–434
Kupfer D, Spiker D, Rossi A, Coble P, Shaw D, Ulrich R (1982) Nortriptyline and EEG sleep in depressed patients. Biol Psychiatry 17(5):535–546
Kupfer DJ, Shipley JM, Perel JM, Pollock B, Coble PA, Spiker DG (1987) Antidepressants and EEG sleep. Search for specificity. In: Dahl M, Gram P, Paul J, Potter W (eds) Clinical pharmacology in psychiatry. Springer, Berlin Heidelberg New York
Kupfer D, Ehlers D, Pollock B, Nathan S, Perel J (1989) Clomipramine and EEG sleep in depression. Psychiatry Res 30:165–180
Kupfer D, Perel J, Pollock B, Nathan R, Grochocinski V (1991a) Fluvoxamine versus desipramine: comparative polysomnographic effects. Biol Psychiatry 29:23–40.
Kupfer D, Pollock B, Perel J, Jarrett D, McEachran A, Miewald J (1991b) Immediate effects of intravenous clomipramine on sleep and sleep-related secretion in depressed patients. Psychiatry Res 36:279–289
Maas JW (1979) Neurotransmitters and depression: too much, too little, or too unstable? Trends Neurosci 2:306–308
Melzacka M, Danek L (1983) Pharmacokinetics of amitriptyline-n-oxide in rats after single prolonged oral administration. Pharmacopsychiatry 16:30–34
Mendels J, Chernik D (1973) The effect of lithium carbonate on the sleep of depressed patients. Int Pharmacopsychiatry 8:184–192
Mendels J, Hawkins D (1967) Sleep and depression A controlled EEG study. Arch Gen Psychiatry 19:445–452

Mirmiran M, Van De Poll N, Corner M, Van Oyen H, Bour H (1981) Suppression of active sleep by chronic treatment with clomipramine during postnatal development: effects upon adult sleep and behavior in the rat. Brain Res 204:129–146

Monti J, Alterwain P, Monti D (1990) The effects of moclobemide on nocturnal sleep of depressed patients. J Affective Dis 20:201–208

Mouret J, Lemoine P, Minuit M, Renardet M (1988) Interet des études polygraphiques de sommeil dans le suivi des traitements antidepresseurs. Application à la trazodone. Psychiatr Psychobiol 3:29–36

Mouret J, Lemoine P, Minuit M, Sanchez P, Taillard J (1989) Sleep polygraphic effects of trimipramine in depressed patients. Drugs 38(1):14–16

Nicholson A, Pascoe P (1988) Studies on the modulation of the sleep wakefulness continuum in man by fluoxetine 5-HT uptake inhibitor. Neuropharmacology 27:597–602

Ogura C, Okuma T, Nakazawa K, Kishimoto A (1976) Treatment of periodic somnolence with lithium carbonate. Arch Neurol 33:143

Oswald I (1969) Sleep and dependence on amphetamine and other drugs. In: Kales A (ed) Sleep physiology and pathology. B Lippincott, Philadelphia, pp 317–330

Oswald I, Adam K (1986) Effects of paroxetine on human sleep. Br J Clin Pharmacol 22:97–99

Poignant J (1979) Revue pharmacologique sur l'amineptine. Encephale V:709–720

Post R (1987) Mechanisms of action of carbamazepine and related anticonvulsants in affective illness. In: Meltzer HY (ed) Psychopharmacology: the third generaton of progress. Raven, New York, pp 567–576

Reynolds CF III, Kupfer DJ (1987) Sleep research in effective illness. State of the art circulation 1988. Sleep 10:199–215

Riemann D Velthaus S, Laubenthal S, Mueller W, Berger M (1990) REM- suppressing effects of amitriptyline and amitriptyline-n-oxide after acute medication in healthy volunteers: results of two uncontrolled pilot trials. Pharmacopsychiatry 23:253–258

Riemann D, Gann H, Fleckenstein P, Hohagen F, Olbrich R, Berger M (1991) Effect of RS 86 on REM latency in schizophrenia. Psychiatry Res 38:89–92

Riemann D, Gann H, Hohagen F, Bahro M, Müller W, Berger M (1993) The effect of carbamazepine on endocrine and sleep EEG variables in a patient with 48-hour rapid cycling, and healthy controls. Neuropsychobiology 27:163–170

Roth E, Smolik P, Soucek K (1980) Kleine-Levin syndrome lithoprophylaxis. Cesk Psychatr 76:156–162

Roy-Byrne P, Uhde T, Post R Joffe R (1984) Relationship of response to sleep deprivation and carbamazepine in depressed patients. Acta Psychiatr Scand 69:379–382

Saletu B (1976) Psychopharmaka und die Schlafqualitaet: quantitative neurophysiologische und subjektive Parameter. Arzeneimittelforschung (Drug Res) 26(6):1042–1047

Saletu B, Frey R, Kruphka M, Anderer P, Gruenberger J, See W (1991) Sleep laboratory studies on the single-dose effects of serotonin reuptake inhibitors paroxetine and fluoxetine on human sleep and awakening qualities. Sleep 14(5):439–447

Scherschlicht R, Polc P, Schneeberger J, Steiner M, Haefely W (1982) Selective suppression of rapid eye movement sleep (REMS) in cats by typical and atypical antidepressants. In: Costa E, Racagni G (eds) Typical and atypical antidepressants: molecular mechanisms. Raven, New York, pp 264–359

Schildkraut J (1965) The catecholamine hypothesis of affective disorders: a review of supporting evidence. Am J Psychiatry 122:509

Sitaram N, Wyatt R, Dawson S, Gillin J (1976) REM sleep induction by physiostigmine infusion during sleep. Science 1991:1281–1283

Sitaram N, Mendelson W, Wyatt R, Gillin J (1977) Time-dependent induction of REM sleep and arousal by physiostigmine infusion during normal human sleep. Brain Res 122:562–567

Snyder F (1972) NIH studies of EEG sleep in affective illness. In: Williams T, Katz M, Shields J (eds) Recent advances in the psychobiology of the depressive illnesses. US Government Printing Office, Washington DC, pp 171–192

Steiger A, Holsboer F, Benkert O (1987) Effects of brofaremine (CGP 11 305A), a short acting, reversible, and selective inhibitor of MAO-A on sleep, nocturnal penile tumescence and nocturnal hormonal secretion in three healthy volunteers. Psychopharmacology 92:110–114

Sugrue M (1980) Chronic antidepressant administration and adaptive changes in central monoaminergic system. In: Enna S, Malick J, Richelson E (eds) Antidepressants: neurochemical behavioral and clinical perspectives. Raven, New York

Sulser F, Mobley PL (1980) Biochemical effects of antidepressants in animals. In: Hoffmeister F, Stille G (eds) Psychotropic agents, part I. Springer, Berlin Heidelberg New York, pp 471–490 (Handbook of experimental pharmacology, vol 55)

Tandon R, Shipley J, Eiser A, Greden J (1988) Association between abnormal REM sleep and negative symptoms in schizophrenia. Psychiatry Res 27:359–361

Vogel G (1975) A review of REM sleep deprivation. Arch Gen Psychiatry 32:749–761

Vogel G (1983) Evidence for REM sleep deprivation as the mechanism of action of antidepressant drugs. Prog Neuropsychopharmacol Biol Psychiatry 7:343–349

Vogel G, Thurmond A, Gibbson P, Sloan K, Boyd M, Walker M (1975) REM sleep reduction effects on depression syndromes. Arch Gen Psychiatry 32:749–761

Waldmeier PC (1981) Noradrenergic transmission in depression: under- or over-function? Pharmacopsychiatry 14:3–9

Ware C, Pittard J (1990) Increased deep sleep after trazodone use: a double-blind placebo-controlled study in healthy young adults. J Clin Psychiatry 51(9):18–22

Ware J, Brown F, Moorad P, Pittard J, Cobert B (1985) Comparison of trimipramine and imipramine in depressed insomniac patients. Sleep Res 14:65

Wehr T, Wirz-Justice A (1982) Circadian rhythm mechanisms in affective illness and in antidepressant drug action. Pharmacopsychiatry 15:31–39

Wiegand M, Berger M (1989) Action of trimipramine on sleep and pituitary hormone secretion. Drugs 1:35–42

Wyatt R, Kupfer D, Scott J, Robinson D, Snyder F (1969) Longitudinal studies of the effect of monoamine oxidase inhibitors on sleep in man. Psychpharmacologia 15:236–224

Wyatt R, Fram D, Kupfer D, Snyder F (1971) Total prolonged drug-induced REM sleep suppression in anxious-depressed patients. Arch Gen Psychiatry 24: 145–155

Yang J, Elphick M, Sharpley A, Cowen P (1989) Effects of carbamazepine on sleep in healthy volunteers. Biol Psychiatry 26:324–328

Zarcone V, Benson K, Berger P (1987) Abnormal rapid eye movement latencies in schizophrenia. Arch Gen Psychiatry 44:45–48

Zung W (1969) Antidepressant drugs and sleep. Exp Med Surg 27:124–137

CHAPTER 18
Anticonvulsant Drugs

A. BARUZZI, F. ALBANI, R. RIVA, E. SFORZA, and E. LUGARESI

A. Introduction

The antiepileptic drugs (AEDs) are a class of chemically different compounds whose major clinical indication is the prevention of spontaneous seizures in humans. Modern therapy of epilepsies began in the second half of the last century with the discovery of the anticonvulsant activity of bromides. Bromides remained the mainstay of antiepileptic therapy until phenobarbital was discovered in 1910. Both bromides and phenobarbital were sedative drugs, and sedation was then considered a prerequisite for antiepileptic activity. Only the introduction of phenytoin in 1938 demonstrated that sedative and antiepileptic effects could be clearly separated. The development of phenytoin also provided a rational approach to the search for new antiepileptics in animal seizure models. Indeed, the majority of the AEDs in use today were identified and developed on the basis of their anticonvulsant activity in experimental models considered predictive of antiepileptic efficacy in humans. A consequence of this empirical approach is that in most cases animal studies have provided a better picture of the potential clinical utilization of the drug than of its basic mechanism of action at molecular and cellular levels.

In this chapter we will discuss in detail the available data on the effects on sleep of the main AEDs, namely phenobarbital (PB), phenytoin (PHT), carbamazepine (CBZ), valproic acid (VPA) and ethosuximide (ESM). Many other AEDs, such as bromides, oxazolidinediones, sulthiame and pheneturide, have been used in the past, but for these data on the effects on sleep are not available. Data on benzodiazepines are omitted as they are discussed in Chaps. 8, 13 and 14 in this volume.

In recent decades new approaches to the search for anticonvulsants have been devised, based on the increasing knowledge of the neurobiology of epilepsy (e.g., the physiopathological role of inhibitory and excitatory amino acids (MELDRUM 1985). New antiepileptic drugs have been developed using this approach (MELDRUM and PORTER 1986), but for them relevant observations on sleep effects are lacking.

Few data on AED effects on sleep parameters are available compared with those on the general sleep-epilepsy relationship. The methodological complexity (and financial cost) of controlled laboratory polygraphic studies may partially explain the paucity of published works. Moreover, available studies differ widely in methodology and experimental design (nocturnal polygraphy vs. 24-h home recording; single-dose vs. chronic treatment) as well as populations studied (healthy volunteers vs. epileptic patients).

A clear definition of the action of AED on sleep is further hampered by the presence of a complex interaction between sleep and epilepsy itself. It is generally accepted that nocturnal sleep in epileptic patients may be disturbed. Several authors have reported that epileptic phenomena (seizures themselves or interictal discharges) and, if any, associated organic brain diseases alter sleep architecture as well as other aspects of sleep physiology (BALDY-MOULINIER 1986). Quality and quantity of the alterations seem related to factors such as the severity of epilepsy and time and frequency of seizures.

Primary and secondary generalized seizures during sleep reduced total sleep time by an increase in wakefulness and in awakenings after the attacks (BALDY-MOULINIER 1986). Non-REM sleep is scarcely affected by seizures: there is an increase in light sleep, i.e., stages 1 and 2, without changes in the amount of stages 3 and 4. REM sleep is most affected, especially when the seizures occur during the first or second cycles of sleep. REM sleep is generally reduced by 50% and associated with an increase in REM latency. Complex partial seizures affect sleep organization and the amount of REM sleep only when the attacks recur throughout the night.

Sleep alterations intrinsic to epilepsy are also present in the absence of nocturnal seizures. Epileptic patients' sleep is characterized by a more marked disruption as indicated by the increase in wakefulness after sleep onset (WASO) and by the increase in arousals and stage shifts. Ambulatory recordings (DRAKE et al. 1990) showed a more marked increase in WASO, decreased total sleep time and prolonged sleep latency in patients with partial epilepsy. Slow-wave sleep and REM sleep were also decreased in patients with partial seizures.

Not only can epilepsy alter sleep, but sleep itself may modulate the manifestation of epileptic events in different ways in different clinical situations. Sleep effects on epilepsy are clinically relevant and have been extensively studied. A discussion of this topic, however, is beyond the scope of this review and interested readers are referred to a recently published book (DEGEN and RODIN 1991). Thus, when AEDs are used for their main clinical application, the effects on sleep must be seen as a component of a more complex interaction between sleep, epilepsy and drugs.

Given the complexity of these interactions and limitations of available studies, we present the data in a descriptive manner under separate headings for each drug.

B. Drug Reviews

I. Phenobarbital

Phenobarbital, a barbiturate with a long elimination half-life, is the oldest AED still in wide use; it is employed especially to control generalized seizures.

As a class of drugs, barbiturates have a broad spectrum of clinical applications in neurology, as hypnotics, anxiolytics, antiepileptics, anaesthetics, etc. (SMITH and RISKIN 1991). These therapeutic actions derive from a variety of effects on neuronal excitability and synaptic transmission, the most important of which are at the postsynaptic level, and include: (a) enhancement of GABA-mediated inhibition; (b) antagonism of the excitatory effect of glutamate; and (c) direct enhancement of membrane chloride ion conductance. At therapeutic concentrations, PB primarily exerts the two first actions, and a much higher concentration is needed to affect chloride conductance, whereas an anesthetic barbiturate such as pentobarbital exerts all three actions at relatively low concentrations. Thus, quantitative differences in the drug concentration needed to elicit the various actions provide a basis for the different clinical uses of barbiturates (FAINGOLD and BROWNING 1987a).

Like other barbiturates, PB has hypnotic and sedative effects, so that some action on human sleep is anticipated. In spite (or perhaps because) of this, only a few studies have examined this aspect of PB pharmacodynamics in humans.

After acute administration in healthy volunteers (100 mg × 2 nights), PB reduced wake and REM time and increased stage 2 sleep (ZUNG 1973). These effects are similar to those observed after acute administration of other barbiturates (LESTER and GUERRERO-FIGUEROA 1966; HARTMANN 1968; KAY et al. 1972; KALES et al. 1975, 1977). WOLF et al. (1984a) reported similar findings in epileptic patients after short-term therapy (i.e., shortly after steady-state plasma levels were reached). In addition, they found that patients fell asleep faster and that arousal awakenings and movement arousals were fewer when patients were on PB therapy than when on placebo. They also observed an interesting difference in PB action on sleep architecture between patients with primary generalized epilepsies (in their sample about half the patients had generalized tonic-clonic seizures on awakening) and patients with partial epilepsies (almost all patients with secondary generalization, half with secondary generalized tonic-clonic seizures during sleep).

During the first non-REM-REM cycle, PB significantly increased non-REM sleep and consequently REM latency in patients with partial seizures (with or without generalization), whereas in patients with primary generalized epilepsies it decreased stage 2 sleep without altering REM latency. In both groups PB significantly increased stage 4 of the first non-REM-REM cycle.

Moreover in patients with generalized epilepsies there was a significant reduction in the number of short REM sleep interruptions. These data suggest a specific interaction between sleep, epilepsy and AEDs.

As previously mentioned, all these effects were observed after short-term treatment (5–10 weeks); but the same author (WOLF 1987) reported that most of the effects were maintained in long-term treatment. The development of tolerance on long-term therapy, however, cannot be discounted, as an adaptation process with disappearance of most of the sleep modifications during chronic administration has been reported for other barbiturates (OSWALD and PRIEST 1965; OGUNREMI et al. 1973; KALES et al. 1975, 1977). Indeed, in a sleep laboratory study on epileptic patients free from seizures, MANNI et al. (1993) found no differences in sleep stability or sleep architecture parameters between subjects treated with PB, or valproate (VPA) (both groups often patients in monotherapy for at least 1 year), and controls. The study was primarily intended to evaluate daytime sleepiness and reported a shorter mean sleep latency time in PB-treated than in VPA-treated patients or controls.

Finally, a recent study (HIRTZ et al. 1993) evaluated sleep patterns in a large group of children treated with PB or placebo to prevent febrile seizure recurrence. Children without febrile seizures were studied as controls. Mean age at entry was about 20 months, and patients were followed up for 2½ years. Sleep was evaluated using a 3-day sleep log kept by parents on different occasions: at entry, 6 weeks later and thereafter every 6 months until completion of the follow-up. The placebo and PB groups did not differ from each other or from controls with regard to total sleep time, night sleep time or nap time. However, children who were poor sleepers at entry showed an increase in night awakenings if treated with PB with respect to placebo.

II. Phenytoin

Phenytoin is a hydantoin derivative used in many countries as a first-choice drug in the treatment of both partial and generalized epilepsies. Apart from its antiepileptic activity, PHT has many potential therapeutic applications, some of them still incompletely defined (FINKEL 1984).

The CNS effects of PHT most likely to be related to its antiepileptic action are those on ion transport and storage. Phenytoin inhibits sodium influx in a use-dependent manner, an effect which probably determines the drug's ability to regulate neuron-sustained repetitive firing. Similarly, PHT inhibits Ca^{2+} influx, particularly when the membrane is in the depolarized state, thus further reducing neuron excitability. However, PHT is also able to reduce Ca^{2+} uptake in nerve terminals following Ca^{2+} entry into the cell, an effect that per se would increase cell excitability. Phenytoin inhibits calcium-calmodulin regulated protein phosphorylation, thus possibly modulating the cyclic nucleotide second messenger system, an action that would affect many cellular processes in the CNS and other body systems (De Lorenzo 1989).

Sleep investigations with PHT share common drawbacks in the small number of subjects studied, with a single exception, and the large variability in study design and sample population that make result comparisons arduous. The early works reported contradictory results. After acute administration in healthy volunteers, Hartman (1970) and Maxion et al. (1975) found a slight decrease in slow-wave (SWS) sleep. After treatment of about 1 week (300 mg/day), Hartman (1970) found the same effect, whereas Zung (1968) reported an increase in sleep stages 1 and 4, and a reduction of stage 3 and of wake phases. In epileptic patients chronically treated with PHT, Maxion et al. (1975) observed that sleep was not different from that of healthy controls, whereas Hartman (1970) studying only four patients found two subjects with very little stage 4 sleep and one with almost no REM sleep.

A more exhaustive study was performed in newly diagnosed epileptic patients starting an AED treatment and followed up for 6 or more months (Wolf et al. 1985; Röder-Wanner et al. 1987). Patients were studied after acute administration (PHT 100 mg), after at least 4 weeks (short term), and after 6 or more months of therapy (long term). Results were reported for 19 patients on acute, 18 on short-term and 12 on long-term treatments. The authors statistically analyzed a number of markers of sleep stability and sleep structure. The most clear-cut findings were: a reduction of sleep latency and an increase in stage 4 during the first non-REM-REM sleep cycle on acute treatment. On short-term therapy (vs. baseline) there was a reduction of sleep latency, confirming the acute data, and a reduction of percentage sleep stage 1, and an increase in stages 3 and 4. With respect to the acute study, no major differences were found, except a further slight decrease in sleep latency. Subjectively, patients more frequently reported evening tiredness and a quietier and more stable sleep. On long-term treatment, most of the effects were reversed, and REM sleep remained unaffected. The only persistent difference with respect to baseline was the reduction of sleep latency. The possible contribution of the *therapeutic* action of PHT in determining these sleep effects is, however, not clear as the clinical status of patients was not described. The observation that some PHT effects on sleep may be time dependent may partially explain the contradictory findings obtained in previous acute and chronic studies.

More recently, Drake et al. (1990) studied 17 epileptic patients treated on monotherapy for at least 6 months with various drugs. Five took PHT, five carbamazepine, five valproic acid and two clonazepam. Patients' sleep was recorded at home using a four-channel Oxford system. The study mainly aimed to compare sleep in patients with partial epilepsies with sleep in those with generalized epilepsies, but data on AED effects were also discussed. Patients on PHT showed increased sleep latency, reduced total sleep time and more wakefulness after sleep onset with respect to subjects treated with valproic acid and clonazepam. Even taking into account the different kind of contrast (between groups), these results are at variance with those of RÖDER-WANNER et al. (1987). The small number of subjects and the fact that statistically significant differences could also be demonstrated for the different epilepsy types (a factor that cannot be separated from drug effects) limit the relevance of these findings.

III. Carbamazepine

Carbamazepine is a tricyclic compound with a chemical structure resembling those of the tricyclic antidepressants. It is effective in the treatment of partial and generalized epilepsies, and trigeminal neuralgia. In recent years it has been introduced in the treatment of some psychiatric disorders (ELPHICK 1988).

The various therapeutic indications of CBZ derive from a wide spectrum of pharmacological actions in the CNS (MACDONALD 1989). The most prominent action seems to be at the neuronal membrane level, where CBZ binds to sodium channels in the inactive state slowing the transition to the closed state from which they can reopen. The consequent reduction of sodium influx diminishes the high-frequency firing capacity of the neuron. Also, CBZ interacts with adenosine receptors and may modify the activity of second messenger AMPc and GMPc. The relevance of CBZ action on adenosine receptors to the antiepileptic effects is still ill defined. However, it may explain some of the different systemic clinical actions of the drug.

The CBZ effects on sleep have been the object of specific studies only in recent years (TOUCHON et al. 1987; YANG et al. 1989; MANNI et al. 1990; DRAKE et al. 1990). In an early work on CBZ therapeutic actions in manic and/or depressed patients, BALLENGER and POST (1980) reported that a 1-week CBZ treatment improved patients' sleep with an increase in total sleep time, primarily due to the increase in SWS. The improvement was maintained during at least 1 month of treatment.

YANG et al. (1989) started from the consideration that CBZ may have serotoninergic activity and, as 5-HT pathways are implicated in the promotion and maintenance of SWS, they speculated that CBZ should also promote SWS. They studied 7 healthy volunteers, treated with CBZ for 10 days at dosages increased from 200 to 700 mg/day in a week. Recordings were made at the subject's home using a Medilog apparatus. Two pairs of consecutive nights were recorded: before and during CBZ treatment (nights 8 and 9). Statistical analysis was conducted on the averages of each pair of nights. The main findings were an increase in percentage SWS (+110%) and a decrease in percentage REM sleep (−14%). In one subject, investigated more intensively during the study, SWS percentage gradually increased over the 1st week of treatment and returned to baseline values 4 days after CBZ withdrawal. The authors postulated that the observed effects may be due to facilitation of 5-HT transmission, in part possibly produced through interaction with adenosine receptors. To place these findings in the correct perspective with respect to epileptic patients, we stress that they were obtained at relatively high CBZ plasma concentrations (a mean of 11.8 g/ml 12 h after drug administration) in drug-naive subjects. In spite of the authors' statement that "CBZ dosage was well tolerated by most subjects," three out of seven reported sedation and difficulty in visual accommodation and one reported nausea.

TOUCHON et al. (1987) studied 15 recently diagnosed epileptic patients with partial complex seizures (temporal lobe epilepsy), who had to start CBZ treatment. Patients were excluded from the study if they had more than two partial seizures or one generalized seizure during the recording night or had CBZ plasma levels below 4 g/ml or above 10 g/ml. Two recording sessions of two nights each were made, before and after 1 month of CBZ treatment, with the first night of each session serving as an adaptation night. CBZ dosage was 800 mg/day, resulting in CBZ plasma levels in the target range. The control group was composed of nonepileptic patients comparable for age and level of anxiety (Hamilton scale), the latter considered by the authors a factor to be controlled for in sleep studies in epileptic patients. The results showed that, before CBZ treatment, sleep architecture in epileptics was substantially similar to that observed in control subjects, with only a slight increase in stage 1 sleep. CBZ normalized this parameter and induced minimal increases in total sleep time and percentage of stages 3 and 4. Although not statistically significant, the latter modifications are worth nothing because they agree with the findings of other authors. Sleep stability, on the other hand, was greatly altered before therapy, as indicated by the frequent arousals and arousal movements, the amount of WASO and the number of shifts to stages 1 and 2. After a 1-month therapy, the number of arousals and arousal movements were reduced, though not "normalized," whereas the WASO and the number of entries in stage 1 were no longer different from those of controls. In summary, untreated patients with complex partial seizures had an altered sleep pattern even if no seizure occurred during sleep. Sleep structure, particularly REM sleep, was relatively spared but sleep stability was profoundly impaired. CBZ reduced sleep fragmentation without relevant alterations of sleep architecture.

In the study already described for PHT, DRAKE et al. (1990) reported observations on five epileptic patients treated with CBZ on monotherapy. Sleep was similar to that of patients on PHT, with increased sleep latency and WASO and decreased total sleep time compared with patients on VPA or clonazepam monotherapy. Moreover, patients on CBZ showed more arousals and slightly less REM sleep. As already mentioned, in this study it is not possible to distinguish the effect of drugs from the effect of the type of epilepsy.

Lastly, MANNI et al. (1990) studied 14 patients with "focal symptomatic epilepsies" in therapy with CBZ for at least 1 year (eight with complete seizure control and normal EEG) and 11 control subjects without drug therapy. Patients who still experienced seizures were recorded at least 48 h after their last seizure. Results showed that sleep structure in epileptics was similar to that of controls, with only a reduction in REM time, but sleep stability was greatly altered (more awakenings, more stage changes, more shifts to stage 1). No significant differences were found between controlled and uncontrolled patients. The authors concluded that the alterations of sleep stability in patients with focal epilepsies were probably related to the pathology and that CBZ did not seem to impair sleep, except perhaps for a small reduction of REM sleep time.

IV. Valproate

Valproic acid and its derivatives are one of the main agents for the treatment of generalized epilepsies, probably the most widely used for the epilepsies of childhood. Valproic acid differs from the other antiepileptic drugs in regard to both the chemical structure, a simple branched fatty acid, and the broad spectrum of antiepileptic activity. The antiepileptic action of VPA is usually explained through its action on CNS GABAergic functions.

In animals, VPA increases whole brain gamma-aminobutyric acid (GABA) levels, apparently by both enhancement of the GABA-synthesizing enzyme, GAD, and inhibition of the GABA-degradative enzymes, SSADH and aldehyde reductase. Most important, there is evidence that VPA increases GABA pools in synaptosomes, the only fraction of GABA involved in neurotransmission. However, some experimental works questioned both the entity and the clinical role of VPA effects on GABA (FARIELLO and SMITH 1989; FAINGOLD and BROWNING 1987b).

As with other antiepileptic drugs, the effects on human sleep have been studied only in a few subjects (patients or healthy volunteers) and results are partially contradictory. In healthy volunteers, SCHNEIDER et al. (1977) found that acute VPA administration (600 mg $\times$ 2 days) reduced sleep latency without altering other sleep characteristics, and that 2 weeks of treatment (900 mg/day) significantly increased percentage stage 2 and reduced percentage SWS, without modifying the number of awakenings. In contrast, HARDING et al. (1985) observed that 2 weeks of treatment (1000 mg/day) in healthy volunteers slightly increased SWS without further sleep modifications. In adolescent and young adult patients with primary generalized epilepsy, RÖDER and WOLF (1981) observed that valproic acid had only minimal effects on EEG sleep, with an increase in stage 1 sleep without reduction of slow-wave or REM sleep. On the other hand, FINDJI and CATANI (1982) in children under a "clinically effective valproate treatment" observed a slight increase in SWS and a more regular organization of sleep cycles. Finally, DECLERCK and WAUQUIER (1991) and MANNI et al. (1993) reported that the nocturnal sleep of patients on valproate treatment did not differ from the sleep of control subjects.

V. Ethosuximide

Ethosuximide a succinimide derivative, has been used extensively in the past for the therapy of absence seizures. Since the introduction of VPA, is use has declined and today it is mainly restricted to the treatment of absence seizures in patients where VPA was ineffective. In spite of its highly selective clinical action, paralleled by a similarly selective activity in experimental seizure models, the definition of the mechanism of action of ESM is as difficult as for other AEDs. The most likely candidate is the ability of ESM to reduce low-threshold calcium current in thalamic neurons. The effect is voltage dependent, being more pronounced with hyperpolarized membrane potentials. A similar property, however, is shared by PB but not by PHT, CBZ and VPA, the latter drug being effective in the treatment of absence seizures. A further hypothesis is enhancement of inhibitory neurotransmitter action, an idea consistent with the ESM general depressant effect on brain activity. However, ESM does not affect GABAergic transmission significantly, and the effect on dopaminergic pathways is still poorly defined, leaving the potentially affected transmitter still to be identified (FERRENDELLI and HOLLAND 1989; ROGAWSKI and PORTER 1990).

Specific studies on the sleep effects of ESM are limited to one investigation (RÖDER and WOLF 1981; WOLF et al. 1984b; WOLF 1987). In eight patients with generalized epilepsies the authors found that an effective therapeutic treatment of at least 4 weeks with ESM increased sleep stage 1 and decreased SWS. Moreover, temporal sleep structure during the night was changed: the first REM phase was longer and SWS shorter, appearing later in the first part of the night. Increase in sleep stage 1 was more prominent around and after the REM phases. Globally these modifications increase sleep instability and susceptibility to disturbances (especially in the early part of the night). Indeed, almost all patients reported that sleep quality was worsened by ESM therapy. The authors concluded that the observed sleep impairment might promote the development of psychiatric disturbances in patients under ESM therapy.

C. Summary

The antiepileptic drugs are a class of chemically different compounds whose major clinical indication is the prevention of spontaneous seizures in humans. A great deal of information is available on the clinical pharmacology of these drugs; however, data on the effects on human sleep parameters are relatively scarce and not homogenously compared, for example, with those on the general relationship between sleep and epilepsy. A clear definition of the action of AEDs on sleep is further hampered by the presence of a complex interaction between sleep and epilepsy itself. Indeed, it is generally accepted that nocturnal sleep in epileptic patients may be disturbed. The quality and quantity of nocturnal sleep seems to be related to factors such as the severity of epilepsy and the time and frequency of seizures. However, sleep alterations intrinsic to epilepsy are also present in the absence of nocturnal seizures.

Phenobarbital shares with other barbiturates hypnotic and sedative properties, and has similar effects on sleep architecture. After acute and short-term administration, phenobarbital reduces sleep latency, wakefulness and REM sleep, and increases stage 2 sleep. Some of these effects may, however, fade during chronic treatment, as happens with other barbiturates. Early reports on the effects of phenytoin on sleep in patients and healthy volunteers presented inconsistent results. More substantial data came from a large investigation in newly diagnosed epileptic patients followed up for up to 6 months or more. Phenytoin was found to reduce sleep latency and slightly increase slow-wave sleep after acute and short-term administration. During chronic treatment only the reduction in sleep latency was still detectable. REM sleep was in general unaffected. Carbamazepine appears to have no negative effects on sleep in treated patients; after initial therapy sleep structure was almost unaltered and sleep stability was actually improved. When compared with untreated subjects, patients on chronic carbamazepine treatment showed only a small reduction in REM sleep time. Data on the effect of valproic acid and ethosuximide are very scarce. Overall, valproic acid appears to have little or no long-term effects on sleep; indeed, studies in patients receiving the drug have variably reported a slight increase in stage 1 or stages 3 and 4 or a sleep structure no different from that of control subjects. In the only investigation available, ethosuximide was found to impair sleep quality, an effect that the authors related to the possible development of psychiatric disturbances during ethosuximide therapy.

References

Ballenger JC, Post RM (1980) Carbamazepine in manic-depressive illness: a new treatment. Am J Psychiatry 137:782–790

Baldy-Moulinier M (1986) Inter-relationships between sleep and epilepsy. In: Pedley TA, Meldrum BS (eds) Recent advances in epilepsy 3. Churchill Livingstone, Edinburgh, p 37

Declerck AC, Wauquier A (1991) Influence of antiepileptic drugs on sleep patterns. In: Degen R, Rodin EA (eds) Epilepsy, sleep and deprivation, 2nd edn. Elsevier Science, Amsterdam, p 153

Degen R, Rodin EA (eds) (1991) Epilepsy, sleep and deprivation, 2nd edn. Elsevier Science, Amsterdam

De Lorenzo JR (1989) Phenytoin: mechanisms of action. In: Levy RH, Dreifuss FE, Mattson RH, Meldrum BS, Penry JK (eds) Antiepileptic drugs, 3rd edn. Raven, New York, p 143

Drake ME, Pakalnis A, Bogner JE, Andrews JM (1990) Outpatient sleep recording during antiepileptic drug monotherapy. Clin Electroencephalogr 21:170–173

Elphick M (1988) The clinical use and pharmacology of carbamazepine in psychiatry. Int Clin Psychopharmacol 3:185–203

Faingold CL, Browning RA (1987a) Mechanisms of anticonvulsant drug action. I. Drugs primarily used for generalized tonic-clonic and partial epilepsies. Eur J Pediatr 146:2–7

Faingold CL, Browning RA (1987b) Mechanisms of anticonvulsant drug action. II. Drugs primarily used for absence epilepsy. Eur J Pediatr 146:8–14

Fariello R, Smith MC (1989) Valproate. Mechanisms of action. In: Levy RH, Dreifuss FE, Mattson RH, Meldrum BS, Penry JK (eds) Antiepileptic drugs, 3rd edn. Raven, New York, p 567

Ferrendelli JA, Holland KD (1989) Ethosuximide: mechanisms of action. In: Levy RH, Dreifuss FE, Mattson RH, Meldrum BS, Penry JK (eds) Antiepileptic drugs, 3rd edn. Raven, New York, p 653

Findji F, Catani P (1982) The effects of valproic acid on sleep parameters in epileptic children: clinical note. In: Sterman MB, Shouse MN, Passouant P (eds) Sleep and epilepsy. Academic, London, p 395

Finkel MJ (1984) Phenytoin revisited. Clin Ther 6:577–591

Harding GFA, Alford CA, Powell TE (1985) The effect of sodium valproate on sleep, reaction times, and visual evoked potentials in normal subjects. Epilepsia 26:597–601

Hartmann E (1968) The effect of four drugs on sleep patterns in man. Psychopharmacology 12:346–353

Hartmann E (1970) The effect of DPH on sleep in man. Psychophysiology 7:316

Hirtz DG, Chen TC, Nelson KB, Sulzbacher S, Farwell JR, Ellenberg JH (1993) Does phenobarbital used for febrile seizures cause sleep disturbances? Pediatr Neurol 9:94–100

Kales A, Kales JD, Bixler EO, Scharf MB (1975) Effectiveness of hypnotic drugs with prolonged use: flurazepam and pentobarbital. Clin Pharmacol Ther 18:356–363

Kales A, Bixler EO, Kales JD, Scharf MB (1977) Comparative effectiveness of nine hypnotic drugs: sleep laboratory studies. J Clin Pharmacol 17:207–213

Kay DC, Jasinski DR, Eingenstein RB, Kelly OA (1972) Quantified human sleep after pentobarbital. Clin Pharmacol Ther 13:221–231

Lester BK, Guerrero-Figueroa R (1966) Effects of some drugs on electroencephalographic fast activity and dream time. Psychophysiology 2:224–236

Macdonald RL (1989) Carbamazepine. Mechanisms of action. In: Levy RH, Dreifuss FE, Mattson RH, Meldrum BS, Penry JK (eds) Antiepileptic drugs, 3rd edn. Raven, New York, p 447

Manni R, Galimberti CA, Zucca C, Parietti L, Tartara A (1990) Sleep patterns in patients with late onset partial epilepsy receiving chronic carbamazepine (CBZ) therapy. Epilepsy Res 7:72–76

Manni R, Ratti MT, Perucca E, Galimberti CA, Tartara A (1993) A multiparametric investigation of daytime sleepiness and psychomotor functions in epileptic patients treated with phenobarbital and sodium valproate: a comparative controlled study. Electroencephalogr Clin Neurophysiol 86:322–328

Maxion H, Jacobi P, Schneider E, Kohler M (1975) Effect of the anticonvulsant drugs primidone and diphenylhydantoin on night sleep in healthy volunteers and epileptic patients. In: Koella WP, Levin P (eds) Sleep 1974. Karger, Basel, p 510

Meldrum BS (1985) GABA and other amino acids. In: Frey H-H, Janz D (eds) Antiepileptic drugs. Springer, Berlin Heidelberg New York, p 153

Meldrum BS, Porter RJ (1986) New anticonvulsant drugs. Libbey, London

Ogunremi OO, Adamson L, Breziova V, Hunter WM, Maclean AW, Oswald I, Percy-Robb IW (1973) Two antianxiety drugs, a psychoendocrine study. Br Med J 2:202–205

Oswald I, Priest RG (1965) Five weeks to escape the sleeping pill habit. Br Med J 2:1093–1095

Röder UU, Wolf P (1981) Effects of treatment with dipropylacetate and ethosuximide on sleep organization in epileptic patients. In: Dam M, Gram L, Penry JK (eds)

Advances in epileptology, XIIth epilepsy international symposium. Raven, New York, p 145
Röder-Wanner UU, Noachtar S, Wolf P (1987) Response of polygraphic sleep to phenytoin treatment for epilepsy. A longitudinal study of immediate, short- and long-term effects. Acta Neurol Scand 76:157–167
Rogawski MA, Porter RJ (1990) Antiepileptic drugs: pharmacological mechanisms and clinical efficacy with consideration of promising developmental stage compounds. Pharmacol Rev 42:223–286
Schneider E, Ziegler B, Maxion H (1977) Gamma-aminobutyric acid (GABA) and sleep. The influence of di-n-propylacetic acid on sleep in man. Eur Neurol 15:146–152
Smith MC, Riskin BJ (1991) The clinical use of barbiturates in neurological disorders. Drugs 42:365–378
Touchon J, Baldy-Moulinier M, Billiard M, Besset A, Valmier J, Cadilhac J (1987) Organisation du sommeil dans l'épilepsie récente du lobe temporal avant et après traitement par carbamazepine. Rev Neurol 143:462–467
Wolf P (1987) Influence of antiepileptic drugs on sleep. In: Wolf P, Dam M, Janz D, Dreifuss FE (eds) Advances in epileptology, XVIth epilepsy international symposium. Raven, New York, p 733
Wolf P, Roder-Wanner UU, Brede M (1984a) Influence of therapeutic phenobarbital and phenytoin medication on the polygraphic sleep of patients with epilepsy. Epilepsia 25:467–475
Wolf P, Inoue Y, Roder-Wanner UU, Tsai J (1984b) Psychiatric complications of absence therapy and their relation to alteration of sleep. Epilepsia 25:S56–S59
Wolf P, Roder-Wanner UU, Brede M, Noachtar S, Sengoku A (1985) Influences of antiepileptic drugs on sleep. In: Da Silva AM (ed) Biorhythms and epilepsy. Raven, New York, p 137
Yang JD, Elphick M, Sharpley AL, Cowen PJ (1989) Effects of carbamazepine on sleep in healthy volunteers. Biol Psychiatry 26:324–328
Zung WWK (1968) Effect of diphenylhydantoin on the sleep dream cycle: an EEG study in normal adults. Psychophysiology 5:206–207
Zung WWK (1973) The effect of placebo and drugs on human sleep. Biol Psychiatry 6:89–92

CHAPTER 19

Mechanisms of Benzodiazepine Drug Dependence

A.N. VGONTZAS and A. KALES

A. Introduction

Classically, two types of drug dependence have been described in the literature. "Psychological dependence" is defined as drug-seeking or drug-taking behavior that is reinforced by the effects of the drug (AMERICAN PSYCHIATRIC ASSOCIATION 1987; GOODMAN and GILMAN 1990; WOODS et al. 1992). This behavior is considered a disorder when the person presents a cluster of symptoms that indicate that he/she has impaired control of the use of the drug and continues use of the drug despite adverse consequences. "Physiological dependence" is the development of symptoms and signs that occur following the discontinuation of drug use (GRIFFITHS and SANNERUD 1987; AMERICAN PSYCHIATRIC ASSOCIATION TASK FORCE 1990; GOODMAN and GILMAN 1990; WOODS et al. 1992). Physiological dependence is manifested by biochemical, physiological and/or behavioral changes.

The American Psychiatric Association, in its *Diagnostic and Statistical Manual of Mental Disorders, Revised Edition (DSM-III-R)*, lists nine criteria to consider for the diagnosis of Psychoactive Substance Dependence, including: (1) taking the substance more often or in larger amounts than intended; (2) unsuccessful efforts to terminate or reduce drug use; (3) large amounts of time spent acquiring or using the drug or recovering from its effects; (4) frequent intoxication or withdrawal symptoms; (5) abandonment of social or occupational activities because of drug use; (6) continued use despite adverse psychological or physical effects; (7) marked tolerance; and (8) frequent use of the drug to relieve withdrawal symptoms (AMERICAN PSYCHIATRIC ASSOCIATION 1987). It should be noted that although symptoms of physiological dependence are included in the diagnostic criteria, they are not required for the diagnosis of substance dependence because there are drugs thought not to produce physical dependence that do elicit drug-seeking behavior just as there are drugs that produce physical dependence without eliciting drug-seeking behavior (GRIFFITHS and SANNERUD 1987).

Benzodiazepines, since their introduction in the 1960s, were found to be associated with frequent discontinuation symptoms (see also Chaps. 13 and 14, this volume). These phenomena were originally believed to occur only when the drugs were prescribed in high doses for long periods (ISBELL et al. 1950; FRASER et al. 1953; HOLLISTER et al. 1961). However, in the late

1970s and early 1980s, withdrawal symptoms following discontinuation of benzodiazepines were reported to occur even after only brief or other relatively short term use of therapeutic doses (Kales et al. 1978). Although the occurrence of these withdrawal phenomena (rebound insomnia and anxiety) is now clearly established, their importance in inducing physical dependence remains controversial. While some investigators believe that these phenomena indicate physiological dependence and potentially may lead to drug-seeking behavior (Kales and Kales 1983; Kales et al. 1991b), others claim that their presence does not imply the existence of physical dependence (Roehrs et al. 1990; Shader and Greenblatt 1993). In turn, this controversy has affected the definition of benzodiazepine dependence.

In recent years, the Task Force on Benzodiazepine Dependency convened by the American Psychiatric Association (American Psychiatric Association Task Force 1990) defined three categories of symptoms that can characterize a "benzodiazepine discontinuance syndrome": recurrence, rebound and withdrawal. Recurrence is a return of symptoms following drug withdrawal, in the same pattern and intensity as the original symptoms. Rebound symptoms are a return of the original symptoms following withdrawal but in a more intense form than before treatment. The term "withdrawal" specifies the development, following drug discontinuance, of new signs and symptoms that were not part of the disorder for which the drugs were originally prescribed as well as a worsening of preexisting symptoms. Withdrawal symptoms defined in this way are considered "a true abstinence syndrome" and suggest that there is physiological change as a consequence of the drug administration.

We find these proposed distinctions unnecessary and potentially confusing. First, it seems inappropriate to characterize the recurrence of the initial condition as a withdrawal reaction because this recurrence may be totally independent of any drug effect and only a reflection of the chronic nature of the disorder itself. For example, the recurrence of hypertension following the discontinuance of antihypertensive medication, e.g., diuretic, cannot be considered a withdrawal symptom of the specific antihypertensive agent. The same applies in the treatment of chronic psychiatric and sleep disorders where the primary original symptom is anxiety, panic or insomnia. Further, the distinction between the exacerbation of one set of symptoms related to the initial condition of the patient for which the medication was prescribed ("rebound" reaction) and the exacerbation of another set of symptoms seemingly unrelated to the original condition ("withdrawal" reaction) is arbitrary and not based on clinical or physiological data. For example, what data support the validity of separating rebound anxiety ("withdrawal symptom") from rebound insomnia ("rebound symptom") in a patient with insomnia and anxiety following the discontinuation of a hypnotic benzodiazepine when it is known that his/her anxiety was the most likely cause of the patient's original insomnia and hypnotic drugs are also associated with anxiolysis? Finally, the development of new symptoms following drug withdrawal can be considered as a more severe manifestation of the withdrawal reaction on a continuum with the original symptom. For example, panic attacks or even seizures can be considered as manifestations of an hyperexcitability phenomenon following anxiolytic/hypnotic drug withdrawal that includes rebound insomnia at one end of the spectrum and panic and seizures at the other end.

Further, we believe the definition of benzodiazepine physiological dependence should be extended to include withdrawal phenomena that occur not only following drug discontinuation but also during drug administration (interdose withdrawal phenomenon). These interdose withdrawal phenomena were first described with short-acting, potent benzodiazepines when used as hypnotics [daytime anxiety (MORGAN and OSWALD 1982) and early morning insomnia (KALES et al. 1983d)]. Later, these findings were reported with short-acting, potent anxiolytics (interdose rebound anxiety) (HERMAN et al. 1987). Although these phenomena appear to be related to the development of tolerance to a drug's efficacy, they also represent true withdrawal reactions because they occur when the drug and its metabolites have mostly been eliminated from the system. Thus, it seems only appropriate to include these interdose withdrawal phenomena in the definition of benzodiazepine physiological dependence.

B. Prevalence of Benzodiazepine Dependence

Most of the surveys of benzodiazepine dependence have been conducted among clinical populations (primarily inpatients in general or psychiatric hospitals), while the information in the general population is sparse. The rates of dependence and/or abuse range from 1.1% to 2.0% (FLEISCHACKER et al. 1986; SCHMIDT et al. 1989; WOLF et al. 1989a) among psychiatric inpatients. Dependent patients tend to be chronic users of low doses (80% used benzodiazepines for more than 1 year) (SCHMIDT et al. 1989). Patients with a history of dependency on other substances, including alcohol, have an increased risk of developing benzodiazepine dependence (WOLF et al. 1989a,b). Also, elderly patients, particularly females, are physiologically dependent in significantly higher proportions (from 18% to 21% among psychiatric inpatients) (FOY et al. 1986; WHITCUP and MILLER 1987), and abrupt discontinuation of benzodiazepines is associated with severe reactions including confusion.

In a survey of a general population, 3.6% were identified as current users of benzodiazepines (DUNBAR et al. 1988). One-third of those reported difficulty when they attempted to discontinue benzodiazepine use. Household surveys of benzodiazepine use reported prevalence of long-term use (daily use for more than 1 year) of benzodiazepines of 1.5% in the United States and 3.0% in Great Britain (BALTER et al. 1984). Long-term users tended to be older females and suffered from physical and psychiatric symptoms

(MELLINGER et al. 1984; CATALAN et al. 1988). Although these figures do not represent rates of chemical dependence, they do provide baseline data for estimating the potential incidence of dependence in the general population (AMERICAN PSYCHIATRIC ASSOCIATION, 1990). It should be noted that the survey data in the United States were collected prior to the introduction of alprazolam and triazolam (the two most widely prescribed benzodiazepines in the United States since 1987), which have a higher propensity for withdrawal problems than other benzodiazepines (KALES et al. 1983b; PECKNOLD et al. 1988; VGONTZAS et al., in press). Thus, these early figures may be an underestimate of the current rate of dependence on benzodiazepines.

The recreational use of benzodiazepines in the general population or in medical and psychiatric patients is rather low. In the United States the annual prevalence of nonmedical use of benzodiazepines in 1990 was 1%–2% (WOODS et al. 1992). This figure is substantially higher among the drug-abusing population who use benzodiazepines as secondary drugs of abuse. There is also evidence that consumers of even moderate quantities of alcohol show greater preference for benzodiazepines than nonalcohol users (WOODS et al. 1992).

C. Description of Benzodiazepine-Withdrawal Symptoms

I. Most Frequent Withdrawal Symptoms Associated with Benzodiazepine Use and Discontinuation

Since the 1960s, when the first benzodiazepines were introduced, there have been hundreds of clinical reports in the psychiatric and medical literature of benzodiazepine withdrawal symptoms occurring after administration of high doses or long-term use of therapeutic doses. The most common withdrawal symptoms include anxiety, insomnia, agitation, diaphoresis and nightmares, while the less common and hence more severe symptoms include confusion, hallucinations and seizures.

In the late 1970s and early 1980s, it was realized that physiological dependence with benzodiazepines was even more frequent than previously thought and that withdrawal symptoms could occur following the use of even therapeutic doses of benzodiazepines for only relatively brief periods (see KALES et al. 1978, 1979, 1983b for reviews). Most of these findings came from studies which were conducted in a sleep laboratory setting which allows objective and precise measurement of sleep/wakefulness patterns. Although these findings on the use and withdrawal of benzodiazepine hypnotics initially became a center of intense controversy, many subsequent studies replicated the initial findings and extended these findings to the use of anxiolytic drugs. In this chapter we emphasize the withdrawal symptoms of benzodiazepine hypnotics and anxiolytics following the use of therapeutic doses for short periods.

1. Daytime Anxiety, Early Morning Insomnia and Other Hyperexcitability Phenomena During Drug Administration

Because withdrawal phenomena during drug administration occur once tolerance has developed, we briefly discuss the development of tolerance among the various benzodiazepines. All benzodiazepines, especially after prolonged use, are associated with development of tolerance (see also Chaps. 13 and 14, this volume). However, it appears there are significant differences among the benzodiazepines in regard to the degree and rapidity of development of tolerance (KALES and KALES 1983; Kales et al. 1985, 1994; KALES and VGONTZAS 1990). Flurazepam, quazepam, diazepam and clonazepam, which are more slowly eliminated, remain efficacious with continued use over a period of up to 1 month (KALES et al. 1982b, 1988, 1991a). In contrast, triazolam, alprazolam, temazepam and lorazepam, which are more rapidly eliminated, show rapid development of tolerance even over a period of continued use of only 1 week (Figs. 1, 2) (KALES et al. 1976c, 1986b, 1987; BIXLER et al. 1978; MORGAN and OSWALD 1982).

Withdrawal phenomena during drug administration which have been classified as "hyperexcitability phenomena" include panic, anxiety, insomnia, agitation, mania and even violence (see also Chaps. 13 and 14, this volume). The best-described symptoms are daytime anxiety and early morning insomnia. In 1982, MORGAN and OSWALD and, subsequently, ADAM and

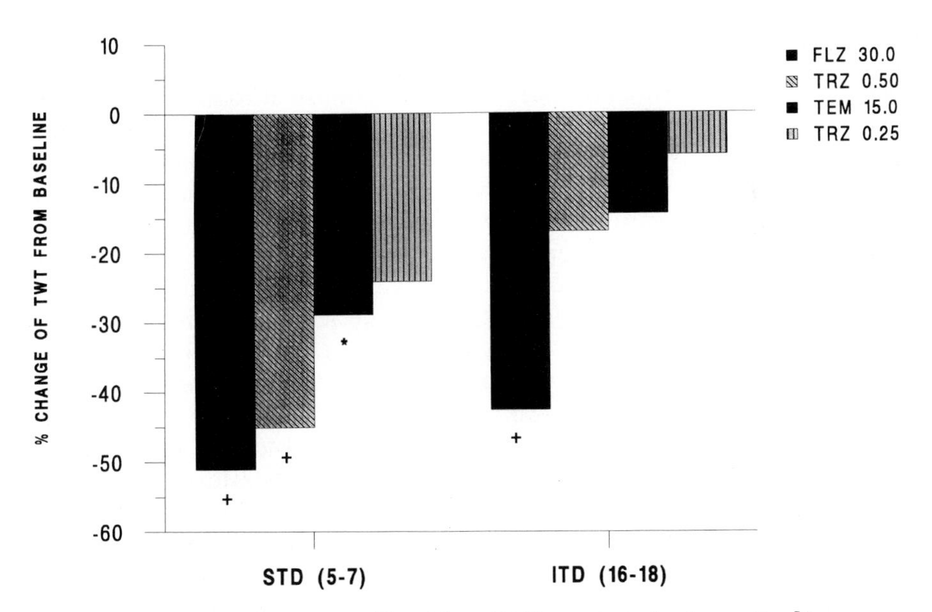

Fig. 1. Efficacy and tolerance profile of three BZPs marketed as hypnotics [flurazepam (*FLZ*) 30 mg, temazepam (*TMZ*) 15 mg and triazolam (*TRZ*) 0.5 and 0.25 mg]. *, $P < 0.05$; +, $P < 0.01$). Comparison of the percentage change of total wake time from baseline during short-term (*STD*) and intermediate-term (*ITD*) drug use

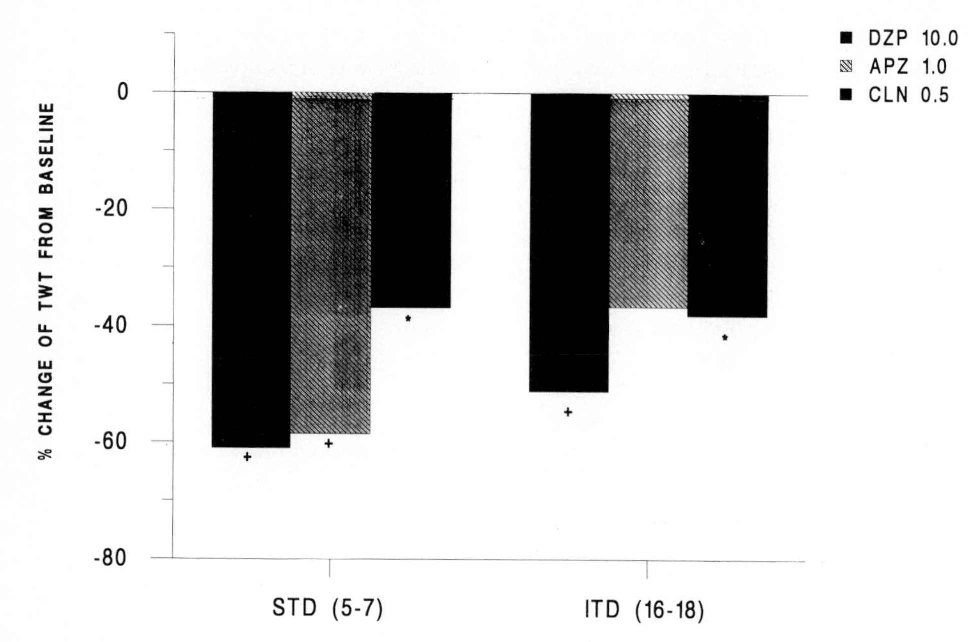

Fig. 2. Efficacy and tolerance profile of three BZPs marketed as anxiolytics [diazepam (*DZP*) 10 mg, alprazolam (*APZ*) 1 mg, and clonazepam (*CLN*) 0.5 mg]. *, $P < 0.05$; +, $P < 0.01$. Comparison of the percentage change of total wake time from baseline during short-term (*STD*) and intermediate-term (ITD) drug use

OSWALD (1989) reported that continued use of triazolam was associated with significantly and progressively increased levels of daytime anxiety. These results were also consistent with the more recent reanalysis of 25 NDA studies of triazolam which showed that 0.5 mg, and even 0.25 mg, triazolam was associated with an increased risk of dropouts due to daytime anxiety with longer drug administration (LAUGHREN and LEE 1992).

In 1983(d) KALES and associates reported that early morning insomnia, a significant increase in wakefulness during the final hours of drug nights, occurred after 1 or 2 weeks of nightly administration of benzodiazepine hypnotics with ultrashort elimination half-lives. In particular, midazolam and triazolam were associated with early morning insomnia, in contrast to flurazepam and quazepam, which remained efficacious during the last 2 h of the nights' recordings. Also, triazolam and midazolam use was associated with increased levels of daytime anxiety and nervousness, confirming the original findings of MORGAN and OSWALD (1982). These results for triazolam were confirmed in large-scale clinical studies, which showed early morning insomnia and daytime anxiety during drug administration (MOON et al. 1985; DE TULLIO et al. 1989).

Subsequently, similar hyperexcitability phenomena (both significant and nonsignificant increases in daytime anxiety, tension and disinhibition) were also reported during the drug administration period with other rapidly eliminated and potent benzodiazepines such as lorazepam (KALES et al. 1986b), flunitrazepam (SCHARF et al. 1979), lormetazepam (KALES et al. 1982a), brotizolam (VELA-BUENO et al. 1983) and alprazolam (KALES et al. 1987) when used as hypnotics. From a clinical standpoint, hyperexcitability phenomena following nightly use of hypnotics can lead to drug-taking behavior and drug dependence. This was clearly shown in a recently published study which reported daytime consumption of triazolam by patients in order to treat daytime anxiety that developed following the nighttime use of this short-acting, highly potent benzodiazepine (MARTINEZ-CANO and VELA-BUENO 1993).

Similar hyperexcitability phenomena have been described with relatively rapidly eliminated and potent benzodiazepines when used as anxiolytics. In particular, alprazolam, another potent triazolobenzodiazepine, has been frequently reported to be associated with hyperexcitability phenomena during drug administration (interdose rebound anxiety, disinhibition) (FRANCE and KRISHNAN 1984; ROSENBAUM et al. 1984; ARANA et al. 1985; GARDNER and COWDRY 1985; PECKNOLD and FLEURY 1986; HERMAN et al. 1987; KALES 1993).

2. Rebound Insomnia and Anxiety Following Drug Withdrawal

It has long been recognized that continued use of large doses of non-benzodiazepine sedative hypnotics for prolonged periods leads to dependence and a fully developed withdrawal syndrome (ISBELL et al. 1950; FRASER et al. 1953; LLOYD and CLARK 1959; JOHNSON and VAN BUREN 1962; SWANSON and OKADA 1963; ESSIG 1964; KALANT et al. 1971; KALES et al. 1974). There have been numerous clinical reports of withdrawal reactions with benzodiazepine anxiolytics and hypnotics, as well (HOLLISTER 1977; PEVNICK et al. 1978; JACOB and SELLERS 1979; KHAN et al. 1980; WINOKUR et al. 1980; HOLLISTER 1981; TYRER et al. 1981; BERLIN and CONNEL 1983; LADER 1983a,b; RICKELS ET AL. 1983; TYRER et al. 1983). In most cases fully developed withdrawal reactions with benzodiazepines have followed their prolonged use in high doses. However, such withdrawal syndromes have been noted after use of even relatively low doses of benzodiazepines, but for extended periods (PEVNICK et al. 1978; JACOB and SELLERS 1979; BERLIN and CONNEL 1983; LADER 1983; RICKELS et al. 1983). These observations have been extended through the striking finding in sleep laboratory studies, previously discussed, that certain withdrawal difficulties (rebound insomnia and anxiety) may occur not only with discontinuation of a low daily dose of a drug but even after relatively short periods of administration of such a low daily dose (see KALES et al. 1978, 1979, 1983b for reviews) (see also Chaps. 13 and 14, this volume).

Separate sleep laboratory studies of 11 benzodiazepine hypnotics and anxiolytics showed that withdrawal of drugs with rapid or intermediate elimination rates, such as lorazepam, triazolam and midazolam, resulted in rebound insomnia at night and, in a number of instances, anxiety during the day (KALES et al. 1976c, 1986b, 1988; SCHARF et al. 1982; and see KALES et al. 1983b; SMITH and WESSON 1985; KALES and VGONTZAS 1990 for reviews). In contrast, benzodiazepines with long half-lives, such as flurazepam and quazepam, were not associated with rebound insomnia (KALES et al. 1982b, 1986b) (Fig. 3).

There is strong agreement that sleep withdrawal difficulties are frequently present, and to a strong degree, with rapidly eliminated drugs (ROTH et al. 1976; VOGEL et al. 1976; OSWALD et al. 1979; MONTI et al. 1982; SCHARF et al. 1982; VELA-BUENO et al. 1983; ADAM et al. 1984; MAMELAK et al. 1984) and infrequently present and, if so, to a milder degree with slowly eliminated drugs (DEMENT et al. 1978; OSWALD et al. 1979; BLIWISE et al. 1984; MAMELAK et al. 1984). In fact, no investigator has demonstrated clear-cut rebound insomnia (statistically significant group mean changes) following the withdrawal of flurazepam or quazepam. In the three studies reporting any sleep disturbance following withdrawal of slowly eliminated drugs, only relatively mild degrees of sleep disturbance have been noted (GREENBLATT et al. 1981; KALES et al. 1982b; MENDELSON et al. 1982).

Of considerable clinical interest was the recent finding with rapidly eliminated benzodiazepines that rebound insomnia can occur even under

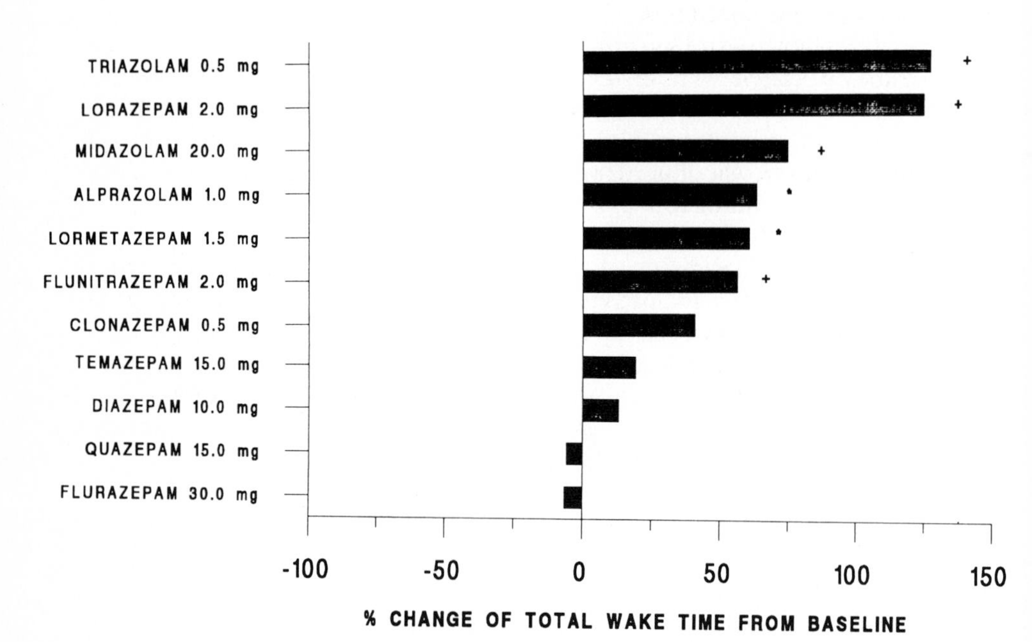

Fig. 3. Comparison of the percentage change of total wake time from baseline during the withdrawal night with the greatest mean percentage of increase in total wake time of 11 BZPs. *, $P < 0.05$; +, $P < 0.001$

conditions of brief intermittent use and withdrawal. One study showed that triazolam is associated with rebound insomnia even after a single night's clinical dose (MAMELAK et al. 1990). In another study, triazolam and, to a much lesser degree, temazepam produced rebound insomnia after only one or two nights of drug administration (KALES et al. 1991b). This side effect, with only brief and intermittent use, predisposes to drug-taking behavior and increases the potential for drug dependence (KALES and KALES 1983). Therefore, it makes the intermittent or prn use (which is the most preferred mode of hypnotic-drug therapy in general) of rapidly eliminated hypnotics, even in recommended low doses, very problematic.

Another withdrawal phenomenon frequently assessed after the cessation of certain benzodiazepine anxiolytics or hypnotics with rapid or intermediate elimination rates is rebound anxiety. This syndrome is an analogue of rebound insomnia and is characterized by an increase in anxiety above baseline levels following termination of the drug (KALES et al. 1983b). Rebound anxiety has been reported following discontinuation of rapidly eliminated benzodiazepines when used as hypnotics, such as triazolam (0.5 mg and 0.25 mg) (KALES et al. 1976c, 1986c), brotizolam (VELA-BUENO et al. 1983), lorazepam (KALES et al. 1986b) and alprazolam (KALES et al. 1987).

Rebound anxiety has also been observed in patients who abruptly or even gradually discontinued benzodiazepines which are rapidly eliminated and potent when used as anxiolytics taken during daytime (FONTAINE et al. 1984, 1985; CHOUINARD 1986; PECKNOLD et al. 1988). In particular, alprazolam has been associated with significant withdrawal difficulties, even with gradual discontinuation of the drug (PECKNOLD et al. 1988). Table 1 summarizes the results of this large-scale multicenter study regarding the prevalence and severity of withdrawal difficulties following gradual discontinuation of alprazolam after 1 month of use.

These findings were confirmed and extended in a more recent study, during which gradual discontinuation of alprazolam after 8 months of use was associated with significant withdrawal (six or more new symptoms reported by a physician) in 96% of patients using alprazolam (RICKELS et al. 1993). These data support the belief that high dose and longer duration increase markedly the risk of developing physical dependence on alprazolam.

Table 1. Withdrawal of alprazolam (data from PECKNOLD et al. 1988)

	Percent
Rebound (increase in original symptoms)	35
Rebound panic	27
Withdrawal (four new symptoms)	35
Either rebound or withdrawal	60

II. Less Frequent but Severe Withdrawal Phenomena Following Benzodiazepine Discontinuation

Severe behavioral and CNS withdrawal reactions following benzodiazepine discontinuation have been published in the form of case reports in the scientific literature and have been reported to the Spontaneous Reporting System of the Food and Drug Administration (FDA). These symptoms include psychosis, confusion, paranoid delusions, hallucinations and seizures (BIXLER et al. 1987; ANELLO 1989; WYSOWSKI and BARASH 1991). The incidence of these withdrawal reactions is rather low, and major predisposing factors include abrupt drug discontinuation, high dose, long-term use and pharmacological properties of the drug. It appears that most of these symptoms occur following abrupt discontinuation of high doses or even therapeutic doses of highly potent, short half-life benzodiazepines (BIXLER et al. 1987; AMERICAN PSYCHIATRIC ASSOCIATION TASK FORCE 1990; WYSOWSKI and BARASH 1991). The fact that common and milder withdrawal reactions, e.g., rebound insomnia, are also associated with short half-life, highly potent benzodiazepines provides further support to the belief that common and mild withdrawal reactions are on a continuum with the less common and more severe reactions and share similar pathophysiological mechanisms.

D. Risk Factors Associated with Benzodiazepine Dependence

I. Dose

Both clinicians and researchers recognized early the role of high doses in benzodiazepine-induced withdrawal reactions. The first withdrawal reactions with benzodiazepines in humans were described following withdrawal of chlordiazepoxide at doses six to ten times higher than those therapeutically recommended (HOLLISTER et al. 1961). Since then, the occurrence of a withdrawal syndrome after discontinuation of high doses of benzodiazepines has been confirmed frequently (SMITH and WESSON 1983, 1985; AMERICAN PSYCHIATRIC ASSOCIATION TASK FORCE 1990; SMITH and LANDRY 1990). High-dose withdrawal reactions have been reported both with long half-life benzodiazepines (HOLLISTER et al. 1961; PRESKORN and DENNER 1977; ALLGULANDER and BORG 1978; DE BARD 1979; ABERNETHY et al. 1981; LADER 1983a) and with short half-life drugs (SELIG 1966; DE LA FUENTE et al. 1980; EINARSON 1980, 1981; BARTON 1981; RATNA 1981; SCHNEIDER et al. 1987; LYNN 1985; NOYES et al. 1985; HERITCH 1987).

However, it soon became clear that withdrawal reactions indicating physiological dependence were present even when benzodiazepines were used at therapeutic doses (BARTEN 1965; COVI et al. 1973; MALETZKY and

Klotter 1976; Rifkin et al. 1976; Dysken and Chan 1977; Agrawal 1978; Allgulander 1978; Pevnick et al. 1978; Van Der Kroef 1979; Rickels et al. 1980; Marks 1981; Laughren et al. 1982; Lader 1984; Murphy et al. 1984; Tien 1985; Patterson 1987). To alleviate clinicians' and patients' concerns, it was suggested that by controlling the duration of drug administration when using the drug within the therapeutic range, one could avoid development of physiological dependence. However, it was Kales and his colleagues (1978) who first showed that within the therapeutic range and when using a drug for only a relatively short period, there are some benzodiazepines which still exhibit a propensity for withdrawal reactions. In 1978 Kales and associates identified "rebound insomnia" as a significant worsening of nocturnal sleep following abrupt withdrawal of short half-life benzodiazepines (triazolam, midazolam) when used at therapeutic doses and for only a relatively short-term period. Since then, numerous studies have documented withdrawal reactions following discontinuation of potent short half-life benzodiazepine hypnotics and anxiolytics used in therapeutic doses for brief periods (Vogel et al. 1975; Kales et al. 1976c; Roth et al. 1976; Vogel et al. 1976; Monti et al. 1982; Gath et al. 1983; Roehrs et al. 1983; Vela-Bueno et al. 1983; Vogel and Vogel 1983; Adam et al. 1984; Mamelak et al. 1984; Mitler et al. 1984; Kales et al. 1986c; Soldatos et al. 1986; Greenblatt et al. 1987; Monti et al. 1987; Lamphere et al. 1990; Mamelak et al. 1990).

Withdrawal difficulties including rebound insomnia have been shown to be dose dependent (Kales et al. 1979, 1981, 1982a; Roehrs et al. 1990) when using doses within the therapeutic range. This has led investigators to suggest that these difficulties can be avoided by using the lowest effective dose (Roehrs et al. 1990). However, it has been shown that doses of short half-life, potent benzodiazepines with even marginal efficacy (e.g., triazolam) can still be associated with rebound phenomena (Kales et al. 1982a, 1983c, 1986c; Scharf et al. 1990). The presence of rebound phenomena even at the level of marginally efficacious doses lowers the benefit-to-risk ratio of a drug and makes its use questionable.

It should be noted that, recently, in contrast to the efforts to control withdrawal problems by lowering the dose, there has been a trend to administer therapeutically high doses of benzodiazepines for the treatment of panic disorders. For example, higher doses of alprazolam than its usual antianxiety dose are recommended to achieve maximum antipanic efficacy. It is believed that the use of these high doses predisposes panic or agoraphobic patients to develop significant physiological dependence. This belief is based on data that showed significant withdrawal phenomena in as many as 60% of patients taking this drug, even after gradual tapering following 4 weeks of drug administration (Pecknold et al. 1988) or 96% following 8 months of drug administration (Rickels et al. 1993). From a practical standpoint, these data call into question the clinical usefulness of trying to gain a better control of panic attacks on a short-term basis by using higher doses while at the same time increasing exponentially the risk of serious withdrawal problems.

II. Duration of Drug Administration

Since the early introduction of benzodiazepines, it has been common experience for clinicians that the longer the administration of a benzodiazepine, the higher the possibility for development of dependence phenomena. However, there is no general agreement about the length of drug administration after which significant withdrawal problems develop. Some authors, based on data from clinical studies with primarily subjective measures, have supported the view that significant withdrawal and rebound phenomena develop after 4 months of regular daily use of therapeutic dose of anxiolytics (AYD 1979; BOWDEN and FISHER 1980; LADER and PETTURSON 1983; RICKELS et al. 1983; BUSTO et al. 1986). More recently, RICKELS et al. (1990) suggested that when duration of use is 1 year or longer, there is no greater degree of dependence. In contrast, MURPHY and TYRER (1991) reported that patients who had been taking benzodiazepines for more than 5 years showed higher withdrawal difficulties than those who had been taking the drugs for less than 5 years. However, these views do not seem to have taken into account the data regarding withdrawal difficulties associated with the use of the newer potent triazolobenzodiazepines. In particular, alprazolam has been associated with significant withdrawal difficulties after even only 1 month of administration (PECKNOLD et al. 1988).

A more accurate assessment of the role of duration of drug administration in the development of withdrawal reactions has been obtained from sleep laboratory studies. In general, benzodiazepine-withdrawal phenomena reported in sleep laboratory studies occur much earlier than those phenomena reported in clinical studies. This difference appears to be due to the fact that the sleep laboratory setting (objective and precise measurements and close monitoring) allows for accurate, objective and, hence, early detection of withdrawal phenomena, in contrast to clinical studies which are based only on subjective estimates (see also Chap. 12, this volume). Rebound insomnia was initially reported following nightly administration of short-acting benzodiazepines for only 2 weeks (KALES et al. 1976c). Since then, many sleep laboratory studies as previously cited have replicated and extended this finding with the use of potent, rapidly eliminated or relatively rapidly eliminated hypnotics and anxiolytics following short-term use. The more potent and the more rapidly eliminated the drug, the shorter the duration needed to develop rebound insomnia. This was clearly demonstrated in recent studies which showed development of rebound insomnia after even only a single night of drug administration of potent, rapidly eliminated benzodiazepines (KALES et al. 1991b; MAMELAK et al. 1990). This was also shown to be the case even with intermittent, brief administration of rapidly eliminated benzodiazepines (KALES et al. 1991b).

Rebound anxiety, which is considered analogous to rebound insomnia, has been described following the use of rapidly eliminated drugs (KALES et al. 1983b). However, in contrast to rebound insomnia, in most of these studies rebound anxiety became significant when the drug was discontinued 2 weeks or more following the start of drug administration (KALES et al. 1986b,c). These findings indicate that, although rebound insomnia may develop immediately, following only a single night's use of a potent, rapidly eliminated hypnotic, the development of rebound anxiety requires at least 2 weeks of drug administration before it is clinically manifested. Finally, the early occurrence of rebound insomnia appears to represent a harbinger of the more severe withdrawal reactions to be developed at a later stage of drug use.

In contrast to the many studies reporting on "withdrawal" reactions and length of prior use, there are relatively few studies reporting on the time required for the development of withdrawal phenomena during drug administration such as the hyperexcitability phenomena of early morning insomnia and daytime anxiety. In general, these phenomena are manifested after a 2-week drug administration period. This was shown clearly in two studies (MORGAN and OSWALD 1982; ADAM and OSWALD 1989) in which daytime anxiety emerged as a distinct symptom in 2 weeks and kept progressively increasing in the following period of these studies. Similar hyperexcitability phenomena (daytime anxiety) reported with other rapidly eliminated and potent benzodiazepines, such as lorazepam (KALES et al. 1986b), flunitrazepam (SCHARF et al. 1979), midazolam (KALES et al. 1983c), lormetazepam (KALES et al. 1982a), brotizolam (VELA-BUENO et al. 1983) and alprazolam (KALES et al. 1987), as well as triazolam 0.25 mg (KALES et al. 1986c), appear also to occur after at least 2 weeks of drug administration. It should be noted that development of tolerance appears to precede the manifestation of interdose withdrawal phenomena.

III. Gradual Versus Abrupt Discontinuation

It has been common clinical practice to withdraw patients dependent on benzodiazepines gradually, rather than abruptly. This is particularly important when patients have been dependent on high doses or even on therapeutic doses for a long time. Also, the pharmacological profile of the drug, i.e., elimination rate, appears to be a significant factor affecting the severity of withdrawal symptoms after abrupt discontinuation. Clinical studies that assessed the effects of gradual versus abrupt withdrawal of short half-life benzodiazepines have shown that abrupt discontinuation produced much more severe withdrawal signs than did gradual withdrawal (BUSTO et al. 1986; GREENBLATT et al. 1987; RICKELS et al. 1990; SCHWEIZER et al. 1990). The results are less consistent when assessing the effects of gradual or abrupt discontinuation of long half-life benzodiazepines. Some authors reported no differences (RICKELS et al. 1990; SCHWEIZER et al. 1990), while others (CANTOPHER et al. 1990) reported more intense withdrawal during abrupt rather than gradual discontinuation of long half-life benzodiazepines. The role of the pharmacological profile of a drug as a significant factor in determining the severity of withdrawal reaction after abrupt discontinuation was shown more clearly in sleep laboratory studies. In these studies, abrupt withdrawal of long half-life benzodiazepines after 4 weeks of administration was not associated with rebound phenomena (KALES et al. 1982b), in contrast to the significant withdrawal phenomena associated with short half-life benzodiazepines, even after a single night's use (KALES et al. 1991b; MAMELAK et al. 1990).

However, it should be noted that gradual discontinuation of short half-life benzodiazepines does not appear to be a panacea, particularly when dealing with the severe and frequent withdrawal phenomena associated with the new potent benzodiazepines. For example, alprazolam has been associated with significant withdrawal difficulties, even with gradual discontinuation spread over a period of 1 month (PECKNOLD et al. 1988; RICKELS et al. 1993). Further, the recommendation for gradual withdrawal appears to be clinically irrelevant for some benzodiazepines because the withdrawal of even the lowest effective dose of certain short-acting benzodiazepines with marginal efficacy, e.g., triazolam, has been associated with significant rebound phenomena. These findings indicate that avoidance of withdrawal problems with the new potent, short-acting benzodiazepines, even with gradual tapering of the dose, is difficult.

IV. Benzodiazepine Dependence and Patient Type

Although it is commonly believed that psychopathology and personality characteristics are important factors in the development of physiological dependence, particularly dependence on high doses, there are few systematic studies evaluating the role of psychopathology. Here we briefly summarize studies on subjects with substance abuse disorders, psychiatric disorders and sleep disorders.

1. Subjects with Substance Abuse Disorders

There is considerable evidence that a history of drug use increases the likelihood of reinforcing effects of benzodiazepines in humans (WOODS et al. 1987). Also, persons who have abused alcohol in the past may be at greater risk for abusing benzodiazepines than are subjects without a history of alcohol abuse (CIRAUOLO et al. 1988; BUSTO et al. 1983). A more interesting finding is that normal subjects with a history of even moderate alcohol consumption appear to be more vulnerable to drug abuse than those with a history of little alcohol use (DE WIT et al. 1989). Finally, the same trend was shown in subjects with a family history of alcoholism (DE WIT 1991).

2. Subjects with Psychiatric Disorders

In two studies by RICKELS et al. (1988, 1990) it was shown that high scores on the "dependence scale" of the MMPI predicted intense withdrawal phenomena. Also, high levels of anxiety or depression were associated with a higher risk for withdrawal reactions. Consistent with his finding are the results of another study (MCCRACKEN et al. 1990), which demonstrated that anxious subjects who are distressed by their anxiety are more likely to develop dependence than those who are not seeking treatment for their anxiety. Finally, MURPHY and TYRER (1991) found that patients with passive-dependent personality disorder had a much higher incidence of withdrawal difficulties than those without such disorders.

3. Subjects with Sleep Disorders

Insomniacs, who constitute the larger group of patients with sleep disorders, tend to use hypnotic benzodiazepines for long periods. This long-term regular use of benzodiazepine hypnotics can produce physiological dependence with definite withdrawal symptoms. It is unclear whether or not these drugs are taken for long periods for their pharmacological effect or for the expectation of their hypnotic properties (placebo effect) or to prevent the development of withdrawal symptoms, i.e., rebound insomnia. Also, there is no information on what percentage of chronic users of benzodiazepine hypnotics tend to escalate the dose as a result of increasing withdrawal difficulties. Clinical experience indicates that this percentage is rather small, and most insomniacs remain on a low therapeutic dose. Supporting evidence to the latter is provided by recent findings of VELA-BUENO et al. (unpublished data), who reported that in a large group of benzodiazepine-dependent subjects who sought treatment for their dependence, the majority were dependent on a relatively low dose. Also, in this study, the chronic benzodiazepine users who tended to escalate the dose had borderline-dependent personality characteristics, in contrast to the low-dose dependent subjects who demonstrated compulsive personality features. It should be noted that the majority of insomniacs present with compulsive personality characteristics (KALES et al. 1976b, 1983a). Finally, the position that most insomniacs are most likely dependent on low doses of benzodiazepines is supported by the findings of LUCKI et al. (1991), who reported that benzodiazepine withdrawal does not increase "craving" for benzodiazepines and that benzodiazepine users appear to be taking the drugs specifically to decrease symptoms that appear whenever they attempt to withdraw.

E. Pharmacological Properties Associated with Benzodiazepine Dependence

I. Absorption Rate

There is no evidence that absorption rate plays a role in the occurrence of withdrawal symptoms and development of dependence in normals or subjects with psychiatric disorders. However, data from surveys among subjects with a history of substance abuse have raised the question whether absorption rate (which determines the onset of action of a drug) is a factor contributing to the nonmedical use of benzodiazepines in this population. It has been reported that diazepam, which has a rapid absorption rate, is preferentially abused by methadone maintenance patients (GRIFFITHS and SANNERUD 1987). Also, studies that compared the reinforcing effects of different benzodiazepines indicated that diazepam has higher liability for abuse than oxazepam (which has a slow absorption rate) (GRIFFITHS and ROACHE 1985). However, another more recent study did not show any difference in terms of "likeability" among substance abusers between lorazepam and diazepam (FUNDERBURK et al. 1988), although lorazepam and diazepam have different absorption rates. The importance of absorption rate as a factor determining the "likeability" of benzodiazepine among substance abusers remains open to further investigation.

II. Elimination Rate

In order to explain the intensity and severity of rebound insomnia found in the early benzodiazepine studies in which only single doses of the drugs were administered nightly, we utilized findings on the metabolism and pharmacokinetics of these drugs as well as information concerning specific benzodiazepine receptors in the brain (KALES et al. 1978; VGONTZAS et al., in press). The intense rebound insomnia that follows the withdrawal of rapidly eliminated benzodiazepines is attributed to the much shorter duration of action of these drugs. Conversely, the absence of rebound insomnia after slowly eliminated benzodiazepines are withdrawn is attributed to their having active metabolites with long half-lives.

The discovery of benzodiazepine receptors in the brain suggested the presence of endogenous benzodiazepine-like molecules whose production would be regulated by concentrations of the circulating molecules or a feedback mechanism (KALES et al. 1978; BRAESTRUP et al. 1982; MOHLER et al. 1982; MÜLLER and STILLBAUER, 1983). Production of endogenous benzodiazepine-like molecules would be decreased if active exogenous benzodiazepine drugs or metabolites were introduced. We hypothesized that abrupt withdrawal of those benzodiazepine drugs with a relative short duration of action results in an intense form of rebound insomnia because of a lag in the production and replacement of endogenous benzodiazepine-like compounds. However, when benzodiazepines with long-acting metabolites are withdrawn, effects on the benzodiazepine receptors are less abrupt because the endogenous benzodiazepine-like compounds may be partially restored before the active metabolites of the exogenously administered drugs are completely eliminated. The ability to produce endogenous benzodiazepine-like compounds within the time these long-acting metabolites are eliminated may also be a function of dosage and the duration of drug administration. Thus, this model does not preclude the possibility that rebound insomnia may develop in response to benzodiazepines with long-acting metabolites that are taken for lengthy periods and/or in high doses.

Since this earlier hypothesis in regard to the underlying mechanism of rebound insomnia, many studies have shown that elimination rate of the parent compound and active metabolites appears to be a critical factor that determines whether a benzodiazepine produces rebound insomnia. Following withdrawal of benzodiazepines with short elimination half-lives, there is a frequent, intense and immediate degree of rebound insomnia (KALES et al. 1976c, 1983b,c, 1986c, 1991b; VOGEL et al. 1976; MONTI et al. 1982; VELA-BUENO et al. 1983; ADAM et al. 1984; GREENBLATT et al. 1990; MAMELAK et al. 1990). After withdrawal of intermediate half-life benzodiazepines, rebound insomnia occurs somewhat less frequently, is of a moderate degree and may appear on a delayed basis (ADAM et al. 1976; BIXLER et al. 1977; BIXLER et al. 1978; ADAM et al. 1984; KALES et al. 1991b). With the longer half-life benzodiazepines, withdrawal sleep disturbances occur even less frequently, are delayed in appearance and are of a milder degree (KALES et al. 1976a, 1978, 1982b, 1986a, 1988; VOGEL et al. 1976; BLIWISE et al. 1984).

III. Receptor-Binding Affinity

With the accumulation of more studies on the phenomenon of rebound insomnia following abrupt withdrawal after short-term use of benzodiazepines, as well as information on the pharmacodynamics of benzodiazepines, i.e., receptor-binding affinity, it appears that although rate of elimination is important in determining a drug's potential for rebound insomnia, other factors, i.e., high receptor-binding affinity, are involved (VGONTZAS et al., in press). Thus, we proposed that rebound insomnia was related to at least two mechanisms, the first, rapid elimination, and the second, high receptor-binding affinity (KALES et al. 1991a). When both factors are present, for example, with triazolam, rebound insomnia is more frequent, immediate and of greater severity (KALES et al. 1976b, 1986c, 1991b). When only one of the two factors is present (for example, temazepam is relatively rapidly eliminated but has a low receptor-binding affinity), rebound insomnia occurs less frequently and is milder in severity (BIXLER et al. 1978; KALES et al. 1986a, 1991b). When neither factor is present, as is the case for both flurazepam and quazepam, which are slowly eliminated and have moderately low receptor-binding affinities, rebound insomnia rarely occurs and any sleep disturbance is infrequent, delayed and mild in severity (KALES et al. 1976a, 1982b).

These two mechanisms (rapid elimination and high binding affinity) appear also to determine to a large extent tolerance and hyperexcitability phenomena during drug administration (MORGAN and OSWALD 1982; KALES et al. 1983d; ADAM and OSWALD 1989). Numerous studies have shown more rapid development of tolerance with benzodiazepines of short-to-intermediate half-life and high-potency (triazolam, brotizolam, midazolam, lorazepam, alprazolam) (KALES et al. 1976c, 1983c, 1986b, 1986c, 1987; VELA-BUENO et al. 1983; MAMELAK et al. 1984). Rapidly eliminated benzodiazepines when given in single nightly doses can result in a daily withdrawal syndrome (early morning insomnia, daytime anxiety), the severity of which is significantly influenced by the potency of the drugs as well as by the duration of drug administration.

At a cellular and molecular level some investigators, based on animal studies, have reported that benzodiazepine tolerance is associated with a "downregulation" of benzodiazepine receptors and that withdrawal and dependence is associated with an "upregulation" of benzodiazepine receptors (MILLER et al. 1987, 1988a,b). However, not all studies have found changes in benzodiazepine receptors following chronic treatment (MOHLER et al. 1978; BRAESTRUP et al. 1979; HENINGER and GALLAGER 1988). Furthermore, the underlying mechanism of these receptor alterations remains unknown (MILLER 1991), while the differential effect on these receptor changes of factors such as elimination rate and binding affinity is also unknown. Given these limited and conflicting findings on receptor changes, no theoretical molecular model can be promoted to explain the differential side effect profile of benzodiazepines.

IV. Benzodiazepine Effects on the LC-NE and HPA Axes

With the introduction of the newer triazolobenzodiazepines, the mechanisms of elimination rate and binding affinity did not appear adequate to explain the intensity and frequency of CNS and psychiatric adverse reactions associated with these drugs, such as daytime anxiety (MORGAN and OSWALD 1982; ADAM and OSWALD 1989) and next-day memory impairment (KALES et al. 1976c; VAN DER KROEF 1979; SCHARF et al. 1988; BIXLER et al. 1991) with triazolam and interdose rebound anxiety (HERMAN et al. 1987), disinhibition (GARDNER and COWDRY 1985) and withdrawal difficulties (PECKNOLD et al. 1988; RICKELS et al. 1993) with alprazolam. Some investigators postulated that triazolam was an anomalous ligand or that it has qualities of an inverse benzodiazepine agonist (ADAM and OSWALD 1989). There is, however, little experimental evidence to support such speculation. Significant differences among benzodiazepines do exist in the quantitative, but not the qualitative, affinity characteristics (GREENBLATT 1992). Other investigators have reported that low doses of alprazolam caused upregulation of benzodiazepine receptors, and they further postulated that this "anomalous" behavior of alprazolam could explain difficulties with tapering the drug (MILLER et al. 1987; LOPEZ et al. 1988). However, this finding cannot explain the initial effectiveness of the drug in small doses, the frequent hyperexcitability phenomena observed in patients adequately dosed, or the fact that the frequency and severity of withdrawal difficulties were directly related to the dose used. Finally, it is unlikely that these differences reflect differential effects on specific receptor subtypes (BZ_1 and BZ_2) because: first, benzodiazepines such as triazolam, temazepam and flurazepam, with very different side-effect profiles, are not known to differ in terms of their effects on their receptor subtypes; and second, benzodiazepines such as quazepam, with known selective effects on those receptors preferentially interacting with BZ_1 (CHUNG et al. 1984), exhibit a side-effect profile (daytime sedation) that can be easily understood from their pharmacokinetics (slow elimination) (KALES 1990).

In the meantime, the accumulation of information about the unique chemical structure of triazolobenzodiazepines and, more specifically, their direct effects on the locus coeruleus-norepinephrine (LC-NE) system and the hypothalamic-pituitary-adrenal (HPA) axis shed more light on understanding the underlying mechanisms of the CNS and psychiatric adverse reactions associated with the use of these drugs.

The first reports on benzodiazepine withdrawal described symptoms such as anxiety, insomnia and other hyperexcitability phenomena which suggested increased adrenergic activation (VGONTZAS et al., in press). These reports prompted investigators to explore the effects of benzodiazepines on NE. Biochemical studies in both humans and animals showed that acute and long-term use of benzodiazepines was associated with decreased levels of NE and its metabolites, while benzodiazepine withdrawal was associated with increased NE levels (CORRODI et al. 1967; TAYLOR and LAVERTY 1969; CORRODI et al. 1971; RASTOGI et al. 1976; RASTOGI et al. 1978; KOZAK et al. 1984; DENNIS et al. 1985; GRANT et al. 1985; DUKA et al. 1986; NUTT and MOLYNEUX 1987; BELL et al. 1988; YANG et al. 1988). Although it is not known exactly how 1,4-benzodiazepines affect NE, it has been proposed that these benzodiazepines affect the regulation of NE release indirectly through GABA, which is the primary inhibiting neurotransmitter system in the brain (SUZDAK and GIANUTSOS 1985a, 1985b; DUKA et al. 1986; KATAOKA et al. 1986).

Also, in both animal and human studies, benzodiazepine use and withdrawal is associated with cortisol suppression and rebound, respectively (BRUNI et al. 1980; ADAM et al. 1984; LOPEZ et al. 1990). Further experimental findings in rats have shown that classic benzodiazepines, e.g., diazepam, suppress central corticotropin-releasing hormone (CRH) secretion, while benzodiazepine inverse agonists are potent stimulators of CRH secretion (CALOGERO et al. 1988).

In addition to this indirect effect of benzodiazepines on the NE system and HPA axis, which is expected to be related to the potency or binding affinity of the drug, triazolobenzodiazepines appear to exert direct effects on the LC-NE system and HPA axis. Triazolobenzodiazepines activate α_2-adrenoreceptors (CHARNEY and HENINGER 1985; FILE and PELLO 1986; ERIKSSON et al. 1986), inhibit platelet-activating factor (PAF) (KORNECKI et al. 1984; CHESNEY et al. 1987; KORNECKI et al. 1987; MIKASHIMA et al. 1987) and do not appear to have total cross-tolerance with other benzodiazepines (ZIPURSKY et al. 1985; VINOGRADOV et al. 1986). In addition, triazolobenzodiazepines have an increased capacity relative to other benzodiazepines to suppress the CRH neuron in the locus coeruleus followed by an overactivation of the CRH neuron during their withdrawal (OWENS et al. 1989; KALOGERAS et al. 1990; OWENS et al. 1991).

α_2-Receptors are mainly located in the brain stem and have an inhibiting effect on the LC-NE system. Also, PAF activates CRH, which in turn has an activating effect on the LC-NE system (CHROUSOS et al. 1988). Therefore, triazolobenzodiazepines by activating α_2-adrenoreceptors, inhibiting PAF and suppressing CRH appear to have a direct suppressant effect on the LC-NE-CRH system. Based on this direct effect on the LC-NE-CRH system, some investigators have attempted to explain both the antipanic and antidepressant effects of the triazolobenzodiazepines. It is likely that the initial dampening or suppression of the LC-NE system and HPA axis is followed by a significant rebound and activation of the LC-NE system and HPA axis, when the drug is rapidly eliminated from the system.

We speculate that this repetitive daily pattern of suppression followed by activation can lead to a chemical overactivation of the LC-NE system and HPA axis and a neurophysiological and behavioral sensitization (kindling phenomenon) of various parts of the limbic system of the brain. This hypothesis can explain the frequency and severity of the major categories of CNS and psychiatric side effects associated with the use of rapidly eliminated, high-potency benzodiazepines and even more frequent and severe adverse reactions with rapidly eliminated, high-potency triazolobenzodiazepines which also have direct effects on the LC-NE system and HPA axis (KALES and VGONTZAS 1990).

Several experimental and clinical findings support our hypothesis for the role of the LC-NE system and HPA axis as significant additional factors in the development of: rapid development of tolerance, daytime anxiety and other hyperexcitability phenomena during drug administration; and withdrawal difficulties (rebound insomnia, anxiety and seizures) following discontinuation even after only relatively short-term use.

1. Tolerance, Daytime Anxiety and Other Hyperexcitability Phenomena During Drug Administration

Animal studies have indicated that brain noradrenergic systems are important prerequisites for developing tolerance to pharmacological effects.

Specifically, an intact brain noradrenergic system seems to be necessary for developing tolerance to the hypnotic and sedative effects of the barbiturates and ethanol in mice (TABAKOFF et al. 1978; TABAKOFF and HOFFMAN 1987). Also, in animals, it has been shown that anxiogenic substances (β-carbolines and inverse benzodiazepine receptor agonists) produce activation of noradrenergic neurons (YOSHISHIGE et al. 1991; YANG et al. 1989).

In humans, anxiety, panic and mania have been associated with increased noradrenergic activity (CHARNEY et al. 1983; CHARNEY and HENINIGER 1985; GORMAN et al. 1989; GOODWIN and JAMISON 1990). In addition to the many biochemical studies assessing the functional state of this system in anxiety, more recently, challenge studies have shown the importance of this system in the genesis of anxiety and panic. Clonidine, which is an α_2-agonist, has anxiolytic-sedative effects, while yohimbine, which is an α_2-antagonist, has anxiogenic effects.

In addition, the HPA axis has shown to be affected in stressful anxiogenic situations. It has been shown that chronic stress and anxiety are associated with hyperactivation of the HPA axis (CHROUSOS et al. 1988). Finally, the administration of CRH in rats is associated with "anxious" behavior opposite to that observed in rats treated with benzodiazepines (CHROUSOS et al. 1988).

2. Withdrawal Difficulties (Rebound Insomnia and Anxiety and Seizures)

Many studies have established the role of the noradrenergic system in the production of a full withdrawal syndrome after long-term use of benzodiazepines (CORRODI et al. 1967; CORRODI et al. 1971; RASTOGI et al. 1976; RASTOGI et al. 1978; DUKA et al. 1986; NUTT and MOLYNEUX 1987; BELL et al. 1988). Also, it has been shown that drugs with a strong direct effect on LC-NE activity have a higher potential for tolerance and withdrawal effects (REDMOND 1987). For example, antiphypertensive medication with direct effects on the NE system, such as clonidine or methyldopa, may, when abruptly withdrawn, cause rebound hypertension and adrenergic hyperactivity (MARTIN et al. 1984). In addition, in animal studies, chemically induced limbic seizures are attenuated with α_2-adrenergic agonists (clonidine), while an α_2-antagonist (yohimbine) potentiates these seizures. It appears that α_2-adrenoreceptors and NE have a significant modulating effect on limbic seizures (BARAN et al. 1985).

Because many of the signs and symptoms observed in animals and humans after abrupt discontinuation of benzodiazepines resemble the stress response, investigators have explored the effects of benzodiazepine withdrawal on CRH neurons and HPA axis (VGONTZAS et al., in press). In fact, experiments in humans and animals have shown a profound activation of CRF neurons and the HPA axis following withdrawal of triazolobenzodiazepines (ADAM et al. 1984; OWENS et al. 1989; OWENS et al. 1991). Furthermore, CRH administered intracerebroventricularly produces seizures (POST et al. 1988). Investigators have conceptualized a continuum of CRH-induced changes depending on dose that progress from a mild increase in aroused behavior to anxiety-like behavior and, with the largest doses, aggressive behavior and ultimately limbic-type seizures.

Finally, the much more severe, frequent and sustained withdrawal reactions (PECKNOLD et al. 1988; WOLF and GRIFFITHS 1991; KALES 1993) associated with even the gradual withdrawal of alprazolam (PECKNOLD et al. 1988; RICKELS et al. 1993) compared with lorazepam (KALES 1993), which has a half-life and binding affinity similar to alprazolam, provide further support for the role of the strong, direct and repetitive effects of alprazolam on both the LC-NE and HPA axes in the production of this drug's withdrawal difficulty.

V. Role of Other Neurotransmitters in Benzodiazepine Dependence

Data on the effects of benzodiazepines on other neurotransmitters and their potential role in the development of benzodiazepine dependence are limited. Cholinergic mechanisms have been implicated in the expression of withdrawal symptoms of barbiturates, ethanol and benzodiazepines (NORDBERG and WAHLSTRÖM 1992). Barbiturates have been shown to bind stereospecifically to muscarinic and nicotinic receptors, but this has not been observed for benzodiazepines (NORDBERG and WAHLSTRÖM 1984). Animal studies have shown that benzodiazepines affect the content of acetylcholine in the brain (CONSOLO et al. 1974) and change the number and affinity of the muscarinic receptors in the brain (POPOVA et al. 1988). Chlordiazepoxide has disruptive effects on learning and memory in mice, which are partially antagonized by physostigmine and scopolamine (NABESHIMA et al. 1990). These data warrant futher exploration of the role of cholinergic mechanisms on the development of dependence and other side effects associated with benzodiazepine use.

Serotonin has been implicated in the manifestation of benzodiazepine withdrawal syndrome. Studies both in animals and humans have shown that following drug withdrawal there is an increased release of 5-HT and its metabolite, 5-HIAA (PETURSSON et al. 1983; ANDREWS and FILE 1993). However, other studies have failed to detect changes in serotonin metabolism during drug withdrawal (COPLAND and BALFOUR 1987; THIEBOT 1986; SHEPHARD 1986). Furthermore, the use of anxiolytics such as buspirone, the primary effects of which are through the serotinergic system, are usually not associated with rebound or dependence phenomena (LADER 1991; MANFREDI et al. 1991). Thus, these data do not appear to support a role of serotonin in the development of benzodiazepine dependence (LADER 1991).

The data on the potential role of other CNS transmitters such as endorphins, dopamine and calcium channel agonists is even more limited and thus inconclusive (WISE 1987; LITTLE 1991; KHAN and SOLIMAN 1993).

VI. Conclusions

In conclusion, rapid elimination, high receptor binding affinity and direct effects on the LC-NE and HPA axes appear to determine to a significant degree the severity and frequency of the major CNS and psychiatric adverse drug reactions associated with the use and withdrawal of benzodiazepines. These factors combined with only an once per day administration of these drugs when used as hypnotics can lead to a repetitive, intermittent activation (kindling or behavioral sensitization phenomenon) of these two systems that increases with continued use of the medication (Post et al. 1988). Empirical support for this model is offered by the finding of progressively increased memory impairment (Bixler et al. 1991; Laughren and Lee 1992) and daytime anxiety (Morgan and Oswald 1982; Adam and Oswald 1989; Laughren and Lee 1992) with continued use of triazolam.

F. Summary

Benzodiazepine physiological dependence is defined as the development of new signs or symptoms and/or a worsening of preexisting symptoms and signs that occur during or following the discontinuation of drug use. One to three percent of the general population are long-term users of benzodiazepines and elderly patients and women are dependent in significantly higher proportions. While, originally, it was believed that withdrawal symptoms following benzodiazepine discontinuation occur only with high doses and/or long-term use, it was the use of the sleep laboratory setting in the 1970s and 1980s, with its precise and objective measurements, that allowed for the early detection of milder withdrawal phenomena associated with the use of hypnotic drugs.

Daytime anxiety, early morning insomnia, rebound insomnia and rebound anxiety are the most frequent withdrawal symptoms associated with benzodiazepine use and discontinuation while psychosis, confusion, hallucinations and seizures are the less frequent but more severe withdrawal phenomena. The first cluster of withdrawal symptoms were described in sleep laboratory studies while the latter symptoms were reported in case reports and post-marketing surveillance studies. The fact that both types of symptoms occur more frequently with the use of high-potency benzodiazepines which are rapidly eliminated supports a common pathophysiological mechanism, which in turn provides additional evidence that both types of phenomena are features of a common withdrawal reaction that varies in its severity.

Dose and duration of administration are risk factors associated with benzodiazepine dependence. However, rebound phenomena may occur even with low therapeutic doses used for only brief periods. Gradual discontinuation of rapidly eliminated benzodiazepines attenuates withdrawal phenomena; however, it does not appear to be a completely effective way of controlling withdrawal problems associated with the use of potent triazolobenzodiazepines. A history of drug use, traits of dependency and high levels of anxiety and depression are associated with a higher risk for withdrawal reactions.

The pharmacological profile of benzodiazepines appears to be a significant factor determining the development of physiological dependence. Rapid elimination was the first characteristic proposed to contribute to the frequency and severity of withdrawal phenomena. The more severe and frequent withdrawal reactions associated with the use of potent benzodiazepines then led to the proposal that binding affinity is an additional factor contributing to the development of physiological dependence. Finally, the unique chemical properties of triazolobenzodiazepines (activation of α_2-adrenoreceptors, inhibition of PAF and suppression of CRH) also appear to contribute to the unique withdrawal difficulties of these drugs. Thus, rapid elimination, high receptor-binding affinity and direct effects on the LC-NE and HPA axes and their interaction appear to determine to a significant degree the severity and frequency of the major CNS and psychiatric adverse reactions associated with the use and withdrawal of benzodiazepines. These factors combined with only once per day administration of these drugs when used as hypnotics can lead to a repetitive, intermittent activation (kindling or behavioral sensitization) of these two systems that increase with continued use of the medication. This theoretical model, which is primarily based on the objective findings of many sleep laboratory studies on hypnotic and anxiolytic drugs, is also supported by behavioral, electrophysiological and biochemical data of many studies conducted in humans and animals.

References

Abernethy DR, Greenblatt DJ, Shader RI (1981) Treatment of diazepam withdrawal syndrome with propranolol. Ann Intern Med 94:354–355

Adam K, Oswald I (1989) Can a rapidly eliminated hypnotic cause daytime anxiety? Pharmacopsychiatry 22:115

Adam K, Adamson L, Brezinova V, Hunter WM (1976) Nitrazepam: lastingly effective but trouble on withdrawal. Br Med J 1:1558–1560

Adam K, Oswald I, Shapiro C (1984) Effects of loprazolam and of triazolam on sleep and overnight urinary cortisol. Psychopharmacology 82:389–394

Agrawal P (1978) Diazepam addiction: a case report. Can Psychiatr Assoc J 23:35–237

Allgulander C (1978) Dependence on sedative and hypnotic drugs: a comparative clinical and social study. Acta Psychiatr Scand Suppl 270S:7–101

Allgulander C, Borg S (1978) A case report: a delirious abstinence syndrome associated with clorazepate (Tranxiline). Br J Addict 73:175–177

American Psychiatric Association (1987) Diagnostic and statistical manual of mental disorders (3rd edn DSM-III-R). American Psychiatric Association, Washington, DC

American Psychiatric Association Task Force (1990) Benzodiazepine dependence, toxicity, and abuse. A task force report of the American Psychiatric Association. American Psychiatric Association, Washington, DC

Andrews N, File SE (1993) Increased 5-HT release mediates the anxiogenic response during benzodiazepine withdrawal: a review of supporting neurochemical and behavioural evidence. Psychopharmacology 112:21–25

Anello C (1989) Presentation on findings from the FDA's spontaneous reporting system to the Psychopharmacological Drugs Advisory Committee meeting. Food and Drug administration, September 22

Arana GW, Pearlman C, Shader RI (1985) Alprazolam-induced mania: two clinical cases. Am J Psychiatry 142:363–369
Ayd F (1979) Benzodiazepines: dependence and withdrawal. JAMA 242:1401
Balter MB, Manheimer DI, Mellinger GE et al. (1984) A cross-national comparison of anti-anxiety/sedative drug use. Curr Med res Opin 8S:5–20
Baran H, Sperk G, Hortnagl H, Sapetschnig G, Hornykiewicz O (1985) α-adrenoceptors modulate kainic acid-induced limbic seizures. Eur J Pharmacol 113:263–2269
Barten HH (1965) Toxic psychosis with transient dysmnesic syndrome following withdrawal from Valium. Am J Psychiatry 121:1210–1211
Barton DF (1981) More on lorazepam withdrawal. Drug Intell Clin Pharm 15:487–488
Bell J, Bickford-Wimer PC, de la Garza R, Egan M, Freedman R (1988) Increased central noradrenergic activity during benzodiazepine withdrawal: an electrophysiological study. Neuropharmacology 27:1187–1190
Berlin RM, Connel LJ (1983) Withdrawal symptoms after long-term treatment with therapeutic doses of flurazepam: a case report. Am J Psychiatry 140:488–490
Bixler EO, Kales A, Soldatos CR, Kales JD (1977) Flunitrazepam, an investigational hypnotic drug: sleep laboratory evaluations. J Clin Pharmacol 17:569–578
Bixler EO, Kales A, Soldatos CR, Scharf MB, Kales JD (1978) Effectiveness of temazepam with short-, intermediate- and long-term use: sleep laboratory evaluation. J Clin Pharmacol 18:110–118
Bixler EO, Kales A, Brubaker BH, Kales JD (1987) Adverse reactions to benzodiazepine hypnotics: spontaneous reporting system. Pharmacology 35:286–300
Bixler EO, Kales A, Manfredi RL, Vgontzas AN, Tyson KL, Kales JD (1991) Next-day memory impairment with triazolam use. Lancet 337:827–831
Bliwise D, Seidel W, Greenblatt DJ, Dement W (1984) Nighttime and daytime efficacy of flurazepam and oxazepam in chronic insomnia. Am J Psychiatry 141:191–195
Bowden CL, Fisher JG (1980) Safety and efficacy of long-term diazepam therapy. South Med J 73:1581–1584
Braestrup C, Nielsen M, Neilsen EB, Lyon M (1979) Benzodiazepine receptors in the brain as affected by different experimental stresses: the changes are small and not unidrectional. Psychopharmacology (Berl) 65:273–277
Braestrup C, Schmiechen R, Nielsen M, Petersen EN (1982) Benzodiazepine receptor ligands, receptor occupancy, pharmacological effect and GABA receptor coupling. In: Usdin E, Skolnick P, Tallman JF, Greenblatt D, Paul SM (eds) Pharmacology of benzodiazepines. MacMillan, London, pp 71–85
Bruni G, Dal Pra P, Dotti MT, Segre G (1980) Plasma ACTH and cortisol levels in benzodiazepine treated rats. Pharmacol Res 12:163–175
Busto U, Simpkins J, Sellers EM, Sisson B, Segal R (1983) Objective determination of benzodiazepine use and abuse in alcoholics. Br J Addict 78:429–435
Busto U, Sellers EM, Naranjo CA, Cappell H, Sanchez-Craig M, Sykova K (1986) Withdrawal reaction after long-term therapeutic use of benzodiazepines. N Engl J Med 315:854–859
Calogero AE, Gallucci WT, Chrousos GP, Gold PW (1988) Interaction between GABAergic neurotransmission and rat hypothalamic corticotropin-releasing hormone secretion in vitro. Brain Res 463:28–36
Cantopher T, Olivieri S, Cleave N, Edwards JG (1990) Chronic benzodiazepine dependence. A comparative study of abrupt withdrawal under propranolol cover versus gradual withdrawal. Br J Psychiatry 156:406–411
Catalan J, Gath DH, Bond A, Martin P (1988) General practice patients on long-term psychotropic drugs: a controlled investigation. Br J Psychiatry 152:399–405
Charney DS, Heninger GR (1985) Noradrenergic function and the mechanism of action of antianxiety treatment I. The effect of long-term alprazolam treatment. Arch Gen Psychiatry 42:458–467

Charney DS, Heninger GR, Redmond D (1983) Yohimbine induced anxiety and increased noradrenergic function in humans: effects of diazepam and clonidine. Life Sci 33:19–29
Chesney CM, Pjifer DD, Cagen LM (1987) Triazolobenzodiazepines competitively inhibit the binding of platelet activating factor (PAF) to human platelets. Biochem Biophys Res Comm 144:359–366
Chouinard G (1986) Rebound anxiety: incidence and relationship to subjective cognitive impairment. J Clin Psychiatry Monograph Series 4:12–16
Chrousos GP, Loriaux L, Gold PW (eds) (1988) Mechanisms of physical and emotional stress. Plenum, New York
Chung M, Hilbert JM, Gural RP, Radwanski B, Symchowicz S, Zampaglione N (1984) Multiple-dose quazepam kinetics. Clin Pharmacol Ther 35:520–524
Ciraulo DA, Sands BF, Shader RI (1988) Critical review of liability for benzodiazepine abuse among alcoholics. Am J Psychiatry 145:1501–1506
Consolo S, Ladinsky H, Peri G, Garattini S (1974) Effect of diazepam on mouse whole brain area acetylcholine and choline levels. Eur J Pharmacology 27:266–2268
Copland AM, Balfour DJK (1987) The effects of diazepam on brain 5-HT and 5-HIAA in stressed and unstressed rats. Pharmacol Biochem Behav 27:619–624
Corrodi H, Fuxe K, Hokfelt T (1967) The effect of some psychoactive drugs on central monoamine neurons. Eur J Pharmacology 1:363–368
Corrodi H, Fuxe K, Lidbrink P, Olson L (1971) Minor tranquilizers, stress and central catecholamine neurons. Brain Res 29:1–16
Covi L, Lipman RS, Pattison JH, Derogatis LR, Uhlenhuth EH (1973) Length of treatment with anxiolytic sedatives and response to their sudden withdrawal. Acta Psychiatr Scand 49:51–64
De Bard ML (1979) Diazepam withdrawal syndrome: a case with psychosis, seizure, and coma. Am J Psychiatry 136:104–105
de la Fuente JR, Rosenbaum AH, Martin HR, Niven RQ (1980) Lorazepam-related withdrawal seizures. Mayo Clin Proc 55:190–192
De Tullio PL, Kirking DM, Zacardelli DK, Kwee P (1989) Evaluation of long-term triazolam use in an ambulatory Veterans Administration medical center population. Ann Pharmacother 23:290–293
De Wit H (1991) Diazepam preference in males with and without an alcoholic first-degree relative. Alcohol Clin Exp Res 15:593–600
De Wit H, Pierri J, Johanson CE (1989) Assessing individual differences in ethanol preference using a cumulative dosing procedure. Psychopharmacology 98:113–2119
Dement WC, Carskadon MA, Mitler MM, Phillips RL, Zarcone VP (1978) Prolonged use of flurazepam: a sleep laboratory study. Behav Med 5:25–31
Dennis T, Curet O, Nishikawa T, Scatton B (1985) Further evidence for, and nature of, the facilitatory GABAergic influence on central noradrenergic transmission. Arch Pharmacol 331:225–234
Duka T, Ackenheil M, Noderer J, Doenicke A, Dorow R (1986) Changes in noradrenaline plasma levels and behavioural responses induced by benzodiazepine agonists with the benzodiazepine antagonist Ro 15-1788. Psychopharmacology 90:351–357
Dunbar GC, Morgan DD, Perera KM (1988) The concurrent use of alcohol, cigarettes and caffeine in British benzodiazepine users as measured by a general population survey. Br J Addict 83:689–694
Dysken MW, Chan CH (1977) Diazepam withdrawal psychosis: a case report. Am J Psychiatry 134:573
Einarson TR (1980) Lorazepam withdrawal seizures. Lancet 1:151
Einarson TR (1981) Oxazepam withdrawal convulsions. Drug Intell Clin Pharm 15:487–488

Eriksson E, Carlsson M, Nilsson C, Soderpalm B (1986) Does alprazolam in contrast to diazepam, activate alpha-2-adrenoreceptors involved in the regulation of rat growth hormone secretion? Life Sci 38:1491–1498

Essig CF (1964) Addiction to nonbarbiturate sedative and tranquilizing drugs. Clin Pharmacol Ther 5:334–343

File SE, Pello S (1986) Triazolobenzodiazepines antagonize the effects of anxiogenic drugs mediated at three different central nervous system sites. Neurosci Lett 61:115–119

Fleischhacker WW, Barnas C, Hackenberg B (1986) Epidemiology of benzodiazepine dependence. Acta Psychiatr Scand 74:80–83

Fontaine R, Chouinard G, Annable A (1984) Rebound anxiety in anxious patients after abrupt withdrawal of benzodiazepine treatment. Am J Psychiatry 141:848–2852

Fontaine R, Chouinard G, Annable A (1985) Efficacy and withdrawal of two potent benzodiazepines: bromazepam and lorazepam. Psychopharmacol Bull 21:91–92

Foy A, Drinkwater V, March S, Mearrick P (1986) Confusion after admission to hospital in elderly patients using benzodiazepines. Br Med J 293:1072

France RD, Krishnan KRR (1984) Alprazolam-induced manic reaction (letter to editor). Am J Psychiatry 141:1127–1128

Fraser HF, Shaver MR, Maxwell ES (1953) Death due to withdrawal of barbiturates. Ann Intern Med 38:1319–1325

Funderburk FR, Griffiths RR, Mcleod DR, Bigelow GE, Mackenzie A, Liebson IA, Nemeth-Coslett R (1988) Relative abuse liability of lorazepam and diazepam: an evaluation in recreational drug users. Drug Alcohol Depend 22:215–222

Gardner DL, Cowdry RW (1985) Alprazolam-induced dyscontrol in borderline personality disorder. Am J Psychiatry 142:98–100

Gath I, Bar-On E, Rogowski Z, Bental E (1983) Automatic scoring of polygraphic sleep recordings: midazolam in insomniacs. Br J Clin Pharmacol 16:89S–96S

Goodman LS, Gilman A (eds) (1990) Goodman and Gilman's the pharmacological basis of therapeutics, 8th edn. Pergamon, New York

Goodwin FK, Jamison KR (1990) Manic-depressive illness. Oxford University Press, New York

Gorman JM, Liebowitz MR, Fyer AJ, Stein J (1989) A neuroanatomical hypothesis for panic disorder. Am J Psychiatry 146:148–161

Grant SJ, Galloway MP, Mayor R, Fenerty JP, Finkelstein MF, Roth RH, Redmond DE (1985) Precipitated diazepam withdrawal elevates noradrenergic metabolism in primate brain. Eur J Pharmacol 107:127–132

Greenblatt DJ (1992) Pharmacology of benzodiazepine hypnotics. J Clin Psychiatry 53:7–13

Greenblatt DJ, Divoll M, Harmatz JS, MacLaughlin DS, Shader RI (1981) Kinetics and clinical effects of flurazepam in young and elderly non-insomniacs. Clin Pharmacol Ther 30:475–486

Greenblatt DJ, Harmatz JS, Zinny MA, Shader RI (1987) Effect of gradual withdrawal on the rebound sleep disorder after discontinuation of triazolam. N Engl J Med 317:722–728

Greenblatt DJ, Miller LG, Shader RI (1990) Neurochemical and pharmacokinetic correlates of the clinical action of benzodiazepine hypnotic drugs. Am J Med 88:18S–24S

Griffiths RR, Roache JD (1985) Abuse liability of benzodiazepines: a review of human studies evaluating subjective and/or reinforcing effects. In: Smith D, Wesson DL (eds) The benzodiazepines: current standards for medical practice. MTP, Lancaster, 209–225

Griffiths RR, Sannerud CA (1987) Abuse of and dependence on benzodiazepines and other anxiolytic/sedative drugs. In: Meltzer HY (ed) Psychopharmacology: the third generation of progress. Raven, New York, pp 1535–1541

Heninger C, Gallager DW (1988) Altered γ-aminobutyric acid/benzodiazepine interaction after chronic diazepam exposure. Neuropharmacology 27:1073–1076
Heritch AJ, Capwell R, Roy-Byrne PP (1987) A case of psychosis and delirium following withdrawal from triazolam. J Clin Psychiatry 48:168–169
Herman JB, Brotman AW, Rosenbaum JF (1987) Rebound anxiety in panic disorder patients treated with shorter-acting benzodiazepines. J Clin Psychiatry 48:22–26
Hollister LE (1977) Withdrawal from benzodiazepine therapy. JAMA 237:1432
Hollister LE, Motzenbacker F, Degan R (1961) Withdrawal reactions from chlordiazepoxide ("Librium"). Psychopharmacology 2:63–68
Hollister LE, Conley FK, Britt RH, Shuer L (1981) Long-term use of diazepam. JAMA 246:1568–1570
Isbell H, Altschul S, Kornetsky CH, Eisenman AG, Flanary HC, Freiser HF (1950) Chronic barbiturate intoxication: an experimental study. AMA Arch Neurol Psychiatry 64:1–28
Jacob MS, Sellers EN (1979) Use of drugs with dependence liability. Can Med Assoc J 121:717–724
Johnson FA, Van Buren HC (1962) Abstinence syndrome following glutethimide intoxication. JAMA 1980:1024–1027
Kalant H, LeBlanc A, Gibbons R (1971) Tolerance to, and dependence on, some non-opiate psychotropic drugs. Pharmacol Rev 23:135–191
Kales A (1990) Evaluations of new drugs. Pharmacotherapy 10:1–12
Kales A (1993) Benefit-to-risk ratio of benzodiazepines. In: World Psychiatric Association, Proceedings of the 9th World Congress of Psychiatry, Brazil, p 39
Kales A, Kales JD (1983) Sleep laboratory studies of hypnotic drugs: efficacy and withdrawal effects. J Clin Psychopharmacol 3:140–150
Kales A, Vgontzas AN (1990) Not all benzodiazepines are alike. In: Stephanis CN, Rabavilas AD, Soldatos CR (eds) Psychiatry: a world perspective, vol 3. Elsevier, Amsterdam, pp 379–384
Kales A, Bixler EO, Tan T-L, Scharf MB, Kales JD (1974) Chronic hypnotic-drug use: ineffectiveness, drug-withdrawal insomnia, and dependence. JAMA 227: 513–517
Kales A, Bixler EO, Scharf MB, Kales JD (1976a) Sleep laboratory studies of flurazepam: a model for evaluating hypnotic drugs. Clin Pharmacol Ther 19: 576–583
Kales A, Caldwell AB, Preston TA, Healey S, Kales JD (1976b) Personality patterns in insomnia, theoretical implications. Arch Gen Psychiatry 33:1128–1134
Kales A, Kales JD, Bixler EO, Scharf MB, Russek E (1976c) Hypnotic efficacy of triazolam: sleep laboratory evaluation of intermediate-term effectiveness. J Clin Pharmacol 16:399–406
Kales A, Scharf MB, Kales JD (1978) Rebound insomnia: a new clinical syndrome. Science 201:1039–1041
Kales A, Scharf MB, Kales JD, Soldatos CR (1979) Rebound insomnia: a potential hazard following withdrawal of certain benzodiazepines. JAMA 241:1692–1695
Kales A, Scharf MB, Bixler EO, Schweitzer PK, Jacoby JA, Soldatos CR (1981) Dose-response studies of quazepam. Clin Pharmacol Ther 30:194–200
Kales A, Bixler EO, Soldatos CR, Mitsky DJ, Kales JD (1982a) Dose-response studies of lormetazepam: efficacy, side effects, and rebound insomnia. J Clin Pharmacol 22:520–530
Kales A, Bixler EO, Soldatos CR, Vela-Bueno, Jacoby JA, Kales JD (1982b) Quazepam and flurazepam: long-term use and extended withdrawal. Clin Pharmacol Ther 32:781–788
Kales A, Caldwell AB, Soldatos CR, Bixler EO, Kales JD (1983a) Biopsychobehavioral correlates of insomnia. II. Pattern specificity and consistency with the Minnesota Multiphasic Personality Inventory. Psychosom Med 45:341–356
Kales A, Soldatos CR, Bixler EO, Kales JD (1983b) Rebound insomnia and rebound anxiety: a review. Pharmacology 26:121–137

Kales A, Soldatos CR, Bixler EO, Goff PJ, Vela-Bueno A (1983c) Midazolam: dose-response studies of effectiveness and rebound insomnia. Pharmacology 26:138–149

Kales A, Soldatos CR, Bixler EO, Kales JD (1983d) Early morning insomnia with rapidly eliminated benzodiazepines. Science 220:95–97

Kales A, Soldatos CR, Vela-Bueno (1985) Clinical comparison of benzodiazepine hypnotics with short and long elimination half-lives. In: Smith DE, Wesson DR (eds) The benzodiazepines, current standards for medical practice. MPT, Boston, pp 121–147

Kales A, Bixler EO, Soldatos CR, Vela-Bueno A, Jacoby JA, Kales JD (1986a) Quazepam and temazepam: effects of short- and intermediate-term use and withdrawal. Clin Pharmacol Ther 39:345–352

Kales A, Bixler EO, Soldatos CR, Jacoby JA, Kales JD (1986b) Lorazepam: effects on sleep and withdrawal phenomena. Pharmacology 32:121–130

Kales A, Bixler EO, Vela-Bueno A, Soldatos CR, Niklaus DE, Manfredi RL (1986c) Comparison of short and long half-life benzodiazepine hypnotics: triazolam and quazepam. Clin Pharmacol Ther 40:378–386

Kales A, Bixler EO, Vela-Bueno A, Soldatos CR, Manfredi RL (1987) Alprazolam: effects on sleep and withdrawal phenomena. J Clin Pharmacol 27:508–515

Kales A, Soldatos CR, Bixler EO, Kales JD, Vela-Bueno A (1988) Diazepam: effects on sleep and withdrawal phenomena. J Clin Psychopharmacol 8:340–346

Kales A, Manfredi RL, Vgontzas AN, Baldassano CF, Kostakos K, Kales JD (1991a) Clonazepam: sleep laboratory study of efficacy and withdrawal. J Clin Psychopharmacol 11:189–193

Kales A, Manfredi RL, Vgontzas AN, Bixler EO, Vela-Bueno A, Fee EC (1991b) Rebound insomnia after only brief and intermittent use of rapidly eliminated benzodiazepines. Clin Pharm Ther 49:468–476

Kales A, Vgontzas AN, Bixler, EO (1994) Not all benzodiazepines are alike: update 1993. In: Biegel A, Lopez Ibor IJ, Costa e Silva JA (eds) Psychiatry: a world perspective, vol II. World Scientific, Singapore, pp 942–946

Kalogeras KT, Calogero AE, Kuribayiashi T, Khan I, Gallucci WT, Kling MA, Chrousos GP, Gold PW (1990) In vitro and in vivo effects of the triazolobenzodiazepine alprazolam on hypothalamic-pituitary-adrenal function: pharmacological and clinical implications. J Clin Endocrinol Metab 70:1462–1471

Kataoka Y, Fujimoto M, Alho H, Guidotti A, Geffard M, Kelly GD, Hanbauer I (1986) Intrinsic gamma aminobutyric acid receptors modulate the release of catecholamine from canine adrenal gland in situ. J Pharmacol Exp Ther 239: 584–590

Khan A, Joyce P, Jones AV (1980) Benzodiazepine withdrawal syndromes. N Z Med J 92:94–96

Khan RM, Soliman MRI (1993) Effects of benzodiazepine agonist, antagonist and inverse agonist on ethanol-induced changes in beta-endorphin levels in specific rat brain regions. Pharmacology 47:337–343

Kornecki E, Ehrlich YH, Lenox RH (1984) Platelet-activating factor-induced aggregation of human platelets specifically inhibited by triazolobenzodiazepines. Science 226:1454–1456

Kornecki E, Lenox RH, Hardwick DH, Bergdahl JA, Ehrlich YH (1987) Interactions of the alkyl-ether-phospholipid activating factor (PAF) with platelets, neural cells, and the psychotropic drugs triazolobenzodiazepines. Adv Exp Med Biol 221:477–488

Kozak W, Valzelli L, Garattini S (1984) Anxiolytic activity on locus coeruleus-mediated suppression of miracidial aggression. Eur J Pharmacology 105:323–326

Lader M (1983a) Benzodiazepine withdrawal states. In: Trimble MR (ed) Benzodiazepine divided. John Wiley, New York, pp 17–32

Lader M (1983b) Dependence on benzodiazepines. J Clin Psychiatry 44:121–127

Lader M (1984) Benzodiazepine dependence. Prog Neuropsychopharmacol Biol Psychiatry 8:85–95

Lader M (1991) Can buspirone induce rebound, dependence or abuse? Br J Psychiatry Suppl 12:45–51

Lader M, Petursson H (1983) Long-term effects of benzodiazepines. Neuropharmacology 22:527–533

Lamphere J, Roehrs T, Vogel G, Koshorek G, Fortier J, Roth T (1990) The chronic efficacy of midazolam. Int Clin Psychopharmacol 5:31–39

Laughren TP, Lee H (1992) Review of adverse events data in Upjohn-sponsored clinical studies of Halcion (triazolam), NDA 17-892. Presented at Food and Drug Administration PDAC Meeting, May 18

Laughren TP, Battey Y, Greenblatt DJ, Harrop DS (1982) A controlled trial of diazepam withdrawal in chronically anxious outpatients. Acta Psychiatr Scand 65:171–179

Little HJ (1991) The role of neuronal calcium channels in dependence of ethanol and other sedatives/hypnotics. Pharmacol Ther 50:347–365

Lloyd EA, Clark LD (1959) Convulsions and delirium incident to glutethimide (Doriden) withdrawal. Dis Nerv Syst 20:524–526

Lopez AL, Kathol RG, Noyes R (1990) Reduction in urinary free cortisol during benzodiazepine treatment of panic disorder. Psychoneuroendocrinology 15:23–28

Lopez F, Miller LG, Greenblatt DJ, Paul SM, Shader RI (1988) Low-dose alprazolam augments motor activity in mice. Pharmacol Biochem Behav 30:511–513

Lucki I, Volpicelli JR, Schwiezer E (1991) Differential craving between abstinent alcohol-dependent subjects and therapeutic users of benzodiazepines. Proceedings of the Annual Meeting of the Committee on Problems of Drug Dependence, June 1990, NIDA Res Monogr 105:322–323

Lynn EJ (1985) Triazolam addiction. Hosp Community Psychiatry 36:779–780

Maletzky BM, Klotter J (1976) Addiction to diazepam. Int J Addict 2:95–115

Mamelak M, Csima A, Price V (1984) A comparative 25-night sleep laboratory study on the effects of quazepam and triazolam on the sleep of chronic insomniacs. J Clin Pharmacol 24:65–75

Mamelak M, Csima A, Price V (1990) The effects of a single night's dosing with triazolam on sleep the following night. J Clin Pharmacol 30:549–555

Manfredi RL, Kales A, Vgontzas AN, Bixler EO, Melda AI, Falcone CM (1991) Buspirone: sedative or stimulant effect? Am J Psychiatry 148:1213–1217

Marks J (1981) Diazepam: the question of long-term therapy with withdrawal reactions. Drug Therapy 11:5–30

Martin PR, Ebert MH, Gordon EK, Weingartner H, Kopin IJ (1984) Catecholamine metabolism during clonidine withdrawal. Psychopharmacology 84:58–63

Martinez-Cano H, Vela-Bueno A (1993) Daytime consumption of triazolam. Acta Psychiatr Scand 88:286–288

McCracken SG, De Wit H, Uhlenhuth EH, Johanson CE (1990) Preference for diazepam in anxious adults. J Clin Psychopharmacol 10:190–196

Mellinger GD, Balter MB, Uhlenhuth EH (1984) Prevalence and correlates of the long-term regular use of anxiolytics. JAMA 251:375–379

Mendelson WB, Weingartner H, Greenblatt DJ, Garnett D, Gillin JC (1982) A clinical study of flurazepam. Sleep 5:350–360

Mikashima H, Takehara S, Muramoto Y, Khomaru T, Terasaawa T, Maruyama Y (1987) An antagonistic activity of estazolam on platelet activating factor (PAF) in vitro effects on platelet aggregation and PAF receptor binding. Jpn J Pharmacol 44:387–391

Miller LG (1991) Chronic benzodiazepine administration: from the patient to the gene. J Clin Pharmacol 31:492–495

Miller LG, Greenblatt DJ, Barnhill JG, Deutsch SI, Shader RI, Paul SM (1987) Benzodiazepine receptor binding of triazolobenzodiazepines in vivo: increased receptor number with low-dose alprazolam. J Neurochem 49:1595–1601

Miller LG, Greenblatt DJ, Barnhill JG, Shader RI (1988a) Chronic benzodiazepine administration: I. Tolerance is associated with benzodiazepine receptor binding downregulation and decreased gamma-aminobutyric $acid_A$ receptor function. J Pharmacol Exp Ther 246:170–176

Miller LG, Greenblatt DJ, Roy BB, Summer WR, Shader RI (1988b) Chronic benzodiazepine administration. II. Discontinuation syndrome is associated with upregulation of gamma-aminobutyric $acid_A$ receptor complex binding and function. J Pharmacol Exp Ther 246:177–182

Mitler MM, Seidel BA, van den Hoed J, Greenblatt DJ, Dement WC (1984) Comparative hypnotic effects of flurazepam, triazolam, and placebo: a long-term simultaneous nighttime and daytime study. J Clin Psychopharamacol 4:2–13

Mohler H, Okada T, Enna SJ (1978) Benzodiazepine and neurotransmitter receptor binding in rat brain after chronic administration of diazepam or phenobarbital. Brain Res 156:391–395

Mohler H, Sieghart W, Polc P, Bonetti EP, Hunkeler W (1982) Benzodiazepine receptors: biochemistry and pharmacology. In: Usdin E, Skolnick P, Tallman JF, Greenblatt D, Paul SM (eds) Pharmacology of benzodiazepines. Macmillan, London, pp 63–70

Monti JM, Debellis J, Gratadoux E, Alterwain P, Altier H, D'Angelo L (1982) Sleep laboratory study of the effects of midazolam in insomniac patients. Eur J Clin Pharmacol 21:479–484

Monti JM, DeBellis J, Alterwain P, D'Angelo L (1987) Midazolam and sleep in insomniac patients. Br J Clin Pharmacol 16:87S–88S

Moon CAL, Ankier SI, Hayes G (1985) Early morning insomnia and daytime anxiety – a multicentre general practice study comparing loprazolam and triazolam. Br J Clin Pract 39:352–358

Morgan K, Oswald I (1982) Anxiety caused by a short half-life hypnotic. Br Med J 284:942

Müller WE, Stillbauer AE (1983) Benzodiazepine hypnotics: time course and potency of benzodiazepine receptor occupation after oral application. Pharmacol Biochem Behav 18:545–549

Murphy SM, Tyrer PJ (1991) A double blind comparison of the effects of gradual withdrawal of lorazepam, diazepam and bromazepam in benzodiazepine dependence. Br J Psychiatry 158:511–516

Murphy SM, Owen RT, Tyrer PJ (1984) Withdrawal symptoms after six weeks' treatment with diazepam. Lancet 2:1389

Nabeshima T, Tohyama K, Ichiara K, Kameyama T (1990) Effect of benzodiazepines on passive avoidance response and latent learning in mice: relationship to benzodiazepine receptors and the cholinergic neuronal system. J Pharmacol Exp Ther 255:789–794

Nordberg A, Walhstrom G (1984) Different interactions of steric isomers of hexobarbital to muscarinic agonist and antagonist binding sites in the brain. Brain Res 310:189–192

Nordberg A, Walhstrom G (1992) Cholinergic mechanisms in physical dependence on barbiturates, ethanol and benzodiazepines. J Neural Transm Suppl [GenSect] 88:199–221

Noyes R, Clancy J, Coryell WH, Crowe RR, Chaudry DR, Domingo DV (1985) A withdrawal syndrome after abrupt discontinuation of alprazolam. Am J Psychiatry 142:114–116

Nutt D, Molyneux S (1987) Benzodiazepines, plasma MHPG and alpha-2-adrenoceptor function in man. Int Clin Psychopharmacol 2:151–157

Oswald I, Adam K, Borrow S, Idzikowski R (1979) The effects of two hypnotics on sleep, subjective feelings and skilled performance. In: Passouant P, Oswald I (eds) Pharmacology of the states of alterness. Pergamon, New York, pp 51–63

Owens MJ, Bissette G, Nemeroff CB (1989) Acute effects of alprazolam and adinazolam on the concentrations of corticotropin-releasing factor in the rat brain. Synapse 4:196–202

Owens MJ, Vargas MA, Knight DL, Nemeroff CB (1991) The effects of alprazolam on corticotropin-releasing factor neurons in the rat brain: acute time course, chronic treatment and abrupt withdrawal. J Pharmacol Exp Ther 258:349–356
Patterson JF (1987) Triazolam syndrome in the elderly. South Med J 80:1425–1426
Pecknold JC, Fleury D (1986) Alprazolam-induced manic episode in two patients with panic disorder. Am J Psychiatry 143:652–653
Pecknold JC, Swinson RP, Kuch K, Lewis CP (1988) Alprazolam in panic disorder: results from a multicenter trial. Arch Gen Psychiatry 45:429–436
Petursson H, Bond PA, Smith B, Lader MH (1983) Monoamine metabolism during chronic benzodiazepine treatment and withdrawal. Biol Psychiatry 18:207–213
Pevnick J, Jasinski D, Haertzen CA (1978) Abrupt withdrawal from therapeutically administered diazepam. Arch Gen Psychiatry 35:995–998
Popova J, Petkov VV, Tokuschieva L (1988) The effect of chronic diazepam and medazepam treatment on the number and affinity of muscarine receptors in different rat brain structures. Gen Pharmacol 19:227–231
Post RM, Weiss SRB, Pert A (1988) Implications of behavioral sensitization and kindling for stress-induced behavioral change. In: Chrousos GP, Loriaux L, Gold PW (eds) Mechanisms of physical and emotional stress. Plenum, New York, pp 441–463
Preskorn SH, Denner LJ (1977) Benzodiazepines and withdrawal psychosis. JAMA 237:36–38
Rastogi RB, Lapierre YD, Singhal RL (1976) Evidence for the role of brain norepinephrine and dopamine in "rebound" phenomenon seen during withdrawal after repeated exposure to benzodiazepines. J Psychiatr Res 13:65–75
Rastogi RB, Lapierre YD, Singhal RL (1978) Some neurochemical correlates of "rebound" phenomenon observed during withdrawal after long-term exposure to 1,4-benzodiazepines. Prog Neuro-Psychopharmacol Biol Psychiatry 2:43–54
Ratna L (1981) Addiction to temazepam. Br Med J 282:1837
Redmond DE (1987) Studies of the nucleus locus coeruleus in monkeys and hypotheses for neuropsychopharmacology. In: Meltzer HY (ed) Psychopharmacology: the third generation of progress. Raven, New York, pp 967–975
Rickels K, Case WG, Diamond L (1980) Relapse after short-term drug therapy in neurotic outpatients. Int Pharmacopsychiat 15:186–192
Rickels K, Case WG, Downing RW, Winokur A (1983) Long-term diazepam therapy and clinical outcome. JAMA 250:767–771
Rickels K, Schweizer E, Case WG, Garcia-Espana F (1988) Benzodiazepine dependence, withdrawal severity, and clinical outcome: effects of personality. Psychopharmacol Bull 24:415–420
Rickels K, Schweizer E, Case WG, Greenblatt DJ (1990) Long-term therapeutic use of benzodiazepines. I. Effects of abrupt discontinuation. Arch Gen Psychiatry 47:899–907
Rickels K, Schweizer E, Weiss S, Zavodnick S (1993) Maintenance drug treatment for panic disorder. II. Short- and long-term outcome after drug taper. Arch Gen Psychiatry 50:61–68
Rifkin A, Quitkin F, Klein DF (1976) Withdrawal reaction to diazepam. JAMA 236:2172–2173
Roehrs T, Zorick F, Koshorek GL, Wittig R, Roth T (1983) The effect of acute administration of brotizolam in subjects with disturbed sleep. Br J Clin Pharmacol 16:371S–376S
Roehrs T, Vogel G, Roth T (1990) Rebound insomnia: its determinants and significance. Am J Med 88:3A–39S
Rosenbaum JF, Woods SW, Groves JE, Klerman GL (1984) Emergence of hostility during alprazolam treatment. Am J Psychiatry 141:792–793
Roth T, Kramer M, Lutz T (1976) Intermediate use of triazolam: a sleep laboratory study. J Int Med Res 4:59–62

Scharf MB, Bixler EO, Kales A, Soldatos CR (1979) Long-term sleep laboratory evaluation of flunitrazepam. Pharmacology 19:173–181
Scharf MB, Kales A, Bixler EO, Jacoby JA, Schweitzer PK (1982) Lorazepam: efficacy, side effects and rebound phenomena. Clin Pharmacol Ther 31:175–179
Scharf MB, Fletcher K, Graham JP (1988) Comparative amnesic effects of benzodiazepine hypnotic agents. J Clin Psychiatry 49:134–137
Scharf MB, Sachais BA, Mayleben DW, Fletcher K, Jennings SW (1990) A polysomnographic comparison of temazepam 15 and 30 mg with triazolam 0.125 and 0.25 mg in chronic insomnia. Curr Ther Res 48:555–567
Schmidt LG, Grohmann R, Mueller-Oerlinghausen B Otto M, Ruether E, Wolf B (1989) Prevalence of benzodiazepine abuse and dependence in psychiatric inpatients with different nosology: an assessment of hospital-based drug surveillance data. Br J Psychiatry 154:839–843
Schneider LS, Syapin PJ, Pawluczyk S (1987) Seizures following triazolam withdrawal despite benzodiazepine treatment. J Clin Psychiatry 48:418–419
Schweizer E, Rickels K, Case WG, Greenblatt DJ (1990) Long-term therapeutic use of benzodiazepines. I. Effects of gradual taper. Arch Gen Psychiatry 47: 908–915
Selig JW Jr (1966) A possible oxazepam abstinence syndrome. JAMA 198:951–952
Shader FI, Greenblatt DJ (1993) Use of benzodiazepines in anxiety disorders. N Engl J Med 328:1398–1405
Shephard RA (1986) Neurotransmitters, anxiety and benzodiazepines: a behavioral review. Neurosci Biobehav Rev 10:449–461
Smith DE, Landry MJ (1990) Benzodiazepine dependency discontinuation: focus on the chemical dependency detoxification setting and benzodiazepine-polydrug abuse. J Psychiatr Res 2:145–156
Smith DE, Wesson DR (1983) Benzodiazepine dependency syndromes. J Psychoactive Drugs 15:85–95
Smith DE, Wesson DR (1985) Benzodiazepine dependency syndromes. In: Smith DE, Wesson DR (eds) The benzodiazepines, current standards for medical practice. MTP, Lancaster, pp 235–248
Soldatos CR, Sakkas PN, Bergiannaki JD, Stenfanis CN (1986) Behavioral side effects of triazolam in psychiatric inpatients: report of five cases. Drug Intell Clin Pharmacol 20:294–297
Suzdak PD, Gianutsos G (1985a) Differential coupling of GABA-A and GABA-B receptors to the noradrenergic system. J Neural Transmission 62:77–89
Suzdak PD, Gianutsos G (1985b) GABA-noradrenergic interaction: evidence for differential sites of action for GABA-A and GABA-B receptors. J Neural Transmission 64:163–172
Swanson LA, Okada T (1963) Death after withdrawal of meprobamate. JAMA 184:780–781
Tabakoff B, Hoffman PL (1987) Ethanol tolerance and dependence. In: Goedde WH, Agarwai DP (eds) Genetics and alcoholism. Liss New York, pp 253–269
Tabakoff B, Yanai J, Ritzmann RF (1978) Brain noradrenergic systems as a prerequisite for developing tolerance to barbiturates. Science 200:449–451
Taylor KM, Laverty R (1969) The effect of chlordiazepoxide, diazepam and nitrazepam on catecholamine metabolism in regions of the rat brain. Eur J Pharmacol 8:296–301
Thiebot MH (1986) Are serotonergic neurons involved in the control of anxiety and in the anxiolytic activity of benzodiazepines? Pharmacol Biochem Beh 24:1471–1477
Tien AY, Gujavarty KS (1985) Seizure following withdrawal from triazolam. Am J Psychiatry 142:1516–1517
Tyrer P, Rutherford D, Huggett T (1981) Benzodiazepine withdrawal symptoms and propranolol. Lancet I:520–522

Tyrer P, Owen R, Dowling S (1983) Gradual withdrawal of diazepam after long-term therapy. Lancet I:1402–1406
van der Kroef C (1979) Reactions to triazolam. Lancet II:526
Vela-Bueno A, Oliveros JC, Dobladez-Blanco B, Arrigain-Igurra S, Soldatos CR, Kales A (1983) Brotizolam: a sleep laboratory evaluation. Eur J Clin Pharmacol 25:53–56
Vgontzas AN, Kales A, Bixler EO (in press) Benzodiazepine side effects: Role of pharmacologic properties. Pharmacology
Vinogradov S, Reiss AL, Csernansky JG (1986) Clonidine therapy in withdrawal from high-dose alprazolam treatment. Am J Psychiatry 143:1188
Vogel GW, Vogel F (1983) Effect of midazolam on sleep of insomniacs. Psychopharmacol Ser 16 [Suppl 1]:103S–108S
Vogel GW, Thurmond A, Gibbons P, Edwards K, Sloan KB, Sexton K (1975) The effect of triazolam on the sleep of insomniacs. Psychopharmacology 41:65–69
Vogel GW, Barker K, Gibbons P, Thurmond A (1976) A comparison of the effects of flurazepam 30 mg and triazolam 0.5 mg on the sleep of insomniacs. Psychopharmacology 47:81–86
Whitcup SM, Miller F (1987) Unrecognized drug dependence in psychiatrically hospitalized elderly patients. J Am Geriatr Soc 35:297–301
Winokur A, Rickels K, Greenblatt DJ, Snyder PJ, Schatz NJ (1980) Withdrawal reaction from long-term, low dosage administration of diazepam. Arch Gen Psychiatry 37:101–105
Wise RA (1987) The role of reward pathways in the development of drug dependence. Pharmacol Ther 35:227–263
Wolf B, Griffiths RR (1991) Physical dependence on benzodiazepines: differences within the class. Drug Alcohol Depend 29:153–156
Wolf B, Brohmann R, Biber D, Brenner PM, Ruther E (1989a) Benzodiazepine abuse and dependence in psychiatric inpatients. Pharmacopsychiatry 22:54–60
Wolf B, Iguchi MY, Griffiths RR (1989b) Sedative/tranquilizer use and abuse in alcoholics currently in outpatient treatment: incidence, pattern and preference. NIDA Res Monogr Ser 95:376–377
Woods JH, Katz JL, Winger G (1987) Abuse liability of benzodiazepines. Pharmacol Rev 39:251–419
Woods JH, Katz JL, Winger G (1992) Benzodiazepines use, abuse, and consequences. Pharmacol Rev 44:151
Wysowski DK, Barash D (1991) Adverse behavioral reactions attributed to triazolam in the Food and Drug Administration's Spontaneous Reporting System. Arch Intern Med 151:2003–2008
Yang X-M, Luo Z-P, Zhou J-H (1988) Behavioral evidence for the role of noradrenaline in putative anxiolytic and sedative effects of benzodiazepines. Psychopharmacology 95:280–286
Yang X-M, Luo Z-P, Zhou J-H (1989) Behavioral evidence for the role of noradrenaline in the putative anxiogenic actions of the inverse benzodiazepine receptor agonist methyl-4-ethyl-6, 7-dimethoxy-β-carboline-3-carboxylate. J Pharmacol Exp Ther 250:358–363
Yoshishege I, Elsworth JD, Roth RH (1991) Anxiogenic β-carboline FG 7142 produces activation of noradrenergic neurons in specific brain regions of rats. Pharmacol Biochem Behav 39:791–793
Zipursky RB, Barker RB, Zimmer B (1985) Alprazolam withdrawal delirium unresponsive to diazepam: case report. J Clin Psychiatry 46:344–345

CHAPTER 20

Sleep Disturbances as Side Effects of Therapeutic Drugs

A. KALES and A.N. VGONTZAS

A. Factors Contributing to a Drug's Capacity to Induce Sleep Disturbances

Sleep disturbances, as well as other CNS side effects, can be caused by a large variety of therapeutic drugs used for the treatment of a wide range of neurological, psychiatric and medical conditions. Disordered sleep can present itself as the only or main side effect or it can be one of several side effects caused by a particular drug.

Side effects can be classified into two main types: those expected and those unexpected (DAVIES 1985) (Table 1). The first type, also called "augmented," arise from the normal pharmacological action of the drug and are an extension of the drug's therapeutic effects. The most typical example is daytime sedation, which can be considered as an extension of the therapeutic action of any CNS depressant drug. The second type of side effect includes those that represent an unexpected, novel or bizarre response to the drug. Often these adverse events are paradoxical, that is, they are opposite to the therapeutic effects of the drug. One example of this second type of side effect is hyperexcitability phenomena associated with the use of a hypnotic with an ultrashort elimination half-life and high-binding affinity.

Different factors can contribute to the occurrence of CNS side effects with any drug: factors related to the drug and factors related to the patient. As for factors related to a drug, the first two to be taken into consideration are the dose and route of administration. Side effects are more likely to occur when high doses are used, even within the therapeutic range. Also, there are differences in a drug's potential to cause side effects depending on the route of administration. For example, when a benzodiazepine is given i.v., the likelihood of producing amnesia increases markedly.

Pharmacokinetic and pharmacodynamic properties of drugs are also important contributing factors that can explain the differences in their potential to cause side effects. Lipophilicity is the property that determines a drug's penetration into the CNS, and, therefore, is a major factor in its capacity to induce CNS-related side effects (GREENBLATT et al. 1982). For example, lipophilicity is considered one of the major factors accounting for the capacity of various β-adrenergic antagonists to cause sleep disturbances, among other side effects. Another property that may have an influence in

Table 1. Sleep disturbances as side effects of therapeutic drugs

Disorder	Therapeutic function	Drug type	Drug
Insomnia	Stimulant		Amphetamine
			Methylphenidate
			Ephedrine
			Theophylline
			Norepinephrine
	Antidepressant	Monoamine oxidase inhibitor	
		Energizing tricyclic	Imipramine
			Protryptiline
		Second-generation	Nomifensine
		Serotonin-reuptake inhibitor	Fluoxetine
			Fluvoxamine
	Hypnotic/anxiolytic	Barbiturate	Secobarbital
		Rapidly eliminated benzodiazepine	Triazolam
			Midazolam
			Brotizolam
			Estazolam
		Nonbenzodiazepine anxiolytic	Buspirone
	Anti-inflammatory	Corticosteroid	
	Antihypertensive	Lipophilic β-blocker	Propranolol
Hypersomnia	Hypnotic/anxiolytic	Barbiturate	Secobarbital
		Slowly eliminated benzodiazepine	Flurazepam
			Quazepam
			Diazepam
	Neuroleptic	Less potent phenothiazine	Chlorpromazine
			Thioridazine
	Antidepressants	Heterocyclic	Doxepin
			Amitryptiline
			Trazodone

	Anticonvulsant	Barbiturate	Phenobarbitol
		Benzodiazepine	Clonazepam
		Deoxybarbiturate	Primidone
		Hydantoin	Phenytoin
	Antihistamine		Diphenhydramine
			Chlopheniramine
Sleep apnea	Narcotics	Opioid	Morphine
			Hydromorphine
	Anesthetics		
	Antihypertensive		
	CNS depressants	Benzodiazepine	
	Hormones		Testosterone
Nocturnal myoclonus (withdrawal)	Antidepressants	Tricyclic	
	Antimanic		Lithium
	Anticonvulsant		
	Hypnotic	Barbiturate	
		Benzodiazepine	
Cataplexy	Antihypertensive	Sedative-adrenoreceptor antagonist	Prazosin
Hypnopompic hallucination	Antidepressant	Tricyclic	Imipramine
		Tetracyclic	Maprotiline
Somnambulism	Antimanic		Lithium
	Hypnotic	Rapidly eliminated benzodiazepine	Triazolam
			Midazolam
REM behavior	Antidepressant	Tricyclic	Clomipramine
		Serotonin-reuptake inhibitor	Fluoxetine
		Monoamine-oxidase inhibitor	
	Hypnotic	Benzodiazepine	Nitrazepam
	Anxiolytic		Meprobamate
Nightmares	Antihypertensive	Lipophilic β-blocker	Propranolol

the appearance of side effects is the amount of a drug that is free at a given point in time which is determined by the amount that binds to plasma protein.

Elimination half-life is one of the most important pharmacokinetic factors in the production of sleep-related side effects (KALES et al. 1985). It is determinant of the capacity a drug has to produce daytime sedation or hyperexcitability states, both during drug administration and following drug withdrawal. For example, benzodiazepines with long elimination half-lives are more likely to accumulate and to cause daytime sleepiness. Those with short elimination half-lives are more likely to cause hyperexcitability states (see also Chap. 13, this volume).

Other factors contributing to the occurrence of side effects are associated with a drug's actions at the receptor level. Thus, the binding affinity of different types of drugs for the receptors of various neurotransmitter systems is closely related to their potential to cause both expected and unexpected side effects. For instance, among those benzodiazepines with long elimination half-lives, clonazepam, the one with the highest receptor-binding affinity, causes more rebound insomnia than other drugs with long elimination half-lives (KALES et al. 1991). Also, many of the unexpected or unusual side effects, such as cognitive impairments and hyperexcitability states, are found with the administration of drugs with high receptor-binding affinity (KALES et al., 1994) (see also Chap. 13, this volume).

The interactions of various types of drugs with various neurotransmitter systems are also factors influencing the occurrence of specific sleep-related side effects (see also Chaps. 4–11, this volume). For example, the action of certain bronchodilators on the adrenergic system underlies the insomnia that occurs with this class of drugs (RALL, 1990a). Another example is the daytime sleepiness caused by the use of antihistamine drugs, sedative antidepressants or neuroleptics, which is related to the effects of these drugs on histamine receptors (BASSUK et al. 1983).

Finally, the individual characteristics of patients can contribute to the occurrence of side effects and to their severity. Factors such as age, gender, health status, concurrent medications and previous experience with a drug may, in isolation or associated together, determine the nature and intensity of side effects.

B. Insomnia

Drugs can cause insomnia either while they are being taken or following their withdrawal. A patient's current drug regimen and previous responses to drugs and their withdrawal are, therefore, crucial aspects of a thorough evaluation and important guides to treatment.

I. Stimulants

Certain drugs' direct effects on the CNS may cause insomnia (see also Chap. 15, this volume). For example, amphetamines and other CNS stimulants that are taken to suppress appetite, control narcolepsy and treat attention-deficit/hyperactivity disorder (ADHD) and other conditions are more likely to cause insomnia if they are taken close to bedtime (LEWIS et al. 1971; NICHOLSON and STONE 1979; SMITH et al. 1979; KALES et al. 1981; RUBINSTEIN et al. 1994). The effects of methylphenidate in children with ADHD have been shown to increase significantly the complaint of insomnia in placebo-controlled evaluations (BARKLEY et al. 1990; AHMANN et al. 1993).

II. Stimulant Antidepressants

Antidepressant drugs that may cause insomnia (see also Chap. 17, this volume) are: monoamine-oxidase inhibitors used in the treatment of depression (SCHOONOVER 1983; ROTHSCHILD 1985; ABRAMOWICZ 1991; BOYSON 1991); energizing tricyclic antidepressants, such as imipramine (KALES et al. 1977) and protriptyline (DAVIS and GLASSMAN 1989); and second-generation antidepressants, such as nomifensine (POHL and GERSHON 1981). Also, the newer antidepressants which act primarily through inhibition of serotonin reuptake are frequently associated with insomnia (ABRAMOWICZ 1991). In one study, fluvoxamine used to treat patients with obsessive compulsive disorder caused insomnia in about 21% (FREEMAN et al. 1994). In another study, fluoxetine was reported to cause activation, including insomnia, in approximately 15% of a sample of 65 patients with major depression (BEASLEY et al. 1991). These clinical findings were confirmed in a polysomnographic study of fluoxetine which showed decreased sleep efficiency and increased stage 1 sleep in seven patients with affective and eating disorders (KECK et al. 1991). Finally, in placebo-controlled clinical trials, about 15% of subjects taking serotonin-reuptake inhibitors reported insomnia (PHYSICIAN'S DESK REFERENCE 1993).

III. Hypnotics and Anxiolytics

Three types of insomnia have been described that relate to the withdrawal of hypnotic drugs that occurs either during drug administration or following drug cessation: drug withdrawal insomnia (KALES et al. 1974); early morning insomnia (KALES et al. 1983b); and rebound insomnia (KALES et al. 1978, 1979, 1983a) (see also Chaps. 13, 14 and 19, this volume).

1. Drug Withdrawal Insomnia

This sleep disturbance is part of the general abstinence syndrome that follows the abrupt withdrawal of older nonbenzodiazepine hypnotics (barbiturates and nonbarbiturates) that have been taken in large nightly doses for a long period of time. Patients with this syndrome have severe difficulty falling asleep, their sleep is fragmented and disrupted, and they have an increase in REM sleep (REM rebound) (KALES et al. 1974).

Drug withdrawal insomnia is the result of both a psychologic process and physiologic changes that are induced by drug withdrawal (KALES et al.

1974). When patients abruptly stop using a nonbenzodiazepine hypnotic drug that has been taken in large nightly doses for a prolonged period, they frequently feel apprehensive about being able to sleep without the drug. They also are affected physically by an abstinence syndrome that includes jitteriness and nervousness. The insomnia that evolves from this combination can be severe, especially in terms of difficulty in falling asleep. Patient's sleeplessness is often further aggravated by sleep and dream alterations induced by hypnotic-drug withdrawal; rebound increases in REM sleep are often associated with intensified and more frequent dreaming, and sometimes nightmares. Finally, patients may experience altered sleep and dream patterns and drug withdrawal insomnia even on nights when they have taken the drug but have slept past its duration of action (KALES et al. 1974).

2. Early Morning Insomnia and Daytime Anxiety

Early morning insomnia is a significant increase in wakefulness during the final hours of the night in which a rapidly eliminated benzodiazepine hypnotic has been taken (KALES et al. 1983b) (see also Chaps. 13 and 19, this volume). This condition occurs even if the drug has been taken in single, nightly doses for relatively short-term periods. Early morning insomnia typically appears after the drug has been taken nightly for 1 or 2 weeks, when tolerance beings to develop and the drug begins to lose its effectiveness. As tolerance develops to the rapidly eliminated drug, sleep generally continues to be improved during the first two-thirds of the night, but it worsens significantly during the last third of the night. Early morning insomnia may not be specific to benzodiazepines because it is known that other rapidly eliminated drugs, such as alcohol, can disrupt sleep in a similar manner (YULES et al. 1967; RUNDELL et al. 1972).

Daytime anxiety is a corollary of early morning insomnia; it consists of an increase in levels of tension and anxiety on days following nights of drug administration (CARSKADON et al. 1982; MORGAN and OSWALD 1982; KALES et al. 1983c; ADAM and OSWALD 1989) (see also Chaps. 13 and 19, this volume). In 1982, MORGAN and OSWALD and, subsequently, ADAM and OSWALD (1989) reported that continued use of triazolam was associated with significantly and progressively increased levels of daytime anxiety during the drug administration period. Similar findings have been reported by KALES et al. (1983b) when triazolam and midazolam use were associated with increased levels of daytime anxiety and nervousness.

3. Rebound Insomnia and Rebound Anxiety

Rebound insomnia is a specific type of insomnia that follows withdrawal of benzodiazepine drugs that have short or intermediate half-lives (KALES et al. 1978, 1979, 1983c) (see also Chaps. 13 and 19, this volume). When these drugs are withdrawn, wakefulness can increase above baseline levels, even if the drug has been taken in single, nightly doses for short-term periods.

The frequency and intensity of rebound insomnia are strongly related to the half-life of the drug. If the benzodiazepine withdrawn has a short half-life [e.g., midazolam (KALES et al. 1983a) and triazolam (VOGEL et al. 1975; KALES et al. 1976b; ROTH et al. 1976; VOGEL et al. 1976; MAMELAK et al. 1984)], rebound insomnia usually occurs and is severe. With benzodiazepines of intermediate half-life, rebound insomnia occurs less often following withdrawal and is of moderate intensity (KALES et al. 1991; MITLER et al. 1979). Withdrawal of benzodiazepines with long half-lives seldom produces withdrawal sleep disturbance, and then only of a mild degree (KALES et al. 1982, 1983c). Because benzodiazepine hypnotics with relatively short half-lives produce frequent and severe rebound phenomena both during administration and especially following withdrawal, these drugs have a greater potential than intermediate and, especially, long half-life benzodiazepine hypnotics for reinforcing drug-taking behavior and producing hypnotic drug dependence.

4. Sleep Disturbances with Newer Anxiolytics

Sleep disturbances as a result of the use of newer anxiolytics, for example, buspirone, have been documented both in clinical trials and sleep laboratory studies. Buspirone has been associated with adverse effects such as headache, nervousness, dizziness and insomnia (NEWTON et al. 1982, 1986). Polysomnographic study of the administration of buspirone in chronic insomniacs showed a moderate increase of wake time after sleep onset (MANFREDI et al. 1991).

IV. Corticosteroids

There have been a number of anecdotal reports on the effects of corticosteroids on sleep (KALES and KALES 1984). In a recent multicenter study in which steroids were used on a short-term basis (14 days), sleep disturbance was one of the most common side effects (CHROUSOS et al. 1993). The frequency and severity of sleep disturbances associated with the chronic use of corticosteroids is unknown. However, the belief that subjects usually develop tolerance to the sleep-disturbing effects of corticosteroids is not supported hy any data. In any event, the physician should assess carefully for sleep-disturbing effects of the chronic use of corticosteroids.

V. Bronchodilators

Because of their stimulant effects on the CNS, bronchodilating drugs, which contain ephedrine, theophylline and norepinephrine, can lead to sleep difficulties (ANDERSSON and PERSSON 1980). Theophylline has a central stimulating effect which should, at least theoretically, be associated with sleep disturbance. However, the available data indicate either no effect (AVITAL et al. 1991; FITZPATRICK et al. 1992) or only a mild stimulation

(RHIND et al. 1985; KAPLAN et al. 1993). These data suggest that stimulatory effects of bronchodilators are mild and, in asthmatics, the enhancement in breathing from the therapeutic effects of the drug may result in an overall improvement in sleep despite stimulatory side effects.

VI. Antihypertensives

Lipophilic β-blockers have been reported to cause more frequent CNS side effects, among which insomnia is the most prevalent, compared with nonlipophilic β-blockers, which do not cross the blood-brain barrier (BETTS 1981; NEIL-DWYER et al. 1981; WOODS and ROBINSON 1981; COVE-SMITH and KIRK 1985; CRUICKSHANK and NEIL-DWYER 1985; FOERSTER et al. 1985; KOELLA 1985; WESTERLUND 1985; SHORE et al. 1987; DIETRICH and HERRMANN 1989) (see also Chap. 4, this volume). In addition, β-blockers with more intrinsic sympathomimetic activity have a greater probability of affecting sleep (BETTS 1981; BETTS and ALFORD 1985). This is consistent with a sleep laboratory finding of increased wakefulness during the night with pindolol (BETTS and ALFORD 1985), a lipophilic drug with partial agonist effect. In contrast, we found that nadolol, a nonlipophilic drug, not only had no sleep-disrupting effects but actually improved sleep efficiency at doses of 80 mg (KALES et al. 1988).

In terms of the effects of β-blockers on sleep stages, both the lipophilic and nonlipophilic drugs have been reported to either suppress REM sleep or cause no significant change in it (DUNLEAVY et al. 1971; WYATT and GILLIN 1976; KAYED and GODTLIBSEN 1977; BENDER et al. 1979; TORMEY et al. 1979; BETTS and ALFORD 1985; MEIER-EWERT et al. 1985). In contrast, nadolol's slight sedative effect, as well as its suppressing effect on slow-wave sleep and enhancing effect on REM sleep, are probably the result of its action outside of the blood-brain barrier (KALES et al. 1988). Sedation from β-blockers has been linked to indirect effects on the CNS, that is, the reticular formation responds to a reduction in blood pressure by decreasing CNS activity (KOELLA 1985). Furthermore, slow-wave sleep suppression and REM sleep enhancement may be related to a serotonin-blocking activity in the area postrema, which thereby affects serotonin's inhibitory influence on the CNS (KOELLA 1985).

VII. Other Drugs

Several medications have been associated with insomnia during their use or following their withdrawal. Methysergide, a serotonin-blocking drug that is used to prevent severe migraine, has been shown to cause overstimulation and insomnia (GRAHAM 1964). Insomnia has been reported in 3.4% of patients taking flunarizine for migraine, a calcium entry blocker (VOLTA et al. 1990). Insomnia has been reported by more than 5% of the patients administered remoxipride, a selective dopamine (D_2)-receptor antagonist used in the long-term treatment of schizophrenia (HOLM et al. 1993). Most narcotic-analgesics have been associated with insomnia on withdrawal (JAFFEE

1990). This association includes potent analgesics considered nonaddictive such as pentazocine (KANE and POKORNY 1975). The use of amantadine has been associated with insomnia in 10 of 295 hospital employees taking the drug (FLAHERTY and BELLUR 1981) and in 49 of 430 Parkinson's disease patients (SCHWAB et al. 1972). Oral use of the antibiotic, ofloxacin, in the treatment of infections due to multiple-resistant bacteria has been reported to cause insomnia in 27% of 99 patients treated, while dose reduction produced an amelioration of this symptom (SCULLY et al. 1991; YEW et al. 1993). Similarly, mefloquine, an antimalaria treatment, has been associated with insomnia as the initial symptom of an acute major depressive disorder of short duration (CAILLON et al. 1992).

C. Excessive Daytime Sleepiness

I. Hypnotics and Anxiolytics

Daytime sedation is an "augmented" (i.e., expected) side effect with any hypnotic drug because it represents a direct extension of the drug's therapeutic effect (see also Chaps. 13 and 14, this volume). Manifestations of CNS depressant effects, such as daytime drowsiness, hangover and dizziness, in general are considered to be more prevalent with benzodiazepines with long elimination half-lives (HINDMARCH 1979; OSWALD et al. 1979). This contention is based primarily on findings from specially designed performance studies which have shown that impairment of daytime psychomotor functioning is usually greater with slowly eliminated benzodiazepines. However, results of clinical studies show that the actual prevalence of hangover and other side effects related to daytime sedation do not vary to a great extent among benzodiazepine hypnotics with different rates of elimination. For example, with flurazepam, a slowly eliminated drug (data on file, Roche Laboratories), the overall incidence of drowsiness was 11.4%, while with triazolam, a rapidly elminated drug (data on file, The Upjohn Company), the incidence of this side effect was 14.0% (SOLDATOS et al. 1985). One explanation for this similarity in incidence of daytime sedation among drugs with varying elimination half-lives is that many of the studies conducted were for only one night. Thus, this type of design minimized the carryover effect of the slowly eliminated benzodiazepines.

A study (BIXLER et al. 1987) of the three benzodiazepine hypnotics marketed in the United States in the 1980s (flurazepam, temazepam and triazolam) reported on data collected through the Spontaneous Reporting System of the Food and Drug Administration. The findings showed that, for the 1st year on the market, the rate of manifestations of daytime sedation was slightly greater with flurazepam than with triazolam, whereas that for temazepam was much lower than for either flurazepam or triazolam. It was noted that temazepam's low rate of reports of daytime sedation may have been a reflection of the drug's reported lack of efficacy, which was much greater than for either of the other two drugs. Further, one explanation of the small difference between the rate of sedation reported for triazolam and for flurazepam could be the fact that daytime sedation with flurazepam is an expected side effect and therefore may be underreported.

Nonetheless, because benzodiazepine hypnotics such as flurazepam accumulate with successive nightly administation, they present a greater potential for producing excessive daytime sedation and decrements in performance (KALES et al. 1976a; OSWALD et al. 1979; CHURCH and JOHNSON 1979; JOHNSON and CHERNICK 1982). In 1976, when we summarized and reviewed a number of our studies with flurazepam, we alerted physicians to both the potential advantages and disadvantages of the drug's accumulation and carryover effectiveness (KALES et al. 1976a).

In separate long-term evaluations of flurazepam 30 mg and quazepam 30 mg and 15 mg, we found that quazepam 30 mg produced more frequent and severe daytime sedation that flurazepam 30 mg (KALES et al. 1982). This is not surprising when one considers that quazepam's three active components (the parent compound and two active metabolites) all have long elimination half-lives and, thus, the drug would be expected to produce a greater carryover effect. Evidence for such a greater degree of carryover effect was the fact that quazepam produced a continued improvement in sleep compared to baseline for a longer period following drug withdrawal than flurazepam. We also reported that quazepam in a 15-mg dose produced less frequent and severe daytime sedation than the 30-mg dose of quazepam.

II. Sedative Neuroleptics

Excessive daytime sedation is a common adverse behavioral change that usually occurs during the first few days after starting an antipsychotic, with some rapidly developing tolerance (ABRAMOWICZ 1991) (see also Chap. 16, this volume). In general, sedative side effects are inversely proportional to the milligram potency of these drugs. Thus, chlorpromazine and thioridazine produce more sedation than fluphenazine, haloperidol, thiothixene or trifluoperazine. When necessary it can be controlled by reducing the dose, switching to a less-sedating agent, or administering the entire dose at bedtime. It should be noted that the usefulness of sedative antipsychotics in controlling agitation and behavioral dyscontrol is not established (RIFKIN and SIRIS 1987).

III. Sedative Antidepressants

Excessive daytime sedation is also a frequent side effect associated with heterocyclic antidepressants (see also Chap. 17, this volume). Daytime drowsiness occurs most often: with doxepin, amitriptyline and trazodone; less with imipramine and maprotiline; and infrequently with protriptyline, nortriptyline and desipramine (DAVIS and GLASSMAN 1989). In one study, trazodone was found to be associated with daytime sedation in 42.6% of 61 patients with major depression (BEASLEY et al. 1991). The pharmacological mechanism for this sedation is not well understood, with 5-HT, H_1, or H_2 receptor blockade all implicated.

IV. Anticonvulsants

Daytime drowsiness is among the major side effects of the anticonvulsants (see also Chap. 18, this volume). Its impact on the quality of life of patients is sometimes marked enough as to interfere with the occupational or learning spheres of the patient's life. However, not all the anticonvulsants have the same potential to cause excessive daytime sedation. Thus, with phenobarbital, sedation is one of the main adverse events (BRENT et al. 1987). Sedation with primidone occurs frequently unless the drug is started in a very low dosage (ABRAMOWICZ 1989). Phenytoin, at serum concentrations higher than 30 μg/ml, often causes drowsiness (ABRAMOWICZ 1989). As mentioned before, drowsiness is among the main adverse events with clonazepam. It may also occur with other anticonvulsants such as carbamazepine and valproate (ABRAMOWICZ 1989). However, with the latter, this symptom usually is mild and temporary.

V. Antihistamines

Excessive daytime sedation is the side effect with the highest incidence among histamine antagonists (BASSUK et al. 1983) (see also Chaps. 6 and 16, this volume). Laboratory studies have shown that this sedation is often associated with impairment of many performance skills (CARRUTHERS et al. 1978; CLARKE and NICHOLSON 1978; HINDMARCH and PARROTT 1978; MATTILA et al. 1986; ROTH et al. 1987; SALETU et al. 1988; NICHOLSON et al. 1991; WALSH et al. 1992). Furthermore, epidemiologic studies have shown that antihistamine users are overrepresented among a group of fatally injured drivers who were deemed responsible for their automobile accidents (STARMER 1985).

Experience with some recently developed antihistamines, such as phenindamine and terfenadine, indicates that these agents have the desired therapeutic effects on the peripheral nervous system without negatively affecting the central nervous system. In a recently completed study (unpublished data), we compared the effects of daytime alertness/sedation of antihistamines with known sedative effects (diphenhydramine, chlorpheniramine) with newer nonsedative antihistamines (phenindamine and terfenadine). Our findings indicated phenindamine produced little difference in terms of daytime sedation from placebo or terfendadine, another antihistamine previously shown to have no sedative effects (WEINER 1982; WOODWARD and MUNRO 1982; NICHOLSON and STONE 1983, 1986; MOSKOWITZ and BURNS 1988). On the other hand, diphenhydramine and chlorpheniramine caused significant daytime sedation, confirming findings from previous laboratory studies (CLARK and NICHOLSON 1978; HINDMARCH and PARROTT 1978; ROTH and ZORICK 1987; MOSKOWITZ and BURNS 1988) and clinical subjective reports.

D. Sleep Apnea

The presence of sleep apnea must be ruled out by history and, if necessary, by the sleep laboratory whenever hypnotics, narcotics or general anesthetics are indicated.

I. CNS Depressants

1. Hypnotics

For over 20 years, benzodiazepines have been the most commonly prescribed hypnotic drugs (SOLOMON et al. 1979; HOLLISTER and CERNANSKY 1990). Benzodiazepines are mild respiratory depressants (see RALL 1990; ROBINSON and ZWILLICH 1994 for reviews). The degree to which benzodiazepines affect respiratory chemosensitivity is not clear (GUILLEMINAULT et al. 1978). This most likely is due to factors such as individual differences (CATCHLOVE and KAFER 1971a,b), route of administration, dosage and specific drug (ROBINSON and ZWILLICH 1994). Oral benzodiazepines appear to cause mild, if any, respiratory depression (DOBSON et al. 1976; LONGBOTTOM and PLEUVY 1984; SKATRUD et al. 1988; RALL 1990). For example, triazolam at both two and three times the clinical dose did not depress respiration (SKATRUD et al. 1988). However, benzodiazepines, when used as anesthetics, can cause a moderate depression of respiration (RALL 1990). For example, diazepam given I.M. decreased hypoxic response without altering hypercapnic response (LAKSHMINARAYAN et al. 1976). Further, midazolam given I.V. depressed hypercapnic response in higher doses (GROSS et al. 1983) and hypoxic response in lower doses (ALEXANDER and GROSS 1988).

In general, hypnotic doses of benzodiazepines have little or no influence on respiration during wakefulness or sleep in asymptomatic subjects (CARSKADON et al. 1982: SKATRUD et al. 1988; BONNETT et al. 1990; POLLACK and KENNY 1993). The likelihood of respiratory depression with orally administered benzodiazepine hypnotics is affected by a number of factors including: presence of respiratory impairment; dose level; and elimination half-life. Several studies have shown a mild degree of impaired respiration with slowly eliminated benzodiazepines including in: patients with mild sleep apnea (DOLLY and BLOCK 1982); patients with chronic obstructive pulmonary disorder (RUDOLF et al. 1978); and elderly asymptomatic subjects (GUILLEMINAULT et al. 1984). On the basis of these findings, MENDELSON (1991) has suggested that short-acting hypnotics should be used in patients with impaired respiration. However, the need for further studies is suggested by the variability of results on respiration with long-acting benzodiazepines in patients with sleep apnea. For example, a reduction in sleep apneic events was associated with the use of clonazepam, a benzodiazepine with relatively slow elimination (GUILLEMINAULT et al. 1988).

There are few studies on the effects of benzodiazepines in patients with obstructive sleep apnea. However, clinical experience indicates that benzodiazepine use in obstructive sleep apnea (OSA) patients is associated with a worsening of breathing abnormalities. This is supported by studies or case reports which showed that apneas become more frequent and oxygen saturation decreases following the use of benzodiazepines in symptomatic apneics (CLARK et al. 1971; MENDELSON et al. 1981; GUILLEMINAULT et al. 1982; CUMMISKEY et al. 1983; BLOCK et al. 1984; MIDGREN et al. 1989; MURCIANO et al. 1993).

2. Narcotics

Narcotics are known to be strong respiratory depressants but their effects on respiration during sleep have not been throughly assessed (see JAFFE and MARTIN 1990; ROBINSON and ZWILLICH 1994 for reviews). Administration of therapeutic doses of opioids to normal subjects or addicts decreases hypoxic and hypercapnic ventilatory response and awake minute ventilation (WEIL et al. 1975; SANTIAGO et al. 1980). Oral administration of lower doses of narcotics do not appear to induce sleep-disordered breathing in asymptomatic subjects (ROBINSON et al. 1987). In two reports, no increase in sleep-disordered breathing events occurred after a low dose of oral hydromorphone or morphine in normal subjects (FORREST and BELLVILLE 1964; ROBINSON et al. 1987). However, there is evidence that larger doses or parenteral use of narcotics may be harmful to breathing during sleep (LAMARCHE et al. 1986; SAMUELS et al. 1986; KEAMY et al. 1987). Further, narcotic analgesia, compared to local anesthetic analgesia, produces considerable sleep-disordered breathing with a higher number of apneic events and a greater degree of oxygen desaturation (CATLEY et al. 1985; CLYBURN et al. 1990). However, a narcotic effect alone cannot be concluded from these results because patients given narcotics also received general anesthetics. Thus an additive effect in producing sleep-disordered breathing may occur in patients given both narcotics and general anesthesia (ROBINSON and ZWILLICH 1994).

The effects of narcotics in snorers and individuals with mild sleep-disordered breathing have not been thoroughly studied. ROBINSON and ZWILLICH (1994) have reviewed several case reports on the occurrence of severe sleep-disordered breathing associated with the use of narcotics perioperatively in patients who subsequently were found to have sleep apnea (LAMARCH et al. 1986; SAMUELS and RABINOV 1986; KEAMY et al. 1987). Also, in a more recent study, preexisting sleep apnea syndrome was shown to be a risk factor associated with the occurrence of respiratory depression following the parenteral use of opioids (ETCHES 1994). In any event, extra caution is recommended in the use of narcotics in patients with sleep apnea.

3. Anesthetics

It has been commonly observed that asymptomatic subjects develop upper airway obstruction during anesthesia (see MARSHALL and LONGNECKER 1990 for review). This is particularly so even with light general anesthesia, if proper positioning of the head and neck is not maintained (Safar et al. 1959). These effects are more pronounced in obese individuals and patients with OSA (Haponik et al. 1983). Even nasal anesthesia may be associated with sleep-disordered breathing in asymptomatic men (White et al. 1985a).

Recognition of the effects of anesthetics and narcotics in patients with sleep apnea has led to the establishment of guidelines for perioperative management of such patients (Bonora et al. 1984; Craddock and Lees 1987; Chan et al. 1989; Wang et al. 1989). Recommendations offered include the following (Robinson and Zwillich 1994). Preoperative sedation should be administered with caution. In those patients administered general anesthesia, careful monitoring in the recovery room for respiratory depression and upper airway obstruction is essential. Further, monitoring in the intensive care unit should be utilized for patients with sleep apnea who have had major surgery, prolonged anesthesia or received narcotic drugs. Whenever possible, local analgesia is preferred over general anesthesia.

II. Antihypertensives

Hypertension is frequently found in patients with sleep apnea (Kales et al. 1985). Also a high prevalence of sleep apnea among study samples of hypertensive patients has been reported by several groups (Kales et al. 1984; Lavie et al. 1984; Fletcher et al. 1985a; Williams et al. 1985). The effects of antihypertensives on sleep apnea have been variable across a number of studies (see Robinson and Zwillich 1994 for a review). Propranolol has been reported to worsen OSA in a report of only two patients (Boudoulas et al. 1983). This was not confirmed in another study of 10 sleep apnea patients (Fletcher et al. 1985b). In another report, metoprolol decreased blood pressure and the number of apneic events in six patients with OSA and hypertension (Mayer et al. 1990). In another study, cilazapril, an angiotensin-converting enzyme inhibitor, produced moderate reductions in blood pressure and number of apneic events in 12 men with both hypertension and sleep apnea (Peter et al. 1989; Mayer and Peter 1991). Finally, in prevalence studies of hypertensive samples, no clear cut evidence has been found between specific drug therapy and severity of sleep apnea (Kales et al. 1984; Fletcher et al. 1985a). Thus, it appears that antihypertensives may be safely administered to patients with sleep apnea.

III. Testosterone

A male gender predisposition for sleep apnea has been reported by most investigators who have evaluated sleep apnea (Guilleminault et al. 1976b; Block et al. 1979; Ancoli-Israel et al. 1987; Leech et al. 1988; Hoch et al. 1990; Metes et al. 1991; Rajala et al. 1991; Young et al. 1993). This strongly implies that sex hormones influence breathing during sleep. Women have been considered to be protected to a large extent from sleep apnea until menopause (Block et al. 1980). The prevalence of sleep apnea in a recent study of 250 obese men and women without a primary sleep complaint who were referred for treatment of their obesity was obtained using polysomnographic methods (Vgontzas et al. 1994). In this study, 40% of the men and only 3% of the women were observed to have sleep apnea severe enough to be treated. Hypogonadal males have been evaluated in two studies. In one study, testosterone replacement was found to increase hypoxic but not hypercapnic ventilatory response (White et al. 1985b) and, in the second study, not to influence upper airway anatomy (Schneider et al. 1986).

There are several anecdotal reports on patients who received testosterone and developed OSA (Sandblom et al. 1983; Johnson et al. 1984). Available data suggest that, in certain individuals, sleep apnea may develop during replacement therapy with testosterone (Scrima et al. 1982; Matsumoto et al. 1985). The most vulnerable to this testoterone effect appear to be those with prior sleep-disordered breathing (Robinson and Zwillich, 1994).

IV. Other Drugs

Sleep apnea has also been reported to be associated with exposure to organic solvents (Teelucksingh et al. 1991; Edling et al. 1993).

E. Nocturnal Myoclonus

Tricyclic antidepressants and lithium carbonate may induce nocturnal myoclonus (Ware et al. 1984; Heiman and Christie 1986), as may withdrawal from a variety of drugs such as anticonvulsants, benzodiazepines, barbiturates and other hypnotics (Thorpy 1990).

Disruptive nocturnal myoclonus has been reported in two patients who were using methsuximide and phenytoin for complex partial and generalized seizures (Drake 1988). Although dopamine agonists are used in the treatment of restless legs syndrome and nocturnal myoclonus, a paradoxical leg restlessness which was treatment-emergent developed in 8 out of 47 patients with L-dopa at bedtime, 7 of whom required daytime medication for relief (Becker et al. 1993).

F. Auxiliary Symptoms of Narcolepsy

Cataplexy was exacerbated in a patient who was taking prazosin, a selective-adrenoceptor antagonist used in the treatment of hypertension (Aldrich and Rogers 1989). This drug had shown a high capacity for inducing cataplectic attacks in narcoleptic dogs (Mignot et al. 1987, 1988).

Hypnopompic hallucinations have been reported in patients with affective disorders treated with antidepressants at doses within the therapeutic range. In one report, the patient was being treated with imipramine (SCHLAUCH 1979) and, in the other, the patients were taking maprotiline (ALBALA et al. 1983).

G. Somnambulism/Night Terrors

I. Lithium

Several reports have documented the occurrence of somnambulistic-like episodes secondary to combined lithium and neuroleptic administration (CHARNEY et al. 1979). Induction of sleepwalking activity with the combined administration of lithium and neuroleptics is consistent with the observation that somnambulism occurs during slow-wave sleep (GASTAUT and BROUGHTON 1965; JACOBSON et al. 1965; KALES et al. 1966; BROUGHTON 1968) and that certain neuroleptics (FEINBERG et al. 1969; LESTER et al. 1971; KAPLAN et al. 1974; JUS et al. 1975) and lithium (CHERNIK and MENDELS 1974; KUPFER et al. 1974) increase slow-wave sleep.

It is not clear how the increase in slow-wave sleep induced by a neuroleptic and lithium carbonate predisposes a patient to sleepwalk. BROUGHTON (1968) has conceived somnambulism as a disorder of arousal in which certain physiological constellations occur during slow-wave sleep that do not appear in normal subjects. For example, sleepwalkers exhibit sleeptalking or complex and gestural movements in association with slow-wave sleep, but independent of the somnambulism itself. They also have more severe and longer confusional episodes following forced arousals from slow-wave sleep (BROUGHTON and GASTAUT 1975). It may be postulated that in patients who experience drug-induced, somnambulistic-like events, the drugs produce a physiologic state during slow-wave sleep which is similar to that in patients with primary somnambulism.

In patients who require neuroleptic and lithium therapy and who have developed sleepwalking or have a past history of sleepwalking, eliminating the evening or bedtime doses of these medications may be helpful. If sleepwalking persists, the daily dosage of the neuroleptic should be reduced whenever possible. The addition of a drug that suppresses slow-wave sleep, such as diazepam (KALES and SCHARF 1973) at bedtime, might provide an additional therapeutic effect, as it did in one of our patients.

II. Hypnotics

There have been several case reports describing somnambulism associated with the use of hypnotics, particularly the more potent, rapidly eliminated benzodiazepines. In one case report, use of 1 mg triazolam precipitated the occurrence of somnambulistic-like episodes in three young patients who did not have a prior history of sleepwalking (Poitras 1980). In another case report, sleepwalking was one of the side effects associated with the use of 0.5 and 1.5 mg triazolam for up to 8 months (Regestein and Reich 1985). Sleepwalking has been reported following the use of either 0.25 mg triazolam or 15 mg midazolam in a patient with restless legs and nocturnal myoclonus syndrome (Laurema 1991). The bedtime doses of various combinations of hypnotics including chloral hydrate, methyprylon, ethchlorvynol, barbiturates and diphenhydramine have also been associated with somnambulism (Christianson and Perry 1956; Luchins et al. 1978; Huapaya 1979; Nadel 1981).

III. Other Drugs

Bedtime administration of single daily doses of neuroleptics or tricyclics have been reported to be associated with somnambulism and night terrors (Flemenbaum 1976; Huapaya 1979). The aggravation or reappearance of somnambulism has been reported with the use of propranolol in patients being treated for migraine (Pradalier et al. 1987). With nonlipophilic β-blockers, such an association has not been reported. Finally, the use of amantadine was associated with night terrors in 10 of 295 hospital employees taking the drug (Flaherty and Bellur 1981).

H. REM Behavior Disorder

This parasomnia basically consists of abnormal movements during REM sleep; these can be excessive, limb or body jerkings, and complex and/or vigorous or violent movements. They occur on the background of certain polygraphic abnormalities, such as excessive augmentation of chin EMG tone or excessive EMG twitching. Altogether the resulting behavior can be disruptive, annoying or even harmful for the patient or the bed partner. Although initially it was described in the chronic form (Schenck et al. 1986), there is also an acute variety (Schenck et al. 1988).

Drugs, both during administration and withdrawal, can be involved in the etiology of both the acute and the chronic presentation. Thus, there have been reports of the acute condition caused by withdrawal of meprobamate (Tachibana et al. 1975), pentazocine (Tanaka et al. 1979) and nitrazepam (Sugano et al. 1980). The administration of tricyclic antidepressants (Passouant et al. 1972; Guilleminault et al. 1976a; Besset 1978; Shimizu et al. 1985; Bental et al. 1979) and monoamine oxidase inhibitors (MAOIs) (Akindele et al. 1970) has been reported to be associated with a lack of REM atonia which may allow the occurrence of the acute form of REM behavior disorder. The chronic condition has been described with the use of clomipramine (Bental et al. 1979) and fluoxetine (Schenck et al. 1992).

I. Nightmares

I. Antihypertensives

Reserpine is an alkaloid extracted from the plant *Rauwolfia serpentina*, which is still widely used in the treatment of hypertension, and has been associated with an increase in reports of nightmares (GOODWIN and BUNNEY 1971).

Nightmares and disturbed sleep quality in some individuals are well-known side effects of β-blockers (BETTS and ALFORD 1985). The potential of β-blockers to induce CNS side effects appears to be related to their degree of lipophilicity (KALES et al. 1989). Among the most common side effects are sleep disturbances, with nightmares among the most frequently reported. In a double-blind, placebo-controlled study, the effects were assessed on dreams of β-blockers with various degrees of lipid solubility: propranolol (high lipophilicity) with nadolol (low lipophilicity). Both drugs were studied at a dose of 40 mg in the sleep laboratory according to an 18-night protocol to evaluate the initial, short-term (night 5–11) and intermediate-term (nights 12–18) effects as compared to baseline (nights 1–4). With short-term administration, propranolol induced significant increases in dream vividness, violence and color and a significant decrease in dream pleasantness. In contrast, nadolol resulted in significant reduction in dream emotionality, pleasantness, color, violence and physical activity.

Nightmares have also been reported with the ophthalmic administration of β-blockers (MORT 1992). Finally, the use of angiotensin-converting enzyme inhibitors such as captopril and enalapril (HAFFNER et al. 1993) has been associated with nightmares.

II. Psychotropic Drugs

Nonbenzodiazephine hypnotic drugs that suppress REM sleep, i.e., barbiturates when abruptly withdrawn, are associated with an increased intensity and frequency of dreaming (KALES and JACOBSON 1967). At times even nightmares occur. These altered dream patterns and nightmares contribute additionally to drug-withdrawal insomnia (KALES et al. 1974). Benzodiazepine hypnotics cause a slight suppression of REM sleep during drug administration while during withdrawal there is a gradual return to baseline. In the many studies we have conducted on the effects of benzodiazepine hypnotics on sleep, we have not reported nightmares as part of the withdrawal reaction. However, the possibility of the occurrence of nightmares as an infrequent withdrawal reaction cannot be ruled out.

Antidepressants and neuroleptics with strong anticholinergic effects, i.e., tricyclics and low-potency phenothiazines, are known to suppress REM sleep (see Chap. 16, this volume). Following drug cessation, there is a rebound phenomenon (KALES et al. 1977). Theoretically, this REM rebound should be associated with intense dreaming and/or nightmares. Although this appears to be the occasional experience of many clinicians, a literature search using Medline from 1966 to the present time did not show any published reports indicating nightmares as a withdrawal symptom following the discontinuation of antidepressants or neuroleptics. This indicates that nightmares following the discontinuation of antidepressants or neuroleptics may not be a frequent phenomenon.

III. Other Drugs

Nightmares have been associated with the use of potent analgesics such as pentazocine (Miller 1975), buprenorphine (MacEvilly and O'Carrol 1989) and naproxen (Bakht and Miller 1991). Nightmares and vivid dreams have also been reported during the administration of the stimulant, fenfluramine (Alvi 1969; Hooper 1971) and following its sudden withdrawal (Oswald 1969). Vivid dreams have been reported in 15% of 88 patients with idiopathic Parkinson's disease treated with levodopa (Moskovitz et al. 1978). Nightmares have also been reported in single cases, with the use of erythromycin (Williams 1988) and the antimalarial drug, mefloquine (Burke 1993).

J. Circadian Rhythm Disorders

Antidepressants (for example, lithium, MAOIs and tricyclics) have been shown to have a direct pharmacologic effect on the circadian pacemaker in the suprachiasmatic nucleus (SCN) by lengthening the period of circadian rhythms (Wirz-Justice 1983). MAOIs and tricyclics given prophylactically may cause a rapid cycling between depression and mania, indicating a drug-induced circadian rhythm disorder (Prien et al. 1973; Quitken et al. 1978; Wehr and Goodwin 1979; Kukopulos et al. 1980). This effect may be accounted for by the finding that, in animals, both MAOIs and tricyclics, in contrast to lithium, have been shown to induce a dissociation of the activity components of the rest-activity cycle from body temperature rhythms (Wirz-Justice 1983). This pattern is strikingly similar to that seen in humans in free-running conditions, as well as in manic depressive patients, when they undergo spontaneous or drug-induced switches of depression into mania. The finding that lithium does not induce this dissociation (Wirz-Justice 1983) may explain why this drug, in contrast to MAOIs and tricyclics, is a good prophylactic agent in the treatment of most bipolar patients (Goodwin 1979).

A shortening of the period of circadian rhythms seems to be caused by hormones (see review in Demet and Chicz-Demet 1987). However, there are no reports of sleep disturbances caused by these drugs.

K. Summary

Sleep disturbances as well as other CNS side effects can be caused by a large variety of therapeutic drugs used for the treatment of a wide range of neurologic, psychiatric and medical conditions. Dose and route of administration as well as pharmacokinetic and pharmacodynamic properties are the major contributing factors that can explain differences in a drug's potential to cause sleep disturbances. Elimination half-life, binding affinity and the interaction of various types of drugs with various neurotransmitter systems appear to be the most important kinetic and dynamic factors in the production of sleep-related side effects. However, sleep-related side effects have been associated on a rather rare basis with the use of a wide variety of drugs, such as antibiotics, which cannot be easily understood from what is known about the pharmacokinetics and dynamics of these drugs.

Insomnia is a frequent side effect of many categories of drugs including those that are used for the treatment of insomnia, i.e., hypnotics. Stimulants used for the treatment of narcolepsy or attention deficit disorder can cause insomnia either while they are being taken or after their withdrawal. Also, stimulant antidepressants including MAOIs, energizing tricyclic antidepressants and the newer serotonin-reuptake inhibitors are often associated with insomnia. Drug withdrawal insomnia is part of the general abstinence syndrome that follows the abrupt withdrawal of older nonbenzodiazepine hypnotics (barbiturates and nonbarbiturates) that have been taken in a large nightly dosage for a long period of time, while early morning insomnia and rebound insomnia are two types of sleep disturbance that have been described following the use and the withdrawal respectively of benzodiazepine hypnotics and anxiolytics with rapid elimination and high potency. Furthermore, nocturnal sleep disturbance has been described with the newer anxiolytics, for example, buspirone and with corticosteroids and lipophilic β-blockers.

Excessive daytime sedation is a frequent side effect of benzodiazepines, particularly those with a long half-life. However, most patients develop tolerance to the side effect in a few days. Sedation is also present during the use of sedative neuroleptics, antidepressants, some anticonvulsants and antihistamines. With neuroleptics, sedative side effects are inversely proportional to the milligram potency of these drugs. With antidepressants, sedation is a frequent side effect associated with the heterocyclics. Among anticonvulsants, phenobarbital is the drug most frequently associated with sedation. Sedation used to be the most troubling side effect of the antihistamines; however, with the introduction of more recently developed antihistamines, such as terfenadine, this side effect has been greatly minimized.

The effects of drugs on nocturnal respiratory function has become of increasing interest because of the epidemiological findings that sleep apnea is a widespread disease, particularly among males. Narcotics and anesthetics are powerful respiratory depressants and can be associated with adverse effects on breathing pattern, particularly in patients with sleep apnea. Benzodiazepines, in contrast, are rather mild respiratory depressants, and in therapeutic dosage they have little or no influence on wakeful and nighttime respiration in asymptomatic subjects. However, there is a suggestion that benzodiazepine use in obstructive sleep apnea patients is associated with a worsening in breathing abnormalities. Information on the effects of other drugs, such as antihypertensives or testosterone, on sleep-breathing patterns is rather inconsistent and indicates a rather mild effect.

Nocturnal myoclonus has been associated with the use of tricyclic antidepressants and lithium carbonate. Auxiliary symptoms of narcolepsy, such as cataplexy or hypnopompic hallucinations, have been rarely reported with drugs such as prazosin and imipramine. Also, somnambulism/night terrors have been associated with the use of a variety of drugs, such as lithium, neuroleptics, propranolol and hypnotics, particularly those with rapid elimination and high binding affinity. REM behavior disorder has been associated with the use or withdrawal of drugs, such as meprobamate, nitrazepam, tricyclic antidepressants and others. Nightmares have been associated with the use of lipophilic antihypertensives, and hypnotic drugs that suppress REM sleep, i.e., barbiturates when abruptly withdrawn. Although antidepressants and neuroleptics, with strong anticholinergic effects, are expected during withdrawal to be associated with REM rebound, few reports of nightmares under these conditions exist in the published literature.

References

Abramowicz M (1989) Drugs for epilepsy. The Med Let 31:1–4

Abramowicz M (1991) Drugs for psychiatric disorders. Med Let 33:43–50

Adam K, Oswald I (1989) Can a rapidly-eliminated hypnotic cause daytime anxiety? Pharmacopsychiatr 22:115–119

Ahmann PA, Waltonen SJ, Olson KA, Theye FW, VanErem AJ, LaPlant RJ (1993) Placebo-controlled evaluation of ritalin side effects. Pediatrics 91:1101–1106

Akindele MO, Evans JI, Oswald I (1970) Mono-amine oxidase inhibitors, sleep and mood. Electroencephalogr Clin Neurophysiol 29:47–56

Albala AA, Weinberg N, Allen SM (1983) Maprotiline-induced hypnopompic hallucinations. J Clin Psychiatry 44:149–150

Aldrich MS, Rogers AE (1989) Exacerbation of human cataplexy by prazosin. Sleep 12(3):254–256

Alexander CM, Gross JB (1988) Sedative doses of midazolam depress hypoxic ventilatory responses in humans. Anesth Analg 67:377–382

Alvi MY (1969) Unusual effects of fenfluramine. Br Med J 4:237

Ancoli-Israel S, Kripke DF, Mason W (1987) Characteristics of obstructive and central apnea in the elderly: an interim report. Biol Psychiatry 22:741–750

Andersson KE, Persson CGA (1980) Extrapulmonary effects of theophylline. Eur J Resp Dis Sup 61:17–28

Avital A, Steljes DG, Pasterkamp H, Kryger M, Sanchez I, Chernick V (1991) Sleep quality in children with asthma treated with theophylline or cromolyn sodium. J Pediatr 119:979–984

Bakht FR, Miller LG (1991) Naproxen-associated nightmares. Southern Med J 84:1271–1273

Barkley RA, McMurray MB, Edelbrock CS, Robbins K (1990) Side effects of methylphenidate in children with attention deficit hyperactivity disorder: a systemic, placebo-controlled evaluation. Pediatrics 86:184–192

Bassuk EL, Schoonover SC, Gelenberg AJ (eds) (1983) The practitioner's guide to psychoactive drugs, 2nd edn. Plenum, New York
Beasley CM, Dornseif BE, Pultz JA, Bosomworth JC, Sayler ME (1991) Fluoxetine versus trazodone: efficacy and activating-sedating effects. J Clin Psychiatry 52:294–299
Becker PM, Jamieson AO, Brown WD (1993) Dopaminergic agents in restless legs syndrome and periodic limb movements of sleep: response and complications of extended treatment in 49 cases. Sleep 16:713–716
Bender W, Greil W, Ruther E, Schnelle K (1979) Effects of the β-adrenoceptor blocking agent sotalol in CNS: sleep, EEG, and psychophysiological parameters. J Clin Pharmacol 19:505–512
Bental E, Lavie P, Sharf B (1979) Severe hypermotility during sleep in treatment of cataplexy with clomipramine. Isr J Med Sci 15:607–609
Besset A (1978) Effect of antidepressants and sleep. Adv Biosci 21:141–148
Betts TA (1981) Adrenoceptor drugs and sleep. In: Wheatley D (ed) Psychopharmacology of sleep. Raven, New York, pp 199–212
Betts TA, Alford C (1985) Beta-blockers and sleep: a controlled trial. Eur J Clin Pharmacol 28:65–68
Bixler EO, Kales A, Brubaker BH, Kales JD (1987) Adverse reactions to benzodiazepine hypnotics: spontaneous reporting system. Pharmacology 35:286–300
Block AJ, Boysen PG, Wynne JW, Hunt LA (1979) Sleep apnea, hypopnea, and oxygen desaturation in normal subjects: a strong male predominance. N Engl J Med 300:513–517
Block AJ, Wynne JW, Boysen PG (1980) Sleep-disordered breathing and nocturnal oxygen desaturation in postmenopausal women. Am J Med 69:75–79
Block AJ, Dolly FR, Slayton PC (1984) Does flurazepam ingestion affect breathing and oxygenation during sleep in patients with chronic obstructive lung disease? Am Rev Respir Dis 129:230–233
Bonnet MH, Dexter JR, Arand DL (1990) The effect of triazolam on arousal and respiration in central sleep apnea patients. Sleep 13:31–41
Bonora M, Shields G, Knuth S, Bartlett D, St. John WM (1984) Selective depression by ethanol of upper airway respiratory motor activity in cats. Am Rev Respir Dis 130:156–161
Boudoulas H, Schmidt H, Geleris P, Clarke RW, Lewis RP (1983) Case reports on deterioration of sleep apnea during therapy with propranolol – preliminary studies. Res Commum Chem Pathol Pharmacol 39:3–10
Boyson SJ (1991) Psychiatric effects of selegiline. Arch Neurol 48:902
Brent DA, Grumbine PK, Yarma RR, Allan M, Allman C (1987) Phenobarbitol treatment and major depressive disorder in children with epilepsy. Pediatrics 80:909–917
Broughton RJ (1968) Sleep disorders: disorders of arousal? Science 159:1070–1078
Broughton RJ, Gastaut H (1975) Recent sleep research on enuresis nocturna, sleepwalking, sleep terrors and confusional arousals. In: Levin P, Koella WP (eds) Sleep 1974. Second European Congress of Sleep Research, Rome 1975, pp 82–91
Burke BM (1993) Mefloquine. Lancet 341:1605–1606
Caillon E, Schmitt L, Moron P (1992) Acute depressive symptoms after mefloquine treatment. Am J Psychiatry 149:712
Carruthers SG, Shoeman DW, Hignite CE, Azarnoff DL (1978) Correlation between plasma diphenhydramine level and sedative and antihistamine effects. Clin Pharmacol Ther 23:375–382
Carskadon MA, Seidel WF, Greenblatt DJ, Dement WC (1982) Daytime carryover of triazolam and flurazepam in elderly insomniacs. Sleep 5:361–371
Catchlove RFH, Kafer ER (1971a) The effects of diazepam on the ventilatory response to carbon dioxide and on steady-state gas exchange. Anesthesiology 34:9–13

Catchlove RFH, Kafer ER (1971b) The effects of diazepam on respiration in patients with obstructive pulmonary disease. Anesthesiology 34:14–18

Catley DM, Thornton C, Jordan C, Lehane JR, Royston D, Jones JG (1985) Pronounced, episodic oxygen desaturation in the postoperative period. Its association with ventilatory pattern and analgesic regimen. Anesthesiology 63:20–28

Chan CS, Grunstein RR, Bye PTP, Woolcock AJ, Sullivan CE (1989) Obstructive sleep apnea with severe chronic airflow limitation. Am Rev Respir Dis 140: 1274–1278

Charney DS, Kales A, Soldatos CR, Nelson JC (1979) Somnambulistic-like episodes secondary to combined lithium-neuroleptic treatment. Br J Psychiatry 135: 418–424

Chernik DA, Mendels J (1974) Longitudinal study of the effects of lithium carbonate on the sleep of hospitalized depressed patients. Biol Psychiatry 9:117–23

Christianson HB, Perry HO (1956) Reactions to chloral hydrate. Arch Dermatol 74:232–240

Chrousos GA, Kattah JC, Beck RW, Cleary PA, and the Optic Neuritis Study Group (1993) Side effects of glucocorticoid treatment. JAMA 269:2110–2112

Church MW, Johnson LC (1979) Mood and performance of poor sleepers during repeated use with flurazepam. Psychopharmacology 61:309–316

Clark TJH, Collins JV, Tong D (1971) Respiratory depression caused by nitrazepam in patients with respiratory failure. Lancet II:737–738

Clarke CH, Nicholson AN (1978) Performance studies with antihistamines. Br J Clin Pharmacol 6:31–36

Clyburn PA, Rosen M, Vickers MD (1990) Comparison of the respiratory effects of IV infusions of morphine and regional analgesia by extradural block. Br J Anaesth 64:446–449

Cove-Smith JR, Kirk CA (1985) CNS-related side effects with metoprolol and atenolol. Eur J Clin Pharmacol 28:69–72

Craddock M, Lees DE (1987) Anesthesia for obstructive sleep apnea patients: risks, precautions, and management. In: Fairbanks DNF (ed) Snoring and obstructive sleep apnea. Raven, New York, pp 235–243

Cruickshank JM, Neil-Dwyer GN (1985) β-Blocker brain concentrations in man. Eur J Clin Pharmacol 28:21–23

Cummiskey J, Guilleminault C, Del Rio G, Silverstri R (1983) The effects of flurazepam on sleep studies in patients with chronic obstructive pulmonary disease. Chest 84:143–147

Davies DM (ed) (1985) Textbook of adverse drug reactions, 3rd edn. Oxford University Press, New York

Davis JM, Glassman AH (1989) Antidepressant drugs. In: Kaplan HI, Sadock BJ (eds) Comprehensive textbook of psychiatry, 5th edn: vol 2. Williams and Wilkins, Maryland, pp 1627–1655

Demet E, Chicz-Demet A (1987) Effects of psychoactive drugs on circadian rhythms. Psychiatric Ann 17:682–688

Dietrich B, Herrmann WM (1989) Influence of cilazapril on memory functions and sleep behaviour in comparison with metoprolol and placebo in healthy subjects. Br J Clin Pharmacol 27:249S–261S

Dobson ME, Yousseff Y, Maddison S, Pleuvry B (1976) Respiratory effects of lorazepam. Br J Anaesth 48:611–612

Dolly FR, Block AJ (1982) Effect of flurazepam on sleep-disordered breathing and nocturnal oxygen desaturation in asymptomatic subjects. Am J Med 73:239–243

Drake ME (1988) Restless legs with antiepileptic drug therapy. Clin Neurol Neurosurg 90:151–154

Dunleavy DLF, MacLean AW, Oswald I (1971) Debrisoquine, guanethidine, propranolol and human sleep. Psychopharmacology (Berl) 21:101–110

Edling C, Lindberg A, Ulfberg J (1993) Occupational exposure to organic solvents as a cause of sleep apnoea. Br J Ind Med 50:276–279
Etches RC (1994) Respiratory depression associated with patient-controlled analgesia: a review of eight cases. Can J Anaesth 41:125–132
Feinberg I, Wender PH, Koresko RL, Gottlieb F, Piehuta JA (1969) Differential effects of chlorpromazine and phenobarbital on EEG sleep patterns. J Psychiatr Res 7:101–109
Fitzpatrick MF, Engleman HM, Boellert F, McHardy R, Shapiro CM, Deary IJ, Douglas NJ (1992) Effect of therapeutic theophylline levels on the sleep quality and daytime cognitive performance of normal subjects. Am Rev Respir Dis 145:1355–1358
Flaherty JA, Bellur SN (1981) Mental side effects of amantadine therapy: its spectrum and characteristics in a normal population. J Clin Psychiatry 42:344–345
Flemenbaum A (1976) Pavor nocturnus: a complication of single daily tricyclic or neuroleptic dosage. Am J Psychiatry 133:570–572
Fletcher EC, DeBehnke RD, Lovoi MS, Gorin MB (1985a) Undiagnosed sleep apnea in patients with essential hypertension. Ann Intern Med 103:190–195
Fletcher EC, Lovoi M, Miller J, Schaaf J, Flournoy DJ (1985b) Propranolol and sleep apnea (abstract). Am Rev Respir Dis 131:A103
Foerster E-Ch, Greminger P, Siegenthaler W, Vetter H, Vetter W (1985) Atenolol versus pindolol: side-effects in hypertension. Eur J Clin Pharmacol 28:89–91
Forrest WH, Bellville JW (1964) The effect of sleep plus morphine on respiratory response to carbon dioxide. Anesthesiology 25:137–141
Freeman CPL, Trimble MR, Deakin JFW, Stokes TM, Ashford JJ (1994) Fluvoxamine versus clomipramine in the treatment of obsessive compulsive disorder: a multicenter, randomized, double-blind, parallel group comparison. J Clin Psychiat 55:301–305
Gastaut H, Broughton R (1965) A clinical and polygraphic study of episodic phenomena during sleep. In: Wortis J (ed) Recent advances in biological psychiatry. Plenum, New York, pp 197–221
Goodwin FK (ed) (1979) The lithium ion: impact on treatment and research. Arch Gen Psychiatry 36:833–916
Goodwin FK, Bunney WE (1971) Depressions following reserpine: a reevaluation. Semin Psychiatry 3:435–448
Graham JR (1964) Methysergide for prevention of headache. N Engl J Med 270: 67–72
Greenblatt DJ, Shader RI, Abernethy DR, Ochs HR, Divoll M, Sellers EM (1982) Benzodiazepines and the challenge of pharmacokinetic taxonomy. In: Usdin E, Skolnick P, Tallman JF Jr, Greenblatt D, Paul SM (eds) Pharmacology of benzodiazepines. MacMillan, London, pp 257–270
Gross JB, Zebrowske ME, Carel WD, Gardner S, Smith TC (1983) Time course of ventilatory depression with thiopental and midazolam in normal subjects and in patients with chronic obstructive pulmonary disease. Anesthesiology 58:540–544
Guilleminault C, Cummiskey J, Silvestri R (1982) Benzodiazepines and respiration during sleep. In: Udsin E, Clarke P, Tellman D, Paul SM (eds) Pharmacology of benzodiazepines. London, Macmillan, pp 229–236
Guilleminault C, Raynal D, Takahaski S, Carskadon M, Dement W (1976a) Evaluation of short-term and long-term treatment of the narcolepsy syndrome with clomipramine hydrochloride. Acta Neurol Scand 54:71–87
Guilleminault C, Tilkian A, Dement WC (1976b) The sleep apnea syndromes. Annu Rev Med 27:465–484
Guilleminault C, Cummiskey J, Silverstri R (1978) Benzodiazepines and respiration during sleep. In: Usdin E, Skolnick P, Tallman JF Jr Greenblatt D, Paul SM (eds) Pharmacology of benzodiazepines. MacMillan, London, pp 229–236

Guillemnault C, Silvestre R, Mondini S, Coburn S (1984) Aging and sleep apnea: action of benzodiazepine, acetazolamide, alcohol and sleep deprivation in a healthy elderly group. J Gerontol 39:655–661

Guillemnault C, Crowe C, Quera-Salva MA, Miles L, Partinen M (1988) Periodic leg movement, sleep fragmentation and central sleep apnea in two cases: reduction with clonazepam. Eur Respir J 1:762–765

Haffner CA, Smith BS, Pepper C (1993) Hallucinations as an adverse effect of angiotensin converting enzyme inhibition. Postgrad Med J 69:240

Haponik EF, Smith PL, Bohlman ME, Allen RP, Goldman SM, Blecker ER (1983) Computerized tomography in obstructive sleep apnea. Am Rev Respir Dis 127:221–226

Heiman EM, Christie M (1986) Lithium-aggravated nocturnal myoclonus and restless legs syndrome (letter). Am J Psychiatry 143:1191–1192

Hindmarch I (1979) Effects of hypnotic and sleep-inducing drugs on objective assessment of human psychomotor performance and subjective appraisals of sleep and early morning behaviour. Br J Pharmacol 8:435–465

Hindmarch I, Parrott AC (1978) A repeated dose comparison of the side effects of five antihistamines on objective assessments of psychomotor performance, central nervous system arousal and subjective appraisals of sleep and early morning behaviour. Arzneimittelforschung 28:483–486

Hoch CC, Reynolds CF, Monk TH, Buysse DJ, Yeager AL, Houck PR, Kupfer DJ (1990) Comparison of sleep-disordered breathing among healthy elderly in the seventh, eighth, and ninth decades of life. Sleep 13:502–511

Hollister LE, Csernansky JG (1990) Clinical pharmacology of psychotherapeutic drugs. Churchill Livingston, New York, p 194

Holm AC, Edsman I, Lundberg T, Odlind B (1993) Tolerability of remoxipride in the long term treatment of schizophrenia. Drug Safety 8:445–456

Hooper AC (1971) Fenfluramine and dreaming. BMJ 3:305

Huapaya LVM (1979) Seven cases of somnambulism induced by drugs. Am J Psychiatry 136:985–986

Jacobson A, Kales A, Lehmann D, Wenner WH (1965) Somnambulism: all-night electroencephalographic studies. Science 148:975–977

Jaffe JH (1990) Drug addiction and drug abuse. In: Goodman LS, Gilman A (eds) Pharmacological basis of therapeutics, 8th edn. Pergamon, New York, pp 522–573

Jaffe JH, Martin WR (1990) Opioid analygesics and antagonists. In: Goodman, Gilman's (eds) The pharmacological basis of therapeutics, 8th edn. Pergamon Press, New York, pp 485–521

Johnson LC, Chernick DA (1982) Sedative-hypnotics and human performance. Psychopharmacology 76:101–113

Johnson MW, Anch AM, Remmers JE (1984) Induction of the obstructive sleep apnea syndrome in a woman by exogenous androgen administration. Am Rev Respir Dis 129:1023–1025

Jus K, Beland C, Jus A, Fontaine P, Brunelle R, Bouchard M (1975) Polygraphic night sleep pattern during chronic single and multiple dose administration. Int Pharmacopsychiatry 10:58–63

Kales A, Jacobson A (1967) Studies of mental activity: Recall somnambulism and effects of REM deprivation and drugs. Exp Neurol 19:81

Kales A, Scharf MB (1973) Sleep laboratory and clinical studies of the effects of benzodiazepines on sleep: flurazepam, diazepam, chlordiazepoxide, and RO 5-4200. In: Garatini S, Mussini E, Randall LO (eds) The benzodiazepines. Raven, New York, pp 577–598

Kales A, Kales JD (1984) Evaluation and treatment of insomnia. Oxford University Press, New York

Kales A, Vgontzas A (1990) Not all benzodiazepines are alike. In: Stefanis CN, Rabavilas AD, Soldatos CR (eds) Psychiatry: a world perspective, vol 3. Elsevier, Amsterdam, pp 379–384

Kales A, Jacobson A, Paulson MJ, Kales JD, Walter RD (1966) Somnambulsim: psychophysiological correlates. I. All-night EEG studies. Arch Gen Psychiatry 14:586–594
Kales A, Bixler EO, Tan TL, Scharf MB, Kales JD (1974) Chronic hypnotic-drug use: ineffectiveness, drug-withdrawal insomnia, and dependence. JAMA 227:513–517
Kales A, Bixler EO, Scharf MB, Kales JD (1976a) Sleep laboratory studies of flurazepam: a model for evaluating hypnotic drugs. Clin Pharmacol Ther 19:576–583
Kales A, Kales JD, Bixler EO, Scharf MB, Russek E (1976b) Hypnotic efficacy of triazolam: sleep laboratory evaluation of intermediate-term effectiveness. J Clin Pharmacol 16:399–406
Kales A, Kales JD, Jacobson A, Humphrey JF II, Soldatos CR (1977) Effects of imipramine on enuretic frequency and sleep stages. Pediatrics 60:431–436
Kales A, Scharf MB, Kales JD (1978) Rebound insomnia: a new clinical sydrome. Science 201:1039–1041
Kales A, Scharf MB, Kales JD, Soldatos CR (1979) Rebound insomnia: a potential hazard following withdrawal of certain benzodiazepines. JAMA 241:1692–1695
Kales A, Soldatos CR, Kales JD (1981) Sleep disorders: evaluation and management in the office setting. In: Arieti S, Brodie HKH (eds) American handbook of psychiatry, 2nd edn: vol 7. Basic Books, New York, pp 423–454
Kales A, Bixler EO, Soldatos CR, Vela-Bueno A, Jacoby J, Kales JD (1982) Quazepam and flurazepam: long-term use and extended withdrawal. Clin Pharmacol Ther 32:781–788
Kales A, Soldatos CR, Bixler EO, Goff PJ, Vela-Bueno A (1983a) Midazolam: dose-response studies of effectiveness and rebound insomnia. Pharmacology 26:138–149
Kales A, Soldatos CR, Bixler EO, Kales JD (1983b) Early morning insomnia with short-acting benzodiazepines. Science 220:95–97
Kales A, Soldatos CR, Bixler EO, Kales JD (1983c) Rebound insomnia and rebound anxiety: a review. Pharmacology 26:121–137
Kales A, Cadieux RJ, Shaw LC, Vela-Bueno A, Bixler EO, Schneck DW, Locke TW, Soldatos CR (1984) Sleep apnea in a hypertensive population. Lancet II:1005–1008
Kales A, Cadieux RJ, Bixler EO, Soldatos CR, Vela-Bueno A, Misoul CA, Locke TW (1985) Severe obstructive sleep apnea-I: onset, clinical course and characteristics. J Chron Dis 38:419–425
Kales A, Soldatos CR, Vela-Bueno A (1985) Clinical comparison of benzodiazepine hypnotics with short and long elimination half-lives. In: Smith DE, Wesson DR (eds) The benzodiazepines: current standards for medical practice. MTP Press, Maine, pp 121–147
Kales A, Bixler EO, Vela-Bueno A, Cadieux RJ, Manfredi RL, Bitzer S, Kanter T (1988) Effects on nadolol on blood pressure, sleep efficiency, and sleep stages. Clin Pharmacol Ther 43:655–662
Kales A, Bixler EO, Vela-Bueno A, Manfredi RL, Soldatos CR (1989) Effects on dreaming of two beta blockers (nadolol and propranolol). Excerpta Medica 899:689
Kales A, Manfredi RL, Vgontzas AN, Baldassano CF, Kostakos K, Kales JD (1991) Clonazepam: sleep laboratory study of efficacy and withdrawal. J Clin Psychopharmacol 11:189–193
Kales A, Vgontzas AN, Bixler EO (1994) Not all benzodiazepines are alike: update 1993. In: Biegel A, Lopez Ibor IJ, Costa e Silva JA (eds) Psychiatry: a world perspective, vol II. World Scientific, Singapore, pp 942–946
Kane FJ, Pokorny A (1975) Mental and emotional disturbance with pentazocine (talwin) use. South Med J 68:808–811
Kaplan J, Dawson S, Vaughan T, Green R, Wyatt RJ (1974) Effect of prolonged chlorpromazine administration on the sleep of chronic schizophrenics. Arch Gen Psychiatry 31:62–66

Kaplan J, Fredrickson PA, Renaux SA, O'Brien PC (1993) Theophylline effect on sleep in normal subjects. Chest 103:193–195

Kayed K, Godtlibsen OB (1977) Effects of the β-adrenoceptor antagonists acebutolol and metoprolol on sleep pattern in normal subjects. Eur J Clin Pharmacol 12:323–326

Keamy MF, Cadieux RJ, Kofke WA, Kales A (1987) The occurrence of obstructive sleep apnea in a recovery room patient. Anesthesiology 66:232–234

Keck PE Jr, Hudson JI, Dorsey CM, Campbell PI (1991) Effect of fluoxetine on sleep. Biol Psychiatry 29:618–625

Koella WP (1985) CNS-related (side-) effects of β-blockers with special reference to mechanisms of action. Eur J Clin Pharmacol 28:55–63

Kukopulos A, Reginaldi D, Laddomada P, Floris G, Serra G, Tondo L (1980) Course of the manic-depressive cycle and change caused by treatments. Pharmako-Psychiatrie Neuropsychopharmakologie 13:156–167

Kupfer D, Reynolds CF, Weiss BL, Foster FG (1974) Lithium carbonate and sleep in affective disorders. Arch Gen Psychiatry 30:79–84

Lakshminarayan S, Sahn SA, Hudson L, Weil JV (1976) Effects of diazepam on ventilatory responses. Clin Pharmacol Ther 20:178–183

Lamarche Y, Martin R, Reiher J, Blaise G (1986) The sleep apnoea syndrome and epidural morphine. Can Anaesth Soc J 33:231–233

Lauerma H (1991) Nocturnal wandering caused by restless legs and short-acting benzodiazepines. Acta Psychiatry Scand 83:492–493

Lavie Ben-Yosef R, Ami-Hai ER (1984) Prevalence of sleep apnea syndrome among patients with essential hypertension. Am Heart J 108:373–376

Leech JA, Onal E, Dulberg C, Lopata MA (1988) A comparison of men and women with occlusive sleep apnea syndrome. Chest 94:983–988

Lester BK, Coulter JD, Cowden LC, Williams HL (1971) Chlorpromazine and human sleep. Psychopharmacologia 20:280–287

Lewis SA, Oswald I, Dunleavy DLF (1971) Chronic fenfluramine administration: some cerebral effects. Br Med J 3:67–70

Longbottom RT, Pleuvry BJ (1984) Respiratory and sedative effects of triazolam in volunteers. Br J Anaesth 56:179–185

Luchins DJ, Sherwood PM, Gillin JC, Mendelson WB, Wyatt RJ (1978) Filicide during psychotropic-induced somnambulism: a case report. Am J Psychiatry 135:1404–1405

MacEvilly M, O'Carroll C (1989) Hallucinations after epidural buprenorphine. Br Med J 298:928–929

Mamelak M, Csima A, Price V (1984) A comparative 25 night sleep laboratory study on the effects of quazepam and triazolam on the sleep of chronic insomniacs. J Clin Pharmacol 24:65–75

Manfredi RL, Kales A, Vgontzas AN, Bixler EO, Isaac MA, Falcone CM (1991) Buspirone: sedative or stimulant effect? Am J Psychiatry 148:1213–1217

Marshall BE, Longnecker DE (1990) General anesthetics. In: Goodman, Gilman's (eds) The pharmacological basis of therapeutics, 8th edn. Pergamon Press, New York, pp 285–310

Matsumoto AM, Sandbloom RE, Schoene RB, Leek A, Giblin EC, Pierson DJ (1985) Testosterone replacements in hypogonadal men: effects on sleep apnea, respiratory drives, and sleep. Clin Endocrinol 22:713–721

Mattila MJ, Mattila M, Konno K (1986) Acute and subacute actions on human performance and interactions with diazepam of temelastine (SK & F93944) and diphenhydramine. Eur J Pharmacol 31:291–298

Mayer J, Peter JH (1991) First experience with cilazapril in the treatment of sleep apnoea related hypertension. Drugs 41:37–37

Mayer J, Weichler U, Herres-Mayer B, Schneider, Marx U, Peter JH (1990) Influence of metoprolol and cilazapril on blood pressure and sleep apnea activity. J Cardiovasc Pharmacol 16:952–961

Meier-Ewert K, Matsubayashi K, Benter L (1985) Propranolol: long-term treatment in narcolepsy-cataplexy. Sleep 8:95–104

Mendelson WB, Garnett D, Gillin JC (1981) Flurazepam induced sleep apnea syndrome in a patient with insomnia and mild sleep-related respiratory changes. J Nerv Ment Dis 169:261–264
Mendelson WB (1991) Safety of short-acting benzodiazepine hypnotics in patients with impaired respiration. Am J Psychiatry 148:1401
Metes A, Ohki M, Cole P, Haight BM, Hoffstein V (1991) Snoring, apnea and nasal resistance in men and women. J Otolaryngol 20:57–61
Midgren B, Hansson L, Skeidsvoll H, Elmqvist D (1989) The effects of nitrazepam and flunitrazepam on oxygen desaturation during sleep in patients with stable hypoxemic nonhypercapnic COPD. Chest 95:765–768
Mignot E, Guilleminault C, Bowersox S, Kilduff T, Dement WC (1987) The effects of alpha-1 adrenoceptor agonist and antagonist in canine narcolepsy. Sleep Res 16:393
Mignot E, Guilleminault C, Bowersox S, Rappaport A, Dement WC (1988) A role of central alpha-1 adrenoceptors in canine narcolepsy. J Clin Invest 82:885–894
Miller RR (1975) Clinical effects of pentazocine in hospitalized medical patients. J Clin Pharmacol 15:198–205
Mitler MM, Carshadon MA, Phillips RL, Sterling WR, Zavcone VP, Spiegel R, Guilleminault C, Dement WC (1979) Hypnotic efficacy of temazepam: a long-term sleep laboratory evaluation. Br J Clin Pharmacol 8:635–685
Morgan K, Oswald I (1982) Anxiety caused by a short-life hypnotic. Br Med J 284:942
Mort JR (1992) Nightmare cessation following alteration of ophthalmic administration of a cholinergic and a beta-blocking agent. Ann Pharmacother 26:914–916
Moskowitz H, Burns M (1988) Effects of terfenadine, diphenhydramine, and placebo on skills performance. Cutis 42:14–18
Moskowitz C, Moses H, Klawans HL (1978) Levodopa-induced psychosis: a kindling phenomenon. Am J Psychiatry 6:669–675
Murciano D, Armengaud MH, Cramer PH, Neveux E, L'Heritier C, Pariente R, Aubier M (1993) Acute effects of zolpidem, triazolam and flunitrazepam on arterial blood gases and control of breathing in severe COPD. Eur Respir J 6:625–629
Nadel C (1981) Somnambulism, bed-time medication and over-eating. Br J Psychiatry 139:79
Neil-Dwyer G, Bartlett J, McAinsh J, Cruickshank JM (1981) β-Adrenoceptor blockers and the blood-brain barrier. Br J Clin Pharmacol 11:549–553
Newton RE, Casten GP, Alms DR, Benes CO, Marunycz JD (1982) The side effect profile of buspirone in comparison to active controls and placebo. J Clin Psychiatry 43:100–102
Newton RE, Marunycz JD, Alderdice MT, Napoliello MJ (1986) Review of the side-effect profile of buspirone. Am J Med 80:17–21
Nicholson AN, Stone BM (1979) Effect of some stimulants on sleep in man. Br J Pharmacol 66:476P
Nicholson AN, Stone BM (1983) The H_1-antagonist mequitazine: studies on performance and visual function. Eur J Clin Pharmacol 25:563–566
Nicholson AN, Stone BM (1986) Antihistamines: impaired performance and the tendency to sleep. Eur J Clin Pharmacol 30:27–32
Nicholson AN, Pascoe PA, Turner C, Ganellin CR, Greengrass PM, Casy AF, Mercer AD (1991) Sedation and histamine H_1-receptor antagonism: studies in man with the enantiomers of chlorpheniramine and dimethindene. Br J Pharmacol 104:270–276
Oswald I (1969) Unusual effect of fenfluramine. Br Med J 4:806
Oswald I, Adam K, Borrow S, Idzikowski G (1979) The effects of two hypnotics on sleep, subjective feelings and skilled performance. In: Passouant J, Oswald I (eds) Pharmacology of the states of awareness. Pergamon, New York, pp 51–63
Passouant P, Cadilhac J, Ribstein M (1972) Les privations de sommeil avec mouvements oculaires par les anti-depresseurs. Rev Neurol 127:173–192

Peter JH, Gassel W, Mayer J, Heurer-Mayer B, Penzel T, Schneider H (1989) Effects of cilazapril in hypertension, sleep, and apnea. Am J Med 87:72S–78S
Physician's Desk Reference (1993) Medical Economics, New Jersey, p 2687
Pohl R, Gershon S (1981) Nomifensine: a new antidepressant. Psychiatr Ann 11:391–395
Poitras R (1980) A propos d'episodes d'amnesies anterogrades associes a l'utilisation du triazolam. Union Med Can 109:427–429
Pradalier A, Giroud M, Dry J (1987) Somnambulism, migraine and propranolol. Headache 27:143–145
Prien RF, Klett CJ, Caffety EM Jr (1973) Lithium carbonate and imipramine in prevention of affective episodes. Arch Gen Psychiatry 29:420–425
Quitkin F, Rifkin A, Kane J (1978) Prophylactic effect of lithium and imipramine in unipolar and bipolar II patients: a preliminary report. Am J Psychiatry 135:570–572
Rajala R, Partinen M, Sane T, Pelkonen R, Huikuri K, Seppalainen A-M (1991) Obstructive sleep apnoea in morbidly obese patients. J Intern Med 230:125–129
Rall TW (1990a) Hypnotics and sedatives; ethanol. In: Goodman, Gilman's (eds) The pharmacological basis of therapeutics, 8th edn. Pergamon Press, New York, pp 345–382
Rall TW (1990b) Drugs used in the treatment of asthma. In: Goodman, Gilman's (eds) The pharmacologic basis of therapeutics, 8th edn. Pergamon Press, New York, pp 618–637
Regestein QR, Reich P (1985) Agitation observed during treatment with newer hypnotic drugs. J Clin Pharmacol 46:280–283
Rhind GB, Connaughton JJ, McFie J, Douglas NJ, Flenley DC (1985) Sustained release choline theophyllinate in nocturnal asthma. Br Med J 291:1605–1607
Rifkin A, Siris S (1987) Drug treatment of acute schizophrenia. In: Meltzer HY (ed) Psychopharmacology. The third generation of progress. Raven, New York, pp 1095–1101
Robinson RW, Zwillich CW, Bixler EO, Cadieux RJ, Kales A, White DP (1987) Effects of oral narcotics in sleep disordered breathing in healthy adults. Chest 91:197–203
Robinson RW, Zwillich CW (1994) Drugs and sleep respiration. In: Kryger MH, Roth T, Dement WC (eds) Principles and practice of sleep medicine, 2nd edn. Saunders, Philadephia, pp 603–620
Roth T, Zorick F (1987) Sedative effects of antihistamines. J Allergy Clin Immunol 80:94–98
Roth T, Kramer M, Lutz T (1976) Intermediate use of triazolam: a sleep laboratory study. J Int Med Res 4:59–62
Roth T, Roehrs T, Koshorek J, Sicklesteel BA, Zorick F (1987) Sedative effects of antihistamines. J Allergy Clin Immunol 80:94–98
Rothschild AJ (1985) Mania after withdrawal of isocarboxazid. J Clin Psychopharmacol 5:340–342
Rubinstein S, Silver LB, Licamele WL (1994) Clonidine for stimulant-related sleep problems. J Am Acad Child Adoless Psychiatry 33:281–282
Rudolf M, Geddes DM, Turner JA, Saunders KF (1978) Depression of central respiratory drive by nitrazepam. Thorax 33:97–100
Rundell OH, Lester BK, Griffiths WJ, Williams HL (1972) Alcohol and sleep in young adults. Psychopharmacologia 26:201–218
Safar P, Escarraga LA, Chang F (1959) Upper airway obstruction in the unconscious patients. J Appl Physiol 14:760–764
Saletu B, Grunberger J, Anderer P, Barbanoj MJ (1988) Pharmacodynamic studies of a combination of lorazepam and diphenhydramine and its single components: electroencephalographic brain mapping and safety evaluation. Curr Ther Res Clin Exp 44:909–937

Samuels SI, Rabinov W (1986) Difficulty reversing drug-induced coma in a patient with sleep apnea. Anesth Analg 65:1222–1224
Sandblom RE, Matsumoto AM, Schoene RB, Lee KA, Giblin EC, Bremner WJ (1983) Obstructive sleep apnea syndrome induced by testosterone administration. N Engl J Med 308:508–510
Santiago TV, Goldblatt K, Winters K, Pugliese AC, Edelman NH (1980) Respiratory consequences of methadone: the response to added resistance to breathing. Am Rev Respire Dis 122:623–628
Schenck CH, Bundlie SR, Smith SA, Ehinger MG, Mahowald MW (1986) REM behavior disorder in a 10 year old girl and aperiodic REM and NREM sleep movements in an 8 year old brother. Sleep Res 15:162
Schenck CH, Hurwitz TD, Mahowald MW (1988) REM sleep behavior disorder. Am J Psychiatry 145:652
Schenck CH, Mahowald MW, Kim SW, O'Conner KA, Hurwitz TD (1992) Prominent eye movements during NREM sleep and REM sleep behavior disorder associated with fluoxetine treatment of depression and obsessive-compulsive disorder. Sleep 15:226–235
Schlauch R (1979) Hypnopompic hallucinations and treatment with imipramine. Am J Psychiatry 136:219–220
Schneider BK, Pickett CK, Zwillich CW, Weil JV, McDermott MT, Santon RJ, Varano LA, White DP (1986) The influence of testosterone on breathing during sleep. J Appl Physiol 61:618–623
Schoonover SC (1983) Depression. In: Bassuk EL, Schoonover SC, Gelenberg AJ (eds) The practitioner's guide to psychoactive drugs, 2nd edn. Plenum, New York, pp 19–77
Schwab RS, Poskanzer DC, England AC, Young RR (1972) Amantadine in Parkinson's disease. JAMA 7:792–795
Scrima L, Broudy M, Nay K, Cohn MA (1982) Increased severity of obstructive sleep apnea after bedtime alcohol ingestion: diagnostic potential, and proposed mechanism of actions. Sleep 5:318–328
Scully BE, Clynes N, Neu HC (1991) Oral ofloxacin therapy of infections due to multiply-resistant bacteria. Diagn Microbiol Infect Dis 14:435–441
Shimizu T, Ookawa M, Iijuma S, Nakamura M (1985) Effect of clomipramine on nocturnal sleep of normal human subjects. Annu Rev Pharmacopsychiat Res Round 16:138
Shore, JH, Fraunfelder FT, Meyer SM (1987) Psychiatric side effect from topical ocular timolol, a beta-adrenergic blocker. J Clin Psychopharmacol 7:264–267
Skatrud JB, Begle RL, Busch M (1988) Ventilatory effects of single, high-dose triazolam in awake human subjects. Clin Pharmacol Ther 44:684–689
Smith DE, Wesson DR, Buxton ME, Seymour RB, Ungerleider JT, Morgan JP, Mandell AJ, Jara G (1979) Amphetamine use, misuse, and abuse. In: Proceedings of the national amphetamine conference, 1978. Hall, Boston, pp 21–27
Soldatos CR, Kales A, Bixler EO, Vela-Bueno A (1985) Behavioral side effects of benzodiazepine hypnotics. Clin Neuropharmacol 8 [Suppl 1]:S112–S117
Solomon F, White C, Parron D, Mendelson WB (1979) Sleeping pills, insomnia, and medical practice. N Engl J Med 300:803–808
Starmer G (1985) Antihistamines and highway safety. Accid Anal Prev 17:311–317
Sugano T, Suenaga K, Endo S, Shimada M (1980) Withdrawal delirium in a patient with nitrazepam addiction. Jpn J EEG EMG 8:34–35
Tachibana M, Tanaka K, Hishikawa Y, Kaneko I (1975) A sleep study of acute psychotic states due to alcohol and meprobamate addiction. Adv Sleep Res 2:177–205
Tanaka K, Kameda H, Sugita Y, Hishikawa Y (1979) A case with pentazocine dependence developing delirium upon withdrawal. Psychiatr Neurol Jpn 81: 289–299
Teelucksingh S, Steer CR, Thompson CJ, Seckl JR, Douglas NJ, Edwards CRW (1991) Hypothalamic syndrome and central sleep apnoea associated with toluene exposure. Q J Med, New Series 78,286:185–190

Thorpy MJ (1990) International classification of sleep disorders: diagnostic and coding manual. American Sleep Disorders Association, Rochester

Tormey WP, Buckley MP, Darragh AS (1979) Propranolol, sleep and the nocturnal release of stress hormones. J Irish Med Assoc 72:450

Vgontzas AN, Tan TL, Bixler EO, Martin LF, Shubert D, Kales A (1994) Sleep apnea and sleep disruption in obese patients. Arch Intern Med 154:1705–1711

Vogel G, Thurmond A, Gibbons P, Edwards K, Sloan KB, Sexton K (1975) The effect of triazolam on the sleep of insomniacs. Psychopharmacologia 41:65–69

Vogel GW, Barker K, Gibbons P, Thurmond A (1976) A comparison of the effects of flurazepam 30 mg and triazolam 0.5 mg on the sleep of insomniacs. Psychopharmacology 47:81–86

Volta GD, Magoni M, Cappa S, Di Monda V (1990) Insomnia and perceptual disturbances during flunarizine treatment. Headche 30:62–63

Walsh JK, Muehlback MJ, Schweitzer PK (1992) Simulated assembly line performance following ingestion of cetirizine or hydroxyzine. Ann Allergy 69:195–200

Wang M, Tu P, Chen T, Hu SY (1989) Anesthetic implications and complications of uvulopalatopharyngoplasty and sleep apnea syndrome. Anaesth Sinica 27:191–196

Ware JC, Brown FW, Moorad PJ, Pittard JT, Murphy M, Franklin D (1984) Nocturnal myoclonus and tricyclic antidepressants. Sleep Res 13:72

Wehr TA, Goodwin FK (1979) Rapid cycling in manic-depressives induced by tricyclic antidepressants. Arch Gen Psychiatry 36:555–559

Weil JV, McCullough RE, Kline JS, Sodal IE (1975) Diminished ventilatory response to hypoxia and hypercapnia after morphine in normal men. N Engl J Med 292:1103–1106

Weiner M (1982) Sedation and antihistamines. Arzneimittelforschung 32:1193–1195

Westerlund A (1985) Central nervous system side effects with hydrophilic and lipophilic β-blockers. Eur J Clin Pharmacol 28:73–76

White DP, Cadieux RJ, Lombard RM, Bixler EO, Kales A, Zwillich CW (1985a) The effects of nasal anesthesia on breathing during sleep. am Rev Respir Dis 132:972–975

White DP, Schneider BK, Santen RJ, McDermott M, Pickett CK, Zwillich C, Weil JV (1985b) Influence of testosterone on ventilation and chemosensitivity in male subjects. J Appl Physiol 59:1452–1457

Williams AJ, Houston D, Finbert S, Lam C, Kinney JL, Santiago S (1985) Sleep apnea syndrome and essential hypertension. Am J Cardiol 55:1019–1022

Williams NR (1988) Erythromycin: a case of nightmares. BMJ 296:214

Wirz-Justice A (1983) Antidepressant drugs: effects on the circadian system. In: Wehr TA, Goodwin FK (eds). Circadian rhythms in psychiatry. Boxwood, Pacific Grove, California, pp 235–264

Woods PB, Robinson ML (1981) An investigation of the comparative liposolubilities of β-adrenoceptor blocking agents. J Pharm Pharmacol 33:172–173

Woodward JK, Munro NL (1982) Terfenadine, the first non-sedating antihistamine. Arzneimittelforschung 32:1154–1156

Wyatt RJ, Gillin JC (1976) Biochemistry and human sleep. In: Williams RL, Karacan I (eds) Pharmacology of sleep. Wiley, New York, pp 239–274

Yew WW, Wong CF, Wong PC, Lee J, Chau CH (1993) Adverse neurological reactions in patients with multidrug-resistant plumonary tuberculosis after coadministration of cycloserine and ofloxacin. Clin Infect Dis 17:288–289

Young T, Palta M, Dempsey J, Skatrud J, Weber S, Badr S (1993) The occurrence of sleep-disordered breathing among middle-aged adults. N Engl J Med 328: 1230–1236

Yules RB, Lippman ME, Freedman DX (1967) Alcohol administration prior to sleep. Arch Gen Psychiatry 16:94–97

Subject Index

Springer-Verlag and the Environment

We at Springer-Verlag firmly believe that an international science publisher has a special obligation to the environment, and our corporate policies consistently reflect this conviction.

We also expect our business partners – paper mills, printers, packaging manufacturers, etc. – to commit themselves to using environmentally friendly materials and production processes.

The paper in this book is made from low- or no-chlorine pulp and is acid free, in conformance with international standards for paper permanency.

Printing: Mercedesdruck, Berlin
Binding: Buchbinderei Lüderitz & Bauer, Berlin